SECOND EDITION

LASER BEAM SHAPING

Theory and Techniques

SECOND EDITION

LASER BEAM SHAPING

Theory and Techniques

EDITED BY

FRED M. DICKEY

CRC Press is an imprint of the
Taylor & Francis Group, an **informa** business

CRC Press
Taylor & Francis Group
6000 Broken Sound Parkway NW, Suite 300
Boca Raton, FL 33487-2742

First issued in paperback 2017

Version Date: 20140402

ISBN 13: 978-1-138-07630-3 (pbk)
ISBN 13: 978-1-4665-6100-7 (hbk)

Library of Congress Cataloging-in-Publication Data

Laser beam shaping : theory and techniques / editor, Fred M. Dickey. -- Second edition.
pages cm
Summary: "This book provides a working understanding of the fundamentals of laser beam shaping techniques, as well as insight into the potential application of laser beam profile shaping in laser system design. It covers the theory and practical application of every important technique for lossless beam shaping, explains when beam shaping is practical and when it is not, offers guidance on when each technique is appropriate, and includes experimental results in most cases. Beam measurement techniques are also treated in considerable detail"-- Provided by publisher.
Includes bibliographical references and index.
ISBN 978-1-4665-6100-7 (hardback)
1. Laser beams. 2. Beam optics. I. Dickey, Fred M., 1941- editor of compilation.

TA1677.L3622 2014
621.36'6--dc23 2014007198

Visit the Taylor & Francis Web site at
http://www.taylorandfrancis.com

and the CRC Press Web site at
http://www.crcpress.com

Dedication

I would like to dedicate this book to Alexander, Ryan, Adrian, Miles, and Leo.

Contents

Preface

At the time of the publication of the first edition of this book, there was a significant interest in laser beam shaping for industrial applications and research. A significant amount of this work was not published for proprietary reasons. At that time, people began to publish their work on beam shaping. The interest in laser beam shaping increased dramatically in the following years. This is due to the increase in the number of laser applications that can benefit from shaping the beam, the increase in research in laser beam shaping techniques and the corresponding increase in the literature, and the advances in optical component fabrication technology. The purpose of this edition is to update the book to include significant developments in laser beam shaping theory and techniques.

After the Introduction chapter, Chapter 2 presents the underlying electromagnetic theory and mathematical techniques applicable to beam shaping. This chapter is very fundamental and has one minor correction or change. Chapter 3 is a new chapter that presents the theory of optimal beam splitting gratings (fan-out gratings). Chapter 4 is a new chapter that addresses the theory and application of vortex beams. Chapter 5 (former Chapter 3) presents the diffraction approach to single-mode Gaussian beam shaping and includes experimental results. The major changes in this chapter are the inclusion of a new section on wavelength dependence of the problem and an expansion of Appendix B. The methods, theory, and application of geometrical optics are discussed in Chapter 6 (former Chapter 4). This chapter is expanded significantly to include the author's research that was not available at the time of the first edition. Optimization-based techniques are presented in Chapter 7 (former Chapter 5). This chapter is greatly revised around the techniques based on the use of current optical software packages. Beam shaping using diffractive diffusers is introduced in Chapter 8 (former Chapter 6). This chapter is significantly revised. Chapter 9 is a new chapter that presents the theory of beam shaping based on the use of microlens diffusers. Multiaperture beam integration systems, including experiment and design, are presented in Chapter 10 (former Chapter 7). The major change in this chapter is the addition of a new section on channel integrators. Chapter 11 is a new chapter that discusses the generation of light ring patterns using axicons. This chapter includes a technique for the generation of rectangular line light patterns. Beam profile measurement technology is addressed in Chapter 12 (former Chapter 9). This chapter is significantly updated. Chapter 13 (former Chapter 8) discusses the application of geometrical optics methods to classical (nonlaser) shaping problems.

The material in these chapters gives the reader a working understanding of the fundamentals of laser beam shaping techniques. It also provides insight into the potential application of laser beam profile shaping in laser system design.

The book is intended primarily for optical engineers, scientists, and students who have a need to apply laser beam shaping techniques to improve laser processes. It should be a valuable asset to someone who researches, designs, procures, or assesses the need for beam shaping with respect to a given application. Due to the broad treatment of theory and practice in the book, we think that it should also appeal to scientists and engineers in other disciplines.

I express my gratitude to the contributing authors whose efforts made the book possible. It was a pleasure working with the staff of Taylor & Francis Group books. Finally, I express my appreciation to the very helpful project coordinator, Laurie Schlags.

Fred M. Dickey
FMD Consulting, LLC

Editor

Fred M. Dickey received his BS (1964) and MS (1965) degrees from Missouri University of Science and Technology, Rolla, Missouri, and his PhD degree (1975) from the University of Kansas, Lawrence, Kansas. He started and chaired the SPIE Laser Beam Shaping Conference, which is in its fifteenth year, for the first 8 years, and is currently a committee member. He heads FMD Consulting, LLC, Springfield, Missouri. He is a fellow of the International Optical Engineering Society (SPIE) and the Optical Society of America, and a senior member of the Institute of Electrical and Electronic Engineers. Dr. Dickey is also the author of over 100 papers and book chapters, and holds 9 patents.

Contributors

Daniel M. Brown
Optosensors Technology, Inc.
Gurley, Alabama

David R. Brown
JENOPTIK Optical System, LLC
Huntsville, Alabama

Jeremiah D. Brown
JENOPTIK Optical Systems, LLC
Huntsville, Alabama

Fred M. Dickey
FMD Consulting, LLC
Springfield, Missouri

Neal C. Evans
PointClear Solutions, Inc.
Innovation Depot
Birmingham, Alabama

Julio C. Gutiérrez-Vega
Photonics and Mathematical Optics Group
Tecnológico de Monterrey
Monterrey, Mexico

John A. Hoffnagle
Picarro, Inc.
Santa Clara, California

Scott C. Holswade
Sandia National Laboratories
Albuquerque, New Mexico

Kevin D. Kirkham
Ophir-Spiricon, LLC
North Logan, Utah

Alexander Laskin
AdlOptica GmbH
Berlin, Germany

Todd E. Lizotte
Via Mechanics USA, Inc.
Londonderry, New Hampshire

Carlos López-Mariscal
US Naval Research Laboratory
Washington, DC

Louis A. Romero
Department of Computational and Applied Mathematics
Sandia National Laboratories
Albuquerque, New Mexico

Carlos B. Roundy
Logan, Utah

Tasso R.M. Sales
RPC Photonics Inc.
Rochester, New York

David L. Shealy
Department of Physics
University of Alabama at Birmingham
Birmingham, Alabama

Louis S. Weichman
Sandia National Laboratories
Albuquerque, New Mexico

1 Introduction

Todd E. Lizotte and Fred M. Dickey

Beam shaping is the process of redistributing the irradiance and phase of a beam of optical radiation. The beam shape is defined by the irradiance distribution. The phase of the shaped beam is a major factor in determining the propagation properties of the beam profile. For example, a reasonably large beam with a uniform phase front will maintain its shape over a considerable propagation distance. Beam shaping technology can be applied to both coherent and incoherent beams.

Arguably, there exists a preferred beam shape (irradiance profile) in any laser application. In industrial applications, the most frequently used profile is a uniform irradiance with steep sides, flat-top beam. This is due to the fact that the same interaction (physics) is accomplished over the illuminated area. Flat-top beams also have applications in laser printing. However, this is not the only profile of interest. Laser disk technology uses a focused beam with minimized side lobes to eliminate cross talk. Other patterns of interest in applications include shaped lines, rings, and array patterns. Some of the major applications of laser beam shaping are discussed in detail in *Laser Beam Shaping Applications*.[1]

Although the laser was invented in 1960, there were only about eight papers on laser beam shaping that appeared in the literature before 1980. A brief history and overview of laser beam shaping is given in the 2003 *Optics & Photonics News* paper "Laser Beam Shaping."[2] The rate of the appearance of laser beam shaping papers grew linearly, but slowly, until about 1995 when the rate increased dramatically. There is evidence that considerable research and development work on laser beam shaping was done in the period before 1995, but was not published for proprietary reasons. Starting in 2000 and continuing to the present, there have been 14 International Society for Optics and Photonics (SPIE) laser beam shaping conferences.[3–16] The history of laser beam shaping is treated by Shealy in Chapter 9 of *Laser Beam Shaping Applications*.

A flat-top laser irradiance profile can be obtained by expanding the beam to obtain a pattern with the desired degree of uniformity. This approach intrudes very large losses in energy throughput. In almost all beam shaping applications, it is desirable to minimize the losses. The two major beam shaping techniques for producing a uniform beam are field mapping and beam integrators (homogenizers). These techniques can be designed to have very low losses.

Field mapping is the technique of using a phase element to map the laser beam into a uniform beam (or other profile) in a given plane. Field mappers are applicable to single-mode (spatially coherent) lasers. Producing vortex beams is also an example of field mapping. The individual lenslets in beam integrators function as field mappers.

Beam integrators break up the input beam into smaller beamlets that are directed to overlap in the output plane with the desired shape. They frequently consist of a lenslet array and a primary lens. Beam integrators can also be implemented using a reflective tube and focusing the laser beam on the input aperture of the tube. This approach is called a channel integrator. Beam integrators are especially applicable to low spatial coherence beams. The low spatial coherence of the input beam reduces the speckle pattern that is inherent in the output of beam integrators. There cases when it is useful to apply beam integrators to spatially coherent beams when the speckle can be tolerated. It is interesting to note that optical configurations that can be considered beam integrators were introduced long before the advent of the laser.[17,18]

The ability to do beam shaping is limited by uncertainty principle of quantum mechanics, or equivalently the time–bandwidth product inequality associated with signal processing. Mathematically, the uncertainty principle is a constraint on the lower limit of the product of the root-mean-square width of a function and its root-mean-square bandwidth. It can be directly applied to the beam shaping problem because of the Fourier transform relation in the Fresnel integral used to describe the beam shaping problem. In fact, the uncertainty principle is generally applicable to diffraction theory. Applying the uncertainty principle to the general diffraction problem associated with laser beam shaping, one obtains a parameter β of the form

$$\beta = C\frac{r_0 y_0}{\lambda z} \tag{1.1}$$

where:

r_0 is the input beam half-width
y_0 is the output beam half-width
C is a constant that depends on the exact definition of beam widths
z is the distance to the output plane

The parameter β is also obtained when applying the method of stationary phase to diffraction problems.

The value of β must be sufficiently large for successful beam shaping to be accomplished. It should be noted that the system designer has some design control over β by specifying r_0 or, possibly, the other three parameters in Equation 1.1. Because of its fundamental nature, β is applicable to field mappers and beam integrators.

It is commonly stated that when β is large the problem can be treated using geometrical optics. This is true for field mapping systems designed to produce flat-top profiles. However, techniques such as diffractive diffusers inherently require the use of diffraction theory in the design process. In addition, diffraction theory is useful in determining some general properties of beam shapers. An example is the wavelength independence of some field mapping configurations (see Chapter 5).

In no case can quality beam shaping be accomplished if β is small. It is suggested that this parameter be considered in the initial stage of any beam shaping system design.

As stated earlier and clearly defined in the subsequent chapters, the theory, calculations, and strategies for designing laser beam shaping systems have come a long way. Although success in the application of beam shaping does not only come from knowing how to calculate a design and understanding the guiding parameters such as β, it simply provides a working baseline of knowledge.

When an engineer or a development team considers the integration of optics into a process, they need to take into account a number of primary as well as ancillary variables that start at the process and work backward through the optical system to the laser source itself. Failure is inevitable if beam shaping is considered simply as an off-the-shelf product that is easy to integrate. Consideration of a new beam shaper design or the purchase of an off-the-shelf beam shaping product must be approached carefully since the beam shaper's performance is dependent on the stability of the laser source itself. Offering a considerable number of challenges due to the dynamic nature of laser processes simple changes to duty cycle often result in pointing instability, divergence shifts, beam intensity distribution, and power fluctuations, to name a few. All of the variables identified need to be scrutinized and prioritized so that the beam shaper can be designed and configured with an appropriate set of preconditioning optics or enough axes of adjustment to provide fine-tuning if required.[19] These items are only part of what it takes to be successful; a willingness to tackle the hardest problems first is the only true guarantee.

Since the introduction of lasers into the industrial marketplace, those of us involved in its application have been in a technology race, whether we like it or not. Driving innovation is the key to success for technologists, but that innovation in many cases is found by simply searching for insight from existing successes within the scientific community or other markets where similar technology is applied. That insight can take many forms such as exposure to existing and past technologies or merely taking calculated risks by blazing a new trail by pulling together various technologies and integrating them into a new solution. Whether applying old or new ideas, innovation of beam shaping technology requires identifying the parameters within the context of a laser process that matter and moving through them systematically to deliver an elegant solution.

Below are two examples that highlight the development of "diffractive and refractive" laser beam shaping technology over the past 20 years and hit upon this theme. From a historical perspective, the examples were selected to illustrate the progression and impact of laser beam shaping on the industrial laser system marketplace. No attempt has been made to select specific technological milestones of equal importance nor should the reader consider these items critical in terms of a grand historical record. These examples are simply moments in time where insight gained by early adopters led to experimentation and the evolution of laser beam shaping within the industrial laser and laser materials processing field. Let us explore a few exemplar beam shaper solutions that were brought to market, which is the point where true innovation ends, as well as demonstrated.

As mentioned earlier, field mappers are phase elements that redistribute laser light to form a new desired irradiance and phase profile. These phase elements commonly take two highly efficient forms, either traditional refractive phase elements, such as custom aspheric lenses (>96% efficient), or diffractive elements, such as diffractive lenses

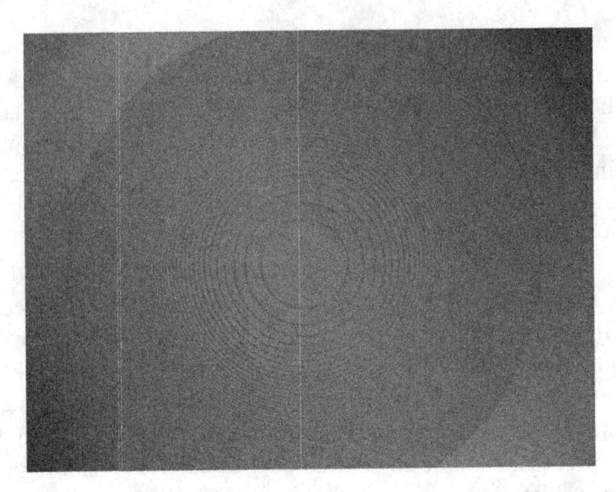

FIGURE 1.1 Gaussian to round flat-top diffractive field mapping beam shaper element.

(>85% efficient), where the phase coefficients are compressed into 2Π surface reliefs. Figure 1.1 shows a Gaussian to round flat-top diffractive field mapping beam shaper element. In many cases, these diffractive optics are based upon an aspheric phase design. It should be stated that fabrication plays a major role in the efficiency of such field mappers and during the early 1990s the costs for fabricating these elements were significant; 20 years later, these elements are now manufactured at scale with competitive pricing that has made them available for lower cost laser marking applications.

Although numerous far- and near-infrared (FIR/NIR) lasers such as CO_2 and Nd:YAG have benefited from more traditional beam shaping and field mapping aspheric systems, frequency tripled and quadrupled diode-pumped solid-state (UV DPSS) lasers (355 and 266 nm, respectively) have benefited from field mappers to a greater degree. As UV DPSS lasers began to be adopted into larger volume laser system markets, it quickly became necessary to begin employing field mappers to improve process performance. Immediate gains in process stability and overall material removal quality were demonstrated. Finding a home within high volume microelectronics packaging manufacturing, industrial ultraviolet (UV) laser tools utilizing field mappers led to the rapid level of microminiaturization of numerous consumer electronics and commercial products. By providing uniform intensities, very small features can be produced.

As field mapper-based beam shapers were employed, the UV DPSS laser system began shifting from a traditional focal point machining technique to a higher precision imaging technique.[20] Optical imaging is the most widely used beam delivery technique in the laser micromachining world. Optical imaging is the preferred method because it offers a high-finesse process that creates accurate and controllable structures into the surfaces of a wide variety of organic and inorganic materials. Defined by an image placed on a mask, the desired structures are optically transferred from the mask to the surface. The design of the beam delivery system is based on the lens imaging equation; however, the calculation of most importance, is the demagnification required to achieve the optimum energy density on target for the material being processed to

FIGURE 1.2 An assembled beam shaping and imaging beam delivery for a UV DPSS marking application installed on an automation system.

produce precision ablated microstructures. The demagnification ratio determines the conjugate distances of the optical imaging system, that is to say the object (mask) distance to the lens and the image distance to the target plane. The optical system reduces the mask design by this factor; therefore, when generating the mask artwork, the feature desired on target is multiplied by the magnification factor. Figure 1.2 shows an assembled beam shaping and imaging beam delivery for a UV DPSS marking application installed on an automation system. The UV beam delivery systems geared for UV DPSS laser micromachining applications have four basic functions once the laser beam exits the laser itself: (1) improve the uniformity of the laser beam, (2) illuminate a fixed aperture or mask plane, (3) reduce/demagnify and project the mask image onto the target material, and (4) control the energy density at the target.

In most cases, the output intensity profile of the UV DPSS laser is Gaussian or TEM_{00} mode. For precision micromachining such as drilling or thin-film patterning, the transformation of the Gaussian laser beam into a flat-top intensity profile was the watershed moment, where the quality and accuracy of laser-based processes truly began to be demonstrated. Microvias that are laser-drilled into microelectronic packaging and laser thin-film patterning for solar cells and flat panel displays were key markets where the benefits of laser beam shaping were initially realized in the late 1990s and continue even today.

Two examples of common designs include field mappers for transforming single-mode Gaussian laser beams into a round or square flat-top intensity distribution profile. Figure 1.3 shows a Gaussian and super Gaussian profile. Within the field of microvia drilling of printed circuit boards (PCBs), the Gaussian output of a UV DPSS laser beam is shaped into a round flat-top intensity distribution and demagnified by using an optical imaging system to achieve the appropriate energy density to either ablate a thin metal film or polymer dielectric layers that make up a PCB assembly. Figure 1.4 shows a 30, 40, and 50 μm blind microvia in a PCB. Conversely, a square flat-top intensity profile can be used to pattern thin films to form structures such as pixel arrays or to precisely cut circuits for repairing advanced displays. Figure 1.5 shows an electrical grounding strap that was laser deleted using a square

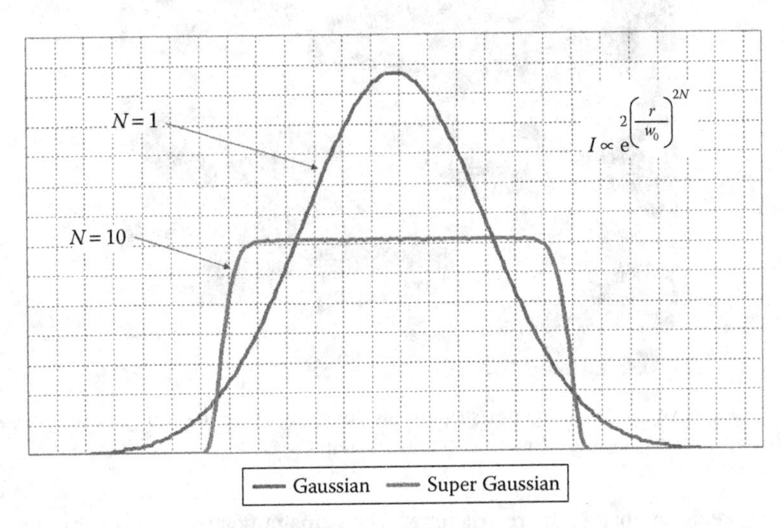

FIGURE 1.3 A Gaussian and super Gaussian profile distribution as an example of the desired transformation for many industrial beam shaping applications.

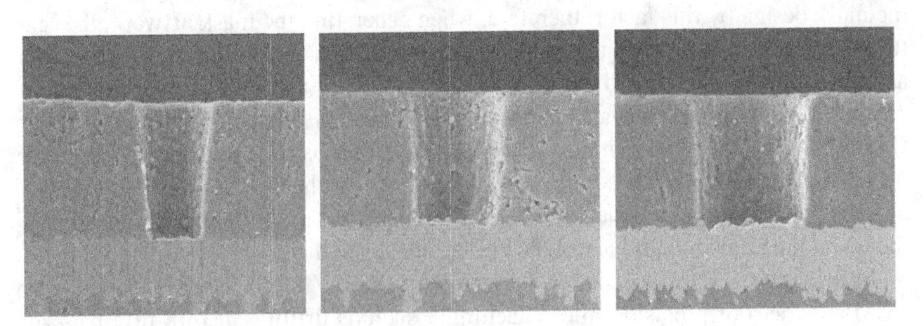

FIGURE 1.4 Microvia hole drilling on a PCB with 30, 40, and 50 μm diameters (left to right). These are possible using a UV DPSS flat-top laser beam.

FIGURE 1.5 A grounding strap deletion using a square flat-top laser beam.

flat-top beam and the energy density was tailored so that only the strap was removed leaving the underlying material undamaged.

Depending on the laser system setup, the imaged beam can precisely drill or pattern materials with such finesse as to minimize or negate any thermal damage to the surrounding material. Therefore, precise and tailored to the material being processed, field mappers have enabled new opportunities for precision micromanufacturing that could not have been possible 20 years ago.

The arrival of integrator technology for laser materials processing was under the radar and developed in earnest when major challenges were encountered with the introduction of excimer lasers into the marketplace. During these early years, integrator development was kept secret until patent filings revealed their implementation. Many of these innovations were championed by laser system developers who were seeking solutions to improve excimer laser beam uniformity for large field size high-precision laser processes. In nearly all applications, excimer laser output uniformity is critical for high-precision applications.[21] During the early 1980s, to create a uniform irradiance beam profile, engineers were limited to either improving the performance of the laser itself, at a hefty price, or utilizing optical techniques. Modification of the laser meant trade-offs; increased uniformity at the sacrifice of pulse energy or power and that still did not guarantee the best uniformity.

At this time the excimer laser was finding new applications within semiconductor processing such as lithography and gaining ground in the promising field of laser micromachining. These early adopters were exposed to similar integrator designs used within illumination systems for lamp-based exposure tools, such as those produced by Oriel Instruments Corporation (Stratford, CT), which were generally simple lens array integrator designs.[22] The earliest and relatively successful example of an industrial excimer laser application that would not have been possible without the use of beam integrators, besides the semiconductor lithography market, was laser annealing of silicon. Patented by XMR, Inc. (Houston, TX) in 1986 and issued in 1988 as

FIGURE 1.6 A beam integrator design from the mid-1980s manufactured by XMR, Inc.

US Patent 4,733,944, the design is one of a handful of instances of an imaging beam integrator for a high-volume industrial laser process requiring absolute stability and uniformity. What was unique about the design at the time was the fact that the spot size produced could be selectively adjusted in size at the working plane. Adjustability allowed variation of the energy density on target, tailoring it to an optimum setting and process area. Figure 1.6 is an example of an XMR design from the actual system.

From this point, the design of laser beam integrators began to take the form of specialized optical configurations with further enhancements and refinements to meet the needs of ever-demanding laser micromachining processes. In 1994 the integration of UV excimer lasers for laser micromachining of fluidic structures was being exploited for consumer and medical device products. One such product was the production of nozzle plates and fluidic channels for inkjet printers. Although personal inkjet printers had entered the marketplace in 1988, it was not until 1991 when inkjet printer manufacturers had begun to seriously consider excimer lasers as a means to reduce the costs of forming precision inkjet nozzle plates, which were upward of $4.00 each to manufacture. Figure 1.7 shows an example of an inkjet nozzle plate with integrated fluidic channels imaged and ablated into polyimide.

At that time, manufacturing nozzle plates to micron accuracies was not an easy task and they were costly to manufacture using traditional lithography and nickel electroforming techniques. It was clear in the early days of process development that the design of excimer beam delivery optics, physical system configuration, methods of optical beam shaping, and laser material interaction would play significant roles in producing inkjet nozzle plates. Due to higher demands for quality and the critical nature of providing consumers with exceptionally reliable products, the excimer processes needed to be robust and repeatable to a 3σ or better level.[23]

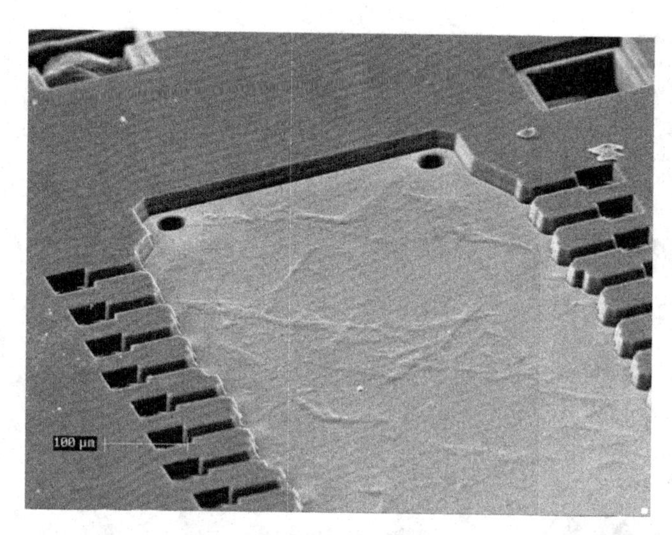

FIGURE 1.7 An inkjet nozzle plate with integrated fluidic channels imaged and ablated into polyimide.

Matching the laser process optics to resolve the nozzle and fluidic channel features to micron tolerances within the desired material was the goal. To complicate matters, inkjet nozzle arrays have dimensional requirements upward of 15 × 2 mm in size (L × W), which made the selection of beam shaping technology to provide uniformity across the entire image plane at a uniform energy density the most critical aspect of the excimer micromachining system. Figure 1.8 shows how the implementation of an imaging lens array beam integrator evolved in less than 10 years (ca. 1994). Larger in size and its zooming feature allowing larger illumination field sizes at the mask plane demonstrates how the integrator design had progressed into a useful and versatile tool for large-field ablation applications. Incorporating beam preconditioning optics, in this case an anamorphic cylindrical lens telescope, the output beam of the excimer was shaped from a rectangular shape into an optimized square configuration for illuminating the zoomable cross-cylindrical lens beam integrator. The beam integrator design was also more advanced than earlier designs, with an adjustable zoom for both axes of the beam, allowing the illumination field at a mask plane to be both uniform in intensity and dimensionally optimum. The shaped beam illuminated a mask that defined the features being produced. The numerical aperture of the beam integrator was designed to match the numerical aperture of the final imaging lens system allowing the mask design to be imaged onto the material to be processed. As the first excimer laser inkjet nozzle drilling system came online, the cost of inkjet nozzle plates had dropped to less than $0.20 a unit—a significant milestone for that industry that would not have been achieved without an integrator-based beam shaper design.

Since that time, further advancements have made it possible to take an excimer beam with a beam size of ~20 × ~10 mm (L × W) and transform it into a uniform beam of ~140 × ~5 mm, another example of how the integrator continues to evolve. Figure 1.9 shows a beam exposed on film at various attenuation factors

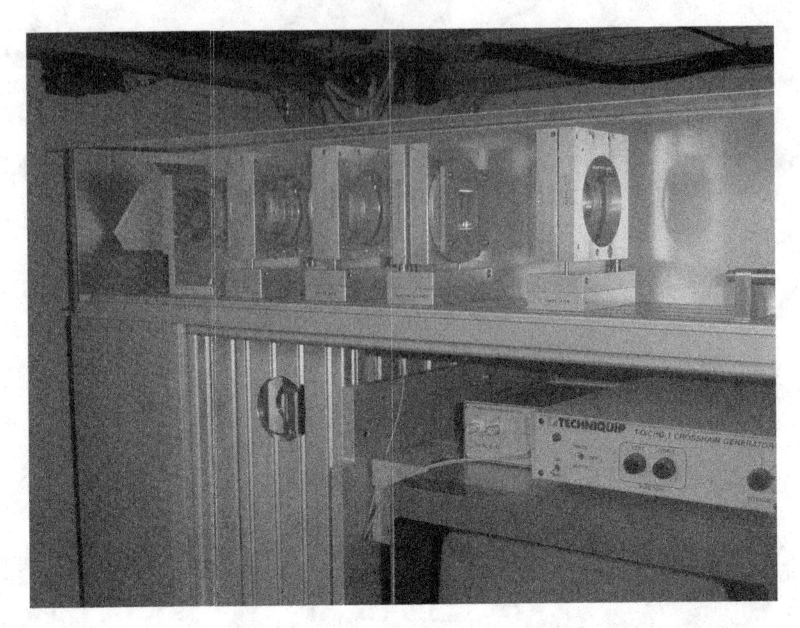

FIGURE 1.8 An advanced imaging excimer laser beam integrator design that is zoomable in the short and long axis (*X*–*Y*), allowing it to be tailored for the illumination field size at the mask plane.

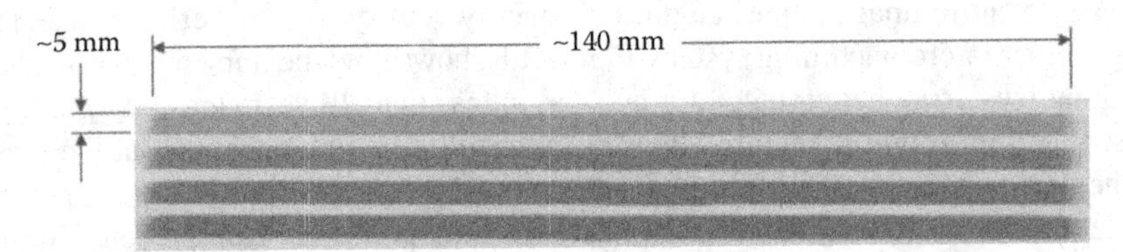

FIGURE 1.9 An integrator-shaped excimer laser beam exposed on a film at various attenuation factors: 25%–100% intensity (top to bottom). The integrator design produces an illumination field size at the mask ~140 × ~5 mm.

25%–100% intensity. This particular design is used in an excimer laser step and scan ablation tool to achieve an on-target ablation field size of ~35 × ~1.25 mm; based on a 4× demagnification the laser beam integrator produced an illumination field size at the mask plane of ~140 × ~5 mm.

When looking for the most dominant market segment that has benefited from beam shaping, you simply need to review history. As a rule of thumb, if you want to understand where technology will be 10 years into the future, you need to only look backward 20 years into the past. Within our optical community we tend to think that beam shaping innovations just came out of nowhere, but in reality there were great challenges that spurred development by pioneers in our field. Each chapter within this book is a testament to those pioneers and represents their continued impact on our laser beam shaping community.

REFERENCES

1. F. M. Dickey, S. C. Holswade, and D. L. Shealy, Editors, *Laser Beam Shaping Applications*, CRC Press, New York, 2005.
2. F. M. Dickey, "Laser beam shaping," *Optics & Photonics News*, 14(4), pp. 30–35, 2003.
3. F. M. Dickey and S. C. Holswade, *Laser Beam Shaping*, *Proceedings of SPIE* 4095, August, San Diego, CA, 2000.
4. F. M. Dickey, S. C. Holswade, and D. L. Shealey, *Laser Beam Shaping II*, *Proceedings of SPIE* 4443, August, San Diego, CA, 2001.
5. F. M. Dickey, S. C. Holswade, and D. L. Shealey, *Laser Beam Shaping III*, *Proceedings of SPIE* 4770, August, Seattle, WA, 2002.
6. F. M. Dickey and D. L. Shealey, *Laser Beam Shaping IV*, *Proceedings of SPIE* 5175, August, San Diego, CA, 2003.
7. F. M. Dickey and D. L. Shealey, *Laser Beam Shaping V*, *Proceedings of SPIE* 5525, August, Denver, CO, 2004.
8. F. M. Dickey and D. L. Shealey, *Laser Beam Shaping VI*, *Proceedings of SPIE* 5876, August, San Diego, CA, 2005.
9. F. M. Dickey and D. L. Shealey, *Laser Beam Shaping VII*, *Proceedings of SPIE* 6290, August, San Diego, CA, 2006.
10. F. M. Dickey and D. L. Shealey, *Laser Beam Shaping VIII*, *Proceedings of SPIE* 6663, August, San Diego, CA, 2007.
11. A. Forbes and T. E. Lizotte, *Laser Beam Shaping IX*, *Proceedings of SPIE* 7062, August, San Diego, CA, 2008.
12. A. Forbes and T. E. Lizotte, *Laser Beam Shaping X*, *Proceedings of SPIE* 7430, August, San Diego, CA, 2009.
13. A. Forbes and T. E. Lizotte, *Laser Beam Shaping XI*, *Proceedings of SPIE* 7789, August, San Diego, CA, 2010.
14. A. Forbes and T. E. Lizotte, *Laser Beam Shaping XII*, *Proceedings of SPIE* 8130, August, San Diego, CA, 2011.
15. A. Forbes and T. E. Lizotte, *Laser Beam Shaping XIII*, *Proceedings of SPIE* 8490, August, San Diego, CA, 2012.
16. A. Forbes and T. E. Lizotte, *Laser Beam Shaping XIV*, *Proceedings of SPIE* 8843, August, San Diego, CA, 2013.
17. J. Mihalyi, "Projection system for color pictures," US Patent Number 1,762,932, June 10, 1930.
18. K. Rantsch, L. Bertlele, H. Sauer, and A. Merz, "Illuminating system," US Patent Number 2,186,123, January 9, 1940.
19. T. Lizotte, F. M. Dickey, and D. Brown, "Laser beam homogenization, splitting and three spot image formation: System design, analysis and testing," *Laser Beam Shaping XIV*, *Proceedings of SPIE* 7789-9, 2010.
20. T. Lizotte and O. Ohar, "UV Nd:YAG laser technology," *Lasers & Optronics*, pp. 17–19, September 1994.
21. K. Weible and T. E. Lizotte, "Adaptation of an existing diffractive mono-mode beam shaping design to compensate a wavelength change," *Laser Beam Shaping X*, *Proceedings of SPIE* 7430, August, San Diego, CA, 2009.
22. T. S. Fahlen, S. B. Hutchison, and T. McNulty, "Optical beam integration system," US Patent Number 4,733,944, March 29, 1988.
23. T. Lizotte and O. Ohar, "Production excimer drilling process for producing micron exit holes in polyimide," *Photon Processing in Microelectronics and Photonics V*, *Proceedings of SPIE* 6106, June, San Jose, CA, 2006.

2 Mathematical and Physical Theory of Lossless Beam Shaping

Louis A. Romero and Fred M. Dickey

CONTENTS

2.1 INTRODUCTION

In this chapter, we present the basic mathematics and physics that are required to understand the theory of lossless beam shaping. Figure 2.1 shows the physical situation that we are concerned with. We assume that a parallel beam of coherent light enters an aperture at the plane $z = 0$. At the aperture, the light gets refracted by a combination of a Fourier transform lens with focal length f and a beam shaping lens. We are interested in the irradiance of the beam at the focal plane $z = f$. The separation of the refractive elements at the aperture into a Fourier transform lens and a beam shaping lens is convenient for our analysis, and sometimes convenient in practice, but it should be emphasized that these two lenses could in fact be combined into a single lens.

The beam shaping problem is concerned with how to choose the beam shaping lens so that we can transform a beam with an initial irradiance distribution at the plane $z = 0$ into a beam with a desired irradiance distribution at the focal plane $z = f$. We assume that the beam shaping lens is lossless. This means that it does not absorb

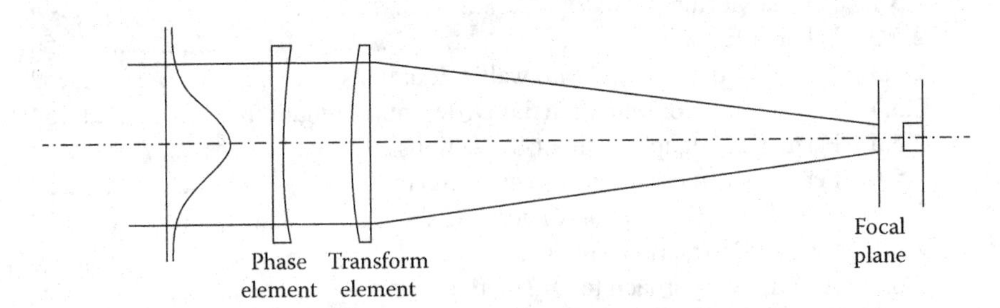

FIGURE 2.1 A schematic of the basic beam shaping system. A parallel beam of light enters the aperture where it encounters a Fourier transform lens, and then the beam shaping element. We choose the beam shaping element so that the output at the focal plane has the desired intensity distribution.

or block out any of the energy of the incoming beam. If we assume that the laws of geometrical optics apply, it is possible to transform any initial distribution into any desired output distribution, provided only that the total energy of the incoming and outgoing beams are the same. When we include the effects of diffraction, it is in general not possible to accomplish our goal exactly.

One of the major themes of this chapter is to determine the scaling properties of beam shaping systems (What happens when we make our system bigger or smaller, or change the wavelength?). In particular, we need to know when the laws of geometrical optics can successfully be applied to designing our system. Due to our emphasis on scaling, we choose to write many of our functions in terms of dimensionless coordinates. For example, if the incoming beam has a radially symmetric Gaussian irradiance distribution, many authors would write the irradiance distribution as

$$I(r) = g(r) \tag{2.1a}$$

where:

$$g(r) = e^{-r^2/R^2} \tag{2.1b}$$

Here, the parameter R determines the basic scale of the irradiance distribution. In this chapter, we would prefer to write this irradiance distribution as

$$I(r) = g(r/R) \tag{2.2a}$$

where:

$$g(\xi) = e^{-\xi^2} \tag{2.2b}$$

It might appear simpler to say that the initial irradiance is given by $g(r)$, rather than saying it is given by $g(r/R)$. However, when we consider the scaling properties, the second form is much more powerful. In particular, if we say that the initial distribution is given by $g(r/R)$, then it will be much clearer how to apply the analysis of a system with distribution $g(r/R_1)$ to a system with distribution $g(r/R_2)$.

This approach is motivated by the practice commonly used in fluid mechanics of writing equations in dimensionless form [1,2]. This approach in fluid mechanics allows us to show that different physical systems will have the same behavior, provided only that certain "dimensionless parameters" are the same. For example, when fluid flows past a sphere, the behavior of the flow depends on the Reynolds number:

$$\mathrm{Re} = \frac{RU_0}{\nu} \tag{2.3}$$

where:

R is the radius of the sphere
U_0 is the velocity far from the sphere
ν is the kinematic viscosity

If two flows have the same Reynolds number, the patterns of fluid flow will be identical, after rescaling our coordinates. However, if the Reynolds numbers are different, the flow patterns will look dramatically different. For example, in one case the flow may be turbulent, and in the other case it may not.

Ideas similar to these can be applied to the theory of beam shaping. Suppose that our initial irradiance distribution is given by $g(x/R,y/R)$, and our desired output irradiance distribution is given by $Q(x/D,y/D)$. The parameter R gives the characteristic length of the incoming beam, and D is the characteristic length of the output beam. If the wavelength of the light is λ, and we are imaging our output at a distance f from the aperture, the dimensionless parameter

$$\beta = \frac{2\pi RD}{\lambda f} \tag{2.4}$$

is very important to understanding beam shaping. In particular, suppose that we design a lens that solves the beam shaping problem in the geometrical optics limit, and now we analyze how this lens works when the wavelength is finite. We will see that the irradiance distributions of two beam shaping systems will be geometrically similar, provided only that they have the same shape functions $g(s,t)$, and $Q(s,t)$, and the parameters β for the two systems are the same. This means that we can transform the irradiance distribution of one system into the irradiance distribution of the other system by merely rescaling our axes. In particular, one system will suffer from diffraction effects if and only if the other system (with identical β) also does.

Geometrical optics is a short wavelength approximation, so it is clear that we would like β to be large in order for geometrical optics to hold. We will see that if β is large it is relatively simple to do beam shaping, but if it is small, the uncertainty principle of signal analysis shows that it is essentially impossible.

Another important feature in determining the difficulty of a beam shaping problem is the continuity of the beam shaping lens. If the surface of the element designed using geometrical optics is infinitely differentiable, we will not need a very high value of β to achieve good results. To be more precise, the effects of diffraction will die down like $1/\beta^2$ as β gets to be large. However, if the lens has a discontinuity in its third derivative, the effects of diffraction will die down like $1/\sqrt{\beta}$ in parts of the image plane, and hence we will need a much larger value of β in order to approach the geometrical optics limit. If the lens has discontinuities in the first or second derivatives, we will need to use even larger values of β before we can ignore the effects of diffraction.

If the input beam is smooth (such as Gaussian), the continuity properties of the lens designed using geometrical optics are controlled by the continuity of the desired output beam. If one has a good understanding of geometrical beam shaping, it is not too difficult to see how the continuity of the desired output beam will affect the continuity of the lens. However, if one is not familiar with this theory, the results can be somewhat surprising. For example, Figure 2.2 shows examples of three desired output beams. One might naively think that all of these beams have abrupt discontinuities in them, so they may all lead to equally difficult beam shaping problems. It turns out, however, that the output in Figure 2.2a will lead to an infinitely differentiable lens, the beam in Figure 2.2b leads to a lens with a discontinuity in the second

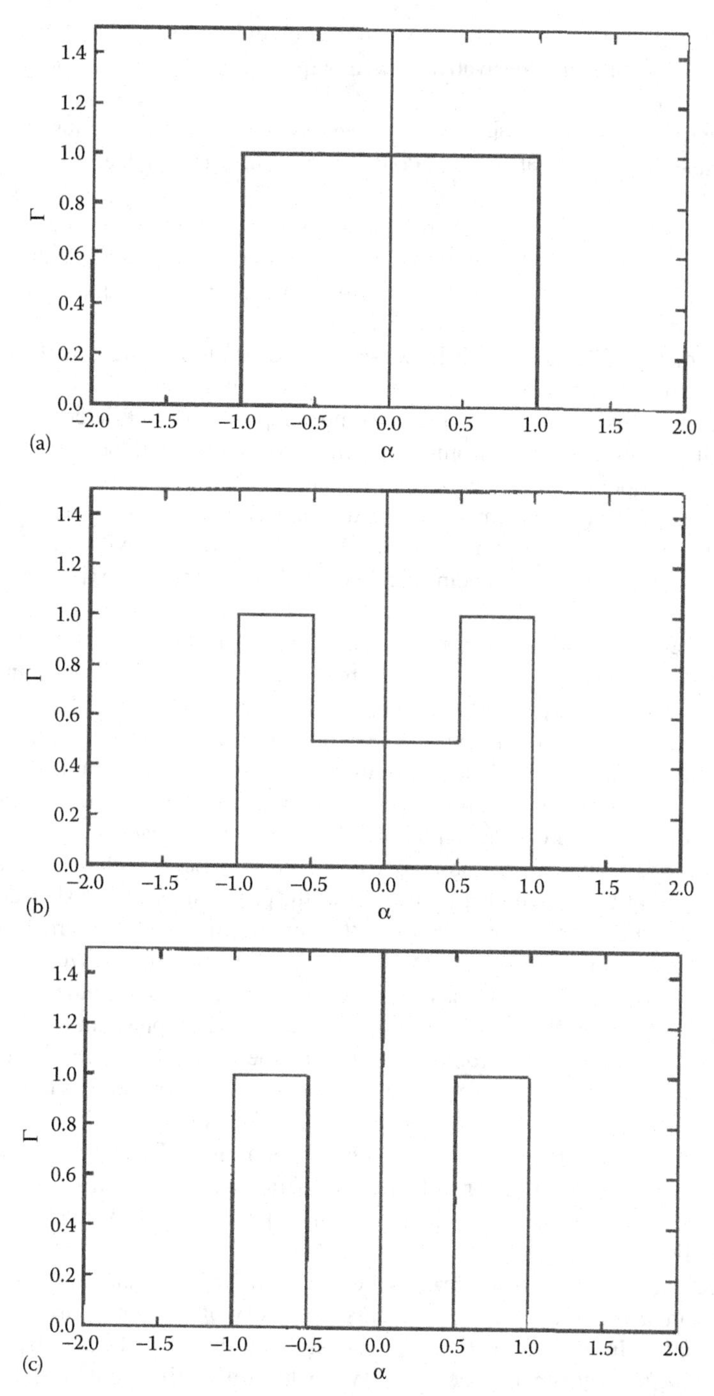

FIGURE 2.2 (a–c) Examples of three desired output distributions. The outputs get progressively harder to achieve when diffraction effects are taken into account.

derivative, and the lens required to produce the output beam in Figure 2.2c will have a discontinuity in the first derivative. These outputs (Figure 2a–c) get progressively harder to achieve.

This entire chapter is devoted to understanding the points we have just discussed. We feel that it is worth writing them down as succinctly as possible.

- In the geometrical optics approximation, it is possible to turn a beam with a given initial distribution into a beam with any desired output distribution, provided only that the total energy of the input and output beams are the same.
- Diffraction effects make it impossible to do beam shaping exactly when we take into account the finite wavelength of light. For given shapes of the input and output beams, the parameter $\beta = 2\pi RD/\lambda f$ determines the difficulty of the beam shaping problem. If β is large, then the laws of geometrical optics will be a good approximation.
- If the surface of the element designed using geometrical optics has discontinuities in its first, second, or third derivatives, then we will need higher values of β in order for geometrical optics to be a good approximation.

In Section 2.2, we discuss some mathematical prerequisites for understanding the theory of beam shaping. After a brief summary of the basics of Fourier transforms, we prove the uncertainty theorem from signal analysis. In Section 2.7, this theorem will be used to show why it is impossible to do a good job of beam shaping when β is small. Section 2.2 also includes a discussion of how to use the Hankel transform in order to obtain radially symmetric Fourier transforms. This is important when analyzing the effects of diffraction on radially symmetric problems.

In Section 2.3, we outline the theory of stationary phase, with an emphasis on how discontinuities in the higher derivatives of the phase function can slow down the convergence. In Section 2.8, we use the method of stationary phase in order to obtain the large β approximation to the diffractive theory of beam shaping. We will see that the first term in the stationary phase approximation is equivalent to the geometrical optics approximation. We also use the method of stationary phase in Section 2.5 in order to analyze the errors introduced by making the Fresnel approximation.

Sections 2.4 through 2.6 discuss the electromagnetic theory necessary to understand beam shaping. Section 2.4 presents a review of Maxwell's equations, Section 2.5 discusses the geometrical optics limit with an emphasis on Fermat's principle, and Section 2.6 discusses the theory of Fresnel diffraction. Fresnel diffraction theory allows us to turn the physical problem of beam shaping into a mathematical problem involving Fourier transforms.

In Sections 2.7 and 2.8, we bring all of our tools together and discuss the theory of beam shaping. Section 2.7 presents the theory of beam shaping in the geometrical optics limit, and Section 2.8 discusses the theory of beam shaping with diffraction effects taken into account. When the diffractive equations for beam shaping are written in dimensionless form, the importance of the parameter β will become evident. We will use the method of stationary phase to analyze the large β limit of the equations. The fact that our geometrical optics solution is based on a stationarity condition (Fermat's principle) and our large β approximation is also based on a stationarity condition (stationary phase), causes these two analyses to look almost identical. We end Section 2.8 by giving some examples that illustrate the principles concerning the importance of β and the smoothness of the shape of the lens.

2.2 MATHEMATICAL PRELIMINARIES

2.2.1 Basic Fourier Analysis

The theory of Fresnel diffraction will allow us to write our beam shaping problem as a problem in Fourier analysis. For this reason, it is impossible to understand our theoretical treatment of beam shaping if one is not familiar with some of the basic concepts from Fourier analysis. We will use both one- and two-dimensional Fourier analyses throughout the text.

There are several definitions of the Fourier transform used in the literature. The differences are very minor, concerning only the sign of the complex exponential and the constant in front of the integral. However, these differences can be annoying when one is using a table of Fourier transforms or applying theorems such as Parseval's equality. The definition used here is probably the most commonly used [3,4].

Definition 1. *The* Fourier transform *of a function f(x) is defined as*

$$F(\omega)=Tf(x)=\int_{-\infty}^{\infty}f(x)\mathrm{e}^{-\mathrm{i}\omega x}\mathrm{d}x \tag{2.5}$$

An almost identical definition holds for two-dimensional functions.

Definition 2. *The* Fourier transform *of a function f(x,y) is defined as*

$$F(\omega_x,\omega_y)=Tf(x,y)=\int_{-\infty}^{\infty}\int_{-\infty}^{\infty}f(x,y)\mathrm{e}^{-\mathrm{i}(\omega_x x+\omega_y y)}\mathrm{d}x\mathrm{d}y \tag{2.6}$$

The following are some well-known theorems in Fourier analysis that will be used throughout this chapter.

Theorem 1. *One-dimensional Fourier inversion theorem—If F(ω) is the Fourier transform of f(x), then*

$$f(x)=T^{-1}\left[F(\omega)\right]=\frac{1}{2\pi}\int_{-\infty}^{\infty}F(\omega)\mathrm{e}^{\mathrm{i}\omega x}\mathrm{d}\omega \tag{2.7}$$

Theorem 2. *Two-dimensional Fourier inversion theorem—If $F(\omega_x,\omega_y)$ is the Fourier transform of $f(x,y)$, then*

$$f(x,y) = T^{-1}\left[F(\omega_x,\omega_y)\right] = \frac{1}{4\pi^2}\int_{-\infty}^{\infty}\int_{-\infty}^{\infty} F(\omega_x,\omega_y)e^{i(\omega_x x+\omega_y y)}d\omega_x d\omega_y \tag{2.8}$$

Theorem 3. *One-dimensional Parseval's equality—A function $f(x)$ and its Fourier transform $F(\omega)$ satisfy*

$$\int_{-\infty}^{\infty} |F(\omega)|^2 d\omega = 2\pi \int_{-\infty}^{\infty} |f(x)|^2\, dx \tag{2.9}$$

Theorem 4. *Two-dimensional Parseval's equality—A function $f(x,y)$ and its Fourier transform $F(\omega_x,\omega_y)$ satisfy*

$$\int_{-\infty}^{\infty}\int_{-\infty}^{\infty} \left|F(\omega_x,\omega_y)\right|^2 d\omega_x d\omega_y = 4\pi^2 \int_{-\infty}^{\infty}\int_{-\infty}^{\infty} |f(x,y)|^2\, dxdy \tag{2.10}$$

Theorem 5. *One-dimensional Fourier convolution theorem—Suppose $F(\omega)$ and $G(\omega)$ are the Fourier transforms of the functions $f(x)$ and $g(x)$, respectively. The inverse Fourier transform of $F(\omega)G(\omega)$ is given by*

$$T^{-1}\left[F(\omega)G(\omega)\right] = \int_{-\infty}^{\infty} f(\xi)g(x-\xi)\, d\xi \tag{2.11}$$

Theorem 6. *Two-dimensional Fourier convolution theorem—Suppose $F(\omega_x,\omega_y)$ and $G(\omega_x,\omega_y)$ are the Fourier transforms of the functions $f(x,y)$ and $g(x,y)$. The inverse Fourier transform of $F(\omega_x,\omega_y)G(\omega_x,\omega_y)$ is given by*

$$T^{-1}\left[F(\omega_x,\omega_y)G(\omega_x,\omega_y)\right] = \int_{-\infty}^{\infty} f(\xi,\eta)g(x-\xi,y-\eta)\, d\xi d\eta \tag{2.12}$$

Theorem 7. *Transforms of derivatives—The Fourier transform of the derivative is given by*

$$T\left(\frac{df}{dx}\right) = i\omega F(\omega) \tag{2.13}$$

Theorem 8. *Transforms of partial derivatives—The Fourier transforms of the partial derivatives are given by*

$$T\left(\frac{\partial f}{\partial x}\right) = i\omega_x F(\omega_x, \omega_y) \tag{2.14}$$

and

$$T\left(\frac{\partial f}{\partial y}\right) = i\omega_y F(\omega_x, \omega_y) \tag{2.15}$$

Although the Cauchy–Schwartz inequality is not really a theorem in Fourier analysis, we will need it in our proof of the uncertainty principle, and hence now state it.

Theorem 9. *The Cauchy–Schwartz inequality (for infinite integrals)—For any function f(x) and g(x), we must have*

$$\left|\int_{-\infty}^{\infty} f(x)\bar{g}(x)\mathrm{d}x\right|^2 \le \int_{-\infty}^{\infty} |f(x)|^2\,\mathrm{d}x \int_{-\infty}^{\infty} |g(x)|^2\,\mathrm{d}x \tag{2.16}$$

The two sides are equal if and only if there is a constant λ such that $f(x) = \lambda g(x)$.

2.2.2 The Uncertainty Principle and the Space Bandwidth Product

In this section, we discuss the space bandwidth product and the uncertainty principle of signal analysis [5]. This discussion is crucial to understanding the theory of beam shaping. As we shall see in later sections, in a beam shaping system, the space bandwidth product is related to the parameter β discussed in Section 2.1. In Section 2.8, we will use the uncertainty principle to show that it is impossible to do a good job of beam shaping if β is small.

The Heisenberg uncertainty principle of quantum mechanics [6] states that the product of the uncertainty in position and the uncertainty in momentum must be greater than $h/2\pi$:

$$\Delta p\,\Delta x > \frac{h}{2\pi} \tag{2.17}$$

To make this precise, we must define precisely what we mean by Δx and Δp. This principle was one of Heisenberg's basic assumptions in his development of matrix mechanics. However, it can also be derived by assuming the wave mechanics of Schrödinger. The derivation of the result depends on the fact that the wave function for momentum is the Fourier transform of the wave function for position, and on the subject of this section, the uncertainty principle from Fourier analysis.

All of our derivations will be limited to one-dimensional functions and their transforms, but almost identical derivations apply for two-dimensional transforms.

Once we have derived the one-dimensional results, we will state the two-dimensional results without proof. We now define the uncertainty in $f(x)$ and $F(\omega)$.

Definition 3. *The* uncertainty *in $f(x)$ and its* transform $F(\omega)$ *are given by*

$$\Delta_f = \sqrt{\frac{\int_{-\infty}^{\infty} x^2 |f(x)|^2 \mathrm{d}x}{\int_{-\infty}^{\infty} |f(x)|^2 \mathrm{d}x}} \tag{2.18}$$

and

$$\Delta_F = \sqrt{\frac{\int_{-\infty}^{\infty} \omega^2 |F(\omega)|^2 \mathrm{d}\omega}{\int_{-\infty}^{\infty} |F(\omega)|^2 \mathrm{d}\omega}} \tag{2.19}$$

The uncertainty principle concerns the product of these two quantities and is simply related to the space bandwidth product.

Definition 4. *The* space bandwidth product *of a function $f(x)$ is defined as*

$$\text{Space bandwith product} = \Delta_f \Delta_F \tag{2.20}$$

It should be noted that the space bandwidth product of a function does not depend on the scaling of the function.

Lemma 1. *For any nonzero constant a and nonzero real number b, the* space *bandwidth product of $af(bx)$ is the same as the space bandwidth product of $f(x)$.*

We are now ready to state the uncertainty principle of signal analysis.

Theorem 10. *One-dimensional uncertainty principle—For any square integrable function $f(x)$, the space bandwidth product must be greater than 1/2. In other words,*

$$\Delta_f \Delta_F \geq \frac{1}{2} \tag{2.21}$$

Proof. The Cauchy–Schwartz inequality implies that

$$\left| \int_{-\infty}^{\infty} (xf) \frac{\mathrm{d}f}{\mathrm{d}x} \mathrm{d}x \right|^2 \leq \int_{-\infty}^{\infty} x^2 |f(x)|^2 \mathrm{d}x \int_{-\infty}^{\infty} \left| \frac{\mathrm{d}f}{\mathrm{d}x} \right|^2 \mathrm{d}x \tag{2.22}$$

Clearly,

$$\left|\int_{-\infty}^{\infty}(x\bar{f})\frac{\mathrm{d}f}{\mathrm{d}x}\,\mathrm{d}x\right|^2 \geq \left|\mathrm{Re}\int_{-\infty}^{\infty}(x\bar{f})\frac{\mathrm{d}f}{\mathrm{d}x}\,\mathrm{d}x\right|^2 \tag{2.23}$$

We can write

$$\mathrm{Re}\int_{-\infty}^{\infty} x\bar{f}\frac{\mathrm{d}f}{\mathrm{d}x}\,\mathrm{d}x = \frac{1}{2}\int_{-\infty}^{\infty} x\left(f\frac{\mathrm{d}\bar{f}}{\mathrm{d}x}+\bar{f}\frac{\mathrm{d}f}{\mathrm{d}x}\right)\mathrm{d}x = -\frac{1}{2}\int_{-\infty}^{\infty}|f(x)|^2\,\mathrm{d}x \tag{2.24}$$

The inequalities (2.22 and 2.23) now imply

$$\frac{1}{4}\left|\int_{-\infty}^{\infty}|f(x)|^2\,\mathrm{d}x\right|^2 \leq \int_{-\infty}^{\infty}x^2|f(x)|^2\,\mathrm{d}x\int_{-\infty}^{\infty}\left|\frac{\mathrm{d}f}{\mathrm{d}x}\right|^2\mathrm{d}x \tag{2.25}$$

Since the Fourier transform of df/dx is i$\omega F(\omega)$, Parseval's equality implies that

$$\int_{-\infty}^{\infty}\left|\frac{\mathrm{d}f}{\mathrm{d}x}\right|^2\mathrm{d}x = \frac{1}{2\pi}\int_{-\infty}^{\infty}\omega^2|F(\omega)|^2\,\mathrm{d}\omega \tag{2.26}$$

The inequality (2.25) can now be written as

$$\frac{1}{4}\left|\int_{-\infty}^{\infty}|f(x)|^2\,\mathrm{d}x\right|^2 \leq \frac{1}{2\pi}\int_{-\infty}^{\infty}|x^2|f(x)|^2\,\mathrm{d}x\int_{-\infty}^{\infty}\omega^2|F(\omega)|^2\,\mathrm{d}\omega \tag{2.27}$$

Using Parseval's equality, we can write this as

$$\begin{aligned}&\frac{1}{4}\frac{1}{2\pi}\int_{-\infty}^{\infty}|F(\omega)|^2\,\mathrm{d}\omega\int_{-\infty}^{\infty}|f(x)|^2\,\mathrm{d}x\\ &\leq\frac{1}{2\pi}\int_{-\infty}^{\infty}x^2|f(x)|^2\,\mathrm{d}x\int_{-\infty}^{\infty}\omega^2|F(\omega)|^2\,\mathrm{d}\omega\end{aligned} \tag{2.28}$$

If we now divide both sides of this inequality by the left-hand side, we arrive at the desired result. QED

Lemma 2. *We have* $\Delta_f\Delta_F = 1/2$ *if and only if the function* $f(x)$ *is a real Gaussian,* $f(x) = Ae^{-\alpha x^2}$ *where* α *is a real number.*

Proof. In order to get an equality in the uncertainty relation, we must have an equality in the Cauchy–Schwartz inequality in Equation 2.22. This implies that df/d$x = -2x\lambda f$, and hence $f(x) = Ae^{-\lambda x^2}$. It is also necessary that we get an equality in Equation 2.23. This will be the case if and only if $\bar{f}(\mathrm{d}f/\mathrm{d}x)$ is real, which will be true if and only if λ is real. QED

Although the space bandwidth product can never be less than 1/2, there is no limitation to how big it can be. For example, the function $f(x) = e^{ix^2}$ has an infinite space bandwidth product.

Suppose we change the phase of the function $f(x)$ by multiplying it by the phase function $e^{iq(x)}$. How should we choose the phase q so that the function $f(x)e^{iq(x)}$ has a minimum space bandwidth product? Note that the phase function does not change the uncertainty in x, but it does change the uncertainty in ω. This question has implications for the depth of the field of a laser beam shaping system. The following theorem gives a very simple answer to this question.

Theorem 11. *The function $q(x)$ that minimizes the space bandwidth product of $f(x)e^{iq(x)}$ is the one that makes the phase of $f(x)e^{iq(x)}$ constant.*

Proof. The only integral in the space bandwidth product that changes with the function $q(x)$ is the integral

$$\int_{-\infty}^{\infty} \omega^2 \mid G(\omega) \mid^2 \mathrm{d}\omega \tag{2.29}$$

where:

$G(\omega)$ is the Fourier transform of $f(x)e^{iq(x)}$

Let $f(x)e^{iq(x)} = A(x)e^{i\psi(x)}$, where $A(x)$ is a positive real function. Parseval's equality and the formula for the Fourier transform of a derivative show that

$$\int_{-\infty}^{\infty} \omega^2 \mid G(\omega) \mid^2 \mathrm{d}\omega = 2\pi \int_{-\infty}^{\infty} \left| \frac{\mathrm{d}}{\mathrm{d}x}\left[A(x)e^{i\psi(x)}\right] \right|^2 \mathrm{d}x \tag{2.30}$$

The last integral can be written as

$$2\pi \int_{-\infty}^{\infty} \left[\left(\frac{\mathrm{d}A}{\mathrm{d}x}\right)^2 + A^2(x)\left(\frac{\mathrm{d}\psi}{\mathrm{d}x}\right)^2 \right] \mathrm{d}x \geq 2\pi \int_{-\infty}^{\infty} \left(\frac{\mathrm{d}A}{\mathrm{d}x}\right)^2 \mathrm{d}x \tag{2.31}$$

This clearly implies that this integral, and hence the space bandwidth product, is minimized by choosing the function ψ so that it is constant. QED

This theorem will be used in Section 3.3 when discussing the collimation of beams.

We now summarize how these results apply for two-dimensional functions. In two dimensions, the uncertainty will be defined as

$$(\Delta_f)^2 = \frac{\int_{-\infty}^{\infty}\int_{-\infty}^{\infty} (x^2 + y^2) \mid f(x,y) \mid^2 \mathrm{d}x\mathrm{d}y}{\int_{-\infty}^{\infty}\int_{-\infty}^{\infty} \mid f(x,y) \mid^2 \mathrm{d}x\mathrm{d}y} \tag{2.32}$$

$$(\Delta_F)^2 = \frac{\int_{-\infty}^{\infty}\int_{-\infty}^{\infty}(\omega_x^2+\omega_y^2)\,|\,F(\omega_x,\omega_y)\,|^2\, d\omega_x d\omega_y}{\int_{-\infty}^{\infty}\int_{-\infty}^{\infty}|\,F(\omega_x,\omega_y)\,|^2\, d\omega_x d\omega_y} \tag{2.33}$$

The space bandwidth product is once again defined as $\Delta_f\Delta_F$. The two-dimensional uncertainty principle gives the following theorem:

Theorem 12. *Two-dimensional uncertainty principle—For any square integrable function f(x,y), the space bandwidth product must be greater than* 1.
In other words,

$$\Delta_f\Delta_F \geq 1 \tag{2.34}$$

2.2.3 Separation of Variables in Cylindrical Coordinates

When you take the Fourier transform of a function $f(x,y)$ that has radial symmetry, you end up with a Fourier transform $F(\omega_x,\omega_y)$ that has radial symmetry in the Fourier domain. That is, if we can write

$$f(x,y) = g(r) \tag{2.35}$$

where:

$$r = \sqrt{x^2+y^2}$$

Then we can write

$$F(\omega_x,\omega_y) = G(\alpha) \tag{2.36}$$

where:

$$\alpha = \sqrt{\omega_x^2+\omega_y^2}$$

The transformation that takes the function $g(r)$ into the function $G(\alpha)$ is known as a Hankel transform [4]. This transform allows us to find the two-dimensional Fourier transform of a radially symmetric function by performing a one-dimensional integral. The Hankel transform can be very useful when analyzing diffraction effects in beam shaping problems with radial symmetry.

To understand Hankel transforms, it is necessary to be familiar with an identity in the theory of Bessel functions. To understand this identity, we begin by considering the reduced wave equation in polar coordinates:

$$\nabla^2 p + k^2 p = \frac{1}{r}\frac{\partial}{\partial r}\left(r\frac{\partial p}{\partial r}\right) + \frac{1}{r^2}\frac{\partial^2 p}{\partial \phi^2} + k^2 p = 0 \tag{2.37}$$

If we assume solutions of the form:

$$p(r,\phi) = f(r)e^{im\phi} \tag{2.38}$$

we find that the function $f(r)$ must satisfy

$$\frac{1}{r}\frac{\mathrm{d}}{\mathrm{d}r}\left(r\frac{\mathrm{d}f}{\mathrm{d}r}\right)-\frac{m^2 f}{r^2}+k^2 f=0 \tag{2.39}$$

If $g(r)$ is a solution to

$$\frac{1}{r}\frac{\mathrm{d}}{\mathrm{d}r}\left(r\frac{\mathrm{d}g}{\mathrm{d}r}\right)-\frac{m^2 g}{r^2}+g=0 \tag{2.40}$$

then $f(r) = g(kr)$ is a solution to Equation 2.39.

Equation 2.40 is known as Bessel's equation. The solutions that are regular at $r = 0$ are called Bessel functions, which can be written as $J_m(r)$. If we were interested in the waves emitted from a circular cylinder, we would not require that the solution was finite at $r = 0$, but that as $r \to \infty$ the solution represented only outgoing waves. In this case, we would use the solution to Bessel's equation, $H_m^1(kr)$. This is known as the Hankel function of the first kind. Our goal is to understand the Hankel transform as a circularly symmetric Fourier transform. For this purpose we only need the regular solutions to Bessel's equation, which means we only need to consider the function $J_n(kr)$, where n is an integer.

One of the most elegant ways of approaching the theory of Bessel functions [7] is through the use of an integral identity, which we will now derive. This identity allows us to derive almost all of the most commonly known properties of Bessel functions such as their asymptotic behavior for large indices, asymptotic behavior for large argument, recursion formulas, and the behavior near the origin. This identity is almost the only property of Bessel functions that will be needed to understand the Hankel transform.

The integral identity can be derived by considering the function

$$F(x,y)=\mathrm{e}^{\mathrm{i}x} \tag{2.41}$$

This clearly satisfies the two-dimensional reduced wave equation

$$\nabla^2 F + F = 0 \tag{2.42}$$

We can express this in terms of polar coordinates, and then expand the function in a Fourier series. If we do this, we find that

$$\mathrm{e}^{ir\cos(\theta)}=\sum_{k=-\infty}^{\infty} a_k(r)\mathrm{e}^{ik\theta} \tag{2.43}$$

From our discussion at the beginning of the section, we know that this last infinite sum will satisfy the reduced wave equation if the functions $a_k(r)$ satisfy Bessel's equation. Due to the rotational symmetry of the reduced wave equation, it can be shown that in order for this infinite sum to satisfy the reduced wave equation it is necessary that each individual term satisfy the reduced wave equation. This means that it is necessary (not just sufficient) that the functions $a_k(r)$ satisfy Bessel's equation. It is also clear that they must be bounded at $r = 0$. It follows that they are multiples of the Bessel functions $J_k(r)$. We will in fact define the Bessel functions so that the multiplicative factor is unity. This gives us the result

$$e^{ir\cos(\theta)} = \sum_{k=-\infty}^{\infty} J_k(r)e^{ik\theta} \tag{2.44}$$

Using the fact that the right-hand side is the Fourier expansion of the function $e^{ir\cos(\theta)}$, we arrive at the identity

$$J_k(r) = \frac{1}{2\pi}\int_{-\pi}^{\pi} e^{i[r\cos(\theta)-k\theta]}d\theta \tag{2.45}$$

2.2.4 Hankel Transforms

The Fourier transform of $f(x,y)$ can be written as

$$F(\omega_x,\omega_y) = \int_{-\infty}^{\infty}\int_{-\infty}^{\infty} e^{-i(\omega_x x+\omega_y y)} f(x,y)dxdy \tag{2.46}$$

Suppose we write both the original function $f(x,y)$ and the Fourier transform in terms of polar coordinates:

$$(x,y) = r[\cos(\theta),\sin(\theta)] \tag{2.47}$$

$$(\omega_x,\omega_y) = \alpha[\cos(\phi),\sin(\phi)] \tag{2.48}$$

The Fourier transform can be written as

$$F(\alpha,\phi) = \int_{0}^{\infty}\int_{-\pi}^{\pi} e^{-ir\alpha\cos(\theta-\phi)} f(r,\theta)r\,drd\theta \tag{2.49}$$

If the function $f(x,y)$ is independent of θ, then the transform $F(\alpha,\phi)$ will be independent of ϕ. It follows that we can write

$$F(\alpha) = \int_{0}^{\infty}\int_{-\pi}^{\pi} e^{-ir\alpha\cos(\theta)} f(r)r\,drd\theta \tag{2.50}$$

If we perform the integral with respect to θ first, and use the integral representation of J_0, we get

$$F(\alpha) = 2\pi\int_{0}^{\infty} J_0(r\alpha)f(r)r\,dr \tag{2.51}$$

The function $F(\alpha)$ is known as the Hankel transform of the function $f(r)$. We can apply the same steps to show that the inverse Hankel transform is given by

$$f(r) = \int_0^{\infty} J_0(kr)F(k)k \, dk \tag{2.52}$$

2.3 METHOD OF STATIONARY PHASE

2.3.1 Basic Idea of Stationary Phase

The method of stationary phase [8] is an asymptotic method, first used by Stokes and Kelvin, for evaluating integrals whose integrands have a very rapidly varying phase. The method is very important in the theory of dispersive wave propagation where it motivates the concept of group velocity [9,10]. In the theory of beam shaping, it can be used to derive the geometrical optics limit from the theory of Fresnel diffraction, and more importantly, it gives us bounds on when the geometrical theory is applicable.

We will now give a brief heuristic derivation of the lowest order term in the approximation. Suppose we have an integral of the form:

$$H(\gamma) = \int_{-\infty}^{\infty} e^{i\gamma q(\xi)} f(\xi) \, d\xi \tag{2.53}$$

and we are interested in evaluating this integral for large values of γ. Intuitively, we expect that intervals where the function $\gamma q(\xi)$ is changing rapidly will give negligible contributions to this integral. If the derivative of q vanishes at $\xi = \xi_0$, we expect the main contribution to come from the region very near ξ_0. To a first approximation we can write

$$H(\gamma) \approx f(\xi_0) e^{i\gamma q(\xi_0)} \int_{-\infty}^{\infty} e^{i\gamma q''(\xi_0)(\xi-\xi_0)^2/2} d\xi \tag{2.54}$$

We have arrived at this expression by assuming that the major contribution comes from a small region around ξ_0, and hence, we have approximated the function $f(\xi)$ as being constant and equal to $f(\xi_0)$. We have also expanded the function $q(\xi)$ in a Taylor series about ξ_0, keeping only the terms up to the quadratic. The integral can now be evaluated analytically to give

$$H(\gamma) \approx f(\xi_0) e^{i\mu\pi/4} e^{i\gamma q(\xi_0)} \sqrt{\frac{2\pi}{\gamma |q''(\xi_0)|}} \tag{2.55}$$

where:

$$\mu = \text{sgn}\left(\left. \frac{d^2 q(\xi)}{d\xi^2} \right|_{\xi_0} \right) \tag{2.56}$$

Here, we have assumed that there is exactly one point where the phase is stationary. If there is more than one point, then we must sum over all points that are stationary in order to get our asymptotic expansion. If there are no stationary points, then the integral will die down exponentially fast with γ provided the functions $q(\xi)$ and $f(\xi)$ are infinitely differentiable, and the function $f(\xi)$ and all of its derivatives decay as $|\xi| \to \infty$. If there is no stationary point, but the function $f(\xi)$ has a discontinuity in it, we are at least guaranteed that the integral dies down like $1/\gamma$ as $\gamma \to \infty$. It is not too difficult to make this heuristic derivation more rigorous.

2.3.2 Rate of Convergence of the Method of Stationary Phase

In our discussion of beam shaping, we will see that the lowest order term in the stationary phase approximation to the diffraction integral gives us the geometrical optics approximation. In this case, the parameter β discussed in Section 2.1 will serve as our large parameter in the phase of the integrand. To understand what sorts of errors are produced when we use the geometrical optics approximation, we need to know the higher order terms in the method of stationary phase. It is not important for us to have exact expressions for the higher order terms, but we need to know how fast they die down with γ.

The subject of how to correct the lowest order term in the method of stationary phase gets somewhat technical, so we feel that it is best if we begin by summarizing the main results. In our analysis of beam shaping, we will have another parameter in our phase function, so our integrals will be of the form:

$$H(x,\gamma) = \int_{-\infty}^{\infty} e^{i\gamma q(\xi,x)} f(\xi) d\xi \tag{2.57}$$

where:

ξ represents a point on the aperture
x represents a point at the focal plane

The function $q(\xi,x)$ will be proportional to the travel time required to get from a point ξ on the aperture to a point x in physical space. In practice, ξ and x will be two-dimensional vectors, but we assume they are scalars here in order to simplify the presentation. This one-dimensional case will be directly relevant for the case where our input and output beams can be written as a direct product of two one-dimensional distributions.

Let $\xi_0(x)$ be the point at which the phase is stationary. We will show that if the functions $q(\xi,x)$ and $f(\xi)$ are infinitely differentiable at the stationary point $\xi_0(x)$, and

$$\left.\frac{\partial^2 q(\xi,x)}{\partial \xi^2}\right|_{\xi_0} \neq 0 \tag{2.58}$$

then the next order correction dies down like $1/\gamma^{3/2}$. This gives us an expression of the form:

$$H(x,\gamma) = \frac{A(x)}{\gamma^{1/2}} + \frac{iB(x)}{\gamma^{3/2}} + \cdots \tag{2.59}$$

In this case, the relative error between the first-order term and the exact solution will die down like $1/\gamma$. If the function $f(\xi)$ is real, then the functions $A(x)$ and $B(x)$ will have the same phase. This implies that the relative error between $|H(x,\gamma)|^2$ and the value predicted by the first term in the method of stationary phase will be $O(1/\gamma^2)$.

This expression will remain valid provided the functions $f(\xi)$ and $\partial^2 q/\partial\xi^2$ are differentiable at $\xi_0(x)$. If these functions are continuous, but not differentiable at some point x^*, then at the point x^*, the next order term in the method will be of the form:

$$H(x^*,\gamma) = \frac{A(x^*)}{\gamma^{1/2}} + \frac{B(x^*)}{\gamma} + \cdots \tag{2.60}$$

We see that in this case, the relative error between the first-order term and the exact solution will die down like $1/\gamma^{1/2}$. Furthermore, in this case, the functions $A(x)$ and $B(x)$ are not phase shifted by 90° when $f(\xi)$ is real. This means that the relative error between $|H(x,\gamma)|^2$ and the term predicted by the first term in the method of stationary phase will be $O(1/\sqrt{\gamma})$. This means that we will need a much larger value of γ before the first-order term is a good approximation. In terms of our beam shaping problem, this will imply that if the surface of the beam shaping lens designed using geometrical optics has a discontinuity in the third derivative, then we will require much larger values of β in order for the results of geometrical optics to be a good approximation.

Suppose that at some point x_0, the function $f(\xi)$ or $\partial^2 q/\partial\xi^2$ is discontinuous at $\xi_0(x_0)$. Since the first-order term in the method of stationary phase requires us to know $f(\xi)$ and $\partial^2 q(\xi,x)/\partial\xi^2$ at $\xi_0(x)$, it is clear that we need to modify the results of the lowest order term in our stationary phase approximation when $x = x_0$. More importantly, the method of stationary phase will hold for values of x near x_0, but the convergence near these points will be dramatically affected. The analysis of this situation is based on the Fresnel integral [11], and we see that this situation is related to the diffraction by a semi-infinite half plane. As with that case, we end up getting oscillations near the point x_0. For the beam shaping problem, this implies that if the surface of the lens designed using geometrical optics has a discontinuity in the second derivative, then we will get even worse convergence, and this will be accompanied by oscillations in the amplitude. When the surface of the lens has a discontinuity in the first derivative, the convergence toward the geometrical optics limit is affected even more dramatically.

Clearly, if the discontinuities are small enough, they will have little effect on the convergence toward the geometrical optics limit. For example, the elements are often manufactured by approximating the element by a piecewise constant element. This should not be any problem if the steps are small enough.

So far, we have assumed that the second derivative of $q(\xi,x)$ does not vanish at the stationary point $\xi_0(x)$. In optics, points where this condition is violated are said to lie on a caustic surface. Suppose we have a point source of light whose rays get refracted by an inhomogeneous medium. It is possible that at certain points in the medium we might have more than one ray arriving from this point source, or possibly none at all. The surfaces separating regions where there are different numbers of rays are known as the *caustic surfaces*. When we analyze the diffraction integral using the method of stationary phase, we find that on the caustic surface the second derivative of $q(\xi,x)$ vanishes.

We are not so much interested in computing the integral for $H(x,\gamma)$ right at a point where the stationary point ξ_0 has a vanishing second derivative. Instead, we are interested in analyzing the integral $H(x,\gamma)$, for x near x_0, where

$$\left.\frac{\partial^2 q(\xi,x_0)}{\partial\xi^2}\right|_{\xi=\xi_0(x_0)} = 0 \tag{2.61}$$

In optics, the point x_0 would be a point on the caustic surface. We would find that for points on one side of x_0, there are no stationary points, and on the other side there are two stationary points. In order to understand the behavior of $H(x,\gamma)$ near such points we need to include cubic terms in the Taylor series expansion of the phase near the stationary point, and this analysis is based on the Airy integral.

We will not give any further discussion of the Airy integral or caustics since when discussing beam shaping we do not present any examples where caustics occur in the classical sense of the word. All of the problems we analyze lead to lenses whose phase functions do not have inflection points. However, in some of the lenses, the phase function grows linearly as we move far away from the center of aperture. This results in a situation where the caustic occurs at a value of $\xi_0 = \infty$.

2.3.3 A Preliminary Transformation

In our analysis of the higher order terms in the method of stationary phase, we will begin by analyzing the situation where $q(\xi) = \xi^2$. This leads us to consider integrals of the form:

$$P(\gamma) = \int_{-\infty}^{\infty} f(\xi)e^{i\gamma\xi^2}\,d\xi \tag{2.62}$$

By making a preliminary transformation, we can transform the analysis of the integral in Equation 2.53

$$H(\gamma) = \int_{-\infty}^{\infty} e^{i\gamma q(\xi)} f(\xi)\,d\xi \tag{2.63}$$

into the analysis of this simpler problem. To do this we assume that $q(\xi)$ has a single stationary point at ξ_0. In this case, we can introduce a new variable s such that

$$s^2 = \mu[q(\xi) - q(\xi_0)] \tag{2.64}$$

$$\mu = \text{sgn}\left[\frac{d^2 q(\xi_0)}{d\xi}\right] \tag{2.65}$$

in the neighborhood of ξ_0. By making the change of variables $s(\xi) = \sqrt{\mu[q(\xi) - q(\xi_0)]}$, we end up with an integral of the form:

$$H(\gamma) = \mathrm{e}^{\mathrm{i}\gamma q(\xi_0)} \int_{-\infty}^{\infty} \mathrm{e}^{\mathrm{i}\mu\gamma s^2} f\big(\xi(s)\big) \frac{\mathrm{d}\xi}{\mathrm{d}s} \mathrm{d}s \tag{2.66}$$

This gives us an integral of the same form as in Equation 2.62, but with the function

$$g(s) = f\big(\xi(s)\big) \frac{\mathrm{d}\xi}{\mathrm{d}s} \tag{2.67}$$

replacing the function $f(\xi)$.

When we apply the method of stationary phase to the integral in Equation 2.62, we see that there is a stationary point at $\xi = 0$. The continuity properties of $f(\xi)$ are important in determining how quickly the first-order term in the method of stationary phase converges toward the exact answer. For the general case, it is important to know the continuity properties of the function $g(s)$. A discontinuity in the kth derivative of $g(s)$ can arise by the kth derivative of $f(\xi)$ being discontinuous at $\xi = \xi_0$, or by the kth derivative of $\mathrm{d}\xi/\mathrm{d}s$ being discontinuous at $s = 0$. The derivatives of $\xi(s)$ depend on the derivatives with respect to ξ of $q(\xi)$ at ξ_0. These derivatives can be calculated using implicit differentiation. In particular, note that

$$2s \frac{\mathrm{d}s}{\mathrm{d}\xi} = \mu \frac{\mathrm{d}q}{\mathrm{d}\xi} \tag{2.68}$$

If we evaluate this at $\xi = \xi_0$, we find that both sides of this equation vanish and we have not determined any derivatives. However, if we differentiate once more with respect to ξ, we get

$$2\left(\frac{\mathrm{d}s}{\mathrm{d}\xi}\right)^2 + 2s \frac{\mathrm{d}^2 s}{\mathrm{d}\xi^2} = \mu \frac{\mathrm{d}^2 q}{\mathrm{d}\xi^2} \tag{2.69}$$

and when we evaluate this at $\xi = \xi_0$, we get

$$2\left[\frac{\mathrm{d}s(\xi_0)}{\mathrm{d}\xi}\right]^2 = \mu \frac{\mathrm{d}^2 q(\xi_0)}{\mathrm{d}\xi^2} \tag{2.70}$$

This gives us two possible values of $\mathrm{d}s(\xi_0)/\mathrm{d}\xi$. We can choose either sign we want to. When we take further derivatives, we find that the $(\mathrm{d}^k/\mathrm{d}\xi^k)s(\xi_0)$ is determined by the derivatives of $q(\xi)$ up to $k + 1$. This means that $(\mathrm{d}^k/\mathrm{d}s^k)\ \xi(0)$ is also determined by these same derivatives. Finally, we see that the derivatives of $g(s)$ up to k will be continuous only if the derivatives of q up to $k + 2$ are continuous. In particular, we see that the function $g(s)$ will have a continuous first derivative if and only if the derivative of $f(\xi)$ and the third derivative of $q(\xi)$ are both continuous.

2.3.4 Generalized Functions

Before discussing the higher order terms in the method of stationary phase, we consider some relevant concepts from the theory of generalized functions [12]. The Dirac delta function and its derivatives are examples of generalized functions. The definition of these functions often arises as infinite integrals whose integrands do not decay at infinity. For example, the delta function is the inverse Fourier transform of a constant, and hence is defined by a divergent integral:

$$\delta(x) = \frac{1}{2\pi}\int_{-\infty}^{\infty} e^{i\omega x}\,d\omega \tag{2.71}$$

One way of thinking of this function is to imagine that it is defined by taking the inverse Fourier transform of $e^{-\alpha\omega^2}$, and then letting $\alpha \to 0$. The function that we get by doing this is the Dirac delta function. Even though it is a rather unusual function, it is extremely useful in practice.

In finding the higher order terms for the method of stationary phase, it will be useful to consider the integrals

$$R_k(\gamma) = \int_{-\infty}^{\infty} e^{i\gamma\xi^2}\xi^k\,d\xi \tag{2.72}$$

These integrals can be confusing since if γ is real then the integrands of these integrals do not approach zero as $\xi \to \infty$. However, if we evaluate these integrals over a finite interval, and let the region of integration go to infinity, we find that these are in fact convergent integrals. Furthermore, if we give γ a very small positive imaginary part, then the integrands approach zero. After evaluating these integrals we could then let the imaginary part go to zero. When we let the imaginary part go to zero we find that all of the integrals R_k are well defined. We could also get the integrals by taking the derivatives of the integral $R_0(\gamma)$ with respect to γ. If we do this, we find that

$$R_{k+2} = i\frac{dR_k}{d\gamma} \tag{2.73}$$

Due to the asymmetry of the integrand, we get

$$R_k = 0 \quad \text{for} \quad k \text{ is odd} \tag{2.74}$$

Carrying out this process, we find that the first few of these integrals are given by

$$R_0(\gamma) = \frac{C_0}{\gamma^{1/2}} \tag{2.75}$$

$$R_1(\gamma) = 0 \tag{2.76}$$

$$R_2(\gamma) = \frac{iC_0}{2\gamma^{3/2}} \tag{2.77}$$

where:

$$C_0 = e^{i\pi/4}\sqrt{\pi} \tag{2.78}$$

We will also be concerned with the integrals

$$S_k(\gamma) = \int_0^{\infty} \xi^k e^{i\gamma\xi^2} d\xi \tag{2.79}$$

We can use the same sort of reasoning on these integrals. If k is even, then

$$S_k(\gamma) = \frac{1}{2} R_k(\gamma) \text{ for } k \text{ even} \tag{2.80}$$

However, unlike R_k these integrals do not vanish when k is odd. In particular, when $k = 1$

$$S_1(\gamma) = \frac{i}{2\gamma} \tag{2.81}$$

The rest of the integrals can be evaluated using

$$S_{k+2}(\gamma) = -i\frac{d}{d\gamma} S_k(\gamma) \tag{2.82}$$

2.3.5 Higher Order Terms in the Method of Stationary Phase

We begin our analysis of the higher order terms in the method of stationary phase by considering the special case

$$H(\gamma) = \int_{-\infty}^{\infty} e^{i\gamma\xi^2} f(\xi) d\xi \tag{2.83}$$

This has the stationary point at $\xi = 0$. To obtain the first term in the method of stationary phase, we argued that the major contribution to this integral came from the region around $\xi = 0$. For this reason we expanded $f(\xi)$ in a Taylor series about $\xi = 0$, and then kept only the first term in the series. It makes sense that we should get more accurate answers if we keep more terms in the Taylor series. For example, if we kept three terms in the Taylor series, this would lead to an approximation of the form:

$$H(\gamma) \approx f(0)R_0(\gamma) + \frac{df(0)}{d\xi} R_1(\gamma) + \frac{1}{2}\frac{d^2 f(0)}{d\xi^2} R_2(\gamma) + \cdots \tag{2.84}$$

where:

$R_k(\gamma)$ are the integrals that are discussed in Section 2.3.4

We conclude that the higher order approximations for $H(\gamma)$ can be written as

$$H(\gamma) \approx e^{i\pi/4}\sqrt{\frac{\pi}{\gamma}}\left[f(0)+\frac{i}{4\gamma}\frac{d^2f(0)}{d\xi}\right]+\cdots \quad (2.85)$$

In this special case, this shows that the next order term in the method of stationary phase dies down like $1/\gamma^{3/2}$, provided that $f(\xi)$ is sufficiently differentiable.

If the derivative of $f(\xi)$ has a discontinuity at $\xi = 0$, then if we keep two terms in our Taylor series about $\xi = 0$, we end up with an expression

$$H(\gamma) \approx f(0)R_0(\gamma)+\frac{df(0_+)}{d\xi}\int_0^\infty e^{i\gamma\xi^2}\xi d\xi+\frac{df(0_-)}{d\xi}\int_{-\infty}^\infty e^{i\gamma\xi^2}\xi d\xi+\cdots \quad (2.86)$$

We can write this as

$$H(x,\gamma) \approx f(0)R_0(\gamma)+S_1(\gamma)\left[\frac{df(0_+)}{d\xi}-\frac{df(0_-)}{d\xi}\right]+\cdots \quad (2.87)$$

Using our values of R_0 and S_1 we get

$$H(\gamma) \approx f(0)e^{i\pi/4}\sqrt{\frac{\pi}{\gamma}}+\frac{i}{2\gamma}\left[\frac{df(0_+)}{d\xi}-\frac{df(0_-)}{d\xi}\right]+\cdots \quad (2.88)$$

We see that if $f(\xi)$ has a discontinuous derivative at $\xi = 0$, then the relative error between the first-order term and the exact answer will die down like $1/\gamma^{1/2}$. This is much slower than when the derivative of $f(\xi)$ is continuous.

As we noted earlier, the general case where $q(\xi)$ is not quadratic can be transformed into the quadratic case, but with the function $f(\xi)$ replacing the function $g(s)$ in Equation 2.67. We saw that the function $g(s)$ will have a discontinuous derivative if the function $f(\xi)$ has a discontinuous derivative, or if $q(\xi)$ has a discontinuous third derivative. It follows that as long as the first derivative of $f(\xi)$ and the third derivative of $q(\xi)$ are continuous, then the next order term in the method of stationary phase will die down like $1/\gamma^{3/2}$. If either of these derivatives is discontinuous, then the next order term will die down like $1/\gamma$. This can be used to justify our earlier statement concerning the effect of a discontinuity in the third derivative of the lens surface on the rate of convergence toward the geometrical optics limit.

2.3.6 Lower Order Discontinuities in the Phase Functions

When the functions $f(\xi)$ or $(d^2/d\xi^2)q(\xi,x)$ are discontinuous, we can get very slow convergence from the first term in the stationary phase approximation. We will begin with a simple example illustrating this point. By suitably changing coordinates, more general problems can in fact be related to this simple example.

Consider the integral

$$V(x,\gamma) = \frac{1}{\sqrt{\pi}} \int_0^\infty e^{i\gamma(\xi - x)^2} d\xi \tag{2.89}$$

This is a special case of our general problem where $q(\xi,x)$ has a quadratic dependence on ξ and $f(\xi) = 1$ for $\xi > 0$ and 0 for $\xi < 0$. This is an example where the function $f(\xi)$ is discontinuous.

If we apply the method of stationary phase to this integral, we see that there is no stationary point for $x < 0$ and that the method predicts that the integral is independent of x for $x > 0$. More specifically, the method predicts

$$V(x,\gamma) \approx 0 \quad \text{for} \quad x < 0 \tag{2.90}$$

$$V(x,\gamma) \approx \sqrt{\frac{1}{\gamma}} e^{i\pi/4} \quad \text{for} \quad x > 0 \tag{2.91}$$

In the stationary phase approximation, the magnitude of $V(x,\gamma)$ is a multiple of the Heaviside function.

$$|V(x,\gamma)|^2 \approx 0 \quad \text{for} \quad x < 0 \tag{2.92}$$

$$|V(x,\gamma)|^2 \approx \frac{1}{\gamma} \quad \text{for} \quad x > 0 \tag{2.93}$$

We now consider this integral in more detail. A simple change of variables allows us to write

$$V(x,\gamma) = \frac{1}{\gamma^{1/2}} \mathrm{Fr}(-x\gamma^{1/2}) \tag{2.94}$$

where:

$$\mathrm{Fr}(s) = \frac{1}{\sqrt{\pi}} \int_s^\infty e^{it^2} dt \tag{2.95}$$

The function $\mathrm{Fr}(s)$ is known as a complex Fresnel integral. Figure 2.3 shows a plot of the intensity $|\mathrm{Fr}(s)|^2$ of the function $\mathrm{Fr}(s)$. This graph shows that for $x < 0$ the function $V(x,\gamma)$ has a monotonic decay toward zero, whereas for $x > 0$ we get an oscillatory approach toward the constant value of 1. We see that if $|x\sqrt{\gamma}| \gg 1$, then $V(x,\gamma)$ will agree very well with the stationary phase solution. The difference between the exact solution and the stationary phase solution is that the stationary phase solution approximates the lower limit of the integrand as being equal to $-\infty$. Since $\gamma \gg 1$, this is usually a good approximation, but when x is close to zero, this is not so good. This is the root of the slow convergence of the method of stationary phase for all problems that have a discontinuity in $f(\xi)$ or $d^2q/d\xi^2$.

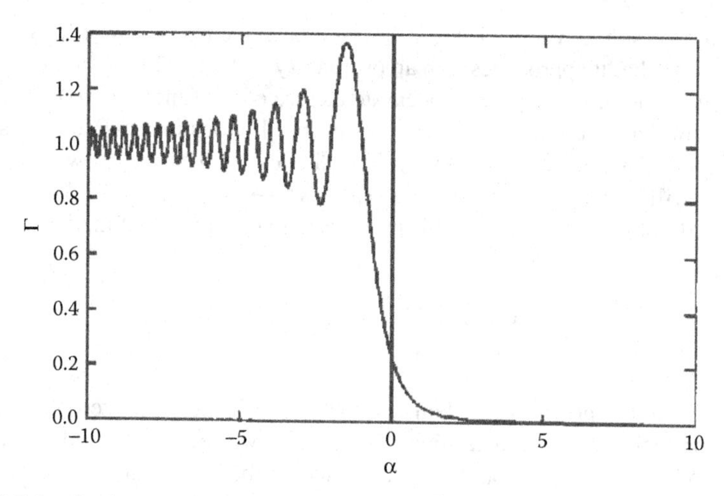

FIGURE 2.3 A plot of the function $|\mathrm{Fr}(s)|^2$ that models the intensity when light gets diffracted by a plane.

We can understand the behavior of Fr(s) by considering the behavior of Fr(s) for large values of $|s|$. For $s \gg 0$, we can integrate by parts to show that

$$\mathrm{Fr}(s) = \frac{1}{\sqrt{\pi}} e^{is^2} \frac{i}{2s} + \frac{1}{\sqrt{\pi}} \int_s^{\infty} \frac{e^{it^2}}{2it} dt \tag{2.96}$$

This shows that

$$\mathrm{Fr}(s) = \frac{1}{\sqrt{\pi}} e^{is^2} \frac{i}{2s} + O\left(\frac{1}{s^2}\right) \tag{2.97}$$

Similarly, for $s = 0$, we can write

$$\mathrm{Fr}(s) = e^{i\pi/4} - \frac{1}{\sqrt{\pi}} \int_{-\infty}^{s} e^{it^2} dt \tag{2.98}$$

An integration by parts now shows that

$$\mathrm{Fr}(s) = e^{i\pi/4} - \frac{1}{\sqrt{\pi}} e^{is^2} \frac{i}{2s} + O\left(\frac{1}{z^2}\right) \tag{2.99}$$

When we compute the amplitude $|\mathrm{Fr}(s)|^2$, we see that

$$|\mathrm{Fr}(s)|^2 = \frac{1}{4\pi s^2} + \cdots \quad \text{for} \quad s \gg 0 \tag{2.100}$$

and

$$|\mathrm{Fr}(s)|^2 = \left[1 - \frac{\sin(s^2 - \pi/4)}{\sqrt{\pi s}}\right] \quad \text{for} \quad s \ll 0 \tag{2.101}$$

We see that the solution approaches its asymptotic value much slower for $s = 0$ than for $s \gg 0$, and that it approaches it in an oscillatory fashion. This is only the asymptotic behavior, but it gives quite an accurate picture of the function Fr(s).

This shows that the function $V(x,\gamma)$ will have oscillations for $x > 0$, and be smooth for $x < 0$. Note that the convergence for large values of γ will be very slow if x is near 0, since $x\sqrt{\gamma}$ will be relatively small in this case as $\gamma \to \infty$.

A very slight generalization of this problem is to consider the function

$$V(x,\gamma) = \int_0^\infty e^{i\gamma(s-x)^2}\,ds + \alpha \int_\infty^0 e^{i\gamma(s-x)^2}\,ds \tag{2.102}$$

This reduces to the previous problem when $\alpha = 0$. When α is nonzero, we can analyze this problem in a similar fashion. In this case, we see that we will get oscillations on both sides of $x = 0$, but the oscillations will be bigger on the side where the stationary phase solutions predict that V is bigger. Generalizing further, we see that when we have an integral of the form:

$$H(x,\gamma) = \int_{-\infty}^\infty e^{i\gamma q(x,\xi)} f(\xi)\,d\xi \tag{2.103}$$

and $f(\xi)$ has a discontinuity at ξ^*, then we need to break the integral up into two parts.

$$H(x,\gamma) = \int_{-\infty}^{\xi^*} e^{i\gamma q(x,\xi)} f(\xi)\,d\xi + \int_{\xi^*}^\infty e^{i\gamma q(x,\xi)} f(\xi)\,d\xi \tag{2.104}$$

Suppose that the function $q(x,\xi)$ has a stationary point at $\xi(x)$, and that $\xi(x^*) = \xi^*$. When we apply the reasoning behind the method of stationary phase to the integral from ξ^* to ∞, we get

$$\int_{\xi^*}^\infty e^{i\gamma q(x,\xi)} f(\xi)\,d\xi \approx e^{i\gamma q[x,\xi(x)]} f\left(\xi(x)\right) \int_{\xi^*}^\infty e^{i\gamma 1/2\left\{d^2 q[x,\xi(x)]/d\xi^2\right\}[\xi-\xi(x)]^2}\,d\xi \tag{2.105}$$

We now make the substitution

$$s^2 = \gamma \frac{1}{2} \frac{d^2 q\left[x,\xi(x)\right]}{d\xi^2} \left[\xi - \xi(x)\right]^2 \tag{2.106}$$

This gives us the approximation

$$\int_{\xi^*}^\infty e^{i\gamma q(x,\xi)} f(\xi)\,d\xi \approx e^{i\gamma q[x,\xi(x)]} f\left(\xi(x)\right) \frac{\sqrt{\pi}}{\sqrt{\gamma/2 \left| \dfrac{d^2 q\left[x,\xi(x)\right]}{d\xi^2} \right|}} \mathrm{Fr}(s^*) \tag{2.107}$$

where:

$$s^* = \left[\xi^* - \xi(x)\right]\sqrt{\frac{\mathrm{d}^2 q\left[x,\xi(x)\right]}{\mathrm{d}\xi^2}\frac{\gamma}{2}} \tag{2.108}$$

If $s^* = 0$, we can make the approximation

$$F(s^*) \approx \mathrm{Fr}(-\infty) = \mathrm{e}^{\mathrm{i}\pi/4} \tag{2.109}$$

and we get back the first term in the method of stationary phase. However, if $\xi(x)$ is too close to ξ^*, this will not be a very good approximation unless γ is extremely large. However, our results should be good if we keep s^* in our expression rather than replacing it by $-\infty$. This is exactly what we did in our analysis of the integral in Equation 2.89.

As in our analysis of Equation 2.89, we can patch up our approximation for points x such that $\xi(x) \approx \xi^*$ by using the Fresnel integral. If we were to do this in the general case, the formulas would get extremely cumbersome. However, it is clear that in this general case we will get the same qualitative behavior as in Equation 2.89.

2.3.7 Method of Stationary Phase in Higher Dimensions

The method of stationary phase carries over to higher dimensional integrals. In particular, suppose we have an integral of the form:

$$H(\gamma) = \int_{-\infty}^{\infty}\int_{-\infty}^{\infty} f(\xi,\eta)\mathrm{e}^{\mathrm{i}\gamma q(\xi,\eta)}\mathrm{d}\xi\mathrm{d}\eta \tag{2.110}$$

Once again if the function $q(\xi, \eta)$ has a stationary point where

$$\nabla q(\xi_0,\eta_0) = 0 \tag{2.111}$$

then the major contribution to the integral will come from points right around this stationary point, and we can approximate the integral by

$$H(\gamma) \approx f(\xi_0,\eta_0)\mathrm{e}^{\mathrm{i}\gamma q(\xi_0,\eta_0)}\int_{-\infty}^{\infty}\int_{-\infty}^{\infty}\mathrm{e}^{\mathrm{i}1/2\gamma Q(\xi,\eta)}\mathrm{d}\xi\mathrm{d}\eta \tag{2.112}$$

where:

$$Q(\xi,\eta) = \frac{\partial^2 q(\xi_0,\eta_0)}{\partial\xi^2}(\xi-\xi_0)^2 + 2\frac{\partial^2 q(\xi_0,\eta_0)}{\partial\xi\,\partial\eta}(\xi-\xi_0)(\eta-\eta_0) + \frac{\partial^2 q(\xi_0,\eta_0)}{\partial\eta^2}(\eta-\eta_0)^2 \tag{2.113}$$

This integral can be evaluated to give

$$H(\gamma) \approx \frac{2\pi i f(\xi_0,\eta_0)}{\gamma\sqrt{J(\xi_0,\eta_0)}} e^{i\gamma q(\xi_0,\eta_0)} \tag{2.114}$$

where:

$$J(\xi_0,\eta_0) = \frac{\partial^2 q(\xi_0,\eta_0)}{\partial \xi^2}\frac{\partial^2 q(\xi_0,\eta_0)}{\partial \eta^2} - \frac{\partial^2 q(\xi_0,\eta_0)}{\partial \xi\,\partial \eta}\frac{\partial^2 q(\xi_0,\eta_0)}{\partial \xi\,\partial \eta} \tag{2.115}$$

2.4 MAXWELL'S EQUATIONS

2.4.1 THE GOVERNING EQUATIONS

The theory of beam shaping is based on diffraction theory, which is itself based on electromagnetic field theory. For this reason we will now review basic electromagnetic theory [13]. The governing equations of electromagnetic field theory are as follows:

$$\nabla \times \mathbf{E} + \frac{1}{c}\frac{\partial \mathbf{B}}{\partial t} = 0 \tag{2.116a}$$

$$\nabla \times \mathbf{B} - \frac{1}{c}\frac{\partial \mathbf{E}}{\partial t} = \frac{4\pi}{c}\mathbf{J} \tag{2.116b}$$

and

$$\nabla \cdot \mathbf{E} = 4\pi\rho \tag{2.117a}$$

$$\nabla \cdot \mathbf{B} = 0 \tag{2.117b}$$

where:

- **E** and **B** are the electric and magnetic fields, respectively
- ρ and **J** are the charge and current densities, respectively

The densities ρ and **J** are related to each other through the law of conservation of charge.

$$\frac{\partial \rho}{\partial t} + \nabla \cdot \mathbf{J} = 0 \tag{2.118}$$

Equation 2.116a is the differential form of Faraday's principle of electromagnetic induction, and Equation 2.116b describes both Ampere's law and Maxwell's displacement current. Equation 2.117a is known as Gauss's law, and Equation 2.117b describes the fact that there is no such thing as a magnetic monopole. It should be noted that the second set of Maxwell's equations almost follows from the first set. If we take the divergence of Equation 2.116a and b, use the fact that the divergence of a curl is zero, and use the law of conservation of charge, we get

$$\frac{\partial}{\partial t}\nabla\cdot\mathbf{B}=0 \tag{2.119}$$

and

$$\frac{\partial}{\partial t}(\nabla\cdot\mathbf{E}-4\pi\rho)=0 \tag{2.120}$$

We see that if the second set of Maxwell's equations is true initially, then the first set requires that they be true for all time.

We are frequently concerned with wave propagation through some medium such as air, water, or glass. In this case, there is an interaction between the charge distributions and the electromagnetic field. This interaction is usually taken into account by assuming that the electric field induces a polarization charge **P** such that the charge density is given by

$$\rho=-\nabla\cdot\mathbf{P} \tag{2.121}$$

Assuming that there are no other charges apart from those induced by the electric field, this gives us the following equation:

$$\nabla\cdot(\mathbf{E}+4\pi\mathbf{P})=0 \tag{2.122}$$

The assumption is typically made that the polarization is given by

$$\mathbf{P}=\chi\mathbf{E} \tag{2.123}$$

Gauss's law can now be written as

$$\nabla\cdot\mathbf{D}=0 \tag{2.124a}$$

where:

$$\mathbf{D}=\varepsilon\mathbf{E} \tag{2.124b}$$

and

$$\varepsilon=1+4\pi\chi \tag{2.124c}$$

The linearity between the polarization and the electric field is usually valid unless the electric field gets to be very large. Here we have also assumed that the medium is isotropic, so there are no preferred directions. In a nonisotropic medium, the polarization is related to the applied electric field by a symmetric second rank tensor. In order to describe the phenomenon of birefringence in crystals, it is necessary to use the general tensor form for the polarization. (This tensor is diagonal for the special case of an isotropic medium.) This relation between the electric field and the polarization also assumes that the polarization depends only on the local value of the electric field. Using such a theory, it is not possible to describe the rotation of the polarization field by an optically active material.

A similar approximation is used to take into account the effect of currents that are produced by the magnetic field. In this case, the currents in the material induce a polarization current **M** such that the current is given by

$$\mathbf{J}_M = c\nabla \times \mathbf{M} \tag{2.125}$$

It follows that the magnetic field must satisfy

$$\nabla \times \mathbf{B} - \frac{1}{c}\frac{\partial \mathbf{E}}{\partial t} = 4\pi\nabla \times \mathbf{M} \tag{2.126}$$

If we introduce the quantity **H** defined by

$$\mathbf{H} = \mathbf{B} - 4\pi\mathbf{M} \tag{2.127}$$

then assuming that the only currents are those arising from the induced current $\mathbf{J}_M$, we can write

$$\nabla \times \mathbf{H} - \frac{1}{c}\frac{\partial \mathbf{E}}{\partial t} = 0 \tag{2.128}$$

In the simplest case, it is assumed that the fields **B** and **H** are linearly proportional to each other:

$$\mathbf{B} = \mu\mathbf{H} \tag{2.129}$$

For most materials that are used in optics the linear relation between the **B** and **H** fields is totally satisfactory. In fact, the constant μ is very nearly equal to unity for most materials of optical interest.

We now collect the macroscopic form of Maxwell's equations in a linear isotropic material:

$$\frac{1}{c}\frac{\partial \mathbf{D}}{\partial t} - \nabla \times \mathbf{H} = 0 \tag{2.130a}$$

$$\frac{1}{c}\frac{\partial \mathbf{B}}{\partial t} + \nabla \times \mathbf{E} = 0 \tag{2.130b}$$

$$\nabla \cdot \mathbf{D} = 0 \tag{2.131a}$$

$$\nabla \cdot \mathbf{B} = 0 \tag{2.131b}$$

where:

$$\mathbf{D} = \varepsilon\mathbf{E} \tag{2.132a}$$

$$\mathbf{B} = \mu\mathbf{H} \tag{2.132b}$$

We have omitted any sources of charges and currents other than those produced by the interaction of the fields with the materials.

2.4.2 Wave Equation

Our analysis of diffraction effects in Section 2.6 is based on the fact that in a linear, homogeneous, and isotropic medium, each component of the electric and magnetic fields satisfies the wave equation. We now give a derivation of this fact. We begin by deriving an equation for **E** that does not assume that ε and μ are constants.

To begin with, we write Equation 2.130b as

$$\frac{1}{c}\frac{\partial \mathbf{H}}{\partial t}+\frac{1}{\mu}\nabla\times\mathbf{E}=0 \tag{2.133}$$

We now take the curl of this equation and use Equation 2.130a to arrive at the result:

$$\frac{\partial^2\mathbf{E}}{\partial t^2}=-c^2\frac{1}{\varepsilon}\nabla\times\left(\frac{1}{\mu}\nabla\times\mathbf{E}\right) \tag{2.134}$$

This is the form of the wave equation for **E** in a medium where ε and μ are not assumed to be constants. If we assume that μ is constant, we can write this equation as

$$\frac{\partial^2\mathbf{E}}{\partial t}=-c^2\frac{1}{\varepsilon\mu}\nabla\times\nabla\times\mathbf{E} \tag{2.135}$$

We can simplify this equation by using the identity $\nabla \times \nabla \times \mathbf{A} = -\nabla^2\mathbf{A} + \nabla\nabla\cdot\mathbf{A}$, along with the fact that $\nabla\cdot\mathbf{E} = 0$ (assuming that ε is constant). For a homogeneous medium, this gives us the equation

$$\frac{\partial^2\mathbf{E}}{\partial t^2}=\frac{c^2}{\varepsilon\mu}\nabla^2\mathbf{E} \tag{2.136}$$

This shows that each component of the electric field satisfies the wave equation. If the fields are time harmonic, with frequency ω, the spatial dependence of the electric field must satisfy

$$k^2\mathbf{E}+\nabla^2\mathbf{E}=0 \tag{2.137}$$

where:

$$k=\frac{\omega\varepsilon\mu}{c} \tag{2.138}$$

We refer to this equation as the reduced wave equation or the Helmholtz equation.

Similar arguments show that the field **H** satisfies

$$\frac{\partial^2\mathbf{H}}{\partial t^3}=-c^2\frac{1}{\mu}\nabla\times\left(\frac{1}{\varepsilon}\nabla\times\mathbf{H}\right) \tag{2.139}$$

Note that this is not quite the same as Equation 2.134 for **E** since we have put ε inside the curl and μ outside the curl. However, if μ and ε are constants, we once again arrive at the conclusion that each component of **H** (and hence **B**) will satisfy the scalar wave equation.

2.4.3 Energy Flux

We will now derive an expression for the flux of energy in an electromagnetic field. If we dot Equation 2.130a with respect to **E**, and Equation 2.130b with respect to **H**, and add the results, we get the following equation:

$$\frac{1}{2c}\frac{\partial}{\partial t}(\mathbf{E}\cdot\mathbf{D}+\mathbf{B}\cdot\mathbf{H})-\mathbf{E}\cdot\nabla\times\mathbf{H}+\mathbf{H}\cdot\nabla\times\mathbf{E}=0 \tag{2.140}$$

If we use the identity

$$\nabla\cdot(\mathbf{A}\times\mathbf{B})=\mathbf{B}\cdot\nabla\times\mathbf{A}-\mathbf{A}\cdot\nabla\times\mathbf{B} \tag{2.141}$$

We see that

$$\frac{1}{2c}\frac{\partial}{\partial t}(\mathbf{E}\cdot\mathbf{D}+\mathbf{B}\cdot\mathbf{H})+\nabla\cdot(\mathbf{E}\times\mathbf{H})=0 \tag{2.142}$$

When putting in integral form, this equation can be written as

$$\frac{1}{2c}\frac{\mathrm{d}}{\mathrm{d}t}\int_V(\mathbf{E}\cdot\mathbf{D}+\mathbf{B}\cdot\mathbf{H})\,\mathrm{d}V=-\int_S(\mathbf{E}\times\mathbf{H})\cdot\mathbf{n}\,\mathrm{d}s \tag{2.143}$$

where:

n is the outward facing normal to the surface

This equation can be interpreted as the fact that the quantity $1/2c(\mathbf{E}\cdot\mathbf{D}+\mathbf{B}\cdot\mathbf{H})$ is the energy density, and $\mathbf{E}\times\mathbf{H}$ is the flux of energy. The interpretation of $1/2c(\mathbf{E}\cdot\mathbf{D}+\mathbf{B}\cdot\mathbf{H})$ as the energy density of the field is actually clearer when we include charges in Maxwell's equations. In this case, we would have to add a term to these equations that would represent the change in kinetic energy of the particles in the system. The vector

$$\mathbf{S}=c\mathbf{E}\times\mathbf{H} \tag{2.144}$$

is referred to as the Poynting vector.

We will use the Poynting vector to justify the fact that the rays in geometrical optics are in fact the direction that energy is begin transported.

2.5 GEOMETRICAL OPTICS

2.5.1 Fermat's Principle

To understand our discussion of beam shaping, it is essential to know how to use the laws of geometrical optics. Although understanding the derivation of the basic principles clearly leads to a deeper understanding, this is not essential for our presentation. For this reason, we begin by stating the basic principles and showing how we use them. Once this is done, we will discuss the derivation of the principles.

Our treatment of geometrical optics is based on Fermat's principle [11]. Fermat's principle is often stated as saying that the ray that gets from point **a** to point **b** will take the path that minimizes the travel time. This is a very concise statement of the principle, but it is not technically correct. Rather than saying that the true path minimizes the travel time, we need to say that the true path is stationary with respect to travel time. In many situations, the travel time is in fact minimized, but it is not always the case.

Before discussing what stationarity means in geometrical optics, we will clarify what we mean by stationarity in a simpler setting. The function $F(x,y) = (x - x_0)^2 + (y - y_0)^2$ has a minimum at $(x,y) = (x_0,y_0)$. A necessary condition that it has a minimum at (x_0,y_0) is that the partial derivatives of F vanish at (x_0,y_0). The vanishing of the partial derivatives is equivalent to saying that the function $F(x,y)$ is stationary at (x_0,y_0). Another way of putting this is to say that if we take any numbers $(\hat{x},\hat{y})$, then

$$F(x_0+\varepsilon\hat{x}, y_0+\varepsilon\hat{y}) = F(x_0,y_0)+O(\varepsilon^2) \quad \text{as} \quad \varepsilon \to 0 \tag{2.145}$$

At a point that is not stationary, we would have $F(x_0+\varepsilon\hat{x}, y_0+\varepsilon\hat{y}) = F(x_0,y_0)+O(\varepsilon)$ as $\varepsilon \to 0$. In order for a function to have a minimum at (x_0,y_0) it must be stationary, but stationarity does not imply that the function is a minimum. For example, the function

$$F(x,y) = -x^2 - y^2 \tag{2.146}$$

is stationary at (0,0), but it has a maximum rather than a minimum. The function

$$F(x,y) = x^2 - y^2 \tag{2.147}$$

is stationary at (0,0), but it has neither a minimum nor a maximum.

Returning to geometrical optics, we will parameterize curves going from point **a** to **b** by a parameter s such that $0 \le s \le 1$. Let $\mathbf{x}(s) = [x(s),y(s),z(s)]$ be a curve such that $\mathbf{x}(0) = \mathbf{a}$ and $\mathbf{x}(1) = \mathbf{b}$. We will denote the travel time along this curve as

$$T\left[\mathbf{x}(s)\right] = \text{Travel time} \tag{2.148}$$

Suppose that $\mathbf{x}_0(s)$ is the true path that a light ray takes to get from **a** to **b**. The stationarity condition implies that for any functions $\hat{\mathbf{x}}(s)$ such that $\hat{\mathbf{x}}(0) = \hat{\mathbf{x}}(1) = 0$,

$$T[\mathbf{x}_0(s)+\varepsilon\hat{\mathbf{x}}(s)] = T[\mathbf{x}_0(s)] + O(\varepsilon^2) \tag{2.149}$$

Fermat's principle applies in an enormous variety of situations. Many times we put constraints on the travel paths. For example, we can use Fermat's principle to show that the angle of incidence equals the angles of reflection for a light ray bouncing off a mirror. In this case, we use the constraint that a ray goes from point **a** to **b** after first touching a surface.

If a light ray gets from point **a** to **b** by first passing through an intermediate point **c**, it can be shown that the paths from **a** to **c** and from **c** to **b** must each be stationary.

We will now give some concrete examples illustrating Fermat's principle. The first few examples that we give are not directly relevant to the beam shaping problem, but the last example is absolutely essential to understanding our discussion of beam shaping.

Example 1

Suppose we have a medium that has a constant speed of light. For a ray to get from point **a** to **b** in the least amount of time, it is clear that it must travel in a straight line. Since the travel path to get from point **a** to **b** is a minimum, it is clear that it is also stationary. It can be shown that in this case, straight lines are the only stationary paths.

Example 2

Suppose the plane $z = 0$ separates medium I with a velocity of c_I from medium II with velocity c_{II}. What path does a light ray take to get from point **a** in medium I to point **b** in medium II? From our last example, we already know that the path must be a straight line in each medium. For simplicity, we will assume that the light ray travels in the plane $y = 0$. Suppose that $\mathbf{a} = (x_1,0,z_1)$ and $\mathbf{b} = (x_2,0,z_2)$. Suppose that in going from point **a** to **b**, the ray goes through $\mathbf{c} = (\xi,0,0)$ on the interface between the two media. We do not know the value of ξ ahead of time, but it can be determined using Fermat's principle. The total travel time to get from point **a** to **b** by going through **c** is

$$T(\xi) = \frac{1}{c_I}\sqrt{(x_1-\xi)^2 + z_1^2} + \frac{1}{c_{II}}\sqrt{(x_2-\xi)^2 + z_2^2} \tag{2.150}$$

In order for the travel time to be stationary, we must have

$$\frac{dT}{d\xi} = 0 \tag{2.151}$$

This implies that

$$\frac{\sin(\theta_I)}{c_I} = \frac{\sin(\theta_{II})}{c_{II}} \tag{2.152}$$

where:

$$\sin(\theta_I) = \frac{\xi - x_1}{\sqrt{(x_1-\xi)^2 + z_1^2}} \tag{2.153}$$

$$\sin(\theta_{II}) = \frac{x_2 - \xi}{\sqrt{(x_2-\xi)^2 + z_2^2}} \tag{2.154}$$

This is equivalent to Snell's law of refraction.

Example 3

Suppose we would like to design a mirror that focuses all of the light rays coming from point **a** to **b**. For simplicity, we will consider this problem to take place in two dimensions. We also assume that the speed of light is constant throughout our medium. Suppose a ray comes from point **a** at an angle of θ with the horizontal. Suppose that this ray bounces off the mirror at a point $\mathbf{p}(\theta)$ and then goes to the point **b**, let $T(\theta)$ be the travel time to get from point **a** to $\mathbf{p}(\theta)$ and then to point **b**. In order for Fermat's principle to hold, we must have

$$\frac{dT}{d\theta} = 0 \tag{2.155}$$

This means that if $\mathbf{q}_1$ is any point on the mirror, then the distance from point **a** to $\mathbf{q}_1$ plus the distance from point $\mathbf{q}_1$ to **b** must be the same as for any other point $\mathbf{q}_2$ on the mirror. This implies that the mirror must in fact have the shape of an ellipse, with foci at points **a** and **b**.

It should be noted that as we move the point **a** off to ∞, this ellipse ends up turning into a parabola. This gives us the solution of how to focus rays coming in from ∞ to a single point **b**.

Example 4

When using Fermat's principle for parallel beams of light, it is necessary to be familiar with the following argument. Suppose we have a parallel beam of rays coming in from ∞. We can think of such rays as coming from a very distant point source. Suppose the point source is at $\mathbf{p} = (-L,0,0)$. Assuming a homogeneous medium, the time to get from point **p** to a point (x,y,z) is given by

$$T(x,y,z) = \frac{1}{c}\sqrt{(x+L)^2 + y^2 + z^2} \tag{2.156}$$

As $L \to \infty$, we can make the approximation

$$T(x,y,z) = \frac{1}{c}\left[L + x + O\left(\frac{1}{L}\right)\right] \tag{2.157}$$

This shows that the travel time to get to any point in space (x,y,z) is independent of y and z. When applying Fermat's principle, the travel time L/c will not matter since it is the same for all paths. We will use this fact whenever we are applying Fermat's principle to rays that are coming in parallel.

Example 5

We now give an example that shows that the ray paths do not always minimize the travel time, but they are still stationary with respect to travel time. Suppose we have a cylindrical mirror (Figure 2.4) whose surface is given by

$$(x,y) = R\left[\cos(\theta), \sin(\theta)\right] - \frac{\pi}{2} \leq \theta \leq \frac{\pi}{2} \tag{2.158}$$

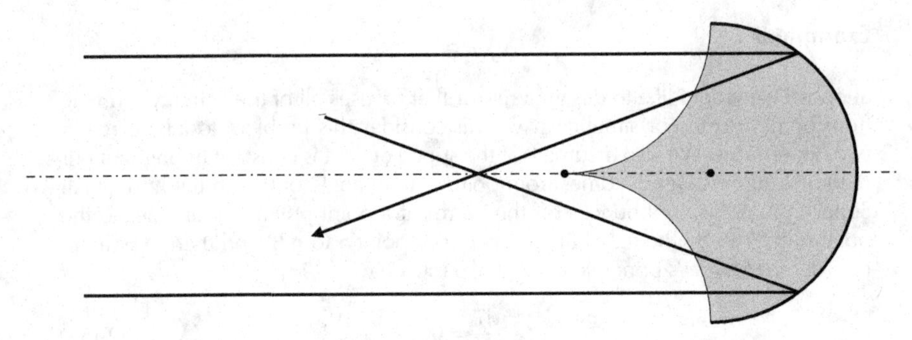

FIGURE 2.4 A schematic of light being reflected by a cylindrical mirror. In the shaded region, there are two rays that reach each point by a single reflection off the mirror. Outside of this region, there is only one ray that reaches each point by a single reflection.

We are interested in finding the paths of rays that are coming in parallel from $x = -\infty$. As we have already mentioned, this can be thought of as rays coming from a distant source at $\mathbf{p} = (-L,0)$ where L is very large. The travel time to get from $\mathbf{p}$ to a point on the surface of the mirror $R[\cos(\theta),\sin(\theta)]$, and then to a point (x,y) is (assuming $L \gg 1$)

$$T(\theta) = \frac{1}{c}\left\{L + R\cos(\theta) + \sqrt{[x - R\cos(\theta)]^2 + [y - R\sin(\theta)]^2}\right\} \tag{2.159}$$

Given a point (x,y) the equation $dT/d\theta = 0$ will determine where the ray that reaches (x,y) reflects off the mirror. For simplicity, we will limit ourselves to points such that $y = 0$. In this case, we can write the stationarity condition as

$$\frac{dT}{d\theta} = \frac{R\sin(\theta)}{c}\left\{-1 + \frac{x}{\sqrt{[x - R\cos(\theta)]^2 + R^2\sin^2(\theta)}}\right\} \tag{2.160}$$

The above equation has the solution

$$\sin(\theta) = 0 \tag{2.161}$$

or

$$x^2 = [x - R\cos(\theta)]^2 + R^2\sin^2(\theta) \tag{2.162}$$

The above equation can be written as

$$x = \frac{R}{2\cos(\theta)} \tag{2.163}$$

For any value of x, the first equation (2.162) gives us the solution $\theta = 0$. However, the second equation (2.163) will have no solutions if $x < R/2$, and will have two solutions if $x > R/2$. When we look throughout the x–y plane, we find that there will be a region that has three reflected rays reaching each point, and another region with only one reflected ray reaching each point. The curve separating the two regions is an example of a caustic surface. These caustic surfaces are easily observed since the irradiance at the surface becomes much larger than at a typical point in the plane. This particular situation can be observed when looking into a cup of tea or a bowl of sugar under the light from a concentrated source such as an incandescent light bulb.

Note that not all of the rays have minimum travel time. If we compute $d^2T/d\theta^2$, we get

$$\frac{d^2T}{d\theta^2} = \frac{R\cos(\theta)}{c}\left\{-1+\frac{x}{\sqrt{[x-R\cos(\theta)]^2+R^2\sin^2(\theta)}}\right\} - \frac{R^2\sin^2(\theta)}{c}\frac{x^2}{\left\{[x-R\cos(\theta)]^2+R^2\sin^2(\theta)\right\}^{3/2}} \tag{2.164}$$

If we restrict our attention to the ray that hits the mirror at $\theta = 0$, we have

$$\frac{d^2T(0)}{d\theta^2} = \frac{R}{c}\left(-1+\frac{x}{|x-R|}\right) \tag{2.165}$$

This shows that if $x > R/2$, then the travel time is a minimum, but if $x < R/2$, the travel time is in fact a local maximum.

Example 6

This example plays a large role in our theory of beam shaping since it shows that our Fourier transform lens has a quadratic time delay, and hence a quadratic phase function. Suppose we would like to place a lens at $x = 0$ that focuses all of the rays coming in from $x = -\infty$ to a single point at $(x,y) = (f,0)$. We will make several approximations. To begin with, we assume that the lens is thin. This means that the rays that enter the lens at $(0,y)$ emerge at very nearly the same point. We can model the effect of the lens by saying that it introduces a time delay of $t_L(y)$. This means that a ray that enters the lens at $(0,y)$ takes a time $t_L(0,y)$ to emerge from the lens. In practice, this time delay can be introduced by making the lens out of a material that has a different index of refraction than the medium that the rays are traveling in, and by varying the thickness of the lens.

We would like to determine the function $t_L(y)$ such that all of the rays from ∞ get focused to the point $(f,0)$. The time for a ray to go from a distant point $(-L,0)$ to a point $(0,y)$ and then to the point $(f,0)$ is approximately

$$t(y) = \frac{1}{c}\left(L+\sqrt{f^2+y^2}\right)+t_L(y) \tag{2.166}$$

Here we have made the approximation that $L \gg 1$, and used the simplified expression for the distance from $(-L,0)$ to $(0,y)$. We will now make the paraxial approximation. This assumes that all points on the lens satisfy $y^2/f^2 \ll 1$. In this case, we can approximate the square root using

$$\sqrt{f^2 + y^2} \approx f + \frac{y^2}{2f} \tag{2.167}$$

Using this approximation, we can write

$$t(y) \approx \frac{1}{c}\left(L + f + \frac{y^2}{2f}\right) + t_L(y) \tag{2.168}$$

Fermat's principle requires that the path that gets from $(-L,0)$ to $(f,0)$ must be stationary with respect to all nearby paths. This implies that

$$\frac{dt}{dy} = 0 \text{ for all } y \tag{2.169}$$

This means that we must have $t(y)$ = constant, and hence

$$t_L(y) = -\frac{y^2}{2fc} \tag{2.170}$$

This shows that in the paraxial approximation, we must use a quadratic lens to focus an initially parallel beam of rays to a point.

2.5.2 Eikonal Equation

The laws of geometrical optics can be derived as a high-frequency approximation to the solutions of Maxwell's equations. The rays of light are related to the direction of energy propagation. There is a very strong connection between Fermat's principle and the method of stationary phase. Both the method of stationary phase and the laws of geometrical optics are high-frequency limits; they both are centered about the phase of the wave field and use a stationarity condition.

Before considering the high-frequency limit of Maxwell's equations, we will begin by considering the high-frequency limit of the scalar wave equation. Suppose we have a solution $p(\mathbf{x},\omega)$ to the following equation:

$$\nabla^2 p + \frac{\omega^2}{c^2(\mathbf{x})} p = 0 \tag{2.171}$$

This is the time-harmonic wave equation, also known as the reduced wave equation. We are interested in determining the behavior of these solutions for large values of ω. In particular, we ask what solutions that are coming from a single point source look like. The theory of Green's functions shows that the general high-frequency limit (not necessarily from a point source) can be built up by integrating over many such point sources. If the velocity is constant, we know that the point source solutions can be written as

$$p(\mathbf{x},\omega) = A\frac{e^{i\omega r/c}}{r} \tag{2.172}$$

Here $r^2 = x^2 + y^2 + z^2$. This solution has a very rapidly varying phase (it varies more rapidly the bigger ω is), and a slowly varying amplitude (that is independent of ω). In the case of variable $c(\mathbf{x})$, we assume that even though the amplitude may not be completely independent of ω, it depends very weakly on ω. Generalizing to the case of nonhomogeneous media, in the high-frequency limit, we will assume that $p(\mathbf{x},\omega)$ can be written as

$$p(\mathbf{x},\omega) = A(\mathbf{x})e^{i\omega\phi(\mathbf{x})} \tag{2.173}$$

This is only the first term in an asymptotic expansion. The general solution needs to include corrections to the amplitude that depend on ω.

We can write

$$\nabla p = \left(\nabla A + Ai\omega\nabla\phi\right)e^{i\omega\phi} \tag{2.174}$$

Using this equation, we can now write

$$\nabla^2 p = \nabla\cdot\nabla p = (\nabla^2 A + 2i\omega\nabla A\cdot\nabla\phi + i\omega A\nabla^2\phi - \omega^2 A\nabla\phi\cdot\nabla\phi)e^{i\omega\phi} \tag{2.175}$$

If we substitute this expression into Equation 2.171 and keep only the highest order term in Ω, we find that

$$|\nabla\phi|^2 = \frac{1}{c(\mathbf{x})^2} \tag{2.176}$$

This equation is usually referred to as the eikonal equation. The next higher order term gives us the following equation:

$$2\nabla A\cdot\nabla\phi + A\nabla^2\phi = 0 \tag{2.177}$$

This equation can be written as

$$\nabla\cdot(A^2\nabla\phi) = 0 \tag{2.178}$$

The fact that this equation can be written in divergence form suggests that the quantity $A^2\nabla\phi$ is the flux of some quantity that is conserved. When we apply these arguments to optical systems, we will see that this quantity is in fact proportional to the flux of energy.

We mentioned that the general high-frequency approximation can be built up by integrating or summing over a family of point sources. As a simple example, if our wave field comes from two point sources, the high-frequency limit of the wave field will look like

$$p(\mathbf{x},\omega) = A_1(\mathbf{x})e^{i\omega\phi_1(\mathbf{x})} + A_2(\mathbf{x})e^{i\omega\phi_2(\mathbf{x})} \tag{2.179}$$

It should be mentioned that even for a single point source, there may be points in space where the high-frequency limit consists of a sum of terms as in the previous equation. This will be the case if the rays are bent so that more than one ray from the same source reaches the same point. The surfaces separating the regions where there are different numbers of rays are the caustic surfaces.

2.5.3 Eikonal Equation and Maxwell's Equations

In the previous section, we derived the eikonal equation from the scalar wave equation. Each component of the electromagnetic wave field satisfies this equation; so it is not surprising that the eikonal equation also arises when considering the high-frequency limit of Maxwell's equations. In this section, we will derive the eikonal equation using Maxwell's equations, and we will see that Poynting's theorem shows that the energy of the electromagnetic field is in fact being propagated normal to the surfaces of constant phase. The derivation of the eikonal equation from Maxwell's equations is almost identical to the analysis of monochromatic plane waves given in most textbooks on electrodynamics [12].

The time-harmonic Maxwell's equations are

$$i\omega\frac{1}{c_0}\mathbf{D} - \nabla\times\mathbf{H} = 0 \tag{2.180a}$$

$$i\omega\frac{1}{c_0}\mathbf{B} + \nabla\times\mathbf{E} = 0 \tag{2.180b}$$

where:

$$\mathbf{D} = \varepsilon(x)\mathbf{E} \tag{2.181a}$$

$$\mathbf{B} = \mu(x)\mathbf{H} \tag{2.181b}$$

Similar to our derivation of the eikonal equation for the scalar wave equation, we assume a solution of the form:

$$\mathbf{E}(\mathbf{x},\omega) = \mathbf{E}_0(\mathbf{x})e^{i\omega\phi(\mathbf{x})} \tag{2.182}$$

$$\mathbf{H}(\mathbf{x},\omega) = \mathbf{H}_0(\mathbf{x})e^{i\omega\phi(\mathbf{x})} \tag{2.183}$$

Substituting this expression into Maxwell's equations and using the vector identity

$$\nabla\times[f(\mathbf{x})\mathbf{A}(\mathbf{x})] = f(\mathbf{x})\nabla\times\mathbf{A}(\mathbf{x}) + \mathbf{A}(\mathbf{x})\times\nabla f(\mathbf{x}) \tag{2.184}$$

we get

$$i\omega\frac{\varepsilon}{c_0}\mathbf{E}_0 - \left(\nabla\times\mathbf{H}_0 + i\omega\mathbf{H}_0\times\nabla\phi\right) = 0 \tag{2.185a}$$

$$i\omega\frac{\mu}{c_0}\mathbf{H}_0 + (\nabla\times\mathbf{E}_0 + i\omega\mathbf{E}_0\times\nabla\phi) = 0 \tag{2.185b}$$

If we only keep the highest order terms in ω in this equation, we get

$$\frac{\varepsilon}{c_0}\mathbf{E}_0 - \mathbf{H}_0 \times \nabla\phi = 0 \tag{2.186a}$$

$$\frac{\mu}{c_0}\mathbf{H}_0 + \mathbf{E}_0 \times \nabla\phi = 0 \tag{2.186b}$$

If we dot each of these equations with $\nabla\phi$, we find that

$$\mathbf{E}_0 \cdot \nabla\phi = 0 \tag{2.187}$$

$$\mathbf{H}_0 \cdot \nabla\phi = 0 \tag{2.188}$$

If we dot Equation 2.186a with respect to $\mathbf{H}_0$ or Equation 2.186b with respect to $\mathbf{E}_0$, we find that

$$\mathbf{E}_0 \cdot \mathbf{H}_0 = 0 \tag{2.189}$$

If we eliminate $\mathbf{H}_0$ from Equation 2.186a and b, we find that

$$\frac{\mu}{c_0}\mathbf{E}_0 + c_0(\mathbf{E}_0 \times \nabla\phi) \times \nabla\phi = 0 \tag{2.190}$$

Using the identity

$$(\mathbf{a} \times \mathbf{b}) \times \mathbf{c} = \mathbf{b}(\mathbf{a} \cdot \mathbf{c}) - \mathbf{a}(\mathbf{b} \cdot \mathbf{c}) \tag{2.191}$$

we find that this can be written as

$$\frac{\varepsilon\mu}{c_0^2}\mathbf{E}_0 - \left(\mathbf{E}_0\nabla\phi \cdot \nabla\phi - \nabla\phi\mathbf{E}_0 \cdot \nabla\phi\right) = 0 \tag{2.192}$$

Using the fact that $\mathbf{E}_0 \cdot \nabla\phi = 0$, this can be written as

$$\mathbf{E}_0\left[\frac{1}{c^2(\mathbf{x})} - \nabla\phi \cdot \nabla\phi\right] = 0 \tag{2.193}$$

where:

$$c^2(\mathbf{x}) = \frac{c_0^2}{\varepsilon\mu} \tag{2.194}$$

We see that we have once again arrived at the eikonal equation.

The direction of energy flux is given by the Poynting vector:

$$\mathbf{S} = c\mathbf{E} \times \mathbf{H} \tag{2.195}$$

Using the fact that $\mathbf{E}$ and $\mathbf{H}$ are orthogonal to each other, and also to $\nabla\phi$, it follows that this vector is in the direction of $\nabla\phi$. We see that the direction of energy flux is in fact normal to the surfaces of constant phase.

2.5.4 First-Order Nonlinear Partial Differential Equations

The theory of ray tracing from the eikonal equation is a special case of the solution of nonlinear first-order partial differential equations [14,15]. In this section, we will give a brief outline of this theory. Suppose we have an equation of the form:

$$F(x,y,z,p,q,r) = 0 \tag{2.196a}$$

where:

$$p = \frac{\partial \phi}{\partial x} \tag{2.196b}$$

$$q = \frac{\partial \phi}{\partial y} \tag{2.196c}$$

$$r = \frac{\partial \phi}{\partial z} \tag{2.196d}$$

For optical applications, we are especially concerned with the case where

$$F(x,y,z,p,q,r) = \frac{1}{2}\left[p^2 + q^2 + r^2 - \frac{1}{c^2(\mathbf{x})}\right] \tag{2.197}$$

which is just the eikonal equation. In this section, we will consider the case for a general function F rather than limiting ourselves to the eikonal equation. Our analysis could be extended to the case where the function F also depends on the function ϕ, but that case is just a bit more complicated, and it never arises in optical applications, so we will not consider it here.

Suppose we know the function ϕ and all of its first derivatives at some point (x_0,y_0,z_0). Is it possible to determine the solution ϕ in the neighborhood of the point (x_0,y_0,z_0)? In particular, is it possible to determine the second derivatives of the function ϕ at the point (x_0,y_0,z_0)? If we take the derivatives of our equation with respect to x, y, and z, we end up with the following equations:

$$\frac{\partial F}{\partial x} + \frac{\partial F}{\partial p}\frac{\partial p}{\partial x} + \frac{\partial F}{\partial q}\frac{\partial q}{\partial x} + \frac{\partial F}{\partial r}\frac{\partial r}{\partial x} = 0 \tag{2.198a}$$

$$\frac{\partial F}{\partial y} + \frac{\partial F}{\partial p}\frac{\partial p}{\partial y} + \frac{\partial F}{\partial q}\frac{\partial q}{\partial y} + \frac{\partial F}{\partial r}\frac{\partial r}{\partial y} = 0 \tag{2.198b}$$

$$\frac{\partial F}{\partial z} + \frac{\partial F}{\partial p}\frac{\partial p}{\partial z} + \frac{\partial F}{\partial q}\frac{\partial q}{\partial z} + \frac{\partial F}{\partial r}\frac{\partial r}{\partial z} = 0 \tag{2.198c}$$

The partial derivatives of p, q, and r can all be expressed in terms of the six second-order partial derivatives of ϕ:

$$\frac{\partial^2 \phi}{\partial x^2}, \frac{\partial^2 \phi}{\partial y^2}, \frac{\partial^2 \phi}{\partial z^2}, \frac{\partial^2 \phi}{\partial x \partial y}, \frac{\partial^2 \phi}{\partial x \partial z}, \frac{\partial^2 \phi}{\partial y \partial z} \tag{2.199}$$

By differentiating our equation $F(x,y,p,q,r) = 0$, we have arrived at three equations for the six second-order partial derivatives of ϕ. Clearly, we do not have enough equations to determine the second-order partial derivatives. The question now arises: Is it possible to determine the derivatives p, q, and r in a particular direction? It turns out that this is in fact possible. To do this, we use the fact that

$$\frac{\partial p}{\partial y} = \frac{\partial^2 \phi}{\partial x \partial y} = \frac{\partial^2 \phi}{\partial t \partial x} = \frac{\partial p}{\partial x} \tag{2.200a}$$

Similarly,

$$\frac{\partial r}{\partial x} = \frac{\partial p}{\partial z} \tag{2.200b}$$

It follows that Equation 2.198a can be written as

$$\frac{\partial F}{\partial x} + \frac{\partial F}{\partial p}\frac{\partial p}{\partial x} + \frac{\partial F}{\partial p}\frac{\partial p}{\partial y} + \frac{\partial F}{\partial r}\frac{\partial p}{\partial z} = 0 \tag{2.201a}$$

Similarly, by switching the order of the other mixed partial derivatives, we can get

$$\frac{\partial F}{\partial y} + \frac{\partial F}{\partial p}\frac{\partial q}{\partial x} + \frac{\partial F}{\partial q}\frac{\partial q}{\partial y} + \frac{\partial F}{\partial r}\frac{\partial q}{\partial z} = 0 \tag{2.201b}$$

and

$$\frac{\partial F}{\partial z} + \frac{\partial F}{\partial p}\frac{\partial r}{\partial x} + \frac{\partial F}{\partial q}\frac{\partial r}{\partial y} + \frac{\partial F}{\partial r}\frac{\partial r}{\partial z} = 0 \tag{2.201c}$$

These equations can be written as

$$\frac{\partial F}{\partial x} + \nabla p \cdot \mathbf{a} = 0 \tag{2.202a}$$

$$\frac{\partial F}{\partial y} + \nabla q \cdot \mathbf{a} = 0 \tag{2.202b}$$

and

$$\frac{\partial F}{\partial z} + \nabla \mathbf{r} \cdot \mathbf{a} = 0 \tag{2.202c}$$

where:

$$\mathbf{a} = \left(\frac{\partial F}{\partial p}, \frac{\partial F}{\partial q}, \frac{\partial F}{\partial r} \right) \tag{2.202d}$$

These equations show that although we do not know the derivatives of p, q, and r in any arbitrary direction, we do know the derivatives in the direction $\mathbf{a} = (\partial F/\partial p, \partial F/\partial q, \partial F/\partial r)$. This suggests that there may be special curves $[x(s),y(s),z(s)]$ such that we can determine $[p(s),q(s),r(s)]$. In particular, if

$$\dot{x} = \frac{\partial F}{\partial p} \tag{2.203}$$

$$\dot{y} = \frac{\partial F}{\partial q} \tag{2.204}$$

$$\dot{z} = \frac{\partial F}{\partial r} \tag{2.205}$$

then we have

$$\dot{p} = -\frac{\partial F}{\partial x} \tag{2.206}$$

$$\dot{q} = -\frac{\partial F}{\partial y} \tag{2.207}$$

$$\dot{r} = -\frac{\partial F}{\partial z} \tag{2.208}$$

The function ϕ changes according to the following equation:

$$\dot{\phi} = \frac{\partial \phi}{\partial x}\dot{x} + \frac{\partial \phi}{\partial y}\dot{y} + \frac{\partial \phi}{\partial z}\dot{z} = p\dot{x} + q\dot{y} + r\dot{z} = p\frac{\partial F}{\partial p} + q\frac{\partial F}{\partial q} + r\frac{\partial F}{\partial r} \tag{2.209}$$

This is a seventh-order ordinary differential equation (ODE) for the unknowns (x,y,z,p,q,r,ϕ). We can solve this system of equations provided we specify initial values of (x,y,z,p,q,r,ϕ). It should be noted that we cannot specify these values arbitrarily, but must require that they satisfy the equation $F(x,y,z,p,q,r) = 0$.

In optics the function ϕ is the phase of our wave field. The curves $[x(s),y(s),z(s)]$ along which we propagate our solution are known as the rays. In optics they are what we intuitively think of as being the rays of light. If we know the phase $\phi(x,y,z)$ on some plane $z = z_0$, then we can parametrically map out the phase in all of space by tracing all the rays from the plane $z = z_0$.

This process of tracing out the phase field assumes that one and only one ray passes from a given point (x,y,z) to the plane $z = z_0$. In practice, it is possible that no rays pass through some points, and multiple rays pass through other points. The surfaces separating regions with different numbers of rays are once again the caustic surfaces.

Example 7

We will now apply this theory to the eikonal equation where

$$F(x,y,z,p,q,r) = \frac{1}{2}\left[p^2 + q^2 + r^2 - \frac{1}{c^2(\mathbf{x})}\right] \tag{2.210}$$

We will use the shorthand notation $\mathbf{p} = (p,q,r)$ and $\mathbf{x} = (x,y,z)$. The theory we have just derived shows that

$$\frac{d\mathbf{x}}{ds} = \mathbf{p} \tag{2.211a}$$

$$\frac{d\mathbf{p}}{ds} = -\frac{\nabla c}{c^3} \tag{2.211b}$$

and

$$\frac{d\phi}{ds} = \frac{1}{c^2(\mathbf{x})} \tag{2.211c}$$

where $\mathbf{p}$ is required to satisfy the initial condition

$$\mathbf{p}.\mathbf{p} = \frac{1}{c^2} \tag{2.212}$$

We can eliminate $\mathbf{p}$ from this equation to get

$$\frac{d^2\mathbf{x}}{ds^2} = -\frac{\nabla c}{c^3} \tag{2.213}$$

Note that the equation for ϕ can be written as

$$\frac{d\phi}{ds} = \frac{|\dot{\mathbf{x}}|}{c(\mathbf{x})} \tag{2.214}$$

This equation can be interpreted that the change in ϕ in going from $\mathbf{x}(s_0)$ to $\mathbf{x}(s_1)$ is the travel time to get from $\mathbf{x}(s_0)$ to $\mathbf{x}(s_1)$ along the curve $\mathbf{x}(s)$.

Example 8

The system of Equation 2.213 is in some ways the simplest set of equations we could write down for the paths of the light rays. However, it suffers from one problem. The equations are not invariant under a change of parameterization. If we parameterize our curves by $\xi = \xi(s)$, we will end up getting a different differential equation for $\mathbf{x}(\xi)$ than for $\mathbf{x}(s)$. The solutions will result in the same curve in physical space, but the differential equations will be different. Unless one is extremely concerned with the aesthetic properties of their equations, this is not a real problem. However, it turns out that the equations that we derive using Fermat's principle will be invariant under a change of parameterization, and hence it will be difficult to compare the two sets of equations unless we write Equation 2.213 so that they are also invariant.

To do this, we note that the first of Equation 2.212 requires that

$$|\dot{\mathbf{x}}| = |\mathbf{p}| = \frac{1}{c} \tag{2.215}$$

It follows that along our solution curve, we have

$$|\dot{\mathbf{x}}|\, c(\mathbf{x}) = 1 \tag{2.216}$$

It follows that we will not change the solutions to Equation 2.211 if we divide the left-hand side by $|\dot{\mathbf{x}}|\,c(\mathbf{x})$. In this case, we get the following equations:

$$\frac{1}{|\dot{\mathbf{x}}|\,c(\mathbf{x})}\dot{\mathbf{x}} = \mathbf{p} \tag{2.217}$$

$$\frac{1}{|\dot{\mathbf{x}}|\,c(\mathbf{x})}\dot{\mathbf{p}} = -\nabla c\frac{1}{c^3} \tag{2.218}$$

Now if we eliminate $\mathbf{p}$ from these equations, we end up with the system of the following equations:

$$\frac{1}{|\dot{\mathbf{x}}|\,c(\mathbf{x})}\frac{d}{ds}\left[\frac{\dot{\mathbf{x}}}{|\dot{\mathbf{x}}|\,c(\mathbf{x})}\right] = -\frac{\nabla c}{c^3} \tag{2.219}$$

This is the final system of equations that we will use to compare to the curves obtained by using Fermat's principle. Note that if we make a change of variables $\xi = \xi(s)$, the differential equation in terms of ξ will be identical to the differential equation in terms of s.

2.5.5 Fermat's Principle without Reflections

In Example 8, we derived the equations for the path $\mathbf{x}(s)$ that a light ray follows in an inhomogeneous medium. We will now show that the path that gets from point $\mathbf{x}_0$ to $\mathbf{x}_1$ is stationary with respect to the travel time between these two points. Suppose we have a curve $\mathbf{x}(s)$ such that $\mathbf{x}(s_0) = \mathbf{x}_0$ and $\mathbf{x}(s_1) = \mathbf{x}_1$. The time to get from point $\mathbf{x}_0$ to point $\mathbf{x}_1$ may be written as

$$T = \int_{s_0}^{s_1} \sqrt{\dot{\mathbf{x}}\cdot\dot{\mathbf{x}}}\,\frac{1}{c(\mathbf{x})}\,ds \tag{2.220}$$

The first variation of this integral may be written as

$$\delta T = \int_{s_0}^{s_1} \frac{\delta\dot{\mathbf{x}}\cdot\dot{\mathbf{x}}}{c(\mathbf{x})\sqrt{\dot{\mathbf{x}}\cdot\dot{\mathbf{x}}}} - |\dot{\mathbf{x}}|\frac{\nabla c\cdot\delta\mathbf{x}}{c^2(\mathbf{x})}\,ds \tag{2.221}$$

If we integrate by parts to get rid of the derivative with respect to $\delta\mathbf{x}$, and if we require that $\delta\mathbf{x}$ vanish at the end points, we find that

$$\delta T = -\int_{s_0}^{s_1} \left\{\frac{d}{ds}\left[\frac{\dot{\mathbf{x}}}{|\dot{\mathbf{x}}|\,c(\mathbf{x})}\right] + |\dot{\mathbf{x}}|\frac{\nabla c}{c^2(\mathbf{x})}\right\}\cdot\delta\mathbf{x}\,ds \tag{2.222}$$

If the path is stationary, then this integral must vanish for all functions $\delta\mathbf{x}$, and hence $\mathbf{x}$ must satisfy the following equation:

$$\frac{1}{c(\mathbf{x})|\dot{\mathbf{x}}|}\frac{d}{ds}\left[\frac{\dot{\mathbf{x}}}{|\dot{\mathbf{x}}|\,c(\mathbf{x})}\right] = -\frac{\nabla c}{c^3} \tag{2.223}$$

This is identical to Equation 2.219, which was derived using the eikonal equation, and requiring that the equation be invariant under a change of parameterization.

2.5.6 Fermat's Principle for Reflecting Surfaces

In our analysis of beam shaping systems, we will not consider any cases where the rays reflect off mirrors. However, since it may sometimes be desirable to use reflecting surfaces in beam shaping systems, we now consider Fermat's principle for reflecting surfaces. Suppose we have a surface S defined parametrically by $\mathbf{x} = \mathbf{f}(\xi_1,\xi_2)$. Suppose that a ray goes from the point $\mathbf{x}_0$ to the point $\mathbf{x}_1$, but first bounces off the surface S. The theory of waves shows that at the point where the rays get reflected by the surface, the following conditions hold:

- The normal to the surface, the incident ray, and the reflected ray all lie in the same plane.
- The incident and reflected rays make the same angle with respect to the normal to the surface.

We now show that these conditions can be derived by assuming that the path from $\mathbf{x}_0$ to $\mathbf{x}_1$ that touches the surface S is stationary with respect to travel time. Suppose we have a path that goes from $\mathbf{x}_0$ to $\mathbf{x}_1$ after first touching some point $\mathbf{q} = \mathbf{f}(\xi_1,\xi_2)$ on the surface S. Clearly, the paths from $\mathbf{x}_0$ to $\mathbf{q}$ and from $\mathbf{q}$ to $\mathbf{x}_1$ must themselves be stationary. It follows that to determine the true path we only need to determine the point $\mathbf{q}$ on the surface S. In particular, suppose $\phi(\mathbf{x},\mathbf{z})$ gives the travel time to get from the point $\mathbf{x}$ to the point $\mathbf{z}$. We have shown that the travel time $\phi(\mathbf{x},\mathbf{z})$ is in fact a solution to the eikonal equation. The total travel time from $\mathbf{x}_0$ to $\mathbf{x}_1$ is given by

$$T = \phi\left[\mathbf{x}_0, \mathbf{f}\left(\xi_1,\xi_2\right)\right] + \phi\left[\mathbf{x}_1, \mathbf{f}\left(\xi_1,\xi_2\right)\right] \tag{2.224}$$

If this travel time is stationary, then we must have

$$\frac{\partial \mathbf{f}}{\partial \xi_k} \cdot (\mathbf{p}_\mathrm{i} + \mathbf{p}_\mathrm{r}) = 0 \quad \text{for} \quad k = 1,2 \tag{2.225}$$

where:

$$\mathbf{p}_\mathrm{i} = \frac{\partial \phi(\mathbf{x}_0,\mathbf{q})}{\partial \mathbf{q}} \tag{2.226a}$$

and

$$\mathbf{p}_\mathrm{r} = \frac{\partial \phi(\mathbf{x}_1,\mathbf{q})}{\partial \mathbf{q}} \tag{2.226b}$$

are the incident and the reflected ray vectors, respectively.

Note that the vectors $\mathbf{p}_\mathrm{i}$ and $\mathbf{p}_\mathrm{r}$ must satisfy $|\mathbf{p}| = 1/c(\mathbf{q})$, and hence we must have $|\mathbf{p}_\mathrm{i}| = |\mathbf{p}_\mathrm{r}|$. Let $\mathbf{n}$ be the normal to the surface S at $\mathbf{f}(\xi_1,\xi_2)$, and let $\mathbf{t}_1$ and $\mathbf{t}_2$ be the two independent tangent vectors to the surface. These vectors can be written as linear combinations of the vectors $\partial \mathbf{f}/\partial \xi_1$ and $\partial \mathbf{f}/\partial \xi_2$, and hence Equation 2.225 shows that the tangential components of $\mathbf{p}_i$ and $\mathbf{p}_r$ must be negatives of each other. That is, if

$$\mathbf{p}_i = a_i\mathbf{n} + b_i\mathbf{t}_1 + c_i\mathbf{t}_2 \tag{2.227}$$

then

$$\mathbf{p}_r = a_r\mathbf{n} - b_i\mathbf{t}_1 - c_i\mathbf{t}_2 \tag{2.228}$$

Furthermore, in order for $\mathbf{p}_r$ and $\mathbf{p}_i$ to have the same magnitude, we must have $a_i = a_r$. This shows that $\mathbf{p}_i$, $\mathbf{p}_r$, and $\mathbf{n}$ all lie in the same plane. Furthermore,

$$\mathbf{p}_i \cdot \mathbf{n} = \mathbf{p}_r \cdot \mathbf{n} \tag{2.229}$$

and hence the vectors $\mathbf{p}_i$ and $\mathbf{p}_r$ make the same angle with respect to $\mathbf{n}$. This is precisely what we wanted to prove.

2.6 FOURIER OPTICS AND DIFFRACTION THEORY

2.6.1 Fresnel Diffraction Theory

Fresnel diffraction theory plays an important role in the theory of beam shaping since it allows us to access the validity of the geometrical optics approximation. Through the theory of Fresnel diffraction, we will be able to turn our physical optics problem into a mathematical problem concerning Fourier transforms. After giving a derivation of the Fresnel approximation, we will outline the conditions necessary for it to be a good approximation.

The Fresnel approximation is concerned with the wave field for $z > 0$ produced when an incoming wave passes through an aperture at $z = 0$. In general, the aperture may contain an optical element that changes the amplitude or phase of the incoming wave. Elementary theories of diffraction usually are concerned with the field far from the aperture, and in a narrow solid angle normal to the aperture. The theory of Fresnel diffraction can be outlined in three basic steps.

1. Write down an exact expression for determining the wave field for all values of $z > 0$ provided one knows the wave field at the plane of the aperture $z = 0$.
2. Use a paraxial approximation that simplifies this expression assuming the observation point is near the axis.
3. Compute the wave field away from the aperture by using the first two steps along with a very simple assumption concerning the field in the plane of the aperture. The assumption is that at the aperture the wave field is equal to the undisturbed incoming wave field (modified by any optical element inside the aperture), and zero everywhere else.

The first step can be carried out rigorously. The second step can be justified quite well using simple asymptotics. The third step is by far the hardest to justify, but it can be argued that it is plausible provided the aperture is large compared to the wavelength of the incoming light.

We begin with a discussion of the Fresnel approximation for the scalar wave equation. Physically, we can think of this equation arising from the equations of acoustics. When we present the vector theory of diffraction, we will see that this theory can be used to determine the various components of the electric field, but a slight error occurs in the component of the field normal to the aperture. This error is not big as long as we are near the axis.

2.6.2 A Fourier Approach to Diffraction Theory

We suppose that the function $u(x,y,z)$ satisfies the Helmholtz equation:

$$\nabla^2 u + k^2 u = 0 \tag{2.230a}$$

where:

$$k = \frac{2\pi}{\lambda} \tag{2.230b}$$

where:
λ is the wavelength

The function u must also satisfy the boundary condition:

$$u(x,y,0) = f(x,y) \tag{2.230c}$$

and

$$u(x,y,z) \text{ has no incoming waves as } z \to \infty \tag{2.230d}$$

This last boundary condition, sometimes referred to as the Sommerfeld radiation condition, is a somewhat subtle condition, but it is quite straightforward to implement when doing analytical work. It requires that as $z \to \infty$ all of the waves will be traveling away from $z = 0$, not toward it.

We choose to solve these equations by spatially Fourier transforming the function $u(x,y,z)$ in the x and y directions. Let

$$U(k_x, k_y, z) = \int_{-\infty}^{\infty}\int_{-\infty}^{\infty} \mathrm{e}^{-\mathrm{i}(k_x x + k_y y)} u(x, y, z)\, \mathrm{d}x\mathrm{d}y \tag{2.231}$$

be the Fourier transform of u, and $F(k_x,k_y)$ be the Fourier transform of $f(x, y)$. The function U must satisfy the following equations:

$$\frac{\mathrm{d}^2 U}{\mathrm{d}z^2} + (k^2 - k_x^2 - k_y^2)\, U = 0 \tag{2.232a}$$

$$U(k_x, k_y, 0) = F(k_x, k_y) \tag{2.232b}$$

$$U(k_x, k_y, z) \text{ has only outgoing waves as } z \to \infty \tag{2.232c}$$

The solution to this set of equations can be written as

$$U(k_x,k_y,z) = F(k_x,k_y)\mathrm{e}^{\mathrm{i}z}\sqrt{k^2-k_x^2-k_y^2} \tag{2.233}$$

The sign of the square root must be chosen so that the field decays as $z \to \infty$, and there are no incoming waves from infinity. To ensure this, we must choose the positive square root for $k^2 - k_x^2 - k_y^2 > 0$ and choose it so that $i\sqrt{k^2 - k_x^2 - k_y^2} < 0$ for $k^2 - k_x^2 - k_y^2 < 0$.

We can now inverse Fourier transform this to get

$$u(x,y,z) = \frac{1}{4\pi^2}\int_{-\infty}^{\infty}\int_{-\infty}^{\infty}\mathrm{e}^{\mathrm{i}(k_x x+k_y y)}F(k_x,k_y)\mathrm{e}^{\mathrm{i}kz\sqrt{1-k_x^2/k^2-k_y^2/k^2}}\,\mathrm{d}k_x\mathrm{d}k_y \tag{2.234}$$

We have now accomplished the first step in deriving the Fresnel approximation; we have derived an exact expression for the field u in terms of its value at $z = 0$.

This form for the field is sometimes used in diffraction theory. However, both analytical and numerical works are usually much simpler if the square root is approximated by

$$\sqrt{1-\frac{k_x^2}{k^2}-\frac{k_y^2}{k^2}} \approx 1 - \frac{k_x^2+k_y^2}{2k^2} \tag{2.235}$$

This approximation is referred to as the paraxial or Fresnel approximation. Using this approximation, we can write

$$U(k_x,k_y,z) = F(k_x,k_y)\mathrm{e}^{\mathrm{i}kz}\mathrm{e}^{-\mathrm{i}z(k_x^2+k_y^2/2k)} \tag{2.236}$$

Assuming the paraxial approximation, the Fourier convolution theorem tells us that the field $u(x,y,z)$ can be written as the convolution of $f(x,y)$ with the inverse Fourier transform of $\mathrm{e}^{ikz}\mathrm{e}^{-\mathrm{i}z(k_x^2+k_y^2)/2k}$. The inverse Fourier transform $\mathrm{e}^{ikz}\mathrm{e}^{-\mathrm{i}z(k_x^2+k_y^2)/2k}$ is $i(k/2\pi z)\mathrm{e}^{ikz}\mathrm{e}^{\mathrm{i}(k/2z)(x^2+y^2)}$. It follows that we can write

$$u(x,y,z) = i\frac{k}{2\pi z}\mathrm{e}^{ikz}\int_{-\infty}^{\infty}\int_{-\infty}^{\infty}f(\xi,\eta)\mathrm{e}^{\mathrm{i}k\left[(x-\xi)^2+(y-\eta)^2\right]/2z}\,\mathrm{d}\xi\,\mathrm{d}\eta \tag{2.237}$$

It is often convenient to write this as

$$u(x,y,z) = \frac{ik}{2\pi z}\mathrm{e}^{ikz}\mathrm{e}^{ik(x^2+y^2)/2z} \times \int_{-\infty}^{\infty}\int_{-\infty}^{\infty}f(\xi,\eta)\mathrm{e}^{ik(\xi^2+\eta^2)/2z}\mathrm{e}^{-ik(x\xi+y\eta)/z}\,\mathrm{d}\xi\,\mathrm{d}\eta \tag{2.238}$$

This is usually referred to as the Fresnel approximation [16]. This approximation can greatly simplify both analytical and numerical calculations.

2.6.3 Fourier Optics

We will now consider what happens when the beam passes through a lens of focal length f at the aperture. We claim that modifying the field at the aperture by the phase factor $e^{-ik(x^2+y^2)/2f}$ is equivalent to passing the beam through a lens with focal length f. Note that if we had a beam of light coming in from infinity, then the field of the incoming light would be constant over the aperture.

$$f(x,k) = Ae^{i\psi} \tag{2.239}$$

At the plane $z = f$, the field would be given by

$$u(x,y,f) = \frac{ik}{2\pi f} Ae^{i\psi} e^{ikf} e^{ik(x^2+y^2)/2f} \int_{-\infty}^{\infty}\int_{-\infty}^{\infty} e^{-ik(x\xi+y\eta)/f} d\xi d\eta \tag{2.240}$$

which can be written as

$$u(x,y,f) = \frac{i2\pi k}{f} e^{ikf} e^{ik(x^2+y^2)/2f} Ae^{i\psi} \delta\left(\frac{kx}{f}, \frac{ky}{f}\right) \tag{2.241}$$

This formula assumes that the aperture is infinitely large, and hence goes beyond the limits of validity of the Fresnel approximation. However, we could consider the case of an aperture of finite diameter, and we would get a more complicated but similar result, namely, that the field at $z = f$ is all concentrated near the origin $(x,y) = (0,0)$. This is exactly what a lens of focal length f would do to an incoming field of this sort.

We now consider the case where the incoming beam is not necessarily constant at the aperture, but is equal to $f(x,y)$. We assume that at the aperture $z = 0$, we have a lens with focal length f, which modifies the phase of the incoming beam by the factor $e^{-ik(x^2+y^2)/2f}$. In this case, the output will be given by

$$u(x,y,z) = \frac{ik}{2\pi z} e^{ikz} e^{ik(x^2+y^2)/2z} \times \int_{-\infty}^{\infty}\int_{-\infty}^{\infty} f(\xi,\eta) e^{ik(\xi^2+\eta^2)/2z} e^{-ik(\xi^2+\eta^2)/2f} e^{-ik(x\xi+y\eta)/z} d\xi d\eta \tag{2.242}$$

The output at the focal plane is given by

$$u(x,y,z) = \frac{ik}{2\pi f} e^{ikz} e^{ik(x^2+y^2)/2f} \int_{-\infty}^{\infty}\int_{-\infty}^{\infty} f(\xi,\eta) e^{-ik(x\xi+y\eta)/f} d\xi d\eta \tag{2.243}$$

which can be written as

$$\frac{ik}{2\pi f} e^{ikf} e^{ik(x^2+y^2)/2f} F\left(\frac{kx}{f}, \frac{ky}{f}\right) \tag{2.244}$$

where:

$F(\omega_x,\omega_y)$ is the Fourier transform of the function $f(x,y)$

We see that except for the term outside of the integral, the field distribution is given by the Fourier transform of the incoming field distribution. Note that the x and y dependence of the term outside of the integral has only a phase dependence. It follows that if we are only concerned with the irradiance distribution, then we can in fact ignore the terms outside of the integral.

2.6.4 Limits of Validity of the Fresnel Approximation

We now comment on the errors introduced by making the Fresnel approximation. We should emphasize that we are only considering the errors introduced in the problem of approximating the field $u(x,y,z)$ assuming we know the field at $z = 0$. In a real diffraction problem, we do not know the field at $z = 0$, but approximate it as being the incoming wave field.

We will now briefly summarize the conditions under which the Fresnel approximation can be assumed to be valid. In what follows R will be the effective dimension of the aperture and λ will be the wavelength of the light. We assume that the aperture lies in the plane $z = 0$, and that we are evaluating the field at a point $[d\cos(\theta), d\sin(\theta), z]$.

The Fresnel approximation always assumes that

$$R \gg \lambda \tag{2.245}$$

Assuming that this restriction holds, a summary of the conditions for the validity of the Fresnel approximation is as follows:

- The Fresnel approximation will be valid for all values of d if

$$N_F = \frac{2\pi R^2}{\lambda z} \text{ is not small} \tag{2.246}$$

- The amplitude of the wave predicted by the Fresnel approximation will be valid even if $N_F \ll 1$, provided

$$d/z \ll 1 \tag{2.247}$$

- Both the phase and amplitude predicted by the Fresnel approximation will be valid when $N_F \ll 1$, if

$$\frac{d^4}{z^4} \ll \frac{1}{kz} \tag{2.248}$$

It should be noted that for the most part we are only concerned with the irradiance of the field, so the phase errors introduced by the Fresnel approximation for large values of z will not be important to us. For this reason, we will be justified in using the Fresnel approximation provided $R \gg \lambda$, and that $d/z \ll 1$.

We will analyze the two-dimensional case where the aperture and field is independent of the y coordinate. The three-dimensional case is conceptually no more difficult, but the notation and the algebraic manipulations are simpler in two dimensions.

We will assume that the incoming wave field is equal to $f(x)$ at the aperture $z = 0$. If $F(k_x)$ is the Fourier transform of $f(x)$, then the field is given by

$$u(x,z) = \frac{1}{2\pi}\int_{-\infty}^{\infty} e^{ik_x x} F(k_x) e^{ikz\sqrt{1-k_x^2/k^2}}\, dk_x \tag{2.249}$$

In analyzing the Fresnel approximation, we find it fruitful to consider a family of problems where the form of the function $f(x)$ stays the same, but the scaling of the function changes. In particular, we will set

$$f(x) = g\left(\frac{x}{R}\right) \tag{2.250}$$

This includes the situation where the function $f(x) = 1$ inside the aperture and 0 elsewhere. In this case, the parameter R would be the characteristic dimension of the aperture. The Fourier transform of $f(x)$ can be written as

$$F(k_x) = RG(k_x R) \tag{2.251}$$

where:

$G(\alpha)$ is the Fourier transform of $g(x)$

The field for $z > 0$ can be written as

$$u(x,z) = R\frac{1}{2\pi}\int_{-\infty}^{\infty} e^{ik_x x} G(Rk_x) e^{ikz\sqrt{1-k_x^2/k^2}}\, dk_x \tag{2.252}$$

If we make the change of variables

$$\xi = Rk_x \tag{2.253}$$

we can write this integral as

$$u(x,z) = \frac{1}{2\pi}\int_{-\infty}^{\infty} e^{i(\xi x/R)}\, G(\xi) e^{ikz\sqrt{1-\xi^2/(Rk)^2}}\, d\xi \tag{2.254}$$

If $g(\xi)$ is a well-behaved function, the Fourier transform $G(\xi)$ goes to zero as $|\xi| \to \infty$. It follows that our answers will not be very sensitive to how we approximate the term $\sqrt{1-\xi^2/(kR)^2}$ when ξ is large. This means that we only need to approximate this well for $\xi = O(1)$. We now make the approximation that

$$kR \gg 1 \tag{2.255}$$

This is the first approximation that will be made when doing the Fresnel approximation. This is equivalent to assuming that the aperture is much bigger than the wavelength, an assumption that will have to hold in order to carry out the general plan of diffraction theory. Under this assumption, it is reasonable to expand the square root in a Taylor series:

$$kz\sqrt{1-\frac{\xi^2}{(kR)^2}} = kz\left[1-\frac{\xi^2}{2(kR)^2}+\frac{\xi^4}{8(kR)^4}+\cdots\right] \tag{2.256}$$

If we ignore the third term and all of the remaining terms in the Taylor series, we will end up with the Fresnel approximation. We will now see when we can ignore these terms, and what sorts of errors we will make when we ignore them. Note that assuming that $kR \gg 1$ and $\xi = O(1)$, these terms will always be small compared to the first two terms. However, it is possible that if $kz \gg 1$, then they will not necessarily be small. For simplicity, we will now keep the first three terms, and see when we can ignore the third term. The conditions for ignoring this term will be the same as for ignoring all of the remaining terms. If we keep the first three terms in the Taylor series, we get

$$u(x,y) = \frac{1}{2\pi}e^{ikz}\int_{-\infty}^{\infty} G(\xi)e^{-i(1/N_F)\left\{-\xi x^* + 1/2\xi^2 - \left[\xi^4/8(Rk)^2\right]\right\}}d\xi \tag{2.257}$$

where:

$$N_F = \frac{kR^2}{z} \tag{2.258}$$

is known as the Fresnel number and

$$x^* = \frac{xkR}{z} \tag{2.259}$$

We see that if

$$\frac{1}{N_F k^2 R^2} = \frac{z}{R}\frac{1}{k^3 R^3} \ll 1 \tag{2.260}$$

then we can ignore the third term in the Taylor series. This means that if we are close to the aperture then the Fresnel approximation will be valid (assuming $kR \gg 1$). In this case, there is no restriction on the value of x. If $1/N_F$ is not large, then the Fresnel approximation will hold. This case is not that interesting, because it is essentially the case when the geometrical optics approximation holds and diffraction effects are unimportant.

We now consider the much more interesting case when N_F is small. In this case, the phase in the integrand is multiplied by the large parameter $1/N_F$, and we can apply the method of stationary phase to the integral.

The phase will be stationary at the point ξ_0 satisfying

$$-x^* + \xi_0 - \frac{\xi_0^3}{2(kR)^2} = 0 \tag{2.261}$$

The method of stationary phase predicts that the field will be given by

$$u(x,z) \approx \frac{1}{2\pi} e^{ikz} e^{i\pi/4} \sqrt{\frac{2N_F \pi}{\psi_0''}} G(\xi_0) e^{-i(1/N_F)\psi_0} \tag{2.262}$$

where:

$$\psi_0 = -x^* \xi_0 + \frac{\xi_0^2}{2} - \frac{\xi_0^4}{8k^2R^2} \tag{2.263}$$

and

$$\psi_0'' = 1 - \frac{3\xi_0^2}{2k^2R^2} \tag{2.264}$$

This is the result predicted by the method of stationary phase assuming that $N_F \ll 1$ when we keep the first three terms in the Taylor series expansion of $\sqrt{1 - k_x^2 / k^2}$. We would like to know how this compares to the answer we would get if we only kept the first two terms (the Fresnel approximation).

In the Fresnel approximation, we would have

$$\xi_0 = x^* \tag{2.265}$$

We will have been justified in ignoring the cubic term in the equation for ξ_0, provided

$$\frac{x^{*2}}{k^2R^2} \ll 1 \tag{2.266}$$

This is equivalent to requiring that

$$\frac{x^2}{z^2} \ll 1 \tag{2.267}$$

This means that the stationary point when we include the higher order term will be nearly the same as the stationary point for the Fresnel approximation, provided the opening angle from the midpoint of the aperture to the point (x,z) is small.

From the form of the answer in Equation 2.262, we see that if we are not concerned with the phase errors, our answer will be accurate provided we have approximated ξ_0 well. This means that the amplitude of the Fresnel approximation will agree with the amplitude of the answer obtained by keeping three terms in the Taylor expansion provided $x^2 / z^2 \ll 1$, and $N_F \ll 1$. However, in order for the phase of the answer predicted by the Fresnel approximation to agree with the more refined answer, it is necessary that we also approximate ψ_0/N_F well. The value ψ_0/N_F predicted by the Fresnel approximation is

$$\frac{\psi_0}{N_F} \approx \frac{1}{N_F}\left(-x^* \xi_0 + \frac{\xi_0^2}{2}\right) \tag{2.268}$$

This will be a good approximation to the more refined answer, provided that

$$\frac{1}{N_F} \frac{x^{*4}}{k^2R^2} \ll 1 \tag{2.269}$$

This can be written as

$$\frac{x^4}{z^4} \ll \frac{1}{kz} \tag{2.270}$$

This agrees with the results we have already summarized concerning the errors in the Fresnel approximation.

2.6.5 Vector Theory of Diffraction

The theory we have presented so far is limited to the scalar wave equation. In optics, we are concerned with the vector fields, the electric and magnetic fields. We begin by outlining the most naive, but nearly correct, approach to the vector theory. We know that each component of the electric and magnetic fields satisfies the scalar wave equation. Just as in the scalar theory, we can assume that the field in the aperture is the same as the incoming field, and that the fields vanish elsewhere in the plane of the aperture. Using the scalar theory, we could compute each component of the electric and magnetic fields.

What are some possible difficulties with this approach? Just because each individual component of the field satisfies the wave equation it does not mean that the vector field satisfies Maxwell's equations. If each component of the field were chosen exactly right at the plane of the aperture, then this would be the case. However, the assumption that we have made for the fields at the aperture are not necessarily consistent with the correct fields. For this reason, we may end up getting inconsistent fields in the far field.

As an example of an inconsistency, suppose that $z = 0$ is the plane of the aperture, and that the incoming field is a plane wave propagating in the z direction. The naive approach to vector diffraction theory would imply that the z components of **E** and **B** vanish at the aperture, and hence vanish everywhere. A thorough analysis of this situation shows that the z components of the fields do not vanish identically.

This example merely shows that the results of scalar diffraction theory cannot be exactly right. However, the theory was never intended to give exact answers. Just because the fields are inconsistent does not necessarily mean that they are worse approximations than a theory where the fields satisfy Maxwell's equations. However, the theory that takes into account the vector nature of the fields is in fact more accurate for large angles.

A more consistent approach to the vector theory can be obtained by noting that it is not possible to arbitrarily specify all three components of **E** and **B** at the aperture. It is only necessary to specify the tangential electric fields at the aperture. We now argue that once these fields are known, we know E_x and E_y for $z > 0$, and we can then determine E_z and **B**.

Clearly both E_x and E_y satisfy the scalar wave equation. It follows that if we know these components at $z = 0$, then we can determine them everywhere for $z > 0$. Once we know E_x and E_y, we can use the following equation to determine E_z up to an arbitrary additive function $f(x,y)$:

$$\frac{\partial E_x}{\partial x} + \frac{\partial E_y}{\partial y} + \frac{\partial E_z}{\partial z} = 0 \tag{2.271}$$

Assuming that we have a finite-sized aperture, the field **E**, and in particular the function E_z, must approach zero as $z \to \infty$. This fact allows us to determine this arbitrary function $f(x,y)$. It follows that we can determine E_z. We can now determine **B** by taking the curl of **E** and using Faraday's law. It follows that we can determine all the components of both **E** and **B** once we specify the tangential components of **E** at the aperture.

The vector theory of diffraction [17] approximates the tangential components of the electric field using scalar diffraction, but then computes the z component based on these fields. We will restrict ourselves to the case where the incoming wave has no z component of the electric field. In this case, the scalar theory of diffraction predicts that the diffracted field will also have no z component of the electric field. We will now show that in this situation the z component of the electric field can be ignored provided we are only interested in small angles, a condition that we have already assumed in making the Fresnel approximation.

Suppose that at the plane of the aperture the tangential components of the electric field are given by

$$\left[E_x(x,y,0),E_y(x,y,0)\right]=\left[g_x\left(\frac{x}{R},\frac{y}{R}\right),g_y\left(\frac{x}{R},\frac{y}{R}\right)\right] \tag{2.272}$$

The x and y components of the electric field each satisfy the scalar wave equation. By Fourier transforming the wave equation, we can conclude that

$$E_x(x,y,z)=\frac{1}{4\pi^2}R^2\int_{-\infty}^{\infty}\int_{-\infty}^{\infty}e^{i(k_x x+k_y y)}G_x(k_xR,k_yR)\times e^{iz\sqrt{k^2-k_x^2-k_y^2}}\,dk_x dk_y \tag{2.273}$$

and

$$E_y(x,y,z)=\frac{1}{4\pi^2}R^2\int_{-\infty}^{\infty}\int_{-\infty}^{\infty}e^{i(k_x x+k_y y)}G_y(k_xR,k_yR)\times e^{iz\sqrt{k^2-k_x^2-k_y^2}}\,dk_x dk_y \tag{2.274}$$

where:

$G_x(k_x,k_y)$ and $G_y(k_x,k_y)$ are the Fourier transforms of $g_x(x, y)$ and $g_y(x, y)$, respectively

Using the fact that $\nabla \cdot E = 0$, we can write the field E_z as

$$E_z(x,y,z)=\frac{1}{4\pi^2}R^2\int_{-\infty}^{\infty}\int_{-\infty}^{\infty}e^{i(k_x x+k_y y)}\Gamma(k_x,k_y)e^{iz\sqrt{k^2-k_x^2-k_y^2}}\,dk_x dk_y \tag{2.275}$$

where:

$$\Gamma\left(k_x,k_y\right)=\frac{1}{\sqrt{k^2-k_x^2-k_y^2}}\left[k_xG_x\left(k_xR,k_yR\right)+k_yG_y\left(k_xR,k_yR\right)\right] \tag{2.276}$$

These are exact expressions assuming that we know the tangential electric field at the plane of the aperture.

The expression for E_z is very similar to the expressions for E_x and E_y except that it has $k_x/\sqrt{k^2 - k_x^2 - k_y^2}$ multiplying G_x, and $k_y/\sqrt{k^2 - k_x^2 - k_y^2}$ multiplying G_y. Under the conditions for the Fresnel approximation, we can make the following approximation:

$$\frac{k_x}{\sqrt{k^2 - k_x^2 - k_y^2}} = \frac{k_x}{k} \tag{2.277}$$

and

$$\frac{k_y}{\sqrt{k^2 - k_x^2 - k_y^2}} = \frac{k_y}{k} \tag{2.278}$$

The Fresnel approximation is based on the assumption that k_x/k and k_y/k are both small in the region of interest. It follows that the factors multiplying G_x and G_y will always make the term E_z negligible compared to E_x and E_y.

For example, if the Fresnel number is small, then we can evaluate these integrals using the method of stationary phase. We could put these integrals in dimensionless form and arrange things so that there was a large parameter multiplying the phase. However, we can take a shortcut and note that in the Fresnel approximation, the phases of the integrands are given by

$$\phi = k_x x + k_y y - z\frac{k_x^2 + k_y^2}{2k} \tag{2.279}$$

The phase will be stationary when

$$x - \frac{k_x z}{k} = 0 \tag{2.280}$$

and

$$y - \frac{k_y z}{k} = 0 \tag{2.281}$$

This shows that when we apply the method of stationary phase, the z component of the electric field can be related to the other two components by

$$E_z(x,y,z) \approx \frac{x}{z}E_x(x,y,z) + \frac{y}{z}E_y(x,y,z) \tag{2.282}$$

This shows that provided $|x/y| \ll 1$, and $|y/z| \ll 1$, the z component of the electric field will be negligible compared to the tangential components. This was based on the assumption that the Fresnel number was small. If the Fresnel number is of order 1, we can show that the z component will be small provided only that $kR \ll 1$.

2.7 GEOMETRICAL THEORY OF BEAM SHAPING

2.7.1 ONE-DIMENSIONAL THEORY

In this section, we present a theory of beam shaping based on geometrical optics. Special cases of this theory may be found in the literature on geometrical beam shaping [18]. The theory we present is not the most general one using geometrical optics since we assume that the rays are moved around continuously, and in a very orderly manner. In the geometrical optics limit, it is possible to accomplish the same goal by moving the rays around in a discontinuous and less orderly manner, but when we analyze beam shaping using diffraction theory we will see that this is very undesirable. We believe that it is very difficult to improve on a beam shaping system designed by the techniques described in this section. However, some systems designed this way will work very well, whereas others will work very poorly. One must go beyond the geometrical theory and use diffraction theory in order to understand why this is so. That will be the subject of Section 2.7.2.

We begin by considering the beam shaping problem in one dimension. This theory is directly applicable to cases where the incoming beam has an irradiance distribution that is the direct product of two one-dimensional distributions. A two-dimensional function $f(x,y)$ is the direct product of one-dimensional functions if

$$f(x,y) = f_1(x)f_2(y) \tag{2.283}$$

If both the input and the desired output can be written as a direct product, then the problem can be decomposed into two one-dimensional beam shaping problems. This is the case when we try to turn a Gaussian beam into a rectangular flat-top beam.

We suppose that an incoming parallel beam of light has an irradiance distribution of $I(x/R)$, and at the plane $z = 0$ the beam passes through a phase element that refracts the beam. We would like to determine the phase element such that the irradiance distribution at the plane $z = f$ is given by $(AR/D)Q(x/D)$, where A is a constant chosen so that the energy of our light beam is conserved.

In our analysis, we assume that the aperture contains a lens of focal length f, plus an additional optical element that allows us to shape the beam. In practice these two optical elements can be combined into a single optical element, but this may not be a desirable feature if one wants to use the same element to shape the beam at several different focal planes. We suppose that our beam shaping element introduces a phase shift of $(RD/fc)\phi(x/R)$ at the plane $z = 0$, where c is the speed of light. The goal of our analysis is to determine the function ϕ such that the beam at the plane $z = f$ has the desired shape. This analysis is carried out in three steps:

- Determine the constant A that determines the irradiance of the output beam. This is accomplished by requiring that the total energy of the output beam is the same as the energy of the incoming beam.
- Determine a function that maps rays at the plane of the aperture into rays at the focal plane. In particular, we determine a function $\alpha(\xi)$ such that a ray that passes through the aperture at $x = R\xi$ passes through the focal plane at $x = D\alpha(\xi)$. This step can be carried out by requiring that the energy of any

bundle of rays that enters the aperture is the same as the energy of the same bundle of rays as they pass through the focal plane.
- Determine the function $\phi(\xi)$ that gives us the phase shift introduced by our beam shaping element. Once we know the function $\alpha(\xi)$, this step can be carried out by requiring that the time for a ray to get from $z = -\infty$ to the focal plane is consistent with Fermat's principle.

At this point, the reader may feel annoyed by our introduction of the lengths R and D. For example, it would be simpler if we said that the input beam had the irradiance $I(x)$ rather than $I(x/R)$. However, the lengths R and D have been included in the definition of our irradiance profiles, our normalization constant A, and our phase shift ϕ in order to bring out certain scaling properties of beam shaping. These scaling properties will be especially important in Section 2.7.2 when we discuss diffraction effects.

To carry out the first step in this process, we note that the energy of the incoming beam can be written as

$$E_{\text{in}} = \int_{-\infty}^{\infty} I\left(\frac{s}{R}\right) \mathrm{d}s = R \int_{-\infty}^{\infty} I(s)\,\mathrm{d}s \tag{2.284}$$

The energy of the outgoing beam can be written as

$$E_{\text{out}} = \frac{AR}{D} \int_{-\infty}^{\infty} Q\left(\frac{s}{D}\right) \mathrm{d}s = AR \int_{-\infty}^{\infty} Q(s)\,\mathrm{d}s \tag{2.285}$$

If we equate these two expressions, we arrive at the result

$$A = \frac{\int_{-\infty}^{\infty} I(s)\,\mathrm{d}s}{\int_{-\infty}^{\infty} Q(s)\,\mathrm{d}s} \tag{2.286}$$

We have accomplished the first of our three steps. We now determine the function $\alpha(\xi)$ using the conservation of energy.

$$\int_{-\infty}^{R\xi} I\left(\frac{s}{R}\right) \mathrm{d}s = A \frac{R}{D} \int_{-\infty}^{D\alpha(\xi)} Q\left(\frac{s}{D}\right) \mathrm{d}s \tag{2.287}$$

This is a mathematical statement of the fact that the energy of all of the rays with initial x coordinates less than $R\xi$ must have the same energy as all of the rays at the focal plane that have x coordinate less than $D\alpha(\xi)$. A simple change of variables gives us the following equation:

$$\int_{-\infty}^{\xi} I(s)\,\mathrm{d}s = A \int_{-\infty}^{\alpha(\xi)} Q(s)\,\mathrm{d}s \tag{2.288}$$

As long as the functions $I(s)$ and $Q(s)$ are both positive, it is clear that the function $\alpha(\xi)$ is uniquely determined by this equation. This follows from the fact that for a given value of ξ we can increase the value α until the integral on the right equals the integral on the left. Since $Q(\xi) > 0$, it is clear that for any value of ξ there is only one value of a such that the two integrals will be equal.

The functions $I(s)$ and $Q(s)$ are both nonnegative, but it is possible that they could vanish on certain intervals. This would be the case if we were trying to transform a beam into a beam that had a core of zero irradiance (such as an annulus). In this case, we could have a whole interval of points α that are assigned to the same point ξ. This degenerate case can be thought of as a limiting case of when the functions $I(s)$ and $Q(s)$ are both positive.

Equation 2.288 determines the functions $\alpha(\xi)$. However, there are a few motivations for differentiating this equation to get

$$AQ(\alpha)\frac{d\alpha}{d\xi} = I(\xi) \tag{2.289}$$

This gives us a differential equation for the function $\alpha(\xi)$. One way of solving this differential equation is to integrate this equation once to get back to Equation 2.288. However, if one needs to solve the equation numerically, it may be more convenient to solve the differential equation than to solve Equation 2.288. Another motivation for writing down the differential equation is that when we make the stationary phase approximation to diffraction theory we end up with this differential equation. Yet another motivation comes from the fact when we consider problems that are neither one dimensional nor radially symmetric; we must revert to a differential equation that is analogous to Equation 2.289.

In the energy equation (2.288), we have assumed that the orientation of the incoming rays is the same as the orientation of the rays at the focal plane $z = f$. By this we mean that incoming rays with $\xi \ll 0$ get mapped into rays with $\alpha \ll 0$ at the focal plane, and incoming rays with $\xi \gg 0$ get mapped into rays with $\alpha \gg 0$ at the focal plane. It is possible to reverse the orientation of the rays so that incoming rays with $\xi \gg 0$ end up at the focal plane with $\alpha \ll 0$ and vice versa. In this case, the energy equation can be written as

$$\int_{-\infty}^{R\xi} I\left(\frac{s}{R}\right)ds = A\frac{R}{D}\int_{D\alpha(\xi)}^{\infty} Q\left(\frac{s}{D}\right)ds \tag{2.290}$$

Changing variables in the integrals gives us the following equation:

$$\int_{-\infty}^{\xi} I(s)\,ds = A\int_{\alpha(\xi)}^{\infty} Q(s)\,ds \tag{2.291}$$

If we differentiate this equation, we get

$$AQ(\alpha)\frac{d\alpha}{d\xi} = -I(\xi) \tag{2.292}$$

These two solutions will give identical irradiance distributions as long as we evaluate the irradiance at the plane $z = f$. However, as we move away from the plane $z = f$, these two solutions have very different properties. When we apply the method of stationary phase, the two different types of solutions appear by choosing different signs of the phase function. These will also be discussed in Section 3.4.5.

The solutions derived using Equation 2.288 or 2.291 are the only ones that allow us to shape the beam so that the rays are moved around in a continuous fashion, and the function $\xi(\alpha)$ [the inverse of $\alpha(\xi)$] is single valued. When we study the effects of diffraction, we will see that beam shaping systems that do not satisfy these requirements will suffer much more from the effects of diffraction than ones that do.

We have now completed the first two steps in our analysis, and we are ready to determine the function $\phi(\xi)$. We assume that the rays that enter the aperture are coming in parallel. For our purposes, it is simpler to assume that rays are coming from a distant point source at $(0,-L)$, and we will then let $L \rightarrow \infty$. The travel time for a ray to get from the point source to a point $(D\alpha, f)$ is of three types:

- The time $t_L(\xi)$ to get from the source at $(0,-L)$ to a point $(R\xi,0)$ on the aperture
- The time $t_{\text{delay}}(\xi)$ to get through the Fourier transform lens and the beam shaping element at $(R\xi, 0)$
- The time $t_f(\xi,\alpha)$ to get from a point $(R\xi,0)$ on the aperture to a point $(D\alpha,f)$ at the focal plane

The total travel time is given by

$$t(\xi,\alpha) = t_L(\xi) + t_{\text{delay}}(\xi) + t_f(\xi,\alpha) \tag{2.293}$$

Fermat's principle requires that the travel time of a ray that starts out at $(0,-L)$, passes through the aperture at $(R\xi,0)$, and ends up at $(D\alpha,f)$ must be stationary. This means that it must be stationary compared with the travel time of any nearby ray. In particular, it will be stationary with respect to the travel time of a ray that goes from $(0,-L)$, passes through the aperture at $(R\xi + Rd\xi,0)$, and then goes straight to the point $(D\alpha,f)$. In order for this to be so, we must have

$$\frac{\partial t(\xi,\alpha)}{\partial \xi} = 0 \tag{2.294}$$

We will now see that this equation allows us to determine $\phi(\xi)$.

The travel time t_L is given by

$$t_L(\xi) = \frac{1}{c}\sqrt{L^2 + \xi^2} \approx \frac{1}{c}\left(L + \frac{\xi^2}{2L}\right) \tag{2.295}$$

In the limit as $L \rightarrow \infty$ we end up with the equation

$$\frac{\partial}{\partial \xi} t_L(\xi) = 0 \tag{2.296}$$

The travel time $t_{\text{delay}}(\xi)$ is given by

$$t_{\text{delay}}(\xi) = -\xi^2 \frac{R^2}{2fc} + \phi(\xi)\frac{RD}{fc} \tag{2.297}$$

The first term on the right gives the time delay introduced by the transform lens, and the second term gives the time delay introduced by the beam shaping element.

Taking the derivative of this, we get

$$\frac{\partial}{\partial \xi} t_{\text{delay}}(\xi) = -\xi \frac{R^2}{fc} + \frac{RD}{fc} \frac{\partial}{\partial \xi} \phi(\xi) \tag{2.298}$$

The travel time $t_f(\xi,\alpha)$ is given by

$$t_f(\xi,\alpha) = \frac{1}{c} \sqrt{f^2 + \left(R\xi - D\alpha\right)^2} \tag{2.299}$$

The paraxial approximation assumes that $D^2\alpha^2/f^2 \ll 1$, and $R^2\xi^2/f^2 \ll 1$ so that we can make the following approximation:

$$\sqrt{\left(\varepsilon^2 + f^2\right)} \approx f + \frac{\varepsilon^2}{2f} \tag{2.300}$$

In this approximation, we get

$$t_f(\xi,\alpha) = \frac{f}{c} + \frac{(D\alpha - R\xi)^2}{2fc} \tag{2.301}$$

and hence

$$\frac{\partial}{\partial \xi} t_f(\xi,\alpha) = \frac{R}{fc} \left(R\xi - D\alpha\right) \tag{2.302}$$

Combining our expressions for $\partial/\partial\xi\left(t_L + t_{\text{delay}} + t_f\right)$, we end up with the very simple equation:

$$\frac{\mathrm{d}\phi}{\mathrm{d}\xi} = \alpha(\xi) \tag{2.303}$$

Assuming we know the function $\alpha(\xi)$, the function $\phi(\xi)$ can be determined by quadrature.

We now collect our beam shaping equations into a single set of equations. Given the functions $I(s)$ and $Q(s)$, the phase function $\phi(\xi)$ is determined by first calculating the constant A

$$A = \frac{\int_{-\infty}^{\infty} I(s)\mathrm{d}s}{\int_{-\infty}^{\infty} Q(s)\mathrm{d}s} \tag{2.304a}$$

and then solving the differential equation to determine $\alpha(\xi)$:

$$AQ(\alpha)\frac{\mathrm{d}\alpha}{\mathrm{d}\xi} = \pm\, I(\xi) \tag{2.304b}$$

The sign in this equation depends on whether we have reversed the orientation of the rays. Finally, the function $\phi(\xi)$ is obtained by solving the differential equation:

$$\frac{d\phi}{d\xi} = \alpha(\xi) \tag{2.304c}$$

A very simple scaling property of these equations will now be pointed out. If we determine a beam shaping system for the lengths D and f, then we can use the same phase function $(RD/fc)\phi(\xi)$ for a new beam shaping system with lengths D_1 and f_1, provided $D_1/f_1 = D/f$. This means that we can change the scale of our system by merely using a different quadratic lens, without changing the optical element determined by ϕ. This follows from the fact that the function $\alpha(\xi)$ is independent of the D, f, and R. It follows that the function $\phi(\xi)$ is also independent of these quantities. Clearly, the function $(RD/fc)\phi(\xi)$ will not change as long as we keep the ratio D/f fixed.

2.7.2 Direct Product Distributions

Once again the theory of the Section 2.7.1 can be applied here when both the input and the desired output can be written as direct products. That is, we can use the theory of Section 2.7.1 if we can write

$$I(x,y) = I_1(x)I_2(y)$$

and

$$Q(x,y) = Q_1(x)Q_2(y)$$

In this case, the phase function of the beam shaping element can also be written as a direct product.

$$\varphi(x,y) = \varphi_1(x)\varphi_2(y)$$

One very important example of this is when the input is a circular Gaussian,

$$I(x,y) = e^{-\left(x^2+y^2\right)/2}$$

and the output is a rectangular flat-top beam:

$$Q(x,y) = \mathrm{Rect}\left(\frac{x}{A}\right)\mathrm{Rect}\left(\frac{y}{B}\right)$$

where:

$$\mathrm{Rect}(x) = 1 \mid x \mid \leq 1$$

$$\mathrm{Rect}(x) = 0 \mid x \mid > 1$$

2.7.3 Radially Symmetric Problems

We now derive a geometrical theory of beam shaping that applies when we are trying to convert a radially symmetric beam with irradiance profile $I(r/R)$ into a radially symmetric beam with irradiance profile that is proportional to $Q(r/D)$. We assume that the desired output beam has the irradiance $(AR^2/D^2)Q(r/D)$. As in the one-dimensional case, we begin by computing the normalization constant A. The total energy of the incoming beam is given by

$$E_{\text{in}} = 2\pi \int_0^\infty I\left(\frac{s}{R}\right) s\,\mathrm{d}s = 2\pi R^2 \int_0^\infty sI(s)\,\mathrm{d}s \tag{2.305}$$

The energy of the output beam is given by

$$E_{\text{out}} = \frac{AR^2}{D^2} \int_0^\infty sQ\left(\frac{s}{D}\right) \mathrm{d}s = AR^2 \int_0^\infty sQ(s)\,\mathrm{d}s \tag{2.306}$$

If we require that the energy of the incoming beam is the same as the outgoing beam, we must have

$$A = \frac{\int_0^\infty sI(s)\,\mathrm{d}s}{\int_0^\infty sQ(s)\,\mathrm{d}s} \tag{2.307}$$

We now determine the function $\alpha(\xi)$ such that a ray that encounters our optical element at $(R\xi,0)$ ends up at $(D\alpha,f)$.

The conservation of energy now implies that

$$\int_R^\infty I\left(\frac{s}{R}\right) s\,\mathrm{d}s = \frac{AR^2}{D^2} \int_{D\alpha(\xi)}^\infty sQ\left(\frac{s}{D}\right) \mathrm{d}s \tag{2.308}$$

This equation is a mathematical statement of the fact that the energy of the rays that encounter the plane $z = 0$ with $r > R\xi$ is the same as the energy of the rays that encounter the focal plane with $r > D\alpha(\xi)$.

A simple change of variables gives us the equations

$$\int_\xi^\infty I(s)s\,\mathrm{d}z = A \int_{\alpha(\xi)}^\infty sQ(s)\,\mathrm{d}s \tag{2.309}$$

Just as in the one-dimensional case, we can argue that Equation 2.309 uniquely determines the function $\alpha(\xi)$. As in the one-dimensional case, it may be convenient to differentiate this equation to get a differential equation for $\alpha(\xi)$.

$$A\alpha Q(\alpha)\frac{\mathrm{d}\alpha}{\mathrm{d}\xi} = \xi I(\xi) \tag{2.310}$$

This equation assumes that the ray that starts at the axis of symmetry ends up at the axis of symmetry at $z = f$. In analogy to the one-dimensional case, we could also consider the case where the ray that started on the axis is sent out infinitely far from the axis when $z = f$. We could devise an optical element that did this, but it would necessarily be quite degenerate and suffer from diffraction effects.

Now that we know the function $\alpha(\xi)$, we can use Fermat's principle to determine the optical thickness $\phi(r/R)$ that can actually accomplish this beam shaping.

Once again, let $-r^2/2fc + RD\phi(r/R)/fc$ be the time delay introduced by our optical element, and $z = f$ be the imaging plane. Fermat's principle requires that

$$\frac{d\phi}{\partial\xi} = \alpha(\xi) \tag{2.311}$$

This is exactly the same equation we used in the one-dimensional case. Since we know the function $\alpha(\xi)$, we can determine the function $\phi(\xi)$ by quadrature.

Once again we can argue that the function ϕ is independent of the parameters D and f, and hence the time delay $(RD/fc)\phi(r/R)$ depends on D and f only through the ratio D/f.

2.7.4 More General Distributions

So far we have considered one-dimensional (applicable to direct product profiles) and radially symmetric beam shaping. In this section, we outline how one would determine an optical element that turns an incoming irradiance profile $I(x/R,y/R)$ into an irradiance distribution that is proportional to $Q(x/D,y/D)$ at the image plane f.

The solution to this problem is much more difficult than the ones we have already encountered. We do not have any firsthand experience in actually doing this, but feel that it is worth writing down the equations that would allow one to solve this problem.

We begin by assuming that the irradiance distribution at the focal plane f is equal to $(AR^2/D^2)Q(x/D,y/D)$. In order for the energy of input beam to be the same as the output beam we must have

$$A = \frac{\int_{-\infty}^{\infty}\int_{-\infty}^{\infty} I(s,t)\,ds\,dt}{\int_{-\infty}^{\infty}\int_{-\infty}^{\infty} Q(s,t)\,ds\,dt} \tag{2.312}$$

We now write down an equation for the conservation of energy of any bundle of rays. Suppose rays that encounter the optical element at $(s,t,0)$ end up at $(x(s,t),y(s,t),f)$. To conserve energy, we must have

$$I(s,t) = \pm\, AQ(x(s,t),y(s,t))J(s,t) \tag{2.313a}$$

where

$$J(s,t) = \frac{\partial x}{\partial s}\frac{\partial y}{\partial t} - \frac{\partial x}{\partial t}\frac{\partial y}{\partial s} \tag{2.313b}$$

This is the generalization of the differential form of the energy equations that we have written down previously. It can be justified by noting that the rays in the area

$s < x < s + ds$, $t < y < t + dt$ get mapped into a region with area $J(s,t)ds\,dt$ at the focal plane.

If the time delay produced by our beam shaping element is given by $(RD/fc)\phi(x/R, y/R)$, then Fermat's principle shows us that the function $\phi(s,t)$ must satisfy

$$\frac{\partial \phi}{\partial s} = x(s,t) \tag{2.314a}$$

$$\frac{\partial \phi}{\partial t} = y(s,t) \tag{2.314b}$$

These two equations can be derived almost identically to the one-dimensional and radial cases. We need two equations because we need to guarantee that the path is stationary with respect to changes in both the x and y directions. Using this last set of equations, we can write our energy equation as

$$I(s,t) = \pm AQ\left(\frac{\partial \phi}{\partial s}, \frac{\partial \phi}{\partial t}\right)\left[\frac{\partial^2 \phi}{\partial s^2}\frac{\partial^2 \phi}{\partial t^2} - \left(\frac{\partial^2 \phi}{\partial s \partial t}\right)^2\right] \tag{2.315}$$

This is a nonlinear partial differential equation for the function $\phi(s,t)$. For the special cases where the profiles are radially symmetric, or can be written as direct products, we end up with our previous results. In general, it is not clear that this equation is enough to determine the function $\phi(s,t)$. In order to get a feel for this equation we consider a linearized version of this equation. We will see that the linearized equations end up giving us an equation that is very similar to Poisson's equation. We will see that the linearized equations give us a well-posed mathematical problem, indicating that the same will likely be true of the full nonlinear equations.

In order to get a linearized system of equations, we suppose that the function $I(s,t)$ is almost identical to the function $Q(x,y)$. This would imply that the function $(x(s,t), y(s,i))$ is very nearly equal to (s, t), and hence

$$\phi(s,t) \approx \frac{s^2 + t^2}{2} \tag{2.316}$$

This means that the function ϕ is merely reversing the phase difference caused by the lens that focuses the beam at $z = f$. We will now assume that

$$Q(x,y) = I(x,y) + \delta P(x,y) \tag{2.317}$$

where:

δ is a very small number

We also assume that

$$A = 1 + \delta a \tag{2.318}$$

and

$$\phi(s,t) = \frac{s^2 + t^2}{2} + \delta\psi(s,t) \tag{2.319}$$

To first order in δ, we can write

$$Q\left(\frac{\partial\phi}{\partial s},\frac{\partial\phi}{\partial t}\right) = I(s,t) + \delta\left[P(s,t) + \nabla\psi(s,t)\cdot\nabla I(s,t)\right] \quad (2.320)$$

$$\frac{\partial^2\phi}{\partial s^2},\frac{\partial^2\phi}{\partial t^2} - \left(\frac{\partial^2\phi}{\partial s\partial t}\right)^2 = I + \delta\left(\frac{\partial^2\psi}{\partial s^2} + \frac{\partial^2\psi}{\partial t^2}\right) \quad (2.321)$$

If we expand Equation 2.315 to first order in δ, we end up with the following equation:

$$P(s,t) + aI(s,t)\left(\frac{\partial^2\psi}{\partial s^2} + \frac{\partial^2\psi}{\partial t^2}\right) + \nabla\psi\cdot\nabla I = 0 \quad (2.322)$$

which can be written as

$$\nabla\cdot\left[I(s,t)\nabla\psi\right] = -P(s,t) - aI(s,t) \quad (2.323)$$

If we integrate these equations over the x–y plane, we find that the left-hand side vanishes (assuming ψ vanishes at ∞), and hence the constant a must be chosen so that

$$\int_{-\infty}^{\infty}\int_{-\infty}^{\infty}\int P(s,t)\mathrm{d}s\mathrm{d}t + a + \int_{-\infty}^{\infty} I(s,t)\mathrm{d}s\mathrm{d}t = 0 \quad (2.324)$$

Once we have chosen a in this way, we can uniquely solve for ψ if we require that ψ vanishes at ∞.

The fact that we can solve the linearized equations is an excellent sign that the nonlinear equation (2.315) will uniquely determine the function ϕ.

2.7.5 Examples

We will now present some concrete examples from the geometrical theory of beam shaping. Some of these examples are important for their own sake, but other examples are presented to illustrate some of the difficulties that can arise when applying the geometrical theory. The difficulties will not appear until we analyze them using diffraction theory.

Example 9: Turning a Gaussian into a Flat-Top Beam—I

Let

$$I(s) = \mathrm{e}^{-s^2} \quad (2.325)$$

and

$$Q(s) = 1 \text{ for } |s| < 1 \quad (2.326)$$

$$Q(s) = 0 \text{ for } |s| > 1 \quad (2.327)$$

The normalization of the energy requires that

$$\int_{-\infty}^{\infty} e^{-s^2} ds = 2A \tag{2.328}$$

or

$$A = \frac{\sqrt{\pi}}{2} \tag{2.329}$$

The function $\alpha(\xi)$ must satisfy

$$Q(\alpha)\frac{d\alpha}{d\xi} = \frac{2}{\sqrt{\pi}} e^{-\xi^2} \tag{2.330}$$

As long as $|\alpha| < 1$, this can be written as

$$\frac{d\alpha}{d\xi} = \frac{2}{\sqrt{\pi}} e^{-\xi^2} \tag{2.331}$$

The solution to the above equation can be written as

$$\alpha(\xi) = \text{erf}(\xi) \tag{2.332}$$

where:

$$\text{erf}(\xi) = \frac{2}{\sqrt{\pi}} \int_{0}^{\xi} e^{-s^2} ds \tag{2.333}$$

Since $|\alpha| < 1$ for $-\infty < \xi < \infty$, we conclude that we do not need to consider the case where $Q(\alpha) = 0$.

We now use the equation

$$\frac{d\phi}{d\xi} = \text{erf}(\xi) \tag{2.334}$$

to find the solution

$$\phi(\xi) = \frac{2}{\sqrt{\pi}}\left[\xi\frac{\sqrt{\pi}}{2}\text{erf}(\xi) + \frac{1}{2}e^{-\xi^2} - \frac{1}{2}\right] \tag{2.335}$$

This example has been presented without any reference to the scalings R and D. If we were trying to turn a beam with the initial distribution $I(x/R)$ into a beam with distribution $Q(x/D)$ at the focal plane f, then our beam shaping element would need to introduce a phase delay of $(RD/fc)\phi(x/R)$.

Example 10: Turning a Gaussian into a Flat-Top Beam—II

We consider the same problem as in Example 9. However, this time we present a solution that reverses the order of the rays.

The function $\alpha(\xi)$ must satisfy

$$Q(\alpha)\frac{d\alpha}{d\xi} = -\frac{2}{\sqrt{\pi}} e^{-\xi^2} \tag{2.336}$$

As long as $|\alpha| < 1$, this can be written as

$$\frac{d\alpha}{d\xi} = -\frac{2}{\sqrt{\pi}} e^{-\xi^2} \tag{2.337}$$

The solution to this equation can be written as

$$\alpha(\xi) = -\operatorname{erf}(\xi) \tag{2.338}$$

We now use the equation

$$\frac{d\phi}{d\xi} = -\operatorname{erf}(\xi) \tag{2.339}$$

to find the solution

$$\phi(\xi) = -\frac{2}{\sqrt{\pi}}\left[\xi\frac{\sqrt{\pi}}{2}\operatorname{erf}(\xi) + \frac{1}{2}e^{-\xi^2} - \frac{1}{2}\right] \tag{2.340}$$

Example 11: Turning a Radial Gaussian into a Radial Flattop

We now consider the problem of turning a radial Gaussian into a radial flattop. In particular, suppose $I(s) = e^{-s^2}$, and

$$Q(s) = 1 \text{ if } s < 1 \tag{2.341}$$

$$Q(s) = 0 \text{ if } s > 1 \tag{2.342}$$

In this case, we must choose the constant A so that

$$A\pi = \int_0^\infty s e^{-s^2} ds \tag{2.343}$$

It follows that

$$A = \frac{1}{2\pi} \tag{2.344}$$

Equation 2.310 implies

$$\alpha\frac{d\alpha}{d\xi} = 2\pi\xi e^{-\xi^2} \tag{2.345}$$

If we require that $\alpha(0) = 0$, this equation implies that

$$\alpha(\xi) = \sqrt{2\pi}\sqrt{1 - e^{-\xi^2}} \tag{2.346}$$

Equation 2.311 for ϕ now implies that

$$\phi(\xi) = \sqrt{2\pi}\int_0^\xi \sqrt{1 - e^{-s^2}}\, ds \tag{2.347}$$

Example 12: Turning a Gaussian into a Stairstep

We consider the case where the input beam is a Gaussian

$$I(s) = e^{-s^2} \tag{2.348}$$

and the desired output beam is a stairstep function.

$$Q(s) = \gamma \; |s| < \alpha_0 \tag{2.349}$$

$$Q(s) = 1 \; \alpha_0 < |s| < 1 \tag{2.350}$$

$$Q(s) = 0 \; |s| > 1 \tag{2.351}$$

This situation is clearly symmetrical, so that the phase function $\phi(\xi) = \phi(-\xi)$, and $\alpha(-\xi) = -\alpha(-\xi)$. For this reason, we will only concern ourselves with finding ϕ and α for $\xi > 0$.

The normalization condition requires that

$$A = \frac{\sqrt{\pi}}{2\alpha_0(\gamma - 1) + 2} \tag{2.352}$$

There will be a point ξ_0 that separates the rays that get sent into the first step from those that get sent into the second step. We do not know this point ahead of time, but must calculate its value given the parameters γ and α_0. The function $\alpha(\xi)$ must satisfy

$$\frac{d\alpha}{d\xi} = \frac{1}{A\gamma} e^{-\xi^2} \text{ for } \xi < \xi_0 \tag{2.353}$$

This equation is valid for $\alpha < \alpha_0$. We also have

$$\frac{d\alpha}{d\xi} = \frac{1}{A} e^{-\xi^2} \text{ for } \xi > \xi_0 \tag{2.354}$$

This equation is valid for $\alpha_0 < \alpha < 1$.

The first of these equations can be integrated from 0 to ξ_0 to give

$$\text{erf}(\xi_0) = \frac{2}{\sqrt{\pi}} A\gamma\alpha_0 \tag{2.355}$$

This is not an explicit expression for ξ_0, but it can very quickly be determined using an iterative method such as Newton's method. Once we have determined ξ_0 and A, we have explicit expressions for $\alpha(\xi)$. We can now determine the function $\phi(\xi)$ by solving the following equations

$$\frac{d^2\phi}{d\xi^2} = \frac{1}{A\gamma} e^{-\xi^2} \text{ for } \xi < \xi_0 \tag{2.356}$$

$$\frac{d^2\phi}{d\xi^2} = \frac{1}{A} e^{-\xi^2} \text{ for } \xi < \xi_0 \tag{2.357}$$

along with the requirements

$$\phi(0) = 0 \tag{2.358}$$

and the requirement that ϕ and its derivative are continuous at ξ_0. These equations are almost identical for those of turning a Gaussian into a flattop. Let $\phi_0(\xi)$ be given by

$$\phi_0(\xi) = \xi\frac{\sqrt{\pi}}{2}\mathrm{erf}(\xi) + \frac{1}{2}e^{-\xi^2} - \frac{1}{2} \tag{2.359a}$$

Then the phase function for the stairstep can be written as

$$\phi(\xi) = \frac{1}{A\gamma}\phi_0(\xi) \text{ for } \xi < \xi_0 \tag{2.359b}$$

and

$$\phi(\xi) = \frac{1}{A\gamma}\left[\gamma\phi_0(\xi) + (1-\gamma)\phi_0(\xi_0) + (\xi - \xi_0)(1-\gamma)\mathrm{erf}(\xi_0)\frac{\sqrt{\pi}}{2}\right] \text{ for } \xi > \xi_0 \tag{2.359c}$$

Example 13: Numerical Solutions for Symmetrical Profiles

There are many situations where it is very cumbersome, or impossible, to obtain closed form analytical solutions for the function $\phi(\xi)$. However, it is not difficult to write a computer code that solves for ϕ. We now consider how to write a code for the special case where both $I(s)$ and $Q(s)$ are symmetric with respect to reflections in s. That is,

$$I(s) = I(-s) \tag{2.360a}$$

and

$$Q(s) = Q(-s) \tag{2.360b}$$

In this case, we can argue that

$$\alpha(-\xi) = -\alpha(\xi) \tag{2.361a}$$

and

$$\phi(-\xi) = \phi(\xi) \tag{2.361b}$$

This means that we can solve for α and ϕ on the interval $\xi > 0$, and this will allow us to determine these functions everywhere.

We now outline how one can use an ODE solver to determine the function ϕ, given the functions $Q(s)$ and $I(s)$. In order to do this we first determine the constant A.

$$A = \frac{\int_{-\infty}^{\infty} I(s)ds}{\int_{-\infty}^{\infty} Q(s)ds} \tag{2.362}$$

In many situations, this constant can be determined analytically, even when the function $\phi(\xi)$ cannot. In these situations, one can analytically compute A. In general, one can use the ODE solver to compute the integrals in both the numerator and the denominator. Once the constant A has been determined, we use the ODE solver to solve the following initial value problems:

$$\frac{d\alpha}{d\xi} = \pm \frac{1}{AQ(\alpha)} I(\xi) \tag{2.363a}$$

$$\frac{d\phi}{d\xi} = \alpha(\xi) \tag{2.363b}$$

Either sign can be taken in the first of these equations. As we have already mentioned, each sign corresponds to a physically different solution.

These initial conditions for these equations can be written as

$$\alpha(0) = 0 \tag{2.364a}$$

$$\phi(0) = 0 \tag{2.364b}$$

These equations can now be integrated out to any value of ξ that you want. A plot of $\phi(\xi)$ can be made by outputting the values as the integration proceeds.

2.8 DIFFRACTIVE THEORY OF LOSSLESS BEAM SHAPING

2.8.1 Scaling Properties

We now present a theory of lossless beam shaping that is based on diffraction theory [19]. In the geometrical theory of beam shaping, it is possible to turn a beam with one irradiance distribution into a beam with any desired irradiance distribution, provided only that the energies of the incoming and outgoing beams are the same. However, when diffraction effects are taken into account, this is no longer possible. The geometrical theory is valid provided the wavelength is small. The major goal of this section is to quantify what we mean by a small wavelength. As in our discussion of geometrical beam shaping, we are interested in turning a beam with an incoming irradiance distribution of $I(x/R,y/R)$ at the plane $z = 0$ into a beam with an irradiance distribution of $Q(x/D,y/D)$ at the plane $z = f$. We will see that the parameter

$$\beta = \frac{2\pi RD}{f\lambda} \tag{2.365}$$

is a dimensionless measure of how small the wavelength λ is. If this parameter is large, then the results from the geometrical theory of beam shaping should be valid. If it is small, then diffractive effects will be important. The parameter β is one, but not the only, measure of how difficult our beam shaping problem is. We will see that the smoothness properties of our input and output beam is another important measure of the difficulty of the beam shaping problem.

Suppose that at the plane $z = 0$ the incoming wave field is given by $g(x/R,y/R)$, and we have an aperture that has a lens with focal length f along with an additional phase element $\psi(x/R,y/R)$. The theory of Fourier optics shows us that the wave field at $z = f$ is given by

$$U(x_f,y_f,f) = \frac{1}{i\lambda f} e^{ikf} e^{ik(x_f^2 + y_f^2)/2f} \times \int_{-\infty}^{\infty}\int_{-\infty}^{\infty} g\left(\frac{x}{R},\frac{y}{R}\right) e^{i\psi(x/R,y/R)} e^{-ik(x_f x + y_f y)/f} \mathrm{d}x\mathrm{d}y \tag{2.366}$$

We would like to determine a function ψ such that the output $U(x_f,y_f)$ satisfies

$$|U(x_f,y_f)|^2 = A\frac{R^2}{D^2} Q\left(\frac{x_f}{D},\frac{y_f}{D}\right) \tag{2.367}$$

where:

The function Q determines the shape of the desired irradiance distribution
D determines the scale of the desired irradiance distribution
A is a scaling factor that guarantees that the energy of the output beam is the same as that of the incoming beam

At this point, our problem has the parameters $\lambda = 2\pi/k$, f, R, and D, and it is not clear what we mean when we say the wavelength is small. We can collect all of our parameters into a single parameter by introducing dimensionless coordinates. In particular, assuming we could choose ψ so that our desired output had exactly the right shape, we would have

$$|G(\omega_x,\omega_y)|^2 = \frac{4\pi^2 A}{\beta^2} Q\left(\frac{\omega_x}{\beta},\frac{\omega_y}{\beta}\right) \tag{2.368a}$$

where:

$$G(\omega_x,\omega_y) = \int_{-\infty}^{\infty}\int_{-\infty}^{\infty} g(\xi,\eta) e^{-i(\omega_x\xi+\omega_y\eta)} e^{i\beta\phi(\xi,\eta)} \mathrm{d}\xi\,\mathrm{d}\eta \tag{2.368b}$$

where:

$$\xi = \frac{x}{R} \tag{2.369a}$$

$$\eta = \frac{y}{R} \tag{2.369b}$$

$$\omega_x = x_f \frac{Rk}{f} \tag{2.369c}$$

$$\omega_y = y_f \frac{Rk}{f} \tag{2.369d}$$

$$\psi(\xi,\eta) = \beta\phi(\xi,\eta) \tag{2.369e}$$

We have chosen to write the phase as $\beta\phi$ rather than as ψ. This will be convenient when we are doing the large β approximation. We will refer to Equation 2.368a and b as the dimensionless beam shaping equation. Given the function g, the function Q, and the parameter β, our goal is to determine a constant A and a function $\phi(\xi,\eta)$ such that Equation 2.368a and b is satisfied. This statement of the beam shaping problem is very nice because we have collected all of our parameters into the single parameter β.

2.8.2 One-Dimensional Beam Shaping

As in the theory of geometrical beam shaping, we now consider problems where the incoming beam $g(\xi,\eta)$ and the desired output $Q(s,t)$ can be written as a direct product. This allows us to separate the beam shaping problem into two one-dimensional problems. In particular, we are trying to find a function ϕ and a constant A such that for a given $g(\xi)$, $Q(s)$, and β, we have

$$|G(\omega)|^2 = A\frac{2\pi}{\beta}Q\left(\frac{\omega}{\beta}\right) \tag{2.370a}$$

where:

$$G(\omega) = \int_{-\infty}^{\infty} g(\xi)e^{-i\omega\xi}e^{i\beta\phi(\xi)}d\xi \tag{2.370b}$$

In general, it is not possible to choose ϕ so that Equation 2.370a and b is satisfied exactly. For example, if β is small, then we would need the Fourier transform of $g(\xi)e^{i\phi(\xi)}$ to be very concentrated around the origin. This would contradict the uncertainty principle. To make this statement more precise, we can apply the uncertainty principle to the function $g(\xi)e^{i\beta\phi(\xi)}$ and its desired Fourier transform to get

$$\Delta_g\Delta_G \geq \frac{1}{4} \tag{2.371}$$

where:

$$\Delta_G = \frac{\int_{-\infty}^{\infty}\omega^2|G(\omega)|^2\,d\omega}{\int_{-\infty}^{\infty}|G(\omega)|^2\,d\omega} \tag{2.372}$$

and

$$\Delta_g = \frac{\int_{-\infty}^{\infty}\xi^2|g(\xi)|^2 d\xi}{\int_{-\infty}^{\infty}|g(\xi)|^2 d\xi} \tag{2.373}$$

If we could choose ϕ so that we accomplished our beam shaping exactly, we would have

$$|G(\omega)|^2 = \frac{A2\pi}{\beta} Q\left(\frac{\omega}{\beta}\right) \tag{2.374}$$

This would imply that

$$\Delta_G = \beta^2 \Delta_Q \tag{2.375}$$

where:

$$\Delta_Q = \frac{\int_{-\infty}^{\infty} \omega^2 |Q(\omega)|^2 \, d\omega}{\int_{-\infty}^{\infty} |Q(\omega)|^2 \, d\omega} \tag{2.376}$$

and hence

$$\beta^2 \Delta_g \Delta_Q \geq \frac{1}{4} \tag{2.377}$$

This inequality cannot be satisfied if β is too small. It should be evident that if β is very small, then it will not even be possible to turn the beam into a profile that is even near the desired profile. This shows that it is not possible to do a good job of beam shaping if the parameter β is small. We now consider the case where β is large, and show that in this case if we choose ϕ to be the function obtained from using geometrical beam shaping, then this will nearly satisfy our beam shaping problem.

We begin our analysis of the beam shaping problem by commenting on our decision to write the phase delay as $\beta\phi(\xi)$. This scaling will allow us to use the method of stationary phase to determine the behavior for large values of β. It should be noted that this scaling predicts that the phase function grows linearly with the frequency of light that we are using, a result that would hold if we designed a lens based on geometrical optics, and kept the same lens for all frequencies of light.

If we use the variable

$$\alpha = \frac{\omega}{\beta} \tag{2.378}$$

our beam shaping problem can be written as follows: given the function g, the function Q, and the parameter β, try to determine the constant A and the function ϕ such that

$$G(\alpha) = \int_{-\infty}^{\infty} g(\xi) e^{i\beta[\phi(\xi) - \alpha\xi]} d\xi \tag{2.379a}$$

$$|G(\alpha)|^2 = \frac{2\pi A}{\beta} Q(\alpha) \tag{2.379b}$$

The integral in Equation 2.379a is in a form that can be evaluated using the method of stationary phase. To lowest order in β, the method of stationary phase shows us that the integral is given by

$$G(\alpha) \approx e^{i\pi/4} e^{i\beta\{\phi[\xi(\alpha)]-\xi(\alpha)\}} \sqrt{2\pi} \frac{g[\xi(\alpha)]}{\sqrt{\beta\phi''[\xi(\alpha)]}} \tag{2.380}$$

where the function $\xi(\alpha)$ is determined implicitly by the following equation:

$$\frac{d}{d\xi}\phi[\xi(\alpha)] - \alpha = 0 \tag{2.381a}$$

If we have chosen ϕ so that the beam has the desired output, then we have

$$AQ(\alpha) = \frac{g^2[\xi(\alpha)]}{\phi''[\xi(\alpha)]} \tag{2.381b}$$

With a little bit of manipulation, we can make these equations identical to the equations for geometrical beam shaping. To do this, we begin by differentiating Equation 2.381a with respect to α. This gives us the following equation:

$$\frac{d^2\phi[\xi(\alpha)]}{d\xi^2}\frac{d\xi(\alpha)}{d\alpha} = 1 \tag{2.382}$$

Using this equation, Equation 2.381b can be written as

$$\frac{d\xi(\alpha)}{d\alpha} g^2(\xi) = AQ(\alpha) \tag{2.383}$$

If we use the fact that the irradiance of the incoming beam is given by $|g(\xi)|^2 = I(\xi)$, we get the system of equations:

$$\frac{d\xi}{d\alpha} I(\xi) = AQ(\alpha) \tag{2.384a}$$

$$\frac{d}{d\xi}\phi[\xi(\alpha)] - \alpha = 0 \tag{2.384b}$$

If we integrate Equation 2.384a from $-\infty$ to ∞, we find the normalization condition:

$$A = \frac{\int_{-\infty}^{\infty} I(\xi)d\xi}{\int_{-\infty}^{\infty} Q(\alpha)d\alpha} \tag{2.384c}$$

These equations are identical to Equation 2.304a–c, derived using the geometrical theory of beam shaping.

2.8.3 Two-Dimensional Beam Shaping

We will now quickly summarize how our results can be extended to apply to arbitrary beam shape problems, that is, ones that are not separable. In general, we want to find a function $\phi(\xi,\eta)$ such that

$$G(x,y) = \int_{-\infty}^{\infty}\int_{-\infty}^{\infty} g(\xi)e^{i\beta[\phi(\xi,\eta) - x\xi - y\eta]}d\xi\, d\eta \tag{2.385a}$$

$$|G(x,y)|^2 = \frac{4\pi^2 A}{\beta^2}Q(x,y) \tag{2.385b}$$

An argument almost identical to that used in the separable case shows that the uncertainty principle requires that

$$\beta^2\Delta_g\Delta_Q \geq 1 \tag{2.386a}$$

where:

$$\Delta_g = \frac{\int_{-\infty}^{\infty}\int_{-\infty}^{\infty}(\xi^2+\eta^2)|g(\xi,\eta)|^2\, d\xi\, d\eta}{\int_{-\infty}^{\infty}\int_{-\infty}^{\infty}|g(\xi,\eta)|^2\, d\xi\, d\eta} \tag{2.386b}$$

and

$$\Delta_Q = \frac{\int_{-\infty}^{\infty}\int_{-\infty}^{\infty}(\omega_x^2+\omega_y^2)|Q(\omega_x,\omega_y)|^2\, d\omega_x d\omega_y}{\int_{-\infty}^{\infty}\int_{-\infty}^{\infty}|Q(\omega_x,\omega_y)|^2\, d\omega_x d\omega_y} \tag{2.386c}$$

As in the separable case, this inequality cannot be satisfied if β is too small. We now consider the limit of the integral in Equation 2.385a as $\beta \to \infty$. Using the two-dimensional method of stationary phase, we find that

$$|G(x,y)|^2 \approx \frac{4\pi^2}{\beta^2 J(\xi_0,\eta_0)}|g(\xi_0,\eta_0)|^2 \tag{2.387}$$

where (ξ_0,η_0) are determined implicitly by the stationarity conditions:

$$\frac{\partial}{\partial\xi}\phi(\xi_0,\eta_0) = x \tag{2.388a}$$

$$\frac{\partial}{\partial\eta}\phi(\xi_0,\eta_0) = y \tag{2.388b}$$

and the function J is defined by

$$J(\xi_0,\eta_0) = \frac{\partial^2\phi(\xi_0,\eta_0)}{\partial\xi^2}\frac{\partial^2\phi(\xi_0,\eta_0)}{\partial\eta^2} - \left[\frac{\partial^2\phi(\xi_0,\eta_0)}{\partial\xi\partial\eta}\right]^2 \tag{2.389}$$

If we use the stationarity conditions, we can write the function J as

$$J(\xi_0,\eta_0) = \frac{\partial x(\xi_0,\eta_0)}{\partial\xi}\frac{\partial y(\xi_0,\eta_0)}{\partial\eta} - \frac{\partial x(\xi_0,\eta_0)}{\partial\eta}\frac{\partial y(\xi_0,\eta_0)}{\partial\xi} \tag{2.390a}$$

If we require that the function $|G(x, y)|^2$ has the desired output, we arrive at the following equation:

$$I(\xi_0,\eta_0) = AQ(x,y)J(\xi_0,\eta_0) \tag{2.390b}$$

These last two equations along with the stationarity conditions in Equation 2.388a and b are identical to the two-dimensional equations that we derived using geometrical optics.

2.8.4 Radially Symmetric Problems

In the section on geometrical beam shaping, we considered problems that have radial symmetry. We now consider how to analyze these problems for the effect of diffraction. Problems with radial symmetry can be considered as a special case of the general theory of two-dimensional beam shaping. These problems are important enough that they deserve some special attention. Suppose both the input beam g and the desired output beam Q have radial symmetry. In this case, the phase function ϕ will also have radial symmetry, and we can replace our two-dimensional Fourier transforms with Hankel transforms (Section 2.2.3). The theory of Hankel transforms shows that our beam shaping problems can be phrased as follows.

Given a function $g(\xi)$, a function $Q(\alpha)$, and a parameter β, find a function $\phi(\xi)$ such that

$$G(\alpha) = 2\pi\int_0^\infty g(\xi)\xi e^{i\beta\phi(\xi)}J_0(\alpha\xi)d\xi \tag{2.391a}$$

satisfies

$$|G(\alpha)|^2 = \frac{4\pi^2 A}{\beta^2}Q\left(\frac{\alpha}{\beta}\right) \tag{2.391b}$$

We already know that a lens designed using the first-order term in the stationary phase approximation gives the same lens as the one designed using geometrical optics. Since radially symmetric problems are special cases of the two-dimensional case, if we design a radially symmetric lens using the large β limit, we should get the same lens as and when we design it using geometrical optics. We conclude that the function $\phi(\xi)$ can be obtained by using the techniques described in the section on the geometrical theory of beam shaping. Once we have obtained this function, we can use Equation 2.391a and b to see how our system performs with a finite value of β. To carry this out in practice, we have used ODE solvers in order to compute the function ϕ and to perform the integration in the definition of the Hankel transform.

2.8.5 Continuity of ϕ

We have seen that the first term in the method of stationary phase is identical to the results obtained using geometrical optics. In order for us to know how well the geometrical optics approximation is working, it is necessary to understand the next order term in the stationary phase approximation. We discussed the higher order terms in the method of stationary phase in Section 2.3. There we saw that if the functions ϕ and g are infinitely differentiable, then the next order term in the method of stationary phase is $1/\beta$ times the size of the first term. However, if the third derivative of ϕ (or g) is discontinuous, then the next order term will only be $1/\sqrt{\beta}$ times smaller than the first-order term. If ϕ has a discontinuity in a lower derivative, we get even worse convergence.

We now consider what class of functions $Q(\alpha)$ will lead to discontinuities in the phase function $\phi(\xi)$ designed by using geometrical optics. We will assume that the function $I(\xi)$ is smooth (such as a Gaussian). Equation 2.384a–c shows that the derivative of ϕ has the same continuity properties as the function $\alpha(\xi)$. If we take the derivative of Equation 2.384a with respect to ξ, we find

$$A\left[\frac{dQ}{d\alpha}\left(\frac{d\alpha}{d\xi}\right)^2 + Q(\alpha)\frac{d^2\alpha}{d\xi^2}\right] = \frac{dI}{d\xi} \tag{2.392}$$

We see that if the function $Q(\alpha)$ has a discontinuous derivative at a point $\alpha = \alpha(\xi_0)$ where $I(\xi_0) \neq 0$, then this will lead to a discontinuity in the second derivative of α with respect to ξ, and hence to a discontinuity in the third derivative of ϕ. It follows that discontinuities in the derivatives of Q or I will slow down the convergence toward the geometrical optics limit.

Note that we excluded the case where the discontinuity in Q occurs at a point where I vanishes. In this case, we must have $d\alpha/d\xi = 0$, and when we look at our expression for the second derivative of a, we find that it does not have a discontinuity. Similar arguments hold for the case where Q itself is discontinuous at a point where I vanishes. A very important example of this is the case where one turns a Gaussian profile into a flat-top beam. In that case, the phase function is infinitely differentiable, even though the function $Q(\alpha)$ has a discontinuity in it. This is because the discontinuity in Q occurs as $\xi \to \infty$, and hence at a point where $I(\xi) = 0$.

For the case where the incoming distribution $I(\xi)$ is a Gaussian, we see that discontinuities in the first derivative of Q will lead to discontinuities in the third derivative of ϕ, unless the discontinuity in Q occurs at an extremity. By an extremity, we mean a point where the rays reaching this point have come from points infinitely far off the axis.

2.8.6 One-Dimensional Examples

In order to illustrate the principles of beam shaping, a computer code was written that allows us to compute the function ϕ as well as the effects of using a finite value of β. In these examples, we calculate

$$G(\alpha) = \int_{-\infty}^{\infty} g(\xi) e^{i\beta[\phi(\xi) - \alpha\xi]} d\xi \tag{2.393}$$

by using an ODE integrator. When an analytical expression for ϕ cannot be found, we compute ϕ with the ODE integrator as we are computing the integral. We output the quantity

$$\Gamma(\alpha,\beta) = \frac{2\pi A}{\beta} |G(\alpha)|^2 \tag{2.394}$$

where:

$$A = \frac{\int_{-\infty}^{\infty} I(\xi) d\xi}{\int_{-\infty}^{\infty} Q(\alpha) d\alpha} \tag{2.395}$$

If the effects of diffraction are negligible, the function $\Gamma(\alpha,\beta)$ should be very close to $Q(\alpha)$.

We could have used a code that computed the function ϕ using the technique described in the section on geometrical beam shaping, and then fed this input into a fast Fourier transform for computing the effects of a finite value of β.

In all of the examples we present, we will use the function

$$g(\xi) = e^{-\xi^2/2} \tag{2.396}$$

and hence

$$I(\xi) = e^{-\xi^2} \tag{2.397}$$

Example 14: Turning a Gaussian into a Flattop

We want to turn the output beam into a flattop with

$$Q(\alpha) = 1 \text{ for } |\alpha| < 1 \tag{2.398}$$

$$Q(\alpha) = 0 \text{ for } |\alpha| > 1 \tag{2.399}$$

We have already considered this example in the section on geometrical beam shaping, where it was shown that the function ϕ is given by

$$\phi(\xi) = \frac{2}{\sqrt{\pi}} \left[\xi \frac{\sqrt{\pi}}{2} \operatorname{erf}(\xi) + \frac{1}{2} e^{-\xi^2} - \frac{1}{2} \right] \tag{2.400}$$

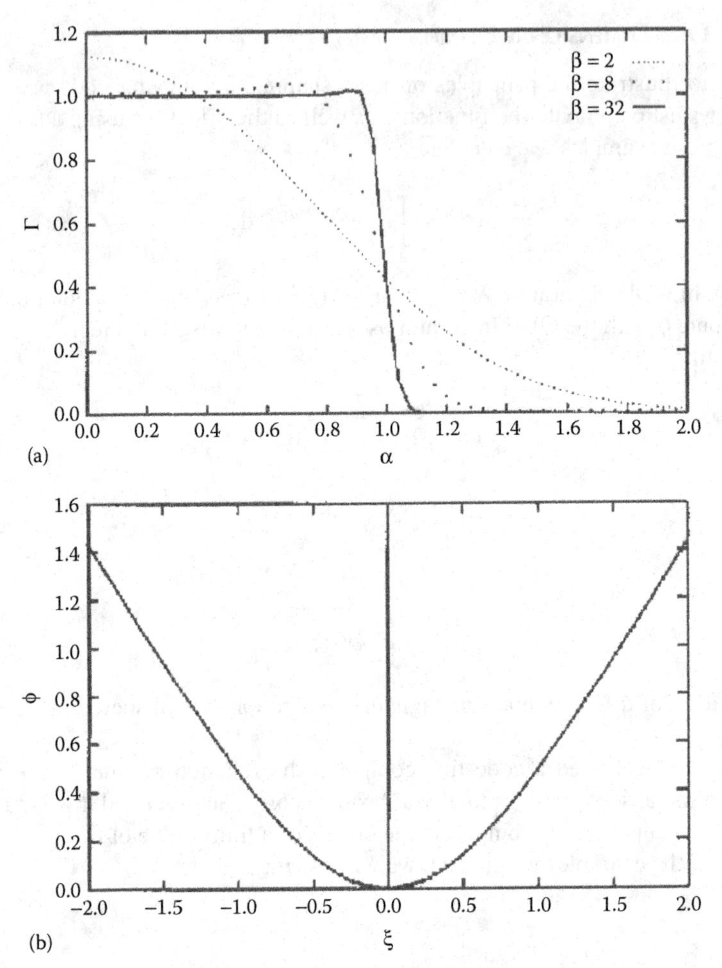

FIGURE 2.5 (a) The intensity distribution for different values of β for the problem of turning a one-dimensional Gaussian into a flat-top beam (Example 14). (b) A plot of the function $\phi(\xi)$ that accomplishes this exactly in the geometrical optics limit.

We will be able to see the effects of having a finite value of β. Figure 2.5a shows plots of $\Gamma(\alpha,\beta)$ for various values of β. We see that for $\beta = 2$ the answer does not look at all like a square pulse, whereas for $\beta = 32$ the answer is starting to look very good.

Figure 2.5b shows a plot of the function $\phi(\xi)$.

Example 15: A Polynomial Output—I

We will now let the output beam be a polynomial that has a hump in it.

$$Q(\alpha) = (1-\alpha^2)(\alpha^2 + \delta) \text{ for } |\alpha| < 1 \tag{2.401}$$

$$Q(\alpha) = 0 \text{ for } |\alpha| > 1 \tag{2.402}$$

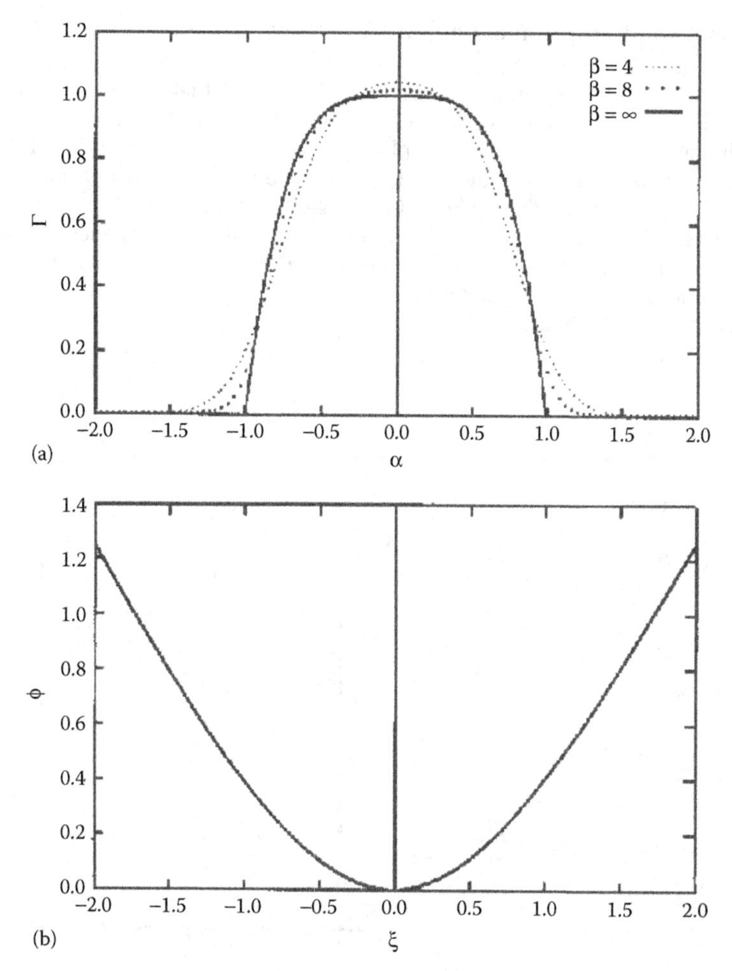

FIGURE 2.6 (a) The intensity distribution for different values of β for the problem of turning a Gaussian into the output $Q(\alpha) = (1 - \alpha^2)(1 + \alpha^2)$ for $|\alpha| < 1$, $Q(\alpha) = 0$ for $|\alpha| > 1$. (b) A plot of the function $\phi(\xi)$ that accomplishes this in the geometrical optics limit.

The constant A is easily computed to be

$$A = \frac{15\sqrt{\pi}}{4 + 20\delta} \tag{2.403}$$

We will choose $\delta = 1$, for this example. Once we know the constant A, we use the ODE solver to compute the function ϕ and the function $\Gamma(\alpha,\beta)$ for various values of β. Figure 2.6a shows plots of $\Gamma(\alpha,\beta)$ for various values of β. Once again, the results are not good for $\beta = 2$, but get progressively better as we increase the value of β. A careful analysis of the data shows that the relative error

$$e(\alpha,\beta) = \frac{\Gamma(\alpha,\beta) - Q(\alpha)}{Q(\alpha)} \tag{2.404}$$

is going to zero like $1/\beta$ everywhere except right at the endpoints $\alpha = \pm 1$.

Figure 2.6b shows a plot of the function $\phi(\xi)$.

Example 16: A Polynomial Output—II

This example is the same as the last example except that we have chosen a value of $\delta = 0.25$ in the function $Q(\alpha)$. This causes the function Q to have two humps in it. Figure 2.7a shows plots of $\Gamma(\alpha,\beta)$ for various values of β, and Figure 2.7b shows a plot of the function $\phi(\xi)$. The relative error is dying down faster than $1/\beta^2$ almost everywhere. Once again right at the ends ($\alpha = \pm 1$), we do not get this behavior, and in the middle ($\alpha = 0$) the convergence is somewhat slower than $1/\beta^2$. The slow convergence at this point does not appear to be illustrating any fundamental principle, but appears to go away if we choose a large enough value of β.

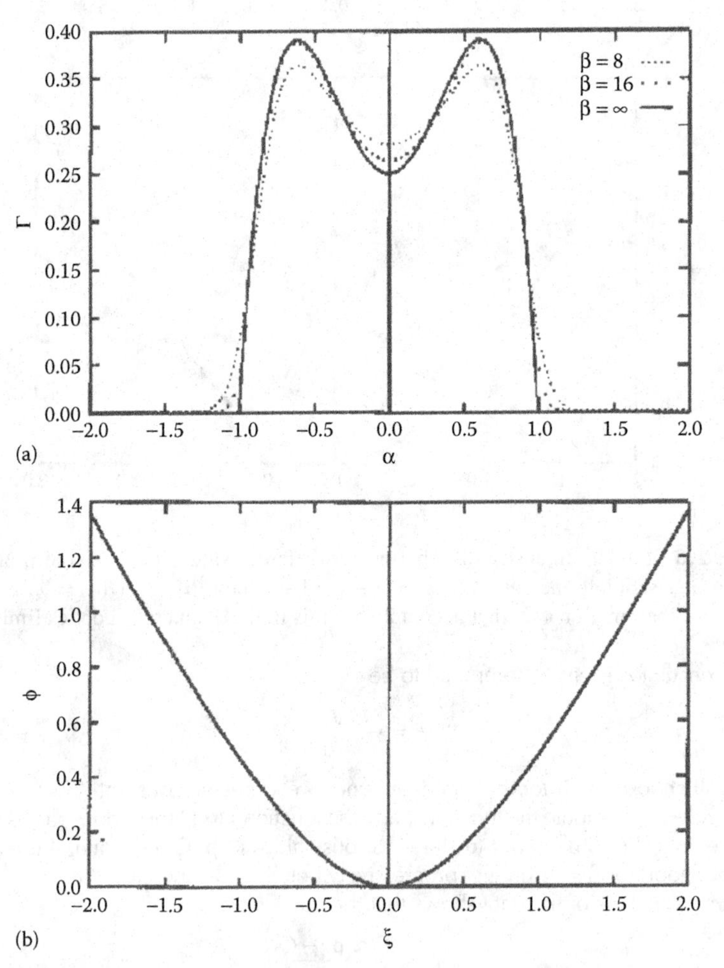

FIGURE 2.7 (a) The intensity distribution for different values of β for the problem of turning a Gaussian into the output $Q(\alpha) = (1 - \alpha^2)(1/4 + \alpha^2)$ for $|\alpha| < 1$ and $Q(\alpha) = 0$ for $|\alpha| > 1$. (b) A plot of the function $\phi(\xi)$ that accomplishes this in the geometrical optics limit.

Example 17: A Triangle Function

Here we consider the case of a triangular function given by

$$Q(\alpha) = 1 - |\alpha| \text{ for } |\alpha| < 1 \tag{2.405}$$

$$Q(\alpha) = 0 \text{ for } |\alpha| > 1 \tag{2.406}$$

This discontinuity in the derivative of the function $Q(\alpha)$ at $\alpha = 0$ causes the function ϕ to have a discontinuity in its third derivative. Figure 2.8a shows plots of the function $\Gamma(\alpha,\beta)$ for various values of β. At the point $\alpha = 0$, the convergence toward the function $Q(\alpha)$ can be seen to be going like $1/\sqrt{\beta}$. Figure 2.8b shows a plot of the function $\phi(\xi)$.

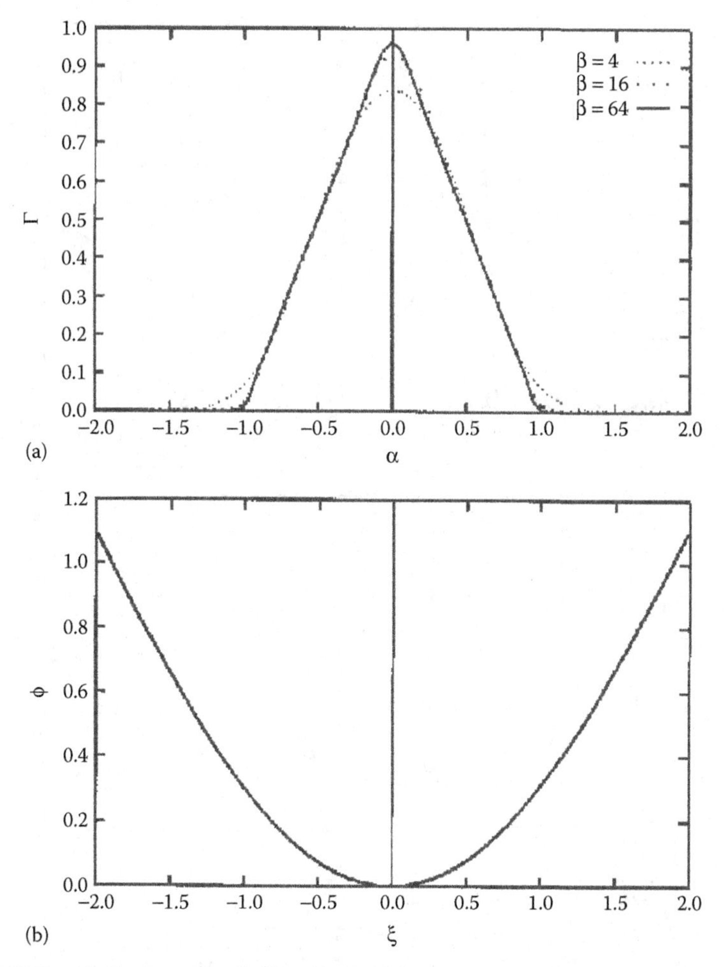

FIGURE 2.8 (a) The intensity distribution for different values of β for the problem of turning a Gaussian into a triangle function $Q(\alpha) = 1 - |\alpha|$ for $|\alpha| < 1$ and $Q(\alpha) = 0$ for $|\alpha| > 1$. (b) A plot of the function $\phi(\xi)$ that accomplishes this in the geometrical optics limit.

Example 18: A Stairstep Function—I

We now consider the case where Q(α) is a stair step function.

$$Q(\alpha) = \gamma \text{ for } |\alpha| < \frac{1}{2} \tag{2.407a}$$

$$Q(\alpha) = 1 \text{ for } \frac{1}{2} < |\alpha| < 1 \tag{2.407b}$$

$$Q(\alpha) = 0 \text{ for } |\alpha| > 1 \tag{2.407c}$$

In this example, we choose $\gamma = 3/4$. The discontinuity in the function Q at $\alpha = \pm 1/2$ causes the function ϕ to have a discontinuity in its second derivative. Figure 2.9a

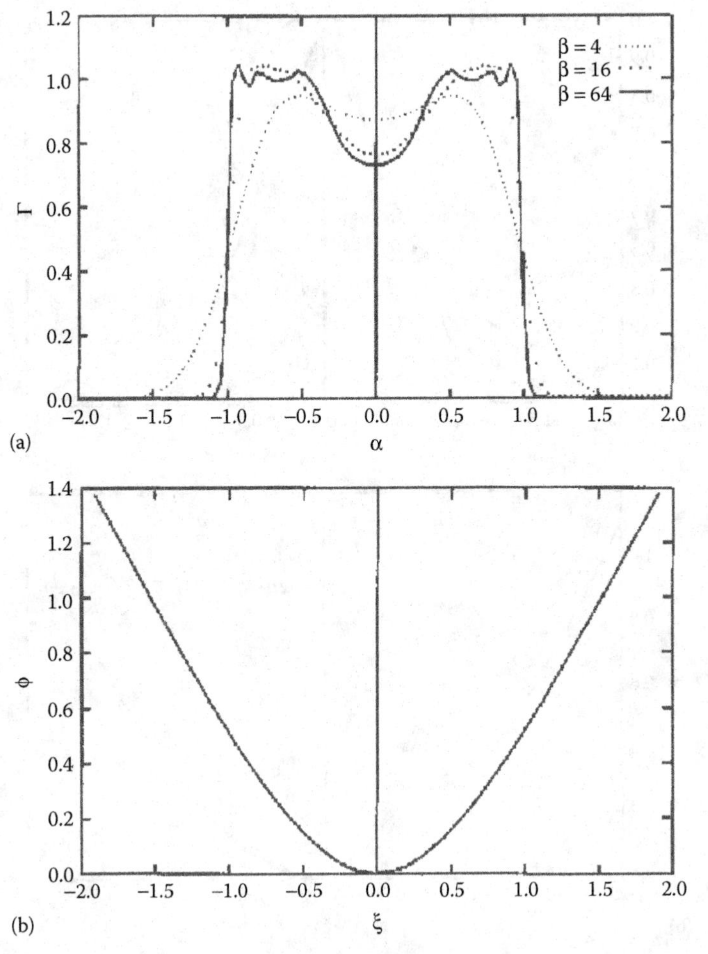

FIGURE 2.9 (a) The intensity distribution for different values of β for the problem of turning a Gaussian into a step function $Q(\alpha) = 3/4$ for $|\alpha| < 1/2$, $Q(\alpha) = 1$ for $1/2 < |\alpha| < 1$, and $Q(\alpha) = 0$ for $|\alpha| > 1$. (b) A plot of the function $\phi(\xi)$ that accomplishes this in the geometrical optics limit.

shows plots of the function $\Gamma(\alpha,\beta)$. The convergences toward the solution $Q(\alpha)$ is extremely slow. Figure 2.9b shows a plot of the function $\phi(\xi)$.

Example 19: A Stairstep Function—II

This is the same as in the last example except we choose the parameter γ in the function Q to be equal to zero. This causes the function Q to have a discontinuity in the first derivative. Figure 2.10a shows plots of the function $\Gamma(\alpha,\beta)$. We see that the convergence toward $Q(\alpha)$ is extremely slow in this case. Figure 2.10b shows a plot of $\phi(\xi)$.

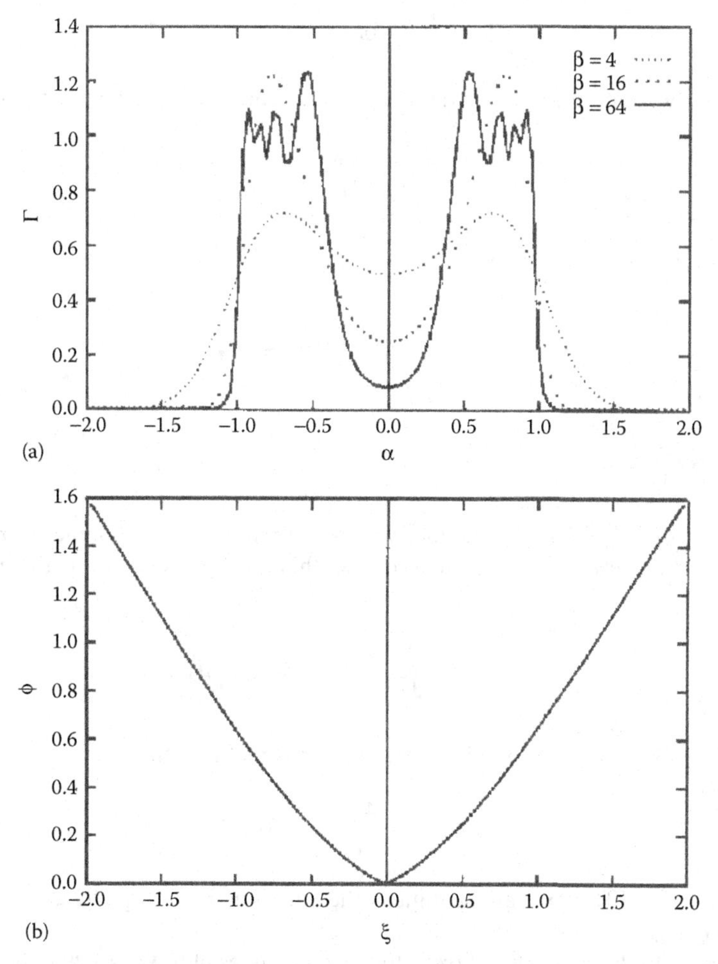

FIGURE 2.10 (a) The intensity distribution for different values of β for the problem of turning a Gaussian into a step function $Q(\alpha) = 0$ for $|\alpha| < 1/2$, $Q(\alpha) = 1$ for $1/2 < |\alpha| < 1$, and $Q(\alpha) = 0$ for $|\alpha| > 1$. (b) A plot of the function $\phi(\xi)$ that accomplishes this in the geometrical optics limit.

2.8.7 An Axisymmetric Example

In Section 2.7.5 on geometric beam shaping, we considered the problem of turning a circular Gaussian beam into an axisymmetric flat-top beam. In this case, the input beam $g(\xi,\eta)$ is given by

$$g(\xi,\eta) = e^{-\xi^2-\eta^2} \tag{2.408}$$

and the desired output is given by

$$Q(x,y) = 1 \text{ for } x^2 + y^2 < 1 \tag{2.409}$$

$$Q(x,y) = 0 \text{ for } x^2 + y^2 > 1 \tag{2.410}$$

The radially symmetric beam shaping equations give us the normalization constant:

$$A = 1 \tag{2.411}$$

The phase function is given by

$$\phi(r) = \int_0^r \sqrt{1-e^{-\xi^2}}\, d\xi \tag{2.412}$$

where:

$r^2 = \xi + \eta^2$

To analyze the effects of diffraction, we compute the radially symmetric Fourier transform. In Section 2.2, we showed that this can be done using the Hankel transform.

$$G(\alpha) = 2\pi \int_0^\infty e^{i\beta\phi(r)} r J_0(\alpha\beta r) g(r) dr \tag{2.413}$$

We are interested in the normalized irradiance of this function.

$$\Gamma(\alpha,\beta) = \frac{4\pi^2}{\beta^2} | G(\alpha)|^2 \tag{2.414}$$

If the effects of diffraction are negligible, then the function $\Gamma(\alpha,\beta)$ should be nearly equal to $Q(\alpha)$.

Figure 2.11a shows a plot of $\Gamma(\alpha,\beta)$ for various values of β. We see that the results are quite similar to the one-dimensional case. Figure 2.11b shows a plot of the function $\phi(\xi)$.

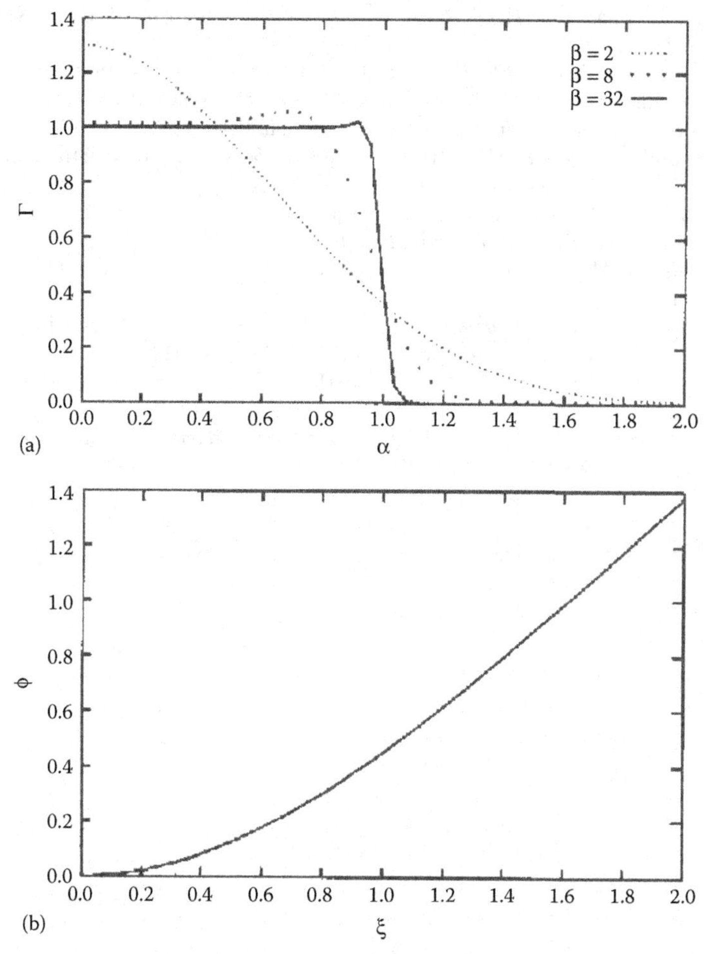

FIGURE 2.11 (a) The intensity distribution for different values of β for the problem of turning a radially symmetric Gaussian into a radially symmetric flat-top beam. (b) A plot of the function $\phi(\xi)$ that accomplishes this in the geometrical optics limit.

REFERENCES

1. G.K. Batchelor. *An Introduction to Fluid Mechanics*. Cambridge: Cambridge University Press, 1967.
2. P.W. Bridgeman. *Dimensional Analysis*. New Haven, CT: Yale University Press, 1922.
3. R.N. Bracewell. *The Fourier Transform and Its Applications*. New York: McGraw-Hill, 1978.
4. R.V. Churchill. *Operational Mathematics*. New York: McGraw-Hill, 1972.
5. L.E. Franks. *Signal Theory*. Englewood Cliffs, NJ: Prentice Hall, 1969.
6. A. Messiah. *Quantum Mechanics*. Amsterdam, the Netherlands: North Holland, 1958.
7. A. Sommerfeld. *Partial Differential Equations in Physics*. New York: Academic Press, 1949.

8. N. Bleistein and R.A. Handelsman. *Asymptotic Expansions of Integrals*. New York: Dover, 1986.
9. M.J. Lighthill. *Waves in Fluids*. Cambridge: Cambridge University Press, 1978.
10. G.B. Whitham. *Linear and Nonlinear Waves*. New York: Wiley, 1974.
11. M. Born and E. Wolf. *Principles of Optics*. London: Permagon Press, 1964.
12. J. Lighthill. *Fourier Analysis and Generalized Functions*. London: Cambridge University Press, 1958.
13. J.D. Jackson. *Classical Electrodynamics*. New York: Wiley, 1962.
14. P.R. Garabedian. *Partial Differential Equations*. New York: Wiley, 1964.
15. N. Bleistein. *Mathematical Methods for Wave Propagation*. Orlando, FL: Academic Press, 1984.
16. J.W. Goodman. *An Introduction to Fourier Optics*. New York: McGraw-Hill, 1968.
17. J.A. Stratton. *Electromagnetic Theory*. New York: McGraw-Hill, 1941.
18. C.C. Aleksoff, K.K. Ellis, and B.D. Neagle. Holographic conversion of a Gaussian beam to a near-field uniform beam. *Optical Engineering* 30:537–543, 1991.
19. L.A. Romero and F.M. Dickey. Lossless laser beam shaping. *Journal of the Optical Society of America A* 13(4):751–760, 1996.

3 Laser Beam Splitting Gratings

Louis A. Romero and Fred M. Dickey

CONTENTS

3.1 INTRODUCTION

Beam splitting gratings are used to split a laser beam into multiple beams for industrial and scientific applications. They are used in a range of applications, including parallel processing in laser machining and material processing, sensor systems, interferometry, communication systems, and image processing and gathering systems. An arbitrary periodic grating will split an incoming beam into a large number of outgoing beams (or orders). For many applications, it is desirable to put as much energy as possible into certain orders, while keeping the energy in all of these orders equal to each other (or more generally in some fixed proportion). If the grating does not absorb any light, it is referred to as a phase grating.

Romero and Dickey have given an overview of the mathematical analysis of laser beam splitting gratings in Reference [1]. In this chapter, we present its condensed version, while including a few new results concerning physical implementation of the theory.

3.2 FOURIER OPTICS

It is well known from books on elementary physics [2,3] that when a light passes through a diffraction grating, it will get split up into diffracted beams traveling in particular directions, and the angular spacing between these diffracted beams depends on the period d of the grating. If λ is the wavelength of the light and d is the period of the grating, the direction cosines of the diffracted beams will be integer multiples of λ/d near the axis. However, few books on elementary physics discuss how the detailed structure of the grating determines the intensity of the diffracted beams. For example, suppose our gratings arise from periodically varying the thickness of the grating. How does the intensity in the diffracted orders depend on the particular periodic function defining the grating?

In almost all theories of beam splitting gratings, it is assumed that the grating is thin and lossless. In this case, the grating can be approximated as an infinitesimally thin surface such that when a light wave passes through the grating, its amplitude, but not its phase, will remain constant. The phase will be changed by $\varphi = \omega\delta\tau$, where ω is the angular frequency of the light and $\delta\tau$ is the time that it takes to pass through the grating (which depends on where we are on the grating). For a line grating, this phase shift will be a function of the distance x in a direction tangent to the surface of the grating. For a thin lossless line grating, the grating can be characterized by giving the phase function $\varphi(x)$ describing the phase shift as one passes through the grating.

For example, if the grating is made out of a material of constant refractive index n and has a variable thickness $h(x)$, the additional time (over the time if the grating were not there) that it takes to pass through the grating at x will be

$$\delta\tau(x) = \frac{h(x)(n-1)}{c} \tag{3.1}$$

and the phase shift will be given by

$$\varphi(x) = \frac{h(x)\omega(n-1)}{c} = \frac{2\pi h(x)(n-1)}{\lambda} \tag{3.2}$$

where:

λ is the wavelength of the light in free space

The theory of Fourier optics [4] shows that the amplitude of the kth diffracted beam will be proportional to the kth Fourier coefficient a_k of the transmission function $t(x) = e^{i\varphi(x)}$ [1,5]:

$$a_k = \frac{1}{d}\int_{-d/2}^{d/2} t(x)e^{-ik\left(\frac{2\pi}{d}\right)x}\,dx \quad k = 0, \pm1, \pm2,... \tag{3.3}$$

where, for a lossless grating, the transmission function $t(x)$ is given by

$$t(x) = e^{i\varphi(x)} \tag{3.4}$$

In particular, the intensity I_k of the kth order is given by

$$I_k = \left|a_k\right|^2 \tag{3.5}$$

More precisely, the diffraction theory predicts that we only get a finite number of these terms since the remaining terms represent the evanescent waves that decay after a few wavelengths of propagation. The upper bound on k is related to the ratio d/λ, where d is the spacing of the grating and λ is the wavelength of the incoming light. The equation for the amplitude of different orders assumes that the illumination is uniform. For the case of beam-like illumination, the output will be the convolution of the plane wave result with the Fourier transform of the beam (see Appendix A of Reference [5]). Assuming that the spatial extent of the beam is large compared to the grating pitch, the uniform beam results are nearly applicable in this case as well.

3.3 THE CONSTRAINED OPTIMIZATION PROBLEM FOR LINE GRATINGS

Suppose that we have a lossless, one-dimensional grating with spatial periodicity d. By a simple rescaling, we can assume that the period of the grating is 2π. Using a coordinate system (x,y,z), we suppose that the grating lies in the plane $z = 0$ and that the transmission function of the grating is independent of the y coordinate. This one-dimensional grating is characterized by a 2π periodic function $\varphi(x) = \varphi(x + 2\pi)$ giving the phase change of an input beam as a function of x at $z = 0$.

In the beam splitting problem under consideration, we choose the transmission function $t(x) = e^{i\varphi(x)}$ to put as much energy as possible into the first $2m + 1$ modes a_k for $k = 0, \pm1, \pm2, \ldots \pm m$. The efficiency of the grating is the ratio of the energy put into these $2m + 1$ orders to the energy in all orders. It is a simple exercise [1] to show that it is not possible to have a lossless grating with 100% efficiency.

After writing down the equations in dimensionless form (so the period of the grating is 2π), we arrive at the following mathematical problem: Choose the real function $\varphi(x)$ so that we maximize the efficiency

$$\eta_{CO} = \frac{\sum_{k=-m}^{m} \left|a_k\right|^2}{\sum_{k=-\infty}^{\infty} \left|a_k\right|^2} \tag{3.6}$$

subject to the constraint that

$$\left|a_k\right|^2 = \left|a_0\right|^2 \quad k = -m,m \tag{3.7}$$

where:

a_k is the kth Fourier coefficient of $e^{i\varphi(x)}$

$$a_k = \frac{1}{2\pi}\int_{-\pi}^{\pi} e^{i\varphi(x)} e^{-ikx} dx \tag{3.8}$$

This is just one example of a beam splitting problem. There are several generalizations of this problem, including the problem of splitting a beam so that the amplitudes of the first $2m + 1$ orders are not all the same, but in some fixed proportion. For example, while trying to split a beam so the mode a_0 has twice as much energy as the other $2m$ modes of interest. Another generalization arises from requiring that we maximize the energy in a different set of modes. The most common variant is to maximize the energy in the modes $[\pm 1, \pm 3, \pm 5, \ldots \pm(2m + 1)]$.

At first sight, this appears to be an infinite-dimensional optimization problem because specifying a function $\varphi(x)$ requires an infinite number of parameters (such as the coefficients in a Fourier series). However, for the case of $m = 1$ (the triplicator), Gori et al. [6] used methods from the calculus of variations to reduce this to a finite-dimensional optimization problem involving a single parameter. Using similar techniques, Borghi et al. [7] solved the problem of splitting a beam into two beams that are not necessarily of the same intensity. Romero and Dickey [1,5,8] generalized this procedure for arbitrary values of m, as well as for two-dimensional gratings. In particular, they showed that the optimal phase function $\varphi(x)$ can always be written as

$$t(x,\boldsymbol{\alpha},\boldsymbol{\mu}) = e^{i\varphi(x)} = \frac{s(x,\boldsymbol{\alpha},\boldsymbol{\mu})}{|s(x,\boldsymbol{\alpha},\boldsymbol{\mu})|} \tag{3.9}$$

where:

$$s(x,\boldsymbol{\alpha},\boldsymbol{\mu}) = \sum_{k=-m}^{m} \mu_k e^{i\alpha_k} e^{ikx} \tag{3.10}$$

where:

$\boldsymbol{\alpha}$ and $\boldsymbol{\mu}$ are vectors containing the parameters α_k and μ_k for $k = -m,m$

The formulas in Equations 3.9 and 3.10 apply to the case in which we are trying to put as much energy as possible into the first $2m + 1$ Fourier coefficients. More generally, if we are trying to put as much energy as possible into a set of modes $k \in K$, we replace Equation 3.10 by

$$s(x,\boldsymbol{\alpha},\boldsymbol{\mu}) = \sum_{k\in K} \mu_k e^{i\alpha_k} e^{ikx} \tag{3.11}$$

If we multiply all of the constants μ_k by the same number, we do not change $e^{i\varphi(x)}$ in Equation 3.9. For this reason, we can arbitrarily set one of the μ_k to unity. Furthermore, for any periodic function $\varphi(x)$, we do not change the magnitude of any of the Fourier coefficients of $e^{i\varphi(x)}$ by adding a constant $\varphi(x)$ or shifting $\varphi(x)$ by any amount x_0. This allows us to arbitrarily set two of the phases α_k. Thus, when solving a beam splitting problem with N beams, we have $2N - 3$ parameters to vary.

In general, if we compute the Fourier coefficients of $t(x,\boldsymbol{\alpha},\boldsymbol{\mu})$ as defined in Equation 3.9, we will get

$$t(x,\alpha,\mu) = \sum_{k \in K} \gamma_k e^{i\beta_k} e^{ikx} + R(x,\alpha,\mu) \tag{3.12}$$

where:

$R(x,\boldsymbol{\alpha},\boldsymbol{\mu})$ gives the other Fourier coefficients that are not of interest to us

In Reference [5], it was shown that for an optimal solution, we must have $\alpha_k = \beta_k$; that is, the phase of the relevant Fourier coefficients of $t(x,\boldsymbol{\alpha},\boldsymbol{\mu})$ must be the same as that of $s(x,\boldsymbol{\alpha},\boldsymbol{\mu})$. If we apply this condition to the $N-2$ unknown phases, and also require that all of the γ_k are equal to each other, this gives us $2N-3$ equations in $2N-3$ unknowns. The parameters $\boldsymbol{\mu}$ and $\boldsymbol{\alpha}$ can be found by solving this system of nonlinear equations using Newton's method. In practice, it is necessary to have a good initial guess for the phase $\boldsymbol{\alpha}$, which can be found by solving the least-squares optimization problem discussed in References [1,5] and Section 3.4.

The fact that the optimal phase grating can be specified by a finite sum (as in Equations 3.9 and 3.10) can simplify the calculation of the optimal phase grating. It also simplifies the communication of results to other workers in the field.

3.4 SOME ALTERNATIVE OPTIMIZATION PROBLEMS

There is a considerable literature on laser beam splitting that does not involve finding the optimal gratings as discussed in Section 3.3. For example, the pioneering work on laser beam splitting was done by Dammann [9,10]. In this and subsequent works [11–16], it was assumed that the gratings were binary, having only two different phases. Extensions to gratings that have a finite number of phases were given in References [17,18]. These gratings with finite numbers of phases are known as Dammann gratings. General reviews of Dammann gratings can be found in References [1,19–21]. Advances in manufacturing techniques have made it possible to manufacture continuous gratings. Hence, in this chapter, we will give no more discussion of Dammann gratings.

Even for the case of continuous gratings, there is a considerable literature devoted to solving other optimization problems that differ from the constrained optimization problem [1] discussed in Section 3.3. We will not present any of the results for these optimization problems, but we will briefly mention what they are and why they are useful.

Romero and Dickey [1] discuss two alternative optimization problems that have been used in laser beam splitting. The least-squares optimization problem is based on the following reasoning. If we could design a grating with 100% efficiency, the transmission function would be given by

$$t(x) = \gamma \sum_{k=-m}^{m} e^{i\alpha_k} e^{ikx} \tag{3.13}$$

where:

γ is a normalization constant and
α_k are the unknown phases

As we have already stated, it is not possible for a function $t(x) = e^{i\varphi(x)}$ to have this form. The least-squares optimization adjusts the parameters γ and α_k and the function $\varphi(x)$ so that we minimize the mean square difference between $e^{i\varphi(x)}$ and the function $t(x)$. In this optimization problem, the optimal phase will typically not produce a solution in which the energy in all of the orders of interest is the same. The solutions to this problem are not as useful as those to the constrained optimization problem. However, the solutions to this problem are easier to find than those to the constrained optimization problem. In particular, the solutions have the same form as in Equation 3.9, except that the coefficients μ_k can all be assumed to be equal to 1. For this reason, the number of parameters that need to be searched is reduced by a factor of 2 (compared to the constrained optimization problem). These efficiencies were tabulated in Reference [22]. They prove that the efficiencies obtained from the least-squares optimization problem are a bound on the efficiencies obtained from the constrained optimization problem [5]. Wyrowski [23] proved a similar result. However, his results were proven in a more general setting, which had an error in them that was corrected in Reference [1].

The third type of optimization problem is called the minimum variance optimization problem [1,24,25]. In this problem, the transmission function of the grating is assumed to have a finite Fourier series as in Equation 3.13, and hence it is not of the form $e^{i\varphi(x)}$. Since the amplitudes of the Fourier coefficients are all the same, the grating puts the exact desired amount of energy in the order of interest. However, the grating is not lossless. The phases of the Fourier coefficients are chosen so that the amplitude of the transmission is as uniform as possible; that is, we are trying to minimize the variance of the transmission function $t(x)$ given by Equation 3.13. As with the least-squares optimization problem, this involves half the number of parameters of the constrained optimization problem and can be useful in finding good initial guesses for solving the constrained optimization problem using Newton's method.

3.5 RESULTS FOR LINE GRATINGS

We begin by showing how the formulas in Equations 3.9 and 3.10 apply to the simple cases of two and three beam splitting. For the case of three beam splitting, we have $m = 1$ in Equation 3.10. We show that the parameters for the optimal solution can be chosen so that $\mu_0 = 1$, $\alpha_0 = 0$, $\mu_1 = \mu_{-1}$, and $\alpha_1 = \alpha_{-1} = \pi/2$ [1,5]. Then Equation 3.10 will become

$$s(x,\alpha,\mu) = 1 + 2i\mu\cos(x) \tag{3.14}$$

Using Equation 3.9, this shows that $\tan[\varphi(x)] = 2\mu\cos(x)$. Hence,

$$\varphi(x) = \tan^{-1}[2\mu\cos(x)] \tag{3.15}$$

The parameter μ must be adjusted to maximize the efficiency. Numerical calculations show that $\mu = 1.32859$ and that the efficiency associated with this parameter is $\eta_{CO} = 0.92556$. This is equivalent to the solution presented in Reference [6] if we shift the solution by $\pi/2$.

For the case of two beam splitting, we use Equation 3.11 where K consists of $k = \pm 1$. In References [1,5], we show that we can choose $\mu_1 = \mu_{-1} = 1$ and $\alpha_1 = \alpha_{-1} = 0$, which gives the following equation:

$$s(x,\alpha,\mu) = \cos(x) \tag{3.16}$$

Hence, using Equation 3.9, we get $e^{i\varphi(x)} = \text{sgn}[\cos(x)]$. This gives us a binary filter, in which the phases are either 0 or π. It is easily shown that this has an efficiency of $\eta_{CO} = 8/\pi^2$.

Table 3.1 shows the results for line gratings that split a beam into an odd number of modes. It gives the coefficients α_k and μ_k specifying the function $e^{i\varphi(x)}$ given in Equations 3.9 and 3.10. The table gives the values of the efficiencies η_{CO} for the constrained optimization problem as well as the efficiencies η_{LS} for the least-squares optimization problem.

Table 3.2 gives the results for splitting a beam into an even number of modes. To understand this table, we note that when splitting a beam into an even number of modes, we use the gratings such that the transmission function $f(x) = e^{i\varphi(x)}$ satisfies the following equation:

$$f(x+\pi) = -f(x) \tag{3.17}$$

TABLE 3.1
Optimum Efficiencies for Splitting a Beam into an Odd Number of Beams

N_{modes}	η_{LS}	η_{CO}	α and μ
3	93.81	92.56	$\alpha = \pi/2$
			$\mu = 1.329$
5	96.28	92.12	$\alpha = (-\pi/2,\pi)$
			$\mu = (0.459,0.899)$
7	97.53	96.84	$\alpha = (-0.984,1.891,0.748)$
			$\mu = (1.289,1.463,1.249)$
9	99.34	99.28	$\alpha = (0.720,5.567,3.033,1.405)$
			$\mu = (0.971,0.964,0.943,1.029)$
11	98.38	97.71	$\alpha = (0.311,4.492,2.847,5.546,4.406)$
			$\mu = (1.207,1.297,1.483,1.427,1.275)$
13	98.57	97.53	$\alpha = (2.308,4.345,1.517,1.692,0.066,6.243)$
			$\mu = (0.912,0.968,0.806,0.923,1.099,1.027)$
15	98.21	97.29	$\alpha = (2.625,4.534,0.970,2.983,3.328,4.070)$
			$\mu = (4.945,1.116,1.463,0.930,1.114,1.466,1.359,1.211)$

Note: We also list the values of α_k and μ_k in Equations 3.9 and 3.10 needed to obtain the following solutions: $\alpha_k = \alpha_{-k}$, $\mu_k = \mu_{-k}$ as well as $\alpha_0 = 0$ and $\mu_0 = 1$. If $N_{modes} = 2M + 1$, the vectors $\boldsymbol{\alpha}$ and $\boldsymbol{\mu}$ contain the values $\boldsymbol{\alpha} = (\alpha_1,\alpha_2,\ldots\alpha_M)$ and $\boldsymbol{\mu} = (\mu_1,\mu_2,\ldots\mu_M)$. The optimal phase functions for $N_{modes} = 3, 11$ are given in Figure 3.1.

TABLE 3.2
Optimum Efficiencies for Splitting a Beam into an Even Number of Beams

N_{modes}	η_{LS}	η_{CO}	α and μ
4	91.94	91.19	$\alpha = 4.438$
			$\mu = 0.523$
6	91.41	88.17	$\alpha = (0.863,3.069)$
			$\mu = (0.274,0.487)$
8	96.12	95.94	$\alpha = (0.724,3.668,5.367)$
			$\mu = (0.560,0.601,0.544)$
10	95.79	92.69	$\alpha = (0.152,4.683,2.681,0.651)$
			$\mu = (0.598,0.412,0.211,0.546)$
12	95.93	95.36	$\alpha = (4.562,3.704,5.465,3.448,1.725)$
			$\mu = (0.523,0.424,0.509,0.586,0.538)$
14	96.80	96.34	$\alpha = (0.235,2.906,1.661,1.521,4.847,2.527)$
			$\mu = (0.430,0.471,0.419,0.505,0.511,0.545)$

Note: Here the modes are given by $k = \pm 2m + 1$, $m = 1$, and M. We also list the values of α_k and μ_k in Equation 3.11 needed to obtain these. Our solutions have $\alpha_k = \alpha_{-k}$, $\mu_k = \mu_{-k}$ as well as $\alpha_1 = 0$ and $\mu_1 = 1$. If $N_{modes} = 2M$, the vectors $\boldsymbol{\alpha}$ and $\boldsymbol{\mu}$ contain the values $\boldsymbol{\alpha} = (\alpha_3, \ldots \alpha_{2M-1})$ and $\boldsymbol{\mu} = (\mu_3, \ldots \mu_{2M-1})$. The optimal phase functions for $N_{modes} = 4$ are given in Figure 3.2.

As discussed in section 2.3 of Reference [1] and in Reference [13], this is a necessary and sufficient condition for all of the even Fourier coefficients of $f(x)$ to vanish. As pointed out in Reference [1], if we solve the problem of putting as much energy as possible into the modes $[\pm 1, \pm 3, \ldots \pm(2m + 1)]$, without requiring that $e^{i\varphi(x)}$ satisfies Equation 3.17, the optimal solution will in fact end up in satisfying Equation 3.17; that is, the optimal solution will be guaranteed of putting no energy into any even mode. Thus, in Table 3.2, the function $s(x,\boldsymbol{\alpha},\boldsymbol{\mu})$ in Equation 3.11 is summing over the odd modes from $-N_{modes} + 1$ to $N_{modes} - 1$.

The theory of laser beam shaping presented in Chapter 2 shows that if the space–bandwidth product is large enough, we can nearly find a phase-only function whose Fourier transform is a flattop. A simple extension of this argument shows that when m is large, we should be able to find a phase-only function such that the Fourier coefficients a_k for $k = 0, \pm 1, \pm 2, \ldots \pm m$ are all nearly equal to each other. Tables 3.1 and 3.2 show that as m grows, it is possible to solve both the even and odd beam splitting problems with high efficiency. However, it should be noted that the efficiency is not monotonic with m.

Figures 3.1 and 3.2 show the phase functions for several of the optimization problems. Note that the optimal phase function for four beam splitting is discontinuous. This is typical of beam splitting problems with an even number of modes. To understand why this happens, note that if we assume that the optimal solution is symmetric (which it is found to be), we will have the following equation:

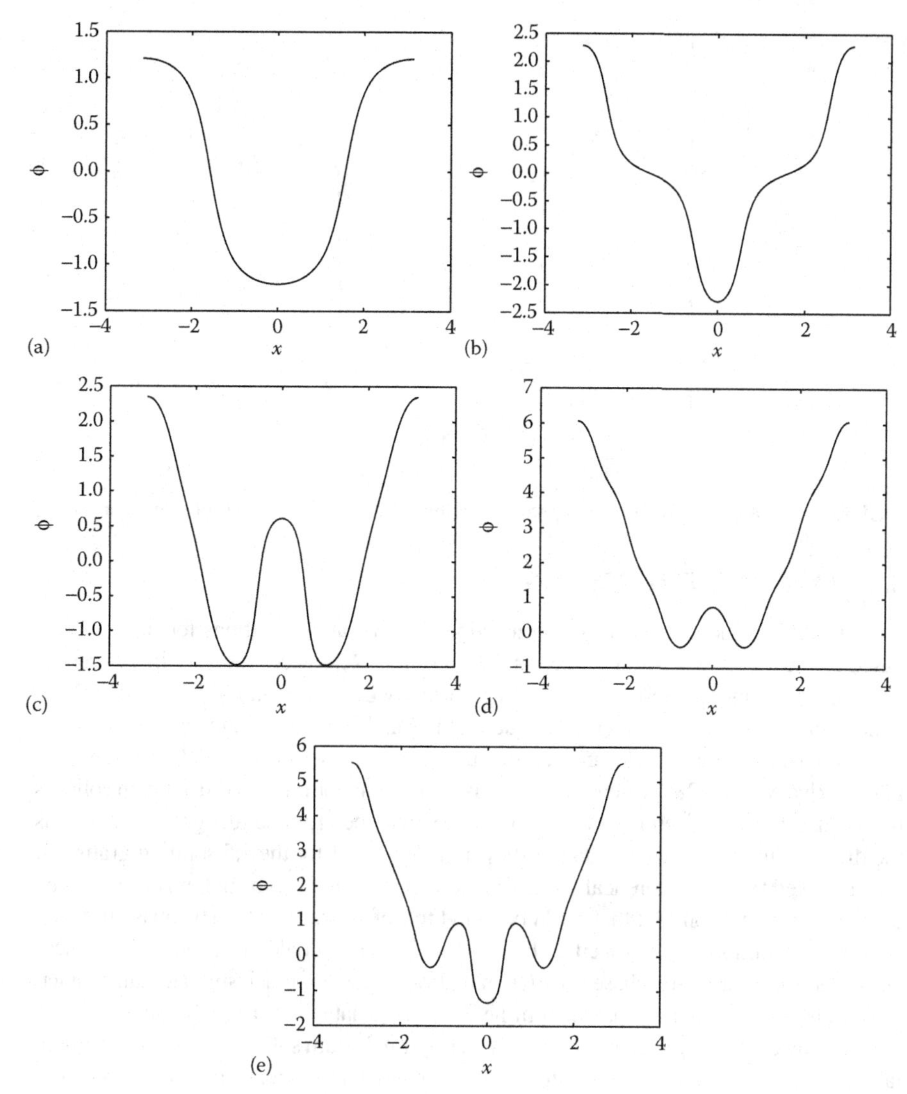

FIGURE 3.1 Plots of the optimal phase functions for one-dimensional beam splitting: (a) $N_{\text{modes}} = 3$; (b) $N_{\text{modes}} = 5$; (c) $N_{\text{modes}} = 7$; (d) $N_{\text{modes}} = 9$; (e) $N_{\text{modes}} = 11$.

$$s(x,\alpha,\mu) = e^{i\alpha_1}\cos(x) + e^{i\alpha_3}\cos(3x) \tag{3.18}$$

This will vanish when $x = \pi/2$. When we compute $e^{i\varphi(x)}$ using Equation 3.9, we will get a different value as we approach $x = \pi/2$ from above and below. This results in a discontinuous phase.

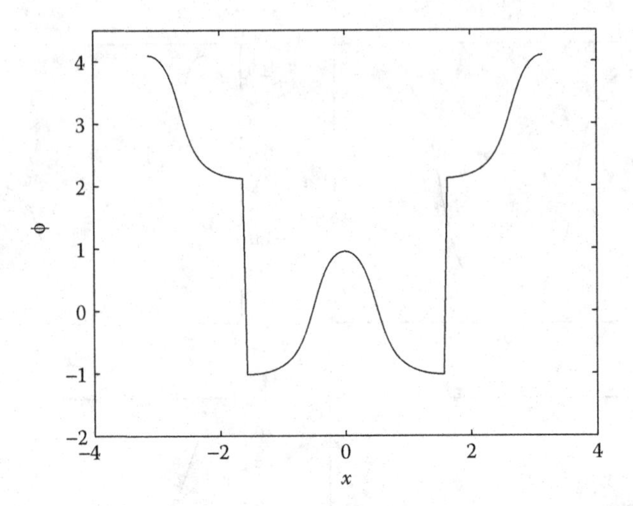

FIGURE 3.2 A plot of the optimal phase function for one-dimensional four beam splitting.

3.6 EXPERIMENTAL RESULTS

Experimental demonstration of the constrained optimization solutions for line gratings based on the theory discussed in Section 3.3 is given in References [26–29]. In References [28,29], the gratings are fabricated using a proprietary Lissotschenko Mikrooptik (LIMO) nonetching material processing technique that is suitable for the manufacturing of high-precision, free programmable, and continuous surface profiles in optical glasses and crystals. Miklyeav et al. [28] present the experimental results for three and five beam splitters based on the results given in Reference [5]. They give the profile of the grating as well as the diffraction pattern. Their data are shown in Figure 3.3 for the 1:5 splitting grating. It may be noted that the theoretical phase function for the three beam splitter is specified by a single numerical constant that could be found in References [1,5,6]. The phase function for the five beam splitter is specified by two constants that could be found in References [1,5]. Communicating the phase function in this way is clearly much simpler than contacting the author of a paper and having him/her send you a data file with the phase function in it. Miklyeav et al. [29] give the results for a 1:11 optimal beam splitter. Further experimental implementation of the theory using spatial light modulators can be found in References [26,27]. Albero et al. present the results for the implementation of a 1:7 splitting grating with both equal and unequal intensities for the orders.

3.7 SQUARE GRATINGS

3.7.1 A Two-Dimensional Beam Splitting Problem

In the original papers on Dammann gratings [9,10], the authors considered two-dimensional gratings, but limited themselves to gratings that were separable, that is, where the transmission function $t(x,y)$ could be written as $t(x,y) = t_1(x)t_2(y)$. In References [30,31], this work was extended to include nonseparable gratings. In these papers, they considered the binary gratings and divided up the unit cell into rectangular blocks, assigning a phase to each block. The papers [32,11,19,33–35] took similar approaches.

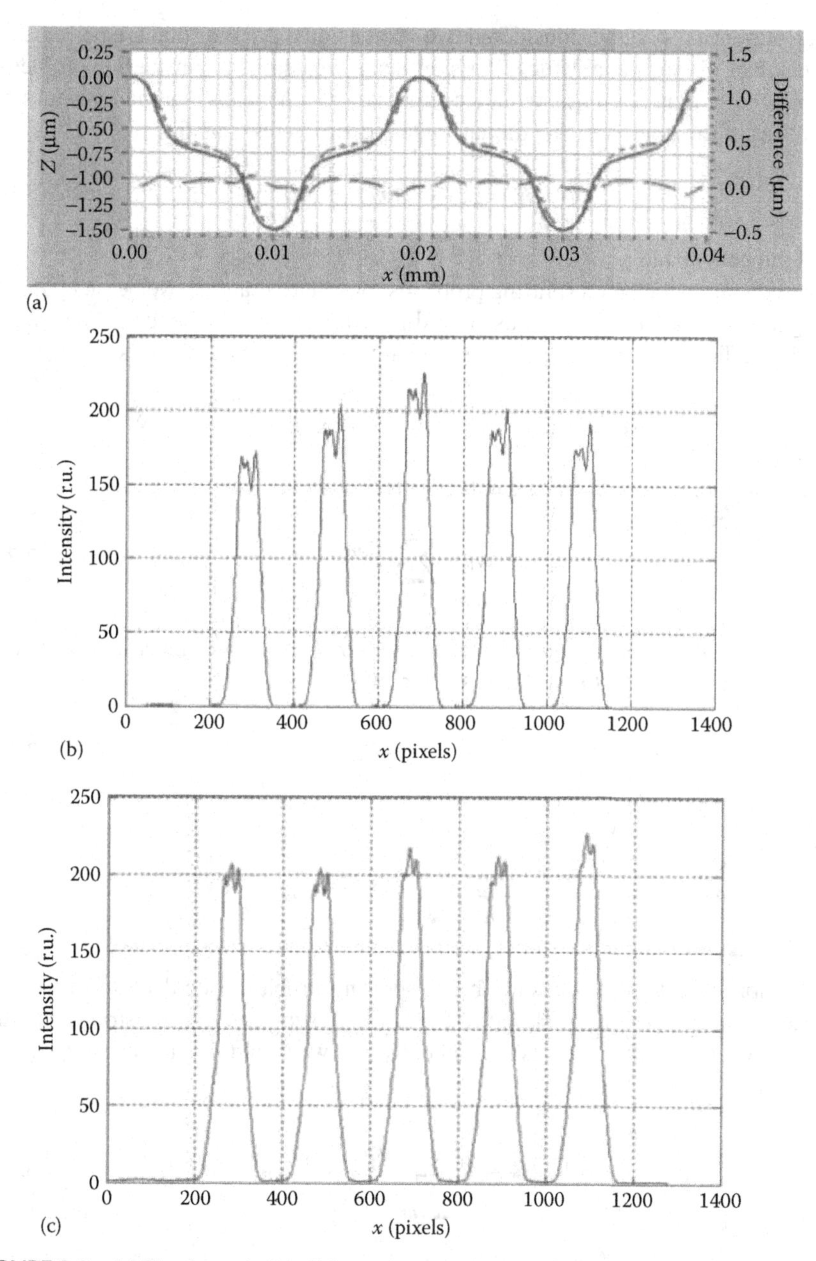

FIGURE 3.3 (a) The theoretical (solid curve) versus the manufactured (dot-dashed curve) profile for a 1:5 beam splitter using the theoretical profile. (From L.A. Romero and F.M. Dickey, *Journal of the Optical Society of America A*, 24, 2280–2295, 2007. With permission.) The period of the grating is 20 m, (b) Cross section of the far-field intensity distribution of 1:5 beam splitter illuminated by multimode Nd:YAG laser. The angle between the beam and the surface is 90°. (c) Same as (b) except the incoming beam is tilted by 15° to correct for sag aspheric fabrication errors. The angle between the beam and the surface is 75°. (Reproduced from Y.V. Miklyeav et al., *Proceedings of SPIE*, 7640, 2010. With permission; Courtesy of LIMO GmbH, Germany.)

In References [1,8], we considered two-dimensional gratings that are periodic on a rectangular or hexagonal lattice. For simplicity, we begin by considering problems in which the grating has a square lattice structure, which means that the transmission coefficient of the grating (using dimensionless coordinates) satisfies

$$\varphi(x,y) = \varphi(x + 2\pi m, y + 2\pi n) \tag{3.19}$$

for all integers m and n.

To understand the beam splitting problems for square gratings, we begin by introducing a bit of notation. We will use two-dimensional vectors of integers to give the Fourier coefficients in two dimensions, which can be written as follows:

$$e^{i\mathbf{m}\cdot\mathbf{x}} = e^{i(mx+ny)},\ \mathbf{m}^T = (m,n),\ \mathbf{x}^T = (x,y) \tag{3.20}$$

In this notation, a Fourier series can be written as

$$f(\mathbf{x}) = \sum_{\mathbf{m}} a_{\mathbf{m}} e^{i\mathbf{m}\cdot\mathbf{x}} \tag{3.21}$$

Here the sum is taken over all pairs of integers $\mathbf{m} = (m,n)$. In this case, the energy diffracted into the $\mathbf{m}^T = (m,n)$ order is given by

$$E_{\mathbf{m}} = |a_{\mathbf{m}}|^2 \tag{3.22}$$

where:

$$a_{\mathbf{m}} = \frac{1}{4\pi^2}\int_{-\pi}^{\pi}\int_{-\pi}^{\pi} e^{-i\mathbf{m}\cdot\mathbf{x}} e^{i\varphi(\mathbf{x})}\,dxdy \tag{3.23}$$

An example of a two-dimensional beam splitting problem would be to maximize the energy in the four orders (1,0),(0,1),(−1,0),(0,−1) subject to the constraint that the energy in each of these orders is the same. Hence, we would like to choose $\varphi(x,y)$ to maximize

$$\eta = \frac{|a_{1,0}|^2 + |a_{0,1}|^2 + |a_{-1,0}|^2 + |a_{0,-1}|^2}{\sum_{\mathbf{m}} |a_{\mathbf{m}}|^2} \tag{3.24}$$

subject to the constraint that

$$|a_{1,0}| = |a_{0,1}| = |a_{-1,0}| = |a_{0,-1}| \tag{3.25}$$

Physically, this would correspond to designing a grating so the diffracted beams (after intersecting them with a plane parallel to the grating) put as much energy as possible onto the four corners of a square.

In the constrained optimization problem for square gratings, we try to maximize the energy in a discrete set of modes $\mathbf{m}_k$ for $k = 1,n$. In References [1,8], we show that if we try to maximize the energy in the modes $a_{\mathbf{m}_k}$ subject to the constraint that the energy in each of these modes is the same, the phase function $\varphi(x,y)$ can be written as

$$t(\mathbf{x},\alpha,\mu) = e^{i\varphi(\mathbf{x})} = \frac{s(\mathbf{x},\alpha,\mu)}{|s(\mathbf{x},\alpha,\mu)|} \tag{3.26}$$

where:

$$s(\mathbf{x},\alpha) = \sum_{k=1}^{n} \mu_k e^{i\alpha_k} e^{i\mathbf{m}_k \cdot \mathbf{x}} \tag{3.27}$$

This is almost identical to the result obtained in one dimensions in that the function $s(\mathbf{x},\alpha,\mu)$ only includes a finite number of terms in the Fourier series. We need to choose the amplitude and phase of each Fourier coefficient so that the Fourier coefficient of each mode $a_{\mathbf{m}_k}$ has the same amplitude, and we are at a maximum of the efficiency subject to that constraint. As with the one-dimensional case, this reduces the infinite-dimensional optimization problem to a finite-dimensional optimization problem.

3.7.2 Separable and Nonseparable Solutions

For a square grating, if we assume that the transmission function is separable

$$e^{i\varphi(x,y)} = e^{i\varphi_1(x)} e^{i\varphi_2(y)} \tag{3.28}$$

we have

$$a_{k,j} = b_k c_j \tag{3.29}$$

where:

$a_{k,j}$ is the (k,j) Fourier component of $e^{i\varphi(x,y)}$
b_k is the kth Fourier coefficient of $e^{i\varphi_1(x)}$
c_j is the jth Fourier coefficient of $e^{i\varphi_2(y)}$

Under the assumption of separability, the optimal solution to the problem stated in Section 4.7.1 of maximizing the energy in the modes $a_{\pm 1,0}$, $a_{0,\pm 1}$ can be obtained by letting $\varphi_1(s) = \varphi_2(s)$ and $\varphi_1(x)$ be the optimal solution for two beam splitting in one dimensions.

In References [1,8], it was shown that the optimal efficiency for this problem under the separability condition is just the square of the efficiency of the one-dimensional two beam splitting problem. This can be shown to be

$$\eta_{\text{sep}} = \frac{64}{\pi^4} \approx 0.658 \tag{3.30}$$

In References [1,8], it is shown that if we drop the assumption of separability, we get an efficiency of

$$\eta \approx 0.9179 \tag{3.31}$$

Clearly dropping the assumption of separability has greatly improved the efficiency of the grating.

The notion of separability is not necessarily related to that of symmetry. However, for all of the simple beam splitting problems on a square lattice, they seem to be related to each other. In particular, the optimal solution assuming a fully symmetric grating is separable. The nonseparable solutions are not fully symmetric. In Section 3.8, we will try to clarify this point.

3.8 THE SYMMETRY OF BEAM SPLITTING PROBLEMS

3.8.1 Symmetry Considerations for One-Dimensional Gratings

It can greatly simplify the search for optimal designs if symmetry is taken into account. However, it is good to be clear on what you may be missing by assuming that your solution is symmetric.

We begin with a simple example from elementary calculus. Suppose we try to find the value of x that minimizes a symmetric function $g(x)$. We assume that $g(x)$ satisfies $g(x) = g(-x)$. If there is a unique global minimum, it must be at $x = 0$ because if $x = x_0$ is the global minimum, $-x_0$ will also be a global minimum since $g(x_0) = g(-x_0)$. Hence, the only way we can have a unique global minimum is to have the minimum occur at $x = 0$.

However, there is no guarantee that the global minimum is unique. For example, the function

$$g(x) = -2x^2 + x^4 \tag{3.32}$$

clearly does not take on its minimum value at $x = 0$, but it takes on its minimum value at $x = \pm 1$.

To connect this example with those that follow, we will restate our results. If a symmetric function $g(x)$ has a unique global minimum, the value of x that minimizes $g(x)$ must be symmetric; that is, we must have $x_0 = -x_0$, and hence $x_0 = 0$. It greatly simplifies our search for the minimum if we assume that the solution is symmetric. In particular, we have no search at all; we just assume that the global minimum occurs at $x = 0$. However, as the example in Equation 3.32 shows, this is not always a valid assumption.

There are many examples in physics in which a physical problem has symmetry, but the physical solution does not. A classic example of such a problem is the buckling of a cantilever beam. If you put a weight on the top of a flagpole, assuming that the flagpole is pointing vertically upward and is circularly symmetric, we would expect on the basis of symmetry that the flagpole will point vertically upward. However, if the flagpole is long enough, or the weight is great enough, this will not be the case. The flagpole will break the symmetry of the problem and will find a particular direction to droop in. This is called symmetry breaking.

In our one-dimensional beam shaping problem, we are looking for a function $\varphi(x)$ that optimizes a functional rather than a single point x_0 that optimizes a function. However, the reasoning is almost identical as in our simple example. In particular, our optimization problem is symmetric in the following sense. If $\varphi(x)$ is a function such that the Fourier coefficients of $e^{i\varphi(x)}$ satisfy the constraints that the energy in the first $2m + 1$ modes is the same, the function $\varphi(-x)$ will also satisfy this constraint. Furthermore, the efficiency of $\varphi(x)$ will be the same as that of $\varphi(-x)$. It follows that if there is a unique global maximum, we must have

$$\varphi(x) = \varphi(-x) \tag{3.33}$$

That is, the function $\varphi(x)$ must be symmetric. For the problem of determining the point x_0 that minimizes $f(x)$, the assumption of symmetry completely eliminates the need for any search at all. For the one-dimensional beam splitting problem, the assumption of symmetry greatly reduces the number of parameters that are needed for a search, but it does not eliminate the need for that search.

For one-dimensional beam splitting problems, we have never found any examples in which the optimal solution is not symmetric [1,5,24]. However, it should be noted that if we impose further restrictions such as assuming that $\varphi(x)$ only takes on two values (we have a binary or Dammann grating [9]), there are situations in which the optimal solution is not symmetric.

In an excellent paper, Morrison [13] considered various aspects of symmetry for one-dimensional gratings. He showed how to apply symmetry arguments to solve the problem where you only want to put energy into the modes with odd orders, as well as how to take into account symmetry in the way we have already discussed. However, he never discusses the possibility of symmetry breaking. For the one-dimensional problems he considered, this turns out not to be an issue. However, one could easily read his paper and draw the wrong conclusion when applying the results to two-dimensional gratings.

3.8.2 Symmetry and Two-Dimensional Gratings

The problem of finding a square grating that maximizes the energy put into the four modes $a_{\pm1,0}$, $a_{0,\pm1}$ has the symmetry of the square. Physically, this problem is trying to create a diffraction pattern that produces four spots of equal intensity on the corners of a square at a plane z = constant. Intuitively, we think of this problem as having the symmetry of a square. To make this mathematically precise, we note that a square centered at $(x,y) = (0,0)$ is left invariant by an eighth order group of linear transformations $\mathbf{G}_k$ for $k = 1,8$. Here the transformations $\mathbf{G}_k$ consist of rotations by 0°, 90°, 180°, and 270°, as well as reflections about the x and y axes, and reflections about the lines $x = y$ and $x = -y$.

We can now state precisely what we mean when we say that our beam splitting problem is invariant under the symmetry group of a square. If $\varphi(\mathbf{x})$ is a phase function such that the Fourier coefficients of $e^{i\varphi(\mathbf{x})}$ satisfy the constraints in Equation 3.25, for any $\mathbf{G}_k$ in our group, the function $\varphi(\mathbf{G}_k\mathbf{x})$ will also satisfy this constraint. Furthermore, the functions $\varphi(\mathbf{x})$ and $\varphi(\mathbf{G}_k\mathbf{x})$ will have the same efficiency.

It follows that if our functional has a unique global maximum, the function maximizing it must be symmetric. Thus, we must have

$$\varphi(\mathbf{x}) = \varphi(\mathbf{G}_k\mathbf{x}) \quad k = 1,8 \tag{3.34}$$

For example, letting $\mathbf{G}_k$ be the reflection about the y axis, we require that $\varphi(x,y) = \varphi(-x,y)$. If $\mathbf{G}_k$ is a rotation by 90°, we get $\varphi(x,y) = \varphi(-y,x)$.

In our discussion, we will ignore the reflections in the symmetry group of the square and only concern ourselves with the rotational symmetry of the function $\varphi(\mathbf{x})$. We can say that the function $\varphi(\mathbf{x})$ has fourfold symmetry if

$$\varphi(\mathbf{x}) = \varphi(\mathbf{R}_{90}\mathbf{x}) \tag{3.35}$$

where:

$\mathbf{R}_{90}$ is a rotation by 90°

That is, the function $\varphi(\mathbf{x})$ has fourfold symmetry if it looks the same when we rotate it by 90°. Similarly, we can say that the function has twofold symmetry if

$$\varphi(\mathbf{x}) = \varphi(\mathbf{R}_{180}\mathbf{x}) \tag{3.36}$$

That is, the function looks the same if we rotate it by 180°. Clearly, if the function has fourfold symmetry, it also has twofold symmetry, but not vice versa.

For example, we consider the problem of splitting a beam into four equal beams using a square grating. In particular, we try to equalize the energy in the four modes $a_{\pm1,0}$, $a_{0,\pm1}$. If we assume that the grating has fourfold symmetry, the phases α_k and μ_k must all be equal. This requires that the function $s(\mathbf{x},\boldsymbol{\alpha},\boldsymbol{\mu})$ used in determining the transmission function $t(\mathbf{x},\boldsymbol{\alpha},\boldsymbol{\mu})$ in Equation 3.26 has the form:

$$s(\mathbf{x},\boldsymbol{\alpha},\boldsymbol{\mu}) = e^{ix} + e^{-ix} + e^{iy} + e^{-iy} = 2\cos(x) + 2\cos(y) \tag{3.37}$$

In this case, we know the phases α_k and the parameters μ_k, and hence there is no need to optimize anything. In Reference [8], it was shown that the efficiency of this grating is given by

$$\eta_{CO} = \frac{64}{\pi^4} \approx 0.658 \tag{3.38}$$

This is the square of the efficiency for one-dimensional two beam splitting. Though the function $s(\mathbf{x},\boldsymbol{\alpha},\boldsymbol{\mu})$ in Equation 3.37 does not appear to be separable, it is separable if we use a coordinate system that is rotated by $\pi/4$ radians. If we had tried to equalize the energy in the modes $a_{\pm1,\pm1}$, we would get the same efficiency, and the transmission function would then clearly be separable in our original coordinate system.

If instead of assuming that the grating has fourfold symmetry, we merely assume that it has twofold symmetry, the Fourier coefficients must satisfy $a_{\mathbf{m}} = a_{-\mathbf{m}}$. This implies that the optimal grating must have the form:

$$s(s,\alpha,\mu) = e^{ix} + e^{-ix} + \mu e^{i(y+\alpha)} + \mu e^{i(-y+\alpha)} \tag{3.39}$$

TABLE 3.3
Efficiencies for Square Gratings

N_{modes}	Symmetry	η_{CO}
4	Fourfold	0.658
4	Twofold	0.9179
5	Fourfold	0.7629
5	Twofold	0.8433
9	Fourfold	0.8456
9	Twofold	0.9327

Note: This table summarizes the various problems we have considered for gratings on a square lattice. For each number of modes, we give the symmetry of the grating and its efficiency η_{CO}.

Here, we have used the fact that we can arbitrarily set the phase of the coefficients of $e^{\pm ix}$ equal to zero and the coefficients μ_k associated with these equal to unity. In Reference [8], it was proven that the optimal solution to this problem is the same as the solution to the least-squares optimization problem. It was shown that the optimal values of α and μ are given by

$$\alpha = \frac{\pi}{2}, \mu = 1 \tag{3.40}$$

It was also shown that the efficiency for this grating is given by

$$\eta_{CO} = \eta_{LS} \approx 0.9179 \tag{3.41}$$

This is considerably greater than the efficiency where we assumed that the grating has fourfold symmetry. This grating is not separable. When it is not assumed that the grating has any rotational symmetry, it appears that the optimal solution is still the grating that has twofold rotational symmetry.

Table 3.3 summarizes the results for other situations involving square gratings. In this table, we present the results for splitting a beam into five beams of equal intensity. If we intersect such a beam with a plane parallel to the grating, we would see four spots on the corners of a square and a spot in the middle of the square. We also give the results for nine beam splitting, where we have a 3×3 array of points. In both of these cases, though the beam shaping problems have fourfold symmetry, the grating with the maximum efficiency has twofold symmetry.

In Appendix A, the formulas for finding the solutions to various beam splitting problems using square gratings are given.

3.9 BEAM SHAPING ON HEXAGONAL GRIDS

In References [1,8], it was shown how to use hexagonal gratings to split beams into patterns that have sixfold symmetry. A hexagonal grating is characterized by the fact that it repeats itself on a hexagonal array. That is, it is possible to overlay a tiling with regular hexagons over the grating such that the grating looks the same inside of each hexagon. An example of hexagonal beam splitting problem is the problem of putting

TABLE 3.4
Efficiencies for Hexagonal Gratings

N_{modes}	Symmetry	η_{CO}
6	Sixfold	0.7107
6	Twofold	0.8338
7	Sixfold	0.8015
7	Twofold	0.9003

Note: This table summarizes the various problems we have considered for Gratings on a hexagonal lattice. For each number of modes, we give the symmetry of the grating and its efficiency η_{CO}.

as much energy as possible into six symmetrical beams (so that there are six spots on the vertices of a regular hexagon on a plane parallel to the grating). Another example is to put as much energy as possible into seven beams so that there are six spots on the vertices of a regular hexagon and a spot in the middle.

Table 3.4 gives the results for these two cases. As with square gratings, the optimal solutions do not have the full symmetry of the beam splitting problem. In both cases, the optimal solutions have twofold symmetry rather than sixfold symmetry. In Appendix B, the formulas for finding the solutions to various beam splitting problems using hexagonal gratings are given.

3.10 SUMMARY

This chapter has given an overview of the mathematical theory of laser beam splitting using phase gratings based on the work in References [1,5,8]. The problem of designing a phase grating to diffract a beam so that more energy is put into a certain set of modes while having the intensity of all of these modes equal has been called the constrained optimization problem. Using the calculus of variations, this can be turned into a finite-dimensional optimization problem in which the phases and amplitudes of the Fourier coefficients of $s(x,\boldsymbol{\alpha},\boldsymbol{\mu})$ (as in Equations 3.9 and 3.10) must be searched for. The theory applies to both one-dimensional line gratings and two-dimensional gratings on square or hexagonal lattices.

The efficiencies for various one-dimensional problems are given in Tables 3.1 and 3.2. Some experimental results for line gratings are presented in Section 3.6. The efficiencies for various problems on square and hexagonal gratings can be found in Tables 3.3 and 3.4. The phases and amplitudes defining these gratings have been given in Appendices A and B.

REFERENCES

1. L.A. Romero and F.M. Dickey. The mathematical theory of laser beam-splitting gratings. *Progress in Optics*, 54:319–386, 2010.
2. R.P. Feynman, R.B. Leighton, and M. Sands. *The Feynman Lectures on Physics.* Redwood City, CA: Addison-Wesley, 1963.

3. D. Halliday and R. Resnick. *Physics: Parts I and II*. New York: Wiley, 1978.
4. J.W. Goodman. *Introduction to Fourier Optics*. New York: McGraw-Hill, 1968.
5. L.A. Romero and F.M. Dickey. Theory of optimal beam splitting by phase gratings. I. One-dimensional gratings. *Journal of the Optical Society of America A*, 24(8):2280–2295, 2007.
6. F. Gori, M. Santarsiero, S. Vicalvi, R. Borghi, G. Cincotti, E. Di Fabrizio, and M. Gentili. Analytical derivation of the optimum triplicator. *Optics Communications*, 157(1–6):13–16, 1998.
7. R. Borghi, G. Cincotti, and M. Santarsiero. Diffractive variable beam splitter: Optimal design. *Journal of the Optical Society of America A*, 17(1):63–67, 2000.
8. L.A. Romero and F.M. Dickey. A theory for optimal beam splitting by phase gratings. II. Square and hexagonal gratings. *Journal of the Optical Society of America A*, 24(8):2296–2312, 2007.
9. G.H. Dammann and K. Gortler. High efficiency in-line multiple imaging by means of multiple phase holograms. *Optics Communications*, 3(5):312–315, 1971.
10. H. Dammann and E. Klotz. Coherent optical generation and inspection of two-dimensional periodic structures. *Optica Acta*, 24(4):505–515, 1977.
11. U. Killat, G. Rabe, and W. Rave. Binary phase gratings for star couplers with high splitting ratio. *Fiber and Integrated Optics*, 4(2):159–167, 1982.
12. U. Krackhardt and N. Streibl. Design of Dammann-gratings for array generation. *Optics Communications*, 74(1/2):31–36, 1989.
13. R.L. Morrison. Symmetries that simplify the design of spot array phase gratings. *Journal of the Optical Society of America A*, 9(3):464–471, 1992.
14. M.R. Feldman and C.C. Guest. Iterative encoding of high-efficiency holograms for generation of spot arrays. *Optics Letters*, 14(10):479–481, 1989.
15. J. Turunen, A. Vasara, J. Westerholm, G. Jin, and A. Salin. Optimization and fabrication of grating beamsplitters. *Journal of Physics D: Applied Physics*, 21:S102–S105, 1988.
16. L.L. Doskolovich, V.A. Soifer, G. Alessandretti, P. Perlo, and P. Repetto. Analytical initial approximation for multiorder binary grating design. *Pure and Applied Optics: Journal of the European Optical Society Part A*, 3:921–930, 1994.
17. J.M. Miller, M.R. Taghizadeh, J. Turunen, and N. Ross. Multilevel-grating array generators: Fabrication error analysis and experiments. *Applied Optics*, 32(14):2519–2525, 1993.
18. S.J. Walker and J. Jahns. Array generation with multilevel phase gratings. *Journal of the Optical Society of America A*, 7(8):1509, 1990.
19. J.N. Mait. Design of binary-phase and multiphase Fourier gratings for array generation. *Journal of the Optical Society of America A*, 7(8):1514, 1990.
20. F. Gori. Diffractive optics: An introduction. In S. Martellucci and A.N. Chester, editors, *Diffractive Optics and Optical Microsystems*. New York: Plenum Press, 1997.
21. J.N. Mait. Fourier array generators. In H.P. Herzig, editor, *Micro-Optics—Elements, Systems and Applications*. London: Taylor & Francis, 1997.
22. U. Krackhardt, J.N. Mait, and N. Streibl. Upper bound on the diffraction efficiency of phase-only fanout elements. *Applied Optics*, 31(1):27–37, 1992.
23. F. Wyrowski. Upper bound of the diffraction efficiency of diffractive phase elements. *Optical Letters*, 16(1915):1917, 1991.
24. D. Prongué, H.P. Herzig, R. Dandliker, and M.T. Gale. Optimized kinoform structures for highly efficient fan-out elements. *Applied Optics*, 31(26):5706–5711, 1992.
25. H.P. Herzig, D. Prongué, and R. Dändliker. Design and fabrication of highly efficient fan-out elements. *Japanese Journal of Applied Physics*, 29(7): L1307–L1309, 1990.
26. J. Albero, I. Moreno, J.A. Davis, D.M. Cottrell, and D. Sand. Generalized phase diffraction gratings with tailored intensity. *Optics Letters*, 37(20):4227–4229, 2012.
27. J. Albero and I. Moreno. Grating beam splitting with liquid crystal adaptive optics. *Journal of Optics*, 14(7):075704, 2012.

28. Y.V. Miklyeav, W. Imgrunt, V.S. Pavelyev, D.G. Kachalov, T. Bizjak, L. Aschke, and V.N. Lissotschenko. Novel continuously shaped diffractive optical elements enable high efficiency beam shaping. *Proceedings of SPIE*, 7640:764023-1, 2010.
29. Y.V. Miklyeav, A. Krasnaberski, M. Ivanenko, A. Mikhailov, W. Imgrunt, L. Aschke, and V.N. Lissorschenko. Efficient diffractive optical elements from glass with continuous surface profiles. *Proceedings of SPIE*, 7913:79130B, 2011.
30. J.N. Mait. Design of Dammann gratings for two-dimensional, nonseparable, noncentrosymmetric responses. *Optics Letters*, 14(4):196–198, 1989.
31. J. Turunen, A. Vasara, J. Westerholm, and A. Salin. Stripe-geometry two-dimensional Dammann gratings. *Optics Communications*, 74(3/4):245–252, 1989.
32. A. Vasara, M.R. Taghizadeh, J. Turunen, J. Westerholm, E. Noponen, H. Ichikawa, J.M. Miller, T. Jaakkola, and S. Kuisma. Binary surface-relief gratings for array illumination in digital optics. *Applied Optics*, 31:3320–3336, 1992.
33. I.M. Barton, P. Blair, and M.R. Taghizadeh. Diffractive phase elements for pattern formation: Phase-encoding geometry considerations. *Applied Optics*, 36(35):9132–9137, 1997.
34. P. Blair, H. Lüpken, M.R. Taghizadeh, and F. Wyrowski. Multilevel phase-only array generators with a trapezoidal phase topology. *Applied Optics*, 36(20):4713–4721, 1997.
35. C. Zhou and L. Liu. Numerical study of Dammann array illuminators. *Applied Optics*, 34(26):5961–5969, 1995.

APPENDIX A

In this appendix, the results for various two-dimensional problems involving square gratings are given. A more thorough discussion of all of these can be found in References [1,8]. In all of the examples in this appendix, it is found that the optimal solution has twofold symmetry, and the efficiency of these solutions is greater than if it is assumed that the grating has fourfold symmetry. For the case of four and nine beam splitting, the optimal solutions assuming fourfold symmetry are identical to those obtained by assuming separability. In References [1,8], the suboptimal solutions assuming fourfold symmetry are given, but these are not given here.

Throughout this appendix, it is assumed that the phase function $\varphi(\mathbf{x})$ can be generated using the two-dimensional equivalent of Equation 3.26 once the function $s(\mathbf{x},\boldsymbol{\alpha},\boldsymbol{\mu})$ is specified. Here, $\mathbf{x} = (x,y)$.

A.1 FOUR BEAM SPLITTING

Suppose we are trying to split a beam into four modes, where each component of $\mathbf{m}$ can take on the values $(\pm 1,0)$ and $(0,\pm 1)$. The most general form for $s(\mathbf{x},\boldsymbol{\alpha},\boldsymbol{\mu})$ that has twofold symmetry is

$$s(x,y,\alpha,\mu) = e^{ix} + e^{-ix} + \mu e^{i(y+\alpha)} + \mu e^{i(-y+\alpha)} \tag{A.1}$$

In Reference [8], it was shown that the values of the parameters that optimize the efficiency in the constrained optimization problem are given by

$$\alpha = \frac{\pi}{2}, \mu = 1 \tag{A.2}$$

It was shown that the efficiency for this grating is given by

$$\eta_{CO} = \eta_{LS} \approx 0.9179 \tag{A.3}$$

A.2 FIVE BEAM SPLITTING

Here we split a beam into five modes, where each component of **m** can take on the values (0,0), (±1,0), and (0,±1). In References [1,8], we show that assuming that the grating has twofold symmetry, we have

$$s(x,y,\alpha,\mu) = \mu_2 e^{i\alpha_2} + e^{ix} + e^{-ix} + \mu_1 e^{i(y+\alpha_1)} + \mu_1 e^{i(-y+\alpha_1)} \tag{A.4}$$

Numerical calculations show that to maximize the efficiency in the constrained optimization problem, we set

$$\mu_1 \approx 1.1928 \tag{A.5a}$$

$$\mu_2 \approx 0.7192 \tag{A.5b}$$

$$\alpha_1 = \frac{\pi}{2} \tag{A.5c}$$

$$\alpha_2 = 0 \tag{A.5d}$$

This gives an efficiency of

$$\eta_{CO} \approx 0.8433 \tag{A.6}$$

A.3 NINE BEAM SPLITTING

We split a beam into nine modes, where each component of **m** can take on the values (−1, 0, 1). In References [1,8], it was shown that the most general solution having twofold symmetry can be written as

$$\begin{aligned} s(x,y,\alpha,\mu) = 1 + \mu_1 g_1(\mathbf{x},\alpha) + \mu_2\left[e^{i(x+y+\alpha_2)} + e^{i(-x-y+\alpha_2)}\right] \\ + \mu_3\left[e^{i(x-y+\alpha_3)} + e^{i(-x+y+\alpha_3)}\right] \end{aligned} \tag{A.7}$$

where:

$$g_1(\mathbf{x},\alpha) = e^{i(x+\alpha_1)} + e^{i(-x+\alpha_1)} + e^{i(y+\alpha_1)} + e^{i(-y+\alpha_1)} \tag{A.8}$$

Numerical calculations show that the parameters maximizing the efficiency in the constrained optimization problem are given by

$$\alpha_1 = 0 \tag{A.9a}$$

$$\alpha_2 = 2.103 \tag{A.9b}$$

$$\alpha_3 = 4.1806 \tag{A.9c}$$

$$\mu_1 = 1.379 \tag{A.9d}$$

$$\mu_2 = 1.111 \tag{A.9e}$$

$$\mu_3 = 1.111 \tag{A.9f}$$

This gives us the efficiency

$$\eta_{CO} = 0.9327 \tag{A.10}$$

APPENDIX B

A discussion of Fourier series on hexagonal lattices can be found in References [1,8]. A function that is periodic on a hexagonal lattice will satisfy

$$f(\mathbf{x} + \mathbf{p_m}) = f(\mathbf{x}) \tag{B.1}$$

for all vectors $\mathbf{p_m}$ of the form:

$$\mathbf{p_m} = m_1\mathbf{p}_1 + m_2\mathbf{p}_2 \tag{B.2}$$

where:
m_1 and m_2 are integers

and

$$\mathbf{p}_1^T = 2\pi\left(\frac{\sqrt{3}}{2}, \frac{-1}{2}\right), \mathbf{p}_2^T = 2\pi\left(\frac{\sqrt{3}}{2}, \frac{1}{2}\right) \tag{B.3}$$

Any function that is periodic on such a lattice can be expanded in a Fourier series of the form:

$$f(\mathbf{x}) = \sum_{\mathbf{m}} a_{\mathbf{m}} e^{i\mathbf{q_m}\cdot\mathbf{x}} \tag{B.4}$$

where:

$$\mathbf{q_m} = m_1\mathbf{q}_1 + m_2\mathbf{q}_2 \tag{B.5}$$

and

$$\mathbf{q}_1^T = \left(\frac{1}{\sqrt{3}}, -1\right), \mathbf{q}_2^T = \left(\frac{1}{\sqrt{3}}, 1\right) \tag{B.6}$$

We will use these formulas in describing our hexagonal beam shaping problems. More details on all of these problems can be found in References [1,8].

B.1 SIX BEAM SPLITTING

We consider the problem of using a hexagonal grating to split an incoming beam into six beams that are on the corners of a hexagon. Though this problem has sixfold symmetry, it was found that the solution to the constrained optimization problem gives a grating that has twofold symmetry. The function $s(\mathbf{x},\boldsymbol{\alpha},\boldsymbol{\mu})$ that gives us the phase function is

$$s(\mathbf{x},\alpha,\mu) = e^{i\mathbf{q}_{1,0}\cdot\mathbf{x}} + e^{i\mathbf{q}_{-1,0}\cdot\mathbf{x}} + \mu_1 e^{i\alpha_1}(e^{i\mathbf{q}_{0,1}\cdot\mathbf{x}} + e^{i\mathbf{q}_{0,-1}\cdot\mathbf{x}}) + \mu_2 e^{i\alpha_2}(e^{i\mathbf{q}_{1,1}\cdot\mathbf{x}} + e^{i\mathbf{q}_{-1,-1}\cdot\mathbf{x}}) \quad \text{(B.7)}$$

This can be written as

$$s(\mathbf{x},\alpha,\mu) = 2\cos(\mathbf{q}_{1,0}\cdot\mathbf{x}) + \mu_1 e^{i\alpha_1}\cos(\mathbf{q}_{0,1}\cdot\mathbf{x}) + \mu_2 e^{i\alpha_2}\cos(\mathbf{q}_{1,1}\cdot\mathbf{x}) \quad \text{(B.8)}$$

Numerical calculations show that the optimal value is given by

$$\alpha_1 = \frac{\pi}{2}, \alpha_2 = 0 \quad \text{(B.9)}$$

$$\mu_1 = 0.5671, \mu_2 = 1 \quad \text{(B.10)}$$

The efficiency is given by

$$\eta_{CO} = 0.8338 \quad \text{(B.11)}$$

B.2 SEVEN BEAM SPLITTING

We now consider the problem of splitting a beam into seven beams that are on the vertices of a hexagon and in the middle of the hexagon. Though this problem has sixfold symmetry, the optimal solution has only twofold symmetry. The solution is given by

$$s(\mathbf{x},\alpha,\mu) = \mu_3 e^{i\alpha_3} + 2\cos(\mathbf{q}_{1,0}\cdot\mathbf{x}) + \mu_2 e^{i\alpha_2}\cos(\mathbf{q}_{1,1}\cdot\mathbf{x}) + \mu_1 e^{i\alpha_1}\cos(\mathbf{q}_{0,1}\cdot\mathbf{x}) \quad \text{(B.12)}$$

Numerically, it is found that

$$(\alpha_1,\alpha_2,\alpha_3) = \left(\frac{\pi}{2},\frac{\pi}{2},0\right) \quad \text{(B.13)}$$

$$(\mu_1,\mu_2,\mu_3) = (1.3368, 1.3368, 0.9811) \quad \text{(B.14)}$$

The efficiency is given by

$$\eta_{CO} = 0.9003 \quad \text{(B.15)}$$

4 Vortex Beam Shaping

Carlos López-Mariscal and Julio C. Gutiérrez-Vega

CONTENTS

4.1 INTRODUCTION

Traditionally, standard approaches to characterizing of optical beams have dealt with parameters that either characterize the beam *as a whole* (e.g., power, beam size, divergence angle) or describe its *shape* using transverse distributions (e.g., intensity pattern, polarization state, wavefront curvature). These approaches are based on *immediately observable* characteristics of the optical beams and have been adequate for most applications. This situation changed dramatically with the recent increased interest in beams carrying angular momentum. It has now been well established that this fundamental mechanical property of light is associated with the energy redistribution inside the beam body under propagation.

Optical phase singularities have become a popular topic in optical physics and beam shaping through their relationship with beams carrying angular momentum [1,2]. Phase singularities are ubiquitous in light fields. They are most easily recognized in spatially coherent monochromatic light, and may be present in many different spatial structures, for example, simple superposition of three plane waves, refracted/diffracted fields, apertured beams, laser beams, random wave fields, and so on. Singularities appear as dark spots in the intensity profile. For partially coherent light, the vortex appears as a diffuse dark core and a zero-intensity point strictly does not occur.

In this chapter, we review the basic theory of vortex beam shaping, the experimental procedures to generate them, and some of their most relevant applications. The theory of singular optics has been built with input and terminology adopted from diverse areas of mathematics and physics. Thus in optics, the terms *singular point*, *phase singularity*, or *wavefront dislocation* are used interchangeably to refer an *optical vortex*. The study of optical vortices and associated localized objects is important from the viewpoint of both fundamental and applied optics.

4.2 ELEMENTS OF SINGULAR OPTICS

Consider the smooth complex amplitude of a scalar monochromatic optical field $U(\mathbf{r},z)\exp(-i\omega t)$, where $\mathbf{r} = (x,y) = (r,\theta)$ denotes the transverse position vector and ω is the angular frequency. The field $U(\mathbf{r},z)$ can be written in terms of either its amplitude and phase (ρ,Φ) or its real and imaginary parts ($f = \rho \cos \Phi$, $g = \rho \sin \Phi$):

$$U = \rho \exp(i\Phi) = f + ig \tag{4.1}$$

where:

all ρ, Φ, f, and g are real functions
Φ is a single-valued modulo 2π

A *singular point* is defined as a point $\mathbf{r}$ where the amplitude is zero

$$\rho(\mathbf{r}) = 0 \tag{4.2}$$

and hence the phase $\Phi(\mathbf{r})$ is undefined. In the neighborhood of the singular point, the whole 2π range of phases occurs. The condition $\rho(\mathbf{r}) = 0$ is fully equivalent to the system of two equations:

$$f(\mathbf{r}) = 0, \quad g(\mathbf{r}) = 0 \tag{4.3}$$

These two conditions imply that phase singularities occur at the crossings of the zero contours of the real functions f and g, as shown in Figure 4.1. Vortices are, usually,

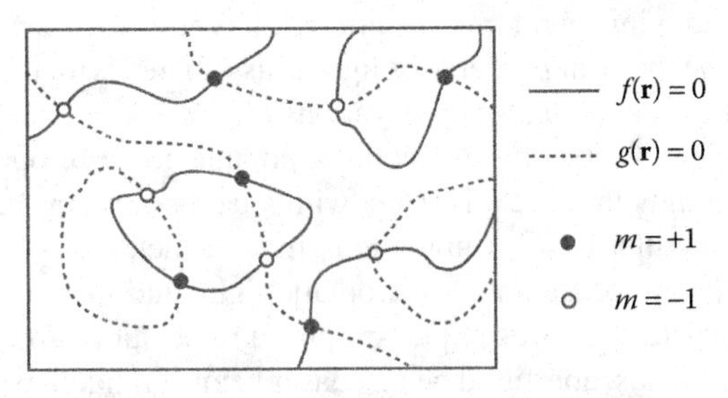

FIGURE 4.1 Zero contours of the real $f(\mathbf{r})$ and imaginary $g(\mathbf{r})$ parts of the complex scalar field. Vortices occur at the intersection points of these sets of contours.

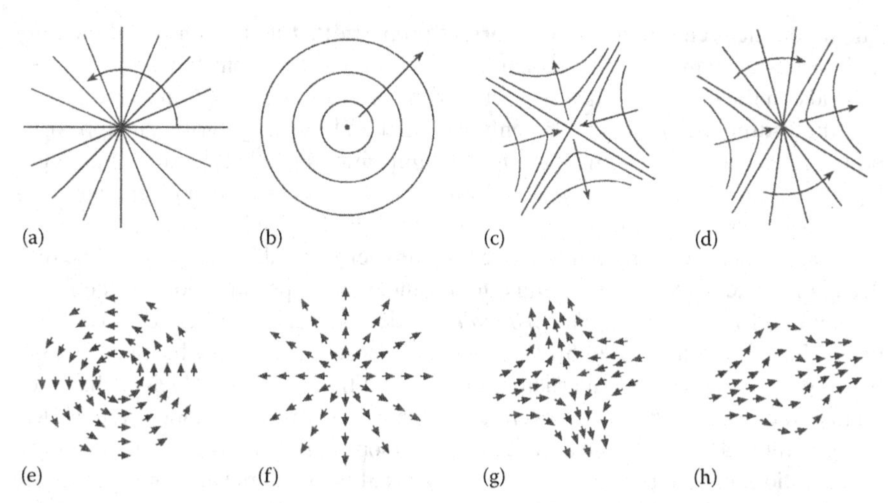

FIGURE 4.2 Equiphase lines of Φ and vector fields $\nabla\Phi$ around a singular point with $s > 0$ (a and e), a stationary point (b and f), a saddle point (c and g), and a monkey saddle (d and h). Arrows point in the direction of increasing phase.

separated points in two dimensions and lines in three-dimensional space. For now, we restrict our attention to the properties of singular points in two dimensions.

Near a phase singularity, the equiphase lines have a starlike structure as shown in Figure 4.2a, and the phase increases or decreases as one moves around the singular point.

The *topological charge* of a phase singularity is the net change of phase modulo 2π around a closed curve C enclosing the singular point:

$$m = \frac{1}{2\pi}\oint_C d\Phi = \frac{1}{2\pi}\oint_C \nabla\Phi \cdot d\mathbf{r} \tag{4.4}$$

Because the phase is continuous on C, the number m is an integer, that is, $m = 0, \pm 1, \pm 2, \ldots$, and its sign is positive or negative if the phase increases or decreases by $m2\pi$ during a positive circulation on the curve C. If $m = \pm 1$, the singularity is called *simple* or *nondegenerate*, otherwise (i.e., $|m| > 1$) *higher order* or *degenerate*. The value of m is independent of the choice of the curve C, as long as the point is the only phase singularity inside C and there are not singularities on C. Finally, the value $m = 0$ implies a nonvortex situation.

Consider now that there are N vortices with topological charges $m_1, \ldots, m_N$ inside C. The total topological charge is given by the sum of the individual charges of the vortices lying inside C:

$$m_{\text{tot}} = \frac{1}{2\pi}\oint_C d\Phi = \sum_{n=1}^{N} m_n \tag{4.5}$$

The topological charge has the important property that it is a conserved quantity under smooth changes of the optical field [3]. In most situations the field smoothly depends on the beam parameters, such as time or the z-axis for a transverse section of a three-dimensional field. The only way that a phase singularity can appear on propagation is together with other phase singularities such that the sum of all topological charges is zero. Likewise, the only way that a vortex can disappear is for it to annihilate with other vortex of opposite sign. The higher order singularities (i.e., $|m| > 1$) are seldom seen because they are very unstable; in other words, they decay in vortices with charges equal to ± 1 under small perturbations of the field.

Figure 4.1 illustrates the *sign principle*: vortices adjacent on a zero contour of $f(\mathbf{r})$ or $g(\mathbf{r})$ have opposite sign [4,5]. If a contour is closed, there must be an even number of phase singularities, alternating in sign, and the overall topological charge on that contour is zero, giving an overall topological neutrality condition: topological charge cannot accumulate on a closed contour loop. This result is consistent with the conservation of the topological charge: as external parameters vary singularities are created or annihilated in pairs of opposite strength as an f contour crosses a g contour. An exception to the sign rule occurs when there is a saddle point on the phase contour line between two vortices: in this case, the vortices have the same strength. An example of this situation is the hollow dark Mathieu beam [6].

In scalar fields $U(\mathbf{r})$ (Equation 4.1), the energy flux vector associated with Φ takes the form [7]:

$$\mathbf{J} = \mathrm{Im} U^* \nabla U = f \nabla g - g \nabla f = \rho^2 \nabla \Phi \tag{4.6}$$

Thus, $\mathbf{J}$ points in the direction of phase change given by the vector field $\nabla\Phi$, and phase singularities are vortices of the energy flux vector flow where the energy circulates around the point. Near a phase singularity, the vector field $\nabla\Phi$ looks like a counterclockwise or clockwise *center* as shown in Figure 4.2e.

A central role is also played by the *vorticity* associated with $\mathbf{J}$:

$$\Omega = \frac{1}{2} \nabla \times \mathbf{J} = \frac{1}{2} \mathrm{Im}(\nabla U^* \times \nabla U) = \nabla f \times \nabla g \tag{4.7}$$

The vector field Ω is important because it points along the vortex line—it is the direction around which the energy flux vector circulates in a right-handed sense. This is because it is perpendicular to the normals to the two surfaces (Equation 4.3). For two-dimensional fields, $\Omega = \omega\hat{\mathbf{z}}$ is in the z direction.

A *stationary point* is defined as a point $\mathbf{r}$ where the phase $\Phi(\mathbf{r})$ is well defined but its gradient vanishes:

$$\nabla \Phi(\mathbf{r}) = 0 \tag{4.8}$$

Stationary points are zeros of the energy flux vector of the field and occur at either local maxima or minima or saddle points of the phase distribution (Figure 4.2). The stationary points may be interpreted as the *sources* and *sinks* of the vector field $\nabla\Phi$. The monkey saddle shown in Figure 4.2d is like a normal saddle, but has three directions in which the phase increases and decreases.

To both singular and stationary points, we can assign a *topological index t* (i.e., the Poincaré index), which is defined as the number of rotations of $\nabla\Phi$ (or **j**) in a closed circuit around the point, that is, signed number of times that the phase lines rotate in the direction in which the circuit is traversed. The Poincaré index t is +1 for nondegenerate positive and negative vortices and also for phase extrema. $t = -1$ for saddle points and −2 for a monkey saddle.

Both topological charge m and index t are quantities that are conserved under smooth variation of the parameters. These conservation laws impose restrictions to the interactions between vortices and stationary points. Typical interactions are the creation, annihilation, and unfolding of vortices. The charge m and index t can only vary via creation and annihilation of multiple singular and/or stationary points. A very common situation is the creation (annihilation) of a positive vortex ($m = 1$, $t = 1$), a negative vortex ($m = -1$, $t = 1$), and two phase saddles ($m = 0$, $t = -1$ for each). Other possibility is the creation of a saddle ($m = 0$, $t = -1$) and a maximum or minimum of phase ($m = 0$, $t = 1$). Of course more complicated reactions are possible [4,5,8,9] and most of them have been experimentally generated [10,11].

4.3 HELICAL VORTEX BEAMS WITH DEPENDENCE exp($im\theta$)

The simplest fields exhibiting a vortex are the helical waves having circularly harmonic phase profiles:

$$U_m(\mathbf{r}) = R_m(r,z)\exp(im\theta) \qquad m = 0, \pm 1, \pm 2, \ldots \tag{4.9}$$

Here, a single vortex of topological charge m coincides with the origin, and thus the complex amplitude $R_m(r,z)$ must vanish at the axis $r = 0$. This is, of course, a consequence that the superposition of all phases along the axis results in perfect destructive interference. An ordinary plane, spherical, or Gaussian wave without any phase singularity is uncharged, that is $m = 0$. In this nonvortex case, the intensity is not required to vanish at the origin.

The circularly symmetric beam (Equation 4.9) may be rewritten as

$$U_m(\mathbf{r}) = |R_m| \exp i[m\theta + kz + \psi(r,z)] \tag{4.10}$$

where:

$\psi(r,z)$ describes the wavefront curvature

The surface of constant phase, $m\theta + kz + \psi(r,z) =$ constant, is helicoidal. Near the optical axis the function $\psi(r,z)$ is approximately constant, and therefore, in that region the helicoid is uniform with constant pitch. The helical wavefront with an axial phase singularity possessing unity charge is depicted in Figure 4.3a. The phase grows linearly around a circular path enclosing the optical vortex and exhibits a phase step (screw dislocation) of size 2π. Due to the phase circulation, vortex beams with dependence exp($im\theta$) carry an intrinsic orbital angular momentum (OAM) content equal to $m\hbar$ per photon, which makes them particularly suitable for multiple applications [12].

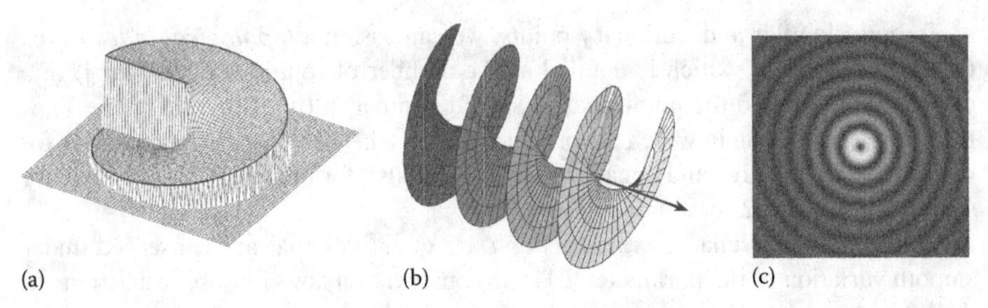

FIGURE 4.3 (a) Mesh plot of the phase function $\Phi = m\theta$ with $m = 1$ where the axial screw dislocation of the circularly harmonic vortex beams can be appreciated. (b) Helical wavefront $kz + m\theta$ = constant of the vortex beams. (c) Transverse intensity pattern of vortex beams is composed by a set of concentric circular rings with a zero on-axis intensity.

An important property of the helical modes is that, in principle, any field $E(r,\theta,z)$ can be decomposed into a series of vortex modes:

$$E(r,\theta,z) = \sum_{m=-\infty}^{\infty} h_m(r,z)\exp(im\theta) \tag{4.11}$$

where:

$h_m(r,z)$ are the radially dependent expansion coefficients of the series

$$h_m(r,z) = \frac{1}{2\pi}\int_{-\pi}^{\pi} E(r,\theta,z)\exp(-im\theta)\mathrm{d}\theta \tag{4.12}$$

In order to represent physical fields, the circularly symmetric functions $U_m(\mathbf{r})$ must themselves satisfy an optical wave equation. For monochromatic fields with time dependence $\exp(-\mathrm{i}\omega t)$, the equation is the time-independent wave equation, or the Helmholtz equation:

$$\nabla^2 U + k^2 U = 0 \tag{4.13}$$

where:

$k = \omega/c$ is the wave number

The known solutions of Equation 4.13 in cylindrical coordinates are the nondiffracting Bessel beams whose transverse intensity pattern remains unchanged on propagation.

In the paraxial domain, the Helmholtz equation can be approximated with the paraxial wave equation for the slowly varying complex amplitude $U(\mathbf{r}) = V(\mathbf{r})\exp(\mathrm{i}kz)$:

$$\nabla_t^2 V + \mathrm{i}2k\frac{\partial V}{\partial z} = 0 \tag{4.14}$$

where:

$\nabla_t^2 = \partial^2/\partial x^2 + \partial^2/\partial y^2$ is the transverse Laplacian

The standard solutions of Equation 4.14 in cylindrical coordinates are the Laguerre–Gaussian (LG) beams. In Sections 4.3.1 and 4.3.2, we will review the Bessel vortex beams and the paraxial vortex beams in more detail.

4.3.1 Bessel Vortex Beams

The separable solutions of the Helmholtz equation (4.13) in cylindrical coordinates are given by the Bessel beams [13,14]:

$$\mathrm{BB}_m(\mathbf{r}) = J_{|m|}(k_t r)\exp(im\theta)\exp(ik_z z) \tag{4.15}$$

where:

$J_{|m|}$ is the mth-order Bessel function

k_t and k_z are the transverse and longitudinal wave numbers, respectively, that satisfy $k^2 = k_t^2 + k_z^2$

The cycle-average power flow of the mth-order Bessel beam is proportional to $J_{|m|}^2(k_t r)$ that it is completely independent of z. That is, at every value of z, the intensity pattern of a Bessel beam has exactly the same (x,y) dependence. Contrary to Gaussian beams, the Bessel beams do not diverge at all. In this sense, they are usually referred to as *nondiffracting beams.*

The transverse intensity pattern of the Bessel beams is composed by a set of concentric circular rings (Figure 4.3b). The zeroth-order beam (i.e., the zero vortex case) is the only one that has a central maximum, whereas all the higher order beams have zero on-axis intensity. As the radius of the inner ring, $r_m = \rho_m/k_t$, is determined by the position ρ_m of the first maximum of the mth-order Bessel function, it increases with the order m. For higher order Bessel vortex beams, the phase at any transverse plane makes a revolution from 0 to $2\pi m$ around the axis $r = 0$. From Equation 4.15, we see that the phase of the Bessel beams is

$$\Phi(\mathbf{r}) = m\theta + k_z z \tag{4.16}$$

Thus, the phase varies linearly with z, and the surfaces of constant phase, the wavefronts, are helicoids with a pitch of $2\pi/k_z$ as shown in Figure 4.3b. As the time progresses, the helicoids move in the z direction with phase velocity ω/k_z.

Far from the origin, the intensity of the Bessel beam $J_{|m|}^2(k_t r)$ is proportional to $\cos^2(k_t r + \delta)/r$. An important consequence of the cosine-squared character of this asymptotic relation is that the energy of a Bessel beam is contained in concentric rings of width given by $k_t r = \pi$, and that the energy is approximately equal in each ring

$$\int_{\text{one ring}} J_{|m|}^2(k_t r) r \mathrm{d}r\,\mathrm{d}\theta \propto \frac{2\pi}{k_t}\int \cos^2(k_t r + \delta)\mathrm{d}r = \frac{\pi}{k_t^2} \tag{4.17}$$

where the integration is carried out over one period of $J_{|m|}$. Thus, just like a single plane wave, a Bessel beam carries an infinite amount of energy in its transverse plane. This result is also in concordance with the fact that the two-dimensional Fourier transform of the Bessel beams is given by a circular delta of radius k_t angularly modulated by the harmonic function $\exp(im\theta)$.

Finally, we should mention that Bessel beams form a complete family of solutions such that any nondiffracting beam $U(r,\theta,z)$ can be expressed as a superposition of Bessel modes with the same transverse wave number k_t:

$$U(\mathbf{r}) = \exp(ik_z z)\sum_{m=-\infty}^{\infty} c_m J_{|m|}(k_t r)\exp(im\theta) \tag{4.18}$$

where:

c_m are the expansion coefficients

4.3.2 Paraxial Vortex Beams

For helical beams of the form $R_m(r,z)\exp(im\theta)$, the paraxial wave equation (Equation 4.14) reduces to the differential equation

$$\frac{1}{r}\frac{\partial}{\partial r}\left(r\frac{\partial R_m}{\partial r}\right) - \frac{m^2}{r^2}R_m + i2k\frac{\partial R_m}{\partial z} = 0 \tag{4.19}$$

for the z-dependent radial function $R_m(r,z)$.

The most general solutions of Equation 4.19 have been obtained and characterized recently in References [15,16]. The corresponding paraxial beams are called *circular beams* and its complex amplitude can be described by either the Whittaker functions or the confluent hypergeometric functions. For special values of their parameters, the circular beams reduce to known families of paraxial vortex beams including the standard, elegant, and generalized LG beams, the Bessel–Gauss (BG) beams [17,18], the hypergeometric beams [19], hypergeometric–Gauss beams [20], the fractional order elegant LG beams [21], the BG beams with quadratic radial dependence [22], and the fractional vortex beams [23]. Among all special cases of the paraxial circular vortex beams, the LG and the BG beams are of particular relevance.

LG beams are the circularly symmetric modes of stable laser resonators with spherical mirrors and form a complete set of modes such that any possible resonator mode can be expanded as a linear combination of LG modes. The normalized LG beam with radial number $n = 0,1,2,\ldots$ and azimuthal number $m = 0,\pm1,\pm2,\ldots$ is written as

$$\begin{aligned}\mathrm{LG}_n^m(\mathbf{r},t) = {} & \frac{A}{w(z)}\left[\frac{\sqrt{2}r}{w(z)}\right]^{|m|} L_n^{|m|}\left[\frac{2r^2}{w(z)^2}\right]\exp\left[\frac{-r^2}{w^2(z)}\right] \\ & \times\exp\left\{i\left[kz + m\theta - \omega t + \frac{kr^2}{2R(z)} - \left(2n+|m|+1\right)\psi_{GS}(z)\right]\right\}\end{aligned} \tag{4.20}$$

where:

$L_n^m(\cdot)$ are the generalized Laguerre polynomials
$w^2(z) = w_0^2(1 + z^2/z_R^2)$ describes the beam width
$R(z) = z + z_R^2/z$ is the radius of curvature of the phase front
$\psi_{GS}(z) = \arctan(z/z_R)$ is the Gouy shift
$z_R = kw_0^2/2$ is the Rayleigh range
w_0 is the beam width at $z = 0$
$A = [2n!/(1 + \delta_{0,m})\pi(n + |m|)!]^{1/2}$ is the normalization constant

The transverse intensity of the LG_n^m beams is composed of a finite set of concentric circular rings. Regardless of the indices, the width of the beam is proportional to $w(z)$, so that as z increases the transverse intensity pattern is affected by the factor $w_0/w(z)$ but otherwise is shape invariant. The Gouy phase shift $(2n + |m| + 1)\psi_{GS}(z)$ is a function of the orders, which means that the phase velocity increases with increasing order numbers. LG_n^m beams with the same number $p = 2n + |m|$ form subsets of beams with the same Gouy shift. Therefore, except for a scale factor, the superpositions of LG_n^m beams belonging to the same subset p preserve the intensity shape on propagation.

In general, the M^2 quality factor of the LG_n^m beams is [16]

$$M^2 = 2n + |m| + 1 \tag{4.21}$$

For a fixed width at the waist plane, a better quality (lower divergence) beam is associated with a lower value of M^2. A minimum value of 1 is reached only for the fundamental nonvortex Gaussian beam LG_0^0.

LG_n^m beams (Equation 4.20) have a higher order vortex of topological charge m at the optical axis. The ringed intensity pattern of the LG beams acquires the form of a single annulus when $n = 0$. For this case, at the plane $z = 0$, Equation 4.20 reduces to

$$LG_0^m(r,\theta, z = 0) \propto r^{|m|}\exp(im\theta)\exp\left(-\frac{r^2}{w_0}\right) = (x + iy)^m \exp\left(-\frac{r^2}{w_0}\right) \tag{4.22}$$

From Equation 4.21, it is clear that the lowest M^2 factor we can get for a vortex LG beam is 2 and corresponds to the one-ringed single vortex with parameters $n = 0$, $m = \pm 1$.

BG beams are solutions of the paraxial wave equation and are finite-energy realizations of the ideal nondiffracting beams in Equation 4.15. The expression of a BG beam carrying an m-charged axial vortex propagating into positive z direction is [18]

$$BG_m(\mathbf{r}) = \exp\left(-i\frac{k_t^2}{2k}\frac{z}{\mu}\right)\frac{1}{\mu}\exp\left(-\frac{r^2}{\mu w_0^2}\right)J_{|m|}\left(\frac{k_t r}{\mu}\right)\exp(im\theta) \tag{4.23}$$

where:

$\mu = \mu(z) = 1 + iz/z_R$, with $z_R = kw_0^2/2$ being the Rayleigh distance

Similar to Bessel beams, the wavefront of the BG_m beams has a helicoidal shape with $m\lambda$ pitch and a slightly varying curvature along the z-axis due to the longitudinal phase shift.

BG_m beams carry a well-defined intrinsic OAM content equal to $m\hbar$ per photon. The M^2 quality factor of the BG_m beams can be calculated in closed form and the expression is [16]

$$M^2 = \sqrt{\left[m+1+\beta\frac{I_{m+1}(\beta)}{I_m(\beta)}\right]^2 - \beta^2} \tag{4.24}$$

where:
$\beta = k_t^2 w_0^2 / 4$
I_m is the modified Bessel function of the first kind

4.3.3 Unwound Vortex Beams: Anisotropic Vortices

An *isotropic* vortex occurs when the phase increases *linearly* from 0 to $2\pi m$ around circles enclosing the singularity (Figure 4.3a), and then the intensity contours are circles centered at the vortex. The field near an isotropic vortex (assumed to be at the origin) may be approximated as follows:

$$U(r,\theta) \simeq r^{|m|}\exp(im\theta) = (x + iy)^m \tag{4.25}$$

The vortices occurring in the helical vortex beams (Equation 4.9) discussed earlier are, obviously, isotropic.

The optical vortices occurring in an optical field are, in general, *anisotropic*. In this case, the intensity contours close to the vortices are elliptical, and around this ellipse, the sectors of equal area sweep out equal intervals of phase. Several parameterizations to describe the vortex anisotropy have been proposed [2,24–28]. Although there are advantages and disadvantages of each parameterization, all of them are fully equivalent. The main conclusion is that in the immediate vicinity of a vortex, the wave function is determined by four independent parameters (six if the vortex is not located at the origin), and the phase structure by three.

The anisotropic vortex can be understood as a geometrical deformation of the isotropic one. Note first that the angle dependence of an isotropic vortex can be decomposed into its real and imaginary parts:

$$h(\theta) = \exp(im\theta) = \cos m\theta + i\sin m\theta \tag{4.26}$$

This relation can be generalized by (1) changing the relative amplitude between the real and imaginary parts, (2) introducing a phase difference between the real and imaginary parts, and (3) introducing an overall rotation. The result is the angular dependence of a general anisotropic vortex:

$$h(\theta) = \cos\varepsilon\cos\left[m(\theta - \theta_0)\right] + i\sin\varepsilon\sin\left[m(\theta - \theta_0 - \sigma)\right] \tag{4.27}$$

where:
ε is the vortex anisotropy
σ is the vortex skewness
θ_0 is the vortex rotation angle

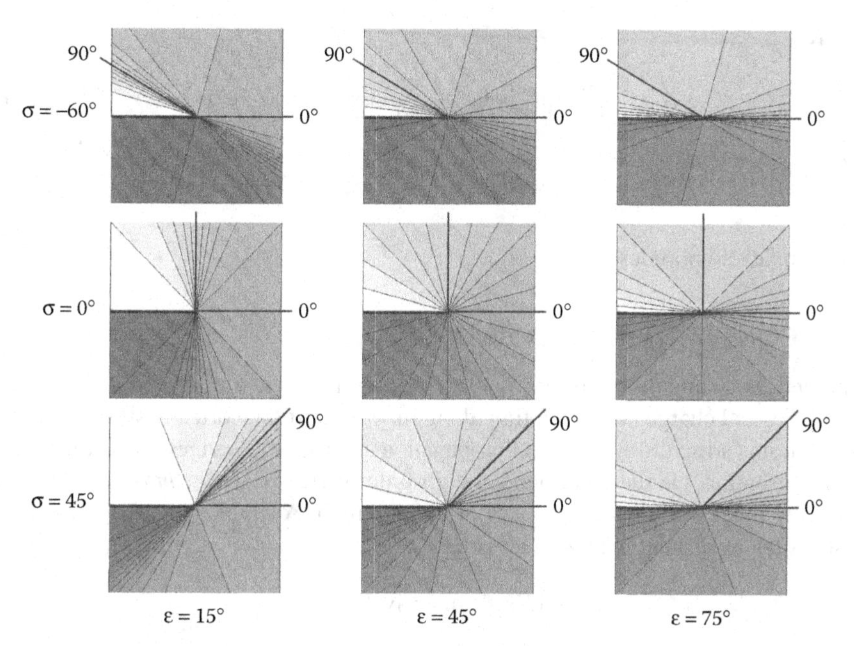

FIGURE 4.4 Phase maps of a single anisotropic vortex illustrating the effects of the anisotropy ε and skewness σ. Contours indicate cophasal lines separated by 15°. Zero phase contours have been oriented along the horizontal axis to facilitate visualization (to do this, we set $\theta_0 = -\sigma$ in all cases).

The phase distribution of the anisotropic vortex is then given by

$$\Phi(\theta) = \arctan\left[\tan\varepsilon \frac{\sin m(\theta - \theta_0 - \sigma)}{\cos m(\theta - \theta_0)}\right] \tag{4.28}$$

and the intensity in the vicinity of the vortex is $|U|^2 \simeq r^{2|m|}|h(\theta)|^2$. It can be easily demonstrated that for unit-strength vortices the contour lines of constant intensity around the singular point are ellipses.

Figure 4.4 illustrates the effects of varying the morphological vortex parameters $(\varepsilon,\sigma,\theta_0)$. As shown, for zero skewness ($\sigma = 0$), the contours $\Phi = 0$ and $\Phi = \pi/2$ remain orthogonal and the lines of constant phase are crowded around the y-axis when $\varepsilon < \pi/4$, and x-axis when $\varepsilon < \pi/4$. The effect of the skewness σ is to change the crossing angle of the contours $\Phi = 0$ and $\Phi = \pi/2$. Changing θ_0 rotates the phase distribution but does not change its shape. For any combination of parameters, the lines of constant phase remain straight lines.

The propagation and properties of optical beams containing anisotropic vortices have been studied in detail by several authors [26–29]. Anisotropic vortices can be constructed by superposing two vortex beams of strength $\pm m$, such as $A(BG_{+m}) + B(BG_{-m})$. Close to the origin, the superposition has the form:

$$U(r,\theta) \simeq r^{|m|}[A\exp(im\theta) + B\exp(-im\theta)] \tag{4.29}$$

where:

$$A = \frac{\cos\varepsilon}{2}\exp(-im\theta_0) + \frac{\sin\varepsilon}{2}\exp\left[-im(\theta_0 + \sigma)\right] \quad (4.30)$$

$$B = \frac{\cos\varepsilon}{2}\exp(im\theta_0) - \frac{\sin\varepsilon}{2}\exp\left[im(\theta_0 + \sigma)\right] \quad (4.31)$$

are complex amplitudes.

4.4 COMPOSITE VORTEX BEAMS

The helical beams discussed in the previous section have an isolated axial vortex of topological charge m propagating along the z-axis. If the vortex is simple, that is, $m = \pm 1$, it is structurally stable on propagation and usually changes its position only on perturbation. On the other hand, a high-order vortex of charge m is unstable and typically breaks up into $|m|$ vortices of unit strength. A simple perturbation can be just adding a constant field with no vortices:

$$U(r) = A(x \pm iy)^m + B \quad (4.32)$$

In this situation, the high-order vortex unfolds into $|m|$ anisotropic vortices evenly spaced on a circle centered on the original vortex. The phase anisotropy of all the perturbed vortices is the same, independent of the strength of the perturbation.

The unfolding of a higher order vortex is an example of a *composite vortex beam*. In general, the field created by the interference of multiple beams may produce composite vortices [30]. A composite vortex beam is a complex structure whose zero-intensity points do not coincide with the zeros of the composing beams. It is even possible to have the situation of free-vortex composing beams that produce composite vortices. A simple example is given by the vortex arrays produced by the interference of three, four, and five plane waves [31]. The location, number, and charge of the composite vortices depend on the relative phase and amplitude of the composing beams. Applications that utilize the properties of optical vortices require the engineering of composite vortices.

As a detailed example of a composite vortex beam, consider the superposition of colinear Bessel beams of integer order to produce Bessel beams with fractional order [32,33]. The composite beam is given by

$$U_\alpha(\mathbf{r}) = \sum_{m=-\infty}^{\infty}\left\{\frac{i^{-(m-\alpha)}\sin\left[\pi(m-\alpha)\right]}{\pi(m-\alpha)}\right\}J_{|m|}(k_0 r)\exp(im\theta) \quad (4.33)$$

where:
α is the continuous order of the Bessel field

Figure 4.5 shows the transverse amplitude and phase distributions of the fractional beam U_α at $z = 0$ for several values of α in the range $5 \le \alpha \le 6$. Noticeably,

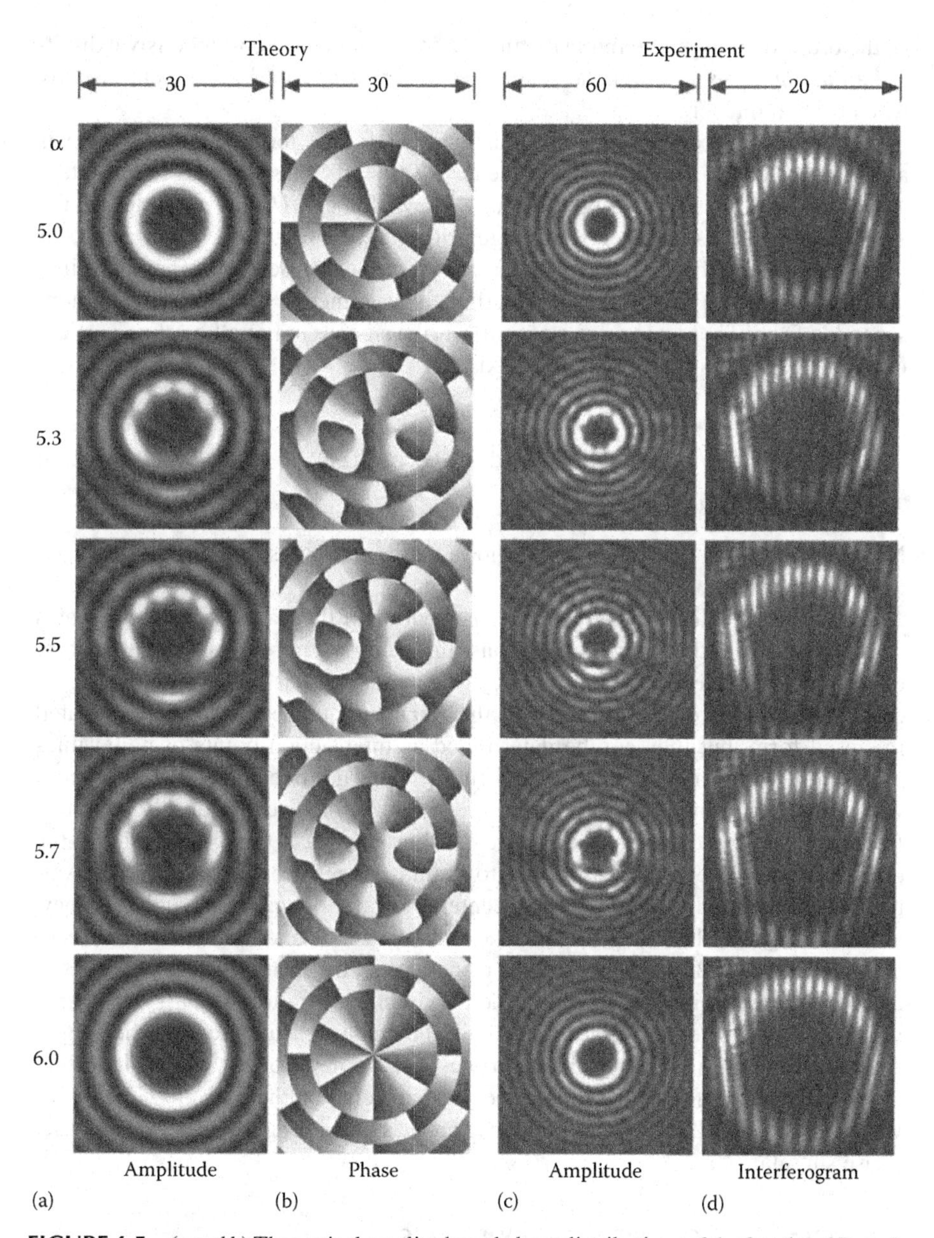

FIGURE 4.5 (a and b) Theoretical amplitude and phase distributions of the fractional Bessel beam U_α for several values of α. (c) Experimental amplitude observations. (d) Interferograms with a plane wave exhibiting the vortex dislocations near the origin. The image columns have been plotted at different transverse scales to show the important details. The image dimensions are in units of k_0^{-1}.

as the order increases, the intensity and phase patterns vary continuously exhibiting an azimuthally asymmetric shape that becomes circularly symmetrical only when α is an integer.

An immediate consequence of Equation 4.33 is that for fractional α there is always at the origin a contribution of the term $J_0(0) = 1$. Thus there is on-axis intensity and no vortex is formed on the z-axis. Now, although U_α has no vortex on the z-axis, it does possess an interesting singularity structure (Figure 4.5). In general, for noninteger orders, the axial vortex of the integer mth-order Bessel beam unfolds into m unit-strength vortices, leading to rotationally asymmetric vortex beam with nonseparable form. The total vortex strength N is the signed sum of all existing vortices threading a large loop including the z-axis:

$$N = \lim_{r\to\infty} \frac{1}{2\pi} \int_0^{2\pi} d\theta \frac{\partial}{\partial\theta}\left[\arg U_\alpha^\pm(r,\theta)\right] \tag{4.34}$$

Numerical evaluation of this integral shows that N converges to the nearest integer, to α, in all cases.

For arbitrary values of α, the phase distribution of the fractional beam deviates significantly from the integer case. Thus, the field $U_\alpha(\mathbf{r})$ presents a rich structure of vortices embedded in the morphology of its transverse field distribution. In general, the positions of the vortices on the transverse plane cannot be calculated in closed form, but they can be determined as intersection points of nodal lines representing the curves $\mathrm{Re}U_\alpha = 0$ and $\mathrm{Im}U_\alpha = 0$. The trajectories of the vortices within the central most region of $U_\alpha = {}_{M+\mu}(\mathbf{r})$ as the order continuously increases from $\alpha = 3$ to 7 are depicted in Figure 4.6. For small $\mu > 0$, we note that (1) the on-axis higher-order vortex unfolds into M unit-strength vortices lying at the vertices of an M-sided regular polygon centered on the optical axis, and (2) a new vortex emerges at the intersection of the negative y-axis and the first circular zero contour of the J_M Bessel function. As μ increases, the M unfolded vortices move away from the origin until reaching the maximum separation when $\mu \sim 1/2$, and, while the new vortex moves further along the negative y-axis toward the origin. As $\mu \to 1$, the original M unfolded vortices and the new vortex tend to the origin forming a new $(M + 1)$-sided regular polygon centered on the optical axis. Finally, when $\alpha = M + 1$, all the vortices collapse on-axis generating the new integer order phase singularity.

4.5 GENERATION OF VORTEX BEAMS

Early work describing the transfer of optical OAM and its connection with the phase circulation characteristic of vortex beams [34] sparked and fueled an increasing interest in the study of optical vortices. The verification of theoretical predictions as well as the development of practical applications acted as the driving forces to devise ways to produce and shape laser beams with embedded vortices. This section summarizes the most prevalent techniques used to achieve this goal.

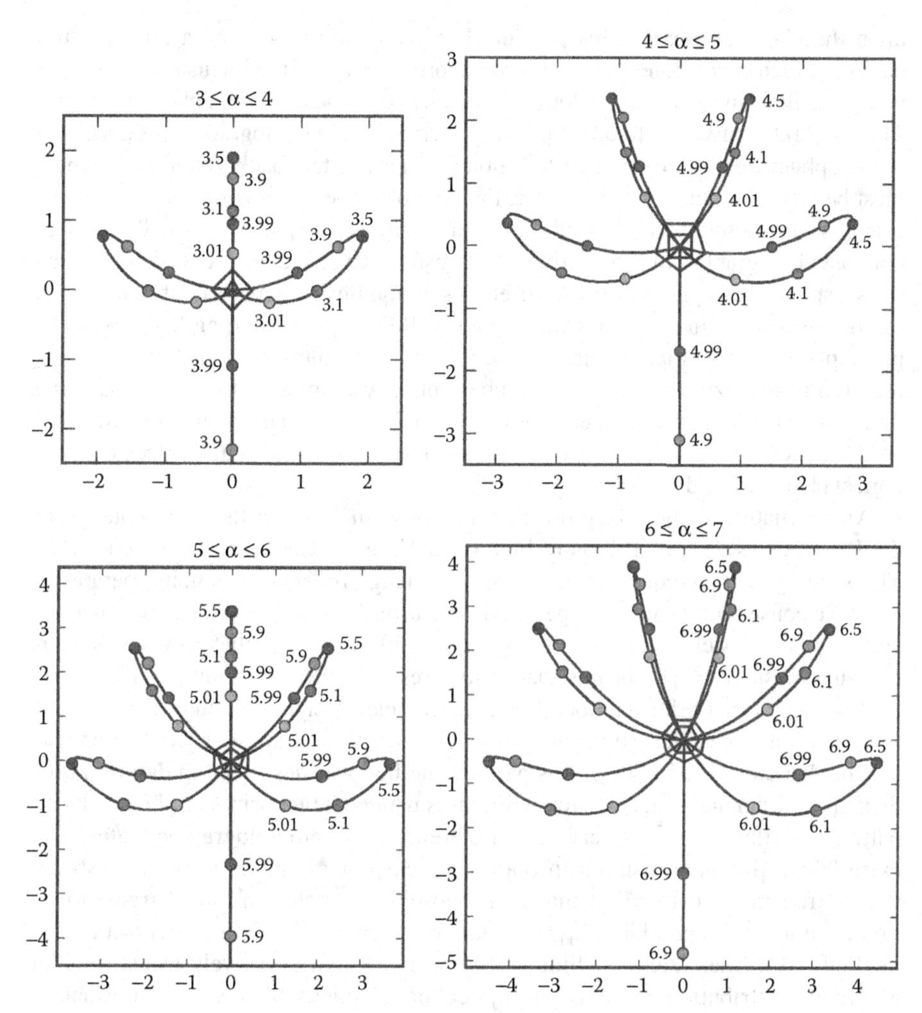

FIGURE 4.6 Theoretical trajectories of the vortices within the central most region of $U_\alpha(\mathbf{r})$ as its order increases continuously from $\alpha = 3$ to 7. The loci of the vortices, as α increases gradually, are given by the circles. The trajectories described by the vortices are represented by the solid lines. Distances are given in units of k_0r.

4.5.1 LG Beams

Arguably, the best-known instance of vortex beam is the doughnut-shaped LG mode LG_0^1. A natural solution of the paraxial wave equation under azimuthal symmetry conditions and an eigenmode of the circular cylindrical laser resonator [35,36] is that the LG_0^1 beam has been extensively observed and used as an OAM-carrying field in numerous experiments [34,37–39].

Some of the earliest approaches to shaping an LG_0^1 beam starting with a fundamental Gaussian beam relied on imprinting the azimuthal phase modulation directly upon the Gaussian beam. This problem has been solved by using a binary spiral pattern, which corresponds to a two-level approximation of the intensity pattern that would result from interfering a doughnut mode and an ideal, coaxial plane wave [40]. This method, equivalent to a rudimentary form of in-line holography, produces the correct phase distribution of a vortex, along with a constant background field, which must be removed with a Fourier filter, for example. The spiral zone plate was fabricated as the photoreduction of a binary black and white opaque original onto high contrast photographic film, resulting in a transmissive optical element often referred to as a spiral phase plate (SPP). A different solution involved an SPP fabricated on a plastic substrate using standard milling tools. Because the resulting features of the phase plate are much larger than the wavelength (and thus the optical path change required in one azimuthal range), the phase plate was immersed in index-matching fluid. By controlling the temperature of the fluid, and in consequence its refractive index, an accurate phase modulation was demonstrated for optical vortices of topological charges 1 and 2 [41].

An alternative method for producing an LG beam relies on the representation of the LG mode as a superposition of high-order Hermite–Gaussian (HG) modes [42]. This shaping method requires a stable source of a high-order HG beam. Its operational principle consists of a pair of properly oriented astigmatic lenses, which introduce the correct complex weight for the mode as well as the Gouy shift that provides the correct superposition of HG beams that in turn result in the corresponding LG beam. A clear advantage of this method, despite its complex design, is its high efficiency.

A seminal paper describing the reliable production of annular LG beams was first published in 1995 [43]. In this paper, a method was described in detail to produce high-efficiency transmission holograms using contact prints of photoreduced patterns similar to those described in Reference [40] onto holographic plates. The extra fabrication step resulted in optically clear holograms on a glass substrate, with a transmission function that more closely resembled the ideal transformation from a fundamental TEM_{00} Gaussian to an annular LG beam. Central to this method is the chemical bleaching of the prints, which effectively transforms an amplitude distribution into a map of optical path values which, when illuminated, produces the desired intensity profile. Bleaching has the effect of rehalogenating the metallic silver grains in the developed emulsion, turning the stored hologram transparent while preserving the thickness variations of the emulsion on the plate. The resulting diffractive optical element has close to ideal transmission efficiency. This particular technique would ultimately become the most successful method for the production of vortex beams until the development of liquid crystal technologies. Furthermore, an adaptation of the key ideas behind this technique would eventually be successfully implemented using spatial light modulators (SLMs) more than a decade later. A detailed recipe for producing bleached holograms on standard photographic film can be found in Reference [44].

Recently, an approach to produce LG beams based on optofluidics has been demonstrated [45]. This approach consists of a two-level hologram fabricated as a fluidic channel through which a fluid can be injected. The capability of varying the refractive index of the fluid results in a tunable optical path upon transmission, which in turn results in a variable phase modulation of the incident light.

4.5.2 Bessel and BG Beams

An intuitive approach to producing a Bessel beam consists of shaping its amplitude profile using one amplitude-only optical element and a different one to imprint the required transverse phase [46]. In this approach, an annular LG beam is first produced with a hologram [40]. The LG beam is then used to illuminate an axicon, which in turn generates the conical wave associated to a Bessel beam. In this way, a reasonable approximation to a Bessel can be produced provided that the beam illuminating the optics is a collimated plane wave. However, when a fundamental Gaussian beam is used as the starting beam, as is most often the case, a BG beam is instead obtained. Milne and colleagues later devised a more elaborate scheme based on this generation design [47], which involved a hollow, fluidic axicon. The axicon, engineered to be filled with a liquid phase of arbitrary refractive index, has thus variable refractive properties.

An alternative to a continuous-phase modulation diffractive optical element, that is, hologram or phase plate, is based on the decomposition of the azimuthal phase in its odd and even parts [48]. In this scheme, only constant phase delays are required to obtain the components of the helical phase profile, which can then be added coherently to produce high-order Bessel beams. The method is inspired in the earliest demonstration of Mathieu beams [49].

4.5.3 Other Vortex Beams

The availability of proven holographic beam shaping techniques led directly to the experimental demonstration of other significant families of beams, namely, Mathieu [50] and parabolic beams [51]. With their intricate vortex structures, both families have now become the subject of countless research efforts.

Known collectively as Helmholtz–Gauss beams, the set of all three beams, BG, parabolic-Gauss, and Mathieu–Gauss beams (together with plane waves), have also been demonstrated experimentally, and their propagation properties calculated and measured to excellent agreement with theory and numerical models [52]. Each member of this set is a family of beams that are eigenmodes of the Helmholtz equation in a cylindrical orthogonal coordinate system. Their demonstration relies on standard holographic techniques already described above. Helmholtz–Gauss beams are separable solutions, which translate into unique and desirable propagation properties, such as the well-known self-healing and extended invariant propagation length of Bessel beams.

4.5.4 Spatial Light Modulators

One of the most significant recent technological advances for the beam-shaping community is without a doubt the development of the modern SLM. SLMs utilize the tunable birefringence of nematic liquid crystal molecules to impart a phase delay onto an incident beam. SLMs are flexible and easy to control with a personal computer and custom software. They are capable of addressing hundreds of thousands of portions of a wavefront and are dynamically addressable to typical video rates.

The advent of current SLMs and their progressively lower cost and increasing availability played a significant role in the advancement of numerical beam shaping algorithms and the development of elaborate applications, as described in Section 4.6. An example of an encoding scheme appropriate for vortex beams in SLMs is provided in Reference [53]. Moreover, recent original work in beam shaping has led to a universal method to design an arbitrary beam profile with the properties of Helmholtz–Gauss beams [54,55]. This work represents outstanding progress in the field of laser beam shaping as it allows for the rapid, efficient sculpting of a self-healing beam of arbitrary transverse profile.

4.6 APPLICATIONS

Soon after their earliest experimental observations, optical vortices became themselves subjects of intensive qualitative and quantitative studies [56]. The ability to generate vortices reliably quickly motivated a growing interest that spread rapidly to several fields of research within the optical sciences, where they have found a diversity of applications. Numerous efforts have been made and countless publications outline different aspects of the optics of optical vortices and their use. Moreover, research in optical vortices is ongoing and particularly fruitful. This section lists only a minute fraction of these efforts due to its relevance to laser beam shaping and does not represent an exhaustive review of the field.

4.6.1 Studies of Vortex Dynamics

An early question regarding the spatial structure of the singularity associated to the phase dislocation of vortex beams was whether a beam with tunable OAM per photon could be created by changing the structure of the vortices within the beam. In fact, tuning the OAM can be achieved by shaping of the odd and even components of the helical phase profile associated to the vortex [28,29]. A relative azimuthal shift of the components results in the harmonic variation of the magnitude of the OAM density of the beam. Using this approach, a vortex is said to be partially unwound as its components are dephased relative to one another. When the vortex is completely unwound, the OAM content vanishes.

One interesting idea with respect to the succession of Bessel functions that represent Bessel beams concerns its connection with the vortex structure of fractional Bessel beams and its resulting OAM content [57]. As the order of the Bessel function is allowed to vary continuously, the degenerate vortex at the optical axis becomes a set of closely spaced vortices within the wave field that describe an apparent coordinated motion that ends in another degenerate vortex of increased order. This dynamic transit of vortices correlate with a continuous variation of the OAM per photon in the sequence of fractional vortices that connect contiguous order Bessel functions [33,57].

However interesting, fractional Bessel functions are not solutions of the Helmholtz equation. In this respect, Gutiérrez-Vega and coworkers have recently demonstrated a new class of vortex beams that connect nondiffracting Bessel beams of successive integer order in a smooth transition while satisfying Helmholtz equation and thus, the long propagation invariance length of the Bessel beam of fundamental order, from which all others can be generated via the application of a raising operator [32].

Also of interest is the connection between the observed rotation rate of the intensity pattern of vortex beams and its OAM content. Recently, the rotation of superpositions of Bessel beams carrying no net OAM has been observed and measured to good agreement with theory that predicted arbitrarily fine control can be achieved over the rotation rates of the intensity profile of the superposition [58]. These rotating fields make an ideal tool for controlled rotation of trapped particles in optical trapping experiments, for example.

4.6.2 Applications of Vortex Beams

A natural avenue of inquiry for vortex beams is the study of the transfer of OAM to material objects [42]. The earliest investigations in this respect used microscopic absorbing particles in conjunction with LG beams in optical tweezers experiments for visual demonstrations of rotational motion [37]. This seminal work distinguished clearly the nature of OAM and its underlying difference with spin, or polarization, angular momentum. To date, countless beam types and shapes have been coupled with optical tweezers to investigate their specific effects on material particles. A few noteworthy examples include high-order Bessel beams [59], helical Mathieu beams [60], and Airy beams [61].

Contemporary work also predicted [62] and resulted in the experimental demonstration by Tabosa et al. of the optical OAM transfer to cold atoms [63]—again, from a LG beam—a milestone that motivated a plethora of subsequent research in the field. In this particular work, four-wave mixing was used as an indirect tool to confirm the transfer of OAM to the atomic medium from an optical pump and again into a probe beam. A similar scheme was used in a later experiment using a colloidal medium, which formed the required nonlinear medium as the colloids aggregated in an organized way in the interference pattern of counterpropagating optical pumps [64]. In another ingenious experiment, the successful relocation of atoms was successfully demonstrated using a vortex beam as a hollow optical duct [65].

Optical vortices have also found application in modern microscopy. Specifically, vortices have been utilized as Fourier filters that produce enhanced edges for an object for small optical paths [66]. An advantage of this innovative application is that for objects with slowly varying topologies, a surface interferogram is readily produced without the need for an interferometric setup.

Observational astronomy has also benefited from the use of optical vortices in the form of an instrument and technique referred to as vortex coronagraph [67]. The central idea behind this concept is the isolation of light that originates in the source or subject of interest, such as a planetoid, by eliminating adjacent light, say from its parent star, by deviating it off the axis using a spiral diffractive optical element. The idea has been tested and it has produced positive results experimentally [68].

Perhaps one of the most wished for applications of vortex beams is the design of tools for advanced cell manipulation. Surgical procedures performed on single cells have been successfully demonstrated with a vortex optical trap [69]. Manipulation with a vortex beam is shown to affect the decreased photodamage on a cell compared to a traditional Gaussian beam, a desirable property when manipulating living organisms.

Vortex beams have also been successfully employed to manipulate and transport aqueous droplets in solution [70]. Droplets with a lower refractive index than their medium rest within the low intensity region near the optical axis while being spatially confined by the annular intensity pattern of the beam.

In recent times, the role of vortex beams in experimental research in quantum optics has become increasingly significant. Three remarkable examples are listed below as a sample of the usefulness of vortex beams in this field of research. Vortex beam pairs made up of entangled photons possess an added degree or dimension of entanglement by virtue of their mutual OAM correlation. Higher dimensional entangled quantum states provide a more secure quantum communication scheme, can potentially carry a higher information content, and offer greater resilience to errors when compared to two-dimensional systems [71]. In Reference [72], the authors demonstrate in a three-dimensional Bell test experiment the suitability of OAM-carrying beams for quantum communication protocols. More recently, McLaren et al. [73] have successfully verified that BG modes can be entangled in the OAM degree of freedom. This experiment specifically establishes the advantage of BG beams in quantum information processes due to the increased number of states that can be utilized and compared to non-OAM-carrying beams, while simultaneously demonstrating high dimensionality in the entangled state [71]. A similar scheme has recently been generalized up to 12 dimensions with the aid of SLMs [74].

In a closely related development, another Bell test was performed using quantum correlations in vortex beams via their OAM in an imaging scheme commonly known as *quantum ghost imaging* [75]. In this scheme, ghost images are formed from the outputs of two detectors: one is a high spatial-resolution detector illuminated by a reference beam and the other is illuminated by light that has interacted with the object but that has no spatial resolution. An image results only by cross-correlating the outputs of both detectors from a sequence of correlated intensity patterns. In this experiment, a phase filter implemented with an SLM in one of the entangled beams results in high coincidence counts that correlate with the imaging objects edges. The enhancement of the edges originates in the nonlocality of the phase filter with respect to the object.

4.7 SUMMARY

The study of the properties and applications of the optical vortices embedded in light fields has become relevant in recent years through their relationship with beams carrying optical angular momentum. This chapter has presented a review of the basic theory of vortex beam shaping, the experimental procedures to generate them, and some of their most important applications. The simplest vortex beams are described by the helical waves having circularly harmonic phase profiles of the form $R_m(r,z)\exp(im\theta)$. The paraxial LG beams and the nondiffracting Bessel beams belong to this class of beams. Suitable superpositions of plane waves can generate optical beams with more complex structures of vortices, namely, composite vortices, that can be specially designed for particular applications.

REFERENCES

1. M.S. Soskin and M.V. Vasnetsov, Singular optics, *Progress in Optics*, 42: 219–276, 2001.
2. M.R. Dennis, K.O.'Holleran, and M.J. Padgett, Singular optics: Optical vortices and polarization singularities, *Progress in Optics*, 53: 293–363, 2009.
3. J.F. Nye and M.V. Berry, Dislocations in wave trains, *Proceedings of the Royal Society of London A*, 336: 45–57, 1984.
4. I. Freund, Saddles, singularities, and extrema in random wave fields, *Physical Review E*, 52: 2348–2360, 1995.
5. I. Freund, Critical-point level-crossing geometry in random wave fields, *Journal of the Optical Society of America A*, 14: 1911–1927, 1997.
6. S. Chávez-Cerda, J.C. Gutiérrez-Vega, and G.H.C. New, Elliptic vortices of electromagnetic wave fields, *Optics Letters*, 26: 1803–1805, 2001.
7. M. Born and E. Wolf, *Principles of Optics*, 7th ed. Cambridge: Cambridge University Press, 1999.
8. J.F. Nye, Unfolding of higher-order wave dislocations, *Journal of the Optical Society of America A*, 15: 1132–1138, 1998.
9. M.V. Berry, Wave dislocation reactions in non-paraxial Gaussian beams, *Journal of Modern Optics*, 45: 1845–1858, 1998.
10. G.P. Karman, M.W. Beijersbergen, A. van Duijl, and J.P. Woerdman, Creation and annihilation of phase singularities in a focal field, *Optics Letters*, 22: 1503–1505, 1997.
11. G.P. Karman, M.W. Beijersbergen, A. van Duijl, D. Bouwmeester, and J.P. Woerdman, Airy pattern reorganization and subwavelength structure in a focus, *Journal of the Optical Society of America A*, 15: 884–899, 1998.
12. L. Allen, S.M. Barnett, and M.J. Padgett, *Orbital Angular Momentum*, Bristol: Institute of Physics Publishing, 2003.
13. J. Durnin, Exact solutions for nondiffracting beams. I. The scalar theory, *Journal of the Optical Society of America A*, 4: 651–654, 1987.
14. J. Durnin, J.J. Micely, Jr., and J.H. Eberly, Diffraction-free beams, *Physical Review Letters*, 58: 1499–1501, 1987.
15. M.A. Bandres and J.C. Gutiérrez-Vega, Circular beams, *Optics Letters*, 33: 177–179, 2008.
16. M.A. Bandres, D. López-Mago, and J.C. Gutiérrez-Vega, Higher-order moments and overlaps of rotationally symmetric beams, *Journal of Optics A: Pure and Applied Optics*, 12: 015706, 2010.
17. F. Gori, G. Guattari, and C. Padovani, Bessel-Gauss beams, *Optics Communications*, 64: 491–495, 1987.
18. J.C. Gutiérrez-Vega and M.A. Bandres, Helmholtz–Gauss waves, *Journal of the Optical Society of America A*, 22: 289–298, 2005.
19. V.V. Kotlyar, R.V. Skidanov, S.N. Khonina, and V.A. Soifer, Hypergeometric modes, *Optics Letters*, 32: 742–744, 2007.
20. E. Karimi, G. Zito, B. Piccirillo, L. Marruci, and E. Santamato, Hypergeometric-Gaussian modes, *Optics Letters*, 32: 3053–3055, 2007.
21. J.C. Gutiérrez-Vega, Fractionalization of optical beams. II. Elegant Laguerre–Gaussian modes, *Optics Express*, 15: 6300–6313, 2007.
22. C.F.R. Caron and R.M. Potvliege, Bessel-modulated Gaussian beams with quadratic radial dependence, *Optics Communications*, 15: 83–93, 1999.
23. M.V. Berry, Optical vortices evolving from helicoidal integer and fractional phase steps, *Journal of Optics A: Pure and Applied Optics*, 6: 259–268, 2004.
24. I. Freund and V. Freilikher, Parameterization of anisotropic vortices, *Journal of the Optical Society of America A*, 14: 1902–1910, 1997.

25. Y.Y. Schechner and J. Shamir, Parameterization and orbital angular momentum of anisotropic dislocations, *Journal of the Optical Society of America A*, 13: 967–973, 1996.
26. J. Masajada and B. Dubik, Optical vortex generation by three plane wave interference, *Optics Communications*, 198: 21–27, 2001.
27. G. Molina-Terriza, E.M. Wright, and L. Torner, Propagation and control of noncanonical optical vortices, *Optics Letters*, 26: 163–165, 2001.
28. C. López-Mariscal and J.C. Gutiérrez-Vega, Vortex beam shaping, *Proceedings of SPIE*, 6290: 62900O, 2006.
29. C. López-Mariscal and J.C. Gutiérrez-Vega, Unwound vortex beam shaping, *Proceedings of SPIE*, 6663: 666306, 2007.
30. I.D. Maleev and G.A. Swartzlander, Jr., Composite optical vortices, *Journal of the Optical Society of America B*, 20: 1169–1176, 2003.
31. K. O'Holleran, M.J. Padgett, and M.R. Dennis, Topology of optical vortex lines formed by the interference of three, four, and five plane waves, *Optics Express*, 14: 3039–3044, 2006.
32. J.C. Gutiérrez-Vega and C. López-Mariscal, Nondiffracting vortex beams with continuous orbital angular momentum order dependence, *Journal of Optics A: Pure and Applied Optics*, 10(015009): 1–8, 2008.
33. C. López-Mariscal, D. Burnham, D. Rudd, D. McGloin, and J.C. Gutiérrez-Vega, Phase dynamics of continuous topological upconversion in vortex beams, *Optics Express*, 16: 11411–11422, 2008.
34. L. Allen, M.W. Beijersbergen, R.J.C. Spreeuw, and J.P. Woerdman, Orbital angular momentum of light and the transformation of Laguerre-Gaussian laser modes, *Physical Review A*, 45: 8185, 1992.
35. C. Tamm and C. Weiss, Bistability and optical switching of spatial patterns in a laser, *Journal of the Optical Society of America B*, 7: 1034–1038, 1990.
36. M. Harris, C.A. Hill, and J.M. Vaughan, Optical helices and spiral interference fringes, *Optics Communications*, 106: 161, 1994.
37. M.E.J. Friese, J. Enger, H. Rubinsztein-Dunlop, and N.R. Heckenberg, Optical angular-momentum transfer to trapped absorbing particles, *Physical Review A*, 54: 1593, 1996.
38. N.B. Simpson, K. Dholakia, L. Allen, and M.J. Padgett, Mechanical equivalence of spin and orbital angular momentum of light: An optical spanner, *Optics Letters*, 22: 52–54, 1997.
39. M.A. Clifford, J. Arlt, J. Courtial, and K. Dholakia, High-order Laguerre-Gaussian laser modes for studies of cold atoms, *Optics Communications*, 156: 300–306, 1998.
40. N. Heckenberg, R. McDuff, C. Smith, and A. White, Generation of optical phase singularities by computer-generated holograms, *Optics Letters*, 17: 221–223, 1992.
41. M.W. Beijersbergen, R.P.C. Coerwinkel, M. Kristensen, and J.P. Woerdman, Helical-wavefront laser beams produced with a spiral phaseplate, *Optics Communications*, 11(5): 321–327, 1994.
42. M.W. Beijersbergen, L. Allen, H.E.L.O. van der Veen, and J.P. Woerdman, Astigmatic laser mode converters and transfer of orbital angular momentum, *Optics Communications*, 96: 123, 1993.
43. H. He, N.R. Heckenberg, and H. Rubinsztein-Dunlop, Optical particle trapping with higher-order doughnut beams produced using high efficiency computer generated holograms, *Journal of Modern Optics*, 42: 217–223, 1995.
44. C. López-Mariscal and J.C. Gutiérrez-Vega, The generation of nondiffracting beams using unexpensive computer-generated holograms, *American Journal of Physics*, 75(1): 36–42, 2007.
45. G. Jeffries, G. Milne, Y. Zhao, C. López-Mariscal, and D. Chiu, Optofluidic generation of Laguerre-Gaussian beams, *Optics Express*, 17: 17555–17562, 2009.
46. J. Arlt and K. Dholakia, Generation of high-order Bessel beams by use of an axicon, *Optics Communications*, 177: 297, 2000.

47. G. Milne, G.D.M. Jeffries, and D.T. Chiu, Tunable generation of Bessel beams with a fluidic axicon, *Applied Physics Letters*, 92: 261101, 2008.
48. C. López-Mariscal, J.C. Gutiérrez-Vega, and S. Chávez-Cerda, Production of high-order Bessel beams with a Mach-Zehnder interferometer, *Applied Optics*, 43: 5060–5063, 2004.
49. J.C. Gutiérrez-Vega, M.D. Iturbe-Castillo, G.A. Ramírez, E. Tepichin, R.M. Rodriguez-Dagnino, S. Chávez-Cerda and G.H.C. New, Experimental demonstration of optical Mathieu beams, *Optics Communications*, 195: 35–40, 2001.
50. S. Chávez-Cerda, M.J. Padgett, I. Allison, G.H.C. New, J.C. Gutiérrez-Vega, A.T. O'Neil, I. MacVicar, and J. Courtial, Holographic generation and orbital angular momentum of high-order Mathieu beams, *Journal of Optics B: Quantum and Semiclassical Optics*, 4: S52–S57, 2002.
51. C. López-Mariscal, M. Bandrés, J.C. Gutiérrez-Vega, and S. Chávez-Cerda, Observation of Parabolic nondiffracting wave fields, *Optics Express*, 13(7): 2364–2369, 2005.
52. C. López-Mariscal, M.A. Bandrés, and J.C. Gutiérrez-Vega, Observation of the experimental propagation properties of Helmholtz-Gauss beams, *Optical Engineering*, 45(6): 068001, 2006.
53. C. López-Mariscal and J.C. Gutiérrez-Vega, Complex scalar fields using amplitude-only spatial light modulators, *Proceedings of SPIE*, 7062: 706209, 2008.
54. C. López-Mariscal and J.C. Gutiérrez-Vega, Numerical calculation of arbitrary Helmholtz-Gauss beams, *Laser Beam Shaping X, Proceedings of SPIE*, 7430: 743009, 2009.
55. C. López-Mariscal and K. Helmerson, Shaped nondiffracting beams, *Optics Letters*, 35: 1215–1217, 2010.
56. I.V. Basistiy, V.Yu. Bazhenov, M.S. Soskin, and M.V. Vasnetsov, Optics of light beams with screw dislocations, *Optics Communications*, 103: 422–428, 1993.
57. J. Leach, E. Yao, and M.J. Padgett, Observation of the vortex structure of a non-integer vortex beam, *New Journal of Physics*, 6: 71, 2004.
58. R. Rop, A. Dudley, C. López-Mariscal, and A. Forbes, Measuring the rotation rates of superpositions of higher-order Bessel beams, *Journal of Modern Optics*, 59(3): 259–267, 2012.
59. K. Volke-Sepulveda, V. Garces-Chavez, S. Chavez-Cerda, J. Arlt, and K. Dholakia, Orbital angular momentum of a high-order Bessel light beam, *Journal of Optics B: Quantum and Semiclassical Optics*, 4: S82, 2002.
60. C. López-Mariscal, J.C. Gutiérrez-Vega, G. Milne, and K. Dholakia, Orbital angular momentum transfer in helical Mathieu beams, *Optics Express*, 14(9): 4182–4187, 2006.
61. J. Baumgartl, T. Cizmr, M. Mazilu, V. Chan, A. Carruthers, B. Capron, W. McNeely, E. Wright, and K. Dholakia, Optical path clearing and enhanced transmission through colloidal suspensions, *Optics Express*, 18: 17130–17140, 2010.
62. M. Babiker, W.L. Power, and L. Allen, Light-induced torque on moving atoms, *Physical Review Letters*, 73: 12391242, 1994.
63. J.W.R. Tabosa and D. Petrov, Optical pumping of orbital angular momentum of light in cold cesium atoms, *Physical Review Letters*, 83: 49674970, 1999.
64. C. López-Mariscal, J. Gutiérrez-Vega, D. McGloin, and K. Dholakia, Direct detection of optical phase conjugation in a colloidal medium, *Optics Express*, 15: 6330–6335, 2007.
65. D.P. Rhodes, D.M. Gherardi, J. Livesey, D. McGloin, H. Melville, T. Freegarde, and K. Dholakia, Atom guiding along holographically generated high order Laguerre-Gaussian light beams, *Journal of Modern Optics*, 53: 547, 2006.
66. S. Frhapter, A. Jesacher, S. Bernet, and M. Ritsch-Marte, Spiral phase contrast imaging in microscopy, *Optics Express*, 13: 689–694, 2005.
67. G. Foo, D. Palacios, and G. Swartzlander, Jr., Optical vortex coronagraph, *Optics Letters*, 30: 3308–3310, 2005.

68. G. Swartzlander, Jr., E. Ford, R. Abdul-Malik, L. Close, M. Peters, D. Palacios, and D. Wilson, Astronomical demonstration of an optical vortex coronagraph, *Optics Express*, 16: 10200–10207, 2008.
69. G.D.M. Jeffries, J.S. Edgar, Y.Q. Zhao, J.P. Shelby, C. Fong, and D.T. Chiu, Using polarization-shaped optical vortex traps for single-cell nanosurgery, *Nano Letters*, 7(2): 415420, 2007.
70. R.M. Lorenz, J.S. Edgar, G.D.M. Jeffries, and D.T. Chiu, Microfluidic and optical systems for the on-demand generation and manipulation of single femtoliter-volume aqueous droplets, *Analytical Chemistry*, 78(18): 6433–6439, 2006.
71. D. Collins, N. Gisin, N. Linden, S. Massar, and S. Popescu, Bell inequalities for arbitrarily high-dimensional systems, *Physical Review Letters*, 88: 040404, 2002.
72. A. Vaziri, G. Weihs, and A. Zeilinger, Experimental two-photon, three-dimensional entanglement for quantum communication, *Physical Review Letters*, 89: 240401, 2002.
73. M. McLaren, M. Agnew, J. Leach, F. Roux, M. Padgett, R. Boyd, and A. Forbes, Entangled Bessel-Gaussian beams, *Optics Express*, 20: 23589–23597, 2012.
74. A.C. Dada, J. Leach, G.S. Buller, M.J. Padgett and E. Andersson, Experimental high-dimensional two-photon entanglement and violations of generalized Bell inequalities, *Nature Physics*, 7: 677680, 2011.
75. B. Jack, J. Leach, J. Romero, S. Franke-Arnold, M. Ritsch-Marte, S.M. Barnett, and M.J. Padgett, Holographic ghost imaging and the violation of a Bell inequality, *Physical Review Letters*, 103: 083602, 2009.

5 Gaussian Beam Shaping: Diffraction Theory and Design

Fred M. Dickey and Scott C. Holswade

CONTENTS

5.1 INTRODUCTION

This chapter describes a diffraction-based method for converting single-mode Gaussian beams into beams with uniform irradiance profiles. The design is based on a Fourier transform relation between the input and output beam functions. This solution can be obtained using geometrical optics methods. However, the diffraction approach introduces a parameter that contains the product of the widths of the input and output beams. This parameter is a significant part of the physical optics solution. The efficacy of the solution is shown to depend on this parameter. The quality of the solution improves asymptotically with increasing value of the parameter.

Many experiments and industrial applications require a laser beam irradiance that is nominally constant over a specified area. Such applications include laser/material processing, laser/material interaction studies, optical data/image processing, and lithography. In many cases, it is desirable, for obvious reasons, that the beam shaping operation conserves energy.

The multifaceted integrator approach to laser beam shaping is especially suitable to laser beams with highly irregular (multimode) irradiance distributions.[1,2] The number and size of the facets is selected to accomplish the required integration or, equivalently, averaging. Doherty[3] has treated the problem of irradiance mapping for laser beams with radial symmetry and regular irradiance distributions. Dickey and O'Neil[4] give a general formulation of the multifaceted beam integrator problem and introduce a configuration that minimizes deleterious diffraction effects.

For single-mode beams with a Gaussian profile, it is possible to map the beam into a uniform intensity profile with steep skirts. This mapping can be accomplished with simpler optics that is more flexible with respect to scaling and does not have the interference patterns inherent in multifaceted beam integration. Several authors address the problem of mapping a Gaussian beam into one with a uniform irradiance distribution. The earliest paper known to the authors that addresses the lossless shaping of a single-mode laser beam is the paper by Frieden.[5] Lee[6] employs an iterative technique to design a phase filter to convert a Gaussian beam into a more uniform irradiance distribution. Veldkamp[7,8] uses an iterative technique to design binary gratings to accomplish the profile shaping. Aleksoff et al.[9] use the geometrical optics approximation to develop a holographic system that maps a Gaussian beam into a rectangularly shaped beam with uniform amplitude and phase. Kosoburd and Kedmi[10] use geometrical optics to design a diffractive system that maps Gaussian beams into beams with uniform irradiance. Eismann et al.[11] apply the Gershberg–Saxton algorithm, or equivalently phase retrieval, to synthesize a two-element design that produces a beam with uniform amplitude and phase. In 1996, Golub et al.[12] presented numerical and experimental results for a diffractive beam shaper based on a geometrically derived phase function.

In this chapter, we give a solution to the problem of mapping a Gaussian laser beam into a beam with uniform irradiance profile. The configuration analyzed exploits the Fourier transform properties of lenses. That is, the output optical field is the Fourier transform of the input optical field and a phase function. This configuration has the advantage of being able to change the output size or the working distance by changing the transform lens. In Section 5.2, we define the general problem of converting a Gaussian beam into a beam with uniform irradiance and give a solution to the problem. Bounds set by the uncertainty principle are also discussed in this section. The problem of creating a collimated beam, a beam with both a uniform irradiance and a uniform phase, is addressed in Section 5.3. Again, the uncertainty principle has implications. It is used to define a generalized Rayleigh range for shaped beams. Wavelength invariance of the beam shaping problem is discussed in Section 5.4. In Section 5.5, considerations associated with designing a shaping element are discussed. Sensitivity to alignment and scaling errors are discussed in Section 5.6. In Section 5.7, we discuss the application of the design methodology to a particular problem. In Section 5.8, we present the results of the design and testing of a prototype system. A summary of the chapter is given in Section 5.9. This chapter is based on the author's papers.[13,14]

5.2 THE ANALYTICAL SOLUTION

The general beam shaping problem is shown schematically in Figure 5.1. In the figure, the beam to be shaped enters the proverbial black box from the left and exits on the right, diffracting to the design irradiance pattern. The black box may contain a single optical element or a combination of several optical components of differing types such as lenses, mirrors, prisms, diffractive optics, and holograms. One approach to solving a beam shaping problem would be to assume an optical configuration and develop a solution around this configuration. An example of this approach would be iterative techniques that provide a solution for a single element diffractive optic. A more general approach would be to obtain a general solution for a shaping function, amplitude and phase, using diffraction theory, and then develop an optical design realizing the shaping function. This approach is usually the most difficult. An approach somewhere between the two is commonly what is taken. Although a given solution might be realizable with a single optical element, it is frequently the case that a more versatile, more practicable, and less expensive design is obtained using multiple elements.

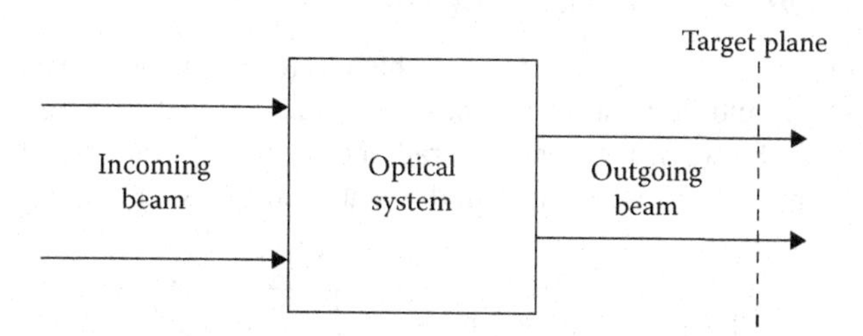

FIGURE 5.1 Schematic of the beam shaping problem.

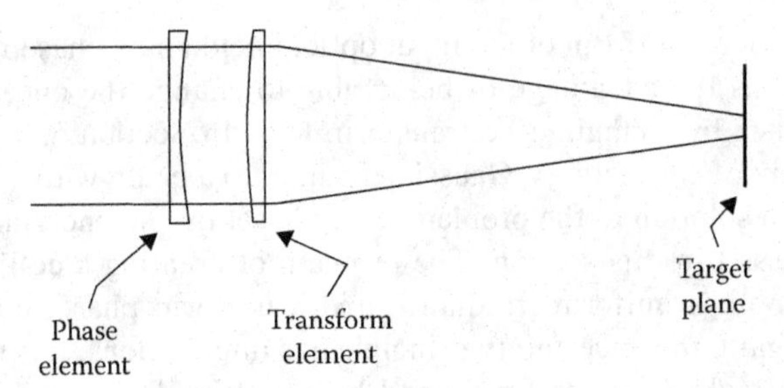

FIGURE 5.2 Fourier transform beam shaping system. (From F.M. Dickey and S.C. Holswade, *Optical Engineering*, 35, 3285–3295, 1996. With permission.)

5.2.1 Optical Configuration

Our approach to lossless beam shaping, illustrated in Figure 5.2, consists of a phase element in conjunction with a Fourier transform lens. The optical field at the focal plane of the transform lens is proportional to the Fourier transform of the product of the input optical field and phase of the phase element.[15] This configuration has several advantages. The phase element can be changed to control both the scale and shape of the output irradiance. The transform lens can be changed to modify the working distance, with a corresponding change in scale of the output. Finally, although the phase element and transform lens could be designed as one optical element, it is generally easier to design and fabricate the two components if their functions are kept separate. It should be noted that using a separate phase element and transform lens to shape the beam produces a more versatile system. The spot size can be changed by changing the focal length of the transform lens and adjusting the working distance correspondingly, or keeping the working distance the same and changing the phase element. Also, it is probably cheaper to fabricate the shaping system as separate elements.

It should be noted that this configuration is more general than it might first appear. Any solution that could be obtained using a Fresnel integral can be obtained using the Fourier transform system shown in Figure 5.2. This can be seen from the fact that the Fresnel integral can be written as a Fourier transform of the product of the input aperture function and a pure (quadratic) phase factor (see Goodman[15] and Equation 4.10). The quadratic phase function becomes part of the beam shaping element, which is discussed in detail in Section 5.4.2.

5.2.2 Minimum Mean Square Error Formulation

Given the configuration of Figure 5.2, the problem is to design the phase element. The direct approach would be to solve for the phase function that minimizes the mean square difference between the desired irradiance and the irradiance produced by the phase element. That is, we want to find ϕ that minimizes an integral of the form

$$R = \int \left| \left| \Im\left[\left(\frac{2}{\sqrt{\pi}}\right)^{1/2} e^{-x^2} e^{i\phi} \right] \right|^2 - \left(\frac{1}{\alpha}\right)^{1/2} \operatorname{rect}\left(\frac{f}{\alpha}\right) \right|^2 df \tag{5.1}$$

where:

$\Im$ denotes a Fourier transform operation
f denotes the corresponding frequency domain variable
α defines the size of the output

The problem is scaled to a unit width ($1/e^2$) Gaussian beam function. Here, the problem is formulated in one dimension, which is appropriate to the separable problem of converting a circular Gaussian beam into a uniform beam with a square cross section. In general, a single variable will be used to represent a one- or two-dimensional variable. Unfortunately, we were not able to obtain a global solution to Equation 5.1. We were able to obtain solutions to the problem using the method of stationary phase. Before presenting the stationary phase solution, it is interesting to discuss the solution to a related problem of requiring both the amplitude and phase of the output to be constant over the region of interest and zero elsewhere.

The solution to the separable uniform amplitude and phase problem can be obtained by determining the phase ϕ that minimizes the functional

$$\begin{aligned} R &= \int \left| \Im\left[\left(\frac{2}{\pi}\right)^{1/2} e^{-x^2} e^{i\phi} \right] - \left(\frac{1}{\alpha}\right)^{1/2} \text{rect}\left(\frac{f}{\alpha}\right) \right|^2 df \\ &= 2 - 2\,\text{Re} \int \Im\left[\left(\frac{2}{\sqrt{\pi}}\right)^{1/2} e^{-x^2} e^{i\phi} \right] \left(\frac{1}{\alpha}\right)^{1/2} \text{rect}\left(\frac{f}{\alpha}\right) df \end{aligned} \tag{5.2}$$

Note that this equation differs from Equation 5.1 in that it involves the differences of fields (complex) functions, while Equation 5.1 is the difference of intensities (magnitude squared) functions. Thus, the problem described in Equation 5.1 is less constrained since the phase of the output is a free parameter. This allows for a broader range of solutions. In Equation 5.2, the input Gaussian function and the output rect function are normalized to unit energy. This ensures that the mean square difference in Equation 5.2 depends on the variation in the shape of the input and output functions and not on any relative amplitude difference between the two functions. As a result, only the integral of the cross terms needs to be evaluated, since the normalization forces the other two integrals to unity.

The solution to Equation 5.2 is readily obtained by applying Parseval's theorem and expanding the integrand. Integrating the magnitude squared terms gives

$$R = 2 - 2\,\text{Re} \int \left(\frac{2\alpha}{\sqrt{\pi}}\right)^{1/2} e^{-\phi}\, \text{sinc}(\alpha x) e^{-x^2} dx \tag{5.3}$$

where:

Re z denotes the real part of z

Clearly, R is minimized if the integral is maximized. This is obtained if ϕ is set equal to the phase of the sine function. Since the phase of the sine function is a binary function with values of 0 and π, the optimum phase function is a binary function. All that remains is to determine the value of α that maximizes the integral in Equation 5.3 with α set equal to the phase of the sine function. This can be evaluated numerically to give $\alpha = 0.710$. The beam irradiance profile for this solution is shown in Figure 5.3.

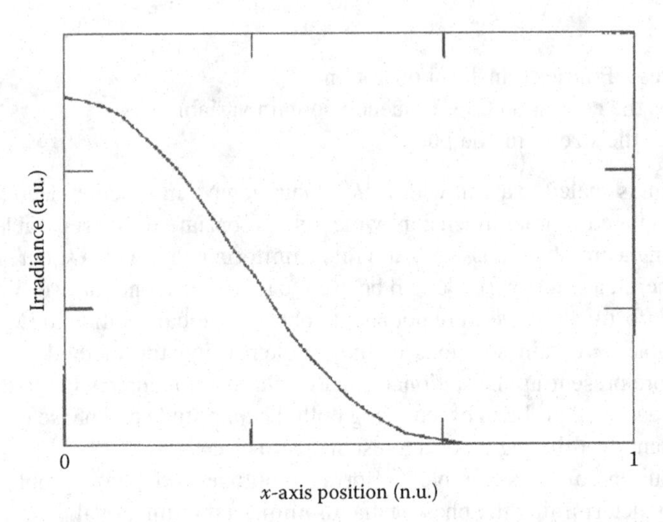

FIGURE 5.3 Beam irradiance profile (right half) for the optimum solution of Equation 5.2. (From F.M. Dickey and S.C. Holswade, *Optical Engineering*, 35, 3285–3295, 1996. With permission.)

This is just the optimum solution, in the sense of Equation 5.2, of the problem posed by Veldkamp.[7] It is interesting to note that the solution is not as flat as might be expected. As mentioned above, α determines the size of the shaped output. In obtaining the minimum mean square error solution, we have let the size of the output to be a free parameter. The value $\alpha = 0.710$ corresponds to small β discussed in the next section. If the output beam size is allowed to increase, one can get a flatter looking output beam. However, the mean square error between the shaped beam and the desired output is larger than that of Figure 5.3. A careful inspection of Equation 5.3 shows that the mean square error increases to a maximum with increasing α (output beam size). This is due to the fact that, with respect to Equation 5.1, Equation 5.2 is overly constrained. That is, in Equation 5.2 both the phase and the amplitude of the output are required to be uniform, whereas only the irradiance of the output in Equation 5.1 is required to be uniform. It can also be noted that this solution corresponds to a value of $R = 1.199$, which is a significant fraction of the maximum of $R = 2$.

5.2.3 The Uncertainty Principle

There are fundamental constraints on the beam shaping problem that can be traced to electromagnetic theory. It is difficult to develop constraints on a problem without a degree of specificity. However, the uncertainty principle of quantum mechanics or, equivalently, the time–bandwidth inequality associated with signal processing can be applied to the beam shaping configuration defined in Figure 5.2. The uncertainty principle is a constraint on the lower limit of the product of the root-mean-square (r.m.s.) width of a function and its r.m.s. bandwidth.[16,17]

$$\Delta_x \Lambda_\upsilon \geq \frac{1}{4\pi} \tag{5.4}$$

The respective widths are defined by

$$(\Delta_x)^2 = \frac{\int_{\infty}^{\infty}\int_{\infty}^{\infty}(x-\bar{x})^2\left|u(x,y)\right|^2 \mathrm{d}x\mathrm{d}y}{\int_{\infty}^{\infty}\int_{\infty}^{\infty}\left|u(x,y)\right|^2 \mathrm{d}x\mathrm{d}y} \tag{5.5}$$

where:

$$\bar{x} = \frac{\int_{\infty}^{\infty}\int_{\infty}^{\infty}x\left|u(x,y)\right|^2 \mathrm{d}x\mathrm{d}y}{\int_{\infty}^{\infty}\int_{\infty}^{\infty}\left|u(x,y)\right|^2 \mathrm{d}x\mathrm{d}y} \tag{5.6}$$

and

$$\left(\Delta_{\upsilon_x}\right)^2 = \frac{\int_{\infty}^{\infty}\int_{\infty}^{\infty}(\upsilon_x-\bar{\upsilon}_x)^2\left|U(\upsilon_x,\upsilon_y)\right|^2 \mathrm{d}\upsilon_x\mathrm{d}\upsilon_y}{\int_{\infty}^{\infty}\int_{\infty}^{\infty}\left|U(\upsilon_x,\upsilon_y)\right|^2 \mathrm{d}\upsilon_x\mathrm{d}\upsilon_y} \tag{5.7}$$

where:

$$\bar{\upsilon}_x = \frac{\int_{\infty}^{\infty}\int_{\infty}^{\infty}\upsilon_x\left|U(\upsilon_x,\upsilon_y)\right|^2 \mathrm{d}\upsilon_x\mathrm{d}\upsilon_y}{\int_{\infty}^{\infty}\int_{\infty}^{\infty}\left|U(\upsilon_x,\upsilon_y)\right|^2 \mathrm{d}\upsilon_x\mathrm{d}\upsilon_y} \tag{5.8}$$

In the last two equations, the upper and lowercase letters denote the field function and its Fourier transform, respectively. The uncertainty principle stated in Equation 5.4 is obtained from Equations 5.5 through 5.8 using the Cauchy–Schwarz inequality, Parseval's theorem, and the Fourier transform correspondence $\partial u/\partial x \leftrightarrow i2\pi\upsilon$.

The field distribution at the focal plane of an ideal lens is proportional to a Fourier transform. The Fourier transform variable is related to the physical variables by the following equation[15]:

$$\upsilon = \frac{x'}{\lambda f} \tag{5.9}$$

where:

- f is the focal length
- λ is the wavelength
- x' is the coordinate in the focal plane

Using Equation 5.9, Equation 5.4 can be written as

$$\frac{\Delta_x \Delta_{x'}}{\lambda f} \geq \frac{1}{4\pi} \tag{5.10}$$

Further, converting the widths on the left-hand side of the above equation to a $1/e^2$ radius for the input Gaussian beam, a full radius for the shaped beam, and multiplying both sides by $2\sqrt{2\pi}$, Equation 5.10 becomes

$$\frac{2\sqrt{2\pi}r_0 y_0}{f\lambda} \geq 0.69 \tag{5.11}$$

where:
r_0 is the Gaussian beam radius
y_0 is the shaped profile radius

The left-hand side of the above inequality is the β defined in Section 5.3.4. It should be noted that this result is strictly true for separable input and output beam functions. In fact, for the nonseparable case, the constant on the right-hand side would be greater because the radii are averaged with respect to the orthogonal coordinate.

Applying the uncertainty principle to the beam shaping problem requires some thought. As derived above, the inequality is strictly applicable to the product of the input and output beam radii. However, it does prohibit focusing a beam to a radius smaller than the lower limit given by the inequality. It is reasonable to expect that good shaping results would not be obtained for beam radii determined by the equality in Equation 5.11. In fact, it is expected that good shaping would not be obtained unless the equality was exceeded by a factor of 3 or more. This is supported by the fact that the equality in Equation 5.4 obtains only when the input and output beam are both Gaussian beams that are Fourier transform pairs.[16] Further, since the beam shaping problem addressed here can be expressed as a convolution in the output plane of the Fourier transform of the Gaussian input beam and the Fourier transform of phase function, one would expect that the uncertainty principle would also be an indication of the ability to achieve steep skirts on the edges of the beam profile. Finally, since the constant on the right-hand side of Equation 5.11 would be greater for a nonseparable function it would be more difficult to produce flat circular beams with steep skirts than it would be to produce corresponding beams with a square cross section. These results are compatible with our numerical modeling of the beam shaping problem (Section 5.5).

5.2.4 Stationary Phase Solution

Solutions to the problem defined in Figure 5.2 and Equation 5.1 can be obtained by the application of the method of stationary phase. Before giving the stationary phase solutions, we present a brief introduction to the one-dimensional stationary phase formula.[18,19] Stamnes[20] provides an extensive discussion of the method of stationary phase and its application to diffraction problems. Walther[21] applies the method of stationary phase to the wave theory of lenses. A treatment of stationary phase is given in Chapter 2.

The method of stationary phase gives an asymptotic approximation to integrals of the form

$$I(\beta) = \int_a^b f(x)e^{i\beta\phi(x)}\,dx \tag{5.12}$$

where:
β is a dimensionless parameter

The first term in the asymptotic phase approximation to the integral in Equation 5.12 is given by

$$I_c(\beta) \sim e^{i[\beta\phi(c)+\mu\pi/4]} f(c)\left[\frac{2\pi}{\beta|\phi''(c)|}\right]^{1/2} \tag{5.13}$$

where:
the double prime symbol (″) denotes the derivatives

$$\mu = \operatorname{sign}\phi''(c) \tag{5.14}$$

and c is a simple stationary point defined by

$$\phi'(c) = 0, \quad \phi''(c) \neq 0 \tag{5.15}$$

Equation 5.13 is commonly referred to as the *stationary phase formula*. Similar results are obtained in two dimensions with $\phi''(c)$ being replaced by the Hessian matrix for ϕ.

The essence of the beam shaping problem is to equate $|I_c(\beta)|^2$ with the desired irradiance in the output plane of Figure 5.2. That is, the magnitude squared of the right-hand side of Equation 5.13 is equal to the desired output irradiance. Using this condition with Equation 5.15 leads to a second-order differential equation for the beam shaping phase function $\phi(x)$. The details of obtaining the explicit form of the differential equation from Equations 5.13 and 5.15 are rather tedious[22] (see Chapter 2). Care must be taken with respect to the absolute value of ϕ'' in the denominator of Equation 5.13. This condition requires that the phase $\phi(x)$ is a convex function, a function whose second derivative is either positive or negative everywhere. This turns out not to be a problem for the case of mapping a Gaussian into a rect function. This can be seen from the geometrical optics representation of the beam shaping problem illustrated in Figure 5.4. In the figure, the input beam consists of collimated rays whose density is accurately scaled to be proportional to a Gaussian irradiance profile. These rays are bent, in the shaping plane, to form a uniform irradiance distribution in the output plane. Near the shaping plane one can form a phase front for the converging beam by integrating the reciprocal of the slope of the rays (wave normals). The curved line in the figure represents the phase front. It can be seen that the slope of the phase front, derivative of the phase function, is a monotonic function giving a phase function with a positive (or negative) second

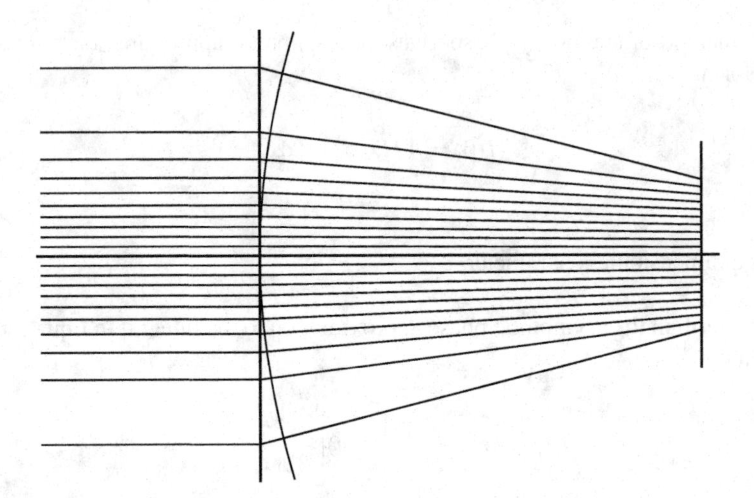

FIGURE 5.4 Geometrical optics representation of the Gaussian to rect function beam shaping problem.

derivative. It may be noted that the construction of Figure 5.4 provides an algorithm for a geometrical optics solution to the beam shaping problem.[9] When the phase of the phase function is not a convex function over the entire input beam, the problem becomes more difficult. In this case, a solution to the problem can be obtained by dividing the input beam into regions over which the phase function is convex and combining the solutions for these regions in a seamless manner. In practice, this may be a very difficult problem.

A very simple and useful application of the method of stationary phase is outlined in Appendix A. In this appendix, we discuss the problem of the lossless mapping of a uniform amplitude and phase beam into a uniform irradiance profile at the focal plane of a transform lens using a phase element as shown in Figure 5.2. In this case, the determination of the differential equation is quite simple. The solution for the phase element is just a quadratic phase function, a thin (ideal) lens. This problem provides the diffraction theory basis for the "fly's eye lens" beam integration system (see Chapter 7).[23,24] If a small array of these elements were placed before the Fourier transform lens, the uniform patterns for each phase element would be superimposed in the focal plane of the Fourier transform lens. This solution is very closely related to the problem if Fourier analysis of chirped or linear frequency modulated signals is occurring in synthetic aperture radar system theory.[25]

In two dimensions, the general form of the equation to be solved is

$$F(\omega_x,\omega_y)\frac{1}{2\pi}\int_{\infty}^{\infty}\int_{-\infty}^{\infty} f(\xi,\eta)\exp\{i[\beta\phi(\xi,\eta)-\xi\omega_x-\eta\omega_y]\}d\xi d\eta \tag{5.16}$$

where:

$\xi = x/r_i$ and $\eta = y/r_i$ are the normalized input variables with r_i defining the length scale

$\omega_x = x_f/r_0$ and $\omega_y = y_f/r_0$ are the normalized output variables in the focal plane of the Fourier transform lens with r_0 defining the length scale

The stationary phase solution improves asymptotically with increasing dimensionless parameter $\beta = 2\pi r_i R_0/f\lambda$, where R_0 is the size of the output beam, λ is the optical wavelength, and f is the focal length of the transform lens.

The stationary phase evaluation of integrals of the type given by Equation 5.16, generally, leads to second-order partial differential equation for the phase function ϕ. The resulting partial differential equation can then be solved for ϕ, subject to an energy boundary condition determined by Parseval's theorem. The partial differential equation reduces to a second-order ordinary differential equation for the separable and circularly symmetric problem. The optical element is then designed to realize $\beta\phi$.

The stationary phase evaluation of Equation 5.16 allows for the mapping of arbitrary single-mode laser beams into arbitrary irradiance profiles using the system in Figure 5.2. However, some irradiance profiles may be more mathematically difficult to realize. Romero and Dickey[22] (see Chapter 2) have obtained solutions for the separable problem of converting circular Gaussian beams to uniform profiles with rectangular cross sections and the problem of converting circular Gaussian beams to uniform beams with circular cross sections.

For a circular Gaussian beam input, the problem of turning a Gaussian beam into a flat-top beam with rectangular cross section is separable. That is, the solution is the product of two one-dimensional solutions. β and $\phi(\xi)$ are thus calculated for each dimension. The phase element will then produce the sum of these phases $[\beta_x\phi_x(x) + \beta_y\phi_y(y)]$. The corresponding one-dimensional solution for ϕ is[22]

$$\phi(\xi) = \frac{\sqrt{\pi}}{2}\xi\,\mathrm{erf}(\xi) + \frac{1}{2}\exp(-\xi^2) - \frac{1}{2} \tag{5.17}$$

where:

$$\xi = \frac{\sqrt{2}x}{r_0} \quad \text{or} \quad \xi = \frac{\sqrt{2}y}{r_0}$$

Here $r_0 = 1/e^2$ is the radius of the incoming Gaussian beam.

The solution for the problem of turning a circular Gaussian beam into a flat-top beam with circular cross section is[22]

$$\phi(\xi) = \frac{\sqrt{\pi}}{2}\int_0^{\xi}\sqrt{1 - \exp(-\rho^2)}\,d\rho \tag{5.18}$$

where:

$$\xi = \frac{\sqrt{2}r}{r_0}$$

Here, r is the radial distance from the optical axis.

As previously mentioned, the quality of these solutions depends strongly on the parameter β. For the two solutions given in Equations 5.17 and 5.18, β is given by

$$\beta = \frac{2\sqrt{2\pi}r_0 y_0}{f\lambda} \tag{5.19}$$

where:

$r_0 = 1/e^2$ is the radius of the incoming Gaussian beam

y_0 is the half-width of desired spot size (the radius for a circular spot, or half the width of a square or rectangular spot)

Examples of the dependence on β will be given later.

5.2.5 Positive and Negative Solutions

An interesting property of the configuration in Figure 5.2 is that if ϕ is any even function solution, then −ϕ is also a solution. If the input and output beams are even functions, then ϕ will be an even function. This is easily demonstrated in one dimension and the development is readily extended to two dimensions. In simplest form, the input or output optical fields shown in Figure 5.2 are related by the Fourier transform:

$$G(\omega) = \int f(x) e^{i\beta\phi(x)} e^{-i\omega x} dx \tag{5.20}$$

where:

$G(\omega)$ is the output field

$f(x)$ is the input field

In this equation, the integrand can be expanded to give

$$\begin{aligned} G(\omega) &= \int f(x)[\cos \beta\phi(x) + i \sin \beta\phi(x)](\cos \omega x - i \sin \omega x)\, dx \\ &= \int f(x) \cos \beta\phi(x) \cos \omega x dx \\ &= \int f(x) \cos[-\beta\phi(x)] \cos \omega x dx \end{aligned} \tag{5.21}$$

The equivalence of the positive and negative solutions follows from the fact that the odd terms in the integrand integrate to zero and the cosine function is an even function. This result has practical implications for system design. The positive solution produces a beam that converges to a small diameter after the output plane, and the negative solution gives a beam that converges to a small diameter before the output plane. It should be noted that this result is independent of the type of solution method such as the method of stationary phase. It depends only on the symmetry assumptions stated at the start of this section. However, the convexity problem discussed following Equation 5.15 is related in that it allows for both a positive and a negative solution.

5.2.6 Quadratic Phase Correction

The solutions described in Section 5.2.4 assume that the input Gaussian beam has a uniform (constant) phase at the beam shaping element. For a Gaussian beam, this condition is obtained at the beam waist, and it is not convenient or practical to always locate the beam waist at the shaping element. One solution is to build into the shaping element a phase conjugate to the input beam phase. A more practicable solution is to exploit the fact that the Gaussian beam phase causes a shift in the location of the output plane. That is, the desired profile is located at a distance from the focal plane of the transform lens. There is also a slight magnification associated with the shift of the output plane. These assertions can be proved using the Fresnel integral and the general form for Gaussian beams.

Gaussian beams propagate with a phase function given by[26]

$$f(x,y) = e^{(-\sigma+i\gamma)(x^2+y^2)} \tag{5.22}$$

where:

σ and γ are functions of the distance from the beam waist and $\gamma = 0$ at the beam waist

The solutions in Section 5.2.4 assume that $\gamma = 0$ and the output is the Fourier transform of the product of a Gaussian and the beam shaping phase function given by

$$U(x_f, y_f) = Ae^{i(k/2f)(x_f^2+y_f^2)} \iint e^{-\sigma(x^2+y^2)} e^{-i\beta\phi} e^{i(2\pi/\lambda f)(xx_f+yy_f)} dxdy \tag{5.23}$$

where:

x and y are the input coordinates

x_f and y_f are the output coordinates in the focal plane of the transform lens

We can arrive at an equivalent expression by applying the Fresnel integral to the field after the lens. If the lens function is given by

$$t_1(x,y) = e^{-i(k/2f)(x^2+y^2)} \tag{5.24}$$

the Fresnel integral gives

$$U(x_0, y_0) = \frac{e^{ikz}}{i\lambda z} e^{i(k/2z)(x_0^2+y_0^2)} \iint e^{(-\sigma+i\gamma)(x^2+y^2)} e^{i\beta\phi} e^{[(ik/2z)-(ik/2f)](x^2+y^2)} \times e^{-i(k/z)(xx_0+yy_0)} dxdy \tag{5.25}$$

where:

x_0 and y_0 are the coordinates in a plane a distance z from the transform lens

If $z = z_0$ is the solution to

$$\gamma - \frac{k}{2f} + \frac{k}{2z} = 0 \tag{5.26}$$

then Equation 5.25 reduces to

$$U(x_0, y_0) = \frac{e^{ikz_0}}{i\lambda z_0} e^{i(k/2z_0)(x_0^2+y_0^2)} \iint e^{-\sigma(x^2+y^2)} e^{i\beta\phi} e^{i(k/z_0)(xx_0+yy_0)}\, dxdy \tag{5.27}$$

The integral in Equation 5.27 is a scaled version of that in Equation 5.23. Thus, both equations produce the same intensity pattern except for scaling and amplitude factors (phase factors do not effect the irradiance pattern).

5.3 COLLIMATED UNIFORM IRRADIANCE BEAMS

A drawback of existing beam shaping systems is the limited depth of field. The uniform profile appears only at the target plane, and the profile quickly degrades beyond it. This is due to the fact that the configuration of Figure 5.2 cannot produce a beam with both a uniform phase and a uniform amplitude without including a loss mechanism (amplitude control). This can readily be seen from the Fourier transform relation between the field in the output plane and the field just before the transform lens (after the shaping element). For example, for the one-dimensional case, if the beam has a uniform amplitude and phase in the output plane it must be a sine function in the input plane. A desirable extension would be to create a uniform beam that could propagate for considerable distances. In other words, besides a uniform profile at the target plane, what is desired is a uniform phase front (see Theorem 11 in Section 2.3.2.2). The uniform profile would then continue to propagate subject only to diffraction effects due to its finite size. In addition, applications such as optical lithography, nonlinear optics, and optical data (image) processing may require beams with uniform phase as well as amplitude.

5.3.1 CONJUGATE PHASE PLATE

A uniform phase and amplitude beam can be obtained by adding a conjugate phase plate at the output plane of the beam shaping system as shown in Figure 5.5. The phase of the conjugate phase plate is designed to cancel the phase of the uniform irradiance beam at the output plane of the beam shaping optics, producing a collimated beam to the right of the output plane. Given the properties of the input beam and the solution for the shaping element, it is theoretically possible to compute the phase of the conjugate phase plate. However, in some cases, it might be more practicable to design the beam shaping system and then measure the phase of the shaped beam. The phase plate would then be designed to give the conjugate of the measured phase.

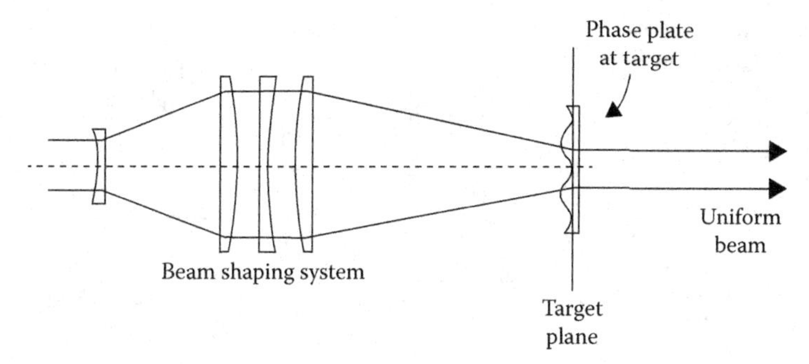

FIGURE 5.5 Optical system for creating a uniform phase beam. (From S.C. Holswade and F.M. Dickey, *Proceedings of SPIE*, 2863, 237–245, 1996. With permission.)

In Appendix B, we show that a beam obeying the scalar wave equation has a minimum r.m.s. radius at a plane of uniform phase, and the beam radius as a function of z (the beam axis coordinate) is a quadratic function given by

$$(\Delta\rho)^2 = a + cz^2 \tag{5.28}$$

where:
  a is the minimum radius squared, $(\Delta\rho_{min})^2$, in the plane $z = 0$

It can be seen from Equation 3.1 that minimizing c minimizes the spread of the beam in the region of the plane of uniform phase. Further, it is shown, using the uncertainty principle, that the Rayleigh range for beams obeying the scalar wave equation is constrained by

$$z_0 \leq \frac{4\pi a}{\lambda} = \frac{4\pi(\Delta\rho_{min})^2}{\lambda} \tag{5.29}$$

The Rayleigh range in this equation is a generalized Rayleigh range defined as the distance (measured from $z = 0$) over which the r.m.s. beam radius increases by a factor of $\sqrt{2}$. Equation 5.29 is just a quantitative statement of the intuitive concept that to obtain a beam with a large depth of field one wants a large beam width with uniform phase. Since the components of the vector wave equation obey the scalar wave equation, these results can be extended to include solutions of the vector wave equation.

5.3.2 Relay Optics

In many applications, especially high-power applications such as material processing, it is not desirable to have a phase plate at the uniform irradiance plane. A solution is to use relay optics to image the beam in the vicinity of the output plane. A relay system for this purpose is illustrated in Figure 5.6. The relay optics configuration in the figure has the additional advantages that it doubles the depth of field and can also be used to magnify the size of the shaped beam. These advantages are, of course, obtained at the expense of additional optics.

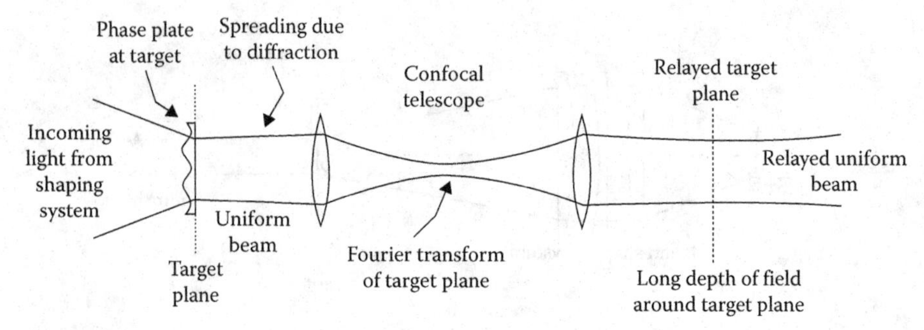

FIGURE 5.6 Relay system for extending the depth of field of uniform irradiance beams. (From S.C. Holswade and F.M. Dickey, *Proceedings of SPIE*, 2863, 237–245, 1996. With permission.)

It should be noted that the relay system in Figure 5.6 consists of a two-lens afocal telescope, that is, the lenses are separated by the sum of their focal lengths. A minimum of two lenses are required to image (relay) the output plane while preserving the uniform phase profile. This can be seen from the fact that an afocal telescope produces a collimated output beam when the input beam is collimated. Another approach is to observe that each lens produces the Fourier transform of the field at its front focal plane at the back focal plane. The result is the Fourier transform of a Fourier transform, which is effectively a Fourier transform followed by an inverse Fourier transform with the coordinates reversed.[15] The output beam is symmetric about the relayed target plane, producing a greater distance where the beam maintains the desired tangential dimensions.

5.4 WAVELENGTH DEPENDENCE OF THE BEAM SHAPING PROBLEM

Forbes et al.[27] have shown the surprising result that the beam shaping problem is independent of wavelength if dispersion of the optical elements is negligible. In the following section, we address the wavelength dependence for the beam shaping problem of Section 5.3. In Section 5.4.1 we extend this result to the general beam shaping problem.

5.4.1 THE FOURIER TRANSFORM OF A PHASE ELEMENT

To address the wavelength dependence of the beam shaping problem, consider the case of the phase element at the transform lens. The problem for a Gaussian to circular flat-top mapping is described by the diffraction integral in Equation 5.30. This is just the Fresnel diffraction integral for the beam shaping problem.

$$u(x,y,z) = \frac{1}{iz\lambda} e^{ik} e^{ik(x^2+y^2)/2z} \int_{-\infty}^{\infty}\int_{-\infty}^{\infty} e^{-\alpha(\xi^2+\pi^2)} e^{\frac{i2\sqrt{2\pi} r_0 y_0}{f\lambda'}\varphi(\xi,\eta)} e^{-i\pi(\xi^2+\eta^2)/f\lambda'} e^{i\pi(\xi^2+\eta^2)/z\lambda} e^{-i2\pi(x\xi+y\eta)/z\lambda} d\xi d\eta \tag{5.30}$$

The first factor in the integrand represents the Gaussian beam, the second factor represents the beam shaping element, the third factor is the phase function for the Fourier transform lens, the fourth factor is the quadratic phase of the Fresnel integral, and the last factor is the Fourier transform kernel. In Equation 5.30, λ' is the beam shaping design wavelength and λ is the applied wavelength.

For beam shaping to be accomplished in the standard way, the integrand in Equation 5.30 must reduce to a Fourier transform of the product of the input beam and the beam shaping function. This is achieved if the two quadratic phase functions cancel each other. This condition is obtained for

$$f\lambda' = z\lambda \tag{5.31}$$

Equation 5.30 then reduces to

$$u(x,y,z) = \frac{k}{i2\pi z}e^{ikz}e^{ik(x^2+y^2)/2z}\int_{-\infty}^{\infty}\int_{-\infty}^{\infty}e^{-\alpha(\xi^2+\eta^2)}e^{\frac{i2\sqrt{2\pi}r_0 y_0}{f\lambda'}\varphi(\xi,\eta)}e^{-i2\pi(x\xi+y\eta)/z\lambda}d\xi d\eta \tag{5.32}$$

Further, if we solve Equation 5.31 for z and substitute the result in Equation 5.32, we obtain

$$u(x,y,z) = \frac{k'}{i2\pi f}e^{ik'f}e^{ik'(x^2+y^2)/2f}\int_{-\infty}^{\infty}\int_{-\infty}^{\infty}e^{-\alpha(\xi^2+\eta^2)}e^{\frac{i2\sqrt{2\pi}r_0 y_0}{f\lambda'}\varphi(\xi,\eta)}e^{-2\pi(x\xi+y\eta)/f\lambda'}d\xi d\eta \tag{5.33}$$

This result is the exact integral for the design beam shaping problem at wavelength λ'; thus, the beam shaping problem is independent of wavelength. The shaped irradiance pattern is obtained at distance z given by Equation 5.31 as

$$z = \frac{f\lambda'}{\lambda} \tag{5.34}$$

Because Equation 5.33 is independent of λ, the spot size does not change with wavelength. This surprising result can be explained by the fact that increasing the wavelength shortens the distance, while longer wavelengths produce a wider diffraction pattern; the two effects cancel. A Gaussian input beam was assumed in the above analysis, but it is easy to see that the result is independent of the input beams shape.

5.4.2 Invariance of the Fresnel Integral

The above argument assumed that beam shaping was implemented as a Fourier transform by the use of a Fourier transform lens. An inspection of the Fresnel integral

$$u(x,y,z) = \frac{1}{iz\lambda}e^{ik}e^{ik(x^2+y^2)/2z}\int_{-\infty}^{\infty}\int_{-\infty}^{\infty}f(\xi,\eta)e^{i\pi(\xi^2+\eta^2)/z\lambda}e^{-i2\pi(x\xi+y\eta)/z\lambda}d\xi d\eta \tag{5.35}$$

shows that the Fresnel integral does not change if the product $z\lambda$ is a constant. This leads to a generalization of the above argument. Consider solving the beam shaping problem starting with the Fresnel integral, that is, we want to solve for ψ in the integral

$$U(x,y,z)=\frac{1}{iz\lambda}e^{ik}e^{ik(x^2+y^2)/2z}\int_{-\infty}^{\infty}\int_{-\infty}^{\infty}e^{-\alpha(\xi^2+\pi^2)}e^{i\psi(\xi,\eta)}e^{i\pi(\xi^2+\eta^2)/z\lambda}e^{-i2\pi(x\xi+y\eta)/z\lambda}d\xi d\eta \quad (5.36)$$

Without loss of generality, we can let

$$\psi'(\xi,\eta)=\psi(\xi,\eta)+\frac{\pi(\xi^2+\eta^2)}{z\lambda} \quad (5.37)$$

We then can solve for ψ' by whatever method one chooses, stationary phase, iterative algorithm, and so on. Once we have done this, we can obtain ψ as

$$\psi(\xi,\eta)=\psi'(\xi,\eta)-\frac{\pi(\xi^2+\eta^2)}{z\lambda} \quad (5.38)$$

Substituting this result in Equation 5.36, we obtain

$$u(x,y,z)=\frac{1}{iz\lambda}e^{ik}e^{ik(x^2+y^2)/2z}\int_{-\infty}^{\infty}\int_{-\infty}^{\infty}e^{-\alpha(\xi^2+\pi^2)}e^{i\psi'(\xi,\eta)}e^{-i2\pi(x\xi+y\eta)/z\lambda}d\xi d\eta \quad (5.39)$$

Thus, we have reduced the solution to the Fourier transform of a phase function. This shows that solving the problem starting with the Fresnel integral is equivalent to solving the problem starting with a Fourier transform. This result is a mathematical statement that for the beam shaping problem using a phase element and a Fourier transform lens, the phase element and lens can be collapsed into one element.

5.5 DESIGN CONSIDERATIONS

A primary design advantage of this lossless beam shaping technique is that the designer can start with a desired target spot quality and determine the optical system required, rather than designing multiple optical systems in the hope of producing an acceptable output. This is because the dimensionless quantity β of Equation 5.19 completely determines the quality of the spot at the target plane. In other words, different optical configurations and wavelengths will produce the same target spot quality if they share the same value of β. Low values of β produce target spots with more rounded sides and wider skirt regions, whereas higher values of β more closely approach the geometric ideal of a uniform intensity profile with infinitely steep sides. As Equation 5.19 implies, the cost of increasing β involves either increasing the size of the Gaussian beam at the phase element, enlarging the target spot, shortening the focal length of the transform lens, or reducing the wavelength. By considering the application and consulting plots of target quality versus β, the designer can determine the minimum value of β that will satisfy the needs and design the most economical system.

Figure 5.7 shows a standard layout for a beam shaping optical system. For most design situations, the size of the target spot and the wavelength will be determined by the application. The focal length of the transform lens may also be determined by

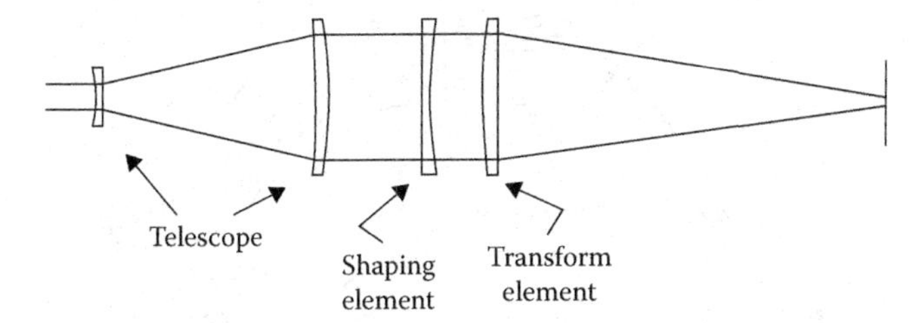

FIGURE 5.7 System optical layout. (From F.M. Dickey and S.C. Holswade, *Optical Engineering*, 35, 3285–3295, 1996. With permission.)

standoff or other considerations, although a minimum focal length will maximize β. The final variable is the Gaussian beam radius at the shaping element. To achieve the desired β, the beam size should be expanded by an afocal telescope, as shown in the figure. With the optical system designed for one target geometry, there are two methods to produce additional target geometries. The first is to change the phase element. With the same expansion and focusing optics, a system could thus produce circular and rectangular beams of several sizes. It should be noted, however, that different target geometries will vary β, and hence spot quality, as determined by Equation 5.19. The second method involves changing the focusing, or transform, lens while leaving the telescope and phase element fixed. This change can vary only the target size, not the geometry, but it has the advantage of maintaining a constant target spot quality. The variation in the focal length changes the spot size proportionally, and thus β remains constant.

Once a target spot quality is determined, the required phase profile imparted on the beam by the phase element is then found by multiplying the phase function of Equation 5.17 or 5.18 by β. This multiplication scales the phase function to the particular geometry of the application. A telescope is then designed to expand the beam to the required value. A transform lens of the required focal length completes the system. The remainder of this section discusses the design considerations in more detail, and also the additional system configurations.

5.5.1 Target Spot Quality

Since β determines target spot quality independently of the circumstances of the design, graphs of the beam shape versus particular values of β are useful. The following simulations include system effects such as beam truncation and lens aberrations. They were calculated for a CO_2 laser system (λ = 10.6 μm) with an *f*/42-to-*f*/21 plano-convex lens. Aperture radii were truncated at $2r_0$ in these simulations, where r_0 was $1/e^2$ radius of the beam. Figure 5.8a shows a square target spot with $\beta = 4$. The profile is fairly rounded. Figure 5.8b shows the square target spot with $\beta = 8$. The skirts of this spot have narrowed considerably. Figure 5.8c shows the square target spot with $\beta = 16$. The skirts of this spot have narrowed further. This system design is beginning to approach the geometric ideal of a uniform profile with infinitely steep sides. The square spot is a special case of the rectangular spot. With a circular input beam, a rectangular output

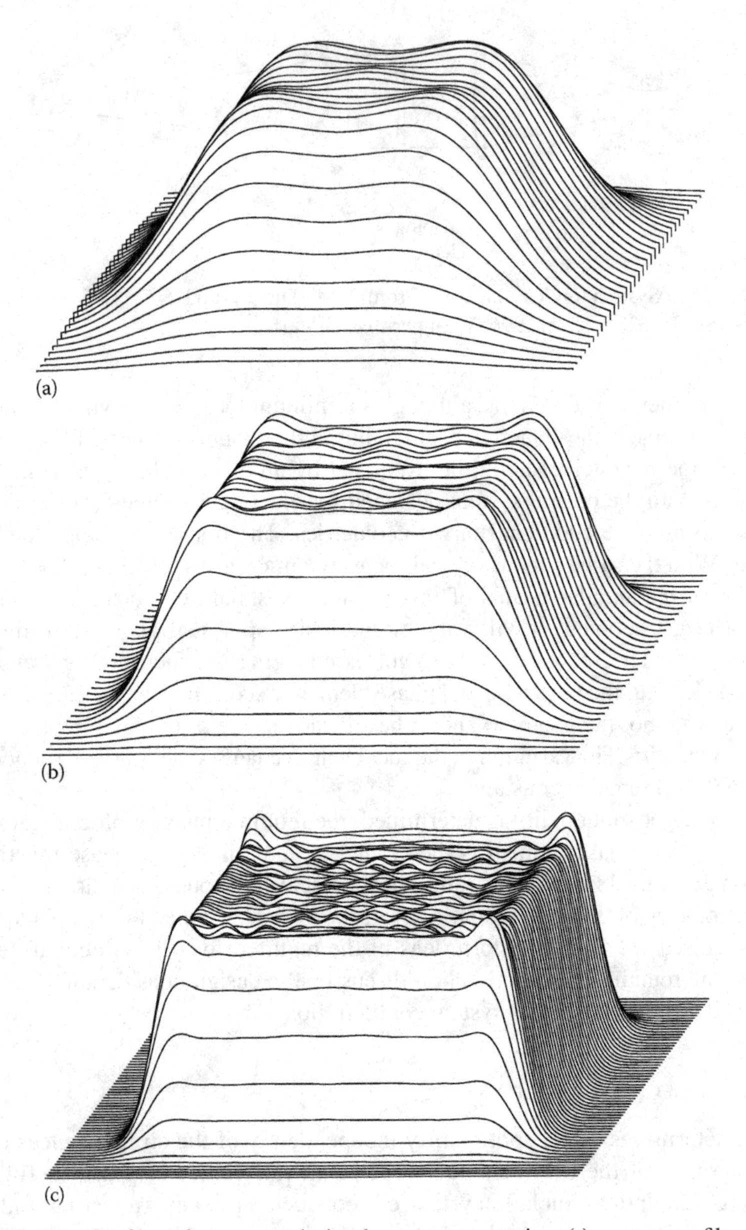

FIGURE 5.8 Profiles of square and circular spot geometries: (a) square profile, $\beta = 4$; (b) square profile, $\beta = 8$; (c) square profile, $\beta = 16$; (d) round profile, $\beta = 8$. (From F.M. Dickey and S.C. Holswade, *Optical Engineering*, 35, 3285–3295, 1996. With permission.)

can be produced by varying β for each axis. For the case of a circular, uniform target spot, Figure 5.8d illustrates the profile for $\beta = 8$ and $3r_0$ truncation. This spot behaves similarly to the square case as β changes. Unlike the square case, however, the circular case exhibits noticeable ripple on the profile as the beam is truncated to $2r_0$.

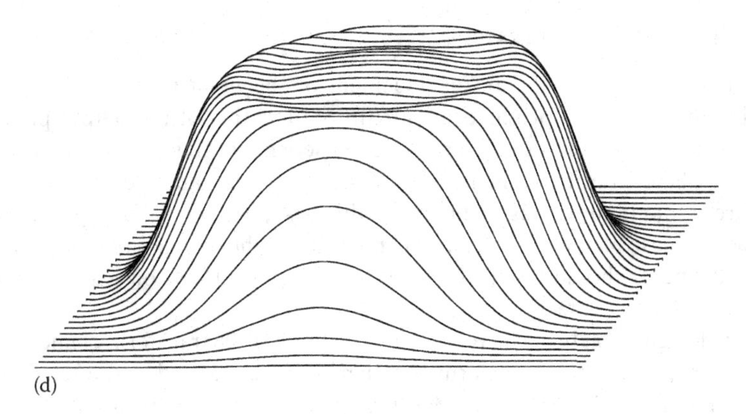

FIGURE 5.8 (Continued)

5.5.2 Modeling System Performance

As was shown in Figure 5.7, the optical system consists of the phase element and three lenses, each of which can contribute aberrations. A complete way to model system performance is thus desirable. Several optical design packages now offer the ability to input surfaces with general aspheric profiles as polynomial functions of x and y coordinates. For small values of β, the phase profile can be well approximated by varying the thickness of the element. In other words, the phase element acts as a "thin" element for small values of β.[15] Some packages allow polynomial phase profiles to be input directly. In either case, the phase functions are fit to a polynomial with appropriate mathematical software, the polynomial multiplied by β, and the phase element inserted into the design package along with the other elements. After tracing a spot diagram with Gaussian apodization of the ray weights, the package then calculates a diffraction-based point spread function for the system, which uses the ray map in the exit pupil. The point spread function provides the diffraction response of the system for a point object. A distant point object produces a planar input wavefront characteristic of a Gaussian beam at its waist. Curvature in the input beam wavefront can be modeled by moving the object point to the appropriate distance. Wavelength variations in the input beam can be modeled with a polychromatic spot diagram and point spread function.

A sufficiently robust design package can model effects due to lens aberrations, beam truncation by optics, beam curvature, and alignment and scaling errors. To avoid aliasing in the point spread function, the spot diagram must sufficiently sample the exit pupil. For βs on the order of 16 or less, the phase profile varies fairly slowly, and most programs can sample it sufficiently. As β increases, however, the phase profile varies more rapidly, and sampling becomes more problematic. However, it should be remembered that β is a measure of how well the system approaches the geometric ideal. For high values of β, therefore, the weighted geometric spot diagram sufficiently models the system performance.

5.5.3 Telescope Considerations

As discussed in Section 5.2.6, displacement of the input Gaussian beam waist from the phase element produces a shift in the location of the output plane and a change in scale of the target spot. The telescope can thus be adjusted to shift the target plane to a different location. The telescope can also compensate for curvature in the input Gaussian beam. In this case, it is adjusted to place a beam waist at the phase element, with a corresponding slight change in input beam size. Both adjustments should generally result in negligible effects on β and target spot geometry.

Phase elements can be located either before or after the transform element.[15] This allows the expansion and transform functions to be combined in the telescope, as shown in Figure 5.9. In some situations, it may be necessary to compensate for tolerances in the incoming laser beam diameter. The phase element may then be located behind the transform element and moved along the beam axis until the beam size matches the design size. This movement will scale the target spot size, but β will remain constant.

5.5.4 Truncation Effects

For standard optical systems, the effects of truncation on Gaussian beams have been reported in the literature.[28] The truncation of the input Gaussian beam by the circular apertures in a beam shaping system will also affect the target profiles. For the square spot, no noticeable degradation is seen for truncation down to $2r_0$. As aperture sizes decrease, however, further ripple becomes apparent. Figure 5.10 illustrates the effects of $1.5r_0$ truncation on the square spot. For the circular spot, ripple becomes apparent at $2r_0$ truncation, as is shown in Figure 5.11. It is interesting to observe why a $2r_0$ circular aperture will affect a circular spot more than the square spot for the same β. For the circular spot, the edge-wave disturbances created by the aperture are symmetric through the system and interfere constructively at the target. For the square spot, however, the disturbances created by a circular aperture are altered by the phase element in a nonsymmetric fashion, and thus do not all constructively interfere. This situation is related to diffraction by circular versus other apertures.[28] For designs producing square or rectangular spots, system apertures of twice the Gaussian beam

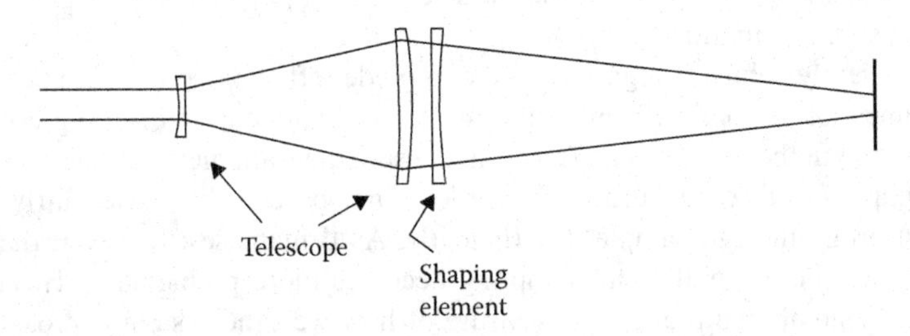

FIGURE 5.9 Combined expansion and transform functions in the telescope. (From F.M. Dickey and S.C. Holswade, *Optical Engineering*, 35, 3285–3295, 1996. With permission.)

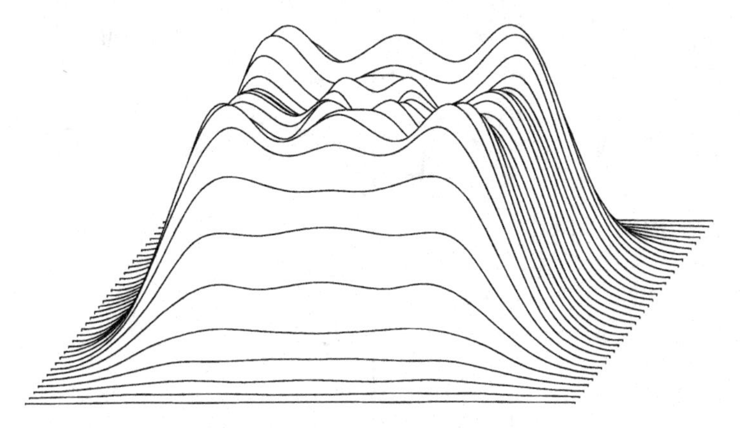

FIGURE 5.10 Square profile, $\beta = 8$, apertures $= 1.5r_0$. (From F.M. Dickey and S.C. Holswade, *Optical Engineering*, 35, 3285–3295, 1996. With permission.)

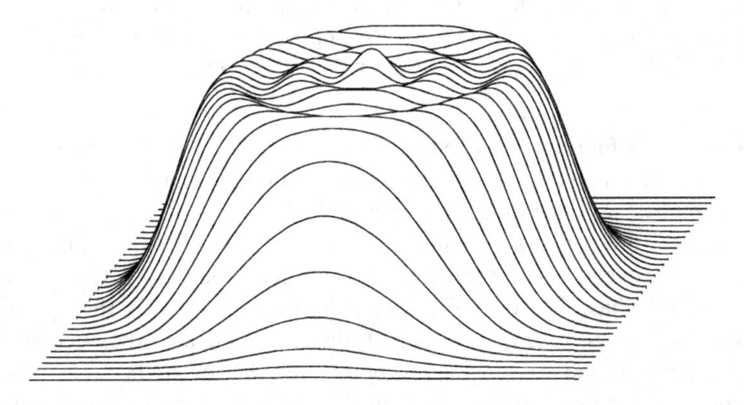

FIGURE 5.11 Round profile, $\beta = 8$, apertures $= 2r_0$. (From F.M. Dickey and S.C. Holswade, *Optical Engineering*, 35, 3285–3295, 1996. With permission.)

radius should provide good performance. For designs producing circular spots, system apertures of three times the Gaussian beam radius will be necessary to avoid ripple effects.

5.5.5 Positive and Negative Phase Functions

As discussed in Section 5.2.5, the phase function has two solutions, positive and negative, for a given configuration. With reference to Equation 5.1, ϕ is the phase delay suffered by a wave in passing through the phase element. This situation is analogous to the phase delays introduced by thin lenses.[15] For a positive phase function ϕ, Equations 5.17 and 5.18 show that the phase delay will vary from zero at the optical axis to increasingly positive values as we move away from the axis. This situation is the same as that for a negative thin lens, and additional insight into the beam shaping system can be gained by viewing the shaping element geometrically,

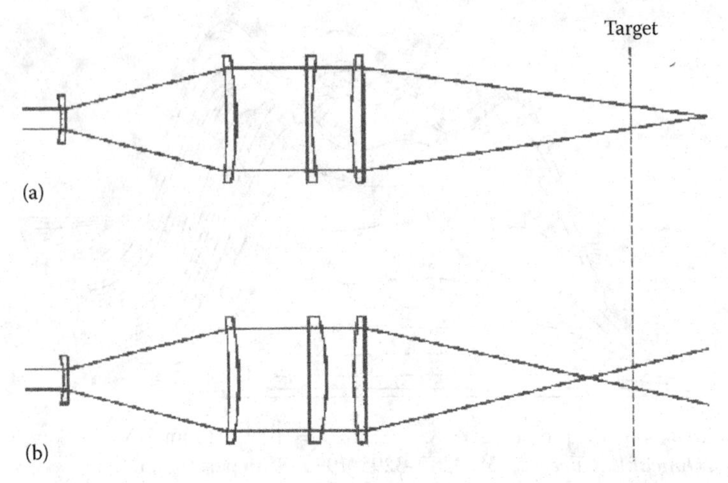

FIGURE 5.12 (a) Element with positive phase function acts geometrically as an aberrated negative lens. (b) Element with negative phase function acts geometrically as an aberrated positive lens. The target plane is the focal plane of the final (transform) lens. (From F.M. Dickey and S.C. Holswade, *Optical Engineering*, 35, 3285–3295, 1996. With permission.)

as is illustrated in Figure 5.12. The target plane is the focal plane of the transform, or focusing, element and this is where the desired target spot appears. For an element with a positive phase function (Figure 5.12a), the beam continues to decrease in size after the target plane. Geometrically, the element has reduced the power of the optical system and behaves much as an aberrated negative lens. For an element with a negative phase function (Figure 5.12b), the beam passes through a minimum diameter before reaching the target plane. The element thus geometrically behaves as an aberrated positive lens. In both cases, the spot at the target plane will be identical. The positive phase function (Figure 5.12a) has an advantage in the depth of field, since the wavefront through the system is closer to planar than for the negative phase function (Figure 5.12b). An analogous geometric explanation is that the marginal rays for Figure 5.12a have smaller angles than Figure 5.12b, thus allowing a larger depth of field. Defocus of the target plane leads to deviations in spot uniformity, and it is treated in the next section. Particular applications may demand that the minimum beam size occur either before or after the target plane. If there is a choice, however, the positive phase function features the least sensitivity to defocus errors.

5.6 ALIGNMENT AND SCALING ERRORS

Unlike the methods based on multifaceted integrators, this lossless beam shaping method is sensitive to alignment errors and variations in the input beam size. Figure 5.13 shows the effect of decentering the Gaussian beam on the phase element by 0.1 r_0 along one of the element axes. For other elements in the system, decentration is most important where it would move the beam on the phase element. For example,

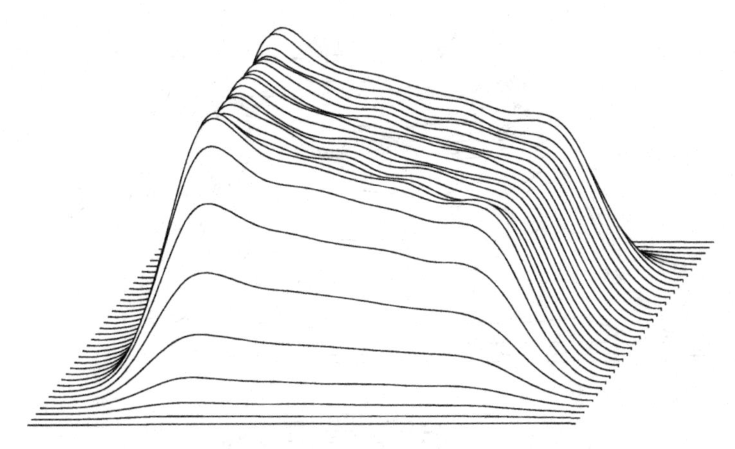

FIGURE 5.13 Input beam decentered along element axis by $0.1r_0$, $\beta = 8$ case. (From F.M. Dickey and S.C. Holswade, *Optical Engineering*, 35, 3285–3295, 1996. With permission.)

decentration of the negative lens in Figure 5.7 would decenter the beam on the phase element and thus produce the effects shown in Figure 5.13. However, decentration of the transform element would have a relatively small impact on the target spot quality. Spot quality is also fairly insensitive to tilt of the phase element, which acts much as a thin plate in this case.

Since the shaping element is designed for a particular input beam size, which in part determines the scaling factor β, it stands to reason that deviations from the design input beam size will affect the target spot. The following cases show the degradation in the square target spot for the $\beta = 8$ design of Figure 5.8b, with a positive phase function. In Figure 5.14a, the input Gaussian beam size is 10% larger than the size used in the design. The target spot shows significantly raised edges. Figure 5.14b shows the target spot for an input beam size that is 10% smaller than that used in the design. The edges of the spot have rounded off. For many applications, this rounding effect is less detrimental than that caused by the raised edges. Thus, if variation in the input beam size is anticipated for systems with a positive phase function, the element should be designed for a beam somewhere near the upper limit of the size range.

Target plane defocus also affects the quality of the target spot for the following reasons. The beam shaping system uses a lens to transform the input beam plus the phase function to the desired shape. The output spot exists at the transform plane of the lens, which is also its focal point. In the derivation of the phase element, the problem was to minimize the difference between the desired irradiance at the target plane and that produced by the system. There were no constraints on the phase of the beam at the target or on the beam irradiance outside the transform plane. Since the phase is generally not uniform at the target plane, the shaped beam will not display the symmetry about the target plane characteristic of Gaussian beams at their waist (Section 5.3). The irradiance of the beam will thus deviate from the desired shape when the target plane is moved away from the focal point of the transform lens.

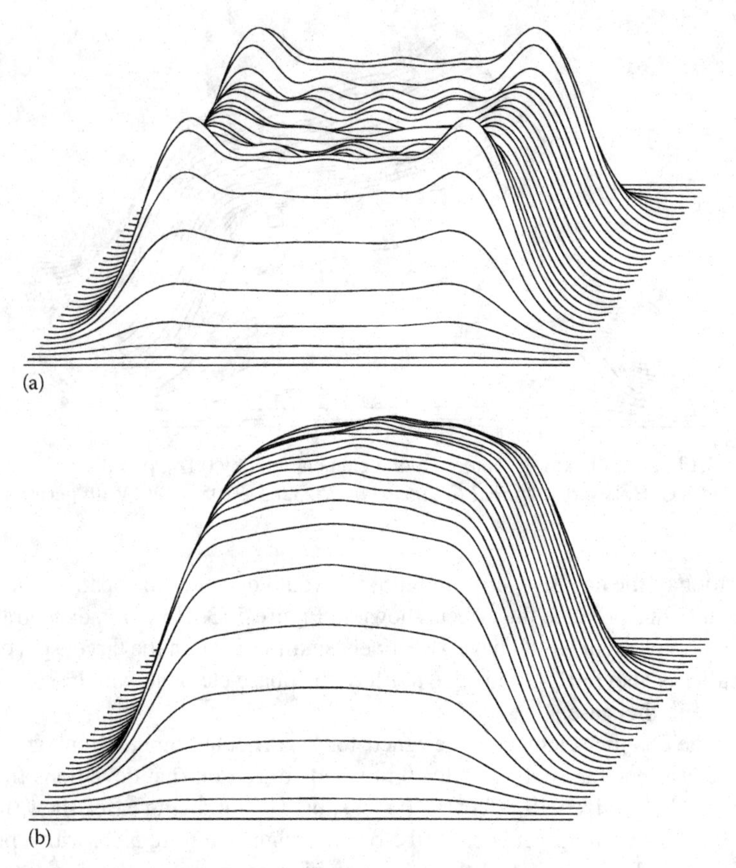

FIGURE 5.14 Effects of deviations in input beam size from design values: (a) input beam 10% larger than design size; (b) input beam 10% smaller than design size. (From F.M. Dickey and S.C. Holswade, *Optical Engineering*, 35, 3285–3295, 1996. With permission.)

The following cases apply to the $\beta = 8$ design of Figure 5.8b, with a positive phase function. Figure 5.15a shows the effect of moving the target plane away from the transform lens by *f*/50, where *f* is the focal length of the lens. The spot decreases in size, increases in average irradiance, and the edge areas rise relative to the center. Figure 5.15b shows the effect of moving the target plane toward the transform lens by *f*/50. Here the spot increases in size and decreases in average irradiance. The spot uniformity remains fairly good, however. Thus, if defocus is anticipated in systems with a positive phase function, the system should be designed for the upper part of the focus range.

As discussed earlier, the beam's phase at the target is unconstrained. Thus, in the system illustrated in Figure 5.7, the target spot cannot simply be collimated with a negative lens to propagate as a flat-top beam. The phase at the target, however, can be computed. A conjugate phase plate placed at the target would cancel these phase differences, and the uniform profile would propagate as a collimated beam, subject to diffraction. However, if the target spot is simply desired at another

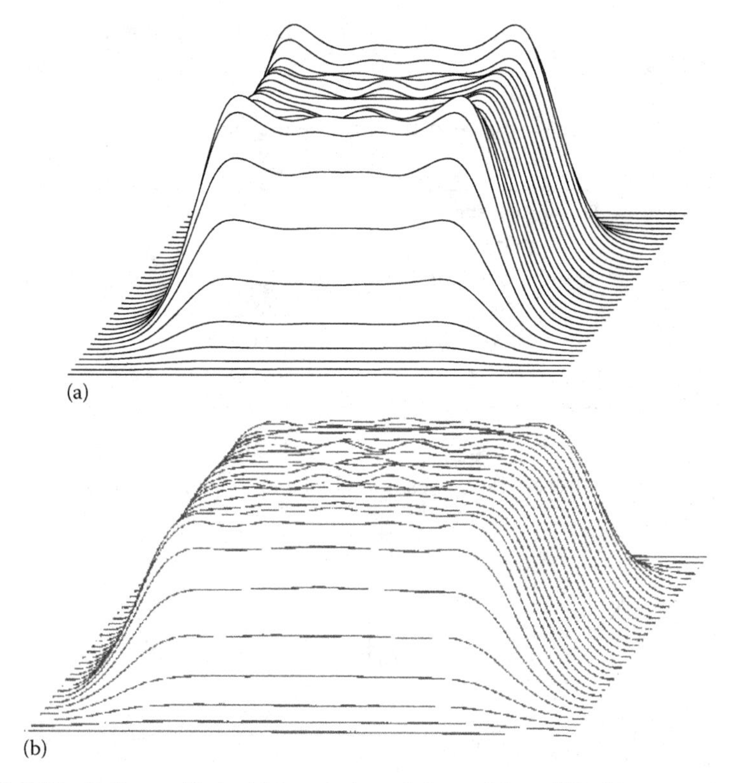

FIGURE 5.15 Defocus effects: (a) target plane defocused by $+f/50$; (b) target plane defocused by $-f/50$. (From F.M. Dickey and S.C. Holswade, *Optical Engineering*, 35, 3285–3295, 1996. With permission.)

location or scale, it can be reimaged with a conventional afocal telescope as discussed in Section 5.3.2.

5.7 METHOD OF DESIGN

As stated earlier, the quality of the target spot can be selected to suit the application, and the necessary optical system parameters calculated directly. In most cases, the size of the Gaussian beam at the phase element will be the free variable that determines β. If the phase element and the optical system are to be studied with an optical design program, the phase function will need to be expressed as a polynomial. With the optical design program, the response of the system to tolerances in beam scaling, beam position, element position, element tilt, and target defocus can be studied. Beam truncation effects can also be modeled if necessary. If tolerances in input beam size or target position are expected in systems with a positive phase function, the target spot will degrade most gracefully if the element is designed for a slightly larger beam than expected with a target plane slightly further away than expected.

To facilitate modeling, the phase functions for rectangular and circular spots have been fitted to 10th-order polynomials. The fits are good to $\xi = 3\sqrt{2}$, which is $3r_0$ at the phase element. The form for the rectangular and circular cross section is

$$\phi(\xi) = a_2\xi^2 + a_4\xi^4 + a_6\xi^6 + a_8\xi^8 + a_{10}\xi^{10} \tag{5.40}$$

Rectangular spot	**Circular spot**
$a_2 = 4.73974 \times 10^{-1}$	$a_2 = 4.31128 \times 10^{-1}$
$a_4 = -5.50034 \times 10^{-2}$	$a_4 = -4.36550 \times 10^{-2}$
$a_6 = 4.99298 \times 10^{-3}$	$a_6 = 3.65204 \times 10^{-3}$
$a_8 = -2.37191 \times 10^{-4}$	$a_8 = -1.65025 \times 10^{-4}$
$a_{10} = 4.41478 \times 10^{-6}$	$a_{10} = 2.97368 \times 10^{-6}$

For the rectangular spot, Figure 5.16a shows the quality of fit to the original function. No difference between the curves is visible, and they have an r.m.s. variation of

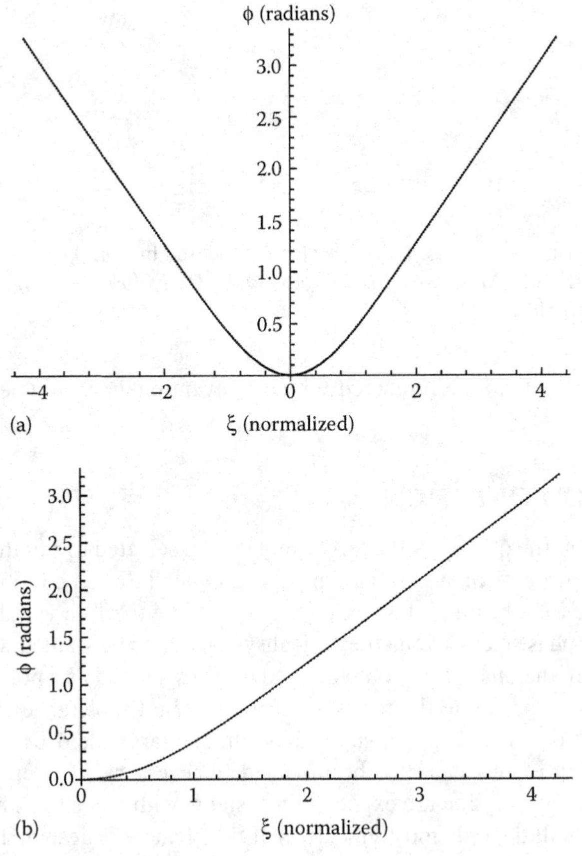

FIGURE 5.16 (a) Rectangular phase function and 10th-order fit. (b) Circular phase function and 10th-order fit. (From F.M. Dickey and S.C. Holswade, *Optical Engineering*, 35, 3285–3295, 1996. With permission.)

0.0046 rad. As discussed earlier, the phase function for each axis is multiplied by β to scale it to the desired geometry, and the dimensionless quantities ξ are replaced by actual coordinates according to Equation 5.17. The scaled phase functions for each axis are then summed to define the complete phase function. For the circular spot, Figure 5.16b shows the quality of fit. The r.m.s. variation between the two curves is 0.0025 rad. ξ is a radial coordinate in this case, and the phase function is radially symmetric.

The following example illustrates the use of the technique to solve an actual problem. Consider the case where a rectangular spot is desired 400 mm away from an optical system. The target spot dimensions are 2 (x-axis) × 4 mm (y-axis). A 10.6 μm laser produces a Gaussian beam with a $1/e^2$ radius of 3 mm, and the optical train is composed of ZnSe ($n = 2.403$). We choose a system layout as shown in Figure 5.7, with an $f = 400$ mm focusing lens. From graphs of target spot quality versus β, we choose $\beta = 8$ as a minimum acceptable value. From Equation 5.19, we see that the beam radius at the phase element, r_0, is the only unconstrained variable. In addition, the x-axis will require the most expansion to produce the required β, since its target dimension is smaller. For $\beta_x = 8$, we obtain $r_0 = 6.76$ mm at the phase element, for an expansion ratio of 2.25 from the telescope. We could anamorphically expand the beam to produce the same β for the y-axis, but we choose a standard radially symmetric telescope for $\beta_y = 16$.

We choose to generate our phase profile by varying the thickness of the phase element. We wish to develop a polynomial that yields the phase element sagitta, or deviations from a plane at the surface vertex and the surface, as a function of distances from the optic axis. We must thus multiply the coefficients of Equation 5.16 by β_x or β_y and convert them to produce sagitta as a function of element coordinates. The following equation gives the sagitta of the phase surface:

$$\begin{aligned} \text{sag}(x,y) = m_2x^2 + n_2y^2 + m_4y^4 + n_4y^4 + m_6x^6 + n_6y^6 + m_8x^8 \\ + n_8y^8 + m_{10}x^{10} + n_{10}y^{10} \end{aligned} \tag{5.41}$$

where:

$$m_i = \frac{a_i\lambda\beta_x\left(\dfrac{\sqrt{2}}{r_0}\right)^i}{2\pi(n-1)} \quad \text{and} \quad n_i = \frac{a_i\lambda\beta_y\left(\dfrac{\sqrt{2}}{r_0}\right)^i}{2\pi(n-1)}$$

The coefficients in this polynomial would be

$$\begin{aligned}
m_2 &= 1.994 \times 10^{-4} & n_2 &= 3.988 \times 10^{-4} \\
m_4 &= -1.0127 \times 10^{-6} & n_4 &= -2.0255 \times 10^{-6} \\
m_6 &= 4.0235 \times 10^{-9} & n_6 &= 8.0471 \times 10^{-9} \\
m_8 &= -8.3652 \times 10^{-12} & n_8 &= -1.673 \times 10^{-11} \\
m_{10} &= 6.8144 \times 10^{-15} & n_{10} &= 1.3629 \times 10^{-14}
\end{aligned}$$

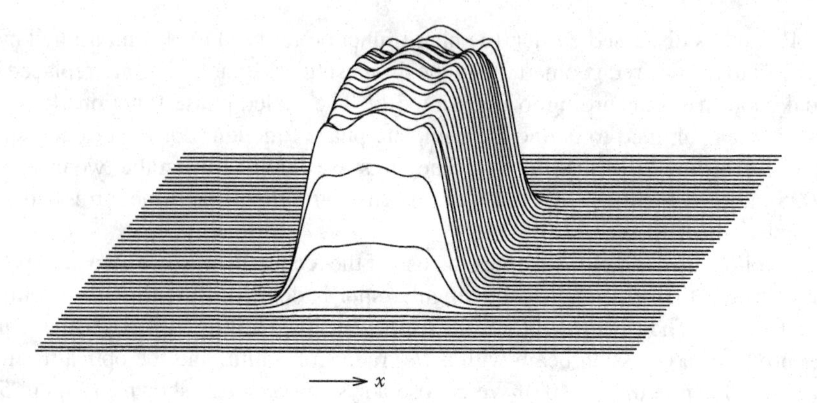

FIGURE 5.17 Simulation of output beam produced by sample problem. $\beta_x = 8$ and $\beta_y = 16$. The patch dimensions are 8 × 8 mm.

If we choose to build an element with a positive phase function, the sign convention on the sagitta would be such that the phase surface had a concave shape. A simulation of the output spot from this example problem appears in Figure 5.17.

5.8 EXPERIMENTAL EVALUATION

A beam shaping system was developed for an application that required a long working distance, limited beam sizes, and operation at 10.6 μm. This resulted in a maximum target spot quality given by $\beta = 4.8$. The phase element was fabricated in ZnSe as a 16-level diffractive optic. Figure 5.18 shows the element profile along the

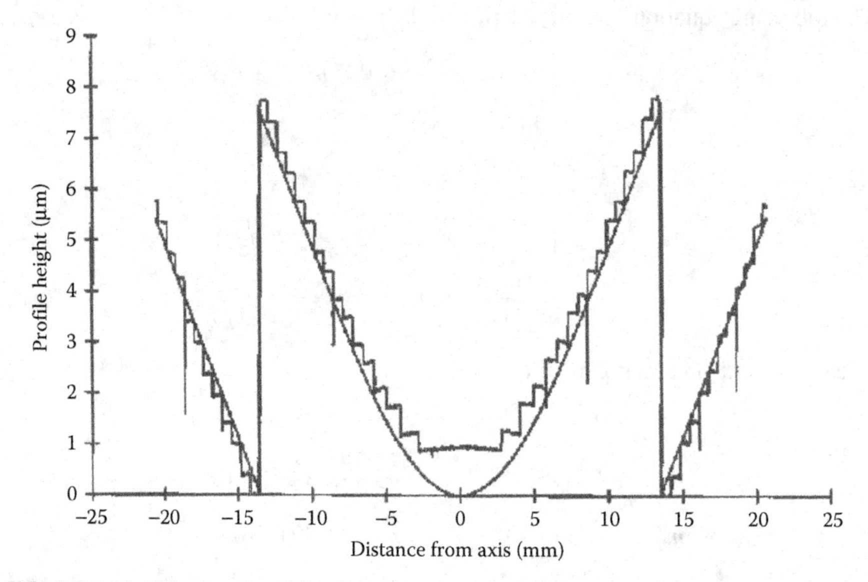

FIGURE 5.18 Measured profile of shaping element along x-axis. It is a 16-level diffractive approximation to the desired profile shown by the smooth curve. (From F.M. Dickey and S.C. Holswade, *Optical Engineering*, 35, 3285–3295, 1996. With permission.)

x-axis as measured by a stylus profilometer. The desired profile in terms of element thickness is shown by the smooth curve overlaid on the measured profile. In order to match the 2π phase shifts of the diffractive optic, 2π phase shifts [or thickness shifts of $\lambda/(n-1)$] were applied to the graph of desired profile as well. Overall, the measured element profile was in reasonable agreement with the desired profile, with the exception of a displacement near the center of the element.

The phase element was tested on a different laser than that for which it was designed, although the measured beam dimensions were within the design goals. The laser beam passed through the beam shaping optical system and on to a target plane. A lens beyond the target plane reimaged and magnified the target spot onto a pyroelectric array camera. The focus and magnification of the reimaging system were set by placing a calibrated pinhole at the target plane and adjusting the lens for a sharp image on the camera. The laser beam dimensions at the input of the optical system were determined with orthogonal scanning knife edges in conjunction with an automated focusing system. This device computed the internal beam waist size and location as well as the beam divergence. It then computed the same parameters for the external laser beam. These quantities then determined the initial beam size at the telescope system.

Before presenting the experimental results, it is instructive to see the predicted spot geometry for the actual beam input parameters. The system was modeled using the computed beam radii for the x and y axes at the first telescope lens. There was a difference of roughly 5% in the computed radius of curvature for the x and y axes, but this was ignored in the modeling. In Figure 5.19, the predicted target spot profiles for the x and y axes are shown, scaled to normalized position units. The optical system was initially aligned using a visible reference beam. Final alignment of the beam shaping element was accomplished by viewing the target image with the pyroelectric

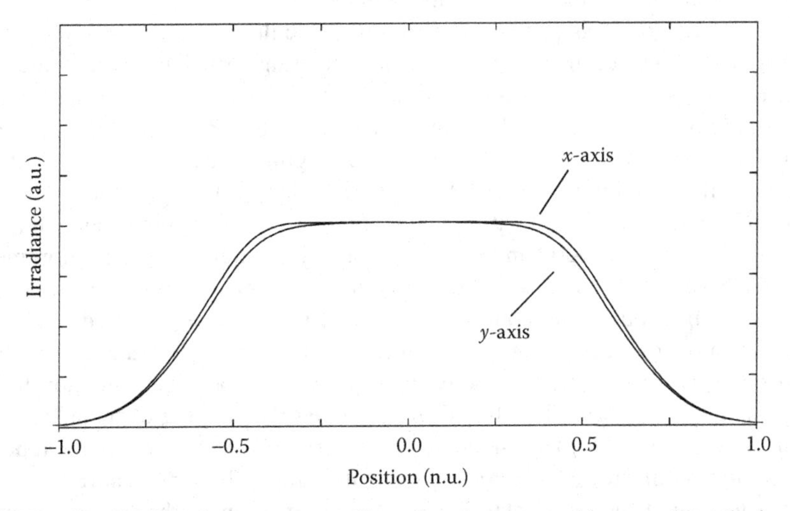

FIGURE 5.19 Predicted target spot profiles using the measured beam radii at the input of the optical system. Units are normalized. (From F.M. Dickey and S.C. Holswade, *Optical Engineering*, 35, 3285–3295, 1996. With permission.)

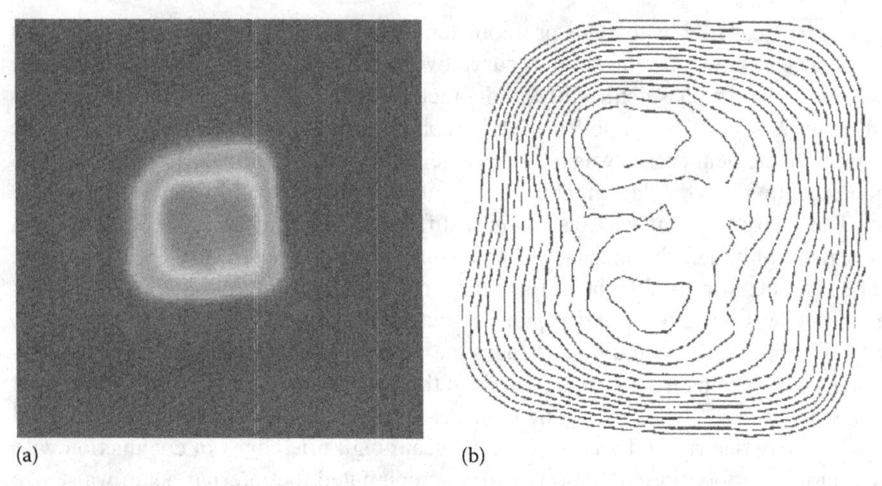

FIGURE 5.20 (a) Image of the target spot for a $\beta/3 = 4.8$ system. (b) Contour plot of the target spot. Contours have equal intervals of 3.2% of the maximum value. (From F.M. Dickey and S.C. Holswade, *Optical Engineering*, 35, 3285–3295, 1996. With permission.)

array camera. Figure 5.20a shows an image of the target spot when the system was aligned. Each change of shade going in toward the center corresponds to an increase of irradiance. The square appearance is evident. Figure 5.20b shows a contour plot of this same target spot. Profiles of this image were extracted and are plotted in Figure 5.21. These profiles use the same normalized position units as the predicted profiles of Figure 5.19. The measured profiles are smaller than predicted, and they deviate somewhat from the desired uniform irradiance. Nevertheless, they show a general agreement with the predicted uniform profiles.

The size difference is partly a function of difficulties in establishing the best target plane. Distance measurements from the transform lens were somewhat inexact, and compensation was necessary for the curvature of the input beam. Also, the reimaging lens and the camera were mounted independently, so that it was difficult to move the target plane once reimaging focus was set. Most likely, the reimaging magnification and focus varied during the alignment process. These uncertainties, coupled with alignment issues for the shaping element, made for difficulties with several independent adjustments during system alignment. It would have been best to be able to mount the camera, reimaging lens, and pinhole together on a common structure, with the pinhole mounted kinematically. When the camera system was used to find the best target plane, the pinhole could be replaced to mark it exactly. In addition, a good approximation to the target plane position could be found by removing the shaping element from the system. The focused beam waist would occur very close to the system focal point, and would account for curvature in the input beam. Telescope separation could be adjusted to put the target plane at the desired location, with fewer subsequent adjustments to make during alignment.

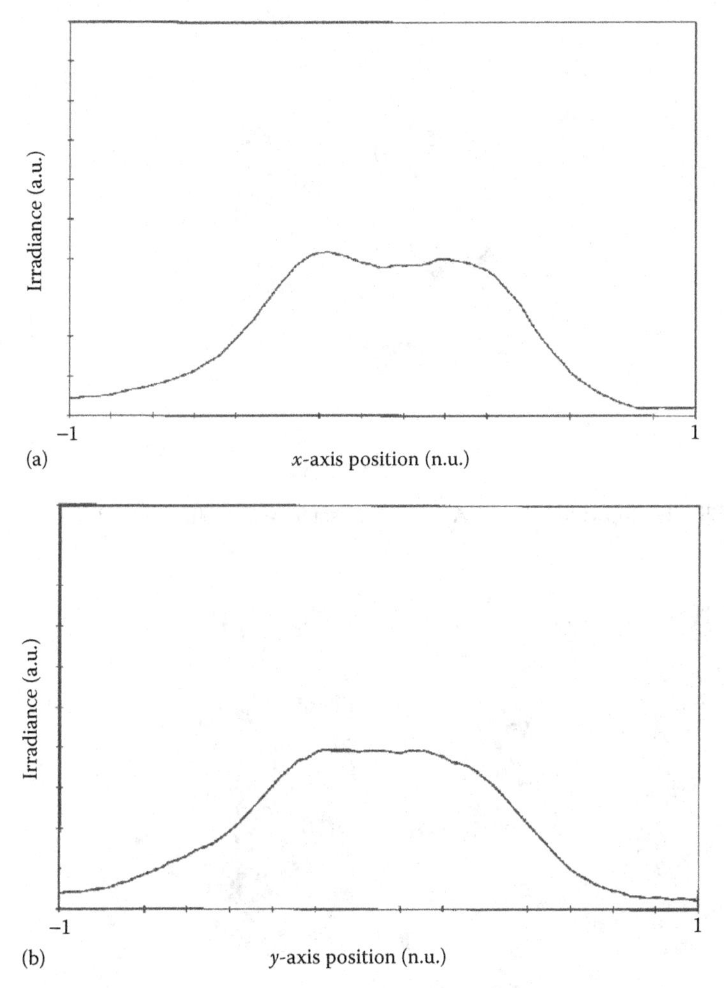

FIGURE 5.21 Measured profiles of the target spot, normalized to the same units as the predicted spot: (a) x-axis; (b) y-axis. (From F.M. Dickey and S.C. Holswade, *Optical Engineering*, 35, 3285–3295, 1996. With permission.)

Differences in the measured profile uniformity are at least partly the result of two factors. First, the test laser displayed near-field deviations from a Gaussian shape due to its unstable resonator configuration. These near-field deviations tend to be masked in far-field beam measurements due to diffraction effects from the focusing lens. Second, the measured element profile shown in Figure 5.18 differed somewhat from the desired profile, leading to a difference in the phase delay applied to the beam.

A common application of beam shaping is for material processing, so this square-shaped target spot was used to burn polymethyl methacrylate (PMMA) material in comparison with a standard Gaussian target spot. Figure 5.22 shows a top view of

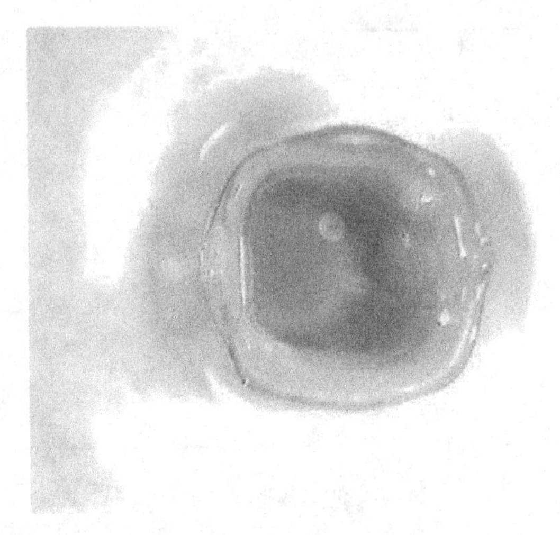

FIGURE 5.22 Top view of PMMA exposed to square-shaped beam for 1 s.

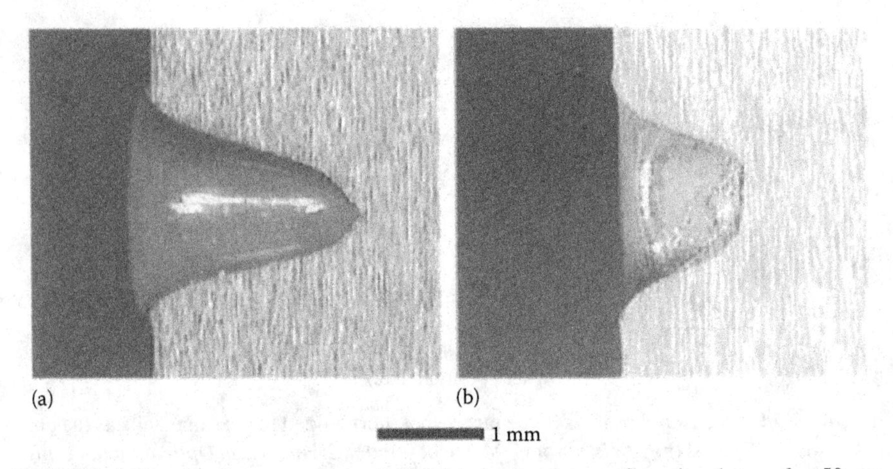

FIGURE 5.23 (a) Cutaway view of PMMA exposed to a Gaussian beam for 50 ms. (b) Cutaway view of PMMA exposed to square-shaped beam for 50 ms.

PMMA material exposed to the square-shaped beam for 1 s. The square shape of the removed material is evident. Figure 5.23 shows side views of PMMA material exposed to both a Gaussian beam and a square-shaped beam for 50 ms. These views were made by cutting the material through the center of the burned spots. The square target beam would be advantageous for situations where a burn needs to be made to a uniform depth. As the figure shows, this target spot could be scanned over the material, and an even burn profile made.

A beam shaping system that produces a circular flat-top profile with $\beta = 20$ has also been fabricated.[29] In this design, the focusing and shaping functions

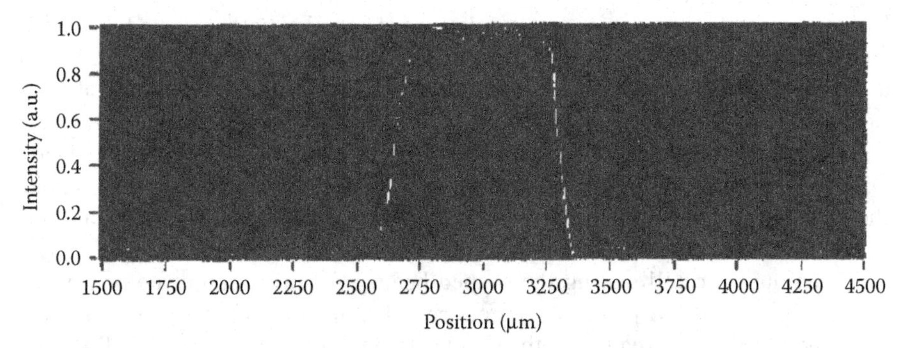

FIGURE 5.24 Experimental results for a combined shaping and focusing element. The system was designed to transform a single-mode Gaussian beam into a uniform, circular irradiance profile at the target with $\beta = 20$. This figure shows a cross section through the center of the irradiance profile. (From X.G. Huang, M.R. Wang, and C. Yu, *Optical Engineering*, 38, 208–213, 1999. With permission.)

were combined in a single diffractive phase element designed for $\lambda = 0.633$ μm. Combining the focusing function implies the addition of a quadratic phase factor to the solution of Equation 5.18. The design was fabricated using laser-assisted chemical etching, which produced a smooth profile with 2π phase discontinuities. The addition of the focusing function increased the slope of the phase function to the point that many more phase jumps were needed than in a shaping-only design. Figure 5.24 illustrates a cross section of the measured target spot for this system; the target spot shows fairly steep skirts and a uniformity error of less than $\pm 5\%$. One likely explanation for the uniformity error is interference caused by the phase discontinuities in the element.

5.9 SUMMARY

Single-mode Gaussian beams can be transformed into circular or rectangular beams with approximately uniform irradiance profiles in a lossless manner by the introduction of an appropriate phase element in conjunction with a Fourier transform lens. This chapter presents a diffraction-theory-based solution for the phase delay obtained using the method of stationary phase. The quality of the target spot was shown to depend on a parameter β that is a function of the input beam size, the target spot size, the focal length of the transform lens, and the wavelength. This dimensionless parameter accounts for diffraction, independently of the particular system. In addition to being a result of applying the method of stationary phase to the evaluation of the diffraction integrals, β was shown to be directly applicable to the general beam shaping problem using the classical uncertainty principle. The use of the β parameter allows the designer to determine the system parameters necessary for a desired target spot quality, rather than iterate through several designs. Once the system parameters are known, either the circular or rectangular phase function can be scaled appropriately. The phase element function can be approximated by a polynomial, which allows standard optics modeling software to predict the effects of system aberrations and tolerances.

A simple mathematical argument is given which shows that the phase-only beam shaping is wavelength invariant. If dispersion in the optical elements is negligible, the output spot size and shape does not change with wavelength, but the axial position of the output does depend on wavelength.

Techniques for collimating the shaped beam by using a conjugate phase element are outlined. The phase plate produces a beam with both uniform phase and amplitude, giving a greater depth of field. Bounds on the generalized Rayleigh range for uniform amplitude and phase beams were derived using the uncertainty principle.

Beam shaping system design techniques, based on the theory, were discussed and several numerical examples were presented to illustrate the range of solutions. The sensitivity of system performance to errors in alignment was discussed and illustrated by numerical simulations.

A particular design was implemented in hardware and tested. Experimental results show that the technique produced a square target spot that was close to the predicted profile. The application of this system to material drilling and ablation is discussed in this chapter.

REFERENCES

1. S. Ream. A convex beam integrator. *Laser Focus* 15(11): 68–71, 1979.
2. D. Dagenais, J. Woodroffe, and I. Itzkan. Optical beam shaping of a high power laser for uniform target illumination. *Applied Optics* 24(5): 671–675, 1985.
3. V.J. Doherty. Design of mirrors with segmented conical surfaces tangent to a discontinuous aspheric base. *Proceedings of SPIE* 399: 263–271, 1983.
4. F.M. Dickey and B.D. O'Neil. Multifaceted laser beam integrators, general formulation and design concepts. *Optical Engineering* 27(11): 999–1007, 1988.
5. B.R. Frieden. Lossless conversion of a plane laser wave to a plane wave of uniform irradiance. *Applied Optics* 4(11): 1400–1403, 1965.
6. W. Lee. Method for converting a Gaussian laser beam into a uniform beam. *Optical Communication* 36(6): 469–471, 1981.
7. W.B. Veldkamp. Laser beam profile shaping with binary diffraction gratings. *Optical Communication* 38(5/6): 381–386, 1981.
8. W.B. Veldkamp. Laser beam profile shaping with interlaced binary diffraction gratings. *Applied Optics* 21(17): 3209–3212, 1982.
9. C.C. Aleksoff, K.K. Ellis, and B.D. Neagle. Holographic conversion of a Gaussian beam to a near-field uniform beam. *Optical Engineering* 30(5): 537–543, 1991.
10. T. Kosoburd and J. Kedmi. Beam shaping with diffractive optical elements. *Proceedings of SPIE* 1971: 390–399, 1993.
11. M.T. Eismann, A.M. Tai, and J.N. Cederquist. Iterative design of a holographic beamformer. *Applied Optics* 28(13): 2641–2650, 1989.
12. M.A. Golub, M. Duparré, E.B. Kley, R. Dowarschik, B. Lüdge, W. Rockstroh, and H.J. Fuchs. New diffractive beam shaper generated with the aid of e-beam lithography. *Optical Engineering* 35(5): 1400–1406, 1996.

13. F.M. Dickey and S.C. Holswade. Gaussian laser beam profile shaping. *Optical Engineering* 35(1): 3285–3295, 1996.
14. S.C. Holswade and F.M. Dickey. Gaussian laser beam shaping: Test and evaluation. *Proceedings of SPIE* 2863: 237–245, 1996.
15. J.W. Goodman. *Introduction to Fourier Optics*. New York: McGraw-Hill, 1968.
16. L.E. Franks. *Signal Theory*. Englewood Cliffs, NJ: Prentice Hall, 1969.
17. R.N. Bracewell. *The Fourier Transform and Its Applications*. New York: McGraw-Hill, 1978.
18. N. Bleistein and R.A. Handelsman. *Asymptotic Expansion of Integrals*. New York: Dover Publications, 1986.
19. N. Bleistein. *Mathematical Methods for Wave Phenomena*. Orlando, FL: Academic Press, 1984.
20. J.J. Stamnes. *Waves in Focal Regions*. Bristol: IOP Publishing, 1986.
21. A. Walther. *The Ray and Wave Theory of Lenses*. Cambridge: Cambridge University Press, 1995.
22. L.A. Romero and F.M. Dickey. Lossless laser beam shaping. *Journal of the Optical Society of America A* 13(4): 751–760, 1996.
23. X. Deng, S. Liang, Z. Chen, W. Yu, and R. Ma. Uniform illumination of large targets using a lens array. *Applied Optics* 25(3): 377–381, 1986.
24. H.-J. Kahlert, U. Sarbach, B. Burghardt, and B. Klimt. Excimer laser illumination and imaging optics for controlled microstructure generation. *Proceedings of SPIE* 1835: 110–118, 1992.
25. R.O. Harger. *Synthetic Aperture Radar Systems: Theory and Design*. Orlando, FL: Academic Press, 1970.
26. B.E.A. Saleh and M.C. Teich. *Fundamentals of Photonics*. New York: Wiley, 1991.
27. A. Forbes, F.M. Dickey, M. DeGama, and A. du Plessis. Wavelength tunable laser beam shaping. *Optics Letters* 37(1): 49–51, 2012.
28. A.E. Siegman. *Lasers*. Mill Valley, CA: University Science Books, 1986.
29. X.G. Huang, M.R. Wang, and C. Yu. High-efficiency flat-top beam shaper fabricated by a nonlithographic technique. *Optical Engineering* 38(2): 208–213, 1999.
30. R.E. Collin and F.J. Zucker, *Antenna Theory: Part 1*. New York: McGraw-Hill, 1969.

APPENDIX A

A simple illustration of the method of stationary phase is given by the problem of mapping a uniform amplitude beam into a uniform irradiance beam. Consider the problem of determining the phase element in Figure 5.2 that maps a uniform amplitude and phase beam into a uniform irradiance beam at the focal plane of the transform lens. If the phase of the input beam is not uniform, it can be corrected by the phase element.

For simplicity, we treat the problem in one dimension. The field at the focal plane of the Fourier transform lens is given by

$$E(\omega) = \int_{\infty}^{\infty} \mathrm{rect}\left(\frac{x}{\alpha}\right) \mathrm{e}^{ik[\phi(x/\alpha) - \omega x/k]}\, \mathrm{d}x \tag{A.1}$$

where:

α is the input beam width

k is an arbitrary parameter (not the wave number)

Letting $\omega/k = \beta$, we can write Equation A.1 as

$$E(\omega) = \int_{\infty}^{\infty} \mathrm{rect}\left(\frac{x}{\alpha}\right) \mathrm{e}^{ik[\phi(x/\alpha)-\beta x]}\, \mathrm{d}x \tag{A.2}$$

The above equation can be approximated by the method of stationary phase, giving good results for large k. It is desired that the intensity of the field at the focal plane approximate a rect function. The stationary phase formula

$$I_c(k) \sim \mathrm{e}^{\mathrm{i}[k\psi(c)+\mu\pi/4]} f(c) \left[\frac{2\pi}{k\left|\psi''(c)\right|}\right]^{1/2} \tag{A.3}$$

gives

$$\left|E(\omega)\right|^2 = \mathrm{rect}(c)\frac{2}{k\phi''(c)} \tag{A.4}$$

where:

$\mu = \mathrm{sign}\, \psi''(c)$

c is determined by $\psi'(c) = 0$

In obtaining Equation A.4, we have used

$$\psi(x) = \phi\left(\frac{x}{\alpha}\right) - \beta x \tag{A.5}$$

Equation A.4 implies that $\phi''(x)$ = a constant. Let

$$\phi\left(\frac{x}{\alpha}\right) = \left(\frac{x}{\alpha}\right)^2 \tag{A.6}$$

and determine c by setting $\mathrm{d}\psi(c)/\mathrm{d}c = 0$, which gives $c = \beta\alpha^2/2$.

The field at the focal plane is then approximated by

$$E \approx \sqrt{\frac{\pi}{k}}\mathrm{e}^{-\mathrm{i}[(\omega^2\alpha^2/4k)-\pi/4]}\mathrm{rect}\left(\frac{\omega\alpha}{2k}\right) \tag{A.7}$$

where:

$\omega = 2\pi x_f/\lambda f$

the output spot width, W, is determined by $\pi W\alpha/k\lambda f = 1$, giving

$$k = \frac{\pi W\alpha}{\lambda f} \tag{A.8}$$

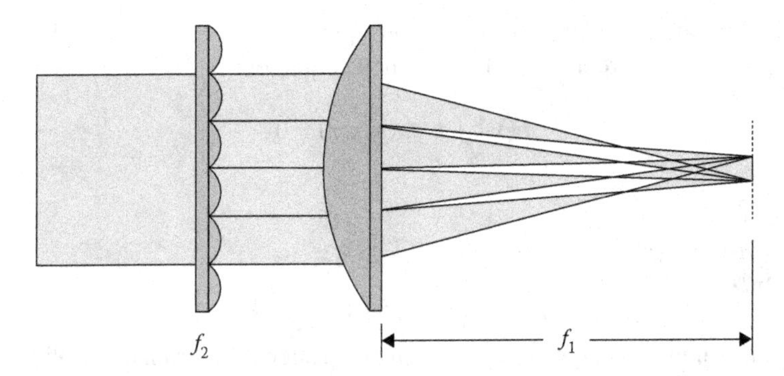

FIGURE 5.25 Optical schematic of the fly's eye lens beam integration system.

The parameter k is similar to β in Section 5.2.4, in that the larger the value of k the better the solution. Note that k is a dimensionless constant.

As in the Gaussian to flat-top solution, the phase element that accomplishes the mapping is $k\phi(x)$ with $\phi(x)$ determined by Equation A.6 and k determined by Equation A.8. It is interesting to note that the phase element is a simple (thin) lens in this case. Within the range of Fourier optics the solution is independent of the spacing between the thin lens phase element and the Fourier transform. If a small array of these elements were placed before the Fourier transform lens, the uniform patterns for each phase element would be superimposed in the focal plane of the Fourier transform lens. This is the diffraction theory basis for the "fly's eye lens" beam integration system.

A schematic diagram of the fly's eye lens beam integration system is shown in Figure 5.25. In the figure, each lenslet array element is focused to a common region by a focusing (Fourier transform lens), effecting beam integration (averaging). Each lens in the array samples the input beam. If the size of the elemental lenses in the array is small enough, the irradiance of each sample will be approximately uniform and, as designed, the output irradiance for each lens element will be approximately uniform. The net output of the beam integration system is a summation of the output from each array element, which should be approximately uniform. However, there will be a fine-structure interference component in the output that will depend on the degree of coherence of the laser system. Beam integration systems are typically used with multimode laser systems. Multielement imaging systems with two lenslet arrays can be designed to eliminate diffraction altogether.[24] A detailed optical analysis of multifaceted beam integration systems is presented in Chapter 10.

APPENDIX B

Assuming the scalar wave equation, it can be shown that the r.m.s. beam width is a quadratic function of distance along the optical axis. Equations 3.1 and 3.2 in

Section 5.3.1 can be obtained as follows. For the case of a monochromatic field, the scalar wave equation reduces to the Helmholtz equation:

$$(\nabla^2 + k^2)u(x,y,z) = 0 \tag{B.1}$$

where:

$$\nabla^2 = \frac{\partial^2}{\partial x^2} + \frac{\partial^2}{\partial y^2} + \frac{\partial^2}{\partial z^2}$$

$$k = 2\pi/\lambda$$

In the Fourier transform domain, the solution to Equation B.1 is given in Equation 5.20 by

$$U(f_x,f_x,z) = U(f_x,f_x,0)e^{iz\sqrt{k^2-(2\pi f_x)^2-(2\pi f_y)^2}} \tag{B.2}$$

where:

the uppercase letters denote Fourier transformed quantities
f_x and f_y are Fourier transform variables
$U(f_x,f_y,0)$ is the Fourier transform of the aperture field (at $z = 0$)

The solution of the wave equation given in Equation B.2 neglects evanescent waves requiring that $(2\pi f_x)^2 + (2\pi f_y)^2 \leq k^2$.

The mean square beam width in the coordinate for a beam with its centroid at the coordinate origin is defined by

$$(\Delta\rho)^2 = \frac{\int_{-\infty}^{\infty}\int_{-\infty}^{\infty} x^2 |u(x,y,z)|^2 \,dxdy}{\int_{-\infty}^{\infty}\int_{-\infty}^{\infty} |u(x,y,z)|^2 \,dxdy}
= \frac{\left(\frac{1}{2\pi}\right)^2 \int_{-\infty}^{\infty}\int_{-\infty}^{\infty} \left|\frac{\partial U(f_x,f_y,z)}{\partial f_x}\right|^2 df_x df_y}{\int_{-\infty}^{\infty}\int_{-\infty}^{\infty} |U(f_x,f_y,z)|^2 \,df_x df_y}. \tag{B.3}$$

The integrand in the numerator of the expression after the last equality sign in Equation B.3 can be expanded using Equation B.2 to obtain

$$\left|\frac{\partial U(f_x,f_y,z)}{\partial f_x}\right|^2 = \left|\frac{\partial U(f_x,f_y,0)}{\partial f_x}\right|^2 - 2\,\mathrm{Im}\left[\frac{(2\pi)^2 z f_x}{\sqrt{k^2-(2\pi f_x)^2-(2\pi f_y)^2}} \times U(f_x,f_y,0)\frac{\partial U^*(f_x,f_y,0)}{\partial f_x}\right] + \frac{(2\pi)^4 z^2 f_x^2}{k^2-(2\pi f_x)^2-(2\pi f_y)^2}|U(f_x,f_y,0)|^2 \tag{B.4}$$

Substituting this result in Equation B.3 gives

$$(\Delta\rho)^2 = \frac{\left(\frac{1}{2\pi}\right)^2 \int_{-\infty}^{\infty}\int_{-\infty}^{\infty} \left|\frac{\partial U(f_x,f_y,0)}{\partial f_x}\right|^2 df_x df_y}{\int_{-\infty}^{\infty}\int_{-\infty}^{\infty} |U(f_x,f_y,0)|^2 df_x df_y}$$

$$-2\frac{z\left(\frac{1}{2\pi}\right)^2 \int_{-\infty}^{\infty}\int_{-\infty}^{\infty} \mathrm{Im}\left[\frac{(2\pi)^2 f_x}{\sqrt{k^2-(2\pi f_x)^2-(2\pi f_y)^2}} \times U(f_x,f_y,0)\frac{\partial U^*(f_x,f_y,0)}{\partial f_x}\right] df_x df_y}{\int_{-\infty}^{\infty}\int_{-\infty}^{\infty} |U(f_x,f_y,0)|^2 df_x df_y} \quad \text{(B.5)}$$

$$+\frac{z^2\left(\frac{1}{2\pi}\right)^2 \int_{-\infty}^{\infty}\int_{-\infty}^{\infty} \frac{(2\pi)^4 f_x^2}{k^2-(2\pi f_x)^2-(2\pi f_y)^2} |U(f_x,f_y,0)| df_x df_y}{\int_{-\infty}^{\infty}\int_{-\infty}^{\infty} |U(f_x,f_y,0)|^2 df_x df_y}$$

The above equation can be written in the simple form:

$$(\Delta\rho)^2 = a - bz + cz^2 \quad \text{(B.6)}$$

where:

a, b, and c are defined as the coefficients of the corresponding powers of z in Equation B.5

The main result expressed in Equation B.5, or equivalently Equation B.6, is that the mean square beam width is a quadratic function z.

Further, if $b = 0$ in Equation B.6, the minimum beam width will be in the plane $z = 0$, and Equation B.6 becomes

$$(\Delta\rho)^2 = a + cz^2 \quad \text{(B.7)}$$

In the above equation, a is, by definition, the minimum mean square beam radius, that is, $a = (\Delta\rho_{min})^2$. Determining the conditions that give $b = 0$ is not a trivial matter. A sufficient, but not necessary, condition is that the term in brackets in the numerator integral defining b is real. This is achieved if $U(f_x,f_y,0)$ is a real function with the possibility of multiplication by a complex constant. This condition includes beams that have purely even or odd amplitude functions and a uniform phase. It would also include, for example, beams modulated by suitably symmetric square waves. If the beam is not circularly symmetric, the above derivation holds for arbitrary orthogonal coordinates. An equation in the form of Equation B.6 is obtained in each coordinate.

To minimize the spread of the beam it is desirable to have c as small as possible. However, a and c are related by the uncertainty principle. To apply the uncertainty principle, we can again use the condition $(2\pi f_x)^2 + (2\pi f_y)^2 \leq k^2$ in the integral defining c to obtain

$$\frac{ck^2}{(2\pi)^2} \geq \frac{\int_{-\infty}^{\infty}\int_{-\infty}^{\infty} f_x^2 \left|U(F_x, f_y, 0)\right|^2 df_x df_y}{\int_{-\infty}^{\infty}\int_{-\infty}^{\infty} \left|U(f_x, f_y, 0)\right|^2 df_x df_y} \tag{B.8}$$

Applying the uncertainty principle (Section 5.2.3), one can obtain

$$ac \geq \frac{\lambda^2}{4(2\pi)^2} \tag{B.9}$$

or equivalently

$$\sqrt{a}\sqrt{c} \geq \frac{\lambda}{4\pi} \tag{B.10}$$

One can define a *Rayleigh range* by the condition $a + cz_0^2 = 2a$, giving

$$z_0 = \frac{\sqrt{a}}{\sqrt{c}} \tag{B.11}$$

This result can be readily checked by comparing it to the Rayleigh range for Gaussian beams. It is well known that a Gaussian function is a minimum uncertainty function,[16] that is, equality obtains in the uncertainty principle. Thus, substituting Equation B.10 in Equation B.11 gives the Rayleigh range for a Gaussian beam as

$$z_0 = \frac{4\pi a}{\lambda} = \frac{4\pi(\Delta\rho_{\min})^2}{\lambda} \tag{B.12}$$

Noting that $\Delta\rho_{\min} = W_0/2$, where W_0 is the $1/e^2$ beam radius, we obtain $z_0 = \pi W_0/\lambda$, which is the standard result for Gaussian beams.[26] Finally, the uncertainty principle gives the inequality as a bound on the Rayleigh range:

$$z_0 \leq \frac{4\pi a}{\lambda} = \frac{4\pi(\Delta\rho_{\min})^2}{\lambda} \tag{B.13}$$

The components of the vector wave equation obey the scalar wave equation. Hence, these results can readily be extended to solutions of the vector wave equation.

It can be seen from Equation B.5 that all beams do not have a finite r.m.s. width. A simple example is the radiation pattern of a small dipole antenna.[30] The radiation pattern of a small dipole (current element) has a finite r.m.s. width in one plane, but the r.m.s. width in an orthogonal plane is infinite. The first term in Equation B.5 diverges if the Fourier transform of the field distribution, $U(f_x, f_y, 0)$, has a discontinuity. An example is a rect function type distribution. This is due to the fact that the Fourier transform of a function with a discontinuity falls off as $1/x$. The second term in Equation B.5 is well behaved.

The third term in Equation B.5 diverges unless $U(f_x,f_y) \rightarrow 0$ in a well-defined way as $(2\pi f_x)^2 + (2\pi f_y)^2$ approaches k^2. For example, if $U(f_x,f_y,0)$ is a Gaussian function the third term will diverge regardless of the width of the Gaussian function. Consider a Gaussian beam with a radius of 5 mm and a wavelength of 1 μm, then $|U(f_x,f_y)|^2$ will be down by a factor of $\exp(-493\times 10^6)$ at $2\pi f_x = k$. Clearly, this is a very small number, but the third term would still diverge. If the beam radius is close to a wavelength the Gaussian function will not have decayed greatly, and the divergence will be fundamental, and the beam does not have a well-defined radius.

The above discussion suggests that Equation B.5 can be used to determine the propagation of the beam radius by truncating the spectrum to avoid the divergence in the third term of Equation B.5. It is simple to truncate the spectrum in a smooth way so that the first term does not diverge. Further, one might define the beam radius based on the energy in the truncated spectrum. For example, the spectrum of the input beam profile could be truncated so the truncated beams contains something like 99.99% of the original beam energy. Clearly, the beam profile associated with the truncated spectrum will be close to the original beam profile.

Using a similar analysis as above, it is simple to show that the centroid of a beam propagates in a straight line. The beam centroid is given by

$$\bar{x} = \frac{\int_{-\infty}^{\infty}\int_{-\infty}^{\infty} x\,|u(x,y,z)|^2\,dxdy}{\int_{-\infty}^{\infty}\int_{-\infty}^{\infty} |u(x,y,z)|^2\,dxdy} = \frac{\left(\frac{i}{2\pi}\right)\int_{-\infty}^{\infty}\int_{-\infty}^{\infty} \frac{[\partial U(f_x,f_y,z)]\,U^*(f_x,f_y,z)}{\partial f_x}\,df_x df_y}{\int_{-\infty}^{\infty}\int_{-\infty}^{\infty} |U^2(f_x,f_y,z)|\,df_x df_y} \tag{B.14}$$

The integrand in the numerator of the expression after the last equality sign in Equation B.14 can be expanded using Equation B.2 to obtain

$$\begin{aligned}\frac{[\partial U(f_x,f_y,z)]\,U^*(f_x,f_y,z)}{\partial f_x} &= \left[\frac{(2\pi i)z f_x}{\sqrt{k^2-(2\pi f_x)^2-(2\pi f_y)^2}}U(f_x,f_y,0) + \frac{\partial U(f_x,f_y,0)}{\partial f_x}\right] \\ &\quad\times [U^*(f_x,f_y,0)] \\ &= \frac{(2\pi i)\,z f_x}{\sqrt{k^2-(2\pi f_x)^2-(2\pi f_y)^2}}|U(f_x,f_y,0)|^2 \\ &\quad+\frac{\partial U(f_x,f_y,0)}{\partial f_x}U^*(f_x,f_y,0)\end{aligned} \tag{B.15}$$

Since the value of the integral in Equation B.14 is real, we can substitute the imaginary part of Equation B.15 in Equation B.14 to obtain

$$\bar{x} = \frac{\left(\frac{z}{2\pi}\right)\int_{-\infty}^{\infty}\int_{-\infty}^{\infty}\frac{(2\pi i)f_x}{\sqrt{k^2-(2\pi f_x)^2-(2\pi f_y)^2}}\left|U(f_x,f_y,0)\right|^2 df_x df_y}{\int_{-\infty}^{\infty}\int_{-\infty}^{\infty}\left|U(f_x,f_y,0)\right|^2 df_x df_y} + \frac{\left(\frac{1}{2\pi}\right)\int_{-\infty}^{\infty}\int_{-\infty}^{\infty}\mathrm{Im}\left[\frac{\partial U(f_x,f_y,0)}{\partial f_x}U^*(f_x,f_y,0)\right]df_x df_y}{\int_{-\infty}^{\infty}\int_{-\infty}^{\infty}\left|U(f_x,f_y,0)\right|^2 df_x df_y} \quad \text{(B.16)}$$

Equation B.16 can be written in the simple form as

$$\bar{x} = az + b \quad \text{(B.17)}$$

where:
a and b are defined as the coefficients of the corresponding powers of z

The main result expressed in Equation B.16, or equivalently Equation B.17, is that the location of the centroid of the beam is a linear function of z, a straight line. The divergence issues discussed above are applicable here. Two orthogonal components are needed to specify the direction of the beam.

6 Geometrical Methods

David L. Shealy and John A. Hoffnagle

CONTENTS

6.1 INTRODUCTION

Laser beam shaping is used in many fields of scientific, engineering, and industrial research and development and has become important for laser-based applications, such as lithography, materials processing, isotope separation, medical and illumination applications, laser printing, optical data storage, spectroscopy, photography, laser fusion, and laboratory research. These applications use a variety of the unique properties of lasers, such as high intensity, coherent, and monochromatic light of lasers. In this chapter, we shall discuss a wide spectrum of geometrical methods that have been used to shape a laser beam profile, which involves application of geometrical optics to solve the optical design problem for the laser beam shaping system.

Bokor and Davidson [1] have identified several criteria to use when evaluating the quality of any beam shaping technique: (1) it forms the desired beam shape with high accuracy and has minimal losses of the total power; (2) the output beam shape is not sensitive to small changes of the input beam irradiance profile or phase distribution; and (3) it has minimal reduction of the beam brightness. However, laser beam shaping techniques have been categorized in different ways by different authors.

For example, Kreuzer [2] described, at the *Symposium on Optical and Electro-Optical Information Processing Technology* in 1964, four methods for creating a uniform irradiance profile for a beam leaving a typical laser cavity:

1. Introduce an aperture into the beam path, which allows only the central portion of the laser beam with more uniform irradiance to propagate, which uses only a small fraction of the input beam power.
2. Use a spatial filter of nonuniform transmission, which attenuates the bright central part of a laser beam more than edges.
3. Integrate the spatial modes with suitable relative phases and amplitudes to provide greater uniformity of irradiance than a Gaussian mode from a confocal laser resonator.
4. Use a pair of aspheric lenses whose surface contours are determined by imposing the conditions of conservation of energy and constant optical path length (OPL) between the input and output beams.

Sinzinger and Jahns [3] describe the use of micro-optics for beam shaping, which they categorize as lateral, axial, and temporal beam shaping outside of the laser cavity for single or multiple laser beams. Further, Sinzinger and Jahns note that the homogenization of a laser beam irradiance profile is one of the most important commercial applications of micro-optical devices. However, Dickey et al. [4] characterize beam shaping based on the methods used to modify the laser beam characteristics, such as attenuators which truncate a Gaussian beam directly with an aperture, neutral density filter, or other electro-optics device with an appropriate transverse transmittance profile; integrators that combine the energy within parts of a light beam to create a more uniform irradiance distribution; and transformers (as field mappers) that perform a one-to-one mapping of the electromagnetic field (amplitude and phase) from an input plane (or surface) to an output plane (or surface). For more information on the history of beam shaping, an extensive history of beam shaping is given in Reference [5].

The process of laser beam shaping discussed in this chapter uses geometrical optics to solve the optical design of components to redistribute the irradiance and phase of the beam. A beam shape generally refers to the irradiance profile of the beam, while the phase of a beam is related to the OPL or distance, which a wavefront of a beam has traveled from a reference location, and affects its propagation characteristics. Specifically, the laws of reflection and refraction are used along with ray tracing, conservation of energy within a bundle of rays, and the constant OPL condition to design laser beam profile shaping optical systems. Interference or diffraction effects are not considered as part of the design process in this chapter. That is, only lenses and mirrors are used for the optical components of the laser beam profile shaping systems and are discussed in this chapter. The characteristics of a laser beam can be changed by using an aperture, apodization, integrator, and/or transformer optical elements, while it is important for beam shaping methods to minimize losses of power and initial beam brightness and to form the desired output beam profile with high accuracy.

McDermit and Horton [6,7] presented a method for designing a rotationally symmetric reflective optical system for illuminating a receiver surface in a prescribed manner using a nonuniform input beam profile. Using their method, two mirrors were designed to allow a laser to uniformly heat a flat surface as part of a material testing procedure. Malyak [8] has designed a two-mirror laser profile shaping system where the second mirror is decentered relative to the first mirror to eliminate the central obscuration present in the axially symmetric design. A set of equations was presented for the mirror surface figure for the nonrotationally symmetric laser profile shaping optical system.

Laser beam shaping has been demonstrated by many other methods during the past 50 years to change the laser beam characteristics when using an aperture, apodization, or integrator optical components [9,10]. Aspheric refractive optics was first proposed by Kreuzer [2] in 1965 for use in a lossless laser beam shaper by using a pair of plano-aspheric lens elements whose surface contours are determined by imposing the conditions of conservation of energy and constant OPL between the input and output beams. For the two plano-aspheric lenses in either the Galilean or the Keplerian configuration, Kreuzer [11] developed and patented a coherent-light optical system, which redistributes the rays of an input laser beam to yield an output beam with a prescribed irradiance distribution, based on the geometrical optics intensity law, while maintaining wavefront shape by using the constant OPL condition between the input and output wavefronts. Kreuzer presented equations for computing the sag of the aspheric surfaces of the two-element refractive system that transforms a collimated input Gaussian beam into a collimated output beam with uniform irradiance. Frieden [12] published independently from Kreuzer a similar method to design a laser beam shaping system. Rhodes and Shealy [13,14] derived a set of differential equations using intensity mapping to calculate the shape of two aspherical surfaces of a lens system that expands and converts a Gaussian laser beam profile into a collimated, uniform irradiance output beam, where these differential equations have come to be known as the beam shaping equations (geometrical optics law of intensity and constant OPL condition). Rhodes and Shealy also analyzed the performance of a two plano-aspheric beam shaper when either element was displaced and/or tilted with respect to its nominal position as part of a tolerance analysis. The design methods of Kreuzer, Rhodes, and Shealy have been shown in Appendix A to be equivalent.

Shafer [15] designed an afocal beam shaper using spherical lenses. Shafer replaces the plano-aspheric elements of Kreuzer's design by an afocal doublet spherical lens with zero net power. By varying the shape of the first group of lenses while keeping the power constant, a large amount of spherical aberration can be introduced into the system which changes the output beam intensity profile. Setting the second group of lenses to have the same, but opposite sign, amount of spherical aberration will result in a collimated output beam with a more uniform intensity profile. Balancing the spherical aberration of the four spherical lenses leads to an output beam with variations of ±2% from uniformity between radial points corresponding to the ($1/e$) power width of the input beam. While seeking to use spherical surfaces for each optical element, Wang and Shealy [16,17] presented a method for designing an expanded, uniform irradiance profile laser beam using two axial gradient-index (GRIN) lenses. Their design procedure yields the GRIN profiles as well as the curvatures and separation of the lens surfaces.

Since efficient beam shaping and high uniformity of the output irradiance require use of aspheric surfaces, there was an interlude of 30 years from the first proposal of beam shaping using aspheric lens pair to the first publication of experimental results by Jiang et al. [18,19] who designed, fabricated, and tested a refractive beam shaper in the Galilean configuration. Single-point diamond lathe fixture was used to create the aspheric surfaces in a CaF_2 substrate. This system was designed to operate at a wavelength of 411.57 nm (HeCd laser), but was also shown to efficiently shape a HeNe laser (632 nm) when the spacing between the two lenses was increased by approximately 2% from its original value [20].

Hoffnagle and Jefferson [21,22] invented an optical system that transforms an input laser beam with an axially symmetric irradiance profile into an output beam with a different axially symmetric irradiance profile with a continuous, sigmoidal irradiance distribution, such as a Fermi–Dirac (FD) distribution. The design method of Kreuzer with convex aspherics in a Keplerian configuration was used to ease fabrication and introduced a continuous roll-off of the output beam profile for more control of the far-field diffraction pattern. Their beam shaper has been assembled, tested, and transferred into a commercially available refractive beam shaper [23] with a large bandwidth from IR to UV. Hoffnagle and Jefferson have analyzed the optical systems for shaping an input Gaussian beam into an FD output beam profile and have shown that the FD function has similar behavior as the super-Gaussian (SG) function. Hoffnagle and Jefferson also provide detailed characterization of a refractive beam shaper at 514 nm, which shows that the output beam irradiance variation is less than 5% and that the root-mean-square (rms) variation of the optical path difference (OPD) over the beam is 13 nm. In addition, Hoffnagle and Jefferson have shown that the flexibility of the two-lens beam shaper is enhanced by its insensitivity to wavelength (deep-UV to near-IR) and by the large depth of field of the output beam [24]. Further,

Hoffnagle and Jefferson have used geometrical optics and the paraxial approximation to derive an expression for the effects that small errors in the slope and curvature of the aspheric lens surfaces have on the output irradiance profile, which leads to setting a tolerance on the lens curvature at the optical axis to be less than 2% variation in the output irradiance. Jefferson and Hoffnagle [25] describe methods to achromatize an aspheric beam transformer using conventional spherical optics.

Cornwell [26,27] introduced nonprojective transformations for use in designing reflective and refractive laser beam profile shaping systems and presented a systematic seven-step procedure for designing laser profile shaping systems with either rectangular or polar symmetry. A number of illustrative examples of using nonprojective transformations to design both reflective and refractive laser profile shaping systems are presented in his work. Annual conferences on laser beam shaping have been held as part of the *SPIE Optics and Photonics Symposium* since 2000 [28–40].

This chapter deals with how beam transformers are designed and used in applications when geometrical optics is valid. A beam transformer or field mapper performs a one-to-one mapping of the optical field (amplitude and phase) of the input laser beam into a specified output beam with a specific irradiance distribution and phase profile. This type of beam shaping is well suited for applications when the input beam has a single mode. One optical element can transform an input beam irradiance profile into one that uniformly illuminates a specific output plane. At least two optical elements are generally required to achieve more general transformations of the irradiance and phase distributions of the input beam. Reflective, refractive, or diffractive optical elements have been used for different configurations of beam transformers. Beam transformers commonly transform a collimated TEM_{00} Gaussian beam into a collimated and more uniform irradiance profile beam, such as top hat, SG, FD, or other flattened distributions.

A general theory of designing and analyzing a laser profile shaping optical systems as applied to a two-lens or two-mirror optical system is presented in Section 6.2. Specific attention is devoted to the application of conservation of energy as expressed by the intensity law of geometrical optics and the constant OPL condition as constraints on the optical surface figure of laser profile shaping systems. Section 6.3 describes the design procedures used to design either refractive or reflective laser beam shapers. A summary of a seven-step procedure for using nonprojective transformation to design laser profile shaping systems is presented in Section 6.3.3. Section 6.4 describes the optical and mechanical tolerances that are important to address when building actual laser beam systems. Section 6.5 describes design, fabrication, and testing of several refractive beam shapers in the Keplerian and Galilean configuration, as well as an axial GRIN laser beam shaper. Section 6.6 describes design and analysis for one- and two-mirror laser beam shapers. Overall conclusions of these efforts to design, build, and test refractive and reflective laser beam shapers are given in Section 6.7.

6.2 THEORY

Geometrical optics has been an effective method for designing laser beam shaping systems when a relationship between beam waist, wavelength, and apertures of the optical elements is such that the effects of diffraction are not important to consider during the design of the beam shaping optics [9,10]. In particular, when the input and output beam profiles are known and when there is a one-to-one mapping between the input and output beams, the geometrical optics intensity law and the constant OPL condition have been used to design reflective and refractive configurations of laser beam shapers [2,8–22]. In addition, a Gaussian to flattened beam shaping system must also be designed to optimize competing constraints, such as efficient use of the beam power, propagation of the desired beam profile across the region of interest, and uniformity of output beam irradiance profile throughout the region of interest. Finally, the aspheric optics must be fabricated and aligned within acceptable cost for the application of interest. This section will highlight the essential results of geometrical optics used in beam shaping and discuss output beam profiles and propagation. This section also addresses the irradiance profiles commonly used in beam shaping applications and the effects of diffraction associated with the finite size of the aperture and of the beam propagation from the optics for flattened profiles using Kirchhoff's diffraction theory with the Fresnel approximation.

6.2.1 Geometrical Optics

The concepts of rays, wavefront, and energy propagation are fundamental to understanding and using geometrical optics for shaping laser beam profiles. A brief overview of these concepts is presented in this section. Then, geometrical optics is used to set up several constraint equations that are used to determine the reflective or refractive surface shape or gradient index profile as part of the optical design of laser beam profile shaping optical systems. There are many discussions of geometrical optics in the literature [41–43]. In order to determine or optimize the illumination within an optical system, the optical field must be determined throughout the system. The optical field is a local plane wave solution of Maxwell's equations for an isotropic, nonconducting, and charge-free medium and is a solution of the scalar wave equation [44,45]:

$$(\nabla^2 + n^2 k_0^2)u(\mathbf{r}) = 0 \tag{6.1}$$

where:

$u(\mathbf{r})$ represents the components of the electric field at any point $\mathbf{r}$
n is the index of refraction at $\mathbf{r}$
$k_0 = \omega/c = 2\pi/\lambda_0$ is the wave number in free space, where ω is the frequency of the wave, c is the speed of light, and λ_0 is the wavelength of light

Assume that a solution to Equation 6.1 can be written as

$$u(\mathbf{r}) = u_0(\mathbf{r}) \exp[ik_0\Phi(\mathbf{r})] \tag{6.2}$$

where:

$u_0(\mathbf{r})$ and $\Phi(\mathbf{r})$ are unknown functions of $\mathbf{r}$

Equation 6.1 leads to the following conditions that must be satisfied by $u_0(\mathbf{r})$ and $\Phi(\mathbf{r})$:

$$(\nabla\Phi)^2 = n^2 \tag{6.3}$$

$$2u_0\nabla\Phi \cdot \nabla u_0 + u_0^2\nabla^2\Phi = 0 \tag{6.4}$$

where the term proportional to $(1/k_0^2)$ has been neglected, or explicitly, ray optics requires for small wavelengths that the following condition be satisfied:

$$\frac{\nabla^2 u(\mathbf{r})}{u(\mathbf{r})} \sim 0 \tag{6.5}$$

Equation 6.3 is known as the eikonal equation and is a basic equation of geometrical optics, which is discussed in more detail in Section 6.2.1.1, where a general solution of the eikonal equation is presented and discussed for the nonsymmetric case and for the rotationally symmetric case. Equation 6.4 is the intensity law of geometrical optics and expresses conservation of radiant energy within a bundle of rays, which will be discussed in Section 6.2.1.2 and is essential for designing laser beam shaping optics.

6.2.1.1 Eikonal Equation

From Equation 6.2, we can identify the function $\Phi(\mathbf{r})$ as the phase of the optical field, $u(\mathbf{r})$. The surfaces of constant phase of the optical field are known as the geometrical wavefronts

$$\Phi(x,y,z) = \text{constant} \tag{6.6}$$

For isotropic media, rays are normal to the wavefront along the direction of the energy propagation. A unit vector normal to the wavefront and along a ray at the point $\mathbf{r}$ is given by

$$\mathbf{A}(\mathbf{r}) = \frac{\nabla\Phi(\mathbf{r})}{|\nabla\Phi(\mathbf{r})|} = \frac{\nabla\Phi(\mathbf{r})}{n(\mathbf{r})} \tag{6.7}$$

where the eikonal equation was used to simplify this expression for the ray vector $\mathbf{A}$. In homogeneous media, it also follows from the eikonal equation that the magnitude of $\nabla\Phi$ is independent of position. Therefore, the position of a wavefront at a later time is parallel to its original position. Thus, rays in homogeneous media are straight lines. It is convenient to define an optical ray vector as $\mathbf{S} \equiv n\mathbf{A}$ with the conventional normalization given by

$$|\mathbf{S}|^2 = u^2 + v^2 + w^2 = n^2 \tag{6.8}$$

where:

(u,v,w) are the optical direction cosines of a ray with respect to the coordinate axes (x,y,z)

The eikonal is also known as Hamilton's characteristic function, and the eikonal equation has also been identified as the analogy of the Hamilton–Jacobi equation. Solutions of the eikonal equation have been very useful in optical design, where a power series solution of the eikonal equation has been used for approximately a century in lens design. The coefficients of the series solution are known as aberrations. Additional properties of rays are given in Section 6.2.1.3.

6.2.1.1.1 General Solution for Nonsymmetric Systems

A general solution of the eikonal equation (Equation 6.3) has been developed by Stavroudis and Fronczek [46,47] for light propagating in a homogeneous medium with a constant index of refraction n in the direction of **S** and to have the form

$$\Phi(\mathbf{r}) = \mathbf{r} \cdot \mathbf{S}(u,v) + k(u,v) \tag{6.9}$$

where the following auxiliary conditions are imposed on the k-function, $k(u,v)$:

$$\left[\frac{\partial k(u,v)}{\partial u}\right]\sqrt{n^2 - u^2 - v^2} = -x\sqrt{n^2 - u^2 - v^2} + uz \tag{6.10a}$$

$$\left[\frac{\partial k(u,v)}{\partial v}\right]\sqrt{n^2 - u^2 - v^2} = -y\sqrt{n^2 - u^2 - v^2} + vz \tag{6.10b}$$

It is convenient to use subscripts on $k(u,v)$ to indicate partial differentiation with respect to the corresponding variables, such as $k_u = \partial k(u,v)/\partial u$. The vector $\mathbf{r} = (x,y,z)$ specifies the coordinates of points on a wavefront at some distance s along the optical axis from the reference origin **O**, and $k(u,v)$ is a general function of two optical direction cosines of the ray vector **S** whose first partial derivatives satisfy Equation 6.10a and b.

A general expression for the k-function for refraction of a plane wave from an arbitrary surface has been presented by Shealy and Hoffnagle [48,49]. Figure 6.1

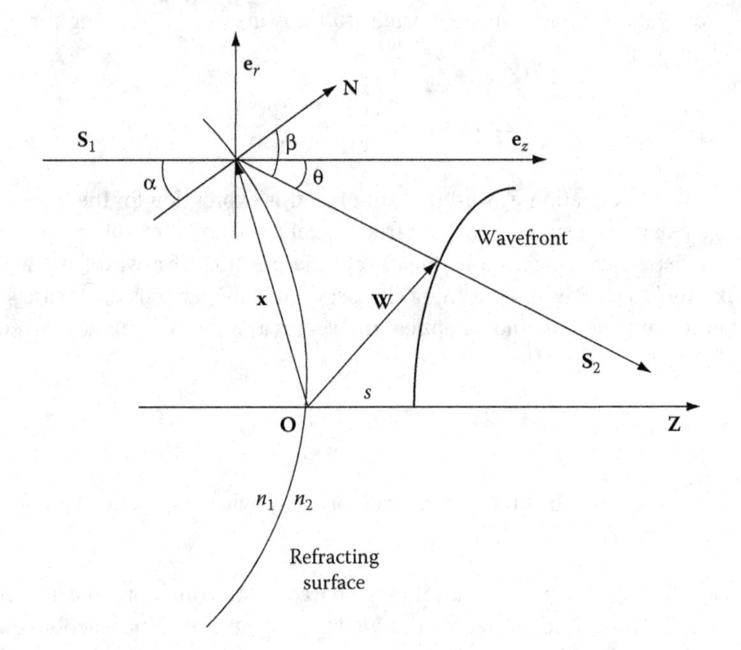

FIGURE 6.1 Schematic diagram of the refracting surface **x**, showing the incident ray $\mathbf{S}_1(= n_1\mathbf{e}_z)$, the refracted ray $\mathbf{S}_2$, and the wavefront **W** when the origin of coordinate system **O** is located at the vertex of refracting surface with the optical axis. (Reproduced from Shealy, D. L. and Hoffnagle, J. A. *J. Opt. Soc. Am. A*, 25, 2370–2380, 2008. With permission.)

illustrates the geometry for refraction of a plane wave from a single surface and the notation used for this problem. The refracting surface separates two regions with uniform indices of refraction n_1 and n_2. A global coordinate system with origin **O** is used such that **x** locates an arbitrary point on the surface separating region 1 and region 2. The normal unit vector to the surface is given by **N**, and the vectors $\mathbf{S}_1$ and $\mathbf{S}_2$ are the optical wave vectors of the incident and refracted rays, respectively. The angles of incidence and refraction are given by α and β, where the deviation of the refracted ray is given by $\theta = \beta - \alpha$. The vector **W** specifies an arbitrary point on refracted ray at **x**. As shown by Shealy and Hoffnagle [48], the k-function in region 2 is given by

$$k(\mathbf{S}_2) = -\Omega \mathbf{x} \cdot \mathbf{N} \tag{6.11}$$

where:

Ω is the factor appearing in relationship between the incident and outgoing wave vectors [50]:

$$\mathbf{S}_2 = \mathbf{S}_1 + \Omega \mathbf{N} \tag{6.12}$$

where:

$$\Omega = -n_1 \cos(\alpha) + n_2 \cos(\beta) \tag{6.13}$$

The k-function given by Equation 6.11 applies to a plane wave propagating along the optical axis and refracting from an arbitrarily shaped surface. In general, the k-function evaluated from Equation 6.11 will be a function of the two optical direction cosines of the refracted rays for point on the input aperture coordinates.

To evaluate the equation of a specific wavefront, we set $\Phi(\mathbf{r})$ in Equation 6.9 equal to a constant, such as

$$\Phi(\mathbf{r}) = n_2 s \tag{6.14}$$

where:

s is the distance along a reference axis, such as the optical axis or the z-axis from a reference origin **O**

Equations 6.9 and 6.10 can be rewritten for the specific wavefront defined by Equation 6.14 as follows:

$$xu + yv + z\sqrt{n_2^2 - u^2 - v^2} + k(u,v) = n_2 s \tag{6.15a}$$

$$-uz + (x + k_u)\sqrt{n_2^2 - u^2 - v^2} = 0 \tag{6.15b}$$

$$-vz + (y + k_v)\sqrt{n_2^2 - u^2 - v^2} = 0 \tag{6.15c}$$

which can be solved for the coordinates (x,y,z) of the position vector **W** on the wavefront defined by Equation 6.14 and whose wavefront equation is given by

$$\mathbf{W}(u,v,s) = q\frac{\mathbf{S}_2}{n_2^2} - \mathbf{K} \tag{6.16}$$

where:

$$q = [n_2 s - k(u,v)] + \mathbf{S}_2 \cdot \mathbf{K} \tag{6.17a}$$

$$\mathbf{K} = (k_u, k_v, 0) \tag{6.17b}$$

From the equation of a wavefront (Equation 6.16), Stavroudis [47] has evaluated the two principal curvatures of the wavefront, $\rho_\pm$, and the associated caustic surfaces:

$$\rho_\pm = \frac{1}{n_2}\left(q - \frac{H \pm \mathcal{S}}{2}\right) \tag{6.18}$$

where:
S is proportional to difference between $\rho_\pm$

$$H = (n_2^2 - u^2)k_{uu} + (n_2^2 - v^2)k_{vv} - 2uvk_{uv} \tag{6.19}$$

$$\mathcal{S}^2 = H^2 - 4n_2^2 w^2 T^2 \tag{6.20}$$

$$T^2 = k_{uu}k_{vv} - (k_{uv})^2 \tag{6.21}$$

From the principal curvatures, one can compute the transformation of irradiance along a ray [42]. Finally, the two caustic surfaces, $\mathbf{C}_\pm$, are described by the following points:

$$\mathbf{C}_\pm = \frac{1}{2n_2^2}(H \pm \mathcal{S})\mathbf{S}_2 - \mathbf{K} \tag{6.22}$$

These methods have been used to evaluate the wavefront, caustic surfaces, and irradiance of the laser beam as it propagates through beam shaping optical systems. In addition, it is well known that the wavefront aberration of a symmetric optical system has been defined as the OPL along a ray between the wavefront in the exit pupil and the ideal reference spherical wavefront centered on the idea (paraxial) image point of the object which generated the given wavefront [51]. Therefore, Equation 6.16 for the wavefront can be used to evaluate the aberrations for an optical system without making a series expansion as illustrated for a plane wave reflected from a spherical mirror [52].

In addition, Hoffnagle and Shealy [49] evaluate an analytical expression for the *k*-function and wavefront reflected from each surface of a Cassegrain telescope. Therefore, in the case, the wave aberration function has been evaluated by calculating the distance from the exact wavefront and the Gaussian reference spherical wavefront which goes through the center of the exit pupil and is centered on the Gaussian image point.

6.2.1.1.2 General Solution for Rotationally Symmetric Case

For systems with rotational symmetry for the incident wavefront, the refracting surface and the refracted wavefront as shown in Figure 6.1, the optical ray vector, $\mathbf{S}_2$, and points on a refracted wavefront, $\mathbf{W}$, are given below in terms of their components along the axial and radial directions:

$$\mathbf{W} = W_r\mathbf{e}_r + W_z\mathbf{e}_z \tag{6.23a}$$

$$\mathbf{S}_2 = -\sqrt{n_2^2 - w^2}\;\mathbf{e}_r + w\mathbf{e}_z \tag{6.23b}$$

Stravroudis and Fronczek [46] have also shown that the wavefront refracted from a rotationally systemic optical system must satisfy the following relationship:

$$n_2 s = -W_r\sqrt{n_2^2 - w^2} + W_z w + k(w) \tag{6.24}$$

where the k-function satisfies the following auxiliary condition:

$$k_w \equiv \frac{\mathrm{d}k(w)}{\mathrm{d}w} = -\frac{W_r w}{\sqrt{n_2^2 - w^2}} - W_z \tag{6.25}$$

Shealy and Hoffnagle [48,49] have established a general expression for the k-function as given by Equation 6.11 which applies at each refracting surface in a multielement optical system. Explicitly, to evaluate the k-function at the second refracting surface of an optical system, then one increments by 1 the indices of each quantity in Equation 6.11, or the k-function for a general refracting surface shown in Figure 6.1 is given by the following equation:

$$k = -\Omega\mathbf{x}\cdot\mathbf{N} \tag{6.26}$$

where:

x and **N** are defined in Figure 6.1
Ω is defined by Equation 6.13

Equation 6.26 provides interesting insight into the physical meaning of the k-function. Since the vector $\Omega\mathbf{N}$ represents the change between the input and refracted optical ray vectors, which is along the direction of the unit normal vector to the refracting surface with a magnitude of Ω, the k-function is the projection of $-\Omega\mathbf{N}$ onto the position vector **x** of the point of refraction. Equation 6.26 is a convenient form for the k-function for evaluating the first and second derivatives of k with respect to w, when w is a function of r, and for evaluating the wavefront and the caustic surfaces.

The first and second derivatives of $k(w)$ with respect to w have been shown to be given by following results:

$$k(w) = r\sqrt{n_2^2 - w^2} + (n_1 - w)z(r) \tag{6.27a}$$

$$k_w = -\frac{rw}{\sqrt{n_2^2 - w^2}} - z(r) \tag{6.27b}$$

$$k_{ww} = \frac{(n_1 w - n_2^2)^2}{z''(n_2^2 - w^2)(w - n_1)^3} - \frac{r n_2^2}{(n_2^2 - w^2)^{3/2}} \tag{6.27c}$$

where the intermediate steps required to obtain Equation 6.27c from Equation 6.27b by taking the derivative of k_w with respect to w are given in the original work [48]. In addition, an equation of the refracted wavefront follows from Equations 6.24 and 6.25 by using the explicit expressions for $k(w)$ from Equation 6.27a and for k_w from Equation 6.27b to obtain

$$\mathbf{W}(r,\phi,s) = \mathbf{x}(r,\phi) + [-n_1 z(r) + n_2 s]\frac{\mathbf{S}_2}{n_2^2} \tag{6.28}$$

which agrees with an earlier result for an equation for the wavefront refracted by a rotationally symmetric surface which was obtained by using a constant OPL argument for the plane wave propagating from a reference plane [53].

Finally, equations for the caustic surfaces for a plane wave refracted by a rotationally symmetric surface [48] for the configuration of the refracting surface shown in Figure 6.1 are given as follows:

$$\mathbf{C}_+ = \mathbf{e}_r\left[r - \frac{\sqrt{n_2^2 - w^2}\,(n_1 w - n_2^2)^2}{z'' n_2^2 (w - n_1)^3}\right] + \mathbf{e}_z\left[z(r) + \frac{w(n_1 w - n_2^2)^2}{z'' n_2^2 (w - n_1)^3}\right] \tag{6.29a}$$

$$\mathbf{C}_- = \mathbf{e}_z\left[z(r) + \frac{rw}{\sqrt{n_2^2 - w^2}}\right] \tag{6.29b}$$

It is interesting to compare Equation 6.29a and b with similar results presented earlier by Shealy and Hoffnagle [53], which were obtained by generalized ray tracing and expressed in terms of the first and second derivatives of the sag function $z(r)$ of the refracting surface with respect to r. Inspection of Equation 6.29a and b suggests that we can write

$$\mathbf{C}_\pm = \mathbf{x} + r_\pm \frac{\mathbf{S}_2}{n_2^2} \tag{6.30}$$

where $r_\pm$ represents the OPL along $\mathbf{S}_2$ from $\mathbf{x}$ to the points on the tangential and sagittal caustic surfaces, which represent the principal radii of curvature of the wavefront just after refraction at $\mathbf{x}$. By inspection of Equations 6.29a through 6.30, we conclude that $r_\pm$ are given by

$$r_+ = \frac{(n_1 w - n_2^2)^2}{z''(w - n_1)^3} \tag{6.31a}$$

$$r_- = \frac{r n_2^2}{\sqrt{n_2^2 - w^2}} \tag{6.31b}$$

Further, we have shown that Equation 6.31a and b is equivalent to Equation 6.23a and b presented by Shealy and Hoffnagle [53] for the tangential and sagittal wavefront curvatures at the point of refraction on the lens.

Shealy and Hoffnagle [53–55] have reported both experimental and computational evaluation of the wavefront propagation and caustic surfaces formed between the two plano-aspheric lenses of a Keplerian configuration of a laser beam shaping device, which was designed to transform a Gaussian beam into a flattened irradiance profile output beam. Equations 6.16 and 6.22 have been used to evaluate the wavefront propagation and the caustic surfaces between the two aspheric lenses of refractive beam shaper, which was designed, built, and tested by Hoffnagle and Jefferson

[21,22,24,56,57]. Figure 6.2a illustrates a cross section of the refracted wavefront near the vertex of the first aspheric lens of this one-to-one Keplerian (1-to-1K) beam shaper and the optical axis, and Figure 6.2b illustrates the wavefront as it propagates through the focal region of the first aspheric lens and on to the second aspheric lens of this 1-to-1K beam shaper. There are cusps and inflection points on the wavefront as it propagates between the two aspheric lenses of the refractive beam shaper, and the wavefront folds back upon itself several times as it comes in contact with the

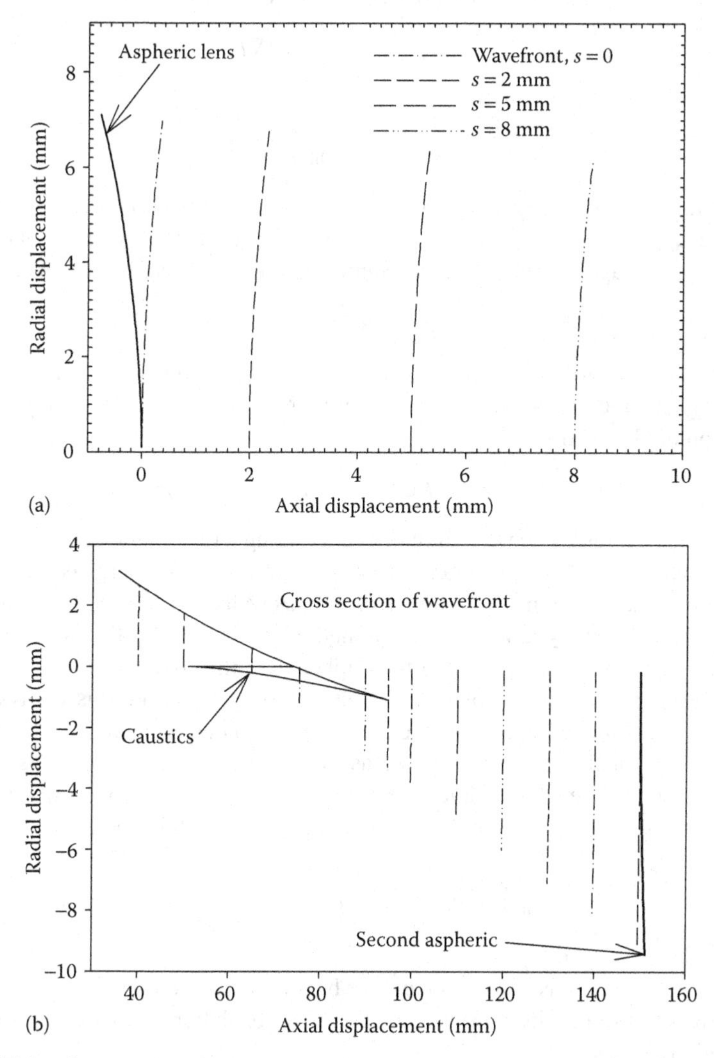

FIGURE 6.2 Cross section of refracted wavefront for (a) $s = 0, 2, 5, 8$ mm and (b) $s = 40, 50, 65, 75.5, 90, 95, 100, 110, 120, 130, 140$, and 150 mm represented by dashed lines. The caustic surfaces of the plano-aspheric lens of a 1-to-1K laser beam shaper with $q = 15$, $n = 1.46071$, $R_{FL} = 3.25$ mm, and $w_0 = 2.366$ mm. (Reproduced from Shealy, D. L. and Hoffnagle, J. A. Wavefront and caustic surfaces of refractive laser beam shaper. In *Novel Optical Systems Design and Optimization X*, Koshel, R. J. and Gregory, G. G., eds., vol. 6668, 666805-1, 2007. With permission.)

tangential and sagittal caustic surfaces as its irradiance is being redistributed over the wavefront according to the optical design condition of the intensity law and the constant OPL condition, which will be discussed in detail in Section 6.3.

6.2.1.2 Intensity Equation

Equation 6.4 is equivalent to the conservation of radiant energy within a bundle of rays and leads to the geometrical optics intensity law for propagation of a bundle of rays as illustrated in Reference [58]. Using the vector identity,

$$\nabla \cdot (f\mathbf{v}) = f\nabla \cdot \mathbf{v} + \mathrm{v} \cdot \nabla f \tag{6.32}$$

Equation 6.4 can be rewritten as

$$\nabla \cdot (u_0^2 \nabla \Phi) = \nabla \cdot (u_0^2 n\mathbf{A}) = 0 \tag{6.33}$$

Recognizing that the energy density of a field is proportional to the square of the field amplitude u_0^2 and that the intensity I is equal to the energy density of the field times the speed of propagation within the medium, Equation 6.33 can be written as

$$\nabla \cdot (I\mathbf{A}) = 0 \tag{6.34}$$

Equation 6.34 expresses conservation of radiant energy for nonconducting medium. Integrating Equation 6.34 over a tube surrounding a bundle of rays [42] gives after application of Gauss' theorem:

$$I_1 \mathrm{d}A_1 = I_2 \mathrm{d}A_2 \tag{6.35}$$

Equation 6.35 expresses conservation of energy along a ray bundle between any two surfaces intersecting the beam and is a basic equation used to design laser beam profile shaping optical systems. Born and Wolf [42] as well as Shealy [59] used the geometrical optics intensity law to provide a method for evaluating the intensity along any ray path based on the integral of the Laplacian of the eikonal.

Integrating Equation 6.35 over the input and output apertures is equivalent to applying conservation of energy between the input and output planes of a laser beam shaper. For an ideal optical system, we assume that none of the power is blocked from the beam as a result of the lens apertures. Therefore, we assume that both the input and output beams contain the same total power, which is normalized to unity

$$2\pi \int_0^\infty I_{\mathrm{in}}(r) r \mathrm{d}r = 2\pi \int_0^\infty I_{\mathrm{out}}(R) R \mathrm{d}R = 1 \tag{6.36}$$

where the input beam is often a Gaussian beam, and the output beam profile has been considered in the literature to have a top-hat or flattened irradiance profile as discussed in Section 6.2.2.

6.2.1.3 Ray Optics

According to geometrical optics, the phase and amplitude of the optical field are evaluated independently. First, the ray paths are evaluated throughout the optical system, which enables computing the phase in terms of the OPL of rays passing through the system. Next, the amplitude of the optical field is determined by monitoring the intensity variations along each ray [60–62]. This approach for evaluating the phase and amplitude of the optical field is in contrast to the more rigorous wave optics or electromagnetic theory approach, which involves solving coupled partial differential equations for the complex electromagnetic fields where the phase and amplitudes are interdependent.

Rays generally characterize the direction of the flow or propagation of radiant energy, except near foci or the edge of a shadow where interference and diffraction takes place. Thus, a ray is a mathematical construct rather than a physical entity. Snell's law relates the direction of incident and refracted rays at an interface between media of different indices of refraction, which can be written in vector form:

$$n_1(\mathbf{a}\times\mathbf{N}) = n_2(\mathbf{A}\times\mathbf{N}) \tag{6.37}$$

where:

a and **A** are the unit vectors along the incident and refracted rays, respectively

N is a unit vector along the normal to the interface surface with the general orientation of the incident ray

(n_1,n_2) are the indices of refraction of the incident and refracting media

The vector nature of this equation ensures coplanarity of rays and interface surface normal, as required by electromagnetic theory. For ray tracing, it is convenient to vector multiply Equation 6.37 by **N** and simplify the resulting triple vector product into the form

$$n_2\mathbf{A} = n_1\mathbf{a} + (n_2 \cos\beta - n_1 \cos\alpha)\mathbf{N} \tag{6.38}$$

where:

$$\cos\beta = \mathbf{A}\cdot\mathbf{N} \quad \text{and} \quad \cos\alpha = \mathbf{a}\cdot\mathbf{N} \tag{6.39}$$

and (α,β) are the angles of incidence and refraction. When mirrors are involved, the refraction ray equations can be used for reflection by setting $n_2 = -n_1$ and using the optics sign convention. Explicitly, a unit vector **A** along a reflected ray is given by the following equation:

$$\mathbf{A} = \mathbf{a} - 2\mathbf{N}(\mathbf{a}\cdot\mathbf{N}) \tag{6.40}$$

where:

N is a unit normal vector at the point of reflections

a is unit vector along the incident ray

Rays may also be defined as lines normal to the geometrical wavefront. Wave propagation is commonly described by wavefronts. A wavefront is a surface of constant phase of the wave or OPL from the source or reference surface. In electromagnetic theory, the direction of radiant energy propagation is given by the Poynting vector or cross product of the electric and magnetic fields.

Each ray generally follows the path of shortest time through the optical system according to Fermat's principle which states that a ray from points P to Q is the curve C connecting these two points such that the integral

$$\text{OPL(C)} = \int_C n(x,y,z)\mathrm{d}s \tag{6.41}$$

is an extremum (maximum, minimum, or stationary). The quantity $n(x,y,z)$ is the index of refraction of the medium and ds is the infinitesimal arc length of the curve. For a homogeneous medium, the OPL between P and Q is the geometrical path length between the two points times the index of refraction of the medium. In general, the OPL divided by the speed of light in free space, c, gives the time for light to travel from point P to Q along the ray path C. The ray path C can be determined using the calculus of variations [50]. It can be shown [45] that when the index of refraction, $n(\mathbf{r})$, is a smooth function, the ray path C satisfies the differential equation:

$$\frac{\mathrm{d}}{\mathrm{d}s}\left[n(\mathbf{r})\frac{\mathrm{d}\mathbf{r}}{\mathrm{d}s}\right] = \nabla n(\mathbf{r}) \tag{6.42}$$

where:
$\mathbf{r}$ is the position vector of any point on the ray

Equation 6.42 is known as the ray equation and is difficult to solve in many cases. For homogeneous medium (n = constant), the ray path is represented by a straight line:

$$\mathbf{r}(s) = \mathbf{A}s + \mathbf{b} \tag{6.43}$$

where:
$\mathbf{A}$ and $\mathbf{b}$ are constant vectors
s is the ray OPL

Constant index of refraction materials are used in many optical systems. When the index of refraction is a function of position, such as the axial distance, z, the ray paths are curved. Section 6.5.3 will illustrate how laser profile shaping systems are designed using materials with a gradient index of refraction, $n(z)$. Now, the ray trace equations, the intensity law which expresses conservation of energy along ray bundle, and the constant OPL condition are used to design optical systems for shaping laser beam profiles.

6.2.2 Irradiance Profiles

For this work, we consider collimated, axially symmetric beam propagating along the optical axis, which we consider to be the z-axis. We label the input and output irradiance distributions by $I_{in}(r)$ and $I_{out}(r)$. In Section 6.2.2.1, we describe general properties of flattened irradiance profiles and how flattened profiles are used in practical applications.

6.2.2.1 Flattened Irradiance Profiles

To avoid diffraction effects in regions of sharp changes in the irradiance profile, several authors have reported using analytic functions for nearly flat or flattened beam profiles, including the SG [63], the flattened Gaussian (FG) [64], the FD [21], the super-Lorentzian (SL) [65], and the flattened Lorentzian (FL) [66] distributions. There have been a number of studies on the propagation characteristics and the far-field intensity pattern of these distributions [65,67–69].

A general flattened output irradiance profile can be written as a product of its normalization constant, $I_0(\xi,R_0)$, which can be determined by requiring the profile be normalized according to Equation 6.36, and its functional dependence $f(\xi,R/R_0)$, which depends on the shape parameter ξ and the ratio of radial coordinate, R, to the beam width, R_0, of each profile:

$$I_{\text{out}}\left(\xi,\frac{R}{R_0}\right)=I_0(\xi,R_0)\times f\left(\xi,\frac{R}{R_0}\right) \tag{6.44}$$

The width parameter, R_0, defines a length scale over which the profile decreases to some significant value for the specific profile, such as half or e^{-2} of its axial value. The shape parameter, ξ, specifies the shape of the profile function, such as the power of the radial coordinate of a SG profile.

Using the irradiance functions and normalization constants given in Table 6.1, we have evaluated the irradiance functions for these flattened profiles, which are presented in Figures 6.3 through 6.5. For example, Figure 6.3a presents the plots of the

TABLE 6.1
Irradiance Profile Normalization Constants, $I_0(\sigma,R_0)$, and Functions, $f\left(\sigma,\frac{R}{R_0}\right)$, for SG, FG, FD, SL, and FL Distributions

Profile	$I_0(\sigma,R_0)$	$f\left(\sigma,\frac{R}{R_0}\right)$
SG	$\dfrac{4^{1/p}p}{2\pi R_{\text{SG}}^2\Gamma\left(\dfrac{2}{p}\right)}$	$\exp\left[-2\left(\dfrac{R}{R_{\text{SG}}}\right)^p\right]$
FG	$\dfrac{2(N+1)}{\pi R_{\text{FG}}^2}\sum_{m,n=0}^{N}\left[\dfrac{m!n!2^{(m+n)}}{(m+n)!}\right]$	$\exp\left[-2(N+1)\left(\dfrac{R}{R_{\text{FG}}}\right)^2\right]\sum_{m,n=0}^{N}\dfrac{1}{n!m!}\left[(N+1)\left(\dfrac{R}{R_{\text{FG}}}\right)^2\right]^{n+m}$
FD	$\dfrac{3\beta^2}{\pi R_{\text{FD}}^2}\{3\beta^2+6\text{dilog}[1+\exp(-\beta)]+\pi^2\}^{-1}$	$\left\{1+\exp\left[\beta\left(\dfrac{R}{R_{\text{FD}}}-1\right)\right]\right\}^{-1}$
SL	$q\sin\left(\dfrac{2\pi}{q}\right)(2\pi^2R_{\text{SL}}^2)^{-1}$	$\left[1+\left(\dfrac{R}{R_{\text{SL}}}\right)^q\right]^{-1}$
FL	$(\pi R_{\text{FL}}^2)^{-1}$	$\left[1+\left(\dfrac{R}{R_{\text{FL}}}\right)^q\right]^{-\left(1+\frac{2}{q}\right)}$

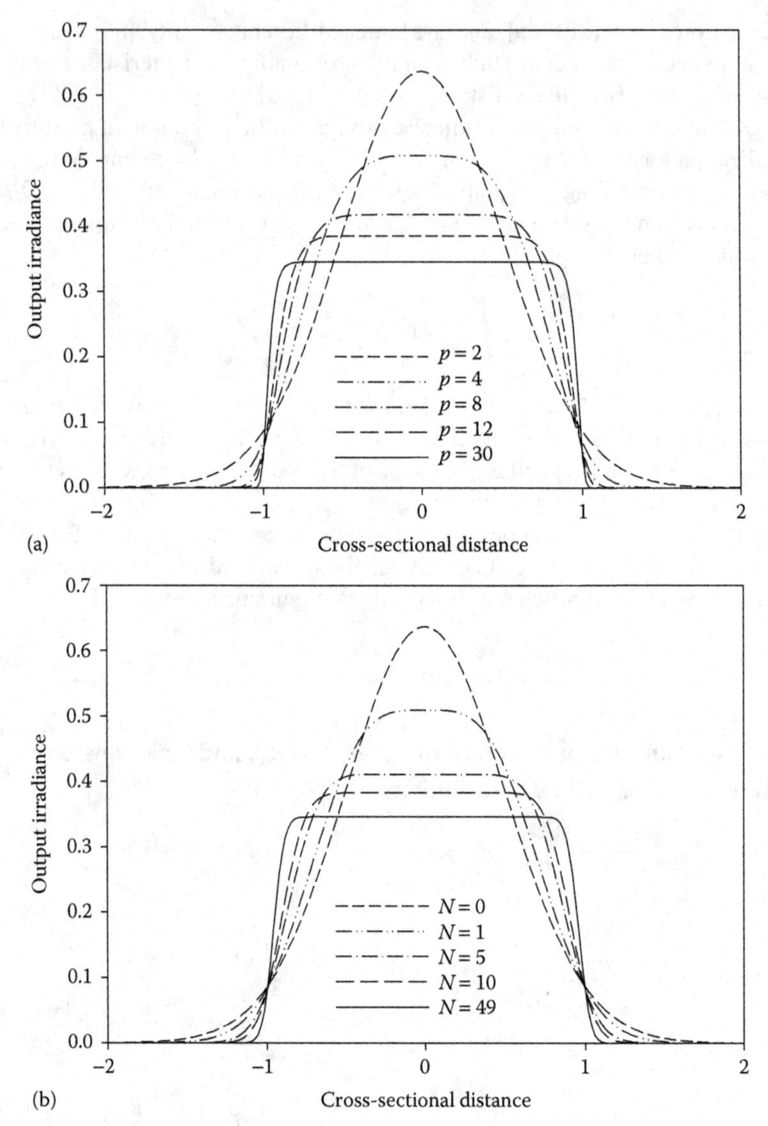

FIGURE 6.3 (a) SG and (b) FG output irradiance for shape parameters p and N. All profiles are normalized such that the total energy contained within each irradiance distribution over all space is equal to unity, and the beam width parameters are equal to unity. The profiles for $p = 2$ and $N = 0$ reduce to the Gaussian profile with axial irradiance equal to 0.6366. The axial irradiance of the SG profiles for $p = 4$, 8, 12, and 30 are equal to 0.5079, 0.4126, 0.3851, and 0.3452, respectively, and the FG profiles with the axial irradiance for $N = 1$, 5, 10, and 49 are equal to 0.5093, 0.4110, 0.3827, and 0.3458, respectively. (Reproduced from Shealy, D. L. and Hoffnagle, J. A. Beam shaping profiles and propagation. In *Laser Beam Shaping VI*, Dickey, F. M. and Shealy, D. L., eds., vol. 5876, 58760D-1, 2005. With permission.)

irradiance of the SG profile for selected shaper parameters p, where the corresponding axial irradiance values are given within the caption of this figure. Figure 6.3b displays the irradiance associated with the FG profile for selected shape parameters, which allows for some comparison of SG and FG profiles for a range of respective shape parameters. Figure 6.4a displays the output irradiance associated with the FD

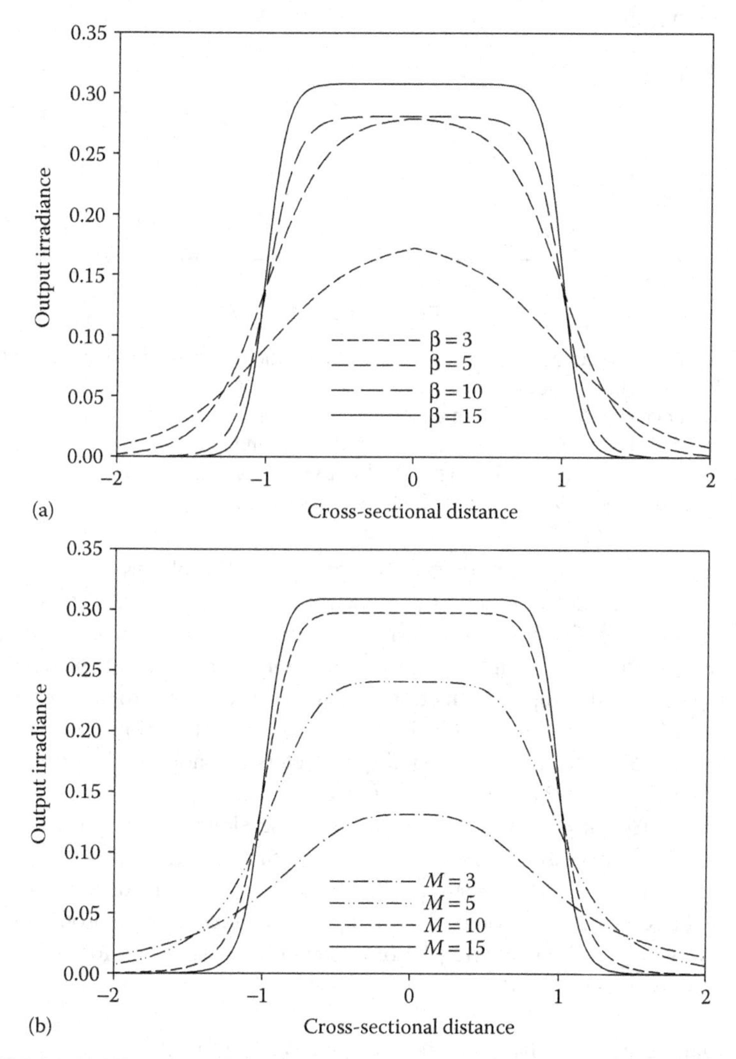

FIGURE 6.4 (a) FD and (b) SL output irradiance for shape parameters β and M. All profiles are normalized such that the total energy contained within each irradiance distribution over all space is equal to unity, as well as the beam width parameters are also equal to unity. The cross-sectional distance is expressed in the normalized units of the beam width. (Reproduced from Shealy, D. L. and Hoffnagle, J. A. Beam shaping profiles and propagation. In *Laser Beam Shaping VI*, Dickey, F. M. and Shealy, D. L., eds., vol. 5876, 58760D-1, 2005. With permission.)

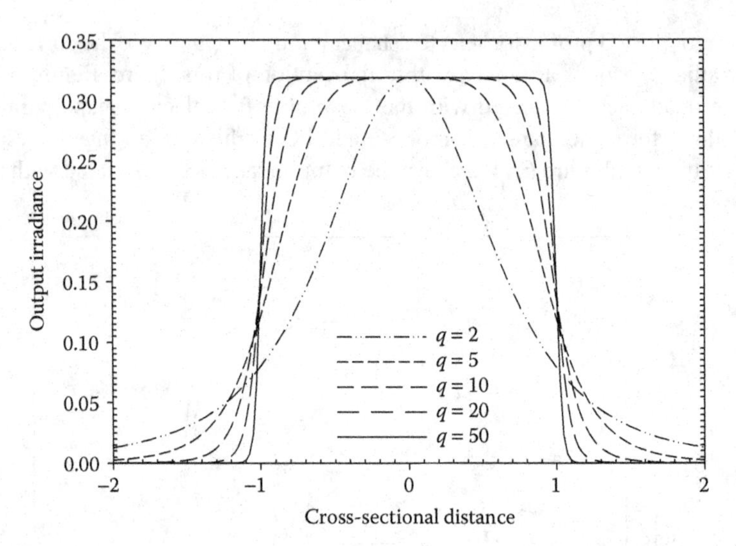

FIGURE 6.5 Cross-sectional plot of the irradiance for the FL profile with the shape parameters $q = 2, 5, 10, 20, 50$ and the width parameter $R_{FL} = 1$. All profiles are normalized such that the total energy contained within each irradiance distribution over all space is equal to unity. (Reproduced from Shealy, D. L., Hoffnagle, J. A., and Brenner, K.-H. Analytic beam shaping for flattened output irradiance profile. In *Laser Beam Shaping VII*, Dickey, F. M. and Shealy, D. L., eds., vol. 6290, 629006-1, 2006. With permission.)

function for different shape parameters β. Qualitatively, it follows from Figure 6.4a that as β increases the FD function approaches a top-hat function, except that the irradiance varies continuously from zero to its uniform axial value as one moves across the aperture. In the limit as β approaches infinity, one can prove that FD profile approaches the top-hat function. Figure 6.4b displays the SL profile that approaches a flat-top profile for $M > 15$. Figure 6.5 presents a plot of the irradiance of the FL profile, where the beam width parameters are not equal to the radius at which the profile is equal to half of its axial value.

Shealy and Hoffnagle [70] have noticed that the slope of the SG, FD, and SL profiles at the half-height point is linearly proportional to their corresponding shape parameters, which raises the question if any of these flattened irradiance profiles can effectively be represented by other flattened profiles under suitable profile matching conditions. Shealy and Hoffnagle [71] have shown that when the following profile matching conditions are used:

- Equal radius when the irradiance is equal to half of its axial value
- Equal slope of the irradiance of the radius of the half-height point

then one can determine the shape and width parameters of all these flattened profiles, since the shape and width parameters of one of the flattened profiles are assumed to be known. Further, Shealy and Hoffnagle have evaluated the implicit functional dependence of the shape and width parameters of the SG, FD, and SL profiles on the FG profile shape N as a result of the two profile matching conditions

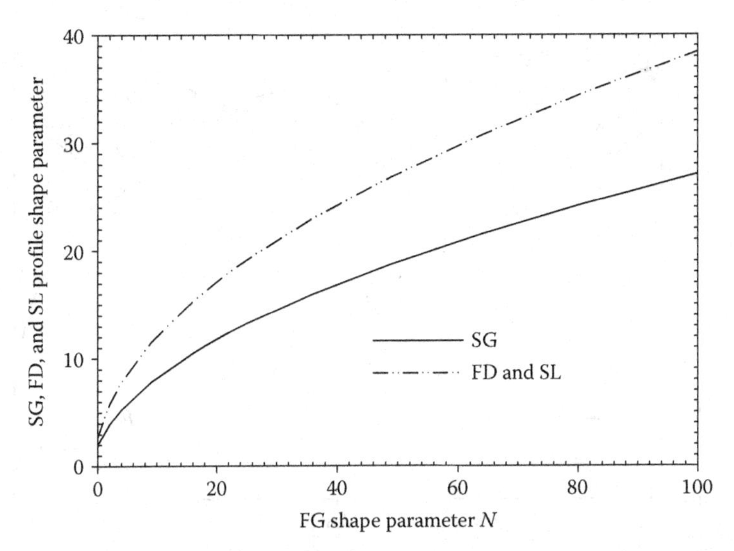

FIGURE 6.6 Comparison between shape parameters of the SG, FD, and SL shape parameters as a function of the FG shape parameters for profiles with matched parameters. The FD shape parameter β and the SL shape parameter q have the same functional dependence on N for the scale of this figure and have been represented by one curve. (Reproduced from Shealy, D. L. and Hoffnagle, J. A. *Appl. Opt.*, 45, 5118–5131, 2006. With permission.)

given in this section. Figure 6.6 presents the dependence of the shape parameters p[SG], β[FD], and q[SL] on N[FD] and allows for comparison of beam shapers that were designed to produce different flattened irradiance profiles. Therefore, explicit beam shaping results produced by flattened profiles in this chapter will be expressed for historical reasons in terms of either an FD profile or an FL profile, where one can use results from Reference [71] to convert results produced for an FD to FL profile or vice versa.

6.2.2.2 Practical Applications

For practical applications, it is important to understand the relationship between uniformity and efficiency for the output beam profile. Consider a beam that illuminates a circular region of space of radius R_{max}, and define the peak-to-peak uniformity of the beam over this region in terms of the extrema of the irradiance within this region:

$$U_{PP}(R_{max}) = \frac{I_{out}(\min)}{I_{out}(\max)} = \frac{f(R_{max}/R_0)}{f(0)} \tag{6.45}$$

where:

$f(R)$ is the output irradiance function as given in Table 6.1

The fraction of the total beam power within the region $R < R_{max}$ represents a measure of the efficiency of the output beam profile shape, since the total beam power has been normalized to unity, and is given by

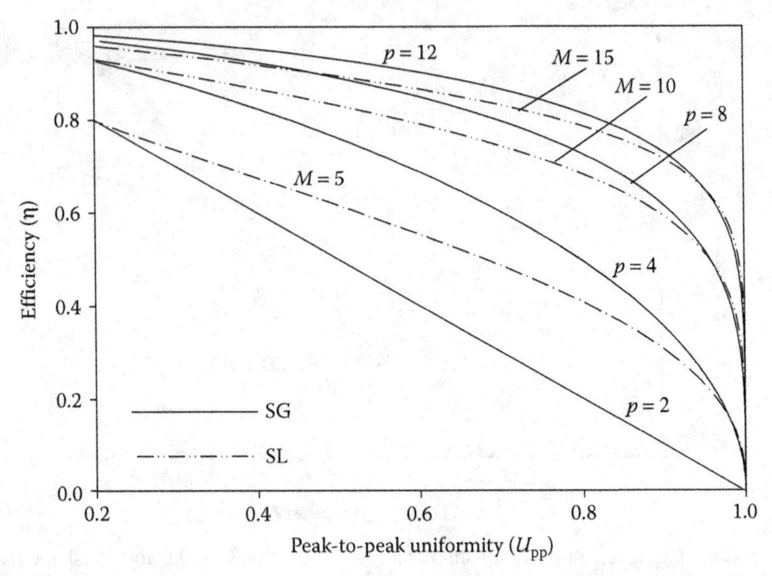

FIGURE 6.7 Relationship between the efficiency and the peak-to-peak uniformity for the SG and the SL profiles for different shape parameters. (Reproduced from Shealy, D. L. and Hoffnagle, J. A. Aspheric optics for laser beam shaping. In *Encyclopedia of Optical Engineering*, Driggers, R., ed., Taylor & Francis, doi:10.1081/E-EOE-120029768, 2006. With permission.)

$$\eta(R_{max}) = 2\pi \int_0^{R_{max}} I_{out}(R) R \mathrm{d}R \tag{6.46}$$

The beam efficiency and uniformity vary from 0 to 1. Equations 6.45 and 6.46 implicitly define a relationship between uniformity and efficiency, which only depends on the beam shape. For a Gaussian beam, $U_{PP}(r_{max}) = \exp(-2r_{max}^2/w_0^2)$ and $\eta(r_{max}) = 1 - \exp(-2r_{max}^2/w_0^2)$. Hence, $\eta = 1 - U_{PP}$ for a Gaussian beam. For other beam shapes there are nonlinear relationships between efficiency and uniformity. A key point of all beam shaping is to balance the trade-off between uniformity of the illumination over some extended region with the efficiency of use of laser beam power. Figure 6.7 illustrates the relationship between efficiency and peak-to-peak uniformity for a family of SL and SG beam profiles [21].

In some applications, it is more useful to consider the rms variation rather than the peak-to-peak variation of the irradiance. The average value of the irradiance leaving the output aperture of radius R_{max} is

$$\langle I_{out}(R_{max})\rangle = \frac{2}{R_{max}^2} \int_0^{R_{max}} I_{out}(R) R \mathrm{d}R \tag{6.47}$$

and its variance is

$$\sigma^2(R_{max}) = \frac{2}{R_{max}^2} \int_0^{R_{max}} \left[I_{out}(r) - \langle I_{out}(R_{max})\rangle\right]^2 R \mathrm{d}R \tag{6.48}$$

The relative rms variation is then defined as

$$V_{\mathrm{rms}}(R_{\max}) = \frac{\sigma(R_{\max})}{\langle I_{\mathrm{out}}(R_{\max})\rangle} \tag{6.49}$$

The implicit relationship between V_{rms} and η depends only on the beam shape in similar manner as between U_{pp} and η.

6.2.3 Propagation

In Section 6.3, we will show how to design optical systems that use the principles of geometric optics to perform an arbitrary transformation of the irradiance profile of a rotationally symmetric laser beam. The transformation of irradiance requires a redistribution of rays; after this has been accomplished, it is possible to include an element that restores the direction of all the rays to that of the incoming beam, thus producing a collimated, shaped beam with a planar wavefront. The purpose of recollimating the beam is to enable it to propagate over some useful distance, but in order to describe the propagation of laser beams it is necessary to go beyond simple ray optics and include the effects of diffraction. While it is possible to generate laser beams, most importantly those with the Gaussian profile,

$$I_{\mathrm{G}}(r) = \frac{2}{\pi w_0^2}\exp\left[-2\left(\frac{r}{w_0}\right)^2\right] \tag{6.50}$$

which retain the same form as they propagate, diffraction generally modifies the shape of a propagating beam. For instance, it is well known that a collimated beam with a "top-hat" profile

$$I_{\mathrm{TH}}(r) = \begin{cases} 1/\pi R_0^2 & \text{if } r \le R_0 \\ 0 & \text{otherwise} \end{cases} \tag{6.51}$$

transforms in the far field to an Airy profile proportional to $[J_1(r)/r]^2$, where $J_1(r)$ is the Bessel function of order 1. For suitable choices of the shape parameter, the flat-top profiles of Section 6.2.2.1 closely resemble the top hat, so it is to be expected that they evolve in the far field to have shapes similar to the Airy profile. This change in shape does not prevent flat-top or other shaped beams from being useful, since their shapes may change only slightly over significant distances in the near field. As part of the design of laser beam shaping optics, it is important to understand the effect that diffraction has on the shape of the output beam and the range of propagation over which the beam can be considered to retain its shape, to within some required tolerance. Propagation effects for the family of FG beams have been studied in detail by Gori and coworkers [64,68,72,73], who introduced quantitative measures of shape-invariance error and shape-invariance range. These works made use of the fact that FG beams are a sum of ordinary Gaussian beams, which evolve in a simple way as they propagate. Here we follow the approach of Campbell and DeShazer [74], which uses the Huygens–Fresnel principle [42] and the Fresnel approximation.

To calculate diffraction effects in the context of the Huygens–Fresnel principle, we assume that the beam can be described by a scalar optical field $u(\mathbf{r})$ which obeys the Helmholtz equation and has the property that the irradiance is proportional to $|u(\mathbf{r})|^2$.

Assuming that at the output aperture of the beam shaper the beam is collimated and rotationally symmetric, with an optical field $u_0(r)$, the Kirchhoff diffraction integral with the Fresnel approximation yields the following expression for $u(r,D)$, the optical field as a function of r, after propagating a distance D from the output aperture

$$u(r,D) = e^{i\varphi} \frac{k}{D} \int_0^{R_{max}} u_0(\rho) J_0\left(\frac{k\rho r}{D}\right) \exp\left(\frac{ik\rho^2}{2D}\right) \rho \, d\rho \tag{6.52}$$

where:

$k = 2\pi/\lambda$ is the wave number of the light

R_{max} is the radius of the output aperture

$J_0(x)$ is the Bessel function of order 0

φ is a phase term that includes the period λ oscillation of the phase of the optical wave and the wavefront curvature due to diffraction, but which does not affect the irradiance calculation

The one-dimensional diffraction integral that describes propagation of a rotationally symmetric beam can easily be evaluated numerically for output beam profile [71]. When considering the propagation of flat-top profiles, it is convenient to combine the three important length scales, λ, D, and R_0 (where R_0 designates any of the parameters used for the radius of the flat-top beam) into a single dimensionless parameter called the Fresnel number

$$N_F = \frac{R_0^2}{\lambda D} \tag{6.53}$$

which determines the change in shape of the propagating beam independently of the individual dimensional parameters. This can be seen by introducing the scaled transverse variables

$$\alpha = \frac{R_{max}}{R_0} \tag{6.54}$$

$$\xi = \frac{r}{R_0} \tag{6.55}$$

$$\tau = \frac{\rho}{R_0} \tag{6.56}$$

and the functions v and v_0 for the optical field in terms of the scaled coordinates

$$v(\xi, N_F) = u(r,D) \tag{6.57}$$

$$v_0(\tau) = u_0(\rho) \tag{6.58}$$

Then Equation 6.52 transforms to

$$v(\xi, N_F) = 2\pi N_F e^{i\varphi} \int_0^{\alpha} v_0(\tau) J_0\left(2\pi N_F \xi \tau\right) \exp(i\pi N_F \tau^2) \tau \, d\tau \tag{6.59}$$

which relates the shape of the beam after propagation to the shape of the beam in the output aperture and the Fresnel number.

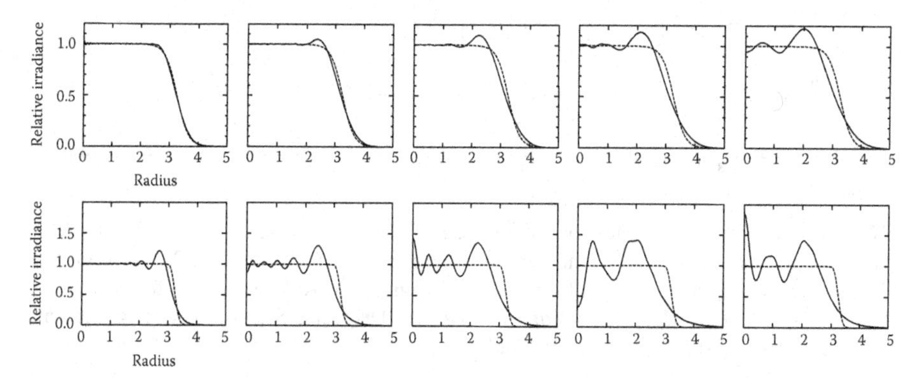

FIGURE 6.8 Propagation of FD beams with $\beta = 16.25$ (top row) and $\beta = 50$ (bottom row). The five plots in each row show the shape of the beam for $N_F = 31$, 15, 10, 8, and 6, from left to right. In each plot, the broken curve represents the original irradiance distribution and the solid curve is the irradiance after propagation. (Reprinted from Shealy, D. L. and Hoffnagle, J. A. Aspheric optics for laser beam shaping. In *Encyclopedia of Optical Engineering*, Driggers, R., ed., Taylor & Francis, doi:10.1081/E-EOE-120029768, 2006. With permission.)

As an illustration of how flat-top beams change shape as they propagate, Figure 6.8 follows the evolution of two beams with FD profiles differing in the beam shape parameter, β. For given wavelength and beam diameter, N_F is inversely proportional to propagation distance. From Figure 6.8, we conclude that beams with a steeper falloff at the edges propagate for much shorter distances before being seriously distorted by diffraction than do beams with more rounded edges. Other authors have come to similar conclusions after studying the effects of diffraction on SG [67] and FG [68] beams. Shealy and Hoffnagle [71] compared the propagation of beams with different functional forms (e.g., SG vs. FG) to see whether the choice of the profile function has any important influence on the propagation beams that have the same size and edge steepness. To make a quantitative comparison between beams with profiles of different functional forms, two beams were considered "comparable" if the normalized irradiance profiles had the same radius at half-maximum and the same derivative of irradiance with respect to radius, evaluated at the half-maximum point. The conclusion, not surprisingly, was that for beams with the same size and edge steepness, the exact functional form of the irradiance profile has very little impact on the propagation of the beam.

The propagation of shaped beams depends not only on the choice of beam shape but also on the aperture of the beam shaping optics, as can be seen from the appearance of the scaled aperture α in Equation 6.59. The functions in Section 6.2.2.1 are defined for all positive radii, but in designing real optics a finite aperture must be chosen, and there is an incentive for the designer to keep the aperture as small as possible, to limit the cost of figuring and testing aspheric surfaces. If the aperture is too small, however, the truncation of the beam can severely limit the propagation range of the beam. This is shown in Figure 6.9, which is obtained by evaluating Equation 6.59 for an FD profile with $\beta = 16.25$, $N_F = 31$, and a range of α from 1.08 to 1.54.

Finite-aperture effects can be surprisingly large. For instance, the aperture corresponding to Figure 6.9b transmits more than 99% of the total power in the shaped beam, yet it seriously degrades the uniformity of the beam after propagation, as can

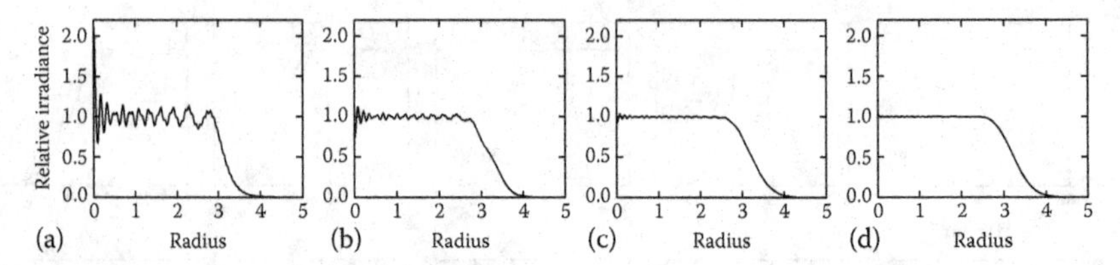

FIGURE 6.9 Effect of truncation on FD beams with $\beta = 16.25$ after propagation with a Fresnel number of 31. The normalized beam aperture is (a) 1.08, (b) 1.23, (c) 1.38, and (d) 1.54. (Reprinted from Shealy, D. L. and Hoffnagle, J. A. Aspheric optics for laser beam shaping. In *Encyclopedia of Optical Engineering*, Driggers, R., ed., Taylor & Francis, doi:10.1081/E-EOE-120029768, 2006. With permission.)

be seen by comparing Figure 6.9b with Figure 6.8a, which shows the profile of the same output beam at the same Fresnel number, but without truncation.

The diffraction integral (Equation 6.52 or 6.59) shows how the beam shape changes as the beam propagates, which is very useful when choosing the appropriate output profile for a beam shaper, but it does not yield a simple figure-of-merit for the propagation of the shaped beam. There is no single number that summarizes the propagation properties of a beam, but one quantity that has found widespread use in this context is the M^2 factor introduced by Siegman [75]. This factor describes how much more rapidly the diameter of an arbitrary beam diverges than a Gaussian beam of the same diameter, and it has the advantage of being relatively easy to evaluate, both in theory and by irradiance measurements of real beams. Parent et al. [67] found an analytic expression for the M^2 factor of SG beams, and Bagini et al. [68] did the same for FG beams. Shealy and Hoffnagle [71] added analytic expressions for M^2 of FD and SL beams and compared the propagation properties of flat-top beams with various profiles. The results are summarized in Figure 6.10, which shows how M^2

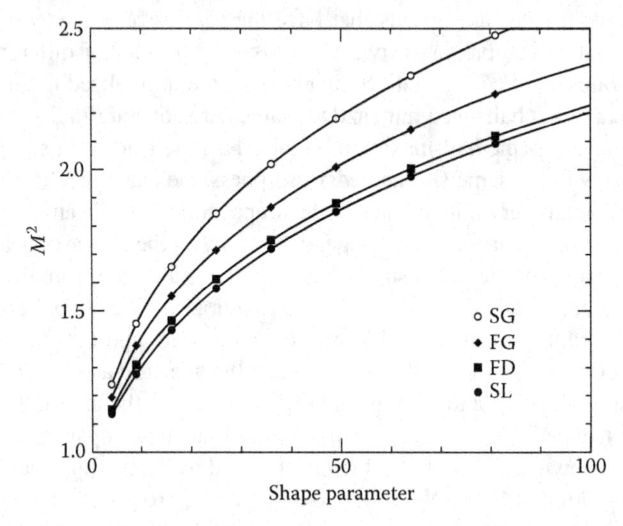

FIGURE 6.10 M^2 as a function of beam shape parameter for the four families of flat-top profiles introduced in Section 6.2.2.1. (Reproduced from Shealy, D. L. and Hoffnagle, J. A. *Appl. Opt.*, 45, 5118–5131, 2006. With permission.)

varies as a function of the shape parameter for the profiles in Section 6.2.2.1. As the shape parameter increases, indicating a steeper roll off at the edge of the beam, M^2 increases. In the limit of infinite shape parameter the profiles considered here all approach the top-hat profile, and M^2 diverges. More details and a summary of all the analytic expressions can be found in Reference [71].

6.3 DESIGN

Cornwell [26] notes that the first element of a beam shaper creates sufficient aberrations in the wavefront to redistribute the irradiance of the beam after propagation by a specified distance and that the second element of the beam shaper restores the wavefront to its original shape. By imposing the geometrical optics intensity law on a bundle of rays passing through the beam shaper, the process of transforming an input Gaussian beam profile into a more uniform output irradiance profile can be described by a nonlinear, one-to-one mapping of the input to the output aperture coordinates. The constant OPL condition is used to ensure that the input and output wavefronts have the same shape. Newman and Oliker [76] have used geometrical optics to understand how an input irradiance distribution is redistributed after reflection or refraction from one optical surface. Later, Oliker [77,78] developed a method for the design of a free-form two-mirror or two-lens optical system without any symmetry to transform an input laser beam into an output beam with a prescribed irradiance profile.

Ries and coworkers [79–82] have developed numerical solutions to the illumination design or tailoring of a reflecting or refracting surface which will transform a specified light source into a desired irradiance distribution on a specified target surface. Tailoring of a two-dimensional system leads to solving an ordinary differential equation for optical surface shape, whereas tailoring a three-dimensional illumination system requires one to solve a nonlinear partial differential equation for the shape of the optical surface. However, Ries [79] asserts that solving either an ordinary or partial differential equation to determine the optical surface shape is the superior method of optical design of illumination system when compared to optimization methods.

To achieve the overall beam shaping, two optical elements are required. These optical elements may be reflective, refractive, or diffractive. A goal of the optical design of laser beam shapers is to define the optical components sufficiently so that the beam shaping system can be analyzed, fabricated, and tested. For reflective and refractive beam shaping, the surface sag, spacing, and index must be determined for all media. Initially, we consider the two plano-aspheric laser beam shaping systems in the Galilean or Keplerian configuration shown in Figure 6.11a and b.

In Section 6.3.1, we describe how to use the geometrical optics law of intensity to obtain the ray mapping function, which describes how an input Gaussian beam is transformed by the first element into a difference irradiance distribution over the second element of the optical system, which may be a detector or another component of a beam shaping system. There are two cases to consider: (1) When one seeks only to illuminate the detector surface with a prescribed irradiance distribution, then the ray mapping function can be used with the slope of the reflected or refracted ray from the first element to obtain a first-order, ordinary differential equation for the

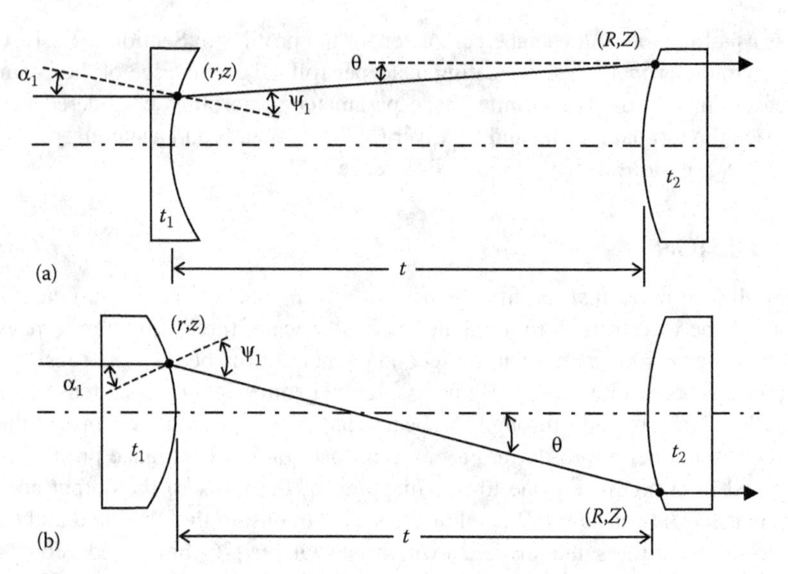

FIGURE 6.11 Ray plots of (a) Galilean and (b) Keplerian configurations of a two plano-aspheric lens laser beam shaper.

sag of the first surface to the desired illumination of the detector. See Equations 6.66 and 6.67 for an example of designing a refractive illumination system. (2) When the second surface is an optical element that is used to create a desired wavefront shape with a specific irradiance distribution, then the sag of the second element is determined by the condition used to fix the shape of the output wavefront. In Section 6.3.2, we describe how the constant OPL is used to determine the sag of the second element of the beam shaper.

When the intensity law and constant OPL condition are satisfied, the beam shaper will transform an input Gaussian beam into a output plane wave with the prescribed irradiance distribution. In Section 6.3.3, a general formalism known in the literature as nonprojective transforms is described.

6.3.1 Geometrical Optics Law of Intensity

The geometrical optics law of intensity requires the intensity times a cross-sectional area of bundle of rays be constant along the beam, which is given for a rotationally symmetric beam by

$$\int_0^r 2\pi I_G\left(\frac{r'}{w_0}\right)r'\mathrm{d}r' = \int_0^R 2\pi I_{FL}\left(\frac{R'}{R_{FL}}\right)R'\mathrm{d}R' \tag{6.60}$$

For a particular Gaussian to FL profile transformation, Shealy et al. [66] integrated Equation 6.60 to obtain

$$1-\exp\left[-2\left(\frac{r}{w_0}\right)^2\right] = \left[1+\left(\frac{R}{R_{FL}}\right)^{-q}\right]^{-(2/q)} \tag{6.61}$$

and then solved the above equation for $r(R)$ to obtain

$$h^{-1}(R) \equiv \epsilon r = \epsilon w_0 \sqrt{-\frac{1}{2}\ln\left\{1-\left[1+\left(\frac{R}{R_{FL}}\right)^{-q}\right]^{-(2/q)}\right\}} \tag{6.62}$$

where:

$$\epsilon = \begin{cases} +1 & \text{Galilean configuration} \\ -1 & \text{Keplerian configuration} \end{cases} \tag{6.63}$$

Now, solve for $R(r)$ by taking the square root of Equation 6.61, and then, raise both sides of the equation to the qth power and solve for $(R/R_{FL})^q$. Next, the ray mapping function, $h(r)$, of the Gaussian to the FL profile transform follows after taking the qth root:

$$h(r) \equiv \epsilon R = \frac{\epsilon R_{FL}\sqrt{1-\exp\left[-2\left(\frac{r}{w_0}\right)^2\right]}}{\sqrt[q]{1-\left\{1-\exp\left[-2\left(\frac{r}{w_0}\right)^2\right]\right\}^{\frac{q}{2}}}} \tag{6.64}$$

Figure 6.12a presents a plot of the ray mapping function of a Gaussian to FL profile, normalized by R_{FL}, as a function of the input aperture radius, normalized by w_0, for several shape parameters q of the FL profile. Figure 6.12b gives a plot of the fraction of total input beam energy within the input aperture of radius ranging from 0 to $2w_0$. The ray mapping function and its inverse contain information about how the input Gaussian beam is changed by the first lens into a more uniform output beam.

As an example of using Figure 6.12a and b to determine the first-order aperture radii of a 1-to-1K beam shaper which transforms more than 99% of the incident energy, assume that the input Gaussian beam waist is $w_0 = 2.366$ mm and that the output FL beam shape parameter is required to be $q = 15$. From Figure 6.12b, we conclude that more than 99% of the input Gaussian beam passes into the beam shaper when the aperture radius of the first lens is equal to or greater than $1.5w_0$ or $r_{max} = 3.546$ mm for this design. From Figure 6.12a, we estimate that $(R/R_{FL}) = 1.18$ for $q = 15$ and input aperture radius of $1.5w_0$. For this 1-to-1K beam shaper, we require $r_{max} = R_{max}$. Then, we can solve for the FL beam width parameter to be $R_{FL} = R_{max}/1.18 = 3.0$ mm for this set of parameters. To ensure higher energy transform efficiency, one can increase the input aperture to $2w_0 = 4.732$ mm, for example. This means in this case that both r_{max} and R_{max} are equal to 4.732 mm, but from Figure 6.12a, we estimate that $(R/R_{FL}) = 1.48$ in this case, and the FL output beam width parameter is 3.2 mm for a 1-to-1K configuration. A detailed analysis of slope and curvatures of aspheric lenses must also be done before finalizing the

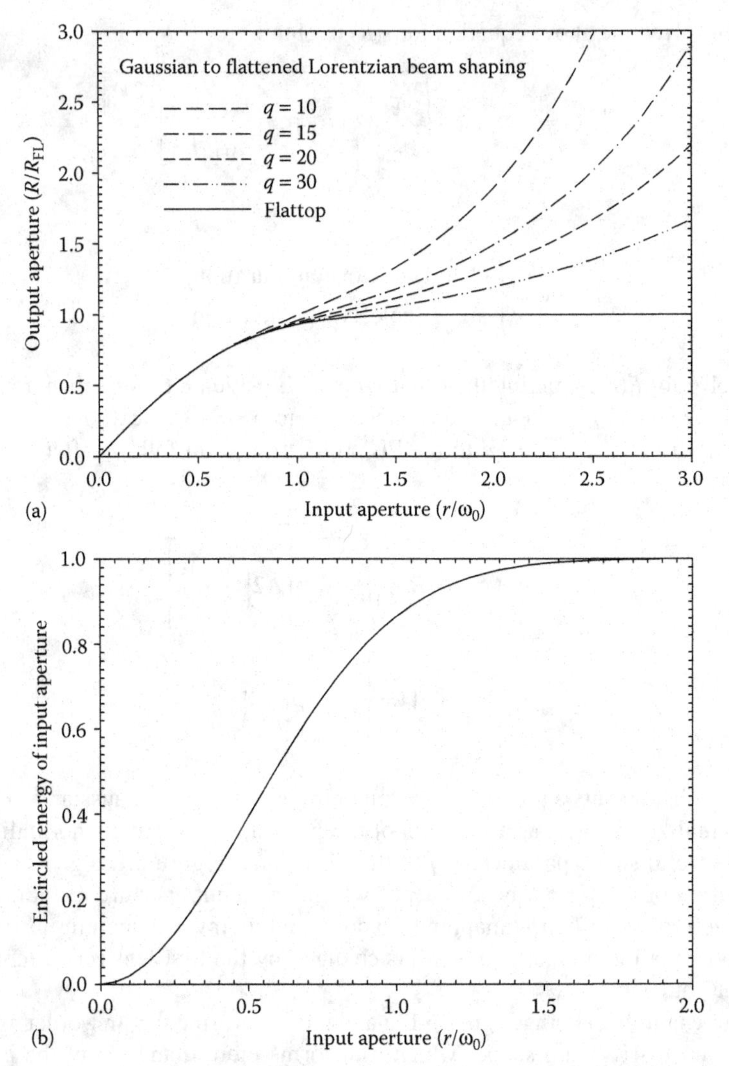

FIGURE 6.12 (a) Plot of the ray mapping function as shown in Equation 6.64, normalized by R_{FL}, as a function of the input aperture radius, normalized by the input beam waist ω_0, for the output FL profiles with shape parameters of q = 10, 15, 20, 30 and for the flat-top profile. (b) Plot of the fraction of input beam energy that is contained within an input aperture radius r. (Reproduced from Shealy, D. L. and Hoffnagle, J. A. Wavefront and caustic surfaces of refractive laser beam shaper. In *Novel Optical Systems Design and Optimization X*, Koshel, R. J. and Gregory, G. G., eds., vol. 6668, 666805-1, 2007. With permission.)

input and output aperture sizes, as well as a detailed assembly and fabrication analysis.

As a second example, Hoffnagle [83] noted that the geometrical optics law of intensity can be used to design a plano-aspheric lens that can be used to uniformly illuminate a plane located a distance Z from this plano-aspheric lens when illuminated with a Gaussian beam. The ray mapping function for this application is given by

$$h(r) = \pm R_0 \sqrt{1 - \exp\left(-\frac{2r^2}{w_0}\right)} \quad (6.65)$$

where:

R_0 is the radius of the top-hat profile on the illumination plane defined by Z = constant

± sign distinguishes between two classes of solutions to this beam shaping problem as described in more detail by Hoffnagle

Further, using elementary trigonometry, Hoffnagle has shown that the slope of aspheric surface must satisfy the following first-order, ordinary differential equation:

$$\frac{dz}{dr} = \frac{h(r) - r}{n\ell - z(r) - Z} \quad (6.66)$$

where:

n is the index of refraction of the lens

ℓ is given by

$$\ell = \sqrt{[h(r) - r]^2 + [Z - z(r)]^2} \quad (6.67)$$

Using the boundary condition of $z(0) = 0$, one can numerically integrate Equation 6.66 for $z(r)$ to obtain an accurate description of the aspheric lens surface.

6.3.2 Constant OPL Condition

Kreuzer [11] first expressed the constant OPL condition of a two plano-aspheric beam shaper in the form given below by Equation 6.75, which is convenient for combining with Snell's law to obtain a differential equation for the sag of the aspheric surfaces. Shealy and Chao [84] incorporated the constant OPL condition into the design of beam shapers by using a generalized, vector-based ray tracing method, which has been applied to the design of systems with circular, elliptical, or rectangular symmetry and has also been shown to be equivalent to the approach used by Kreuzer. For the rotationally symmetric systems considered in this chapter, the trigonometric approach of Kreuzer is presented. Following the work of Kreuzer [2,11], Shealy and Hoffnagle [85] evaluated the sag of the first aspheric lens, $z(r)$, of a refractive beam shaper as a function of the input ray height, r, and expressed in terms of the ray mapping function, $h(r)$, lens spacing, t, and index of refraction, n, of each lens as described in this section. The constant OPL condition is imposed for all rays passing through a laser beam shaper to obtain a functional relationship between the sag functions of each aspheric and the direction cosines of the refracted ray from the first aspheric. The OPL of a ray passing along the optical axis and through the beam shaper shown in Figure 6.11a is given by

$$(\text{OPL})_0 = nt_1 + t + nt_2 \tag{6.68}$$

and the OPL of a ray at a radial height r is given by

$$(\text{OPL})_r = n(t_1 + z) + \ell + n(t_2 - Z) \tag{6.69}$$

where:
ℓ is the distance along the ray from (r,z) to (R,Z)

The sag functions $z(r)$ and $Z(R)$ are defined relative to a local coordinate system with their origin located at the vertex of the surface with the optical axis. The constant OPL condition for this system can be written as

$$\ell = n(t + Z - z) + t(1 - n) \tag{6.70}$$

where nt was added and subtracted to the right-hand side of Equation 6.70 for convenience in a later simplification. The constant OPL condition (Equation 6.70) is expressed in terms of the unknown sag functions $z(r)$ and $Z(R)$, whereas the ray mapping function (Equation 6.64) is an explicit relationship between r and R. To determine the sag functions, it is necessary to introduce Snell's law into the derivation of a physically meaningful expression for the slope of the aspheric surfaces. This can be achieved by transforming Equation 6.70 into an equation relating t, n, R, r, and the angle θ which a ray from (r,z) to (R,Z) makes with the optical axis. Referring to Figure 6.11a and the right triangle formed between the points (r,z) and (R,Z) with ℓ as the hypotenuse that forms an angle θ with the horizontal side that is parallel to the optical axis, the trigonometric relations hold

$$\ell \sin\theta = R - r \tag{6.71}$$

$$\ell \cos\theta = t + Z - z \tag{6.72}$$

Combining Equations 6.70 and 6.72 gives

$$\ell(1 - n\cos\theta) = t(1 - n) \tag{6.73}$$

which can further be simplified by eliminating ℓ by dividing Equation 6.73 by 6.71 to obtain

$$\frac{t(1-n)}{(R-r)} = \frac{1 - n\cos\theta}{\sin\theta} \tag{6.74}$$

Squaring the above equation and adding $(n^2 - 1)$ to both sides leads to the following simplification for the constant OPL condition:

$$\left[\frac{t(1-n)}{(R-r)}\right]^2 + (n^2 - 1) = \left(\frac{\cos\theta - n}{\sin\theta}\right)^2 \tag{6.75}$$

6.3.2.1 Sag of First Aspheric Surface

In this section, an equation for the sag of the first aspheric surface will be obtained by combining the ray mapping function and the constant OPL condition with results obtained by applying Snell's law for refraction of rays at the aspheric surfaces. Specifically, Snell's law at the first aspheric surface is

$$n\sin\alpha_1 = \sin\psi_1 = \sin(\alpha_1 + \theta) = \sin\alpha_1 \cos\theta + \cos\alpha_1 \sin\theta \tag{6.76}$$

where:

α_1 is the angle of incidence
ψ_1 is the angle of refraction
$\psi_1 = \alpha_1 + \theta$ from the geometry of the beam shaper, as shown in Figure 6.11a

Dividing Equation 6.76 by $\cos\alpha_1$ gives after collecting terms

$$\tan\alpha_1 = \frac{\sin\theta}{n - \cos\theta} \tag{6.77}$$

Since the input and output beams are parallel to the optical axis, it follows that the slope of each aspheric surface for radial points, which satisfy the ray mapping function, is equal to the tangent of the angle of incidence:

$$\frac{dz(r)}{dr} = \tan\alpha_1 = \frac{dZ(R)}{dR} \tag{6.78}$$

Combining Equations 6.75 and 6.77 with the above equation leads to an expression for the slope of the first aspheric surface

$$z'(r) = \frac{1}{\sqrt{(n^2 - 1) + \dfrac{t^2(1-n)^2}{(R-\tau)^2}}} \tag{6.79}$$

where Equation 6.64 is used for evaluating R as a function of r when evaluating the sag of the first aspheric surface and the prime denotes differentiation with respect to its argument. Equation 6.79 was first derived by Kreuzer [11]. When computing the sag of the first aspheric surface, it is convenient to rewrite Equation 6.79 with the term $(R - r)$ in the numerator

$$z(r) = \int_0^r \frac{[h(\tau) - \tau]\,d\tau}{\sqrt{t^2(1-n)^2 + (n^2 - 1)[h(\tau) - \tau]^2}} \tag{6.80}$$

By expanding the square root in the denominator of Equation 6.80, the slope of the first aspheric surface can be evaluated from the series

$$z'(r) = \frac{[h(r) - r]}{t(n-1)}\left\{1 - \frac{(n+1)}{2(n-1)}\left[\frac{h(r) - r}{t}\right]^2 + \cdots\right\} \tag{6.81}$$

For points near the optical axis, the slope can be approximated by the first term of this series

$$z'(r) = \frac{[h(r) - r]}{t(n-1)} \quad \text{for small } r \tag{6.82}$$

The second derivative of the $z(r)$ with respect to r can be evaluated from Equation 6.82 for small r

$$z''(r) = \frac{[h'(r) - 1]}{t(n-1)} \quad \text{for small } r \tag{6.83}$$

Since the input and output irradiance are constant for the point near the optical axis, the geometrical optics intensity law implies

$$\lim_{r \to 0} h(r) = \epsilon r \sqrt{\frac{I_{\text{in}}(0)}{I_{\text{out}}(0)}} \tag{6.84}$$

$$\lim_{r \to 0} h'(r) = \epsilon \sqrt{\frac{I_{\text{in}}(0)}{I_{\text{out}}(0)}} \tag{6.85}$$

Equation 6.83 can be written for small r as

$$z''(0) = \frac{\epsilon \sqrt{I_{\text{in}}(0)/I_{\text{out}}(0)} - 1}{t(n-1)} = c_r(0) \tag{6.86}$$

where:
$c_r(0)$ is the axial curvature of the first aspheric surface

6.3.2.2 Sag of Second Aspheric Surface

Kreuzer [11] notes that it is apparent from Figure 6.11 that the slope on the first aspheric surface at the point (r,z) is equal to the slope of the second along the ray at the point (R,Z). Combining Equations 6.75 and 6.77 with Equation 6.78 and integrating leads to an expression for the sag of the second aspheric surface

$$Z(R) = \int_0^R \frac{\{\rho - [h^{-1}(\rho)]\}d\rho}{\sqrt{t^2(1-n)^2 + (n^2-1)\{\rho - [h^{-1}(\rho)]\}^2}} \tag{6.87}$$

where Equation 6.62 for $h^{-1}(R) = \epsilon r$ is used to evaluate the integral while computing the sag of the second aspheric surface. In a similar manner as presented in Section 6.3.2.1, the curvature of the second aspheric surface on axis is given by

$$Z''(0) = \frac{\left[1 - \epsilon \sqrt{I_{\text{out}}(0)/I_{\text{in}}(0)}\right]}{t(n-1)} = c_R(0) \tag{6.88}$$

where:
$c_R(0)$ is the axial curvature of the second aspheric surface

6.3.3 Nonprojective Transforms

For a projective transformation in optics, a point in image space can be expressed as a linear function of the coordinates of the object point. Perfect imaging systems, such as Maxwell's "fish-eye" lens or stigmatic imaging of surfaces, are examples of projective transformations in optics. In practice, aberrations are present in many optical systems, and point-to-point imaging is not possible, except to the first-order or paraxial approximation. Cornwell [26,27] notes that all real optical systems perform nonprojective transformations to some extent. That is, there is a nonlinear dependence between input (or object) and output (or image) coordinates, as illustrated by the ray mapping function of a Gaussian to FL beam shaping transformation, which describes the nonlinear relationship between the input and output aperture coordinates as given by Equation 6.64 and illustrated in Figure 6.12a. Therefore, the geometrical methods of Section 6.3 for designing a laser profile shaping system are an example of a nonprojective transformation in optics. Cornwell notes that the first element of a laser beam profile shaping (nonprojective transform) system creates sufficient aberrations in the wavefront to restructure the intensity of the beam after propagation of the wavefront over a specified distance. Then, the second element of a laser beam profile shaping system has suitable contour to restore the original wavefront shape of the beam. If the purpose of a laser beam profile shaping system is to uniformly illuminate a surface, then the second element is not needed. Symbolically, a laser beam profile shaping system may be considered to be a "black box" that transforms an input laser beam (plane wave) with a Gaussian intensity distribution into an output beam (plane wave) with uniform intensity distribution. It is also convenient to consider that the input and output beams have radii r and R, which are related by the ray mapping function that characterizes the beam transformation, such as Equation 6.64 that describes the ray mapping function of a Gaussian to FL beam transformation. References [26,27] present extensive discussion of many types of laser beam profile shaping systems and draw some interesting and general conclusions. In particular, Cornwell provides a seven-step recipe for designing two-element systems, which perform nonprojective transformations, such as laser profile shaping systems. Since the contents of Refs. [26,27] are not widely available in the optics literature to the knowledge of this author, these seven steps are summarized as follows:

1. Write out differential power expressions for the intensity distributions over the input and output planes.
   Rectangular coordinates

$$I_{\text{in}}(x,y)\text{d}x\text{d}y = I_{\text{out}}(X,Y)\text{d}X\text{d}Y \tag{6.89}$$

   Polar coordinates

$$I_{\text{in}}(r)r\text{d}r = I_{\text{out}}(R)R\text{d}R \tag{6.90}$$

2. Use the conservation of energy to relate the input and output beam parameters.
   Rectangular coordinates

$$\int_{\text{Input aperture}} I_{\text{in}}(x,y)\text{d}x\text{d}y = \int_{\text{Output aperture}} I_{\text{out}}(X,Y)\text{d}X\text{d}Y \tag{6.91}$$

Polar coordinates

$$\int_{\text{Input aperture}} I_{\text{in}}(r) r \mathrm{d}r = \int_{\text{Output aperture}} I_{\text{out}}(R) R \mathrm{d}R \tag{6.92}$$

3. Determine the magnification relating the input and output ray heights.
Rectangular coordinates: Assume the intensity functions are separable:

$$I_{\text{in}}(x,y) = a_x(x) a_y(y) \tag{6.93}$$

$$I_{\text{out}}(X,Y) = A_X(X) A_Y(Y) \tag{6.94}$$

Allowing for nonuniform shaping of a laser beam profile in two orthogonal directions, $X = m_x(x)x$ and $Y = m_y(y)y$, the rectangular magnifications follow from combining Equations 6.89, 6.93, and 6.94:

$$m_x(x) = \frac{1}{x}\left\{ C_1 \int_0^x \frac{a_x(u)\mathrm{d}u}{A_X[u m_x(u)]} + C_2 \right\} \tag{6.95}$$

$$m_y(y) = \frac{1}{y}\left\{ \frac{1}{C_1} \int_0^y \frac{a_y(v)\mathrm{d}v}{A_Y[v m_y(v)]} + C_3 \right\} \tag{6.96}$$

where:

C_i are constants determined by boundary conditions, such as the magnification for a rim ray

Polar coordinates

$$R = m(r) r \tag{6.97}$$

$$m(r) = \frac{1}{r}\left\{ 2 \int_0^r \frac{I_{\text{in}}(r) r \mathrm{d}r}{I_{\text{out}}[m(r) r]} + C \right\} \tag{6.98}$$

4. Express the OPL between input and output reference surfaces of an arbitrary ray in terms of the OPL of a reference ray.
5. Determine the sag $z(r)$ of the first element.
6. Determine the inverse magnification relating the ray coordinates at the first and second elements.
7. Determine the sag $Z(R)$ of the second element.

6.4 OPTICAL AND MECHANICAL TOLERANCES

We have shown how the principles of geometrical optics can be applied to design reflective and refractive systems that transform the irradiance profiles of collimated laser beams. For the important special case of rotationally symmetric input and output profiles, the problem of designing the optical elements can be reduced to a pair of integral equations with exactly two solutions, such as Equations 6.80 and 6.87 for the refractive beam shaper. It is not difficult to solve these equations to any desired numerical precision, thus obtaining a prescription for a set of lenses or mirrors that exactly solves the beam shaping problem.

However, real optics fabricated by an optical shop inevitably departs to some degree from their ideal mathematical prescription. Physical lens or mirror surfaces do not exactly obey Equations 6.80 and 6.87, and the positioning of the beam shaping optics with respect to each other and to the input beam is never absolutely perfect. The input beam itself, in reality, only approximately follows an idealized functional form such as Equation 6.50. This section considers the performance of a laser beam shaper that differs slightly from the exact solution derived in Section 6.3. It relates to the important practical considerations of optical and mechanical tolerances. When fabricating real optics, an optical shop needs to know how precisely the surfaces must be figured. Likewise, mechanical engineers and machine shops need to know the demands on the mounting hardware. The answers to these questions depend on the properties of the optical system that deviates slightly from an ideal configuration. The issue of alignment tolerances in beam shaping optics was first raised by Rhodes and Shealy [13], who performed numerical calculations for a two-lens refractive reshaper. Hoffnagle [86] published a limited treatment of figure error, but Oliker [78,87] published the first complete analysis of the effects of figure error on the output irradiance of a beam shaper. The discussion that follows is based on the work of Shealy and Hoffnagle [85], analyzing the optomechanical tolerances of the two-element refractive beam shaper.

The description of real, imperfect optics is more complex than the derivation of the perfect solution to the beam shaping problem, because it requires consideration of a multitude of ways in which real optics can depart from the ideal solution. One approach to this study would be to use the power of numerical ray tracing to model a large number of optical systems obtained by varying the prescription obtained from the exact, ideal solution. This chapter relies mainly on a more limited but analytic approach, making use of the fact that an exact ideal solution is known and also imposing two simplifying conditions. The first condition is that the deviations of the optics from the exact solution are assumed to be small enough so that the rays in the perturbed system are the same as the rays in the ideal system except for a perturbation that we need only to compute the first order in the quantity describing the perturbation. The second condition is that we consider individual fabrication and alignment errors in isolation, for instance we consider optics that is ideal except that the axis of the beam shaper is translated from the axis of the input beam. Breaking down the many sources of error in this way simplifies the calculations, and is intended also to provide some insight into the properties of imperfect optics that can be useful when testing and aligning practical beam shaping systems. Tolerances are considered in the following order: first, the effects of imperfect figuring of the surfaces; second, the effects of incorrect placement of the optical surfaces with respect to each other; and third, the consequences of the input beam not being matched properly to the beam shaping optics. In all cases, it is important to consider both the irradiance and the wavefront of the beam at the output aperture. The discussion in this section is limited to the special case of a rotationally symmetric, two-element refractive beam shaper, for which the ideal prescription has been derived in Section 6.3.2. Clearly, analogous calculations could be carried out for beam shapers using reflective or GRIN optics.

6.4.1 Figure Error

Figure error refers to imperfect shaping of the lens or mirror surfaces, which causes errors in both the irradiance and the wavefront generated by the beam shaper. The irradiance at the output aperture of a two-element beam shaper is determined by the first element, so the output irradiance is distorted by error in its figure. The first, incomplete discussion of the effect of figure error on a beam shaper was given by Hoffnagle [86], considering only rotationally symmetric perturbations of a refractive system using plano-aspheric lenses. While the assumption of rotational symmetry is too restrictive for a realistic treatment of the imperfections that arise in the grinding and polishing processes, it is relevant to one consideration of practical importance, namely, the error that can be inadvertently introduced to the optical surfaces by approximating the exact solution of the design equations by a polynomial. Equations 6.80 and 6.87 can be solved numerically to determine the sag function $z(r)$ to any desired precision at any radius r. However, software for optical design and analysis often requires that aspheric surfaces be expressed in the form of a power series. (Conic sections are also generally allowed, but since conic sections have no special significance for beam shaping optics it is simpler to speak just of power series.) A rotationally symmetric sag function $z(r)$ can be described exactly by an infinite power series, but in practice the series must be truncated after some finite number of terms, and this introduces error into the description of the surface. As an example, Reference [86] considers the effect of approximating the exact solution for the first surface of a refractive Gaussian to flat-top beam shaper by an eighth-order polynomial. In the specific case considered, the deviation of the polynomial from the exact solution was at most 50 nm, and the effect of this figure error was to introduce unwanted radial variations of ±2%–3% into the output irradiance profile. This level of nonuniformity is very noticeable when a flat-top output is desired, underscoring the importance of keeping numerical errors small when one uses a power series to describe the surfaces of beam shaping optics.

The first complete treatment of the effect of figure error on the irradiance profile generated by a refractive beam shaping lens was published by Oliker [78,87]. His method analyzes the perturbation of the known, ideal lens shape, using a formalism that has no constraints (such as symmetry) on the shape of the reshaping optics or the input and output profiles. To summarize his analysis in the notation of this review, suppose that an ideal lens with rotationally symmetric sag function $z(r)$ transforms the input irradiance profile $I_{in}(r)$ to an output profile $I_{out}(R)$, where r and R are related by the ray mapping function $R = h(r)$. Next consider an imperfect lens with the sag function (not necessarily rotationally symmetric) $z(r) + \tilde{z}(r,\phi)$, where the perturbation $\tilde{z}$ is assumed to be small enough that we only need to consider the lowest order effects on the slope and curvature of the wavefront after the lens. In addition, define the quantity

$$M(z) = \sqrt{1 + (1 - n^2)z_r^2} \tag{6.99}$$

where the subscript denotes differentiation, that is, $z_r \equiv \partial z/\partial r$. Then Oliker finds the perturbed irradiance distribution to be related to the ideal distribution by

$$\frac{\tilde{I}_{\text{out}}(R,\phi)}{I_{\text{out}}(R)} = \frac{(n-1)t}{M(z)}$$

$$\left\{ \frac{\tilde{z}_{rr}}{M^2(z)h_r(r)} + \frac{\tilde{z}_{\phi\phi}}{rh(r)} + \frac{1}{M^2(r)}\left[\frac{1}{h(r)} + \frac{3(n^2-1)z_r z_r r}{M^2(z)h_r(r)}\right]\tilde{z}_r \right\} \tag{6.100}$$

This expression describes the perturbed output irradiance in terms of the error in the figure of the first lens at the point of refraction on the first lens. Oliker also derives a more general expression for the case in which the ideal optical system is not rotationally symmetric, which can be found in Reference [78].

6.4.2 Error in Relative Location of the Optical Surfaces

A two-element beam shaper is most commonly constructed with mounting hardware to hold the individual elements in their prescribed positions. Mechanical imperfections in the alignment of the elements with respect to each other can be broken down into three possibilities: incorrect lens separation, centration offset, and tilt between the symmetry axes of the lenses. This section considers the associated tolerances, assuming that the optical system is otherwise perfect; the input beam is assumed to have its ideal form and be perfectly aligned to the first lens. This means that the irradiance at the nominal position of the second lens has its ideal form, but because the second lens is positioned incorrectly it does not perfectly perform its function of generating a collimated output beam. Consequently, the class of errors considered in this section affects the output wavefront but not the irradiance profile at the output aperture.

First, we consider lens separation error, which is particularly simple to analyze because it preserves the rotational symmetry of the optical system. Rhodes and Shealy [13] analyzed this tolerance by numerically computing the irradiance after the output beam propagated for some distance; here we prefer to describe the aberration of the wavefront at the output aperture. The geometry of the optics is shown schematically in Figure 6.13.

By construction, when the lens separation is correct the angle of incidence of each ray is the same at the first and second lenses, so that after two refractions the ray is parallel to the optical axis. When the lens separation is incorrect, the ray intersects the second lens at a different radius than designed, at which point the angle of incidence on the lens surface is different from the design value, leading ultimately to the ray exiting the system with a nonvanishing angle δ to the optical axis, as shown in an exaggerated way in Figure 6.13b. The connection between lens spacing error and error in the slope at the second lens is illustrated in Figure 6.14.

The first lens deviates the ray by the angle

$$\theta_1 = \psi_1 - \alpha_1 \tag{6.101}$$

where:

$\alpha_1 = \arctan(\mathrm{d}z/\mathrm{d}r)$

$\psi_1 = \arcsin[n\sin(\alpha_1)]$

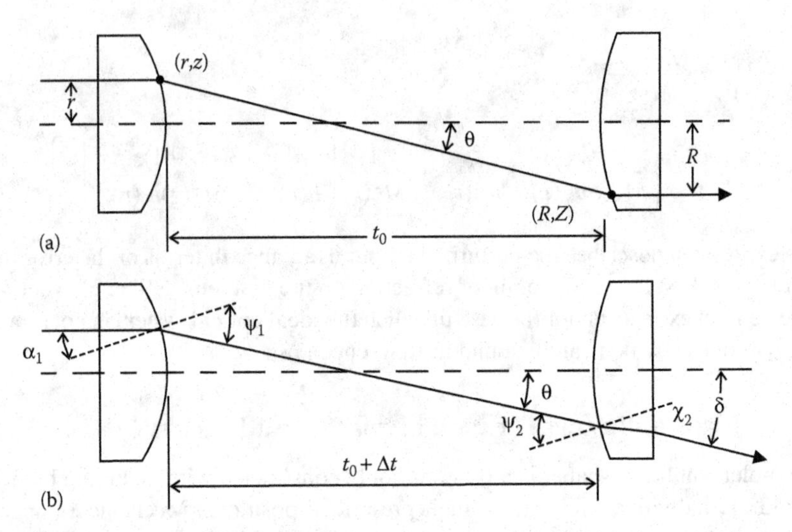

FIGURE 6.13 Ray paths through a refractive beam shaper with (a) lenses spaced correctly and (b) lenses spaced incorrectly.

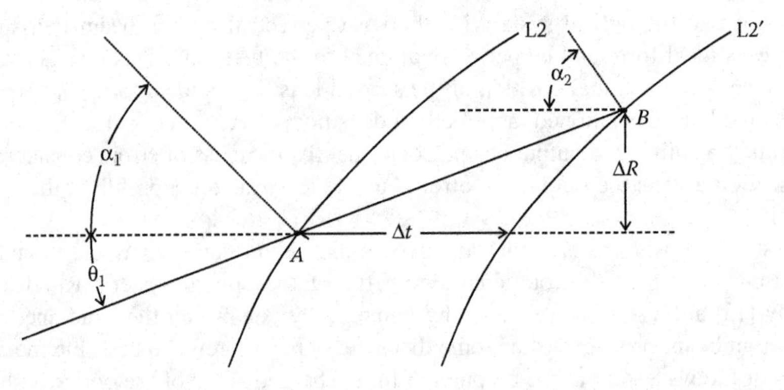

FIGURE 6.14 Geometry of a ray in the neighborhood of the second lens surface. Translating the surface from its design location L2 to L2′ changes the radius at which the ray intersects the lens and thereby the slope of the lens at the point of intersection. The dotted line is drawn parallel to the optical axis of the lens system. (From Shealy, D. L. and Hoffnagle, J. A. Aspheric optics for laser beam shaping. In *Encyclopedia of Optical Engineering*, Driggers, R., ed., Taylor & Francis, doi:10.1081/E-EOE-120029768, 2006. With permission.)

Ideally, the refracted ray intersects the second lens at the point A, where the lens slope $Z'(R) = \mathrm{d}Z(R)/\mathrm{d}R$ is equal to α_1 by design. A lens spacing error Δt causes a radial displacement ΔR in the intersection, which is now at the point B. If the alignment error is small, the lens surfaces can be approximated by a Taylor series in the neighborhood of A. In this approximation, we can solve for the intersection point to lowest order in Δt, obtaining

$$\Delta R = \frac{\Delta t \sin(\theta)\cos(\alpha_1)}{\cos(\alpha_1 + \theta_1)} \tag{6.102}$$

The slope of the second lens at point B is given by $Z'(R + \Delta R)$ and the angle α_2 shown in Figures 6.13 and 6.14 is given by $\alpha_2 = \arctan[Z'(R + \Delta R)]$.

From the angle α_2 it is simple to compute the error in collimation δ. Referring to Figure 6.13, the angle of incidence at the second refracting surface is

$$\psi_2 = \theta_1 + \alpha_2 \tag{6.103}$$

and the corresponding angle of refraction is

$$\chi_2 = \arcsin\left[\frac{\sin(\psi_2)}{n}\right] \tag{6.104}$$

The angle between the ray internal to the second lens and the optical axis is then given by $\delta_g = \chi_2 - \alpha_2$. A final refraction at the plano surface of the second lens gives

$$\delta = \arcsin[n\sin(\delta_g)] \tag{6.105}$$

If the lens spacing error is small, we can approximate $\delta = n\delta_g$, then combining the relations above gives the result

$$\delta = n\left(\arcsin\left\{\sin\frac{[\arcsin(n\sin\alpha_1) + \alpha_2 - \alpha_1]}{n}\right\} - \alpha_2\right) \tag{6.106}$$

The collimation error described by δ is equivalent to an error in the phase of the output wavefront, φ. For a beam with rotational symmetry, as here

$$\delta = \frac{\lambda}{2\pi}\frac{d\varphi}{dR} \tag{6.107}$$

Since δ can be calculated as a function of R, we can integrate to get $\varphi(R)$.

These expressions simplify greatly for rays close enough to the optical axis so that the sine and tangent functions can be approximated by their arguments. Then Equation 6.106 reduces to

$$\delta = (1 - n)\Delta Z' \tag{6.108}$$

The small-angle approximation for θ yields

$$\Delta R = \Delta t(n - 1)Z' \tag{6.109}$$

Expanding ΔZ as a Taylor series in ΔR and taking only the first term gives the approximation

$$\Delta Z' = \Delta R\, Z'' \tag{6.110}$$

The slope of the lens surface near the optical axis, also to the lowest order, is

$$Z'(R) = R\, Z''(0) \tag{6.111}$$

where:

$Z''(0)$ is the curvature of the second lens on axis, given explicitly by Equation 6.88

Combining Equations 6.108 through 6.111 gives the following expression for the wavefront aberration as a function of R, valid to the lowest order in R,

$$\delta(R) = -\Delta t[(n - 1)Z''(0)]^2 R \tag{6.112}$$

This describes a spherical wavefront with curvature c_{out} (equal to the inverse of the radius of curvature) equal to

$$c_{out} = -[(n-1)Z''(0)]^2 \Delta t \tag{6.113}$$

Consequently, the lowest order aberration caused by a lens spacing error is wavefront curvature. Considering larger values of R, away from the optical axis, one finds that spherical aberration and higher order, rotationally symmetric terms appear. For a specific beam shaper design, it is easy to find all the aberrations by evaluating Equation 6.106 numerically.

Next, we consider centration error, by which we mean a lateral displacement of the second element of the beam shaper from its ideal position. Since the ideal geometry is rotationally symmetric, there is no loss of generality in taking the direction of the centration error to be in the x-axis. Denote the displacement of the lens center by Δx. Centration error destroys the overall rotational symmetry of the beam shaper, which complicates the ray tracing. However, for rays along the x-axis the geometry is still relatively simple and a geometrical construction similar to that of Figure 6.14 leads to an expression for the deviation of the output rays from the optical axis. When both Δx and R are small, the lowest order effect due to lens centration error can be shown to be a constant angular deviation of the output beam

$$\delta = \Delta x(1-n)Z''(0) \tag{6.114}$$

equivalent to tilt in the output wavefront. Higher order aberrations are those that are odd in the x-coordinate, principally coma.

Finally, we consider the situation where the second lens is tilted by the angle ω with respect to the first lens. As with centration error, rotational symmetry of the complete optical system is broken. The geometry is more complicated than for the previous two cases. Additional details can be found in Reference [85]. Here we simply state the result that the lowest-order aberration is third-order coma.

6.4.3 Error in Matching the Beam Shaper to the Input Beam

This section considers how the performance of the beam shaper is affected if the aspheric optics are improperly matched to the input beam. First, we consider mechanical alignments errors, namely, an offset or tilt of the beam shaping optics with respect to the input laser beam axis. Next, we consider the possibility that the beam shaping optics are built and assembled as designed, but the actual input beam does not satisfy the assumptions made in the design. If the beam waist parameter w differs from the design value or the beam is not collimated, then the output beam will differ from the design. Finally, the design of the two-element refractive beam shaper depends on the index of refraction of the lenses. Because of dispersion, the beam shaper performs as designed only for one wavelength. We conclude this section with a discussion of the effects of dispersion on the refractive beam shaper. The results presented in this section are of practical use when designing the optomechanical fixturing for a system that uses beam shaping optics, and also for the process of aligning the beam shaping optics.

To begin, we consider the case of a pure offset of the beam shaping optics from the optical axis of the input beam. Choose the offset to be by the amount Δx along the x-axis. This is a special case of a more general problem: For a beam shaper designed to transform the collimated input beam with irradiance $I_{in}(r)$ to a collimated beam with irradiance $I_{out}(R)$, what is the output profile when the input irradiance has nonideal profile, $\tilde{I}_{in}(x,y)$? The answer is given by the ray mapping function and the intensity law of geometrical optics. The beam shaping optics transform a point (r,φ) in the input aperture to the point (R,Φ) in the output aperture, where R is equal to $\epsilon h(r)$ and Φ is equal to either φ, for the Galilean geometry, or $\varphi - \pi$, for the Keplerian geometry. The optics also transforms surface area elements in the neighborhood of the ray that enters at (r,φ) and exits at (R,Φ) by the factor $I_{in}(r)/I_{out}(R)$. This transformation applies to any collimated input beam, regardless of the irradiance profile; consequently, the input irradiance $\tilde{I}_{in}(x,y)$ is transformed to

$$\tilde{I}_{out}(X,Y) = \frac{I_{out}(R)\tilde{I}_{in}(r)}{I_{in}(x,y)} \tag{6.115}$$

where the pairs of points (x,y) and (X,Y) are related by the ray mapping function, and the coordinates (x,y) and (r,φ) are understood to describe the same point in a Cartesian or polar representation, respectively. Applied to the case of a Gaussian input beam that is offset from the axis of the beam shaper, the output irradiance is found to be

$$\tilde{I}_{out}(R,\Phi) = I_{out}(R)\exp\left[\frac{4x\Delta x - 2(\Delta x)^2}{w_0^2}\right] \tag{6.116}$$

where:

$x = r\cos\varphi$

This equation was first derived by Romero and Dickey [88] using physical optics considerations of a phase plate and Fourier transform lens designed to implement a one-dimensional Gaussian to top-hat beam transformation. Here, we see how it arises as a general consequence of the intensity law of geometrical optics.

For the case of a Gaussian to flat-top beam shaper, Equation 6.116 can be used to find a tolerance on the acceptable offset. Assuming $\Delta x \ll w_0$, the exponential can be expanded to give

$$\tilde{I}_{out}(R,\Phi) = I_{out}(R)\left[1 + 4\frac{\Delta x}{w_0}\frac{x}{w_0} + \mathcal{O}\left(\frac{\Delta x}{w_0}\right)^2\right] \tag{6.117}$$

where:

$\mathcal{O}$ indicates that the omitted terms in the series are proportional to the square and higher orders of $\Delta x/w_0$

Near the center of the output beam, this expression implies a linear irradiance variation equal to $4(\Delta x/w_0)(x/w_0)$ times the nominal output irradiance. If the relative invariance error over the input aperture a is required to be less than ϵ, then

$$\frac{\Delta x}{w_0} < \frac{\epsilon}{4(a/w_0)} \tag{6.118}$$

describes the tolerance on acceptable beam offset.

Aside from a transverse offset, the other mechanical misalignment of the beam shaper assembly with respect to the input beam is tilt. All the treatments of the effects of tilt on beam shaper performance have been based on numerical ray tracing for a specific beam shaper design. If the tilt angle is small, then the irradiance at the output aperture is hardly affected, but the wavefront is aberrated. Rhodes and Shealy [13] considered the effect of tilt on a Galilean beam shaper by computing the irradiance profile of the output beam at a moderately large distance after the output aperture of the beam shaper, where the wavefront aberrations distort the desired output profile. Similar calculations for a Keplerian beam shaper were reported in Reference [85]. A more direct way to quantify the effect of tilt on the performance of the beam shaper is to compute the primary wavefront aberrations as a function of the tilt angle; this calculation was carried out for a Keplerian beam shaper by Hoffnagle and Jefferson [89]. For small tilt angles, third- and fifth-order comas, which both increase linearly with angle, are the most important aberrations. The coma introduced by tilting the beam shaper can severely distort the output beam at propagation distances that are small enough so that pure diffraction effects would be negligible. To illustrate this effect, Figure 6.15 reproduces the calculations reported in Reference [85].

Having looked at the issues of mechanical misalignment of a beam shaper with respect to the input beams, we proceed to consider the output of a properly constructed and aligned beam shaper when the input beam does not satisfy the assumptions used for the beam shaper design. One possibility has already been handled above: If the collimated input beam has an irradiance distribution $\tilde{I}_{in}$ different from the nominal input beam profile, then the output profile deviates from the designed profile according to Equation 6.115. A special case of this formula that turns up often enough to deserve explicit mention is for a Gaussian to flat-top beam shaper that is designed to

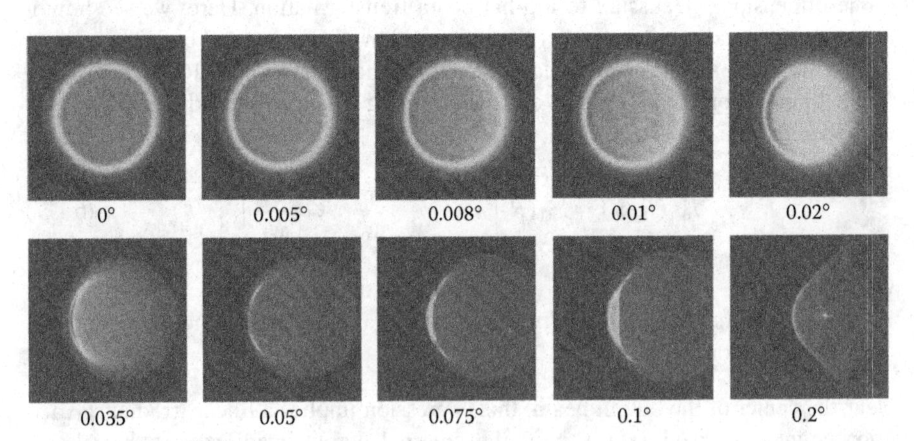

FIGURE 6.15 Output beam irradiance in a plane 1 m from the output aperture of a Gaussian to FD beam shaper for different tilt angles between the input beam and the beam shaper assembly, as indicated by the labels underneath the images. The numerical calculations of irradiance were made by Michael Jefferson. (Reproduced from Shealy, D. L. and Hoffnagle, J. A. Aspheric optics for laser beam shaping. In *Encyclopedia of Optical Engineering*, Driggers, R., ed., Taylor & Francis, doi:10.1081/E-EOE-120029768, 2006. With permission.)

accept an input beam with waist w_0 but actually receives a beam with waist w_1. In this case, the entire optical system is rotationally symmetric and we can write

$$\tilde{I}_{\text{out}}(R) = \frac{I_{\text{out}}(R)\tilde{I}_{\text{in}}(r)}{I_{\text{in}}(r)} \tag{6.119}$$

where r and R are related by the ray mapping function as always. Inserting the Gaussian input profiles gives [90]

$$\tilde{I}_{\text{out}}(R) = I_{\text{out}}(R)\left(\frac{w_0}{w_1}\right)^2 \exp[-2r^2(w_1^{-2} - w_0^{-2})] \tag{6.120}$$

If a given degree of output beam uniformity is required, this expression gives rise to a tolerance on the input beam size. Roughly speaking, the relative tolerance on w is about half the tolerable irradiance nonuniformity; for example, a 5% limit on irradiance variation over the usable aperture requires that w be within ±2.5% of the design value, which is important to bear in mind when designing the integrated system of laser and beam shaper.

Wavefront aberration is another way in which a real input beam can deviate from the idealization that is used to design the beam shaper. By construction, the beam shaping optics transforms a collimated input beam (i.e., plane wavefront) to a collimated output beam. If the input beam has a nonplanar wavefront, then the wavefront variations in the input aperture will be transformed to the output aperture with exactly the same amplitude, because the optics are constructed to have equal OPL for all rays. Consequently, the peak-to-valley variation of the input and output wavefronts will be exactly the same. However, the nonlinearity of the ray mapping function implies that the spatial distribution of the wavefront variations will be different in the input and output planes. So, for instance, if the input wavefront is perfectly spherical, the output wavefront will have not only curvature but also spherical aberration and higher order aberration corresponding to the higher order, rotationally symmetric Zernicke polynomials.

Finally, we consider the effect that glass dispersion has on the performance of a two-element, refractive beam shaper. Because Equations 6.80 and 6.87 contain the index of refraction of the glass, the lenses designed using these equations operate perfectly only at the design wavelength. Operation of the refractive beam shaper with wavelengths other than the design wavelength has been analyzed by Hoffnalge and Shealy [91] and we summarize the results here.

To start, we note that dispersion mainly affects the wavefront of the output beam. If the lenses are made of low-dispersion glass, then the wavelength can be varied over hundreds of nanometers with very little change in the shape of the irradiance profile. Over the same wavelength range, the output wavefront can acquire many waves of aberration [92]. Next, we distinguish two possible classes of beam shaping applications, which both can be called "multi-wavelength applications" but which have different consequences for the wavefront. The first application we consider is the one in which the beam shaper is designed for use at wavelength λ, with corresponding glass index n but used with a monochromatic laser source at wavelength $\tilde{\lambda}$ and glass index $\tilde{n}$. The second application is for simultaneous operation at multiple wavelengths such as with a broadband source or a short-pulse laser. In the first case, monochromatic operation at a wavelength other than the design wavelength, good wavefront quality can be achieved over a wide wavelength range provided that (1) the lens spacing is chosen to be large enough so that the deviation angle, θ in Figure 6.11, is much smaller than the radian everywhere and (2) the beam shaper is constructed so that the lens spacing can be adjusted for each wavelength. The smallness of θ means that the paraxial approximation $\theta \approx (n - 1)\arctan[z(r)]$ is a good approximation everywhere, so that the factor $(n - 1)$ is, to a good approximation, a global scaling factor for the deviation of all rays. Consequently, the ray mapping function is nearly the same when n is changed to $\tilde{n}$ and the lens spacing is changed from t to [21]

$$\tilde{t} = \frac{t(n-1)}{(\tilde{n}-1)} \tag{6.121}$$

This adjustment removes all the wavefront curvature and most of the chromatic-spherical aberration. The possibility of using a single refractive beam shaper at multiple wavelengths has been noted by Jiang et al. [18] and Hoffnagle and Jefferson [56]. Hoffnagle and Shealy [91] showed that a fused silica beam shaper designed for use at $\lambda = 532$ nm could be used from about 400–2000 nm and still have a wavefront that varies over the flat-top region of the beam by less than $\lambda/100$, provided that the lens spacing can be optimized at each wavelength. It is a useful feature that one prescription can be used over a very wide wavelength range, nearly equal to the transparency range of the glass, and limited in practice by the bandwidth of antireflection coatings.

For true multi-wavelength applications, in which the lens spacing cannot be adjusted for each wavelength, the two-element beam shaper exhibits wavefront curvature and chromatic-spherical aberration for wavelengths other than the design wavelength. The calculation of collimation error for a ray with wavelength $\tilde{\lambda}$ is exactly the same as for lens spacing error in Section 6.4.2. The angle between an output ray and the optical axis of the beam shaper is given by Equation 6.106 with n replaced by $\tilde{n}$. Applying the paraxial approximation, one finds that the lowest order aberration is wavefront curvature

$$c_{\text{out}} = [Z''(0)]^2(\tilde{n}-1)(n-\tilde{n})t \tag{6.122}$$

One way to deal with this wavefront aberration in true multi-wavelength applications is to add achromatizing optics to the beam shaper. Jefferson and Hoffnagle [92] presented an example, designed with conventional optical design software, that compensates the aberrations of a two-element refractive beam shaper and generates a beam with less than 0.03λ wavefront error over the wavelength range of 450–650 nm.

6.5 REFRACTIVE BEAM SHAPERS

Laser beam shaping optics is well suited for applications whose overall efficiency increases when the irradiance over the detector (or substrate) is uniform, such as in compact holographic projector systems [93–95]. These compact holographic projection systems have been reported to offer a practical way to make a highly corrected mesh or grid pattern over curved surfaces where the pattern can range in size from submicron to multimicron. The laser profile shaping optics within a holographic projection system enables uniform features to be written over substrates of several centimeters in diameter [96].

To understand this increase in system efficiency when using laser beam shaping optics, note that for a Gaussian beam with irradiance given by Equation 6.50, the intensity of the beam decreases to $1/e^2$ of its axial value at the beam radius equal to its waist. The effect of this variation in beam intensity over a Gaussian beam is illustrated in Figure 6.16. Figure 6.16a shows significant variation in pattern densities at the center and edge of the beam for the same substrate (film) and exposure time when laser profile shaping optics is not part of the system. Figure 6.16b shows almost uniform pattern densities at the center and edge of the beam when laser profile shaping optics is part of the system.

FIGURE 6.16 Interference patterns produced by a four-beam holographic projection processing system when illuminated with a Gaussian beam. The image on the left side of the figure was taken near the center of the beam, and the image on the right side of the figure was taken near the edge of the beam. The images (a) were taken when the laser beam shaping optics was not part of the system, and the images (b) were taken when the laser beam shaping optics was part of the projection system. (Reproduced from Jiang, W. Application of a laser beam profile reshaper to enhance performance of holographic projection systems. PhD thesis, University of Alabama at Birmingham, Birmingham, AL, 1993; Shealy, D. L. Geometrical methods. In *Laser Beam Shaping—Theory and Techniques*, Dickey, F. M. and Holswade, S. C., eds., Marcel Dekker, New York, 163–213, 2000. With permission.)

Therefore, when beam shaping optics is introduced into a holographic projection processing system, the detector substrate will be uniformly illuminated, and photochemical reactions take place at the same rate over the entire substrate area, thus enabling the full beam diameter to be available for material processing. Introducing laser shaping optics into holographic projection processing systems has led to a significant increase in quality of micro-optics fabricated over the substrate.

In Section 6.5.1, a detailed discussion of the design, analysis, fabrication, and testing of the Galilean configuration of a two-lens beam shaper is presented. Section 6.5.2 describes the design, analysis, and testing of the Keplerian configuration of a two-lens beam shaper. Section 6.5.3 describes using axial GRIN materials to design a laser beam profile shaping optical system with spherical lens surfaces.

6.5.1 Galilean Configuration

The first report of experimental results in the literature of laser beam shaping by two plano-aspheric lenses appeared in 1993 [18], almost 30 years after Kreuzer first proposed using plano-aspheric lenses for laser beam shaping in 1965 [2] as a result of technical limitations in making the plano-aspheric lenses required by Kreuzer's designs. Jiang et al. [18] reported the design, fabrication, and testing of an expanding beam shaper in the Galilean configuration, as illustrated in Figure 6.11a. In this section, we describe a Galilean configuration that expands the input beam by the factor of 1.6× and was used to reshape a HeCd laser beam operating at a wavelength of 441.57 nm. The aspheric optics was fabricated by diamond turning of CaF_2. The parameters of this beam shaper are given in the column labeled 1.6× G of Table 6.2.

TABLE 6.2
Parameters for the Galilean (1.6× G) and Keplerian (1-to-1K) Configurations of Laser Beam Shaping Design Examples Discussed in Sections 6.5.1 and 6.5.2, Respectively

Parameters	1.6× G	1-to-1K
Input beam waist (mm)	8.0	2.366
Input aperture radius, r_{max} (mm)	8.0	4.05
Input lens diameter (mm)	30.0	8.1
Output profile shape	TH	FD
Output beam radius, R_0 (mm)	12.5	3.25
Output shape parameter	–	16.25
Output aperture radius, R_{max} (mm)	12.5	4.05
Output lens diameter (mm)	30.0	8.1
Design wavelength (nm)	442	532
Refractive index	1.43916	1.46071
Lens spacing (mm)	150.0	150.0
Geometrical parameter, ϵ	+1	−1

Source: Shealy, D. L. and Hoffnagle, J. A. Aspheric optics for laser beam shaping. In *Encyclopedia of Optical Engineering*, Driggers, R., ed., Taylor & Francis, 2006.

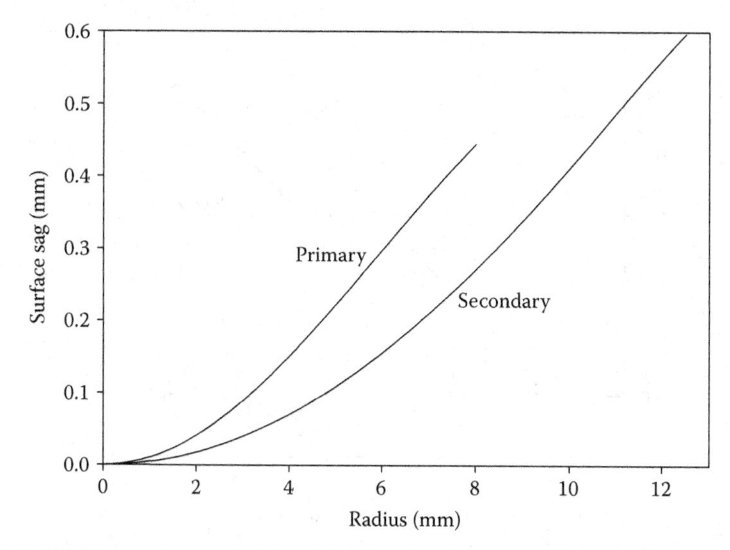

FIGURE 6.17 The surface sag of the primary and secondary aspheric surfaces is shown as a function of the aperture radius for the expanding (1.6×) Galilean configuration of a laser beam shaper. (Reproduced from Shealy, D. L. and Hoffnagle, J. A. Aspheric optics for laser beam shaping. In *Encyclopedia of Optical Engineering*, Driggers, R., ed., Taylor & Francis, 2006. With permission.)

Using the results presented in Section 6.3 to compute the sag and slope of the aspheric surfaces, Figure 6.17 displays the sag of the primary and secondary aspheric surfaces of the 1.6× expanding Galilean laser beam shaper. From analysis of the computed data for the surface sag, slope, and curvature of these two aspheric lenses, the following results were obtained:

- The maximum surface slope is 0.076383.
- The surface curvature on axis for the primary is equal to 0.020895 mm^{-1}.
- The surface curvature on axis for the secondary is equal to 0.008792 mm^{-1}.

It is interesting to explore the meaning and limitations of the optical design equations for laser beam shapers by considering the effect of increasing the primary lens aperture radius to r_{max} = 15 mm for the 1.6× G beam shaper, while keeping other system parameters as listed in Table 6.2. Figure 6.18a presents the results for the surface sag and slope data for this case, where one sees that the sag of the primary aspheric surface has an inflection point off-axis. To understand the implication of an off-axis inflection point in the sag of the primary aspheric surface in this case, we analyze the slope of primary aspheric surface and the plot of $[h(r) - r]$ and $h(r)$ versus r as shown in Figure 6.18b, from which it is clear that the slope of the primary is equal to zero at r = 12.46 mm, as well as on-axis, and the slope of the primary aspheric surface changes sign at the radial point r = 12.46 mm, as well as on-axis. Fabrication of aspheric surfaces with multiple inflection points for their sag is difficult and expensive. However, Kasinski and Burnham [97] have fabricated and tested aspheric surfaces with an inflection point. When the slope of primary aspheric

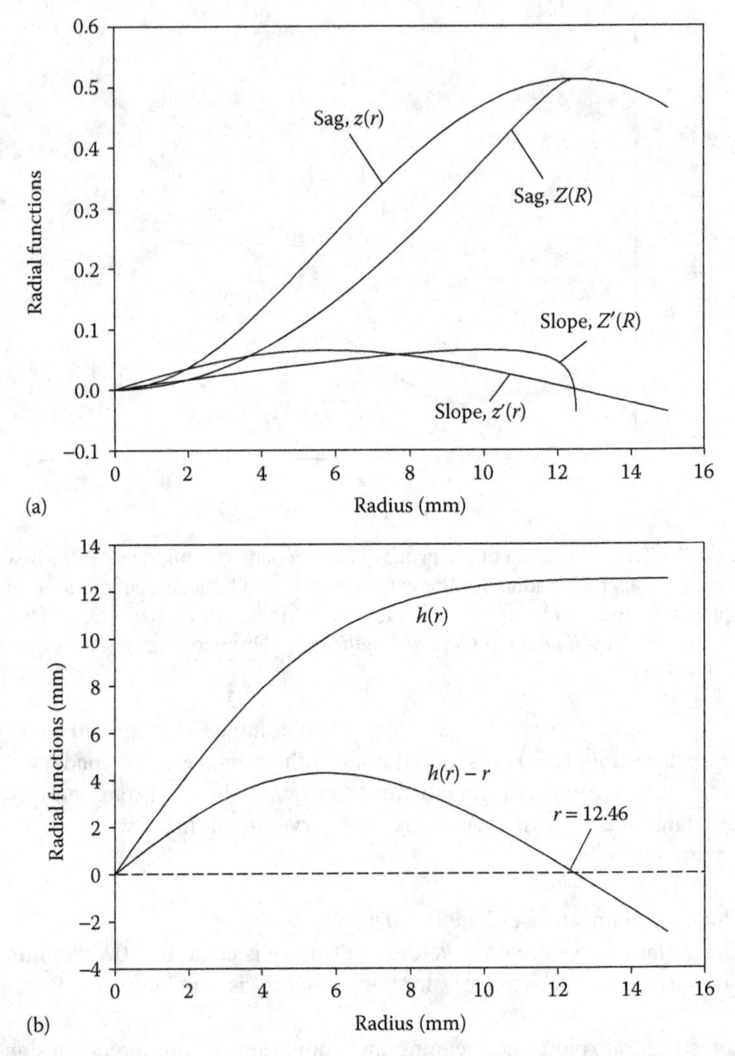

FIGURE 6.18 (a) Plot of the surface sag and slope of the primary and secondary aspheric surfaces as a function of the aperture radius for the expanding (1.6×) Galilean configuration of a laser beam shaper. (b) Plot of the ray mapping function $h(r)$ and $[h(r) - r]$ versus the input aperture radius r for the 1.6× Galilean beam shaper. (Reproduced from Shealy, D. L. and Hoffnagle, J. A. Aspheric optics for laser beam shaping. In *Encyclopedia of Optical Engineering*, Driggers, R., ed., Taylor & Francis, 2006. With permission.)

surface is equal to zero for off-axis points, then the incident ray will leave the first lens of beam shaper parallel to the optical axis, which seems to occur when the input lens aperture is larger than the output lens aperture. Therefore, if one must reduce the diameter of a laser beam after shaping, it seems advisable to use a beam shaper with a one-to-one ratio of the input and output beams, and then to change the diameter of laser beam after reshaping by using conventional optics.

6.5.1.1 Analysis and Simulation Results

The shape of the two aspheric refracting surfaces of the 1.6× Galilean beam shaper are defined by an array of data for the primary (r,z) and the secondary (R,Z), which can be computed from Equations 6.80 and 6.87 as accurately as desired. However, fabrication vendors of aspheric lenses have requested surface data to be provided in terms of coefficients of the conventional optics surface equation

$$z(r) = \frac{cr^2}{1+\sqrt{1-(1+\kappa)c^2r^2}} + \sum_{i=1}^{N} A_{2i} r 2i \qquad (6.123)$$

where:

- c is the vertex curvature
- κ is the conic constant
- A_{2i} are the coefficients of the polynomial deformation terms

For a refracting plano-aspheric beam shaper, the coefficients of the conventional optics surface equation must be evaluated by a nonlinear fitting program for each aspheric lens surface. A nonlinear least squares fitting program based on the simplex method has been successfully used to represent the optical surfaces of a laser profile shaping system. Table 6.3 gives the surface parameters for the primary and secondary surfaces of the 1.6× Galilean laser beam shaper used as part of the holographic projection system developed by Jiang [18,19,96,98].

Conventional optical design software has been used to ray trace and analyze the performance of the 1.6× Galilean laser beam shaper. The OPD of the output beam for this Galilean laser beam shaper has been evaluated over the aperture. The maximum

TABLE 6.3
Surface Parameters of a HeCd (441.57 mm) Laser Beam Shaper System Where the Distance between the Primary and Secondary Lenses Is Equal to 100 mm

Lens Parameters	Primary	Secondary
Diameter (mm)	30.0	30.0
Vertex radius (mm)	47.861445	113.64905
Index of refraction	1.43916 (CaF_2)	1.43916 (CaF_2)
Thickness (mm)	10.0	10.0
Conic constant, κ	−1.1143607	−1.4877144
A_4 (mm)$^{-3}$	$-7.1532887 \times 10^{-5}$	$-2.6322455 \times 10^{-6}$
A_6 (mm)$^{-5}$	3.3729843×10^{-7}	9.4058758×10^{-9}
A_8 (mm)$^{-7}$	$-1.4916816 \times 10^{-9}$	$-2.3096843 \times 10^{-10}$
A_{10} (mm)$^{-9}$	$5.9836543 \times 10^{-12}$	$1.5839557 \times 10^{-12}$
A_{12} (mm)$^{-11}$	$-1.5166511 \times 10^{-14}$	$-4.8438745 \times 10^{-15}$

Source: Shealy, D. L. and Hoffnagle, J. A. Aspheric optics for laser beam shaping. In *Encyclopedia of Optical Engineering*, Driggers, R., ed., Taylor & Francis, 2006.

OPD for this system was 0.0017λ, which corresponds to an absolute OPD of 0.75 nm for the HeCd laser. This demonstrates that the shape of the output wavefront was very close to the same shape as the input wavefront as required by the constant OPL design condition of this laser beam shaper. The flux flow equation [61] has been used to compute the irradiance along a ray as it propagates through the optical system [13,14]. Specific results for computing the irradiance over an output surface of a laser shaping system will be presented and discussed in Section 6.5.1.2.

6.5.1.2 Experimental Results

The two plano-aspheric lenses of Table 6.3 were fabricated by Janos Technology, Inc. of Townshend, Vermont, using a single-point diamond lathe; CaF_2 was used as the lens material. A scanning video system was used to measure the input and output irradiance profiles. A Panasonic TV camera (Model WV-1800) was used to sample the laser beam before and after passing through the laser beam profile shaping optics. The camera was mounted on a translation stage behind a pinhole. By scanning across the beam, it was possible to use the same region of the detector for measuring the intensity of all parts of input and output beams. The image processing software, NIH Image v. 1.44 [99], was used to acquire, display, edit, enhance, analyze, and print images. Reference [19] provides a full discussion of this testing procedure as well as tolerance analysis and other results not summarized in this section.

The input and output beam profiles using the HeCd laser are shown in Figure 6.19. The open diamond symbols are the measured intensity of input beam, and the solid curve is a Gaussian profile fitted to the input beam data. The solid diamond symbols

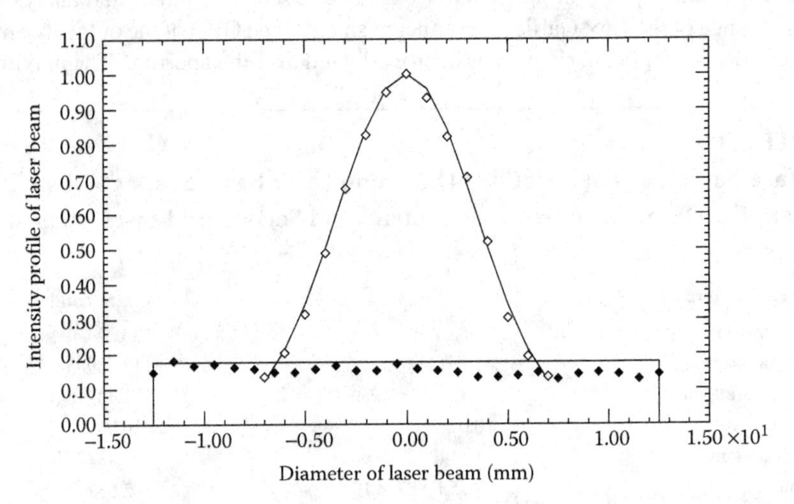

FIGURE 6.19 The input and output irradiance profiles of a HeCd laser beam shaper. The open diamond symbols are measured data points of the input beam, which is fit to a Gaussian curve shown as a solid curve. The solid diamond symbols are measured data points of the output beam. (Reproduced from Jiang, W., Shealy, D. L., and Martin, J. C. Design and testing of a refractive reshaping system. In *Current Developments in Optical Design and Optical Engineering III*, Fischer, R. E. and Smith, W. J., eds., vol. 2000, 64–75, 1993; Shealy, D. L. and Hoffnagle, J. A. Aspheric optics for laser beam shaping. In *Encyclopedia of Optical Engineering*, Driggers, R., ed., Taylor & Francis, doi:10.1081/E-EOE-120029768, 2006. With permission.)

are the measured intensity of the output beam, and the horizontal solid line is the designed output beam intensity. These results clearly show that the input beam has been transformed into a more uniform beam. However, manufacturing and alignment errors can be linked to the variations of the output beam profile from its theoretical value. In addition, a HeNe laser was used to illuminate these beam shaping optics. When using a HeNe laser with 632.8 nm radiation, the lens spacing was increased to 152.2 mm according to the predictions of Equation 6.121.

6.5.2 Keplerian Configuration

Hoffnagle and Jefferson [21,22] first reported the solution to design and fabrication of a Keplerian configuration of a refractive laser beam shaper, where extensive details of the design, analysis, and testing of this beam shaper are given in the literature. Table 6.2 gives the design parameters for the Keplerian configuration, which is labeled by 1-to-1K.

It is instructive to view the ray plots of a 1-to-1K beam shaper as shown in Figure 6.20a, which illustrates the reshaping of a Gaussian beam into a FL beam, where each ray path is associated with equal power. Recall, Shealy and Hoffnagle [71] have shown that when using the matched profile conditions, any flattened irradiance profile has very similar near- and far-field diffraction patterns. Therefore, the conclusions of Hoffnagle and Jefferson [21,22] on the design, analysis, performance, and testing of a 1-to-1K configuration for shaping a Gaussian input beam into an output beam with a FD irradiance profile can be generalized to apply to any flattened irradiance profile with matching profile parameters. Figure 6.20b presents the sag of the primary and secondary aspheric optics when designed to transform a Gaussian beam into a top-hat profile or into a FD profile. It is interesting to note from Figure 6.20b that the sag at the periphery of the primary or secondary aspheric lens surface is smaller by approximately 20–30 μm for a beam shaper designed to transform an input Gaussian beam into a flattened irradiance profile than one designed to shape a Gaussian beam into a top-hat profile. The rate of change of the slope and curvatures of the primary and secondary aspheric surfaces change in a nonlinear manner as a function of the radial coordinate, which has been studied and documented by Shealy and Hoffnagle [85]. Hoffnagle and Jefferson also observed that the surfaces of a Keplerian beam shaper are always convex, that is, the sign of the curvature is the same over the entire lens surface [21]. This property can be advantageous for practical fabrication of the aspheric optical surfaces, because it means that every point on the surface is accessible to the polishing tool. In contrast, no general statement can be made about convexity of the surfaces of a Galilean beam shaper—they may be concave, convex, or neither, depending on the details of the profile transformation. Oliker generalized this result to any reflective [77] or refractive [78] beam shaping system. In the absence of rotational symmetry, one cannot speak of a Galilean or Keplerian telescope, but nevertheless Oliker showed that the beam shaping optics always has one convex solution and one solution for which no general statement about convexity is possible [78].

6.5.2.1 Analysis and Simulation Results

The wavefront propagation through the focal region of this 1-to-1K beam shaper is illustrated in Figure 6.2, which was evaluated from Equation 6.28 for different values

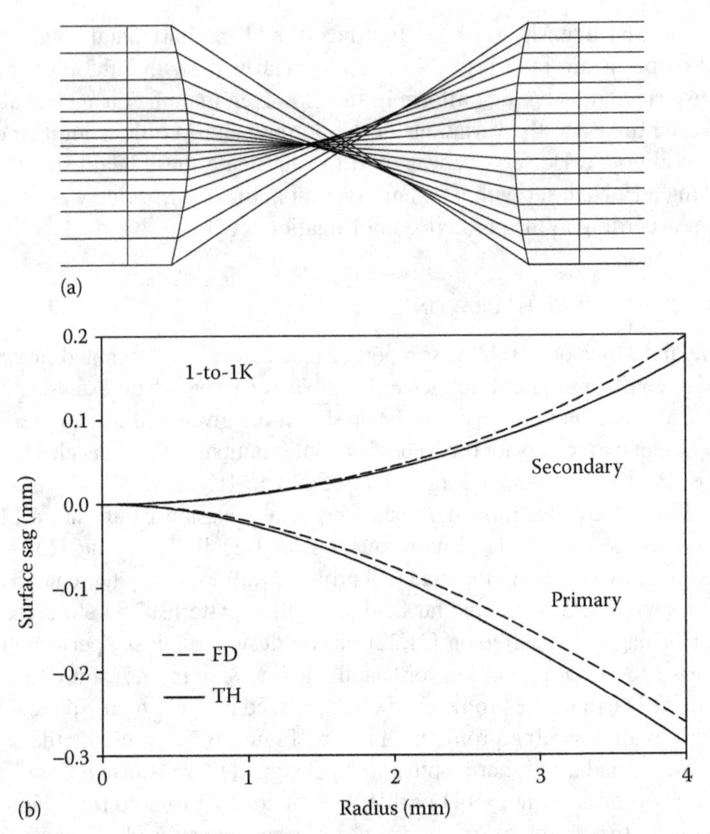

FIGURE 6.20 (a) Illustration of ray plot of the reshaping of a Gaussian input beam into a flattened irradiance output beam by a 1-to-1K configuration of a refractive beam shaper with $q = 15, n = 1.46071, t = 150$ mm, $R_{FL} = 3.25$ mm, and $w_0 = 2.366$ mm. (Reproduced from Shealy, D. L. and Hoffnagle, J. A. Wavefront and caustic surfaces of refractive laser beam shaper. In *Novel Optical Systems Design and Optimization X*, Koshel, R. J. and Gregory, G. G., eds., vol. 6668, 666805-1, 2007. With permission.) (b) Comparison of the sag of the primary and secondary aspheric surfaces of a 1-to-1K beam shapers that have been designed to transform a Gaussian profile into either a top-hat or a flattened irradiance FD profile as designed by Hoffnagle and Jefferson. (Reproduced from Shealy, D. L. and Hoffnagle, J. A. Aspheric optics for laser beam shaping. In *Encyclopedia of Optical Engineering*, Driggers, R., ed., Taylor & Francis, 2006. With permission.)

of the parameter s. Figure 6.2 presents a cross section of the refracted wavefront (solid lines) for $s = 0, 10, \ldots, 150$ mm and the caustic surfaces (dashed lines). For clarity and simplicity, only half of the wavefront (positive x-coordinates for input aperture) is plotted in Figure 6.2. Shealy and Hoffnagle [48,53] examined the behavior of the wavefront in more detail for the region from $z = 35$ to 95 mm where the wavefront passes through its caustic surfaces. For the scale used in Figure 6.2, it is difficult to understand how the wavefront and caustic surfaces interact, but it appears that the caustic surfaces represent a boundary or envelope for the wavefront $s = 40$ to 70 mm.

Before addressing the detailed interactions between the wavefront and its caustics within the focal region, it is helpful to understand the overall meridional and sagittal wavefront curvatures as the wavefront leaves the first aspheric lens and forms its caustic surfaces. As noted, Equation 6.31a and b give the principal radii of curvature of the wavefront as it has been refracted from the first plano-aspheric lens of the laser beam shaper. To better understand the changing curvature of the wavefront after refraction from the aspheric lens, the principal curvatures of the aspheric lens and refracted wavefront are presented in Figure 6.21. It is interesting to note from

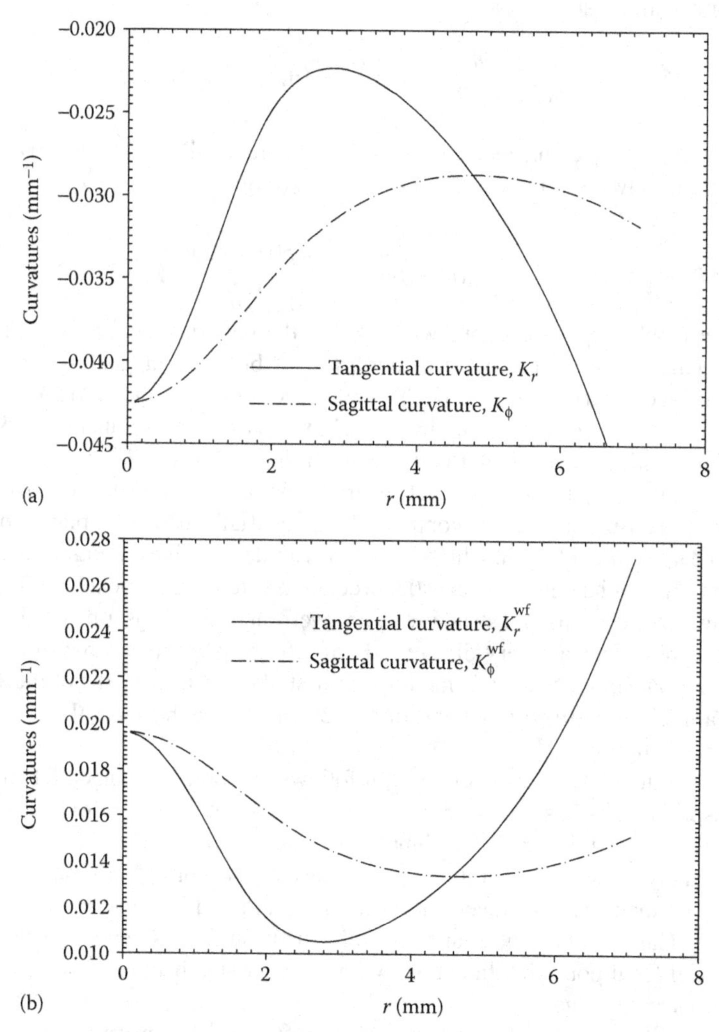

FIGURE 6.21 Tangential and sagittal curvatures of (a) first aspheric lens surface, evaluated in the meridional plane for the Keplerian laser beam shaper and (b) refracted wavefront from the first aspheric lens. (Reproduced from Shealy, D. L. and Hoffnagle, J. A. Wavefront and caustic surfaces of refractive laser beam shaper. In *Novel Optical Systems Design and Optimization X*, Koshel, R. J. and Gregory, G. G., eds., vol. 6668, 666805-1, 2007. With permission.)

Figure 6.21 that the axial curvature of the aspheric lens surface is approximately equal to $-0.0425\ \text{mm}^{-1}$, which can be understood from the following analysis. From Equation 6.86, we calculate the axial curvature of the first aspheric lens of the beam shaper, where we use the parameters $n_1 = 1.46072$, $n_2 = 1$, $t = 150$ mm, $q = 15$, $R_{FL} = 3.25$ mm, and $\epsilon = -1$ to obtain the following:

$$c_r(0) = -0.042579\ \text{mm}^{-1} \tag{6.124}$$

From paraxial optics and the lens maker formula, we then can calculate the paraxial focal point of the first aspheric

$$\frac{n_1}{o} + \frac{n_2}{z_c(0)} = \frac{n_2 - n_1}{R} = (n_2 - n_1)c_r(0) \tag{6.125}$$

where we set the object distance o equal to some large number or allow it to approach ∞ for a plane wave. Solving Equation 6.125 for $z_c(0)$ gives

$$z_c(0) = \frac{n_2}{(n_2 - n_1)c_r(0)} = 50.975\ \text{mm} \tag{6.126}$$

which we label as $z_c(0)$, since we will see that the paraxial focal point of the first aspheric lens is also equal to the z-coordinate of both the tangential and sagittal caustic surfaces on axis (Equation 6.29a and b). For on-axis incident light, the sagittal caustic is a spike or line along the optical axis, as shown in Equation 6.29b. The tangential caustic is typically in the shape of a horn, as shown in Figure 6.22a, where it folds back upon itself for the 1-to-1K beam shaper and has its horn region directed away from the aspheric lens. In contrast, the tangential caustic of a plano-spherical lens of the same focal length, which is shown with dashed lines in Figure 6.22a, has the horn region of its tangential caustic directed toward the spherical lens. Thus, one concludes that the shape and structure of a wavefront used for laser beam shaping or imaging applications are very different. Figure 6.22b presents a three-dimensional view of the tangential caustic formed by the first plano-aspheric lens of the 1-to-1K laser beam shaper with an input aperture of 5 mm, where the sagittal caustic is not visible in this figure.

Returning to analysis of Figure 6.21, it follows that both the tangential curvature of the aspheric lens has a maximum and the tangential curvature of the refracted wavefront has a minimum for an input Gaussian beam at the same ray height of approximately 2.8 mm. In addition, at the input ray height of 4.6 mm, the sagittal curvature of the refracted wavefront has a minimum and is equal to the tangential curvature. Thus, the tangential and sagittal caustic surfaces intersect as they do at the paraxial focal point, but the power within a Gaussian beam at $r = 2w_0$ is much less than its axial value.

To better illustrate how the wavefront and caustic surfaces interact, a significantly expanded scale has been presented in Figure 6.9a–d for $s = 60$, 70, 80, and 90 mm [48]. Shealy and Hoffnagle have identified that as the wavefront passes through a region where there is either a sagittal or tangential caustic surface, then the wavefront develops an inflection point when it touches the tangential caustic and folds back

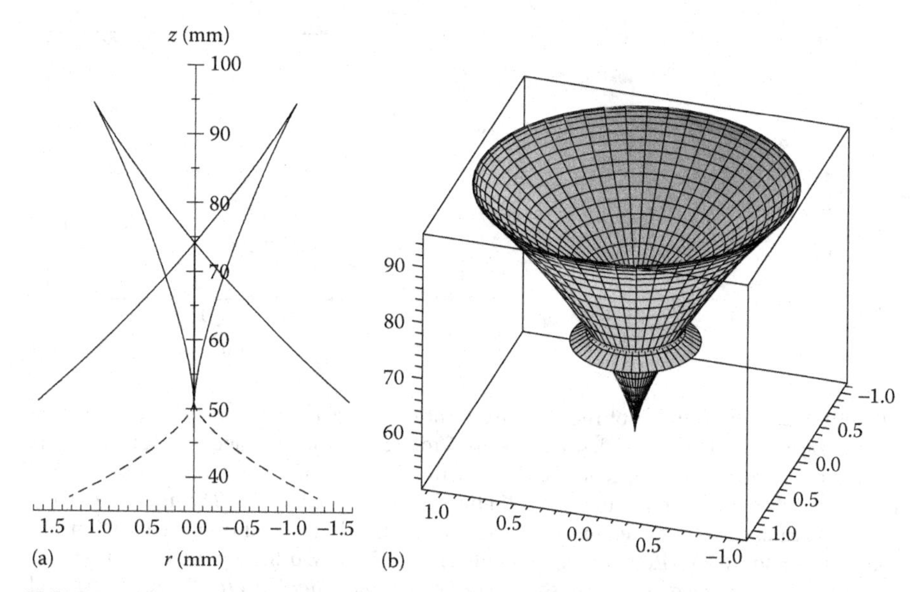

FIGURE 6.22 (a) Cross-sectional view of the tangential and sagittal caustic surfaces of a plano-spherical lens (dashed line) with the same focal length as the first plano-aspheric lens of a Keplerian laser beam shaper (solid line) described in Figure 6.20. The input aperture radius of the plano-aspheric lens is 6 mm, where the z-coordinate of the tangential caustic is approximately equal to the paraxial focal length. (b) **(See color insert.)** Three-dimensional view of the tangential caustic surface of the first plano-aspheric lens of a Keplerian laser beam shaper with an input aperture radius of 5 mm. The sagittal caustic spike is not visible in this view. (Reproduced from Shealy, D. L. and Hoffnagle, J. A. *J. Opt. Soc. Am. A*, 25, 2370–2380, 2008. With permission.)

upon itself with increasing input aperture x-coordinates until the wavefront touches a second part of the tangential caustic where the wavefront develops a second inflection point and folds back upon itself with further increasing input aperture coordinates. As the wavefront propagates through s = 70 to 80 mm, we see that the distance between the inflection points of the wavefront decreases until the wavefront passes the tangential caustic at a distance of 95 mm from the first aspheric lens as it continues to expand its size and becomes more planar until it refracts at the second aspheric lens where final phase corrections are introduced so that the wavefront leaves the beam shaper as a plane wave with the designed irradiance profile. This behavior suggests that there is a significant amount of interference happening within the focal region, which has been confirmed experimentally [54], and will be summarized below.

In addition, when the wavefront touches the caustic surfaces as described earlier, then the wavefront folds back upon itself, which has been discussed in more detail in Reference [48]. Since rays are always perpendicular to optical wavefronts, this means that the ray paths also become entangled among themselves. There are regions of image space where multiple rays from different regions of the aspheric surface actually intersect and interfere constructively or destructively. By carefully following wavefronts and their rays through the focal region, it is possible to identify regions

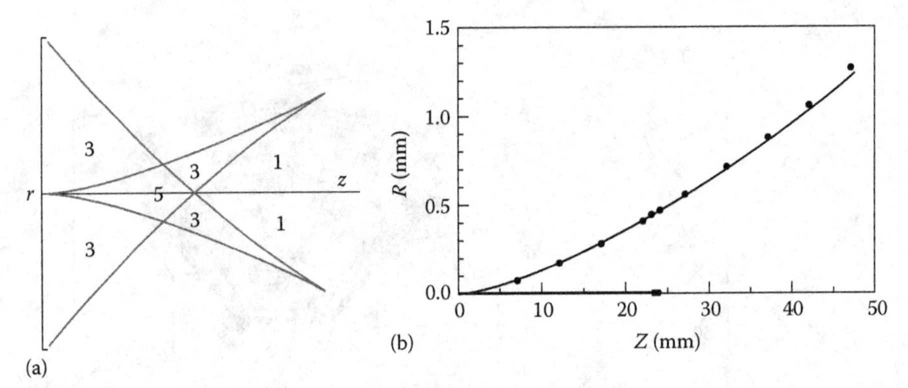

FIGURE 6.23 (a) Number of roots of inverse of generalized ray mapping function within regions of caustic surfaces of first lens of the 1-to-1K laser beam shaper. (b) Radius of the caustic rings, that is, the tangential caustic of the first lens of 1-to-1K laser beam shaper. Filled circles are measured results and solid lines are calculated results. The filled box on the *z*-axis indicates the range over which the sagittal caustic disappeared from the CCD images shown in Figure 6.25. (Reproduced from Hoffnagle, J. A. and Shealy, D. L. Caustic surfaces of a keplerian two-lens beam shaper. In *Laser Beam Shaping VIII*, Dickey, F. M. and Shealy, D. L., eds., vol. 6663, 666304-1, 2007. With permission.)

around and inside of the tangential caustic which have either one or three or five ray intersections, where all of these rays have phases that are likely different, and thus, there is constructive or destructive interference. Figure 6.23a displays the number of ray intersections within the regions of the tangential caustic surface of the first lens of the 1-to-1K beam shaper.

6.5.2.2 Experimental Results

The optics described in this section has been fabricated from fused silica and the aspheric surface figured by magneto-rheological figuring. The accuracy of the polished surfaces, measured interferometrically with the use of a computer-generated hologram, was approximately 25 nm. The irradiance after the beam shaper was measured with a charge-coupled device (CCD) sensor array of 1024 × 1024 pixels (12 μm square) placed directly in the output beam. The rms uniformity, as defined in Equation 6.49, was measured to be less than 5% in the central region enclosing 78% of the incident beam power [21]. Cross sections of the input and output beams—single rows of the CCD output going through the center of the beam—are presented in Figure 6.24, together with the theoretical Gaussian and FD profiles for this reshaper.

The departure of the output wavefront from a plane wave was 0.26λ peak-to-valley and 0.025λ rms [22,57], as measured with a Shack–Hartmann wavefront sensor, with $\lambda = 514$ nm for these measurements. The measured M^2 of the shaped beam was 1.8, which is approximately 25% larger than the value for an ideal, plane-wave FD beam, as shown in Figure 6.10. Measurements of the irradiance profile at several planes after the output aperture of the beam shaper are also presented in Reference [21]. The FD beam had negligible change in profile over a range of approximately 0.5 m, which was an adequate range for its intended use, but nevertheless considerably shorter than what one would predict for an ideal FD beam, based on the diffraction

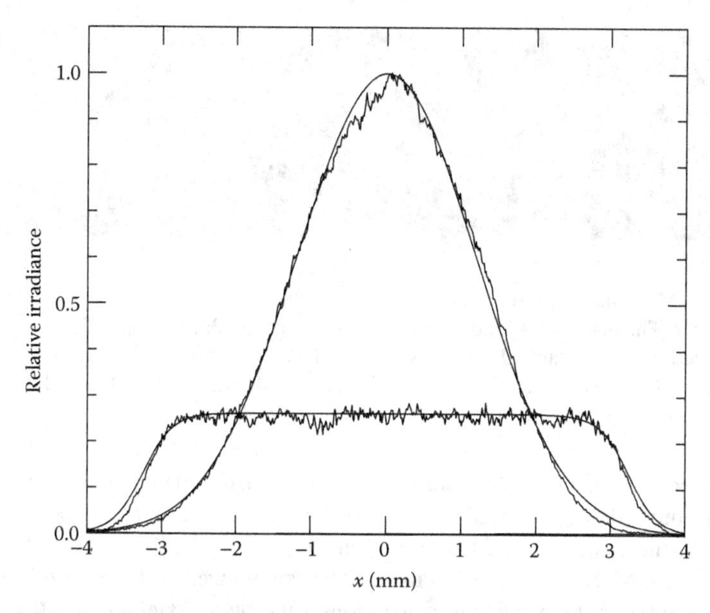

FIGURE 6.24 Irradiance profiles for a slice through the center of a laser beam before (Gaussian profile) and after (flat-top profile) the Keplerian beam shaper described in the text. The rough curves are measured pixel values from a CCD camera and the smooth curves are the ideal Gaussian and FD functions for which the beam shaper was designed. (Reproduced from Hoffnagle, J. A. and Jefferson, C. M. *Appl. Opt.*, 39, 5488–5499, 2000. With permission.)

integral alone as described in Section 6.2.3. Reshaping of a 257 nm beam was also demonstrated with the optics that was designed for a nominal wavelength of 514 nm when the lens spacing was adjusted according to Equation 6.121.

The results presented in Section 6.5.2.1 for the refracted wavefront and caustic surfaces, which are formed after an incident plane wave with a Gaussian irradiance profile has been refracted by the first aspheric lens of a Keplerian beam shaper, represent a complete geometrical solution of this problem. As discussed earlier, the caustic surfaces formed by the first aspheric lens of the Keplerian beam shaper have multiple inflection points, where both the tangential and sagittal caustics fold back upon themselves and intersect axial rays as well as off-axis rays near the full aperture of the lens. The refracted wavefront also develops inflection points when it touches some caustic points and folds back on itself as the irradiance is being redistributed over the wavefront. It has also been shown experimentally in this section that rays from different areas of the input aperture intersect over planes within the caustic region and create strong interference rings.

To test the geometrical theory of the evolution of the wavefront and irradiance, Hoffnagle and Shealy [54] analyzed the irradiance distribution at several planes after the first plano-aspheric lens of a 1-to-1K beam shaper similar to the one described above, where the 514 nm beam from an Ar-ion laser was used after being spatially filtered, collimated, and expanded to match the waist of $w_0 = 2.366$ mm as required by the design. Using a Shack–Hartman sensor, the wavefront was monitored and observed to be planar within an accuracy better than $\lambda/5$ for these measurements.

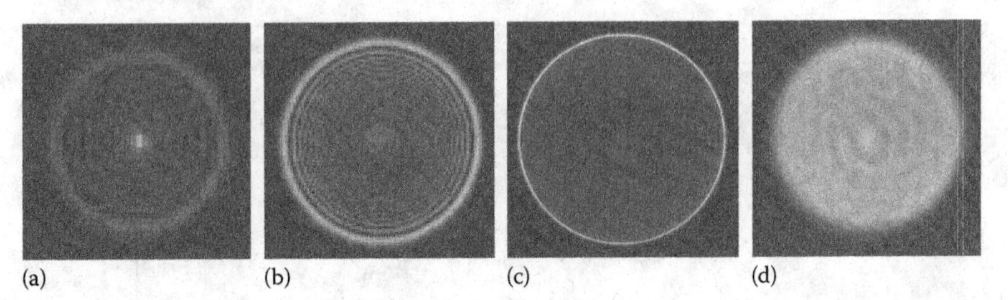

FIGURE 6.25 Images of the beam over four planes after the first lens of the 1-to-1K beam shaper. The distance behind the paraxial focus is (a) 12, (b) 23, (c) 42, and (d) 67 mm. (Reproduced from Hoffnagle, J. A. and Shealy, D. L. Caustic surfaces of a keplerian two-lens beam shaper. In *Laser Beam Shaping VIII*, Dickey, F. M. and Shealy, D. L., eds., vol. 6663, 666304-1, 2007. With permission.)

The 1024 × 1024 pixel CCD sensor was used to measure the irradiance after a thin-film aluminum filter which reduced the laser power and prevented sensor saturation. The CCD sensor was mounted on a micrometer-driven translation stage for precise measurements of the relative translations. The raw images of the shaped laser beam are shown in Figure 6.25 at different locations after the paraxial focus of the first lens of the beam shaper. Figure 6.25a shows the irradiance over a plane that is located just behind the paraxial focus of lens where the sagittal caustic spike is intense, while the outer ring represents the tangential caustic at this location; Figure 6.25b shows the irradiance over a plane where the axial caustic is weak and tangential caustic ring is strong; Figure 6.25c shows the irradiance over a plane near the end of the tangential caustic ring; and Figure 6.25d shows the irradiance over a plane outside of the caustic region as the irradiance profile is becoming uniform. Within the caustic region, there are oscillations of the irradiance, which are visible in Figure 6.25a and b and which are due to interference and diffraction of light within the focal region. From the CCD images, one can directly measure the diameter of the caustic rings and compare to the simulated results given in Figure 6.22a. Figure 6.23b presents a comparison between computed diameter of the tangential caustic to the experimentally measured diameter, where there is very good agreement between computed and measured locations of caustic surface from Figure 6.25.

6.5.3 Axial GRIN Lens Configuration

GRIN glasses have been shown to be able to provide additional degrees of freedom for designing optical systems. Sands [100] has shown that the contributions of an axial GRIN to the third-order aberrations of an optical system are equivalent to those of an aspheric surface. This suggests that the aspheric surfaces of the laser beam profile shaping (Section 6.5.1) can be replaced by axial GRIN lenses with spherical surfaces. Wang and Shealy [101] have demonstrated, without taking into explicit account the functional dependence of the index of refraction on the wavelength of light, that it is possible to design axial GRIN laser beam profile shaping systems with realistic materials and spherical surfaces (see Reference [101] for a more detailed discussion of the results presented in this section).

6.5.3.1 Mathematical Developments

The ray equations for propagation of light through GRIN materials follow from Fermat's principle and Equation 6.42. For an axial-GRIN material with the symmetry axis along the z-axis, the ray equations can be written in the following form [102, Chapter 5]:

$$x = x_0 + K_0 \int_{z_0}^{z} \frac{dz}{M} \tag{6.127}$$

$$y = y_0 + L_0 \int_{z_0}^{z} \frac{dz}{M} \tag{6.128}$$

with

$$M = \sqrt{n^2(z) - K^2 - L^2} = \sqrt{n^2(z) - K_0^2 - L_0^2} \tag{6.129}$$

where:

K, L, and M are the three optical direction cosines
K_0, L_0, and M are the initial values of the three optical direction cosines of ray within the GRIN material
x_0, y_0, and z_0 are the initial coordinates of the ray within the GRIN material

The geometrical configuration of a two-lens GRIN laser beam profile shaping system is shown in Figure 6.26. The optical axis is also the symmetry axis of both GRIN lenses. The input laser beam will not be deflected by the plano-surface of the first lens, but will diverge from the spherical surface S_1. The GRIN profile of the first lens will cause the rays at different heights to refract in such a way as to convert the input Gaussian intensity profile into a uniform intensity profile at the second lens.

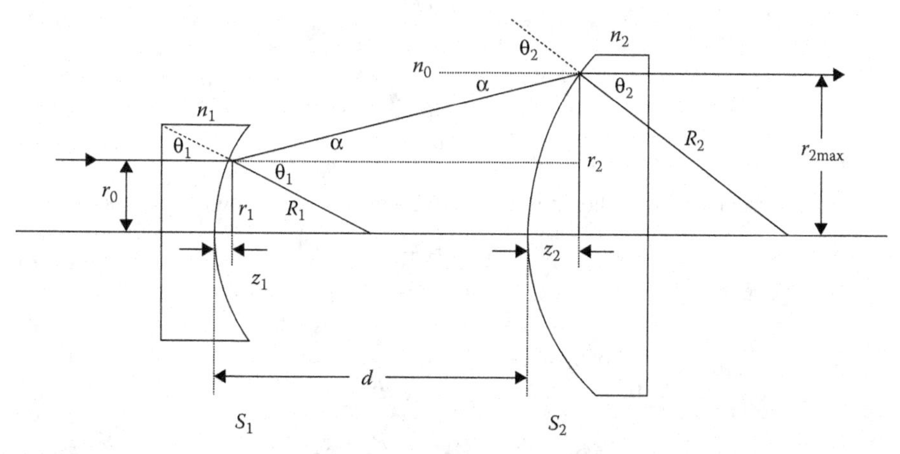

FIGURE 6.26 Geometrical configuration of a two-GRIN lens laser beam shaper. (Reproduced from Wang, C. and Shealy, D. *Appl. Opt.*, 32, 4763–4769, 1993. With permission.)

The spherical surface S_2 of the second lens will refract the rays so that the output beam will be parallel to the optical axis.

Following the discussion in Section 6.2.1.2 on the geometrical optics intensity equation, one can use Equations 6.36, 6.50, and 6.51 to evaluate the energy collected within a circle of radius r_1 given by

$$E_{\text{in}}(r_1) = 2\pi \int_0^{r_1} I_G(r')r'\text{d}r' = 1 - \exp\left[-2\left(\frac{r_1}{w_0}\right)^2\right] \tag{6.130}$$

If the beam reaches a uniform intensity of the top-hat irradiance profile of $\Sigma \equiv (\pi R_0^2)^{-1}$ within a circle of radius $r_2 < R_0$ on the second lens, then applying conservation of energy between the input and output beam gives

$$r_2 = R_0\sqrt{1 - \exp\left[-2\left(\frac{r_1}{w_0}\right)^2\right]} \tag{6.131}$$

As shown in Figure 6.26, the surfaces S_1 and S_2 are two spherical surfaces, whose surface equations are given by Equation 6.123 when using the following parameters for this GRIN application: $\kappa = 0 = A_{2i}$ and $c_i = R_i$; R_1 and R_2 are the radii of curvature of the spherical GRIN lenses. Then, the surface sag z_1 and z_2 can be written as

$$z_1 = \frac{r_1^2/R_1}{1 + \sqrt{1 - r_1^2/R_1^2}} \tag{6.132}$$

or

$$r_1^2 = 2z_1R_1 - z_1^2 \tag{6.133}$$

and

$$z_2 = \frac{r_2^2/R_2}{1 + \sqrt{1 - r_2^2/R_2^2}} \tag{6.134}$$

or

$$r_2^2 = 2z_2R_2 - z_2^2 \tag{6.135}$$

The geometrical relations shown in Figure 6.26 justify the following expressions:

$$\tan\alpha = \frac{r_2 - r_1}{d - z_1 + z_2} \tag{6.136}$$

$$\cot\theta_1 = \frac{R_1 - z_1}{r_1} = \frac{R_1 - z_1}{\sqrt{2R_1z_1 - z_1^2}} \tag{6.137}$$

$$\cot\theta_2 = \frac{R_2 - z_2}{r_2} = \frac{R_2 - z_2}{\sqrt{2R_2z_2 - z_2^2}} \tag{6.138}$$

Applying Snell's law at surfaces S_1 and S_2 gives

$$n_1(z_1)\sin\theta_1 = n_0\sin(\theta_1+\alpha) \tag{6.139}$$

$$n_2(z_2)\sin\theta_2 = n_0\sin(\theta_2+\alpha) \tag{6.140}$$

The GRIN profiles of these lenses can be determined from Equations 6.139 and 6.140. Applying the sum of two angles trigonometric formula, these equations can be written as

$$n_1(z_1) = n_0(\cot\theta_1\sin\alpha + \cos\alpha) \tag{6.141}$$

$$n_2(z_2) = n_0(\cot\theta_2\sin\alpha + \cos\alpha) \tag{6.142}$$

To evaluate the GRIN profile of the first lens, the right-hand side of Equation 6.141 needs to be expressed as a function of z_1. Combining Equations 6.131 and 6.133 gives

$$r_2 = R_0\sqrt{1-\exp\left[-2\frac{(2z_1R_1 - z_1^2)}{w_0^2}\right]} \tag{6.143}$$

Solving Equations 6.143 and 6.134 for z_2 as a function of z_1 leads to the following:

$$z_2 = \frac{R_0^2(1-\exp\{-2[(2z_1R_1 - z_1^2)/r_1^2]\}/R_2)}{1+\sqrt{1-R_0^2(1-\exp\{-2[(2z_1R_1 - z_1^2)/w_0^2]\})/R_2^2}} \tag{6.144}$$

The resulting expression for GRIN function of the first lens is

$$n_1(z_1) = n_0\left\{\left[\frac{R_1 - z_1}{(2R_1z_1 - z_1^2)^{1/2}}\right]\sin\alpha(z_1) + \cos\alpha(z_1)\right\} \tag{6.145}$$

In a similar fashion, the GRIN function for the second lens can be written in terms of z_2 to give

$$n_2(z_2) = n_0\left\{\left[\frac{R_2 - z_2}{(2R_2z_2 - z_2^2)^{1/2}}\right]\sin\alpha(z_2) + \cos\alpha(z_2)\right\} \tag{6.146}$$

Equations 6.145 and 6.146 are the formulas for the GRIN profiles of the two plano-spherical lenses in a refractive laser profile shaping system. These results are based on geometrical optics, energy conversion along a tube of rays, and the constant OPL condition. Now, these results will be used to design two GRIN laser beam profile shaper systems.

6.5.3.2 Using GRIN Lenses

The use of GRIN materials in optical systems has been limited by fabrication capabilities of these materials. Considerable progress has been made toward better controlling the GRIN profile while also increasing the change of the index of refraction and the depth of the gradient of the index [103–106] within the material. Until the recent development of the GRADIUM™ GSF glass family by LightPath Technologies, Orlando, FL [106], it has been difficult to obtain an overall index change larger than 0.08 and a depth of the gradient greater than 5 mm. It is possible to obtain linear and near-parabolic GRIN profiles. For the designs presented in this section [101], these constraints have been used—the maximum overall index of refraction change is 0.08 and the depth of the GRIN gradient is 5 mm. As GRIN technology improves, optical designs for laser profile shaping systems will be able to use a broader spectrum of materials to reduce size and cost of laser beam profile shaping systems. For example, the GRADIUM™ GSF glass family [107] has an overall index of refraction change ranging from 0.04 to 0.14 for a thickness of the gradient ranging from 6 to 26 mm. With this expanded range of parameters, more versatile GRIN laser beam profile shaping systems can be developed.

Using Equations 6.145 and 6.146, the GRIN profiles of a two-lens laser beam profile shaping system shown in Figure 6.26 can be computed for any set of layout parameters including the lens spacing d and the spherical lens surface vertex radii R_1 and R_2. However, only GRIN materials with realistic GRIN profiles can be used when building a laser beam profile shaping optical system. For a particular case, when the input and output beam radii, r_0 and r_{2max}, are given, the depth of the gradient Δz is completely determined by the vertex radius of a spherical lens. The relationship between the depth of the index gradient and the vertex radius is illustrated in Figure 6.27.

In summary, smaller vertex radii yield larger depth of the index gradient. From the equation (Equation 6.133) of a spherical surface S_1, it follows that if the depth of the index gradient Δz_1 has a specific value, then the lower limit on choosing R_1 will be given by

$$R_1 \geq \frac{r_0^2 + (\Delta z_1)^2}{2(\Delta z_1)} \tag{6.147}$$

It is interesting to note that the refraction of rays at different heights from the optical axis are determined by the vertex radius and the GRIN distribution across the beam diameter. The more planar a spherical surface is (larger vertex radius), then the greater the gradient of the index across the surfaces S_1 and S_2 will be for achieving the same deflection of the rays. That is, the overall index change Δn, which can be fabricated, sets the upper limit for choosing the vertex radii R_1 and R_2.

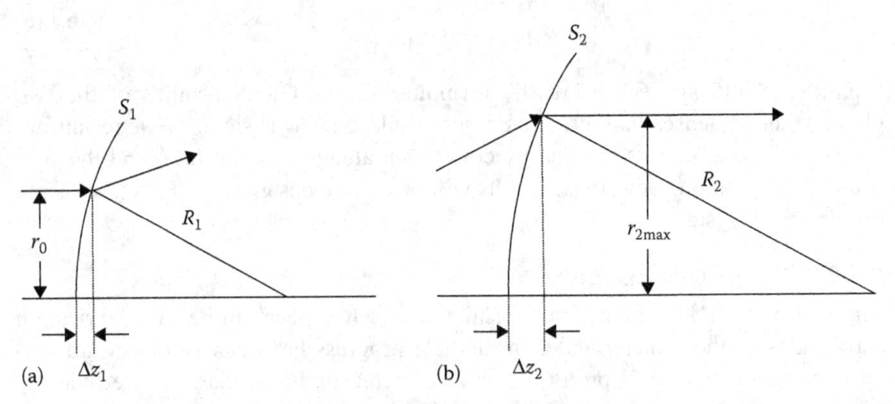

FIGURE 6.27 Relationship between the vertex radius and the depth of the GRIN material with (a) for the primary lens, S_1, and (b) for the secondary lens, S_2. (Reproduced from Wang, C. and Shealy, D. *Appl. Opt.*, 32, 4763–4769, 1993. With permission.)

The spacing d between the two lenses also affects the GRIN profile. Shorter systems require stronger index gradients to achieve the same redistribution of the laser beam profile. Selection of the layout parameters R_1, R_2, and d needs to be guided by the current GRIN fabrication technology. The relationship between GRIN characteristics (Δn and Δz_1) and system layout parameters (R_1 and d) of the primary lens for the GRIN laser beam profile shaping system is illustrated in Figure 6.28.

A similar plot for the secondary lens can be constructed. Then, for a given beam waist r_0 and manufacturing specifications for the GRIN material (Δn and Δz_1), the system layout parameters (R_1, R_2, and d) can be determined. Two specific laser profile shaping optical systems using GRIN lenses will be discussed in more detail in this section. The first system transforms the input Gaussian beam profile into a uniform output beam profile of the same diameter as the input beam. The second system expands the input beam by a factor of 2 while also transforming the input Gaussian beam to a uniform output beam profile.

Consider a laser profile shaping system with layout parameters given in Table 6.4 and illustrated in Figure 6.29. For this system, the input and output beams have the same diameter, which means that the marginal rays (displaced a distance r_0 from the optical axis) must not be deflected by the optical components. According to Snell's law, a ray will not be deflected by a surface when the index of refraction is same on both sides of the interface. This means that there must be a dense material of index n_0 connecting the two lenses. The connector in this design is a glass bar with the

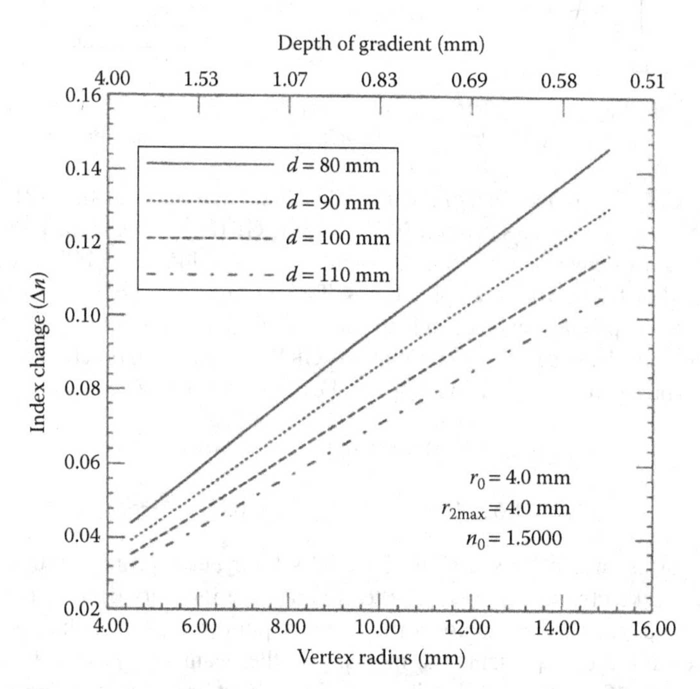

FIGURE 6.28 Relationship between GRIN characteristics and laser profile shaping optical system design parameters for the primary lens S_1. (Reproduced from Wang, C. and Shealy, D. *Appl. Opt.*, 32, 4763–4769, 1993. With permission.)

TABLE 6.4
Layout Parameters for a Laser Beam Shaping System with No Expansion of the Diameter of the Input Beam

System Variables	Values
Primary lens vertex radius, R_1	5.0 mm
Secondary lens vertex radius, R_2	5.0 mm
Spacing between lenses, d	100.0 mm
Incident beam waist (radius), r_0	114.0 mm
Exiting beam radius, r_{2max}	4.0 mm
Index of connector, n_0	1.5

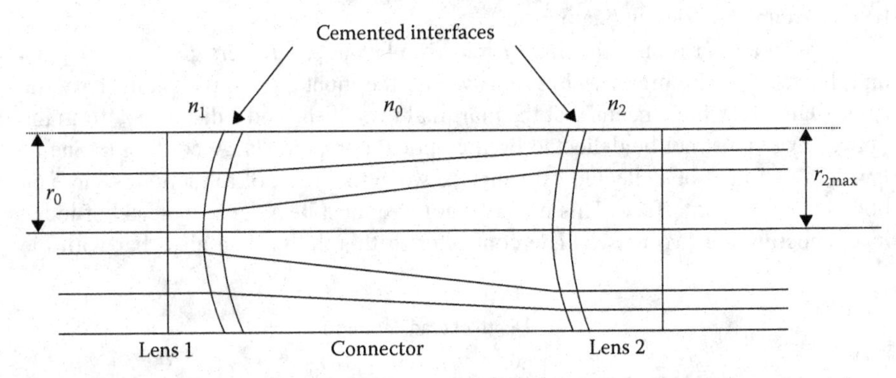

FIGURE 6.29 Layout of a GRIN 1-to-1 laser beam shaping system. (Reproduced from Wang, C. and Shealy, D. *Appl. Opt.*, 32, 4763–4769, 1993. With permission.)

same index of refraction as that of the base glasses used to fabricate the GRIN lenses. Using the layout parameters given in Table 6.4, the GRIN profiles of the primary and secondary lenses have been computed from Equations 6.145 and 6.146 as a function of the radial distance from the optical axis. These results are shown in Figures 6.30 and 6.31 for the primary and secondary lens materials, respectively.

Fitting with a least-squares technique, the GRIN profiles as a function of the sag z of each surface give the following empirical expressions for $n_1(z_1)$ and $n_2(z_2)$:

$$n_1 = 1.537910 - 0.036171z_1 + 0.008827z_1^2 \tag{6.148}$$

$$n_2 = 1.525456 - 0.010882z_2 - 0.000801z_2^2 \tag{6.149}$$

The self-consistency of this design of a GRIN laser beam profile shaping system has been checked by doing a ray trace to compute the intensity of the output beam. A grid of Gaussian distribution over the entrance pupil was used so that the number of rays per unit area represents the intensity of the beam as it passes through this optical system. The intensity distribution of the input and output beams is shown in Figure 6.32. It is clear from these results that the input Gaussian beam has been transformed into a uniform intensity output beam.

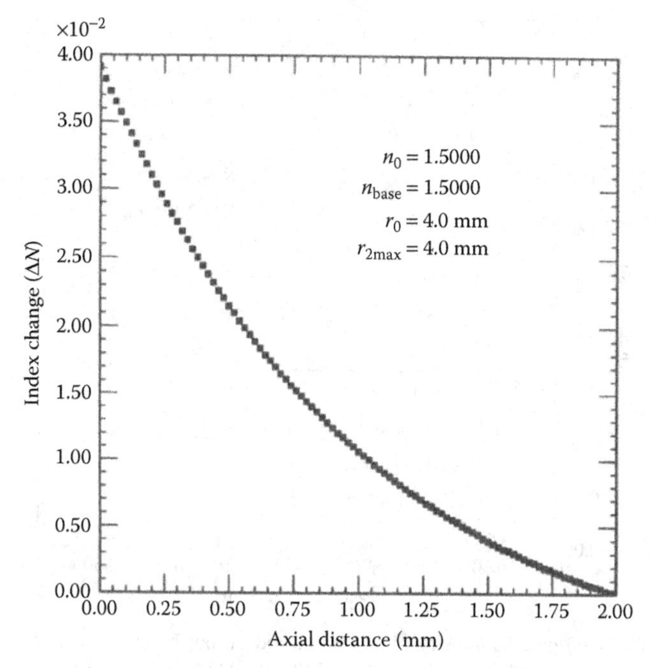

FIGURE 6.30 GRIN profile of the primary lens of a 1-to-1 laser beam shaper. (Reproduced from Wang, C. and Shealy, D. *Appl. Opt.*, 32, 4763–4769, 1993. With permission.)

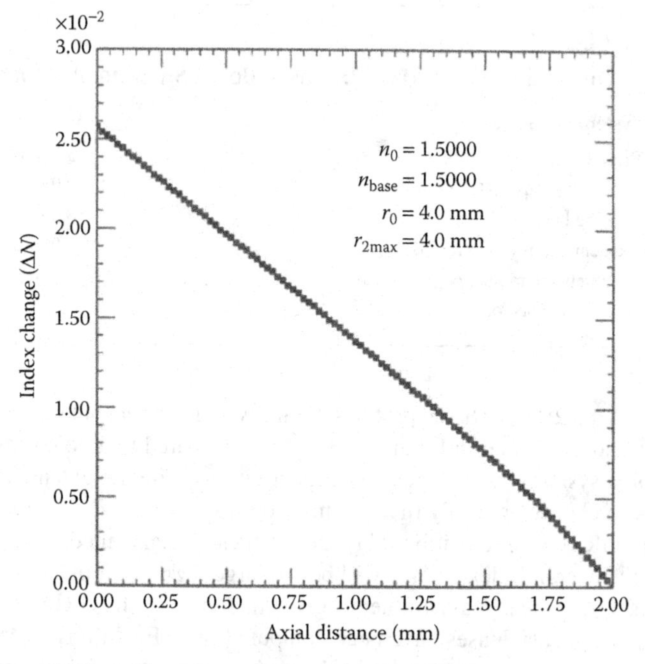

FIGURE 6.31 GRIN profile of the secondary lens of a 1-to-1 laser beam shaper. (Reproduced from Wang, C. and Shealy, D. *Appl. Opt.*, 32, 4763–4769, 1993. With permission.)

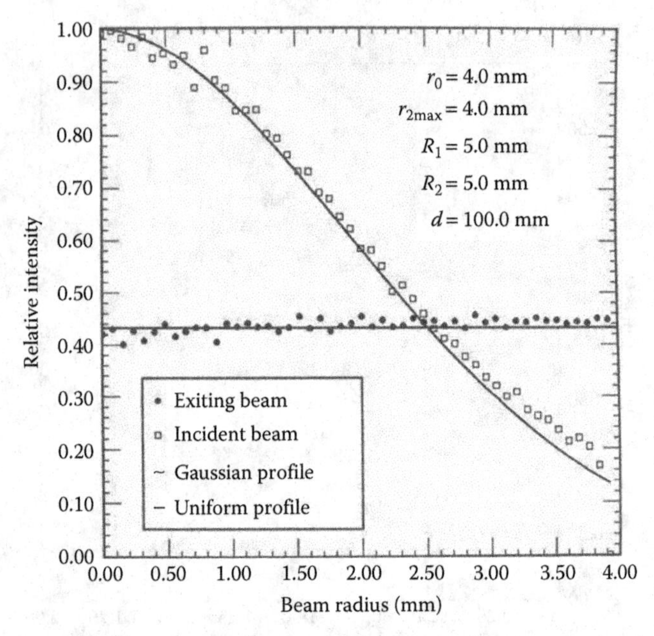

FIGURE 6.32 Computed intensity of the input and output beams of a 1-to-1 GRIN laser beam shaper. (Reproduced from Wang, C. and Shealy, D. *Appl. Opt.*, 32, 4763–4769, 1993. With permission.)

TABLE 6.5
Layout Parameters for a 2× Laser Beam Shaping System

System Variables	Values
Primary lens vertex radius, R_1	5.0 mm
Secondary lens vertex radius, R_2	10.0 mm
Spacing between lenses, d	150.0 mm
Incident beam waist (radius), r_0	4.0 mm
Exiting beam radius, r_{2max}	8.0 mm
Index of connector, n_0	1.5

Now consider a 2× laser beam profile shaping system with layout parameters given in Table 6.5 and illustrated in Figure 6.33. When compared to a nonexpanding laser profile shaping system, beam expanders deflect rays to a greater extent. Therefore, it is important to choose carefully the system layout parameters (R_1, R_2, and d) so that the resulting GRIN profile can be fabricated. Unless a dense medium connects the two lenses, the overall index change will be too large for current material fabrication technologies. Using the layout parameters given in Table 6.5, the GRIN profiles of the primary and secondary lenses have been computed from Equations 6.145 and 6.146 as a function of the radial distance from the optical axis. These results are shown in Figures 6.34 and 6.35 for the primary and secondary lens materials, respectively.

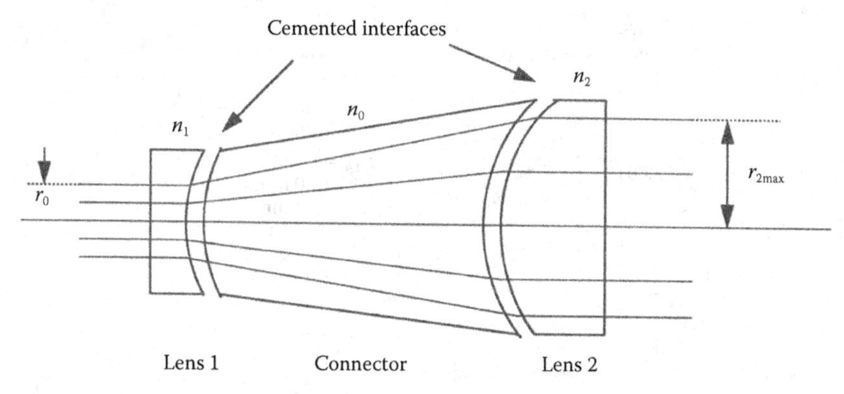

FIGURE 6.33 Layout of a GRIN 2× expander and laser beam shaping system. (Reproduced from Wang, C. and Shealy, D. *Appl. Opt.*, 32, 4763–4769, 1993. With permission.)

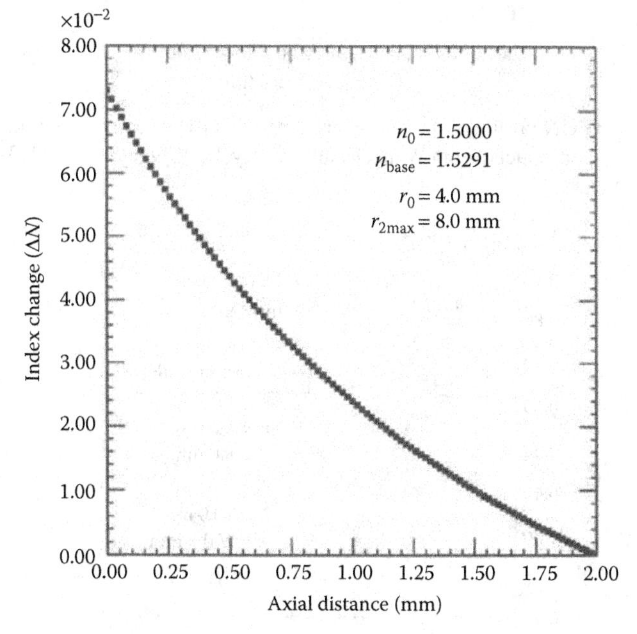

FIGURE 6.34 GRIN profile of the primary lens of a GRIN 2× expander and laser beam shaping system. (Reproduced from Wang, C. and Shealy, D. *Appl. Opt.*, 32, 4763–4769, 1993. With permission.)

Fitting with a least-squares technique, the GRIN profiles as a function of the sag z of each surface give the following empirical expressions for $n_1(z_1)$ and $n_2(z_2)$:

$$n_1(z_1) = 1.600350 - 0.059840z_1 + 0.012423z_1^2 \tag{6.150}$$

$$n_2(z_2) = 1.567088 - 0.009410z_2 \tag{6.151}$$

Similarly, the self-consistency of this design of a GRIN laser beam profile shaping system has been checked by doing a ray trace to compute the intensity of the output

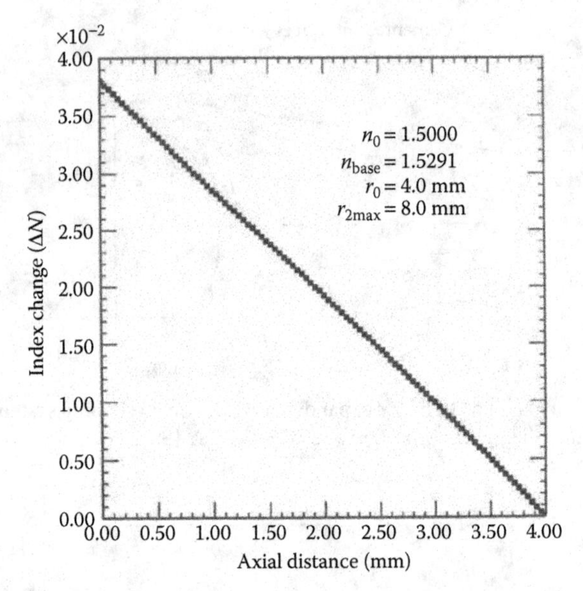

FIGURE 6.35 GRIN profile of the secondary lens of a GRIN 2× expander and laser beam shaping system. (Reproduced from Wang, C. and Shealy, D. *Appl. Opt.*, 32, 4763–4769, 1993. With permission.)

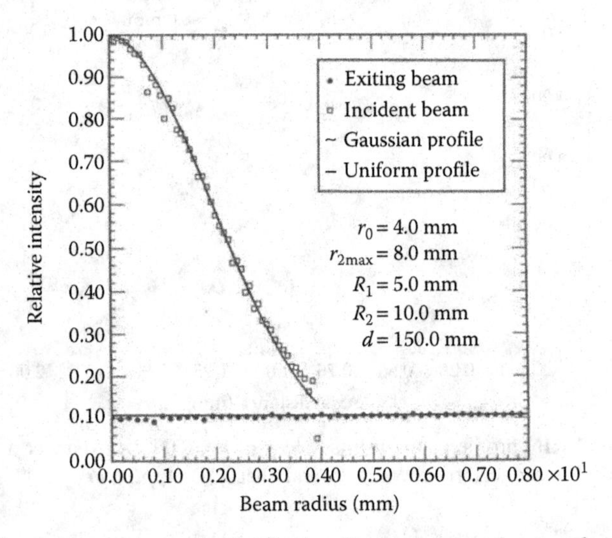

FIGURE 6.36 Computed intensity distribution for the input and output laser beams of a GRIN 2× expander and laser beam shaping system. (Reproduced from Wang, C. and Shealy, D. *Appl. Opt.*, 32, 4763–4769, 1993. With permission.)

beam. A grid of Gaussian distribution over the entrance pupil was used so that the number of rays per unit area represents the intensity of the beam as it passes through this optical system. The intensity distribution of the input and output beams are shown in Figure 6.36. It is evident that the Gaussian input beam has been transformed into a uniform output beam.

6.5.3.3 Summary of GRIN Profile Shaping System

The theory and design procedures for using axial GRIN plano-convex lenses with spherical surfaces have been presented. This is in contrast to the aspheric surfaces required for constant index materials, as presented in Section 6.5. Two axial GRIN laser beam profile shaping systems—2× expander and nonexpander—have been designed, analyzed, and shown via simulations to transform a Gaussian input beam into a uniform intensity profile output beam. Least-squares fitting techniques have shown that the index of refraction of the required GRIN materials is either a linear or near-parabolic functions of the axial distance, where the depth of the gradient and the overall index change are within current GRIN fabrication techniques. As GRIN fabrication technologies improve as illustrated by the LightPath Technologies development of the GRADIUM™ GSF glass family, there will be new opportunities for building compact and versatile laser profile shaping systems.

6.6 REFLECTIVE BEAM SHAPERS

Applying geometrical optics to the design of reflective laser beam shapers follows similar procedures used for the design of refractive laser beam shapers as described in Section 6.8.2. These reflective systems may or may not have a central obscuration. One- and two-mirror systems with central obscuration have been used for shaping the irradiance profile of laser beams [6,7], where solving differential equations has been used to determine the shape of the mirror surfaces which achieve the desired redistribution of the irradiance. Cornwell [26,27] has developed a general approach using nonprojective transformations to design two-mirror laser beam profile shaping systems with either rectangular or polar symmetry. Malyak [8] has designed a two-mirror unobscured optical system using rotationally symmetric aspheric to convert an input Gaussian beam into a uniform intensity output beam.

In Section 6.6.1, the differential equation approach of McDermit and Horton [6,7] for designing a one-mirror system to transform a collimated input beam profile into prescribed illumination of a receiver surface will be summarized. Next, in Section 6.3.3, the nonprojective transformations of Cornwell [26,27] and the differential equation approach of Malyak [8] have been used to describe the design of two-mirror laser profile shaping systems with either rectangular or polar symmetry.

6.6.1 One-Mirror Profile Shaping Systems

Consider the geometrical configuration of a one-mirror laser beam profile shaping system shown in Figure 6.37. The input radiation is collimated (parallel to optical axis) with a known intensity profile. The receiving surface is illuminated with a prescribed intensity distribution while the output beam is not collimated. Unit vectors along the input and output beams are given by

$$\mathbf{a} = \mathbf{k} \tag{6.152}$$

$$\mathbf{A} = \mathbf{a} - 2\mathbf{n}_1(\mathbf{a}\cdot\mathbf{n}_1) = \frac{2z'\mathbf{r} - (1 - z'^2)\mathbf{k}}{(1 + z'^2)} \tag{6.153}$$

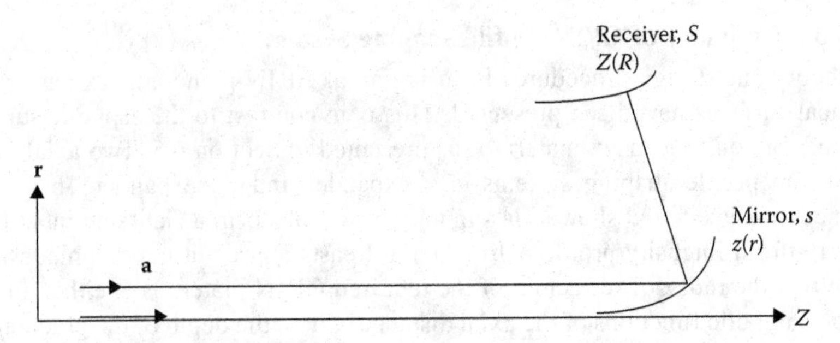

FIGURE 6.37 Geometrical configuration of a one-mirror laser beam profile shaping system.

where Equations 6.40 and A.3 were used. The ray trace equation connecting the mirror surface s with the receiving surface S in the r–z plane is given by

$$\frac{R-r}{Z(R)-z(r)}=\frac{A_r}{A_z}=\frac{2z'(r)}{-[1-z'^2(r)]} \tag{6.154}$$

where:
$z = z(r)$ represents the unknown equation of the mirror surface s
$Z = Z(R)$ represents the known equation of the receiving surface S

Equation 6.154 can be written as

$$-(R-r)z'^2+2(Z-z)z'+(R-r)=0 \tag{6.155}$$

Applying the differential energy balance equation (Equation 6.35) to this problem gives

$$I_1(r)2\pi r\,\mathrm{d}r = I_2(R)2\pi R[\mathrm{d}R^2+\mathrm{d}Z^2]^{1/2} \tag{6.156}$$

where:
$I_1(r)$ is the beam intensity incident upon the first mirror surface
$I_2(R)$ is the intensity incident upon the second mirror surface

Equation 6.156 can be rearranged into the form:

$$\frac{\mathrm{d}Z}{\mathrm{d}r}=\frac{I_1(r)}{I_2(R)}\frac{r}{R}\frac{2}{\left[1+\left(\frac{\mathrm{d}R}{\mathrm{d}Z}\right)^2\right]^{1/2}} \tag{6.157}$$

where:
$I_1(r)$, $I_2(R)$, and $Z(R)$ are the known functions of their respective variables
$z(r)$ is an unknown function at this point of the analysis

In addition, note that the ray trace equation (Equation 6.154) expresses a mapping between surfaces s and S:

$$(r,z)\Rightarrow(R,Z) \tag{6.158}$$

which implies that R is a function of r, $R(r)$. From the chain rule for differentiation of a function of function, the term (dZ/dr) in Equation 6.157 can be written as

$$\frac{dZ}{dr} = \frac{dZ}{dR}\frac{dR}{dr} = Z'(R)\frac{dR}{dr} \tag{6.159}$$

where:

$Z'(R) = dZ(R)/dR$ can be evaluated directly from the equation of the surface S
dR/dr can be evaluated from the ray trace equation (Equation 6.155)

Differentiating Equation 6.155 with respect to r gives

$$-2z'''(R-r) - z'^2\left(\frac{dR}{dr} - 1\right) + 2z''(Z-z) + 2z'\left(\frac{dZ}{dr} - z'\right) + \left(\frac{dR}{dr} - 1\right) = 0 \tag{6.160}$$

Combining Equations 6.159 and 6.160 leads to

$$2z''[-z'(R-r) + (Z-z)] + \frac{dZ}{dr}\left[(1-z'^2)\left(\frac{dR}{dZ}\right) + 2z'\right] - (1+z'^2) = 0 \tag{6.161}$$

However, rewriting Equation 6.155 gives the relationship

$$(Z-z) = \frac{(R-r)}{2z'}(z'^2 - 1) \tag{6.162}$$

which has been used to express Equation 6.161 in the following form:

$$\frac{z''}{z'} = \frac{1}{(R-r)}\left\{\left(\frac{dZ}{dr}\right)\left[\left(\frac{dR}{dZ}\right)\frac{(1-z'^2)}{(1-z'^2)} + \frac{2z'}{(1-z'^2)}\right] - 1\right\} \tag{6.163}$$

Replacing the term (dZ/dr) in Equation 6.163 with the right-hand side of Equation 6.157 gives the following differential equation for the mirror surface in terms of known functions:

$$\frac{z''}{z'} = \frac{1}{(R-1)}\left\{\frac{I_1(r)}{I_2(R)}\left(\frac{r}{R}\right)\frac{\left[\left(\frac{dR}{dZ}\right)\left(\frac{1-z'^2}{1-z'^2}\right) + \frac{2z'}{(1+z'^2)}\right]}{\sqrt{1+\left(\frac{dR}{dZ}\right)^2}} - 1\right\} \tag{6.164}$$

The above equation is equivalent to Equation 3.14 [6] and Equation 13 [7]. When appropriate boundary conditions are given, then Equation 6.164 can be solved for the shape of the mirror surface which will illuminate the receiver surface S with a prescribed intensity $I_2(R)$ for a given source intensity profile $I_1(r)$. McDermit and Horton [6,7] develop an extension of this analysis to two-mirror intensity profile shaping systems. A number of specific solutions for both one- and two-mirror systems are given in Refs. [6,7] including two laser beam profile shaping systems:

1. Uniform illumination of a plane perpendicular to the incident beam using a one-mirror system for an input Gaussian beam (Figure 7.3 [6])
2. Uniform illumination of a plane on the optical axis with a two-mirror system for an input Gaussian beam (Figure 7.9 [6])

In Section 6.6.2, the nonprojective transformations of Cornwell [26,27] will be presented as a geometrical method for designing two-mirror laser beam profile shaping system for either rectangular or polar symmetry, which are described in Section 6.6.

6.6.2 Two-Mirror Laser Beam Shapers

In this section, the design equations of a two-mirror intensity profile shaping system will be developed. Results in both rectangular and polar coordinate systems will be presented. For more details and applications of these results, the interested reader is encouraged to see Refs. [8,26]. Reference [27] is the original source of the development of nonprojective transformations in optics used to develop the material presented in this section. Development of the design equations for rectangular and polar coordinate systems will follow the seven-step recipe of Cornwell [26,27] for nonprojective transformation summarized in Section 6.3.3. The geometrical configuration of nonprojective transformations is illustrated in Figure 6.38. The nonprojective transformations represent a mapping between the input plane and the output beam which takes into account conservation of energy, constant OPL of wavefront between the input and output planes, and the ray trace equations. The input and output beams will be assumed to be collimated and parallel to the optical axis.

6.6.2.1 Systems with Rectangular Symmetry

The relationship between the element of areas on the input and output planes is illustrated in Figure 6.39 and can be written in the following form:

$$I_{in}(x,y)\mathrm{d}x\mathrm{d}y = I_{out}(X,Y)\mathrm{d}X\mathrm{d}Y \tag{6.165}$$

where:

I_{in} and I_{out} are the input and output intensity profiles of the laser beam, respectively

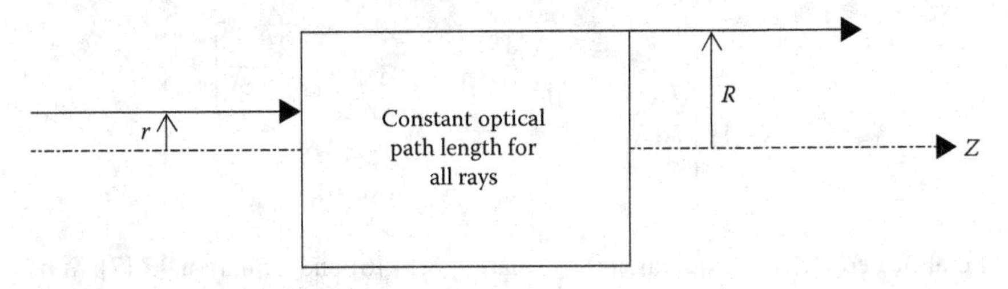

FIGURE 6.38 Geometry of the input plane and output plane for a nonprojective transformation. Either rectangular or polar coordinate systems can be used, depending on the symmetry of the laser beam profiles and optical system. (Reproduced from Shealy, D. L. and Chao, S.-H. *Opt. Eng.*, 42, 3123–3138, 2003. With permission.)

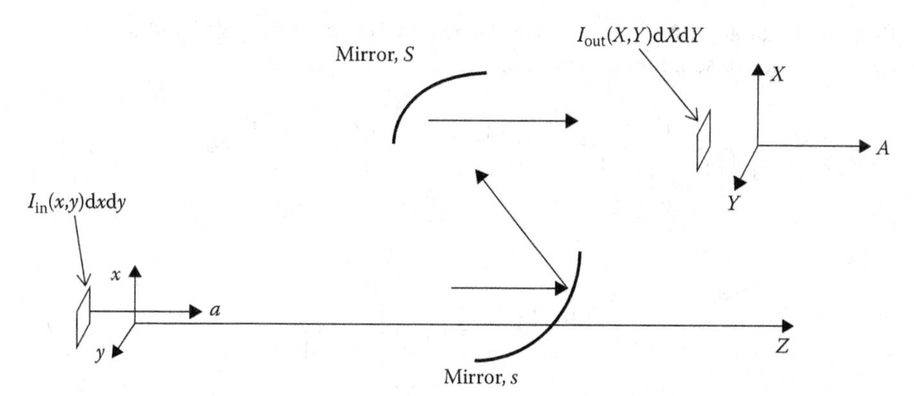

FIGURE 6.39 Symbolic relationship between the input and output elements of the area in the rectangular coordinates. Conservation of energy within the corresponding element of areas on the input and output planes is one principle used to determine the optical surface shapes of the laser beam shapers. The constant OPL condition of all rays passing from the input to the output plane is the second principle used to determine the optical surface shapes of laser beam shaper systems. For two-mirror systems, the axial OPL is the geometrical path length L. (Reproduced from Shealy, D. L. and Chao, S.-H. *Opt. Eng.*, 42, 3123–3138, 2003. With permission.)

The total energy must also be conserved which is represented by integrating Equation 6.165 over the full aperture of the input and output planes:

$$\int_{\text{Full input aperture}} I_{\text{in}}(x,y)\text{d}x\text{d}y = \int_{\text{Full output aperture}} I_{\text{out}}(X,Y)\text{d}X\text{d}Y \tag{6.166}$$

For laser beam profile shaping systems with rectangular symmetry, assume that the input and output intensity profiles can be separated into a product of one-dimensional amplitude functions, as illustrated in Section 6.3.3:

$$I_{\text{in}}(x,y) = a_x(x)a_y(y) \tag{6.167}$$

$$I_{\text{out}}(X,Y) = A_X(X)A_Y(Y) \tag{6.168}$$

Allowing for nonuniform shaping of a laser beam profile in two orthogonal directions, assume that there is an independent and nonuniform magnification of the x and y ray coordinates between the input and output planes:

$$X = m_x(x)x \tag{6.169}$$

$$Y = m_y(y)y \tag{6.170}$$

The rectangular magnifications $m_x(x)$ and $m_y(y)$ can be determined by imposing the incremental expression of conservation of energy (Equation 6.165) for the separated intensity functions (Equations 6.167 and 6.168):

$$a_x(x)a_y(y)\text{d}x\text{d}y = A_X(X)A_Y(Y)\left\{\frac{\partial[xm_x(x)]}{\partial x}\right\}\text{d}x\left\{\frac{\partial[ym_y(y)]}{\partial y}\right\}\text{d}y \tag{6.171}$$

Rewriting Equation 6.171 with terms depending on x on the left-hand side of the equation leads to separation of variables:

$$\frac{a_x(x)}{A_X[m_x(x)]}\frac{1}{\left\{\frac{\partial[xm_x(x)]}{\partial x}\right\}}=\frac{A_Y[m_y(y)]}{a_y(y)}\left\{\frac{\partial[ym_y(y)]}{\partial y}\right\}=\frac{1}{C_1},\text{ constant} \tag{6.172}$$

or

$$\left\{\frac{\partial[xm_x(x)]}{\partial x}\right\}=C_1\frac{a_x(x)}{A_X[m_x(x)]} \tag{6.173}$$

$$\left\{\frac{\partial[ym_y(y)]}{\partial y}\right\}=\frac{1}{C_1}\frac{a_y(y)}{A_Y[m_y(y)]} \tag{6.174}$$

The constant C_1 is determined from the boundary conditions, such as the magnifications at the edge of the beam. Integrating Equations 6.173 and 6.174 gives

$$m_x(x)=\frac{1}{x}\left\{C_1\int_0^x\frac{a_x(u)du}{A_X[um_x(u)]}+C_2\right\} \tag{6.175}$$

$$m_y(y)=\frac{1}{y}\left\{\frac{1}{C_1}\int_0^y\frac{a_y(v)dv}{A_Y[vm_y(v)]}+C_3\right\} \tag{6.176}$$

where:

C_2 and C_3 are constants determined by boundary conditions, such as the magnification of a rim ray, as described in detail by Shealy and Chao [84]

Equations 6.175 and 6.176 are integral equations for the x–y ray magnifications. For many applications using laser beam profile shaping systems, the output intensity profile is uniform and solution to either the differential equations (Equations 6.173 and 6.174) or the integral equations (Equations 6.175 and 6.176) is straightforward.

Now, the ray trace equations connecting points on the input plane to the output plane will be developed. The geometrical configuration of a two-mirror laser beam profile shaping system is illustrated in Figure 6.40. The unit ray vector $\mathbf{A}_{1\to2}$ connecting the two mirror surfaces $s[x,y,z(x,y)]$ and $S[X,Y,Z(X,Y)]$ along a ray path can be written in the following form:

$$\mathbf{A}_{1\to2}=\frac{(X-x)\mathbf{i}+(Y-y-h)\mathbf{j}+(L+Z-z)\mathbf{k}}{\sqrt{(X-x)^2(Y-y)^2+(L+Z-z)^2}} \tag{6.177}$$

where:

L is the distance along the z-axis separating the local coordinate system on each mirror

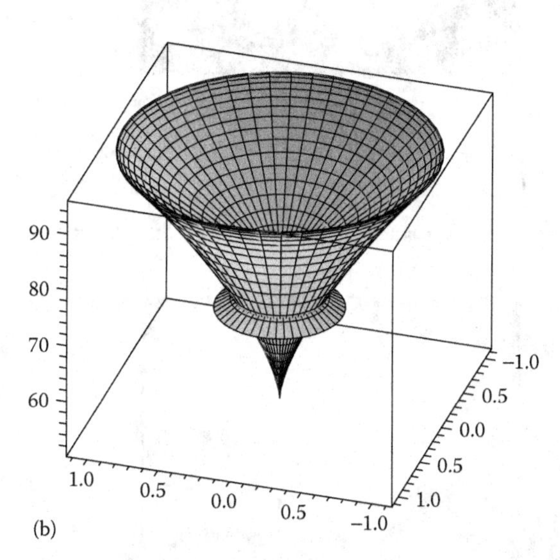

FIGURE 6.22 (b) Three-dimensional view of the tangential caustic surface of the first plano-aspheric lens of a Keplerian laser beam shaper with an input aperture radius of 5 mm. The sagittal caustic spike is not visible in this view. (Reproduced from Shealy, D. L. and Hoffnagle, J. A. *J. Opt. Soc. Am. A*, 25, 2370–2380, 2008. With permission.)

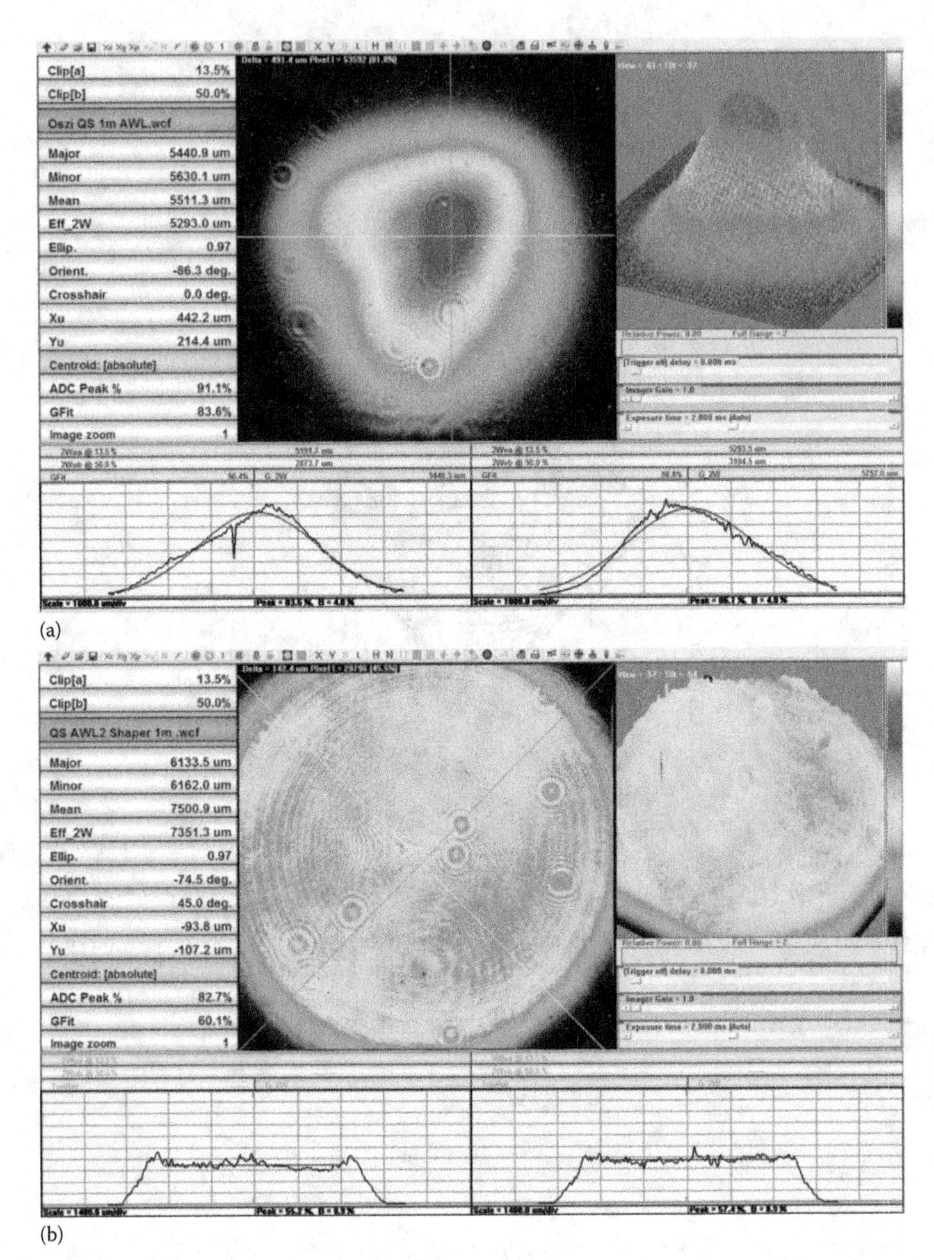

FIGURE 7.22 Example of beam shaping with πShaper: (a) input TEM_{00}; (b) output from πShaper. (Courtesy of InnoLas Laser GmbH; Reproduced from Laskin, A. and Laskin, V., Variable beam shaping with using the same field mapping refractive beam shaper, in *Laser Resonators and Beam Control XIV*, Kudryashov, A. V., Paxton, A. H., and Ilchenko, V. S., eds., SPIE, Bellingham, WA, 2011. With permission.)

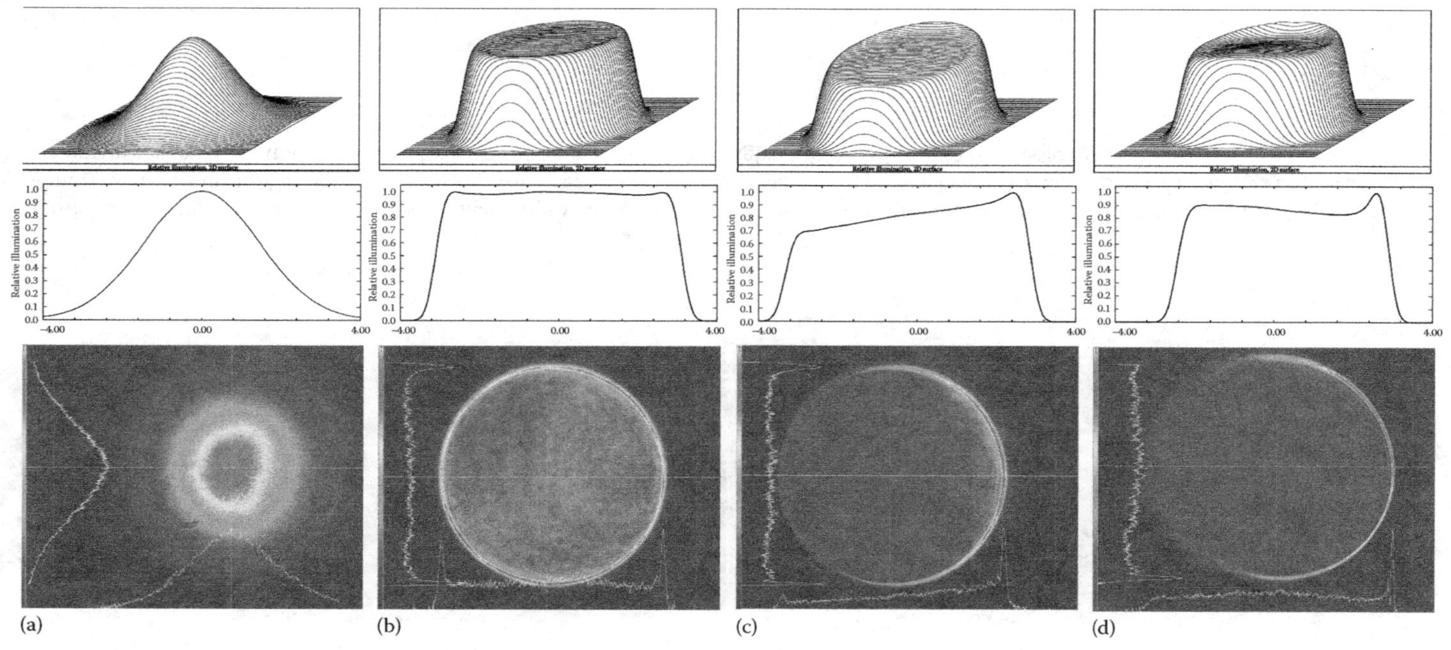

FIGURE 7.23 Evaluation of πShaper to sensitivity of misalignments, theoretical and experimental intensity profiles: (a) input TEM_{00} beam; (b) output beam with perfect alignment; (c) output with lateral shift of 0.5 mm; (d) output with tilt of 1°. (Reproduced from Laskin, A. and Laskin, V., Variable beam shaping with using the same field mapping refractive beam shaper, in *Laser Resonators and Beam Control XIV*, Kudryashov, A. V., Paxton, A. H., and Ilchenko, V. S., eds., SPIE, Bellingham, WA, 2011. With permission.)

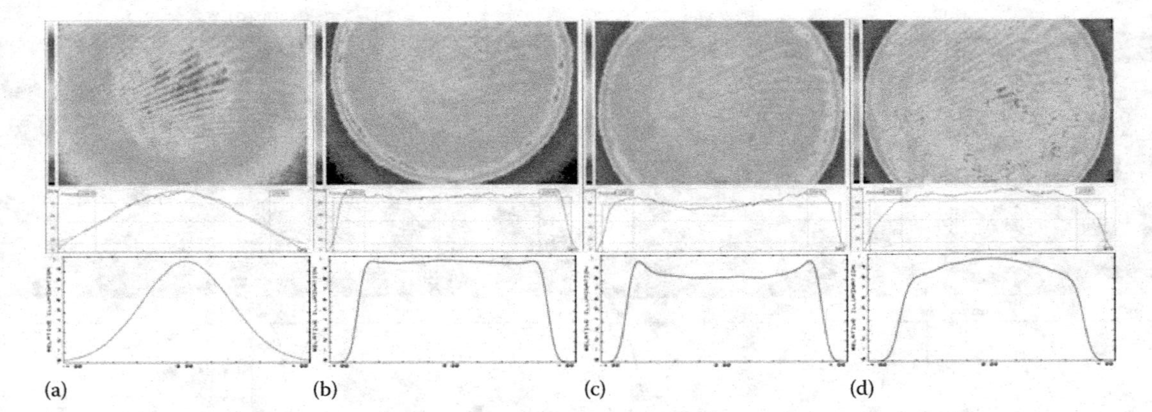

FIGURE 7.26 Experimental and theoretical intensity profiles as described within the text. (Figure was provided by IPG Photonics; Reproduced from Laskin, A. and Laskin, V., Variable beam shaping with using the same field mapping refractive beam shaper, in *Laser Resonators and Beam Control XIV*, Kudryashov, A. V., Paxton, A. H., and Ilchenko, V. S., eds., SPIE, Bellingham, WA, 2011. With permission.)

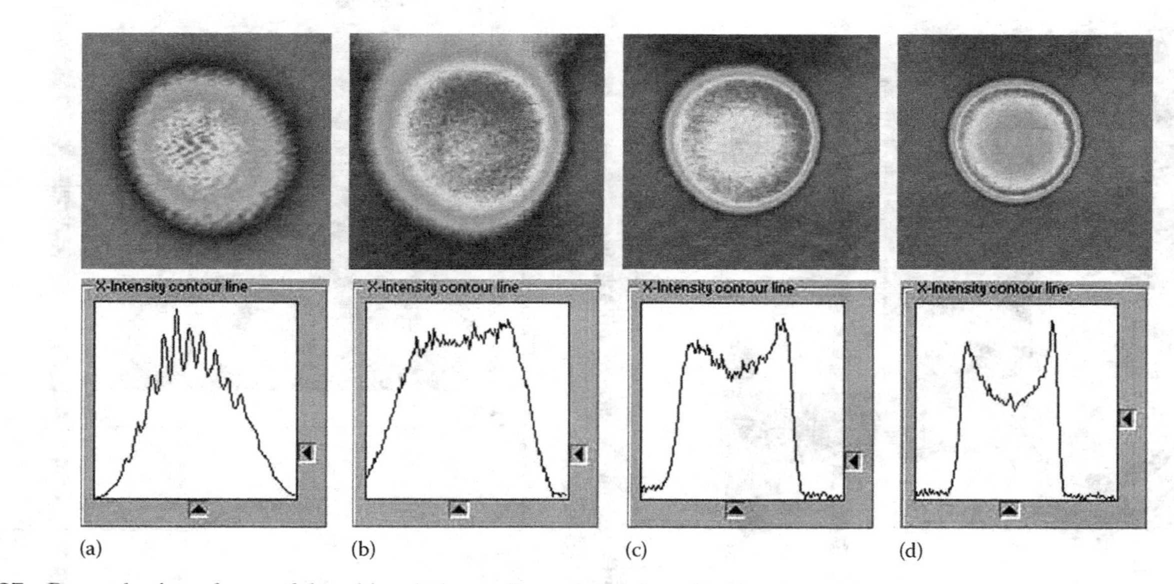

FIGURE 7.27 Beam shaping of powerful multimode laser. (Reproduced from Laskin, A. and Laskin, V., Refractive beam shapers for material processing with high power single mode and multimode lasers, in *Laser Resonators and Beam Control XV*, Kudryashov, A. V., Paxton, A. H., and Ilchenko, V. S., eds., SPIE, Bellingham, WA, 2013; Laskin, A. and Laskin, V., *Proceedings of the ICALEO*, 707, 2012. With permission.)

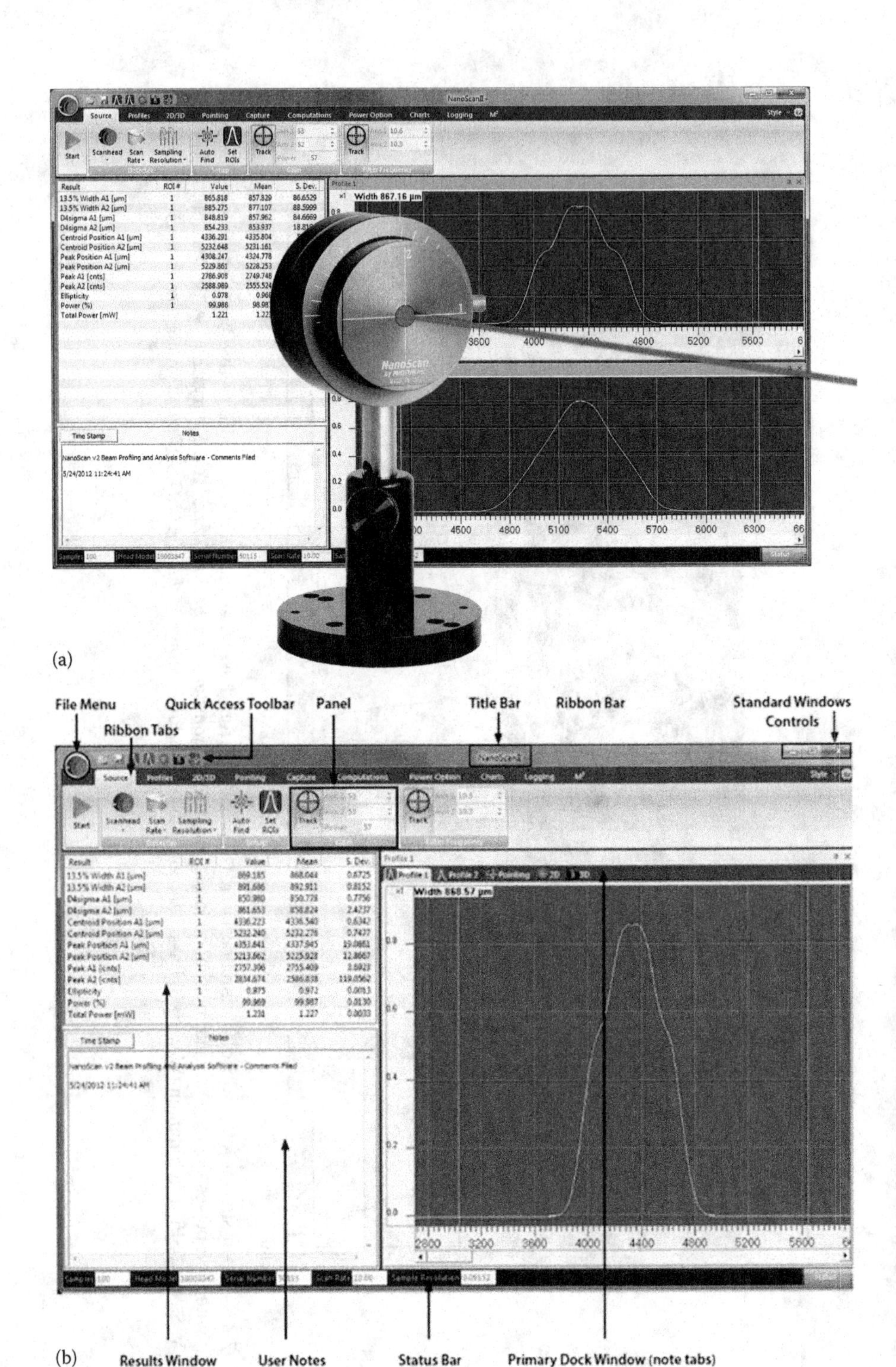

FIGURE 12.10 (a) Commercial knife scanner; (b) Windows PC display.

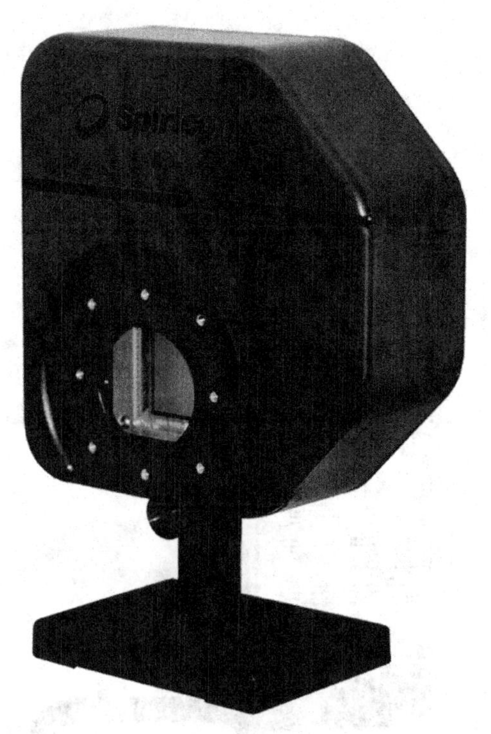

FIGURE 12.14 Pyroelectric camera video graphics array (VGA) output of CO_2 laser.

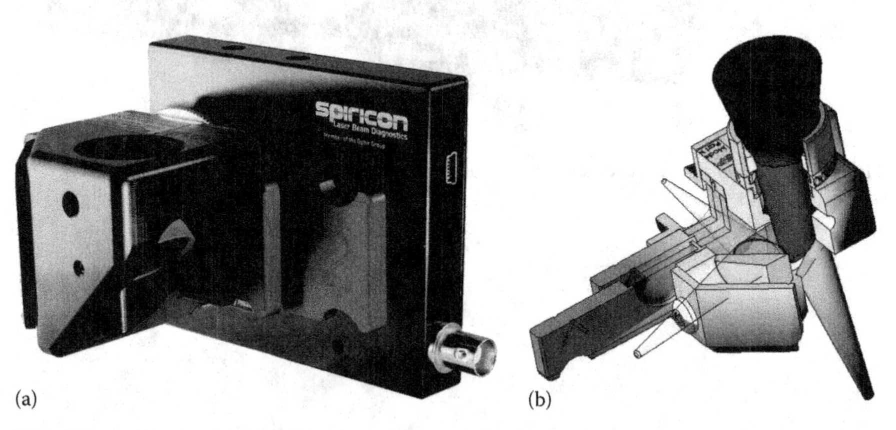

FIGURE 12.16 (a) Combined beam splitter and ND filter holder; (b) mechanical diagram of combined beam splitter and ND filter holder.

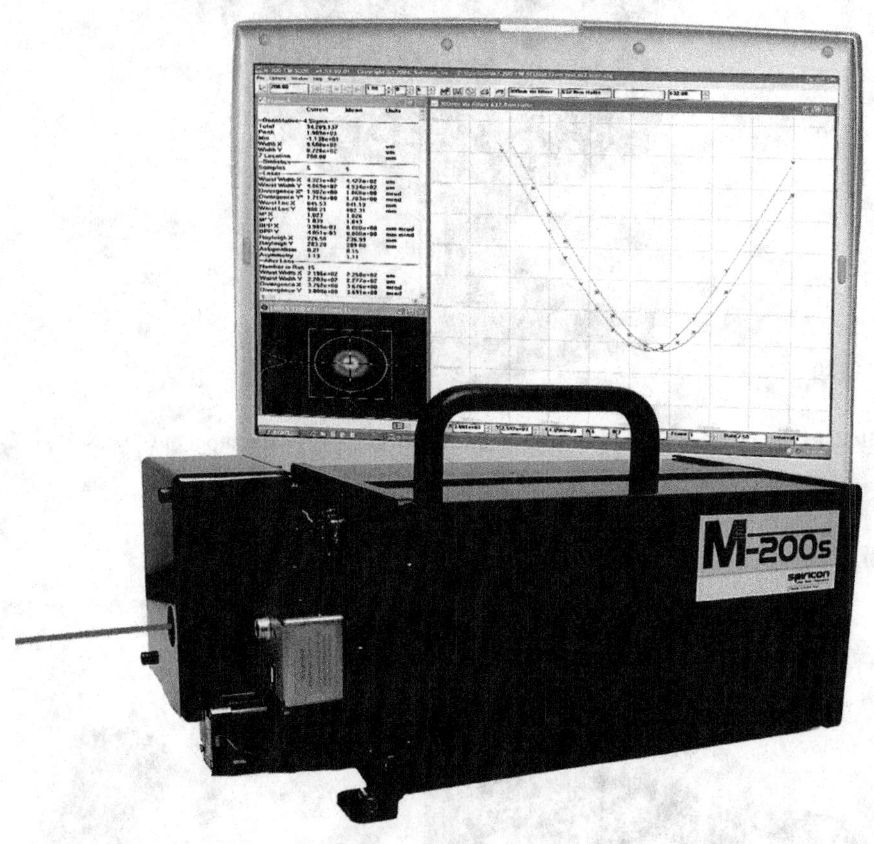

FIGURE 12.33 M^2 measurement display and calculation.

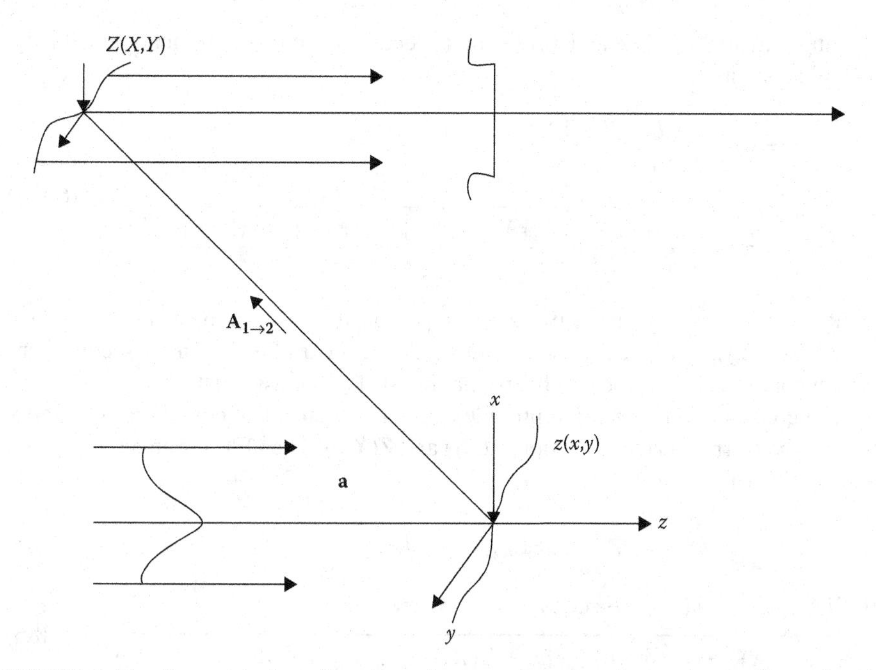

FIGURE 6.40 Geometrical configuration of a two-mirror laser beam shaper. (Reproduced from Shealy, D. L. and Chao, S.-H. *Opt. Eng.*, 42, 3123–3138, 2003. With permission.)

The functions $z(x,y)$ and $Z(X,Y)$ are the sag of each mirror in the local coordinate system of each mirror. Using Equation 6.40, the ray vector connecting each mirror may also be written in terms of the slope of the first mirror:

$$\mathbf{A}_{1\to 2} = \mathbf{a} - 2\mathbf{N}(\mathbf{a} \cdot \mathbf{N}) = \frac{-2z_x\mathbf{i} - 2z_y\mathbf{j} + (1 - z_x^2 - z_y^2)\mathbf{k}}{(1 + z_x^2 + z_y^2)} \tag{6.178}$$

where:

$$\mathbf{a} = \mathbf{k} \tag{6.179}$$

$$\mathbf{n} = \frac{-z_x\mathbf{i} - z_y\mathbf{j} + \mathbf{k}}{\sqrt{1 + z_x^2 + z_y^2}} \tag{6.180}$$

$$z_x = \frac{\partial z(x,y)}{\partial x} \text{ and } z_y = \frac{\partial z(x,y)}{\partial y} \tag{6.181}$$

Equating the x–y and x–z components of the ray vector $\mathbf{A}_{1\to 2}$ from Equation 6.178 to the corresponding coordinates on the ray vector leads to the following ray path equations:

$$\frac{z_x}{X - x} = \frac{z_y}{Y - y - h} \tag{6.182}$$

$$\frac{-2zx}{X - x} = \frac{1 - z_x^2 - z_y^2}{L + Z - z} \tag{6.183}$$

Solving Equations 6.182 and 6.183 for z_x leads to a quadratic equation with the following solutions:

$$z_x = \frac{(X-x)[L+Z(X,Y)-z(x,y)]}{(X-x)^2+(Y-y-h)^2} \pm \frac{(X-x)\sqrt{[L+Z(X,Y)-z(x,y)]^2+(X-x)^2+(Y-y-h)^2}}{(X-x)^2+(Y-y-h)^2} \quad (6.184)$$

Equation 6.184 is a partial differential equation for the unknown mirror surface functions $z(x,y)$ and $Z(X,Y)$, which can be determined after adding a second constraint condition that these two beam shaping surfaces must satisfy.

The constant OPL condition provides another independent condition to be satisfied by the mirror surface functions $z(x,y)$ and $Z(X,Y)$. The OPL for an axial ray and a general ray is given by

$$\mathrm{OPL}_{\mathrm{Axial\ ray}} = \sqrt{L^2 + \mathrm{h}^2} \quad (6.185)$$

$$(\mathrm{OPL})_{\mathrm{General\ ray}} = Z(X,Y) - z(x,y) + \sqrt{(X-x)^2 + (Y-y-h)^2 + [L + Z(X,Y) - z(x,y)]^2 + Z(X,Y) - z(x,y)} \quad (6.186)$$

However, the OPL is a constant for all rays. Therefore, equating the right-hand side of Equations 6.185 and 6.186 leads to the following expression for $Z(X,Y) - z(x,y)$:

$$Z(X,Y) - z(x,y) = \frac{-(x-X) - (y-Y+h)^2 + h^2}{2(L+l_0)} \quad (6.187)$$

where:

$$l_0 = \sqrt{L^2 + h^2}$$

Using the negative sign in Equation 6.184 as the physically meaningful solution, it has been shown in Reference [84] that combining Equations 6.182, 6.184, and 6.187 leads to the following expressions for $z_x(x,y)$ and $z_y(x,y)$:

$$z_x = \frac{x - X}{L + l_0} \quad (6.188)$$

$$z_y = \frac{y - Y + h}{L + l_0} \quad (6.189)$$

Assuming the sag of the first mirror can be written in the form:

$$z(x,y) = \int z_x(x,y)\mathrm{d}x + \int z_y(x,y)\,\mathrm{d}y \quad (6.190)$$

where:

$z_x(x,y)$ and $z_y(x,y)$ are given by Equations 6.188 and 6.189

Then, the sag of the first mirror can be written as

$$z(x,y) = \frac{1}{L + l_0}\left\{ \int_0^x u[1 - m_x(u)]\,du + \frac{1}{L + l_0}\int_0^y v[1 - m_y(v)]\,dv + C_4 \right\} \tag{6.191}$$

where:

C_4 is a constant of integration

Equations 6.169 and 6.170 have been used in the above equation. Expressions for m_x and m_y in Equation 6.191 are found from Equations 6.173 through 6.176. Finally, an expression for the sag of the second mirror follows from Equation 6.187:

$$Z(X,Y) = \left[z(x,y) + \frac{-(x - Y)^2 - (y - Y + h)^2 + h^2}{2(L + l_0)} \right]_{\substack{x=m_x^{-1}(X)X \\ y=m_y^{-1}(Y)Y}} \tag{6.192}$$

where the x and y terms are eliminated from Equation 6.192 by solving for the inverse of the magnifications.

A number of interesting applications of these results are presented in Refs. [8,27,84]. These applications include transformation of a linear ramp beam profile of one slope to another slope with different offset distances from the x or y axes and development of an unobscured two-mirror laser profile shaping system.

6.6.2.2 Systems with Polar Symmetry

A general solution for the shapes of the two mirror surfaces of a rotationally symmetric laser profile shaping system is given in Refs. [26,27]. In this section, the results developed in the previous section for rectangular symmetry will be used to obtain equations for the sag of the two mirror surfaces for a laser profile shaping system with polar symmetry.

Assuming both the input and output beam profiles have rotational symmetry, then a ray entering the system at a polar angle $\theta = \theta_0$ with respect to the x–y coordinate system of the first mirror will leave the system at a polar angle $\Theta = \Theta_0$ with respect to the X–Y coordinate system of the second mirror. Converting to polar coordinates and setting $\theta = \Theta$ as a result of radial symmetry of the beams, then the partial derivatives of the surface sag of the first mirror are given by

$$g_r = g_x\frac{\partial x}{\partial r} + g_y\frac{\partial y}{\partial r} = \frac{1}{L + l_0}(r - R + hr\sin\theta) \tag{6.193}$$

$$g_\theta = g_x\frac{\partial x}{\partial \theta} + g_y\frac{\partial y}{\partial \theta} = \frac{1}{L + l_0}(hr\cos\theta) \tag{6.194}$$

where:

$$x = r\cos\theta;\;\; y = r\sin\theta;\;\; X = R\cos\Theta;\; Y = R\sin\Theta \tag{6.195}$$

Integrating Equations 6.193 and 6.194 gives

$$g(r,\theta) = \frac{1}{L + l_0}\left\{\int_0^r r'[1 - m(r')]\,dr' + hr\sin\theta\right\} \tag{6.196}$$

where $m(r)$ is given by Equation 6.98. An equation for the sag of the second mirror, $G(R,\Theta)$, can be determined by substituting Equation 6.196 into Equation 6.192. Alternatively, an expression for $G(R,\Theta)$ follows from geometric considerations when the input and output beams are parallel to the z-axis. As a result of symmetry, the surface of the first mirror at the point (r,θ) is parallel to the surface of the second mirror at the point (R,Θ), where (r,θ) and (R,Θ) are connected by a ray. Thus, the partial derivatives of g and G are equal, that is, $G_R = g_r$ and $G_\Theta = g_\theta$. Following a similar derivation leading to Equation 6.196 gives

$$G(R,\Theta) = \frac{1}{L + l_0}\left\{\int_0^R R'[m^{-1}(R') - 1]\,dR' + hR\sin\Theta\right\} \tag{6.197}$$

Recall that beam shaping places a constraint on the ray heights as expressed through the radial magnification given by Equations 6.97 and 6.98.

This completes the analysis of a two-mirror system. To design a specific two-mirror laser beam shaping system, the radial magnification function $m(r)$ must be determined. Then, the mirror surface sag functions are computed from Equations 6.196 and 6.197. Malyak [8] presents specific examples of a Gaussian input beam being transformed into a uniform intensity output beam with a smaller diameter. Integral equations for the sag of the first- and second-mirror surfaces are given, and several numerical results for the mirror surfaces are also presented in Refs. [8,84].

6.7 CONCLUSION

The ability to control the irradiance profile of a laser beam is an important asset to the optical designer. We have shown how the basic laws of geometrical optics, the eikonal equation, and the intensity transformation law allow one to analyze a beam shaping optical system, including wavefronts, irradiance profiles, and caustics. A nonimaging (anamorphic) transformation of a bundle of parallel rays to another bundle of parallel rays, which we describe by a ray mapping function, results in a transformation of irradiance profiles. Such a transformation can be implemented using many optical technologies; we have considered refractive, reflective, and gradient index systems in detail.

The most straightforward implementations of laser beam shaping optics require two aspheric reflecting or refracting surfaces, one to redirect the incident rays so as to satisfy the desired ray mapping function and the other to ensure that the output beam has a planar wavefront (collimated rays). Points on these surfaces must satisfy three constraints imposed by the ray mapping function, the equal OPL condition, and the law of reflection or refraction at the surfaces. These three conditions define exactly two solutions for the beam shaping mirrors or lenses. For the special case of rotationally symmetric laser beams, by far the most important practical situation, the constraints on the beam shaping surfaces can be combined to give exact design equations in the form of ordinary differential equations that can be solved numerically to any desired precision. More specifically, for a rotationally symmetric two-lens refractive beam shaper the aspheric surfaces are each described by a simple integral equation, first derived by Kreuzer in 1964.

In practice, the most common application of laser beam shaping optics has been the transformation of a Gaussian to a flat-top profile. Many lasers emit in the Gaussian TEM_{00} mode and non-Gaussian laser beams can often be made Gaussian or nearly so by the simple expedient of spatially filtering with an aperture or a single-mode fiber. As an output profile, uniform illumination is ideal for many important applications, including illumination, lithography, and materials processing (such as annealing, cutting, and drilling). We have described several families of functions that are suitable for flat-top laser beam irradiance profiles. This discussion necessarily extends beyond geometrical optics per se to include the effects of diffraction on flat-top profiles. For a beam that approaches a top-hat profile, diffraction severely limits the range over which the beam exhibits an approximately uniform irradiance. The propagation of such flat-top beams can be described by Siegman's M^2 parameter or by numerical integration of the diffraction equation.

In addition to the equations for the design of beam shaping optics, we have also discussed optical and mechanical tolerances, which are essential considerations for anyone involved in the practical applications of fabrication, mounting, and alignment of beam shaping optics. The effects of dispersion on the aspheric lens doublet also naturally fit into this discussion.

Finally, we have described the measured performance of fabricated refractive beam shapers of both Galilean and Keplerian designs. Irradiance profiles, wavefronts, caustic surfaces (in the case of the Keplerian telescope), and wavelength-dependent operation are all in good agreement with optical theory. Laser beam shaping based on the principles of geometrical optics is thus established as an important piece of the optical designer's toolkit.

REFERENCES

1. Bokor, N. and Davidson, N. Anamorphic, adiabatic beam shaping of diffuse light using a tapered reflective tube. *Opt. Commun.* 201: 243, 2002.
2. Kreuzer, J. Laser light redistribution in illuminating optical signal processing systems. In *Optical and Electro-Optical Information Processing*, Tippett, J. T., Berkowitz, D. A., Clapp, L. C., Koester, C. J., and Vanderburgh, A. J., eds., Cambridge, MA: Massachusetts Institute of Technology Press, p. 365, 1965.
3. Sinzinger, S. and Jahn, J. *Microoptics*. Weinheim, Germany: Wiley-VCH, 2003.
4. Dickey, F. M., Weichman, L., and Shagam, R. Laser beam shaping techniques. In *High-Power Laser Ablation III*, Phipps, C., ed., Bellingham, WA: SPIE, p. 338, 2000.
5. Shealy, D. L. History of beam shaping. *Opt. Sci. Engin.* 102: 307–347, 2006.
6. McDermit, J. H. Curved reflective surfaces for obtaining prescribed irradiation distributions. PhD thesis, University of Mississippi, Mississippi, 1972.
7. McDermit, J. H. and Horton, T. E. Reflective optics for obtaining prescribed irradiative distribution from collimated sources. *Appl. Opt.* 13: 1444–1450, 1974.
8. Malyak, P. W. Two-mirror unobscured optical system for reshaping the irradiance distribution of a laser beam. *Appl. Opt.* 31: 4377–4383, 1992.

9. Dickey, F. M. and Holswade, S. C. *Laser Beam Shaping: Theory and Techniques.* New York: Marcel Dekker, 2000.
10. Dickey, F. M., Holswade, S., and Shealy, D. L. *Laser Beam Shaping Applications.* Boca Raton, FL: Taylor & Francis, 2005.
11. Kreuzer, J. L. Coherent light optical system yielding an output beam of desired intensity distribution at a desired equiphase surface. US Patent 3,476,463, November 4, 1969.
12. Frieden, B. R. Lossless conversion of a plane laser wave to a plane wave of uniform irradiance. *Appl. Opt.* 4: 1400–1403, 1965.
13. Rhodes, P. W. and Shealy, D. L. Refractive optical systems for irradiance redistribution of collimated radiation: Their design and analysis. *Appl. Opt.* 19: 3545–3553, 1980.
14. Design and analysis of refractive optical systems for irradiance redistribution of collimated radiation. MS Thesis, Department of Physics, University of Alabama at Birmingham, Birmingham, AL, 1979.
15. Shafer, D. Gaussian to flat-top intensity distributing lens. *Opt. Laser Technol.* 14: 159–160, 1982.
16. Wang, C. and Shealy, D. L. Design of gradient-index lens systems for laser beam reshaping. *Appl. Opt.* 32: 4763–4769, 1993.
17. Wang, C. and Shealy, D. L. Multi-mirror anastigmat design. In *Current Developments in Optical Design and Optical Engineering III*, Fisher, R. E. and Smith, W. J., eds., Bellingham, WA: SPIE, pp. 28–33, 1993.
18. Jiang, W., Shealy, D. L., and Martin, J. C. Design and testing of a refractive reshaping system. In *Current Developments in Optical Design and Optical Engineering III*, Fischer, R. E. and Smith, W. J., eds., Bellingham, WA: SPIE, pp. 64–75, 1993.
19. Jiang, W. Application of a laser beam profile reshaper to enhance performance of holographic projection systems. PhD thesis, University of Alabama at Birmingham, Birmingham, AL, 1993.
20. Jiang, W. and Shealy, D. L. Development and testing of a refractive laser beam shaping system. In *Laser Beam Shaping*, Dickey, F. M. and Holswade, S. C., eds., New York: Marcel Dekker, pp. 165–175, 2000.
21. Hoffnagle, J. A. and Jefferson, C. M. Design and performance of a refractive optical system that converts a gaussian to a flattop beam. *Appl. Opt.* 39: 5488–5499, 2000.
22. Hoffnagle, J. A. and Jefferson, C. Refractive optical system that converts a laser beam to a collimated flat-top beam. US Patent 6,295,168, September 25, 2001.
23. Newport Corporation. GBS Series Beam Shaper, http://www.newport.com/.
24. Hoffnagle, J. A. Sensitivity of a refractive beam reshaper to figure error. In *Laser Beam Shaping III*, Dickey, F. M., Holswade, S., and Shealy, D. L., eds., Bellingham, WA: SPIE, p. 67, 2002.
25. Jefferson, C. M. and Hoffnagle, J. A. An achromatic refractive laser beam reshaper. In *Laser Beam Shaping IV*, Dickey, F. M. and Shealy, D. L., eds., Bellingham, WA: SPIE, pp. 1–11, 2003.
26. Cornwell, D. F. Nonprojective transformations in optics. In *New Methods for Optical, Quasi-Optical, Acoustic, and Electromagnetic Synthesis*, Stone, W. R., ed., Bellingham, WA: SPIE, pp. 62–72, 1981.
27. Cornwell, D. F. Non-projective transformations in optics. PhD thesis, University of Miami, Florida, 1980.
28. Dickey, F. M. and Holswade, S. C., eds. *Laser Beam Shaping I.* Boca Raton, FL: Taylor & Francis, 2000.
29. Dickey, F. M., Holswade, S. C., and Shealy, D. L., eds. *Laser Beam Shaping II.* Bellingham, WA: SPIE, 2001.
30. Dickey, F. M. and Shealy, D. L., eds. *Laser Beam Shaping III.* Bellingham, WA: SPIE, 2002.
31. Dickey, F. M. and Shealy, D. L., eds. *Laser Beam Shaping IV.* Bellingham, WA: SPIE, 2003.

32. Dickey, F. M. and Shealy, D. L., eds. *Laser Beam Shaping V.* Bellingham, WA: SPIE, 2004.
33. Dickey, F. M. and Shealy, D. L., eds. *Laser Beam Shaping VI.* Bellingham, WA: SPIE, 2005.
34. Dickey, F. M. and Shealy, D. L., eds. *Laser Beam Shaping VII.* Bellingham, WA: SPIE, 2006.
35. Dickey, F. M. and Shealy, D. L., eds. *Laser Beam Shaping VIII.* Bellingham, WA: SPIE, 2007.
36. Forbes, A. and Lizotte, T. E., eds. *Laser Beam Shaping IX.* Bellingham, WA: SPIE, 2008.
37. Forbes, A. and Lizotte, T. E., eds. *Laser Beam Shaping X.* Bellingham, WA: SPIE, 2009.
38. Forbes, A. and Lizotte, T. E., eds. *Laser Beam Shaping XI.* Bellingham, WA: SPIE, 2010.
39. Forbes, A. and Lizotte, T. E., eds. *Laser Beam Shaping XII.* Bellingham, WA: SPIE, 2011.
40. Forbes, A. and Lizotte, T. E., eds. *Laser Beam Shaping XIII.* Bellingham, WA: SPIE, 2012.
41. Mouroulis, P. and MacDonald, J. *Geometrical Optics and Optical Design.* New York: Oxford University Press, 1997.
42. Born, M. and Wolf, E. *Principles of Optics.* 7th edn. Cambridge: Cambridge University Press, 1999.
43. Welford, W. T. *Aberrations of the Symmetrical Optical System.* New York: Academic Press, 1974.
44. Solimeno, S., Crosignani, B., and DiPorto, P. *Guiding, Diffraction, and Confinement of Opical Radiation.* New York: Academic Press, 1986.
45. Ghatak, A. K. and Thyagarajan, K. *Contemporary Optics.* New York: Plenum Press, 1980.
46. Stavroudis, O. N. and Fronczek, R. C. Caustic surfaces and the structure of the geometrical image. *J. Opt. Soc. Am.* 66: 795–800, 1976.
47. Stavroudis, O. N. *The Mathematics of Geometrical and Physical Optics.* Weinheim, Germany: Wiley-VCH Verlag, 2006.
48. Shealy, D. L. and Hoffnagle, J. A. Wavefront and caustics of a plane wave refracted by an arbitary surface. *J. Opt. Soc. Am. A* 25: 2370–2380, 2008.
49. Hoffnagle, J. A. and Shealy, D. L. Refracting the k-function: Stavroudis's solution to the eikonal equation for multielement optical systems. *J. Opt. Soc. Am. A* 28: 1312–1321, 2011.
50. Stavroudis, O. N. *The Optics of Rays, Wavefronts, and Caustics.* New York: Academic Press, 1972.
51. Welford, W. T. *Aberrations of Optical System.* Bristol: Adam Hilger, 1986.
52. Malacara, D. and Malacara, Z. *Handbook of Optical Design, Second Edition.* New York: Marcel Dekker, 2004.
53. Shealy, D. L. and Hoffnagle, J. A. Wavefront and caustic surfaces of refractive laser beam shaper. In *Novel Optical Systems Design and Optimization X,* Koshel, R. J. and Gregory, G. G., eds., Bellingham, WA: SPIE, vol. 6668, 666805-1, 2007.
54. Hoffnagle, J. A. and Shealy, D. L. Caustic surfaces of a Keplerian two-lens beam shaper. In *Laser Beam Shaping VIII,* Dickey, F. M. and Shealy, D. L., eds., Bellingham, WA: SPIE, 2007.
55. Shealy, D. L. and Hoffnagle, J. A. Review: Design and analysis of plano-aspheric laser beam shapers. In *Laser Beam Shaping XIII,* Forbes, A. and Lizotte, T. E., eds., Bellingham, WA: SPIE, pp. 8490–8492, 2012.
56. Hoffnagle, J. A. and Jefferson, C. M. Measured performance of a refractive gauss-to-flattop reshaper for deep-UV through near-IR wavelengths. In *Laser Beam Shaping II,* Dickey, F. M., Holswade, S., and Shealy, D., eds., Bellingham, WA: SPIE, pp. 115–124, 2001.

57. Hoffnagle, J. A. and Jefferson, C. Beam shaping with a plano-aspheric lens pair. *Opt. Eng.* 42(11): 3090–3099, 2003.
58. Deschamps, G. A. Ray techniques in electromagnetics. *Proc. IEEE* 60: 1022–1035, 1972.
59. Shealy, D. L. Classical (non-laser) methods. In *Laser Beam Shaping: Theory and Techniques*, Dickey, F. M. and Holswade, S. C., eds., Bellingham, WA: SPIE, pp. 313–348, 2000.
60. Koch, D. G. Simplified irradiance/illuminance calculations in optical systems. *Int. Symp. Opt. Syst. Des.* 1780: 226–240, 1993.
61. Burkhard, D. and Shealy, D. L. Simplified formula for the illuminance in an optical system. *Appl. Optics* 20(5): 897–909, 1981.
62. Burkhard, D. G. and Shealy, D. L. A different approach to lighting and imaging: Formulas for flux density, exact lens and mirror equations and caustic surfaces in terms of the differential geometry of surfaces. In *Materials and Optics for Solar Energy Conversion and Advanced Lighting Technology*, Lampert, C. M. and Holly, S., eds., Bellingham, WA: SPIE, pp. 248–272, 1986.
63. Silvestri, S. D., Laporta, P., Magni, V., Svelto, O., and Majocchi, B. Unstable laser resonators with super-gaussian mirrors. *Opt. Lett.* 13(3): 201–203, 1988.
64. Gori, F. Flattened Gaussian beams. *Opt. Commun.* 107: 335–341, 1994.
65. Li, Y. Light beams with flat-topped profiles. *Opt. Lett.* 27(12): 1007–1009, 2002.
66. Shealy, D. L., Hoffnagle, J. A., and Brenner, K.-H. Analytic beam shaping for flattened output irradiance profile. In *Laser Beam Shaping VII*, Dickey, F. M. and Shealy, D. L., eds., Bellingham, WA: SPIE, vol. 6290, 629006-1, 2006.
67. Parent, A., Morin, M., and Lavigne, P. Propagation of super-Gaussian field distributions. *Opt. Quantum Electron.* 24: S1071–S1079, 1992.
68. Bagini, V., Borghi, R., Gori, F., Pacileo, A., Santarsiero, M., Ambroshini, D., and Schirripa-Spagnolo, G. Propagation of axially symmetric flattened gaussian beams. *J. Opt. Soc. Am. A* 13: 1385–1394, 1996.
69. Li, Y. New expressions for flat-topped light beams. *Opt. Commun.* 206: 225–234, 2002.
70. Shealy, D. L. and Hoffnagle, J. A. Beam shaping profiles and propagation. In *Laser Beam Shaping VI*, Dickey, F. M. and Shealy, D. L., eds., Bellingham, WA: SPIE, vol. 5876, 58760D-1, 2005.
71. Shealy, D. L. and Hoffnagle, J. A. Laser beam shaping profiles and propagation. *Appl. Opt.* 45: 5118–5131, 2006.
72. Gori, F., Vicalvi, S., Santarsiero, M., and Borghi, R. Shape-invariance range of a light beam. *Opt. Lett.* 21: 1205–1207, 1996.
73. Vicalvi, S., Borghi, R., Santarsiero, M., and Gori, F. Shape-invariance error for axially symmetric light beams. *IEEE J. Quant. Electr.* QE-34: 2109–2116, 1998.
74. Campbell, J. P. and DeShazer, L. G. Near fields of truncated-Gaussian apertures. *J. Opt. Soc. Am.* 59: 1427–1429, 1969.
75. Siegman, A. E. New developments in laser resonators. In *Optical Resonators*, Holmes, D. A., ed., Bellingham, WA: SPIE, pp. 2–14, 1990.
76. Newman, E. and Oliker, V. Determining the intensities produced by reflected and refracted wave fronts in geometrical optics. *J. Opt. Soc. Am. A* 12: 784–793, 1995.
77. Oliker, V. Optical design of freeform two-mirror beam-shaping systems. *J. Opt. Soc. Am. A* 24(12): 3741–3752, 2007.
78. Oliker, V. On design of free-form refractive beam shapers, sensitivity to figure error, and convexity of lenses. *J. Opt. Soc. Am. A* 25: 3067–3076, 2008.
79. Ries, H. and Muschaweck, J. Tailored freeform optical surfaces. *J. Opt. Soc. Am. A* 19: 590–595, 2002.
80. Ries, H. Boundary conditions for balancing light in tailoring freeform surfaces. *Proc. SPIE* 8170: 817002-1–817002-5, 2011.

81. Ries, H. and Rabl, A. Edge-ray principle of nonimaging optics. *J. Opt. Soc. Am. A* 11: 2627–2632, 1994.
82. Ries, H., Muschaweck, J., and Timinger, A. New methods of reflector design. *Opt. Photonics News* 12(8): 46–49, 2001.
83. Hoffnagle, J. A. A new derivation of the Dickey-Romero-Holswade phase function. In *Laser Beam Shaping VI*, Dickey, F. M. and Shealy, D. L., eds., Bellingham, WA: SPIE, 2005.
84. Shealy, D. L. and Chao, S.-H. Geometric optics-based design of laser beam shapers. *Opt. Eng.* 42: 3123–3138, 2003.
85. Shealy, D. L. and Hoffnagle, J. A. Aspheric optics for laser beam shaping. In *Encyclopedia of Optical Engineering*, Driggers, R., ed., New York: Taylor & Francis, 2006.
86. Hoffnagle, J. A. Sensitivity of a refractive beam reshaper to figure error. In *Laser Beam Shaping III*, Dickey, F. M., Holswade, S., and Shealy, D. L., eds., Bellingham, WA: SPIE, p. 67, 2002.
87. Oliker, V. Sensitivity to figure error of a freeform refractive beam shaper. In *Nonimaging Optics: Efficient Design for Illumination and Solar Concentration VI*, Winston, R. and Gordon, J. M., eds., Bellingham, WA: SPIE, p. 742304, 2009.
88. Romero, L. A. and Dickey, F. M. Lossless laser beam shaping. *J. Opt. Soc. Am.* 13: 751–760, 1996.
89. Hoffnagle, J. A. and Jefferson, C. M. Beam shaping with a plano-aspheric lens pair. *Opt. Eng.* 42: 3090–3099, 2003.
90. Hoffnagle, J. A. and Jefferson, C. M. Design and performance of a refractive optical system that converts a Gaussian to a flattop beam. *Appl. Opt.* 39: 5488–5499, 2000.
91. Hoffnagle, J. A. and Shealy, D. L. Effects of dispersion on the performance of a refractive laser beam shaper. *Proc. SPIE* 5876: 128–137, 2005.
92. Jefferson, C. M. and Hoffnagle, J. A. An achromatic refractive laser beam reshaper. In *Laser Beam Shaping IV*, Dickey, F. M. and Shealy, D. L., eds., Bellingham, WA: SPIE, p. 1, 2003.
93. Baker, K., Shealy, D. L., and Jiang, W. Directional light filters: Three-dimensional azo-dye-formed microhoneycomb images within optical resins. *Proc. SPIE* 2404: 144, 1995.
94. Baker, K. Highly corrected submicrometer grid patterning on curved surfaces. *Appl. Opt.* 38: 339–351, 1999.
95. Baker, K. Highly corrected close-packed microlens arrays and moth-eye structuring on curved surfaces. *Appl. Opt.* 38: 339–351, 1999.
96. Jiang, W., Shealy, D. L., and Baker, K. Optical design and testing of a holographic projection system. *Proc. SPIE* 2152: 244, 1994.
97. Kasinski, J. J. and Burnham, R. L. Near-diffraction-limited laser beam shaping with diamond-turned aspheric optics. *Opt. Lett.* 22: 1062–1064, 1997.
98. Jiang, W. and Shealy, D. L. Development and testing of a refractive laser beam shaping system. In *Laser Beam Shaping*, Dickey, F. M. and Holswade, S. C., eds., Boca Raton, FL: Taylor & Francis, pp. 165–175, 2000.
99. Image processing and analysis program for Macintosh computers. http://rsb.info.nih.gov/ij/.
100. Sands, P. J. Inhomogeneous lenses. IV. Aberrations of lenses with axial index distributions. *J. Opt. Soc. Am.* 61: 1086–1091, 1971.
101. Wang, C. and Shealy, D. L. Design of gradient-index lens systems for laser beam reshaping. *Appl. Opt.* 32: 4763–4769, 1993.
102. Marchand, E. W. *Gradient Index Optics*. New York: Academic Press, 1978.
103. Shingyouchi, K. and Konishi, S. Gradient-index doped silica rod lenses produced by a solgel method. *Appl. Opt.* 29: 4061–4063, 1990.
104. Samuel, J. E. and Moore, D. T. Gradient-index profile control from mixed molten salt baths. *Appl. Opt.* 29: 4042–4050, 1990.

105. Kindred, D. S., Bentley, J., and Moore, D. T. Axial and radial gradient-index lens. *Appl. Opt.* 29: 4036–4041, 1990.
106. Koike, Y., Hidaka, H., and Ohtsuka, Y. Plastic axial gradient-index lens. *Appl. Opt.* 24: 4321–4325, 1985.
107. LightPath Technologies, http://www.lightpath.com.
108. Shealy, D. L. Geometrical methods. In *Laser Beam Shaping: Theory and Techniques*, Dickey, F. M. and Holswade, S. C., eds., New York: Marcel Dekker, pp. 163–213, 2000.

APPENDIX: EQUIVALENCE OF SOLUTIONS TO THE BEAM SHAPING PROBLEM

The literature of beam shaping includes two independently derived solutions for the surfaces of a two-element refractive beam shaper, one formulated as an integral equation [11] and the other as an ordinary differential equation [13]. Although they appear superficially different, the same physical principles apply in both cases, so the solutions must be equivalent. This appendix shows explicitly that the solutions of Refs. [11,13] are identical.

A ray entering the first lens parallel to the z-axis is refracted by the first aspheric surface and emerges in the direction **A**, defined in Equation 6.7. Rhodes and Shealy [13] use the vectorial formulation of Snell's law presented in Section 6.2.1.3 and rearrange Equation 6.40 to express the unit vector in the direction of the refracted ray as

$$\mathbf{A} = n\mathbf{e_z} + \Omega\mathbf{N} \tag{A.1}$$

where:

$$\Omega = \frac{-n + \sqrt{1 + z'^2(1 - n^2)}}{\sqrt{1 + z'^2}} \tag{A.2}$$

and

$$\mathbf{N} = \frac{-z'\mathbf{e_r} + \mathbf{e_z}}{1 + z'^2} \tag{A.3}$$

Writing the components of **A** explicitly,

$$\mathbf{A} = \mathbf{e_r}\left[\frac{nz' - z'\sqrt{1 - (n^2 - 1)z'^2}}{1 + z'^2}\right] + \mathbf{e_z}\left[\frac{nz'^2 + \sqrt{1 - (n^2 - 1)z'^2}}{1 + z'^2}\right] \tag{A.4}$$

which leads to an equation relating the points of refraction on the first and second lenses as follows:

$$\frac{(R - r)}{(Z - z)} = \frac{nz' - z'\sqrt{1 - (n^2 - 1)z'^2}}{nz'^2 + \sqrt{1 - (n^2 - 1)z'^2}} \tag{A.5}$$

where:

r and R are related by the ray mapping function

z and Z are the axial coordinates of the first and second points of refraction, respectively

Cross multiplying Equation A.5 yields, after squaring and collecting terms in power of z',

$$z'^4[n^2(R-r)^2-(1-n^2)(Z-z)]-2z'^3(R-r)(Z-z)$$
$$-z'^2(n^2-1)[(R-r)^2+(Z-z)^2]-2z'(R-r)(Z-z)-(R-r)^2=0 \quad \text{(A.6)}$$

Factoring Equation A.6 into a quadratic equation in z' times, the factor $(1+z'^2)$ gives the following result:

$$(1+z'^2)\{az'^2+bz'^2+c\}=0 \quad \text{(A.7)}$$

where:

$$a=n^2(R-r)^2+(n^2-1)(Z-z)^2 \quad \text{(A.8a)}$$

$$b=-2(R-r)(Z-z) \quad \text{(A.8b)}$$

$$c=-(R-r)^2 \quad \text{(A.8c)}$$

Canceling the term $(1+z'^2)$ in Equation A.7 and solving the quadratic equation in z' yields

$$z'(r)=\frac{-(R-r)(Z-z)\pm n(R-r)\sqrt{(R-r)^2+(Z-z)^2}}{(1-n^2)(Z-z)^2-n^2(R-r)^2} \quad \text{(A.9)}$$

In fact, only the negative sign in Equation A.9 is physically meaningful. This can be seen by considering the limit $(R-r)\to 0$, in which case $(Z-z)\to t$, where t denotes the axial separation of the lenses. Then the limiting case of the equation for the slope is

$$z'\to -(R-r)\frac{1\pm n}{t(1+n)(1-n)} \quad \text{(A.10)}$$

Physically, the signs of $R-r$ and z' are the same if $n>1$ and opposite if $n<1$. This means that the correct limiting case is $z'\to (R-r)/(n-1)t$, which requires the positive sign in Equation A.10, corresponding to the negative sign in Equation A.9.

Equation A.9 is a consequence of ray tracing alone. To complete the solution for z', we need to add the constant OPL condition. In Reference [13], it is expressed as

$$\sqrt{(R-r)^2+(Z-z)^2}=n(Z-z)-t(n-1) \quad \text{(A.11)}$$

Squaring Equation A.11 and substituting the resulting expression in Equation A.9 allows one to eliminate Z, resulting in a first-order ordinary differential equation for $z(r)$, which is Rhodes and Shealy's solution. To prove that it is the same as Kreuzer's solution [11], we go back to Equation A.9, with the negative sign, and write it as

$$z'(r)=-(R-r)\frac{(Z-z)+n\sqrt{(R-r)^2+(Z-z)^2}}{(Z-z)^2-n^2[(R-r)^2+(Z-z)^2]} \quad \text{(A.12)}$$

Substituting Equation A.11 in this equation gives

$$z'(r) = -(R-r)\frac{(Z-z) + n[n(Z-z) - t(n-1)]}{(Z-z)^2 - n^2[n(Z-z) - t(n-1)]} \tag{A.13}$$

Rearranging the numerator, and factorizing and rearranging the denominator leads to

$$z'(r) = -(R-r)\frac{(Z-z)(1+n^2) - tn(n-1)}{[(Z-z)(1+n^2) - tn(n-1)][(Z-z)(1-n^2) + tn(n-1)]} \tag{A.14}$$

or

$$z'(r) = \frac{-(R-r)}{(Z-z)(1-n^2) + tn(n-1)} \tag{A.15}$$

Squaring Equation A.11, collecting terms, and solving the resulting quadratic equation result in an equivalent relation:

$$(Z-z) = \frac{n(n-1)t + [(n-1)^2t^2 + (n^2-1)(R-r)^2]^{1/2}}{n^2-1} \tag{A.16}$$

Substituting this expression in the denominator of Equation A.15 gives

$$z'(r) = \frac{R-r}{[(n-1)^2t^2 + (n^2-1)(R-r)^2]^{1/2}} \tag{A.17}$$

which is easily rearranged to yield Kreuzer's solution [11] (Equation 6.79).

7 Optimization-Based Designs

Alexander Laskin, David L. Shealy, and Neal C. Evans

CONTENTS

7.1 INTRODUCTION

Recently, the application of optical design and optimization techniques to laser beam shapers has led to new applications and development. These methods have shown great promise for shaping virtually any stable laser beam profile as well as single and multimode laser beams and for using achromatic beam shaping designs. Laser beam shaping techniques have also been used in holographic and interferometric nanomanufacturing, in controlling beam intensity profiles for task in laser materials processing, and for creating round and square flat-top laser spots in microprocessing systems with scanning optics.

Kreuzer [1] first described, in 1964 at the Symposium on Optical and Electro-Optical Processing Technology, four methods for creating a uniform irradiance profile for a laser beam, which are given in Section 6.1. Later in 1969, Kreuzer [2] developed and patented a coherent light optical system to transform the rays of a laser beam into a collimated beam with a uniform irradiance profile, which is based on (1) the intensity law of geometrical optics and (2) the constant optical path length condition between the input and output apertures of the laser beam shaper system. Kreuzer's patent presents the equations for calculating the sag of two plano-aspheric lenses required to accomplish the specified beam shaping. Further, Shafer [3] used optical design methods to develop a laser beam shaping design with two afocal doublet spherical lenses with zero total optical power. By varying the shape of the first doublet while keeping the total power of the first doublet constant, a large amount of spherical aberration is introduced into the input beam, which redistributes the irradiance of the input beam. Then, the second doublet is given the same amount of spherical aberration, but with opposite sign as the first doublet, which produces a collimated output beam with moderate uniformity of its irradiance.

Evans and Shealy [4] developed a genetic algorithm optimization method for use in the optical design of laser beam shapers during the 1990s, where their optimization methods were applied to designs that contain both discrete and continuous variables within the solution space. Hoffnagle and Jefferson [5,6] developed an optical system that transforms an input laser beam with axially symmetric irradiance profile into an output profile with a different irradiance profile, such as a Fermi–Dirac or other distribution. A Keplerian configuration of two plano-aspheric lenses was used to simplify fabrication and introduce a continuous roll-off of the output beam irradiance distribution for more control of the far-field diffraction pattern. Their Keplerian configuration of beam shaper has been fabricated, assembled, tested, and transferred into a commercially available refractive beam shaper with a large bandwidth from IR to UV [7]. Hoffnagle and Jefferson provide detailed characterization of a refractive beam shaper at 514 nm, which gives an output beam with less than 5% variation in the irradiance and with the root-mean-square variation of the optical path difference (OPD) over the beam of 13 nm. Jefferson and Hoffnagle report a technique to achromatize an aspheric laser beam shaper using conventional spherical optics [8].

Laskin [9,10] uses optical design and optimization methods to extend refractive laser beam shaping optical systems to the goals of (1) a fully achromatic operation with zero or very small total wave aberration and (2) ease of adjustment of optics to accommodate a range of input beams which may be collimated or divergent. This goal was achieved by an achromatic field mapping laser beam shaper system [11]. The achromatic optical unit consists of at least two lens groups with zero total optical power, which is realized by requiring the two optical groups to satisfy the intensity transformation law and the constant optical path length condition, while also satisfying the achromatic conditions over a specific spectral range. The achromatic feature is satisfied by using lens materials with different dispersion characteristics for the lenses of the system. The condition of achromatization has been formulated as equations linking the optical power of the lenses, distances between the lenses, and the Abbe numbers of the lens materials.

Control of the laser beam irradiance profile, particularly providing uniform distribution, is of great importance for various holography and interferometry applications in research and industry. The flatness of the phase front of a laser beam should be conserved during any irradiance profile transformation of both the flatness of the phase front and the irradiance distribution. There are several beam shaping techniques applied in modern laser technologies, such as the integration systems based on arrays of microlenses, micromirrors, or prisms, which cannot be applied since their physical principle implies destroying the beam structure and, hence, leads to loss of spatial coherence. Other techniques include truncation of a beam by an aperture or attenuation by apodizing filters yield acceptable homogeneity of irradiance profile in some cases, but an evident disadvantage of these techniques is the essential loss of the costly laser energy. To meet the demands of holography and interferometry, it is suggested to apply beam shaping systems built on the base of field mapping refractive beam shapers, such as the πShaper,* whose operational principle implies almost lossless transformation of laser irradiance distribution from Gaussian to flattop, conserving of beam consistency, flatness of output phase front, low divergence of collimated output beam, high transmittance, extended depth of field (EDOF), capability to operate with TEM_{00} or multimode lasers, and implementations as telescopes or collimators. This chapter discusses the basic principles and important features of refractive beam shapers as well as some optical layouts that can be built on their base to meet requirements of modern laser technologies.

In Section 7.1.1, we summarize the scope of the many applications of laser beam shaping as a result of the work of Laskin and coworkers, who have moved laser beam shaping into the well-established domain of optical design and optimization for all laser-based optical design technologies and applications. Section 7.1.2 presents a method for computing the irradiance during ray tracing by applications of the geometrical optics intensity transformation law [12]. An overview of the general theory of optimization-based methods for design of achromatic laser beam shapers is presented in Section 7.2. Specific attention is given to the use of optical design and optimization methods to formulate components and specifications of refractive laser beam shapers which satisfy the following: (1) the geometrical optics intensity transformation law; (2) the constant optical path length condition to give zero wave aberration; and (3) the condition of achromatization in the form of combined equations linking focal power of the lenses, distances between lenses, and the Abbe numbers of the lens materials. Section 7.3 summarizes several specific procedures discussed in Section 7.2 to realize applications of refractive, achromatic laser beam shapers in many topical areas of scientific, engineering and technology research and development, such as micro-machining, interferometry, holography, laser design, scientific applications, industrial technologies, and life sciences. In addition, Laskin and coworkers have shown that (1) shaping the spatial (transverse) profile is highly desirable when building optical systems for high-power lasers and their applications [13]; (2) using anamorphic optical element either before or after the rotationally symmetric refractive beam shaper generates a wide collection of spot shapes and intensity profiles [14]; (3) using a refractive laser beam shaper to improve the output radiation of the photocathode of a free-electron laser (FEL) [15]; and (4) using refractive laser beam shaper to also improve operation efficient of spatial light modulators (SLMs) [16]. Overall conclusions of this work are given in Section 7.4.

* πShaper is a registered trademark of AdlOptica GmbH, Berlin, Germany, http://www.adloptica.com.

7.1.1 Scope of Applications

Laser beam shaping is an important component of many industrial processes and has been described as being an enabling technology [17]. There have been annual topical conferences on laser beam shaping since 1999 [17], and there is a wealth of literature published in the laser and optics literature on laser beam shaping. This field has matured and developed very specific applications in many topical areas:

- In micromachining, where specific laser beam shaping solutions are used for drilling blind vias in and cutting of printed circuit board (PCB), repair of display pixels, microwelding, trimming
- Industrial applications of welding, cladding, hardening, and illumination with SLMs
- Interferometry applications of measuring devices and generation of periodic structures
- Holography applications of security holograms, multicolor Denisyuk holography, and holographic data storage
- Life sciences applications of flow cytometry, illumination in confocal microscopy, and homogenizing the radiation of several lasers in fluorescence techniques; scientific applications of flying plate technique in spectroscopy, laser heating in geophysics, various laser ablation techniques, and basic research
- Laser design of pumping of ultra-short pulse solid-state lasers, optimizing master oscillator power amplifier (MOPA) laser design, and irradiation of photocathodes in FELs [19]

7.1.2 Irradiance Calculations via Ray-Tracing Methods

Fundamental to laser beam shaping computations is a fast, accurate means of determining irradiance (energy per unit area per unit time) profiles at different locations within a system. To do this, one must start with first principles: energy must be conserved in a nondissipative optical system. This principle is mathematically expressed in the form of the energy conservation law. The energy conservation law has broad application, from designing reflective beam shapers via analytical differential equation methods to the development of finite-element mesh methods for the design of beam shaping holograms [20–22]. To employ energy conservation, one starts by describing the irradiance profile of a bundle of rays striking the input pupil of a beam shaping system by a radially symmetric function $\sigma(\rho)$. Consider the optical configuration illustrated in Figure 7.1. These rays propagate through the beam shaping system and exit to strike the output surface, where the irradiance distribution on the output surface is represented by the function $u(P)$. Assuming that no energy is dissipated by the system, the energy conservation law can be expressed as

$$E = \int_I \sigma(\rho)\, \mathrm{d}a = \int_O u(P)\, \mathrm{d}A \tag{7.1}$$

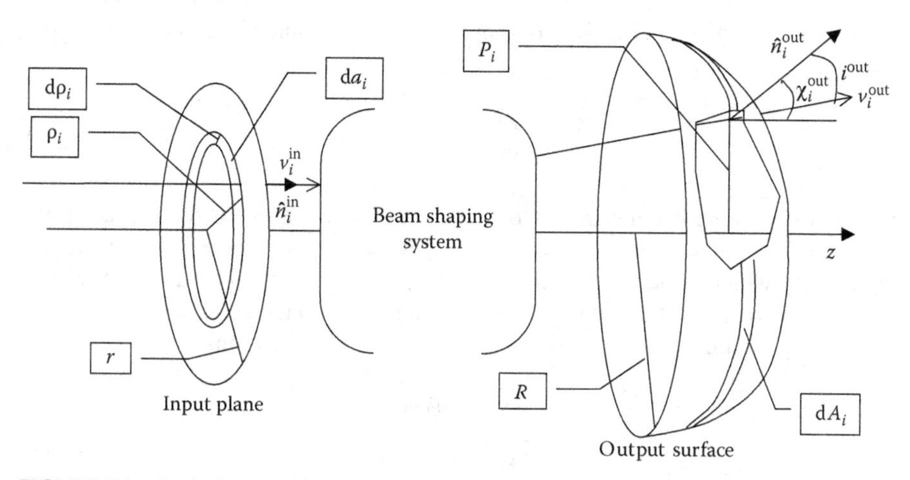

FIGURE 7.1 Basic layout of beam expander with input and output surfaces.

where:

E is the total energy entering the system

I and O are the input plane and output surfaces, respectively, over which the respective integrations occur

Now, we use Equation 7.1 to develop a simple, efficient, numerical method for calculating the intensity that is incident upon the output surface, which can be expressed as

$$u(P) = \sigma(\rho)\frac{da}{dA} \tag{7.2}$$

where expressions for da and dA are derived below.

In the examples presented in this chapter, the input irradiance is assumed to be Gaussian, measured in units of rays per unit area:

$$\sigma(\rho_i) = \exp(-\alpha\rho_i^2) \tag{7.3}$$

where:

α is a dimensionless quantity given by $2/\rho_N^3$

Here, the beam waist of the incoming beam is expressed by ρ_N and is defined as the radius of the circle where the irradiance drops to $1/e^2$ of the central irradiance. The N rays that are traced through the system are distributed uniformly over the input plane accordingly. Though it need not be, a Gaussian input profile is chosen because it describes typical laser profiles when the laser is in the fundamental mode TEM_{00}.

To determine the ratio da/dA in Equation 7.2, we start by tracing N rays through the system, where N is a reasonably large number. For this application, we have used N = 200, which gives adequate resolution for our input and output irradiance profiles. Each ray enters parallel to the optical axis at a specified height, ρ_i, where the set of ρ_i are distributed equally across the radius of the input plane according to the following function:

$$\rho_i = \left(\frac{r}{N}\right) i, \quad i = 0,\ldots,N \tag{7.4}$$

Each ray leaves the beam shaper and strikes a point of the output surface as shown in Figure 7.1. At the point where the ray strikes the output surface, we measure the distance P_i, which is the distance from the optical axis, and χ_i, the angle between the unit normal vector of the output surface at the point of interception and the optical axis. Using Figure 7.1, we can also write the following expressions:

$$da_i = 2\pi\rho_i \, d\rho_i \tag{7.5a}$$

$$d\rho_i = \rho_i - \rho_{i-1} \tag{7.5b}$$

Here, defining $d\rho_i$ in this manner is arbitrary, since definitions such as $d\rho_i = \rho_{i+1} - \rho_i$ or $d\rho_i = \rho_{i+1} - \rho_{i-1}$ would be just as effective. Furthermore, the subscript i is introduced to emphasize the numerical nature of the solution to the now discrete function in Equation 7.2. Calculation of dA_i is somewhat more complicated, since the output surface is not necessarily flat like the input plane. In general, dA_i is given by $2\pi P_i \, dS_i$:

$$dS_i = \frac{dP_i}{\cos \chi_i^{\text{out}}} \tag{7.6}$$

Combining these observations with Equations 7.2 through 7.6, one has

$$u(P_i) = \sigma(\rho_i)\left[\frac{\cos(i_i^{\text{in}})\rho_i(\rho_i - \rho_{i-1})\cos(\chi_i^{\text{out}})}{\cos(i_i^{\text{out}})P_i(P_i - P_{i-1})}\right] \tag{7.7}$$

Equation 7.7 expresses the output beam irradiance in terms of the input beam irradiance times a ratio of areas expressing the beam expansion as a result of ray propagation through the optical system. Equations 7.7 and 7.3, along with the ray trace array, provide an accurate means of calculating the beam profile over any reasonable surface. The accuracy of this method has been verified by calculating the profiles for several benchmark systems [23,24]. Calculations of output beam profiles using Equation 7.7 are in close agreement with the profiles given in the benchmark papers. Now, a merit function can be developed based on Equation 7.7.

7.2 THEORY

Generally, the idea behind optimization is that one has some function f which may be evaluated computationally. This function is expressed in terms of several variables which may be discrete or continuous in nature. One wishes to find the values of these variables that make f assume either its maximum or minimum value. The difficulty of the problem is related to whether one is searching for local extrema, of which there may be many, or the global extrema, which represent the absolute best solutions. The complexity of the problem is related to the number of variables that make up f, in addition to the ease with which f can be calculated. The greater the complexity of the problem, the longer it takes to arrive at a solution. Thus, search algorithms that arrive at solutions quickly are to be coveted, which is evident by the voluminous amount of research regarding the subject present in the literature [25].

It is well known that several commercial optical design and analysis packages implement these techniques in their optimization routines to varying degrees. ZEMAX* and CODE V† also contain proprietary global optimization methods. The problem with these implementations, among other things, is that the merit functions in these packages are oriented toward imaging systems (ZEMAX, however, allows for user-defined merit functions computed by macros or an external programming interface), limiting one's ability to manipulate the merit function for one's own purpose. Also, since the makers of these packages keep their optimization codes proprietary, one has limited ability to modify core optical design routines.

7.3 APPLICATIONS

The scope of applications of laser beam shapers has significantly expanded during the past 5 years after the developments by Laskin et al. [9–11,13–16,26] of using optical design codes and optimization techniques as provided by standard optical design codes, such as ZEMAX and Code V for implementation of achromatic laser beam shapers within a wide range of applications. The general areas of these applications have been summarized in Section 7.1.1. In Section 7.3.1, we discuss how optical design and optimization techniques have been used to obtain similar results as analytical solutions to laser beam shaping design equations [2,22,27]. In Section 7.3.2, we discuss how to use optical design and optimization techniques to obtain achromatic laser beam shapers.

7.3.1 Optimization of Plano-Aspheric Lens Laser Beam Shaper

As was earlier considered [2,22,27], a refractive beam shaping system can be realized in the form of two-lens system with aspheric surfaces, whose analytical expressions are rather complicated and not convenient for practical use. Therefore, numerical calculations are applied. It is also very promising and fruitful to approximate lens surfaces by aspheric terms of high order, such as $A_i r^{2i}$, which gives an analytical description (Equation 6.123) [28] and is widely used by optical designers and by manufacturers of aspheric optics.

There may be different ways to approximate the design of two-lens laser beam shapers. In this section, we shall use an engineering approach toward approximating the aspheric surfaces of the refractive laser beam shaper, which implies finding the

* ZEMAX is a registered trademark of ZEMAX Development Corporation, San Diego, CA, http://www.zemax.com.

† CODE V is a registered trademark of SYNOPSYS, San Francisco, CA, http://www.synopsys.com/.

parameters of an optical system by using the theory of third-order aberrations with further correction of those parameters by using well-developed algorithms of optimization in optical system design software, such as ZEMAX and Code V. A primary task of optimization is finding the global minimum of a merit function, and it is very important to define a set of initial parameters of a system close to that global minimum. Evidently, the task of defining these initial data is not a new one in optics, since different approaches have been developed and used. One approach is making a series expansion of the aberration functions of a system, where one analyzes low-order terms of the series expansion, which are the third-order terms of the transverse aberration. This engineering method was developed in the beginning of the twentieth century, and it works well for systems of low numerical aperture and allows one to define the following parameters of optical systems: radii of curvature, indices of refraction, thicknesses or air gaps, and conic constants of second-order aspheric surfaces.

One should note that the majority of beam shaping systems used in practice have a rather low numerical apertures, so that using third-order aberration theory is justified. A remarkable feature of the third-order aberration theory is using the paraxial variables of an optical system to express coefficients of the series expansion of aberration. In terms of the third-order aberration theory, we consider surfaces of the second-order spherical and conic sections, which include ellipsoid, hyperboloid, paraboloid, and spheroid. Therefore, as a first step of this approximation, we shall model the two-lens beam shaping system by using these aspheric terms of second order. Then, defining of high-order polynomial terms of even-order radial aspheric terms can be used during the further stages of the optimization by using specialized optical system design software.

We shall consider a two-lens beam shaper of the Keplerian design providing theoretically flat phase front and flat-top irradiance profile of the output beam. The third-order aberration theory is used to define initial set of parameters of the system which is used as a starting point for further optimization by using commercial optical designing software.

7.3.1.1 Theoretical Considerations

The beam shaping effect in refractive field mapping systems is achieved through application of lenses with aspheric surfaces; most often these systems are implemented as telescopes of the Keplerian or Galilean type [6,29]. Now, consider a one-to-one Keplerian (1-to-1K) telescopic two-lens beam shaper by defining the system parameters by using the third-order aberration theory and subsequent computer optimization. Later, this method can be also extended in Section 7.3.2 to design achromatic beam shaping systems.

7.3.1.1.1 Defining the Aberration to Be Introduced by First Lens

Consider a 1-to-1K two-lens beam shaper as illustrated in Figure 7.2 and also considered by Shealy and Hoffnagle [27] with the following parameters:

$$w_0 = 2.366\,\text{mm};\ R_{max} = 4.05\,\text{mm};\ r_{max} = 4.05\,\text{mm};\ d = 150\,\text{mm};\ n = 1.46071 \quad (7.8)$$

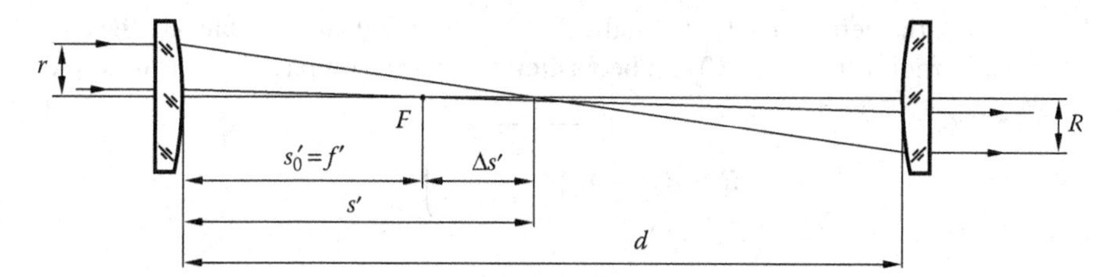

FIGURE 7.2 Basic layout of two aspheric lens laser beam shaping system.

where:

w_0 is the radius of the input Gaussian beam waist
n is the index of refraction of lenses

The first plano-convex lens focuses the laser beam and is responsible for introduction of the wave aberration in order to redistribute the beam irradiance, while the second plano-convex lens compensates for this wave aberration and collimates the beam so that the output beam has a flat phase front and flat-top irradiance profile.

Detailed theory of designing refractive beam shapers is presented in Chapter 6 [30] and in Reference [27]. Here, some important conclusions are used to develop the optimization methods for designing more complex laser beam shapers than were considered in Chapter 6. As a result of imposing the geometrical law of intensity [12] between the input and output apertures of a refractive beam shaper shown in Figure 7.2, one obtains the following ray mapping function [27] for an input Gaussian beam and output top-hat beam profile:

$$R^2 = R_{\max}^2 \left[\frac{1 - \exp(-2r^2 / w_0^2)}{1 - \exp(-2r_{\max}^2 / w_0^2)} \right] \tag{7.9}$$

Since r and R are radial coordinates that are always positive, one must be careful when taking the square root of Equation 7.9 to obtain a valid expression for R for either Keplerian or Galilean configuration of a refractive laser beam shaper. Using the same notation introduced by Shealy and Hoffnagle [27], the output radial coordinate, R, of a refractive beam shaper is given by the following result:

$$R = \epsilon R_{\max} \sqrt{\frac{1 - \exp[-2(r / w_0)^2]}{1 - \exp[-2(r_{\max} / w_0)^2]}} \tag{7.10}$$

where:

ϵ is defined by

$$\epsilon = \begin{cases} +1 & \text{Galilean configuration} \\ -1 & \text{Keplerian configuration} \end{cases} \tag{7.11}$$

As a result of the definition of ϵ, the radical in Equation 7.10 always refers to the positive square root. Equation 7.10 can be written in a more compact form as follows:

$$R = B\sqrt{1-\exp\left[-2\left(\frac{r}{w_0}\right)^2\right]} \tag{7.12}$$

where:

$$B = \frac{\epsilon R_{max}}{\sqrt{1-\exp[-2(r_{max}/w_0)^2]}} \tag{7.13}$$

From the geometry of the input and output rays shown in Figure 7.2, we can write the following expression:

$$\tan\alpha = \frac{r-R}{d} = \frac{r}{s'} \tag{7.14}$$

or

$$s' = \frac{rd}{r-R} \tag{7.15}$$

where:

α is the angle that a refracted ray from the first lens makes with the optical axis

Substituting Equation 7.12 for Keplerian beam shaper with $\epsilon = -1$ into Equation 7.15 gives

$$s' = \frac{d}{1-(B/r)\sqrt{1-\exp(-2r^2/w_0^2)}} \tag{7.16}$$

where it should be noted that for a Keplerian configuration we have $\epsilon = -1$, and thus, $B < 0$ from Equation 7.13. The location s_0' of the paraxial focus F of first plano-aspheric lens can be found by considering the limit as r approaches zero:

$$s_0' = \lim_{r\to 0} s' \tag{7.17}$$

To define this value, it is convenient to use the Taylor series expansion of the exponential function; the transformations give the result as follows:

$$s_0' = \frac{d}{1-\sqrt{2}B/w_0} \tag{7.18}$$

Then, the longitudinal aberration (LA) $\Delta s'$ to be introduced by the first aspheric lens is given by

$$\Delta s' = s' - s_0' = \frac{d}{1-(B/r)\sqrt{1-\exp(-2r^2/w_0^2)}} - \frac{d}{1-\sqrt{2}B/w_0} \tag{7.19}$$

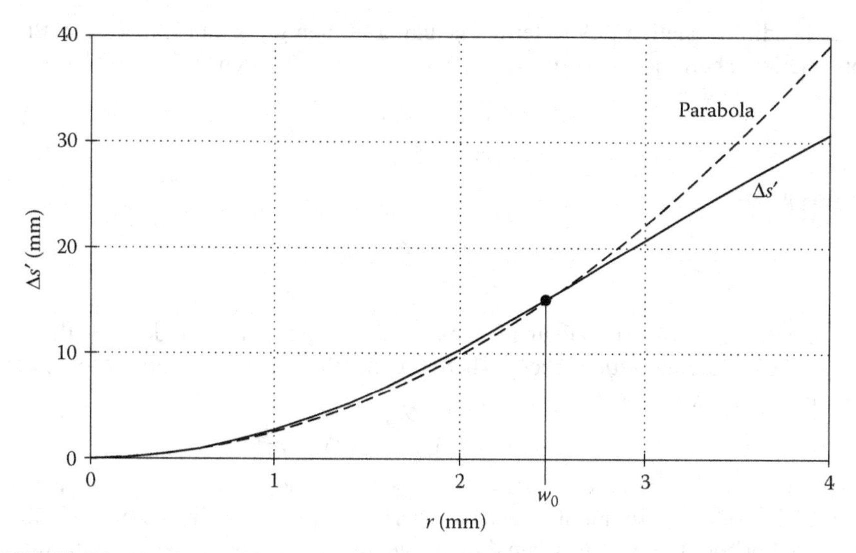

FIGURE 7.3 Analysis of LA: solid curve represents aberration; dashed curve represents approximate parabola.

The character of the aberration function $\Delta s'$ can be understood by evaluating Equation 7.19 for the 1-to-1K example [27] with the parameters given by Equation 7.8. Then, the distance s'_0 to the paraxial focus is equal to 43.81 mm, and the LA $\Delta s'$ as the function of the input beam radius r is shown in Figure 7.3.

In contrast, Shealy and Hoffnagle evaluated the axial curvature of the first plano-aspheric lens of 1-to-1K beam shaper by using the intensity law of geometrical optics to be equal to -0.042656mm^{-1} or an axial distance of 50.767 mm from the vertex of the first lens with the optical axis to the axial focal point [31,32]. Therefore, in contrast to imaging applications, the paraxial focus is less important in beam shaping applications than the overall ray mapping function that describes how the input rays are mapped to the output aperture as required by the intensity law for beam shapers.

There are several important conclusions. First, it is important to note that the LA required to transform an input Gaussian beam into a flat-top beam is reached over several tens of millimeters, which corresponds to a very strong wave aberration, up to 100λ, existing between the components of this laser beam shaping system. The aberration is positive, which means that it is not possible to introduce this level of aberration by a simple spherical surface, since the positive spherical lenses have negative spherical aberration [33]. Therefore, the surface profile of the laser beam shaper must be aspheric. Another important feature is that in the range of input beam radii from 0 to w_0, the function of the LA can be approximated by parabolic function with very high precision, which is shown in Figure 7.3, where the approximate parabola is shown by the dashed line. This parabola and function of LA are crossing at the point equal to the waist radius w_0. Majority of the energy of the Gaussian laser beam, about 87%, is concentrated within a circle of radius of w_0. Therefore, any calculations based on parabolic approximation of the aberration function would be valid for evaluating the performance of a laser beam shaper system.

Expanding Equation 7.19 to terms including r^2 then gives an explicit expression for spherical aberration. The function of the LA can be written as follows:

$$\text{LA} = \Delta s' \approx Ar^2 \tag{7.20}$$

where:

A is given by

$$A = \frac{Bd}{\sqrt{2}(w_0 - \sqrt{2}B)^2} \tag{7.21}$$

As we will see, a similar mathematical expression has been used to describe the LA in the third-order aberration theory. Therefore, the third-order aberration theory can be used to design laser beam shapers.

7.3.1.1.2 Third-Order Aberrations of Two-Lens System

The theory of the third-order aberrations is a powerful tool for analysis of optical systems and developing an initial design that can be used as a starting point for further optimization. For the systems when the aberrations can be approximated by the third-order series approximation, this theory allows calculating of the system parameters being close to global minimum solution. Hence, the further optimization process is very short and gives optimum system parameter values after few iterations. There are adequate descriptions in the literature of several approaches of using third-order approximation of aberrations of optical system on the basis of Seidel sums, where we are using the formulations given in the literature [33–35]. An essential feature of this theory is in calculating the third-order coefficients of the aberration series expansion by using paraxial values of the parameters and specifications of an optical system. Now, this approach is convenient for using in the designing process.

A beam shaping effect is always achieved through introducing certain aberration. In the particular case of the 1-to-1K beam shaping system, the transformation of its irradiance distribution is achieved through introducing just spherical aberration by the first lens. Since these systems work typically in narrow angular field, it is sufficient to consider relationships for spherical aberration only.

According to the third-order aberration theory [36–38], for an infinitely remote object the transverse spherical aberration $\Delta y'_{\text{III}}$ of an optical system with focal length f' is written as

$$\Delta y'_{\text{III}} = -\left[\frac{f'}{2r}\right] S_{\text{I}} \tag{7.22}$$

where:

r is the height of a ray

S_{I} is the sum of the first Seidel coefficient

Then the LA is given by

$$\Delta s'_{\text{III}} = \frac{f'}{r} \Delta y'_{\text{III}} = -\left[\frac{(f')^2}{2r^2}\right] S_{\text{I}} \tag{7.23}$$

The Seidel coefficients have to be calculated for each surface of an optical system by using special formulas described in Refs. [33–35] for the particular surface parameters: curvature radius or focal length, refractive index, air gap or lens thickness, conic constant for the second-order surfaces. In the case of the system illustrated in Figure 7.2, we assume that the convex surfaces are aspheric. In terms of the third-order aberration theory, they can be presented as the second-order surfaces. Then the Seidel coefficients S_{I1} and S_{I2} for correspondingly first and second aspheric surfaces can be written as

$$S_{I1} = \frac{r^4(n^2 + k_1)}{f_1'^3(n-1)^2} \tag{7.24}$$

$$S_{I2} = -\frac{R^4(n^2 + k_2)}{f_2'^3(n-1)^2} \tag{7.25}$$

where:

n is the refractive index of the material of lenses

k_1 and k_2 are the conic constants of the second-order surfaces

Based on these relationships, we now calculate the basic parameters of the beam shaping optical system.

7.3.1.1.3 Parameters of Second-Order Aspheric Lens

The expression for S_{I1} is used to evaluate the parameters of the first lens, which introduces the spherical aberration sufficient to create the necessary ray mapping function. Combining Equations 7.23 and 7.24, we obtain the third-order LA of the first lens:

$$\Delta s'_{III1} = -\frac{(n^2 + k_1)}{2f_1'(n-1)^2} r^2 \tag{7.26}$$

Clearly, the aberration of the first aspheric lens represents a parabolic function analogous to Equation 7.20 and can be used to approximate the ray mapping function. Now, we assume

$$\Delta s' = \Delta s'_{III1} \tag{7.27}$$

and take into account Equation 7.26. Then, we obtain the following expression for the conic constant of the first aspheric surface, k_1:

$$k_1 = \frac{-2\Delta s' f_1'(n-1)^2}{r^2} - n^2 \tag{7.28}$$

As discussed above and shown in Figure 7.3, it is convenient to consider the LA $\Delta s'_{w0}$ corresponding to the ray height at the beam waist of radius w_0. Taking this into account and also noticing that for a plano-convex lens,

$$f_1' = s_0' \tag{7.29}$$

then we can write a final expression for the conic constant as

$$k_1 = \frac{-2\Delta s'_{w0} s'_0 (n-1)^2}{w_0^2} - n^2 \tag{7.30}$$

Since the first lens is plano-convex, the vertex radius r_{c1} of its aspheric surface is

$$r_{c1} = f'_1(n-1) = s'_0(n-1) \tag{7.31}$$

The second basic condition of a laser beam shaper is for there to be no wave aberration present in the plane wavefront as it leaves the system. This means that the spherical aberration of the complete laser beam shaper system must be equal to zero. The corresponding condition of the third-order aberration theory implies that the total sum S_I of first Seidel coefficients for all optical surfaces is equal to zero, which can be expressed by the following relationship:

$$S_\mathrm{I} = S_{\mathrm{I}1} + S_{\mathrm{I}2} = 0 \tag{7.32}$$

This expression is convenient to calculate the aspheric parameters, conic constant k_2 and vertex radius r_{c2}, for the second lens. Combining Equations 7.24, 7.25, and 7.32 gives

$$k_2 = \frac{s'_0 k_1 + n^2 d}{s'_0 - d} \tag{7.33}$$

$$r_{c2} = (n-1)(d - s'_0) \tag{7.34}$$

Thus, all parameters of the plano-aspheric lens pair beam shaping system are defined.

7.3.1.2 Example of Designing the Beam Shaper

Now, we carry out the calculations of the lens parameters on the example of a beam shaper analogous to the 1-to-1K system described by Shealy and Hoffnagle [27]. The initial data are assumed to be given by Equation 7.8, where n = 1.46071 for fused silica when $\lambda = 532$ nm.

7.3.1.2.1 Calculations of the Parameters for Second-Order Aspheres

Using Equations 7.13, 7.18, 7.19, 7.30, 7.31, 7.33, and 7.34 to calculate the second-order parameters gives the following results:

$$s'_0 = 43.81\,\mathrm{mm} \quad \Delta s'_{\omega 0} = 14.02\,\mathrm{mm} \tag{7.35}$$

$$r_{c1} = -20.182\,\mathrm{mm} \quad k_1 = -48.71 \tag{7.36}$$

$$r_{c2} = 48.925\,\mathrm{mm} \quad k_2 = 17.08 \tag{7.37}$$

where the radii signs are consistent with the optics sign convention [33]. The optical system layout is shown in Figure 7.4, where the design data in the form adopted in practice of optical system designing are given in Table 7.1.

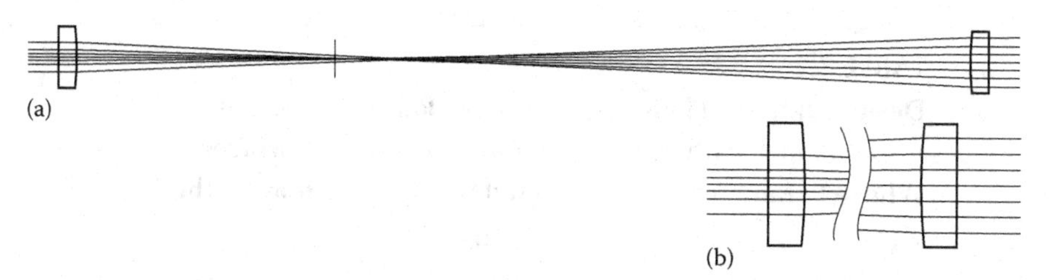

FIGURE 7.4 (a) Layout of plano-aspheric lens pair and Keplerian laser beam shaper; (b) expanded view of layout.

TABLE 7.1
Design Data for Plano-Aspheric Lens Pair of Keplerian Beam Shaper Calculated Based on the Third-Order Aberration Theory

No.	r_c	t_c	Glass	k	n_{532}
		Infinity			1
1	Infinity	3	Fused silica		1.46071
2	−20.182	150		−48.71	1
3	48.925	3	Fused silica	17.08	1.46071
4	Infinity				1

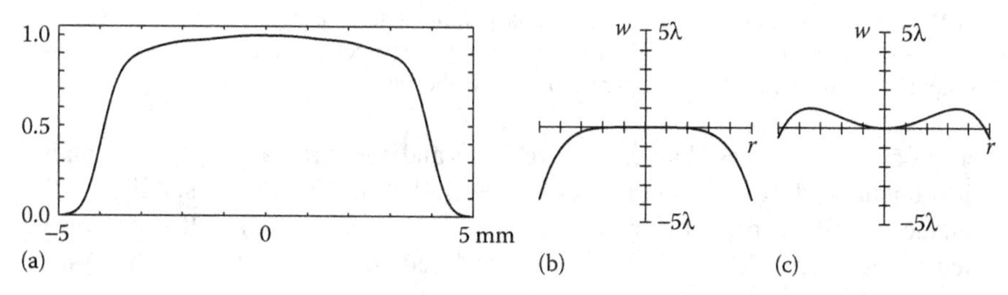

FIGURE 7.5 Performance data for two-lens plano-aspheric laser beam shaper when using the second-order aspheric surfaces calculated based on the third-order aberration theory: (a) output irradiance distribution; (b) residual wave aberration; (c) aberration after the air gap correction.

Modeling of this system by using the optical design software ZEMAX and calculations of output irradiance distribution and wave aberration gives the results presented in Figure 7.5, where the input beam is characterized by Gaussian irradiance profile with a waist radius of 2.37 mm. This system demonstrates good performance, and one can see that the resulting irradiance distribution is very close to uniformity and that the residual wave aberration does not exceed 4λ. Figure 7.5b is a good result when taking into account that wave aberration between the lenses is of order of 100λ. Therefore, it is clear that using the third-order aberration theory for calculating the beam shaper parameters gives a good starting point for further optimization. The aberration correction can be improved by changing the air gap d between the lenses by approximately 1% to a distance of 151.5 mm. Then, the residual wave

TABLE 7.2
Design Data for Plano-Aspheric Lens Pair of Keplerian Beam Shaper with the Second-Order Aspheric Surfaces Whose Parameters Are Corrected by Optimization Method

No.	r_c	t_c	Glass	k	n_{532}
		Infinity			1
1	Infinity	3	Fused silica		1.46071
2	−20.182	150		−54.8	1
3	48,925	3	Fused silica	29.5	1.46071
4	Infinity				1

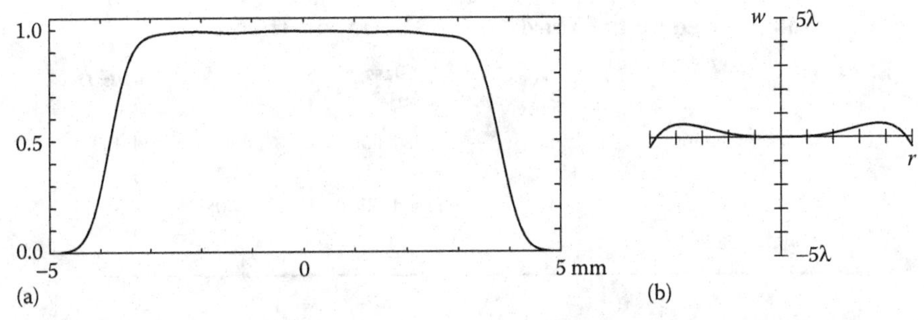

FIGURE 7.6 Performance data for two-lens plano-aspheric laser beam shaper with the second-order aspheric surfaces when parameters are corrected by optimization method: (a) output irradiance distribution; (b) residual wave aberration.

aberration becomes less than λ (Figure 7.5c) and the irradiance profile remains almost unchanged, which is an interesting result that indicates good stability of the irradiance profile as provided by a beam shaper when there are small variations of system parameters. This feature gives some freedom in correction of the system parameters and brings reliability into the operation of a beam shaper.

Performance of the optical system can be improved through optimization of the parameters when using the well-developed mathematical algorithms in modern optical designing software, such as ZEMAX, which was used in this work. As a first step, optimization is suggested to correct the parameters of the above considered system with the second-order aspheric surfaces. By modeling the ray mapping function, setting condition of minimizing the spherical wave aberration and using radii of curvature and conic constants of aspherics as variable parameters, one obtains the optical system described in Table 7.2, where the corresponding results of the irradiance profile and wave aberration are shown in Figure 7.6.

Optimization requires 5–10 iterations and takes several seconds. Evidently, the resulting intensity distribution is almost flat; the deviation from uniformity is less than ±3% which can be considered as a perfect result for many practical applications. After optimization, the residual wave aberration does not exceed ±λ/3, which is acceptable for many practical applications. This example demonstrates a very important feature: The refractive beam shaper can be implemented as a pair of lenses with the second-order aspheric surfaces for which the manufacturing and testing techniques are well developed and widely used in optical industry. The ray mapping function, which is required for irradiance transformation of Gaussian to flat-top beam, can be realized with a hyperbolic aspheric surface ($k < -1$), while to achieve the aberration correction level, being acceptable for some applications, it is sufficient to use a spheroid surface ($k > 0$). Many applications based on using the TEM_{00} lasers, such as, interferometry and holography, require much higher levels of aberration correction. Therefore, further optimization should focus on defining parameters of higher order aspheric terms.

7.3.1.2.2 Optimization Using Higher Order Aspheres

Since the second-order aspheric surface is sufficient to realize a required ray mapping function, one can consider the higher order aspheric surface for the second lens only. It is convenient to apply the optical design equation for aspheric expressions in terms of even orders of the radial coordinates [33], which is realized in modern optical design software as a standard form for optical surfaces. Now, we consider that the second aspheric surface has the fourth-order radial term in its surface equation, and then, we optimize the 1-to-1K system with the following variables: radii of curvature, conic constants of both aspheric surfaces, and the coefficient of the fourth-order polynomial term of second aspheric surface. Results of calculations are presented in Table 7.3 and Figure 7.7.

This optimization process ends after about 10 iterations with a duration of several seconds. The resulting irradiance profile is practically perfect, the deviation from uniformity is within 1%, and the residual wave aberration does not exceed $\pm\lambda/15$. The achieved aberration correction is acceptable for many of real applications, and the residual deviation of phase front is comparable to results which can be associated with manufacturing tolerances while machining and testing of aspheric surfaces.

In summary, a beam shaping optical system composed from plano-convex lens with the second-order aspheric surface and plano-convex lens with the fourth-order

TABLE 7.3
Design Data for Plano-Aspheric Lens Pair of Keplerian Beam Shaper Where First Aspheric Has the Second-Order and the Second Aspheric Has the Fourth Order

No.	r_c	t_c	Glass	Asphere Coefficients	n_{532}
		Infinity			1
1	Infinity	3	Fused silica		1.46071
2	−20.1	150		$k = -55.62$ $A_4 = -6.27 \times 10^{-5}$	1
3	48.75	3	Fused silica	$K = 67.22$	1.46071
4	Infinity				1

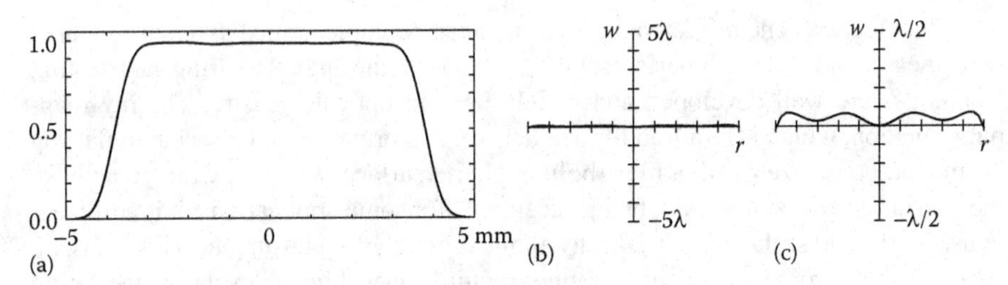

FIGURE 7.7 Performance data for two-lens plano-aspheric laser beam shaper when first aspheric has second order and the second aspheric has fourth order: (a) output irradiance distribution; (b) residual wave aberration; (c) wave aberration using enlarged scale.

TABLE 7.4
Design Data for Plano-Aspheric Lens Pair of Keplerian Beam Shaper When the First Aspheric Has the Second Order and the Second Aspheric Has the Sixth Order

No.	r_c	t_c	Glass	Asphere Coefficients	n_{532}
		Infinity			1
1	Infinity	3	Fused silica		1.46071
2	−20.1	150		$k = -55.6$ $A_4 = -6.27 \times 10^{-5}$ $A_6 = -2.06 \times 10^{-6}$	1
3	49.11	3	Fused silica	$k = 86.42$	1.46071
4	Infinity				1

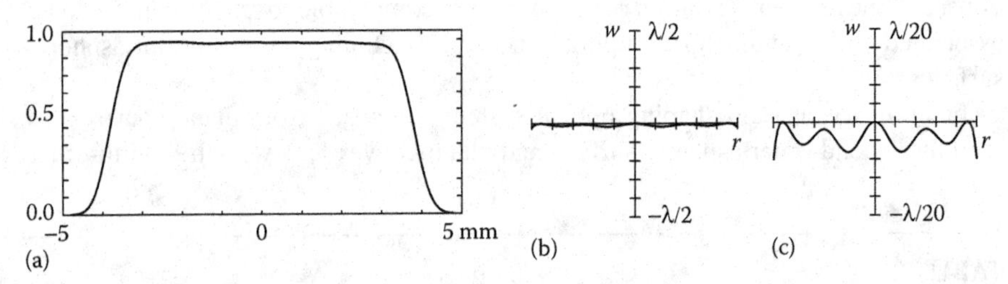

FIGURE 7.8 Performance data for two-lens plano-aspheric laser beam shaper when the first aspheric has second order and the second aspheric has sixth order: (a) output irradiance distribution; (b) residual wave aberration; (c) wave aberration using enlarged scale.

aspheric surface solves the task of irradiance profile redistribution from Gaussian to flat-top profile for many real laser-based applications. Further improvement of the system can be achieved through increasing the degree or power of the second-order aspheric. Optimization of the parameters of the system with the sixth-order aspheric surface leads to the results presented in Table 7.4 and Figure 7.8. The optimization process ends after about 20 iterations within several seconds. The resulting irradiance profile is practically the same as in the previous design. The residual wave aberration does not exceed $\pm\lambda/100$, which presents practically an ideal solution, since this aberration is order of magnitude less than phase front deviations issued by manufacturing errors while machining and testing of aspheric surfaces.

7.3.1.3 Conclusions

We have presented an engineering method for the design of a refractive laser beam shaping system which redistributes an input irradiance Gaussian distribution into a more uniform output irradiance profile. The core technique of this approach is the third-order aberration theory that allows calculating a good approximate solution in the form of set of parameters of an optical system that can be improved by subsequent optimization algorithms of modern optical system designing software. In the case of a plano-aspheric lens pair implementation, the first lens with an aspheric surface of second order allows realizing the ray mapping function required for redistribution of a Gaussian irradiance profile to a uniform profile. The resulting uniform profile demonstrates high stability, which gives some freedom in correction of the system parameters and brings reliability in operation of a beam shaper. Correction of the wave aberration to provide a flat wavefront can be achieved by increasing the power of radial terms of second aspheric surface.

7.3.2 Building Achromatic Beam Shaper Based on the Two-Lens System

In Section 7.3.1, we discussed an optical design and optimization-based method for designing a plano-aspheric laser beam shaping system, which realizes the irradiance profile transformation at a particular wavelength. When low dispersive materials for lenses are applied, then the two-lens laser beam shaper demonstrates acceptable performance within a certain spectral band. For example, when using fused silica, there is an opportunity to achieve a bandwidth up to ± 4 nm in the visible spectrum where the residual wave aberration is less than $\lambda/10$. At the same time, there are a number of laser applications where ultra-short pulse lasers are applied: micromachining with using femtosecond pulses, irradiating photocathodes of FEL, and various ablation techniques. A specific feature of these lasers is the broad spectrum, up to ± 100 nm. However, there are widely used applications such as confocal microscopy and multi-color holography, and various fluorescence techniques in life sciences where several laser sources in the spectrum of several hundreds of nanometers are simultaneously applied, for example, in visible spectrum spanning from 405 to 650 nm. All of these applications will benefit from transformation of Gaussian laser beam into a beam of uniform intensity distribution. Hence, providing an inherently achromatic design of the field mapping beam shapers is of great importance.

Now, we shall present a method of achromatizing the plano-aspheric lens pair beam shaper by using the optical designing technique, known as *chromatic radius* that is widely used when designing broad-spectrum visual optics. Then, the achromatic design of the beam shaper is further developed to provide an optimum solution for a particular application optical system.

7.3.2.1 Theory

There are several ways to develop a beam shaping optical system which operates over a wide spectral bandwidth by satisfying (1) the geometrical optics law of intensity for the irradiance transformation and (2) the constant optical path length. For example, it is possible to apply together with a beam shaper an auxiliary optical component whose main function is achromatizing of entire optical system [8], but this approach leads to more complexity and difficulty in using these systems. A more fruitful approach is to build an inherently achromatic laser beam shaper by achromatizing each optical component of a beam shaping system while retaining the optical functionality of each component, where the first optical component transforms the irradiance distribution by introducing spherical aberration and the second component compensates for the wave aberration as the beam leaves the system.

For the case of a telescopic beam shaper, achromatizing implies fulfilling of the same conditions of beam profile transformation as well as providing zero optical power for extreme wavelengths of a given spectral band. To achieve this goal, we apply the buried achromatizing surface, which is also known as *chromatic radius* designing technique; the basic idea of this is discussed in Section 7.3.2.1.2. Then, optical design software is used to further optimize the performance of an achromatic laser beam shaper for a particular optical application, where a system with air-spaced lenses is considered.

7.3.2.1.1 Evaluating the Chromatic Aberration of a Two-Plano Aspheric Lens Laser Beam Shaper

Now, consider the behavior of the beam shaping system described in Table 7.4, which uses low dispersive fused silica for the spectrum around the design wavelength 532 nm. Calculations of the chromatic aberrations in the spectral range 532±70 nm give results presented in Figure 7.9, where Figure 7.9b gives the wave spherical aberration at central wavelength of 532 nm, as well as the *red* (602 nm) and *blue* (462 nm) regions of the spectrum. Figure 7.9a demonstrates dependence of the optical power on wavelength, which means that evaluation of just optical power is more convenient for the telescopic beam shaper. This is analogous to calculations of chromatic focal shifts for focusing systems like objective lenses.

A plot of the wave aberration at 532 nm is almost null when using the scale ±5λ, while the wave aberration in the red and blue ends of spectrum reaches values of several units of λ. Since the chromatic aberrations of the optical power vary from positive to negative values, the wavefront has converging and diverging components within the beam,

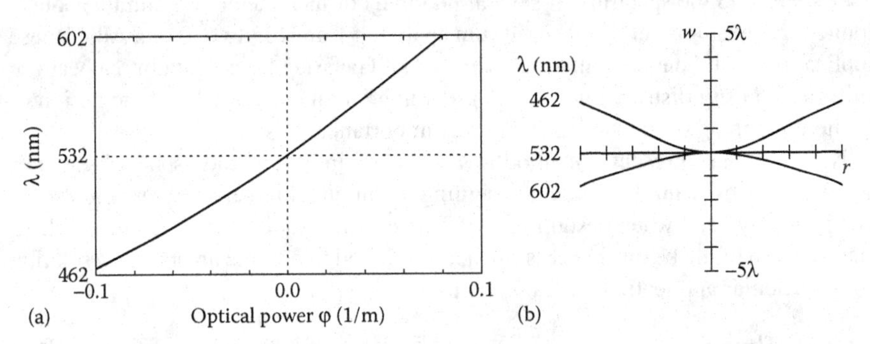

FIGURE 7.9 Chromatic aberration of the laser beam shaper using fused silica glass for the wavelength region of 532 ± 70 nm: (a) optical power variation versus spectrum; (b) wave aberration. Parameters for this system are given in Table 7.4.

and the optical power variation over this spectral band is approximately equal to 0.2m^{-1} which implies that a beam with 8 mm diameter corresponds to a divergence of approximately 1.6 mrad, which is 19 times larger than natural divergence of a 532 nm laser beam with the same beam waist diameter. It is well known that the chromatic aberration can be reduced when low dispersion materials such as CaF_2 are used for optical elements. However, even in this case, the chromatic wave aberration is essential and prevents successful applying of beam shapers with ultra-short pulse or multiwavelength laser sources.

7.3.2.1.2 Designing Technique Using the Chromatic Radius

Achromatization of an optical system implies using optical glasses (or other refractive materials) with different dispersion properties. Most often, the dispersion of a material is characterized by the Abbe number [33,34], which can be written as

$$\nu = \frac{n_{532} - 1}{n_{462} - n_{602}} \tag{7.38}$$

where:

n_{532}, n_{462}, and n_{602} are the indices of refractive at the corresponding wavelengths

Since the main design idea of a field mapping beam shaper optical system is to introduce relatively strong spherical aberration between the beam shaper components, it is very important to avoid or at least minimize spherochromatism (chromatic variation of spherical aberration). These objectives can be realized when each lens component of a beam shaper is inherently achromatic. Historically, the beam shaper components [2,5,6,23] have been implemented as plano-aspheric lenses. The simplest way to achromatize this two-lens refractive beam shaper is to use a pair of lenses from different glasses: either cemented doublets or air-spaced doublets.

Before developing of an achromatic design, it is convenient to apply the so-called chromatic radius technique which is widely used for designing of broadband visual optics. The basic idea of the chromatic radius technique is illustrated in Figure 7.10, which can be described as follows:

- The central wavelength of a given spectrum is defined.
- There is a chosen pair of optical glasses having the same refractive index but different Abbe numbers at the central wavelength, so-called chromatic glasses pair.
- There is designed an optical system with correction of monochromatic aberrations at the central wavelength; in the considered research it should be just a beam shaping system as a pair of plano-aspheric lenses.
- Each lens of the monochromatic design is split virtually into a doublet, whose lenses are made from glasses of earlier chosen chromatic pair, and the air gap between lenses is zero.
- The inner surface radius of the doublets, the so-called chromatic radius, is defined by the condition of compensation of chromatic difference of the focal lengths for the red (λ_{max}) and blue (λ_{min}) ends of the given spectrum (see $\Delta s'_{chrom}$ in Figure 7.10).
- Later, the air gap between the lenses is to be enlarged to realize a real optical design.

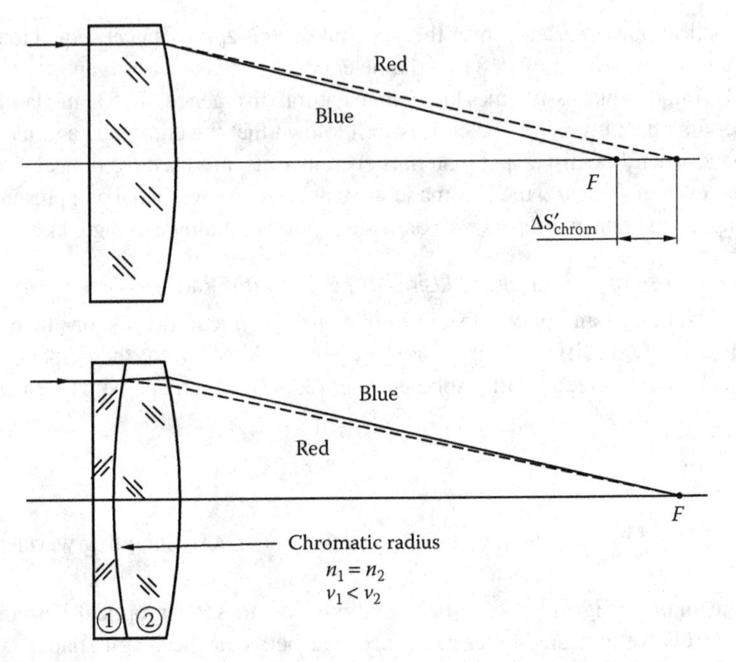

FIGURE 7.10 Description of the chromatic radius concept used for designing achromatic laser beam shaper.

Obviously, this approach guarantees that the monochromatic aberration stays unchanged for whole spectral band. At the same time, the chromatic aberration is eliminated for the extreme wavelengths of that band which exactly meets the needs of achromatizing of a beam shaper.

To define the chromatic radius, we use the techniques for designing of achromatic doublets described in the literature [33,34]. The conditions for achromatizing a thin-lens doublet are given:

$$\varphi = \varphi_1 + \varphi_2 \tag{7.39}$$

$$\frac{\varphi_1}{\nu_1} + \frac{\varphi_2}{\nu_2} = 0 \tag{7.40}$$

where:

φ is the optical power of the full system

φ_1 and φ_2 are the optical power of the first and second lenses, respectively

ν_1 and ν_2 are the Abbe numbers of the corresponding lens glasses

Applying these formulas for the plano-convex lens shown in Figure 7.10 for the chromatic pair glasses chosen earlier, it is easy to get the following formula for calculation of chromatic radius r_{chrom1}:

$$r_{chrom1} = \frac{\nu_1 - \nu_2}{\nu_1} r_{c1} \tag{7.41}$$

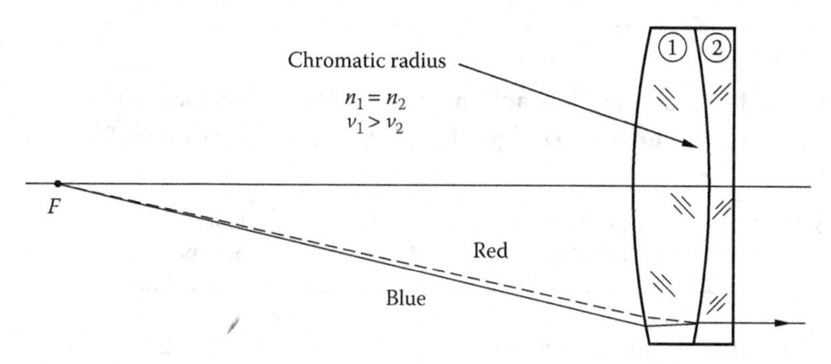

FIGURE 7.11 Defining the chromatic radius for the second beam shaper element.

where:

r_{c1} is the vertex curvature radius of the last surface that is just the asphere of first lens component of the beam shaping system considered in this example

By analogy, it is possible to define the chromatic radius r_{chrom2} for the second plano-convex aspheric lens component of the two-lens beam shaping system shown in Figure 7.11:

$$r_{chrom2} = \frac{\nu_2 - \nu_1}{\nu_2} r_{c2} \tag{7.42}$$

where:

r_{c2} is the vertex curvature radius of the aspheric surface

Thus, the design parameters of a two-component achromatic beam shaper with aspheric surfaces are defined. As a rule, it is recommended to choose glasses with essential difference of Abbe numbers: the more difference, the less curvature of the inner *chromatic* surface responsible for correction of chromatic aberrations and, hence the smoother optical design, less high-order aberrations, and more capabilities to correct chromatism in wider spectrum.

7.3.2.2 Application

Now, consider designing of the achromatic system following a similar approach described above:

- Monochromatic design to be analogous to the beam shaping system presented in Section 7.3.1.2.2
- Achromatization of the system to be done by using the chromatic radius technique
- Realizing a real air-spaced achromatic beam shaper by simulation and optimizing the system by using the optical design software ZEMAX

The initial data for designing of $\lambda = 532 \pm 70$ nm have been presented in Equation 7.8. Despite the popularity of fused silica as a refractive material in laser applications, it is very difficult to use a matching *chromatic pair* of materials. Therefore, we use

TABLE 7.5
List of the Indices of Refraction and Abbe Numbers for Two Glasses Used for This Design of an Achromatic Laser Beam Shaper

Glass	n_{462}	n_{532}	n_{602}	ν
S-BSM16	1.630007	1.623827	1.619669	60.34
S-TIM2	1.636446	1.625702	1.618838	35.54

glasses from the Ohara catalog, crown S-BSM16 and flint S-TIM2, where the index of refraction and the Abbe numbers, ν, can be calculated from Equation 7.38 and are given in Table 7.5.

The refractive indices of the glasses are almost equal at $\lambda = 532$ nm and have different Abbe numbers. Thus, these glasses present a chromatic pair for a given spectral band. Note that the index of refraction of S-TIM2 is larger at $\lambda = 462$ nm and smaller at $\lambda = 602$ nm than corresponding index of S-BSM16.

7.3.2.2.1 Monochromatic Design of a Plano-Convex Aspheric Laser Beam Shaper from S-BSM16

As a first step in the design, it is necessary to repeat calculations of the pair of plano-aspheric lenses made from one of the glasses of the chromatic pair, such as for S-BSM16. Calculations with Equations 7.13, 7.18, 7.19, 7.30, 7.31, 7.33, and 7.34 give the following results:

$$s_0' = 43.81\,\text{mm} \quad \Delta s_{W0}' = 14.02\ \text{mm} \tag{7.43}$$

$$r_{c1} = -27.327\ \text{mm} \quad k_1 = -88.31 \tag{7.44}$$

$$r_{c2} = 66.247\ \text{mm} \quad k_2 = 32.588 \tag{7.45}$$

Further optimization of this system by analogy with the process considered in Section 7.3.1 gives a set of system parameters presented in Table 7.6. Results of

TABLE 7.6
Optimization of the System Defined by the Parameters Given by Equations 7.43 through 7.45

No.	r_c	t_c	Glass	Asphere Coefficients	n_{532}
		Infinity			
1	Infinity	3	S-BSM16		1.623827
2	−27.39	150		$k = -97.85$	
3	66.314	3	S-BSM16	$k = 154.2$	1.623827
				$A_4 = -4.53 \times 10^{-5}$	
				$A_6 = -1.4 \times 10^{-6}$	
4	Infinity				

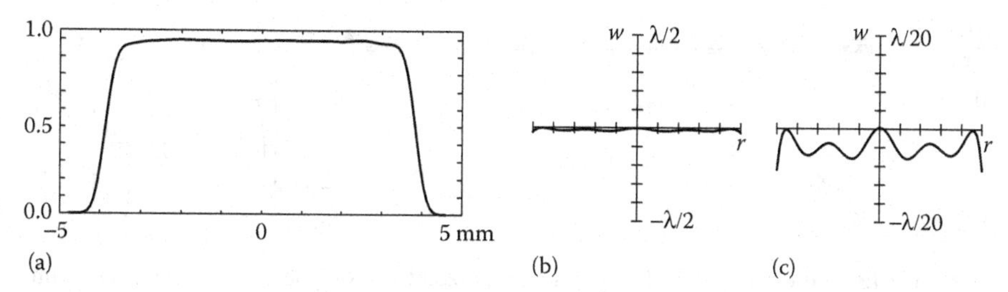

FIGURE 7.12 Performance data for two-lens laser beam shaper using Ohara glass S-BSM16 when the first aspheric has second order and second aspheric has sixth order: (a) output irradiance distribution; (b) residual wave aberration; (c) wave aberration using enlarged scale.

calculations in ZEMAX of output irradiance distribution and wave aberration, characterizing the system performance, are shown in Figure 7.12. Evidently, performance of this system is analogous to the one designed in Section 7.3.1 and presented in Table 7.4 and Figure 7.8. The resulting irradiance profile is practically uniform, and the residual wave aberration does not exceed $\pm\lambda/100$, which presents practically an ideal solution for real manufacturing technologies and laser applications. The irradiance profiles of this beam shaping system in further developments are almost identical to that in Figure 7.12. Therefore, these data will be omitted while presenting the data characterizing system performance.

7.3.2.2.2 Achromatizing by Using the Chromatic Radius

The design technique of the chromatic radius, which is discussed in Section 7.3.2.1.2, is used to achromatize the beam shaping system in a given spectrum. Calculations using Equations 7.41 and 7.42 give the system presented in Table 7.7 and Figure 7.13. Modeling of the system described in Table 7.7 with ZEMAX leads to the results presented in Figure 7.14.

TABLE 7.7
Design Data for Achromatic Beam Shaper Shown in Figures 7.13 and 7.14

No.	r_c	t_c	Glass	Asphere Coefficients	n_{532}	ν
		Infinity				
1	Infinity	1	S-TIM2		1.625702	35.54
2	19.11	2	S-BSM16		1.623827	60.34
3	−27.39	150.8		$k = -97.85$		
4	66.314	2	S-BSM16	$k = 154.2$ $A_4 = -4.53 \times 10^{-5}$ $A_6 = -1.4 \times 10^{-6}$	1.623827	60.34
5	−46.27	1	S-TIM2		1.625702	35.54
6	Infinity					

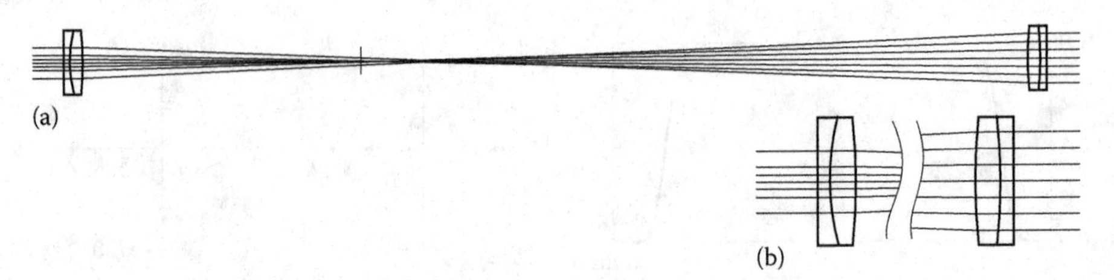

FIGURE 7.13 (a) Optical layout of the achromatized two-component aspheric laser beam shaper; (b) enlarged view of the input and output components.

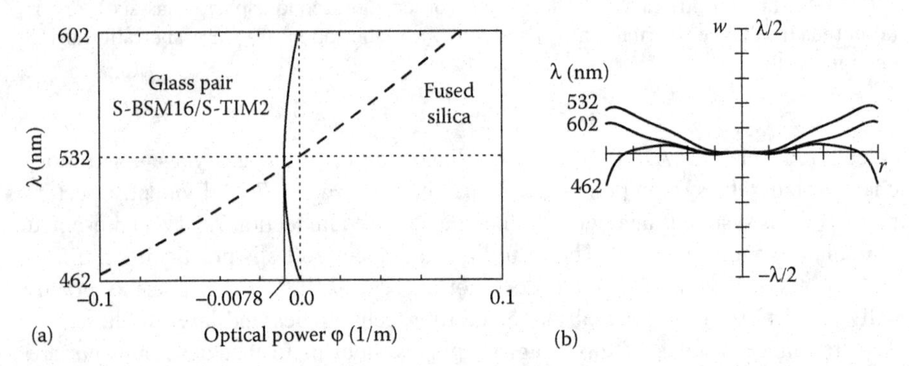

FIGURE 7.14 (a) Optical power variation versus the spectrum of the achromatized two-component laser beam shaper. The dashed curve gives the data for the fused silica system. (b) Wave aberration for the system, which is shown on an expanded scale.

Clearly, application of chromatic radii approach leads quickly to a near-workable solution:

- The residual wave aberration over the given spectrum is approximately equal to $\pm\lambda/7$, as shown in Figure 7.14 is much smaller when comparing to analogous results for the beam shaper made with fused silica which is shown in Figure 7.9.
- The system is exactly afocal in paraxial approximation for the red and blue ends of spectrum.
- The optical power for the central wavelength is equal to -0.0078m^{-1}, which is called the secondary spectrum [33,34] and which is an order of magnitude smaller than the performance obtained for a fused silica design. For the beam diameter of 8 mm, this gives a full divergence angle of 0.064 mrad which corresponds to approximately $\pm 1/3$ of the natural divergence of the comparable TEM_{00} laser beam.
- The difference in the index of refraction of the glasses at the central wavelength of 532 nm is easily compensated by slightly changing the distance between the beam shaper components, such as using 150.8 mm instead of 150 mm for the inter-lens spacing.

Further improvement of optical design performance is achieved by using ZEMAX software to obtain the system layout described in Table 7.8, where the results of the aberration calculations are presented in Figure 7.15.

TABLE 7.8
Design Data for System Obtained by Optimizing with ZEMAX Shown in Figure 7.15

No.	r_c	t_c	Glass	Asphere Coefficients	n_{532}	N
		Infinity				
1	Infinity	1	S-TIM2		1.625702	35.54
2	22.6	1	S-BSM16		1.623827	60.34
3	−27.26	150.8		$k = -100.0$		
4	66.92	2	S-BSM16	$k = 173.7$ $A_4 = -4.1 \times 10^{-5}$ $A_6 = -2.4 \times 10^{-6}$	1.623827	60.34
5	−44.36	1	S-TIM2		1.625702	35.54
6	Infinity					

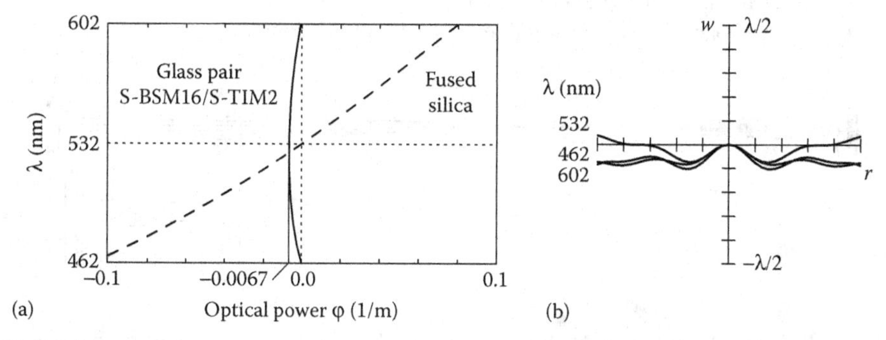

FIGURE 7.15 (a) Optical power variation versus spectrum for the system given in Table 7.8; (b) wave aberration.

Using the optimization algorithms allows for an increase in the system performance:

- The residual wave aberration over whole spectrum is approximately equal to $\pm\lambda/20$.
- The secondary spectrum of the system is further reduced, which implies that the output beam divergence due to chromatism becomes much smaller than natural divergence of a laser beam.
- The functions of spherical aberration for red (602 nm) and blue (462 nm) wavelengths are almost coincident with each other, while the function for the central wavelength (532 nm) shows certain differences. Evidently, this behavior is the logic for operation of an achromatic optical system.

From the point of view of efficiency of beam irradiance profile transformation as well as aberration correction level, this optical system demonstrates performance that meets the practical needs of almost any laser application with a wide spectrum. The last improvement to be done is providing a totally air-spaced design, which is considered in the following section.

7.3.2.2.3 Air-Spaced Achromatic Beam Shaper

In practice for laser applications, the optical design of a beam shaper should be implemented in such a way that its components are air-spaced doublets with *realistic* distances between the lenses. Therefore, the next step in developing the beam shaper system is to increase the distance between the lenses and optimization by using the optical design software. The resulting optical layout is shown in Figure 7.16, where the parameters are given in Table 7.9. The results of aberration calculations are presented in Figure 7.17.

This layout can be considered as a final one:

- It demonstrates near the same performance as the design of the previous system.
- The irradiance profile is practically uniform, where the deviation from uniformity does not exceed ±1% as shown in Figure 7.12.
- The system parameters are realizable by using modern technologies of optics manufacturing and assembly.
- The air-spaced design is suitable for laser applications.

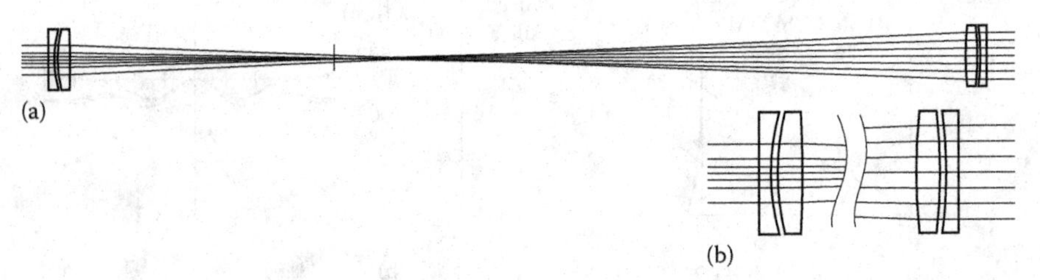

FIGURE 7.16 (a) Optical layout of the achromatized air-spaced aspheric laser beam shaper; (b) enlarged view of the input and output components.

TABLE 7.9
Parameters for the Air-Spaced System Obtained by Optimizing System with ZEMAX Associated with Figures 7.16 and 7.17

No.	r_c	t_c	Glass	Asphere Coefficients	n_{532}	ν
		Infinity				
1	Infinity	1	S-TIM2		1.625702	35.54
2	19.5	0.5				
3	19.95	2	S-BSM16		1.623827	60.34
4	−27.2	151.2		$k = -100.0$		
5	69.23	2	S-BSM16	$k = 173.7$ $A_4 = -4.1 \times 10^{-5}$ $A_6 = -2.4 \times 10^{-6}$	1.623827	60.34
6	−46.21	0.5				
7	−47.0	1	S-TIM2		1.625702	35.54
8	Infinity					

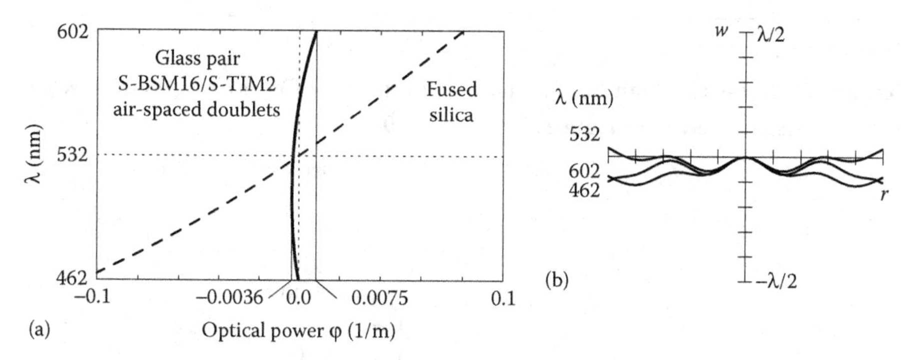

FIGURE 7.17 (a) Optical power variation versus spectrum for the system given in Table 7.9; (b) wave aberration.

It should be noted that the parameters of the final optical system are very close to the corresponding parameters calculated by using the third-order formulas, which confirm applicability of third-order aberration theory and popular techniques of achromatizing the design of a refractive beam shaper, which can be used as a starting point for relatively quick optimization with popular optical design software. It is possible to continue optimizing the system parameters through applying other glasses with other indices of refraction and Abbe numbers, which are close to those of the above considered as a chromatic pair. Of course, in designing of real systems for particular applications, it is possible to apply not only optical glasses but also other refractive materials such as crystals, polymers, or combinations of materials.

7.3.2.2.4 Example of Achromatic Beam Shaper without Internal Focusing

The design methods used are based on the third-order aberration theory, which can also be applied to designing of an achromatic beam shaper configured as a Galilean telescope. These beam shapers [9,10] have no internal focusing of a beam that is very important when working with high peak power short-pulse lasers. An example of the layout of such a system with the initial design data—$r_0 = 1.7$ mm, $R_{max} = 3.0$ mm, and $\lambda = 633–1064$ nm (850 ± 210 nm)—is illustrated in Figure 7.18, where the design parameters of this system are presented in Table 7.10. The performance results of the aberration calculations are presented in Figure 7.19.

While developing these results, the design techniques discussed in Chapter 7 were used, and the optimization of the final system was obtained by using the ZEMAX software. The glasses applied are very close to a chromatic pair. It should be noted that the Galilean design implies combining of optical components with negative

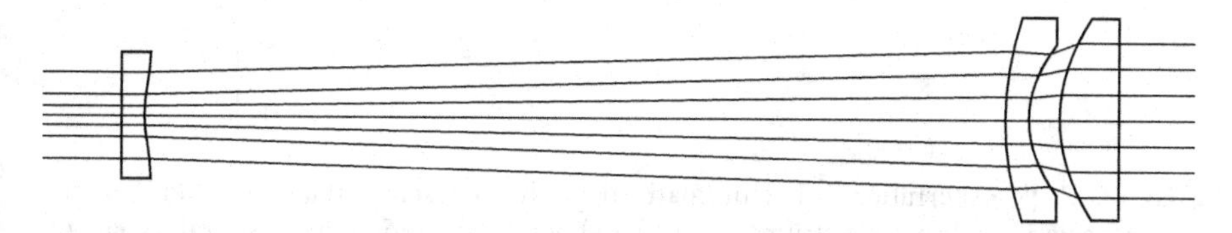

FIGURE 7.18 Layout of achromatic laser beam shaper in Galilean configuration. (Adapted from Laskin, A. US Patent No. 8,023,206 B2, September 20, 2011.)

TABLE 7.10

Design Data for Galilean Beam Shaper Obtained by Optimizing System with ZEMAX Associated with Figures 7.18 and 7.19

No.	r_c	t_c	Glass	Asphere Coefficient	n_d	ν_d
		Infinity				
1	Infinity	1.0	Schott-F5		1.60342	38.03
2	5.859	35.0		$k = -0.05$ $A_2 = -5.0 \times 10^{-2}$ $A_4 = -6.27 \times 10^{-3}$ $A_6 = 5.5 \times 10^{-4}$ $A_8 = -2.44 \times 10^{-5}$		
3	12.328	1.0	Schott-F5		1.60342	38.03
4	4.664	1.3		$k = 5.38$ $A_2 = 4.282 \times 10^{-2}$ $A_4 = 4.77 \times 10^{-4}$ $A_6 = 1.9 \times 10^{-5}$ $A_8 = -1.23 \times 10^{-6}$		
5	18.364	2.4	Schott-SSK3		1.614837	51.16
6	Infinity					

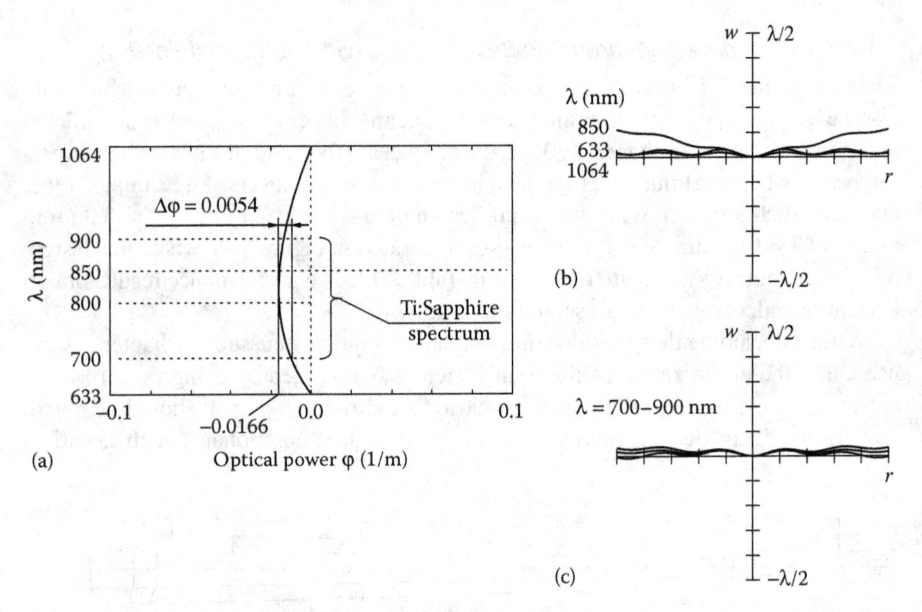

FIGURE 7.19 Aberrations of achromatized Galilean laser beam shaper: (a) optical power variation versus spectrum; (b) residual wave aberration for spectral range of 633–1064 nm; (c) wave aberration in spectral range of ultra-short pulse Ti:Sapphire laser of 700–900 nm.

and positive optical power. Therefore, there are additional capabilities to compensate the chromatic aberrations and real design can be built from three lenses.

From the point of view of aberrations, the system demonstrates very good performance:

- There is a relatively wide spectrum from 633 nm (He-Ne) to 1064 nm (Nd:YAG), which is approximately 850 ± 210 nm.
- The residual wave aberration over whole spectrum is approximately equal to $\pm\lambda/20$.
- The residual wave aberration over spectral range is widely used with ultra-short pulse Ti:Sapphire lasers, 700–900 nm (800 ± 100 nm), and is approximately equal to $\pm\lambda/50$.
- The secondary spectrum in the range of 700–930 nm is characterized by an optical power 0.0054 m^{-1} which corresponds to approximately ±1/10 of the natural divergence of comparable TEM_{00} laser beam with a 6 mm $1/e^2$ diameter at 800 nm.

Since the divergence due to the system chromatism is an order of magnitude smaller than the natural divergence of a laser beam, it is negligible in real applications. This has been confirmed in practice by using the beam shapers built based on this design approach.

7.3.2.3 Conclusions on Achromatic Laser Beam Shapers

The achromatic laser beam shaper discussed in Section 7.3.2 has grown to meet many diverse applications in scientific and technology applications in research and industry and has expanded laser applications significantly in areas such as control of beam profiles; EDOF in laser imaging systems; effective beam shaping of high-power lasers and multimode fiber or fiber-coupled lasers; imaging with flat-top beams; enhanced laser operations in holography and interferometry, SLM, and FEL; and ultra-short pulse lasers, multicolor fluorescence life science techniques, and confocal microscopy [9,10].

7.3.3 Applications of Achromatic Refractive Laser Beam Shaper

In this section, we shall describe an example of a real implementation of a refractive laser beam shaper from the series of devices known as a πShaper,* which has been commercially deployed by AdlOptica GmbH, Berlin, Germany. Here, we shall emphasize the design features and capabilities of this beam shaping technology while presenting some examples from real applications.

7.3.3.1 Optical Design Features

The optical design principles of the field mapping refractive beam shaper, which were discussed in Sections 7.3.1 and 7.3.2, have been realized in series of refractive beam shapers, which are finding various applications in research, industrial

* πShaper is a registered trademark of AdlOptica GmbH, Berlin, Germany, http://www.adloptica.com.

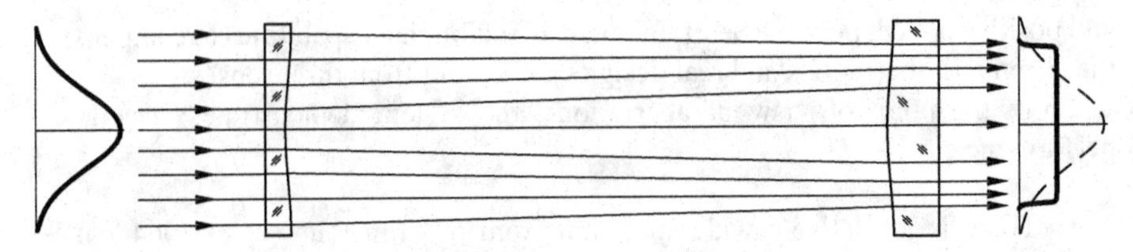

FIGURE 7.20 Configuration of refractive field mapping beam shaper, πShaper. (Reproduced from Laskin, A., *Proceedings of SPIE*, 7430, 2009. With permission.)

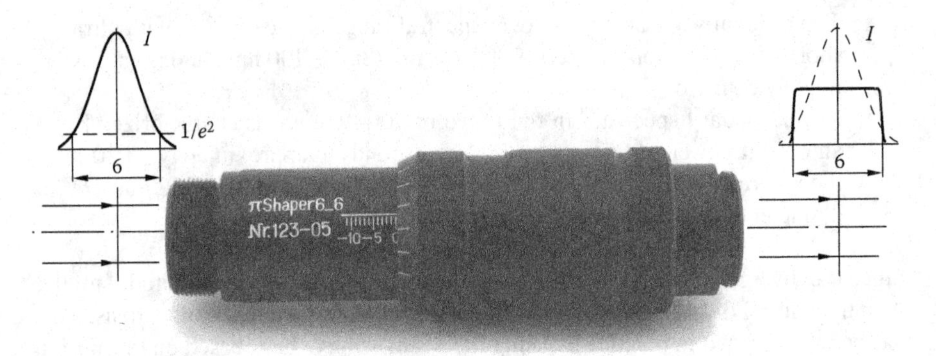

FIGURE 7.21 Refractive field mapping beam shaper, πShaper. (Reproduced from Laskin, A., Williams, G., and Demidovich, A., Applying of refractive beam shapers in creating spots of uniform intensity and various shapes, in *Laser Resonators and Beam Control XII*, Kudryashov, A. V., Paxton, A. H., and Ilchenko, V. S., eds., SPIE, Bellingham, WA, 2010. With permission.)

technologies, and medical instruments. Often these devices are implemented as a Galilean telescopic system with two optical components, where the output wavefronts are planar or flat, and the irradiance profile is transformed from a Gaussian to uniform profile in a controlled manner, by accurately introducing wave aberration by the first lens and further its compensation by the second lens, as illustrated in Figures 7.20 and 7.21. Thus, the resulting collimated output beam has a uniform irradiance profile and a flat wavefront, which has a low divergence—almost the same like one of the input beam. In other words, the field mappers transform the irradiance distribution without deterioration of the wavefront shape or increasing of the beam divergence. A summary of the main features of this refractive field mapper is given as follows:

- Refractive optical system transforms the Gaussian input profile into an output flat-top (top-hat, uniform) irradiance distribution.
- Transformation is through controlled phase front manipulation, where the first optical component introduces spherical aberration required to redistribute the beam irradiance and the second optical component compensates the spherical aberration.
- The output beam is free of aberrations where the phase profile remains flat with a low output divergence.

- This refractive beam shaper works equally well with TEM_{00} lasers and multimode beams with Gaussian-like irradiance profiles.
- Output beam is collimated, its divergence is defined by the natural divergence of the input beam.
- Resulting beam profile is stable over a large distance.
- Implementations as telescopic or collimating optical systems
- Some beam shaper models have achromatic optical design, where the beam shaping effect is provided for a certain spectral range simultaneously.
- All beam shaper models have Galilean design with no internal focusing.

An example of beam shaping for the Nd:YAG laser is presented in Figure 7.22 for beam shaping for an input TEM_{00} beam and output uniform beam profile, where images have been provided by InnoLas Laser GmbH, Krailling, Germany. These measured profiles show that the beam shaper not only converts the irradiance profile but also improves the spot shape, where one can see that the slightly distorted input beam is transformed into a flat-top output beam with regular round spot shape.

In contrast to many other beam shaping techniques, the physical principle of operation of the refractive field mappers does not require the input beam to be a TEM_{00} mode, or to have a common phase front for input and output beams. The refractive beam shapers work well with multiple, stable modes of laser beams. The only condition is that the irradiance distribution of input beam should be similar to the Gaussian function, where the irradiance has a peak in the center of beam and decreasing toward the periphery. For high-power solid-state lasers, the input beam profile may have a parabolic, Gaussian-like irradiance distribution. Capability to work simultaneously with TEM_{00} and other stable mode lasers allows for switching easily from one laser source to another.

7.3.3.1.1 Sensitivity to Tilt and Decenter

Any beam shaping technique implies introduction of aberrations in a certain way and, therefore, requires fulfillment of some predetermined conditions for proper transformation of a laser irradiance distribution. As in other beam shaping techniques, such as the refractive field mapping beam shaper, it is necessary to take into account the input beam size, its irradiance profile, and proper alignment of a beam shaper. Now, we shall evaluate the influence of misalignments in case of refractive field mapping beam shapers. Figure 7.23 presents results of mathematical simulations and measurements of real profiles for the refractive beam shaper in three cases: perfectly aligned, lateral shift of a beam, and angular tilt of the beam shaper.

For proper operation in real applications, such as in industrial equipment, the refractive beam shaper should provide certain tolerances for probable misalignments, such as spatial shifts or tilting. Therefore, realistic designs should provide the same aberration correction level not only for the clear aperture of a system but also in certain extent outside of the clear aperture. The practice of building real beam shaping systems shows that the aberration correction should be provided for diameter at least 1.6 times larger than $1/e^2$ diameter of an input Gaussian beam. Therefore a small, up to about ±20% of diameter, lateral shift of a beam with respect to the beam shaper, or vice versa, does not lead to aberration but allows for interesting beam shaping effects,

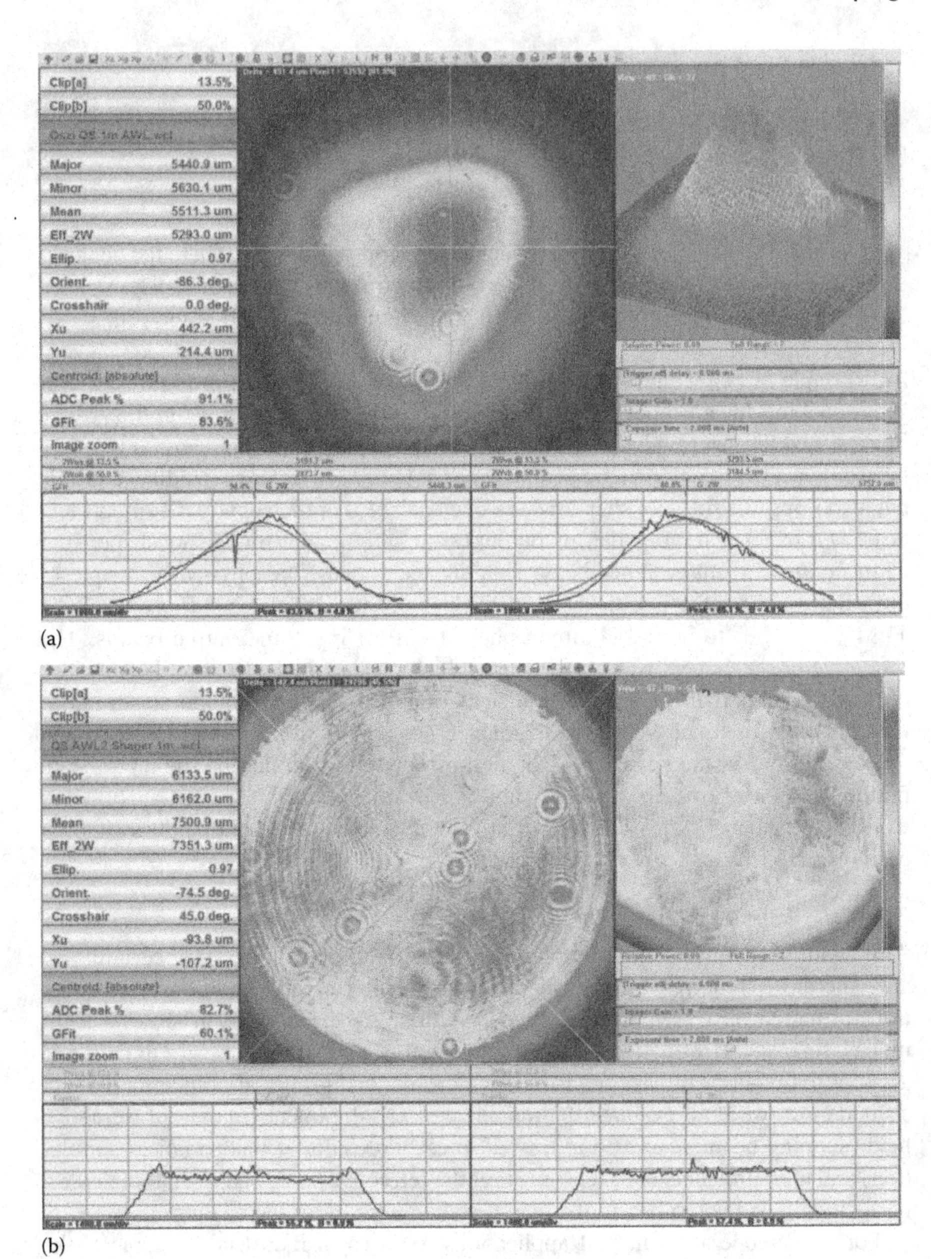

FIGURE 7.22 (See color insert.) Example of beam shaping with πShaper: (a) input TEM_{00}; (b) output from πShaper. (Courtesy of InnoLas Laser GmbH; Reproduced from Laskin, A. and Laskin, V., Variable beam shaping with using the same field mapping refractive beam shaper, in *Laser Resonators and Beam Control XIV*, Kudryashov, A. V., Paxton, A. H., and Ilchenko, V. S., eds., SPIE, Bellingham, WA, 2011. With permission.)

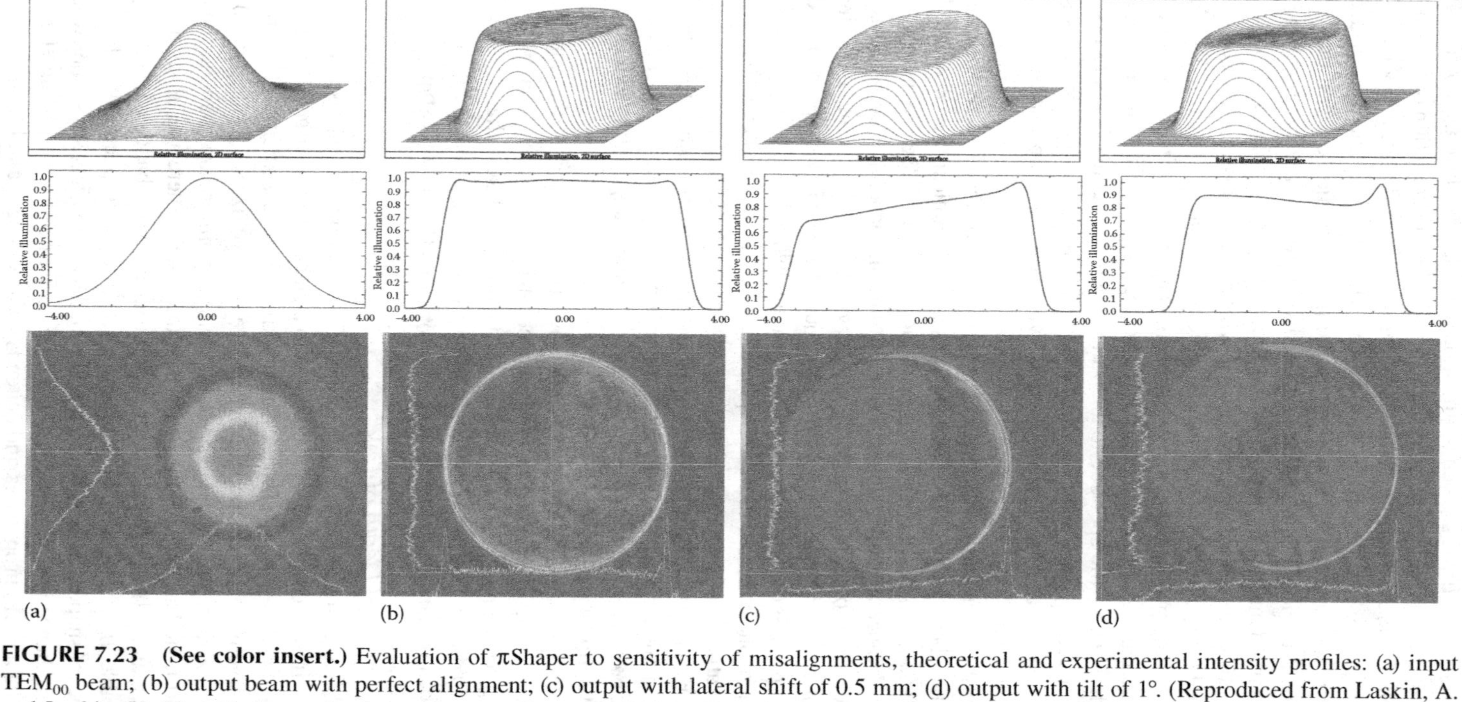

FIGURE 7.23 **(See color insert.)** Evaluation of πShaper to sensitivity of misalignments, theoretical and experimental intensity profiles: (a) input TEM_{00} beam; (b) output beam with perfect alignment; (c) output with lateral shift of 0.5 mm; (d) output with tilt of 1°. (Reproduced from Laskin, A. and Laskin, V., Variable beam shaping with using the same field mapping refractive beam shaper, in *Laser Resonators and Beam Control XIV*, Kudryashov, A. V., Paxton, A. H., and Ilchenko, V. S., eds., SPIE, Bellingham, WA, 2011. With permission.)

such as the output profile is skewed in direction of the lateral shift, as shown in Figure 7.23c. The intensity profile itself stays flat but is tilted in the direction of the shift, and a remarkable feature is that the beam itself stays collimated with a low divergence. This skewed profile can be used in applications where a steady increase or decrease of intensity is required, such as to compensate the attenuation of acoustic waves in acousto-optical devices. This profile can be also useful in hardening techniques to sustain a desired temperature profile on a device within a movable laser spot.

As an optical system designed to work with axial beams, the refractive beam shaper operates in relatively narrow angular field. The data in Figure 7.23d demonstrate the intensity profile behavior by the beam shaper when tilted by 1°. The intensity profile remains stable, but there is a visible degradation of the quality on the left and the right sides of the spot due to aberrations, which is coma. It should be noted that if the refractive beam shaper is tilted by 1° with respect to the optical axis, then there is a 2 mm lateral shift of one of its ends, which can be compensated by ordinary opto-mechanical mounts.

The data show that the misalignments have influence on the refractive beam shaper operation, but the sensitivity to misalignments is not significant, even with a lateral shift up to 0.5 mm and a tilt up to 1°, the resulting profiles are close to a flattop. In other words, the tolerance of positioning of a refractive beam shaper in accommodating and misalignments can be compensated by ordinary opto-mechanical mounts. Since the influence of a tilt on wave aberration of the output beam is very pronounced, it is advisable to pay more attention to angular alignment while configuring beam shapers.

In practice, one can state that the requirements of a refractive beam shaper are not difficult to satisfy. For example, alignment of a refractive beam shaper with a tolerance of 0.1 mm for lateral shift and about 10 arc minute for tilt, while the tolerance of input beam diameter is about 10%. Evidently, proper alignment of a refractive beam shaper can be done by using ordinary opto-mechanical alignment devices, such as the 4-axis tilt/tip mounts.

Another important feature of the refractive beam shaper is the capability to compensate for the divergence or convergence of input beam by changing the air gap spacing between components and easy adaptation to lasers that deviate from a Gaussian irradiance profile. All of these features are of great practical importance. The input beam size can be provided by widely used zoom beam expanders.

7.3.3.1.2 Extended Depth of Field

It is common to characterize beam shaping optics by their working distance, which is the distance from the last optical component to a plane where a target irradiance profile, flattop or another one, is created. The working distance is an important specification for diffractive beam shapers and refractive homogenizers (or integrators) based on multilens arrays. But in the case of the field mapping beam shapers, the output beam is collimated and, hence, instead of a definite plane where a resulting irradiance profile is created, there exists certain space after a beam shaper where the profile remains stable. In other words, the working distance is not a specification for the field mapping beam shapers. It is better to specify the DOF after a beam shaper where the resulting irradiance profile is stable. This DOF is defined by diffraction effects happening while a beam propagates and depends on wavelength and beam size.

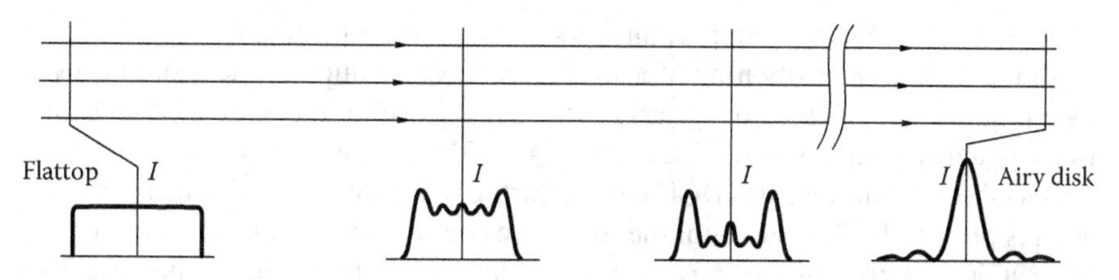

FIGURE 7.24 Intensity profile variation of a flat-top beam during propagation. (Reproduced from Laskin, A. and Laskin, V., *Proceedings of SPIE*, 8490, 2012. With permission.)

When a TEM_{00} laser beam with Gaussian irradiance distribution propagates in space, its size varies due to inherent beam divergence but the irradiance distribution remains stable, which follows from the well-known feature of TEM_{00} beams. When light beams with non-Gaussian irradiance distributions, for example flat-top beams, propagate in space, there is a simultaneous variation of both size and irradiance profile. Suppose a coherent light beam has uniform irradiance profile and flat wavefront (Figure 7.24) which is a popular example considered in diffraction theory [12,33,36], and is also a typical beam created by field mapping refractive beam shapers converting Gaussian to flat-top laser beam.

Due to diffraction of the beam propagating in space, there are variations in the irradiance distribution, where some common profiles are shown in Figure 7.24: At certain distance from initial plane with uniform irradiance distribution, there appears a bright rim that is then transformed to a more complicated ring pattern, and finally at infinity (so-called far field), the irradiance profile is featured with relatively bright central spot and weak diffraction rings—this is the well-known *Airy disk* distribution described mathematically by the formula:

$$I(\rho) = I_0 \left[\frac{J_1(2\pi\rho)}{2\pi\rho} \right]^2 \tag{7.46}$$

where:

$I(\rho)$ is the irradiance
J_1 is the Bessel function of first kind
ρ is the polar radius
I_0 is a constant

The *Airy disk* function is the result of Fourier–Bessel transform for a circular beam of uniform initial irradiance [12,36]. Evidently, even a *pure* theoretical flat-top beam is transformed to a beam with essentially nonuniform irradiance profile. There exists, however, certain propagation length where the profile is relatively stable and where this length is inversely proportional to the wavelength and in square proportion to the beam size. For example, with visible light, single-mode initial beam and flat-top beam of diameter 6 mm after a πShaper 6-6, the distance over which the irradiance does not deviate from uniformity by ±10% is about 200–300 mm, for the 12 mm beam it is approximately equal to 1 m.

There are many laser applications where conserving a uniform irradiance profile over certain distance is required, for example, holography, interferometry; the extended DOF is also very important in various industrial techniques to provide less tough tolerances on positioning of a workpiece. As a solution to the task of providing a required spot size with conserving the flat-top profile over extended DOF, it is useful to apply imaging techniques that are considered in next section.

One should note that the DOF can be further extended when a super-Gaussian beam is provided at the output of the refractive beam shaper. Another way to extend the DOF is to apply magnifying and imaging optics after the refractive beam shaper, which is considered below.

7.3.3.1.3 Beam Shaping for TEM_{00} and Multimode Fiber or Fiber-Coupled Lasers

This section addresses implementations such as telescopes or collimators for applications using high-power fiber lasers and fiber-coupled lasers, cladding, welding, hardening, as well as holography where TEM_{00} fiber-coupled laser sources are very fruitful due to inherent spatial filtering, convenience, and finally reliability.

One of the characteristic trends in modern laser technologies is in expansion of using fiber delivery of laser radiation—both fiber lasers and fiber-coupled diode or solid-state lasers. In addition to convenience of use and reliability while building optical systems, the fiber sources are important for holography and interferometry feature—a TEM_{00} fiber functions as a spatial filter; as a result, the laser radiation emerging from that fiber is characterized by almost perfect Gaussian irradiance profile.

A remarkable feature of field mapping beam shapers is their capability to meet these challenging demands of modern industrial applications and realize collimating optical systems combining the functions of beam shaping and collimation: Divergent Gaussian beam is converted to a collimated flattop, where the collimating beam shaper is shown in Figure 7.25.

7.3.3.1.4 Variable Profiles by Variable Input Beam Size

The feature of field mapping beam shapers where the output beam profile depends on the input beam size can be used as a powerful and convenient tool to vary the resulting intensity distribution by simply changing the laser input beam diameter using an ordinary beam expander before the refractive beam shaper.

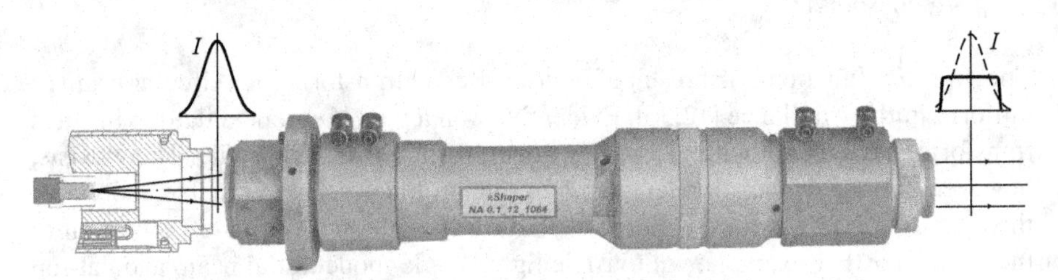

FIGURE 7.25 Collimating πShaper with integrated alignment for powerful fiber lasers. (Reproduced from Laskin, A., *Laser Technik Journal*, 37–40, 2013; Laskin, A. and Laskin, V. Refractive beam shapers for material processing with high power single mode and multimode lasers, in *Laser Resonators and Beam Control XV*, Kudryashov, A. V., Paxton, A. H., and Ilchenko, V. S., eds., SPIE, Bellingham, WA, 2013. With permission.)

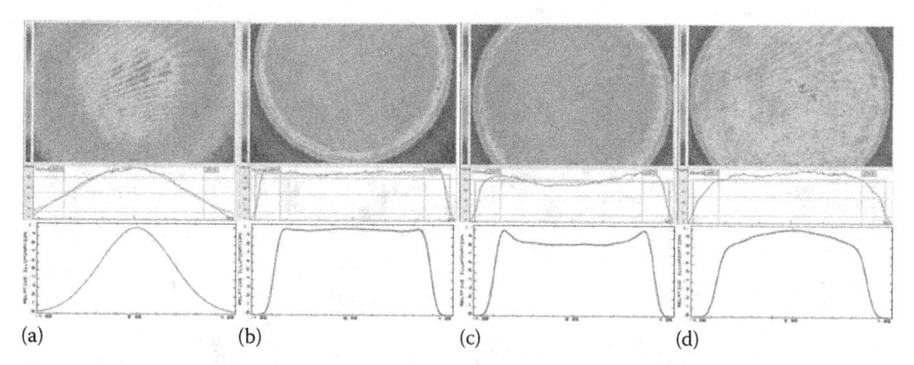

(a) (b) (c) (d)

FIGURE 7.26 **(See color insert.)** Experimental and theoretical intensity profiles as described within the text. (Figure was provided by IPG Photonics; Reproduced from Laskin, A. and Laskin, V., Variable beam shaping with using the same field mapping refractive beam shaper, in *Laser Resonators and Beam Control XIV*, Kudryashov, A. V., Paxton, A. H., and Ilchenko, V. S., eds., SPIE, Bellingham, WA, 2011. With permission.)

This approach is demonstrated in Figure 7.26 where the results of theoretical calculations as well as measured in real experiments beam profiles for TEM_{00} laser are shown. The data relate specifically to the πShaper 6-6, whose design presumes that a perfect Gaussian beam with $1/e^2$ diameter 6 mm to be converted to a beam with uniform intensity (flattop) with FWHM diameter 6.2 mm. When the input beam has a proper size (Figure 7.26a), the resulting beam profile is flattop (Figure 7.26b). Increasing of input beam diameter leads to decreasing of intensity in the center (Figure 7.26c), sometimes, this distribution is called as inverse-Gaussian distribution. Input beam size reduction leads to a convex profile that approximately can be described by super-Gaussian functions (Figure 7.26d).

The considered intensity profiles correspond to about 10% beam size change. Also, for larger changes of input beam size, there are more pronounced variations of the output intensity profile. Another interesting feature of field mapping beam shapers is the stability of the output beam size, where a variation of input beam diameter results in variation of the intensity profile while the output beam diameter stays almost invariable. This is very important in practice and brings element of stability while searching for optimum conditions for a particular laser application.

The next set of profiles in Figure 7.27 demonstrates beam shaping of multimode laser. The radiation from a high-power, solid-state fiber-coupled laser ($\lambda = 1064$ nm, $P = 2$ kW, fiber core diameter 600 μm) was input into the collimating πShaper combining the functions of beam shaping and collimation. The beam emerging from the fiber is divergent and has a profile shown in Figure 7.27a. The output of the πShaper* is a collimated beam with a flat-top intensity distribution (Figure 7.27b). Since the input beam for the πShaper is divergent, there is no possibility to change its size as was done in previous experiments with TEM_{00} laser. However, the collimating πShaper is able to vary the beam size internally, through changing the distance

* πShaper is a registered trademark of AdlOptica GmbH, Berlin, Germany, http://www.adloptica.com.

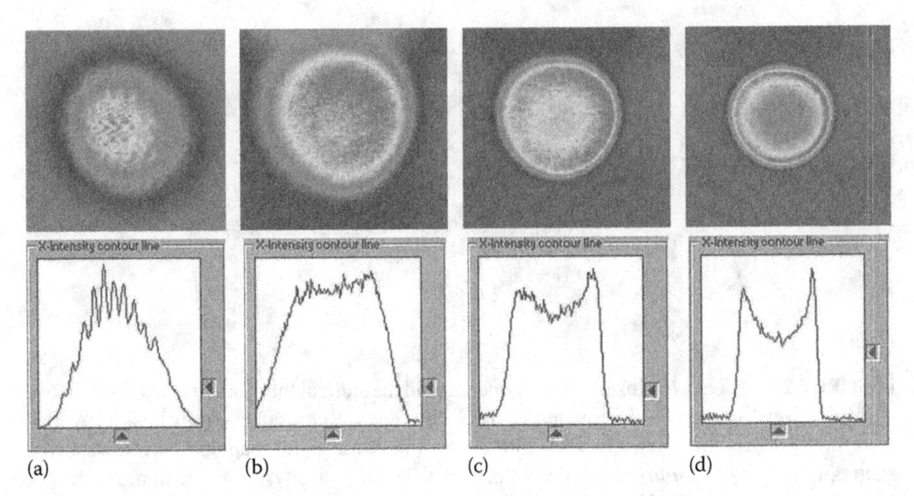

FIGURE 7.27 **(See color insert.)** Beam shaping of powerful multimode laser. (Reproduced from Laskin, A. and Laskin, V., Refractive beam shapers for material processing with high power single mode and multimode lasers, in *Laser Resonators and Beam Control XV*, Kudryashov, A. V., Paxton, A. H., and Ilchenko, V. S., eds., SPIE, Bellingham, WA, 2013; Laskin, A. and Laskin, V., *Proceedings of the ICALEO*, 707, 2012. With permission.)

between its optical components, as was done to realize the inverse-Gaussian profiles presented in Figure 7.27c and d. One can see that the variation of internal parameters of the beam shaper allows varying the resulting profile.

As shown in Figures 7.26 and 7.27, a simple external or internal variation of laser beam size allows generating various profiles with the same beam shaper unit. To vary the beam diameter, the ordinary beam expanders can be used when using the zoom beam expanders, or by choosing the distance between the refractive beam shaper components one can steadily vary the resulting beam profile and choose an optimum one for a particular laser technology.

There are many applications where the variation of intensity profile helps to optimize a laser technology. For example, the welding of plastics, laser heating, or hardening techniques benefit from uniform *temperature profile* on a workpiece, and the inverse-Gaussian intensity distribution is optimum for this purpose. The super-Gaussian distributions are useful in techniques of spectral laser combining, pumping of DPSS lasers such as Ti:Sapphire, and MOPA laser designs.

7.3.3.2 Imaging Methods with a Refractive Laser Beam Shaper

Imaging techniques are powerful tools used to meet the application demands within industrial and scientific laser technologies in combination with a refractive laser beam shaper to obtain various laser spot sizes and shapes as required by each application. For example, implementation of an imaging optical system depends on the specific laser technology being used, such as scanning mirror systems and F-theta lenses in industrial micromachining processes, while telecentric optical systems that conserve phase of the wavefront are desired for interferometry and holography applications. Since a refractive beam shaper has been designed to introduce and remove in a controlled manner accurate amounts of wave aberration, the reshaped beam has low divergence.

7.3.3.2.1 Image Formation

Using a lens to image a uniform intensity laser beam helps in reducing the effects of diffraction and creating a spot of optimum size. The basic optical layout is shown in Figure 7.28, which only shows a singlet lens, but for high-quality imaging, one should use an aplanat or micro-objective lens. Using geometrical optics [33,34] is sufficient to calculate parameters of the imaging system. For example, geometrical optics assumes that each image point is created by a beam of rays emitted by the corresponding object point, where the object and image points are located in the conjugate planes with equal optical path length for all rays of each beam. Also, the real image is created after the lens focus, where the transverse magnification β is defined as a ratio of the distances from the principal planes to the lens to corresponding image and object points:

$$\beta = \frac{-h'}{h} = \frac{-s'}{s} \tag{7.47}$$

Further, the product of the object size h and the aperture angle u or more correctly $\sin(u)$ is constant throughout the optical system:

$$hu = h'u' = \text{constant} \tag{7.48}$$

Further, we assume that the optical system is free of aberrations.

7.3.3.2.2 Imaging with Laser Beams

It is well known that laser beams have a low divergence, such as the full divergence angle of 2Θ for a single-mode laser beam with λ being equal to 532 nm and beam waist 2ω being equal to 6 mm, which is approximately equal to 0.12 mrad or 24 arcseconds. This characteristic of laser beams affects imaging as illustrated in Figure 7.29, which illustrates the behavior of the intensity profile of a low divergence laser beam in an imaging system. According to diffraction theory [12,36], the intensity distribution in a specific plane is formed by interference of light diffracted from the previous plane. As a result of interference, the intensity in the image plane will be similar to the intensity distribution in the object plane, as a result that both planes are optically conjugate of each other. The image size is defined by the transverse

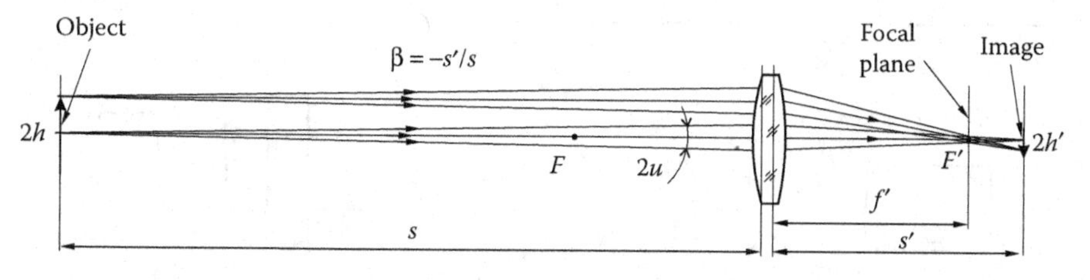

FIGURE 7.28 Image formation with a lens. (Reproduced from Laskin, A. and Laskin, V., *Proceedings of SPIE*, 8490, 2012. With permission.)

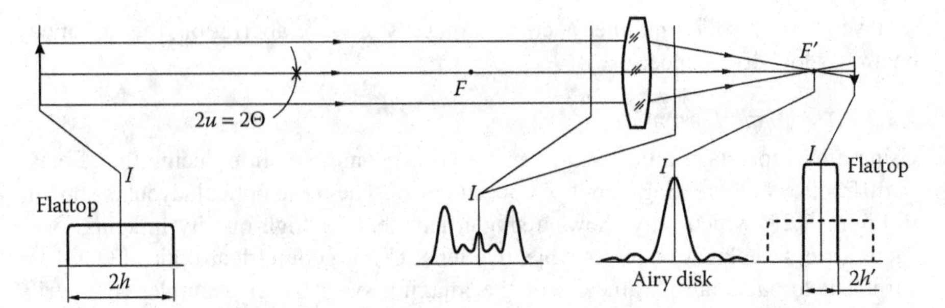

FIGURE 7.29 Irradiance profile transformation of flat-top laser beam in imaging layout. (Reproduced from Laskin, A. and Laskin, V., *Proceedings of SPIE*, 8490, 2012. With permission.)

magnification β. Further, it is well known that a positive lens performs a Fourier transform of the input intensity at the back focal plane of the lens. Thus, the intensity distribution at the back focal plane is the Airy disk distribution. Summarizing, the uniform irradiance distribution of the object is transformed, due to diffraction, to nonuniform, nearby imaging lens, then to Airy disk in its focal plane, and is ultimately restored to be uniform in the image plane.

In addition, there is an extended EDOF associated with imaging of low divergence laser beams, which is illustrated in Figure 7.30. The DOF length can be approximately evaluated by taking into account that the longitudinal magnification of an imaging system is equal to the square of the transverse magnification [33]. Figure 7.30 suggests that the image size within the image space distance of $\Delta s'$ varies and depends on the transverse magnification of each image plane. This feature can be used in some cases to fine-tune the size of spot by shifting the working plane along the optical axis.

7.3.3.2.3 Two-Lens Imaging System

The imaging system can also have two lenses, and the results of Section 7.3.3.2.2 will also apply. A two-lens layout is presented in Figure 7.31, where the object is located in front focal plane of lens 1, and therefore, lens 1 works as a collimator, which produces a collimated beam from each object point. Lens 2 focuses the beams collimated from the object via lens 1 and creates an image on its focal plane. The transverse magnification of this layout is given by the ratio of the focal lengths of these two lenses:

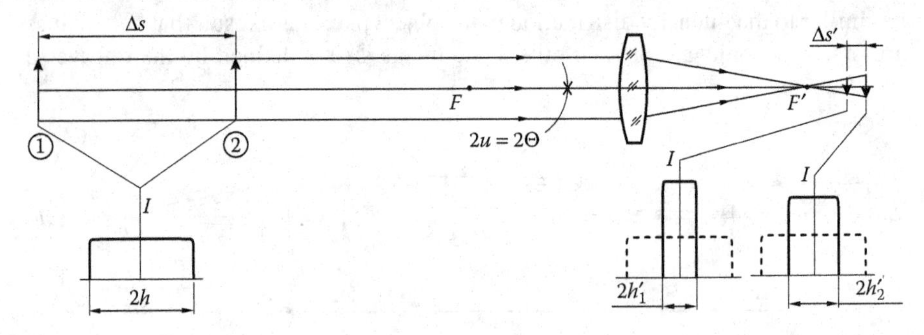

FIGURE 7.30 Evaluation of the DOF within an imaging layout. (Reproduced from Laskin, A. and Laskin, V., *Proceedings of SPIE*, 8490, 2012. With permission.)

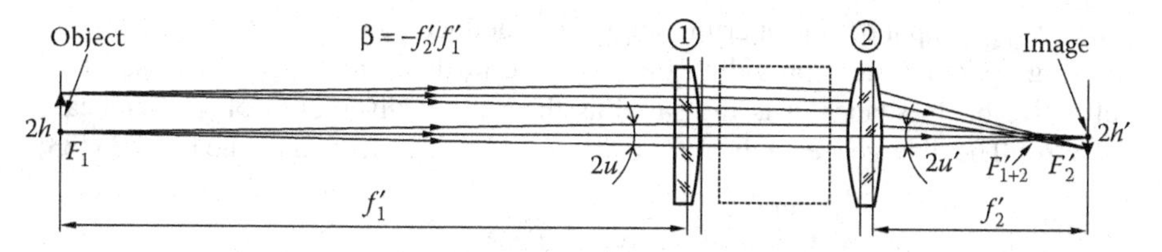

FIGURE 7.31 Two-lens imaging layout. (Reproduced from Laskin, A. and Laskin, V., *Proceedings of SPIE*, 8490, 2012. With permission.)

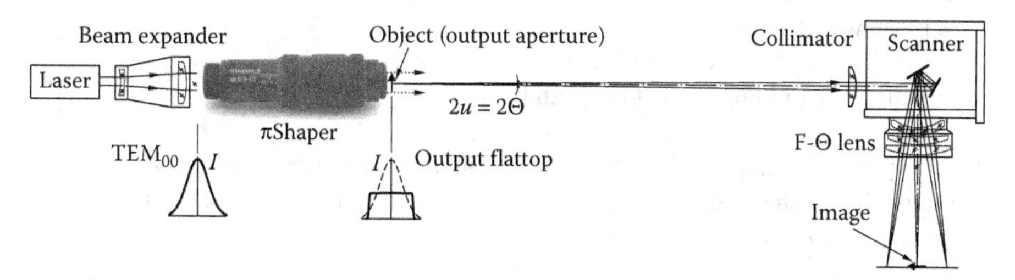

FIGURE 7.32 Combined system with πShaper, collimator, and scanning head with F-theta lens. (Reproduced from Laskin, A. and Laskin, V., *Proceedings of SPIE*, 8490, 2012. With permission.)

$$\beta = \frac{-h'}{h} = \frac{-f_2'}{f_1'} \tag{7.49}$$

Since each object point is collimated by lens 1, the separation between lens 1 and lens 2 is not a critical distance.

A two-lens imaging optical system can be combined with refractive beam shapers in facilities using micromachining technologies to create a demagnified laser spot of required intensity profiles, shapes, and size. Figure 7.32 illustrates an optical layout of a combined optical system including a laser, a beam expander, a beam shaper, and an imaging system consisting of collimator, galvo-mirror scanning head, and F-theta lens. The output of the πShaper is imaged to the working plane which is coincident with the back focal plane of the F-theta lens. The imaging beams from each object point are parallel in the space between the collimator and the F-theta lens. Therefore, the distance between these two elements is not critical, and mirrors of scanning systems can be located in this space between these lenses. The focal length of the F-theta lens is determined by requirements of the laser technology being used.

Now, we wish to evaluate the achievable transverse magnifications for the imaging systems, which are based on widely used industrial optical components. Assume that we are working with a laser with $\lambda = 532$ nm and the imaging optical system is composed of a collimator and an F-theta lens as shown in Figure 7.32. Assume that the focal length of the F-theta lens is $f_2' = 100$ mm and the entrance pupil diameter is $D = 10$ mm, the optical designs of modern F-theta lenses provide diffraction limited image quality over whole working angular field. Evidently, the maximum aperture angle u' for that F-theta lens can be found as the ratio of the pupil diameter D and the focal length f_2':

$$u' = \frac{D}{2f_2'} \tag{7.50}$$

However, the input double aperture angle $2u$ is defined by specifications of the beam shaping optical system providing beam profile in the *object* space. In case of the refractive beam shaper, it is the same as the natural divergence of a laser beam: $2u = 2\Theta$. The divergence angle of a TEM_{00} laser beam is defined by the formula [33]:

$$\Theta = \frac{\lambda M^2}{\pi\omega} \tag{7.51}$$

where:
- λ is the wavelength
- M^2 is the laser beam quality factor
- ω is the waist radius of the Gaussian beam

Transforming Equations 7.48 through 7.51 and taking $\omega = h$, which is valid for refractive beam shapers, one can get a common expression for an achievable transverse magnification:

$$\beta = \frac{-2\lambda M^2 f_2'}{\pi\omega D} \tag{7.52}$$

By substituting the values of the considered example of πShaper 6-6 using $\lambda = 532$ nm,

$$2\omega = 3\,\text{mm}; \quad M^2 = 1 \tag{7.53}$$

$$f_2' = 100\,\text{mm}; \quad D = 10\,\text{mm} \tag{7.54}$$

Then, the calculations for a refractive beam shaper give the magnification down to 1/1000×. In other words, theoretically with ordinary modern off-the-shelf industrial optical components and lasers, it is possible to drastically reduce the output beam of a refractive beam shaper and provide resulting spot sizes of several tens of microns. In practice, compact imaging layouts with transverse magnification down to 1/200× are used.

Since the imaging of the refractive beam shaper output beam is a best way to create demagnified laser spots with uniform intensity profile and high edge steepness, it is typically recommended to be applied in techniques where the required flat-top laser spots are of size below 1 mm diameter, for example, in microwelding, patterning on polymer layers, welding of polymers, laser marking, in some solar cell microprocessing applications such as drilling PCB blind vias and thin-film scribing.

7.3.3.2.4 *Telecentric Imaging of Refractive Beam Shaper Output for Holography and Interferometry*

The holographic and interferometry applications as well as other techniques based on an SLM obtain essential benefits from homogenizing a laser beam. Therefore, beam shaping optics has become more and more popular in these fields. A primary requirement of these techniques is conserving the phase front of a laser beam, which requires simultaneously a flat wavefront and a flat-top (uniform) intensity profile, as shown in Section 7.3.3.1, which describes the refractive field mapping beam shaper.

Most often, these devices are implemented as telescopic systems with two optical components, where it is presumed that the wavefronts at input and output are flat, the transformation of intensity profile from Gaussian to uniform one is realized in a controlled manner by accurate introduction of wave aberration by the first component and further its compensation by the second one [2,9,14]. Thus, the resulting collimated output beam has a uniform intensity and flat wavefront, and it is characterized by low divergence, which is the same as the input beam. In other words, the refractive beam shaper transforms the beam profile without deterioration of the beam consistency and without increasing its divergence.

The holographic and interferometric applications often require expansion of a beam after a refractive beam shaper, for example, to illuminate an SLM or a mask with a collimated laser beam of uniform intensity whose sizes are larger than output beam diameter of a standard πShaper. This is an actual option in techniques such as mastering of security holograms, Denisyuk holography, field illumination in confocal microscopes, interferometric techniques of recording the volume Bragg gratings and periodic structuring, holographic data storage, and many others.

Obviously, this expansion can be realized by using a telescopic beam expander. Popular solutions are beam expanders of Galilean type built from negative and positive lenses that provide system compactness and avoidance of internal focusing; these beam expanders are widely used in industrial applications. However, the Galilean expanders do not overcome the problem of intensity distribution transformation discussed in Section 7.3.3.1.2 and do not create a real image to restore the flat-top intensity distribution like it is realized by imaging optical systems, as discussed in Section 7.3.3.2 and Reference [11].

Therefore, the more advisable solutions for holography and interferometry are beam expanders of Keplerian type, Figure 7.33, built from two positive lenses, whose well-known feature is the capability to create real image. Since the optical power of this telecentric system is zero, we conclude the following:

- The flat phase front in the object space is mapped to the flat phase front in the image space.
- The transverse magnification of the optical system is constant and does not depend on position of the object.
- If the object is located in the front focal plane of the first component, its image is in the back focal plane of the second component.

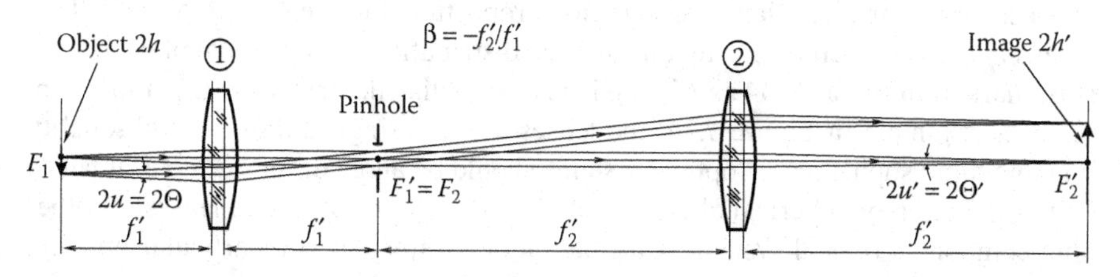

FIGURE 7.33 Telecentric imaging with Keplerian beam expander. (Reproduced from Laskin, A. and Laskin, V., Beam shaping to improve holography techniques based on spatial light modulators, in *Emerging Liquid Crystal Technologies VIII*, Chien, L., Broer, D., Chigrinov, V., and Yoon, T., eds., SPIE, Bellingham, WA, 2013. With permission.)

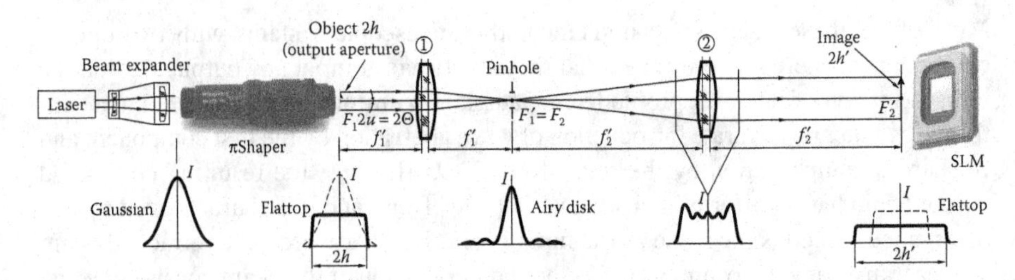

FIGURE 7.34 Layout to illuminate an SLM. (Reproduced from Laskin, A. and Laskin, V., Beam shaping to improve holography techniques based on spatial light modulators, in *Emerging Liquid Crystal Technologies VIII*, Chien, L., Broer, D., Chigrinov, V., and Yoon, T., eds., SPIE, Bellingham, WA, 2013. With permission.)

Now, consider the transformation of an irradiance profile for the example of the optical system illuminating an SLM, as shown in Figure 7.34. The Gaussian beam, TEM_{00} mode from a laser is transformed by the refractive beam shaper to a collimated flat-top beam. We consider the output of the refractive beam shaper as an object plane for the telecentric imaging system. Since the refractive beam shaper conserves low divergence of laser beam, the irradiance profile after it has been transformed as a result of diffraction is similar to the pattern shown in Figure 7.24. As a result, the irradiance distribution is not uniform in the region of the lenses 1 and 2, as shown in Figure 7.34, but the irradiance distribution in this region typically has some diffraction rings, where a particular profile depends on wavelength, beam size, and distance from the object to lenses.

According to the diffraction theory, the irradiance distribution in a certain plane is the result of interference of light diffracted from the previous plane of observation. One of the well-known conclusions from diffraction theory [12,33,36] is the similarity of the irradiance distribution in optically conjugated *Object* and *Image* planes: *If the irradiance distribution is uniform in the Object plane, it is uniform in the Image plane as well.* The profile at the refractive beam shaper output aperture will be repeated in the image plane of that aperture; the resulting spot size is defined by transverse magnification β. Evidently, if an SLM is located in the image plane, the incident radiation will be characterized by flat phase front and flat-top intensity profile. Another well-known conclusion of the diffraction theory is the ability of a positive lens to perform two-dimensional Fourier–Bessel transform and create in its back focal plane irradiance distribution proportional to the one in far field. This means in the considered case that irradiance distribution in back focal plane of first lens, marked in Figure 7.34 as $F_1' = F_2$, is just Airy disk described by Equation 7.46.

In the example in Figure 7.34, the lenses are just singlets, but for high-quality imaging more sophisticated optical systems should be applied, for example, aplanats (with correction of spherical aberration and coma), micro-objective lenses, or other multicomponent optical systems. Calculation of parameters of a particular imaging setup can be done using well-known formulas of geometrical optics, described, for example, in the literature [33,34].

Summarizing results of this example show that the uniform irradiance after the refractive beam shaper, the object plane, is transformed to a nonuniform irradiance distribution in region around the lenses, to essentially nonuniform Airy disk distribution in the back focal plane of first lens, and finally is restored to uniform irradiance profile in the image plane as a result of interference of the diffracted beam. An important conclusion for practice is that it does not matter how the irradiance profile is transformed along the beam path, since the irradiance distribution in the image plane repeats the object plane distribution by taking into account transverse magnification.

There is one useful effect accompanying the magnification of a homogenized beam. Namely, the extended DOF leads to the creation of flat-top intensity profile not only in the area of the image but also practically in whole image space after the second lens. As discussed in Section 7.3.3.1.2, the longitudinal magnification of an imaging system is proportional to the square of the transverse magnification. The transverse magnification of the considered telecentric system is constant and does not depend on the position of an object. Hence, the DOF in image space is proportional to square of transverse magnification as well and can reach large values. Practically, a resulting intensity profile (most often flattop) is restored right after the second component of the telecentric system. An important conclusion for practice of using the beam shaping optics is that when a Keplerian beam expander is applied, the beam of uniform intensity and flat wavefront is created almost right after that expander. Hence, in a real holographic or interferometric installation, a work piece or other optical components can be installed close to the expander, which makes an installation more compact and easier to use.

Since the image is a result of interference of light beams being emitted by the object and diffracted according to physics of light propagation, it is necessary to take care of transmitting full light energy through a system and avoid any beam clipping, to be sure, except the below considered case of spatial filtering with using an enlarged pinhole. The telecentric optical system also has some capabilities for spatial filtering of a beam that is typically required in holographic and interferometric techniques. As a general rule, it is recommended to realize spatial filtering before a refractive beam shaper and do not do this after it to conserve the conditions for interference of diffracted beams. But in some cases in holography, it is strongly advisable to carry out spatial filtering after a refractive beam shaper to eliminate high-frequency modulation of beam intensity happening because of dust or other reasons, and here it is possible to use one trick.

Since the irradiance distribution in the plane of common focuses, $F_1' = F_2$, is just Airy disk, this plane can be used to put a pinhole for the spatial filtering. Using an ordinary pinhole transmitting only the central diffraction spot makes no sense, since this would destroy the beam structure and give an approximate Gaussian intensity profile that is useless for the considered applications. It is possible, however, to apply a pinhole of larger diameter that transmits not only the central spot but also several diffraction rings carrying majority of beam energy. For example, a pinhole, whose diameter is 15 times larger than the that of a typical pinhole for classical spatial filtering, transmits almost 99% of energy. Evidently, when putting such a pinhole and further beam collimation with the second lens, the flat-top intensity profile would be approximately restored; at the same time, that pinhole would filter the high spatial frequency modulation components from the dust or other imperfections. Definitely, the diameter of such a pinhole is a trade-off between the high-frequency modulation to be removed and the diffraction effects appearing due to beam clipping.

7.4 CONCLUSION

This chapter has presented an engineering method for designing optical systems that realize the physical effects of laser beam shaping to redistribute the irradiance profile from a Gaussian distribution into a uniform irradiance profile while retaining the input wavefront nature. The core technique of this approach is the third-order aberration theory that allows for calculating a good approximate solution in the form of a set of parameters of an optical system that can be improved by subsequent optimization algorithms of modern optical design software. Further, an achromatic laser beam shaper has been described, which has grown to meet many diverse applications in scientific and technology applications in research and industry.

Perhaps, the most important conclusion of this chapter has been the effective demonstration of using commercially available optical design codes that lead to many new designs and applications of refractive laser beam shapers in both the Galilean and Keplerian configurations, where the output beam is a plane wave with a uniform irradiance distribution. It is significant to note that the refractive laser beam shapers discussed in this chapter largely preserve the coherence of the input laser beam.

Application of the refractive beam shaper in holography and interferometry makes it possible to provide two basic conditions of illumination with laser beam: flat-top irradiance profile and flat phase front, which are mandatory for computer-generated holography, dot-matrix hologram mastering, multicolor Denisyuk holography, holographic data storage, as well as for interferometric techniques such as volume Bragg gratings recording; these applications obtain essential benefits from homogenized laser beams: high contrast and equal brightness of reproduced images, higher process reliability and efficiency of laser energy usage, and easier mathematical modeling. Availability for various wavelengths, achromatic design, implementations as telescopes and collimators, low divergence, and extended DOF make the refractive beam shaper a unique tool in building holography systems. Collimator versions of the refractive beam shaper perfectly suit the TEM_{00} fiber lasers and fiber-coupled lasers characterized by high quality and *cleanness* of radiation. Telecentric imaging systems expand capabilities of the refractive beam shaper and allow creating image fields of practically unlimited size. Applying spatial filtering with enlarged pinhole allows, simultaneously, providing irradiance uniformity of the image field and suppressing of contrast or eliminating of parasitic patterns from small dust particles.

REFERENCES

1. Kreuzer, J., "Laser light redistribution in illuminating optical signal processing systems." In *Optical and Electro-Optical Information Processing*, Tippett, J. T., Berkowitz, D. A., Clapp, L. C., Koester, C. J., Alexander Vanderburgh, J. (eds.), Massachusetts Institute of Technology Press, Cambridge, MA, p. 365, 1965.
2. Kreuzer, J. L. "Coherent light optical system yielding an output beam of desired intensity distribution at a desired equiphase surface." US Patent No. 3,476,463, November 4, 1969.

3. Shafer, D. "Gaussian to flat-top intensity distributing lens." *Optics & Laser Technology*, 14, 159–160, 1982.
4. Evans, N. and Shealy, D. "Design and optimization of an irradiance profile-shaping system with a genetic algorithm method." *Applied Optics*, 37(22), 5216–5221, 1998.
5. Hoffnagle, J. A. and Jefferson, C. M. "Design and performance of a refractive optical system that converts a Gaussian to a flat-top beam." *Applied Optics*, 39(30), 5488–5499, 2000.
6. Hoffnagle, J. and Jefferson, C. "Refractive optical system that converts a laser beam to a collimated flat-top beam." US Patent No. 6,295,168, September 25, 2001.
7. GBS Series Beam Shaper, Newport Optics, http://www.newport.com/.
8. Hoffnagle, J. and Jefferson, C. "Apparatus for achromatizing optical beams." US Patent No. 6,879,448 B2, April 12, 2005.
9. Laskin, A. Achromatic optical system for beam shaping. US Patent No. 8,023,206 B2, September 20, 2011.
10. Laskin, A. "Achromatic refractive beam shaping optics for broad spectrum laser applications." In Laser Beam Shaping X, *Proceedings of SPIE*, 7430, San Diego, CA, August 21, 2009.
11. Laskin, A. and Laskin, V. "Imaging techniques with refractive beam shaping optics." In Laser Beam Shaping XIII, *Proceedings of SPIE*, 8490, San Diego, CA, August 12, 2012.
12. Born, M. and Wolf, E. *Principles of Optics*. 7th edn., Cambridge University Press, Cambridge, 1999.
13. Laskin, A. and Laskin, V. "Beam shaping in high-power laser systems with using refractive beam shapers." In Laser Sources and Applications, *Proceedings of SPIE*, 8504, San Diego, CA, May 15, 2012.
14. Laskin, A. and Laskin, V. "Applying of refractive beam shapers of circular symmetry to generate non-circular shapes of homogenized laser beams." In Laser Resonators and Beam Control XIII, *Proceedings of SPIE*, 7913, San Francisco, CA, January 22, 2011.
15. Laskin, A. and Laskin, V. "Applying field mapping refractive beam shapers to improve irradiation of photocathode of FEL." In X-Ray Free-Electron Lasers: Beam Diagnostics, Beamline Instrumentation, and Applications, *Proceedings of SPIE*, 8504, San Diego, CA, October 16, 2012.
16. Laskin, A. and Laskin, V. "Refractive beam shaping optics to improve operation of spatial light modulators." In Acquisition, Tracking, Pointing, and Laser Systems Technologies XXVI, *Proceedings of SPIE*, 8395, Baltimore, MD, June 8, 2012.
17. Shealy, D. L. "History of beam shaping." *Optical Science and Engineering*, 102, 307–347, 2006.
18. Dickey, F. M. and Holswade, S. C. (eds.), *Laser Beam Shaping*, CRC Press, Boca Raton, FL, 2000.
19. Dickey, F. M., Holswade, S. C., and Shealy, D. L. (eds.), *Laser Beam Shaping Applications*, CRC Press, Boca Raton, FL, 2006.
20. McDermit, J. H. and Horton, T. E. "Reflective optics for obtaining prescribed irradiative distribution from collimated sources." *Applied Optics*, 13, 1444–1450, 1974.
21. Oliker, V., Prussner, L., Shealy, D., and Mirov, S. "Optical design of a two-mirror asymmetrical reshaping system and its application in superbroadband color center lasers." In Current Developments in Optical Design and Optical Engineering IV, *Proceedings of SPIE*, 2263, San Diego, CA, September 30, 1994.
22. Dickey, F. M. and Holswade, S. C. *Laser Beam Shaping: Theory and Techniques*, Marcel Dekker, New York, 2000.
23. Jiang, W., Shealy, D. L., and Martin, J. C. "Design and testing of a refractive reshaping system." In *Current Developments in Optical Design and Optical Engineering III*, Fischer, R. E. and Smith, W. J. (eds.), SPIE, Bellingham, WA, pp. 64–75, 1993.

24. Wang, C. and Shealy, D. "Design of gradient-index lens systems for laser beam reshaping." *Applied Optics* 32(25), 4763–4769, 1993.
25. Dennis, J. and Schnabel, R. *Numerical Methods for Unconstrained Optimization and Nonlinear Equations*, Prentice-Hall, Englewood Cliffs, NJ, 1983.
26. Laskin, A., Drachenberg, D., Mokhov, S., Venus, G., Glebov, L., and Laskin, V. "Beam combining with using beam shaping." In *Acquisition, Tracking, Pointing, and Laser System Technologies XXVI*, Dubinskii, M. (ed.), SPIE, Bellingham, WA, 2012.
27. Shealy, D. L. and Hoffnagle, J. A. "Aspheric optics for laser beam shaping." In *Encyclopedia of Optical Engineering*, Driggers, R. (ed.), Taylor & Francis, New York, 2006.
28. US Department of Defense. *Military Standardization Handbook on Optical Design,* MIL-HDBK-141, October 5, 1962. US Defense Supply Agency, Washington, DC.
29. Jiang, W., Shealy, D., and Martin, J., "Design and testing of a refractive reshaping system." In Current Developments in Optical Design and Optical Engineering III, *Proceedings of SPIE*, 2000, doi:10.1117/12.163670, San Diego, CA, November 25, 1993.
30. Shealy, D. L. and Hoffnagle, J. A. *Geometrical Methods. Laser Beam Shaping: Theory and Techniques*, Taylor & Francis, New York, 2013.
31. Shealy, D. L. and Hoffnagle, J. A. "Wavefront and caustic surfaces of refractive laser beam shaper." In *Novel Optical Systems Design and Optimization X*, Koshel, R. J. and Gregory, G. G. (eds.), SPIE, Bellingham, WA, pp. 666805–1:5–11, 2007.
32. Hoffnagle, J. A. and Shealy, D. L., Caustic surfaces of a Keplerian two-lens beam shaper. In *Laser Beam Shaping VIII*, Dickey, F. M. and Shealy, D. L. (eds.), SPIE, Bellingham, WA, pp. 666304–1:4–9, 2007.
33. Smith, W. *Modern Optical Engineering*, McGraw-Hill, New York, 2000.
34. Begunov, B., Zakaznov, N., Kiryushin, S., and Kuzichev, V. *Optical Instrumentation: Theory and Design*, MIR Publishers, Moscow, Russia, 1988.
35. Cox, A. *A System of Optical Design*, The Focal Press, London, 1964.
36. Goodman, J. *Introduction to Fourier Optics*, Roberts & Company, Englewood, CO, 2005.
37. Laskin, A., Williams, G., and Demidovich, A., "Applying of refractive beam shapers in creating spots of uniform intensity and various shapes." In *Laser Resonators and Beam Control XII*, Kudryashov, A. V., Paxton, A. H., and Ilchenko, V. S. (eds.), SPIE, Bellingham, WA, 2010.
38. Laskin, A. and Laskin, V. "Variable beam shaping with using the same field mapping refractive beam shaper." In *Laser Resonators and Beam Control XIV*, Kudryashov, A. V., Paxton, A. H., and Ilchenko, V. S. (eds.), SPIE, Bellingham, WA, 2011.
39. Laskin, A. "Solutions for beam shaping." *Laser Technik Journal*, 37–40, WILEY-VCH Verlag GmbH & Co. KGaA Weinheim, Germany, January 14, 2013.
40. Laskin, A. and Laskin, V. "Refractive beam shapers for material processing with high power single mode and multimode lasers." In *Laser Resonators and Beam Control XV*, Kudryashov, A. V., Paxton, A. H., and Ilchenko, V. S. (eds.), SPIE, Bellingham, WA, 2013.
41. Laskin, A. and Laskin, V. "Controllable beam intensity profile for the tasks of laser material processing." *Proceedings of the ICALEO*, p. 707, Anaheim, CA, September 23–27, 2012.
42. Laskin, A. and Laskin, V. "Beam shaping to improve holography techniques based on spatial light modulators." In *Emerging Liquid Crystal Technologies VIII*, Chien, L., Broer, D., Chigrinov, V., and Yoon, T. (eds.), SPIE, Bellingham, WA, 2013.

8 Beam Shaping with Diffractive Diffusers

Jeremiah D. Brown and David R. Brown

CONTENTS

8.1 INTRODUCTION

In this chapter, we discuss an approach to beam shaping that often has a different realm of applications than the more conventional techniques. Specifically, we discuss what is called a band-limited diffuser. Many diffusers, such as ground glass, diffuse light over an angular extent that is often larger and not as well defined as desired. We will see in this chapter that diffractive diffusers offer a technique to diffuse light over a very well-controlled angular spectral band.

Section 8.2 describes the properties of diffusers and differentiates their characteristics from other perhaps more familiar optics. Diffractive diffusers share some properties with both conventional single diffractive order beam shapers and with diffraction gratings. We differentiate between these different classes of diffractive elements and develop the theory used to describe the diffuser.

Sections 8.3 and 8.4 describe the design process and outline the approaches and figures of merit used to specify the optical performance of the structures given realistic fabrication tolerances. Section 8.4 includes a discussion of various fabrication techniques and describes fabrication limitations and considerations. It is meant to give the reader an appreciation for what is possible and which fabrication method is appropriate for a given design. Section 8.5 applies these methods and illustrates the design and tolerancing of a simple ring diffuser.

Section 8.6 summarizes the major negative aspect to this beam shaping technique: speckle. In it, we derive the size of speckle and discuss a few methods for reducing its impact.

The final section of this chapter outlines a few of the possible applications for diffractive diffusers. Section 8.7 also notes when and when not to use diffusers for beam shaping.

8.2 PROPERTIES OF DIFFRACTIVE DIFFUSERS

To help define the properties of beam shaping with diffusers, it is useful to describe the differences between it and the other beam shaping techniques. There are two general categories of beam shapers, the first of which is a near-field or remapping beam shaper.

8.2.1 Remapping Beam Shapers

In general, diffractive optics, such as gratings, can utilize many diffractive orders in combination to generate the desired output pattern. Remapping diffractive optics conversely use a single diffractive order to produce the desired optical effect. A simple example of a diffractive optic that uses only one diffractive order is a lens. Remapping beam shapers are much like a complex aberrated lens that performs a functional mapping of the incident beam's energy distribution to the desired shape at a specified output plane. If the specified output plane is in the near-field, the shaped beam will exist only at this predefined plane, though it is possible to relay the output with an imaging lens or use a shaper designed for an image plane at infinity in conjunction with a Fourier transform lens to transfer the shaped beam to a new plane. Because of their relatively small depths of focus, remapping beam shapers are sometimes referred to as near-field beam shapers.

It is possible to introduce a second optic to correct the phase in the beam as shown in Figure 8.1. The resulting phase of a near-field beam shaper can be canceled to produce a collimated beam, which is then allowed to diverge to give a shaped beam over an extended but finite range. The corrected shaped beam will experience diffraction and will degrade as the beam propagates. The diffraction of the corrected shaped beam will be as if the beam originated from an aperture function that is the same as the shaped beam. To minimize the diffraction of the edges, it is often advantageous to design the desired shape of the beam to have soft or smooth edges.[1] The function

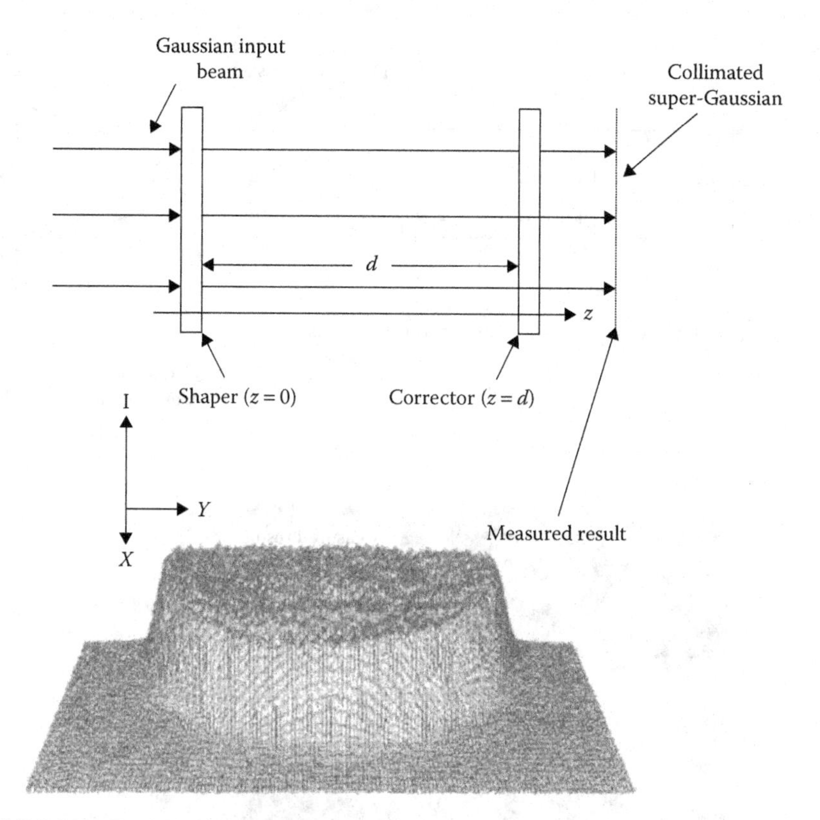

FIGURE 8.1 Typical system layout of a near-field beam shaper. Only the first optic is required to shape the beam at plane $z = d$. To extend the range at which the top hat will exist requires a second optic to correct for aberrations in the phase of the beam. The empirical result shown has a nonuniformity standard deviation of 5.6%.

that is used to describe the soft edge can have many forms. One such soft aperture function is a high-order Gaussian or super-Gaussian of the following form:

$$I \propto e^{-2(r/\omega)^{2N}} \tag{8.1}$$

where:

- I is the intensity
- r is the radius
- ω is the waist radius
- N is an integer

As the value of N increases, the closer the function approximates a true top-hat function.

It is possible to extend the range of a near-field beam shaper by optically taking the Fourier transform[2] of the output as shown in Figure 8.2. The lens transforms the shaped beam into its Fourier transform at the back focal plane of the lens. As the field propagates beyond the back focal plane, the diffraction caused by the propagation

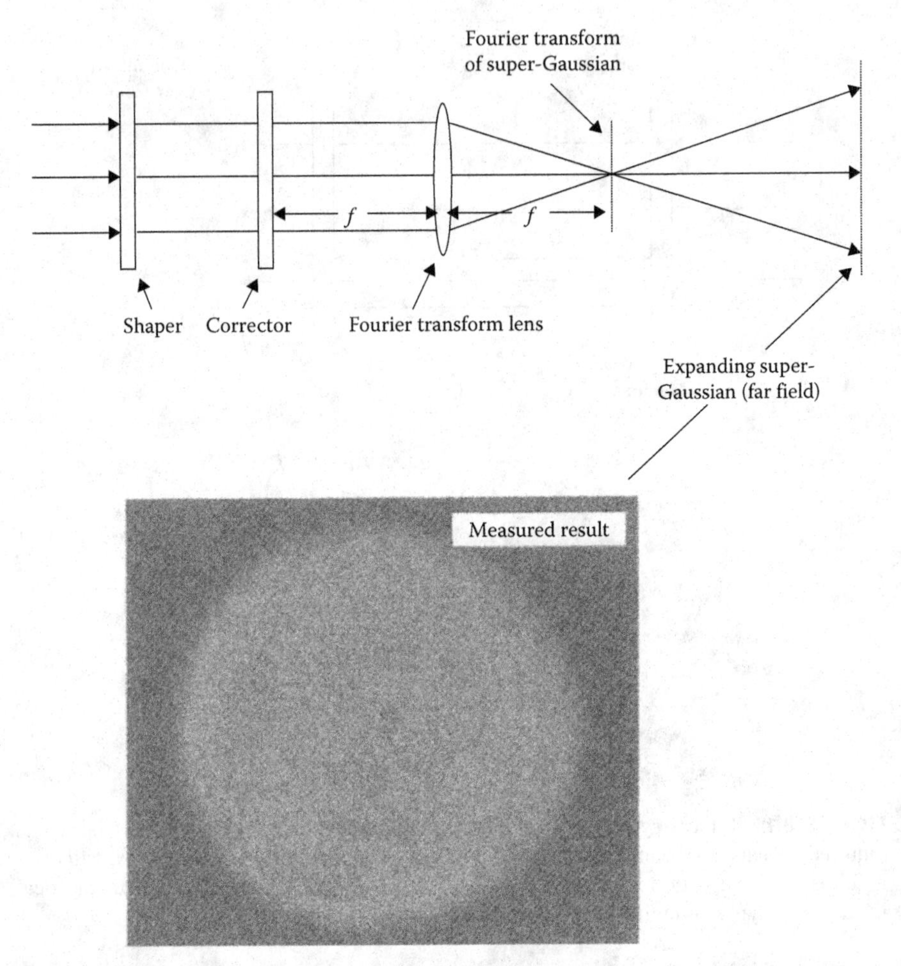

FIGURE 8.2 System to transfer the shaped beam into the far field. A simple lens with the appropriate focal length can be added after this system to recollimate the beam. The measured result shown was taken approximately 300 mm beyond the Fourier transform lens and is approximately 1 cm in diameter. The structure in the beam is from multiple reflections within the system.

transforms the field back into the shaped beam with a spherically diverging phase. This creates a diverging cone of light whose intensity envelope has the desired shape. Experimental results of this setup for a round super-Gaussian are shown in Figure 8.2. The structure that is observed in the measured result is caused by multiple reflections within the system due to optics that do not have antireflection coatings. The beam can then be collimated at any point by selecting the appropriate lens. This also allows one to size the output beam. It should also be noted that it is possible to specify the output plane at a sufficiently large distance from the beam shaper such that the shaped beam will propagate undistorted in the far-field.

Due to the remapping nature of a near-field beam shaper, the output is highly sensitive to the intensity and phase of the input beam. Any deviation in the input beam size, shape, or location relative to the near-field beam shaper will cause degradation

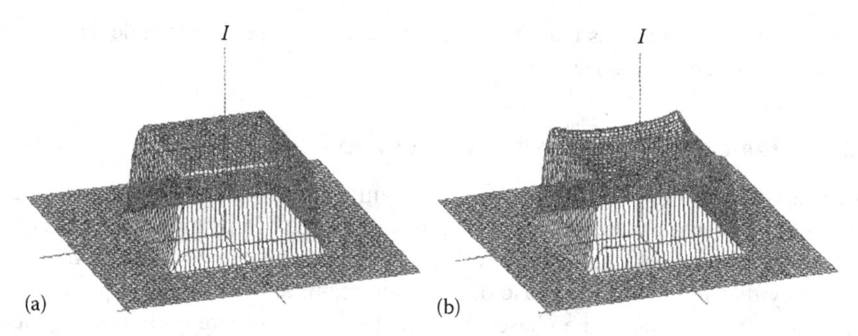

FIGURE 8.3 Simulation results of the output intensity of a Gaussian to square super-Gaussian top-hat beam shaper. (a) The result with the perfect input beam. The output is >99% efficient and the peak-to-valley nonuniformity is <2%. (b) The result with an input beam that is 5% too large. The peak-to-peak nonuniformity is ~18%.

to the resulting output. Figure 8.3 shows the output intensity of a simulation of a Gaussian to square top-hat beam shaper. There are several methods for designing the beam shaping diffractive optic.[2,3] Commercially available ray-tracing computer codes such as ZEMAX and Code V, as well as physical optics design software such as LightTrans VirtualLab, can be used to design the beam shaping optic if one is careful to include diffraction effects. The results in Figure 8.3 were simulated using a scalar wave propagation computer code which accurately models diffraction effects. Figure 8.3a is a plot of the output intensity with a perfect input beam. The peak-to-valley nonuniformity of the intensity of this top hat is less than 2% and the simulated diffraction efficiency is better than 99%. Figure 8.3b shows the output of the same optic with the input beam 5% larger than the designed beam. The peak-to-valley nonuniformity of the intensity is now about 18%. From this, we see the sensitivity of the near-field beam shaper to input beam variations. In general, the desired intensity footprint is maintained over a fairly large range of variations in input beam. The uniformity of the output intensity is however very sensitive to the input beam. However, with care, extremely good results can be obtained. In Figure 8.4, experimental results for a UV beam shaper for a lithography application are shown. The nonuniformity

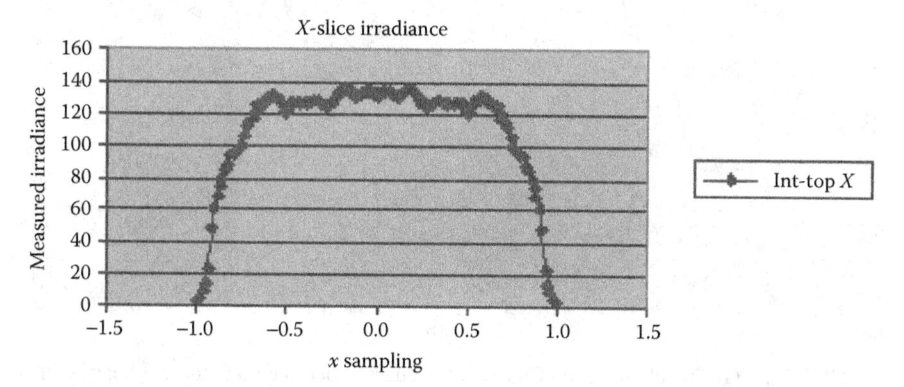

FIGURE 8.4 Experimental results of a beam shaper for a Coherent Innova Sabre-7 UV laser (363.8 nm). Nonuniformity (σ/μ) was measured to be less than 3%.

(σ/μ) was measured at less than 3%, where σ and μ are the standard deviation and mean of the intensity, respectively.

8.2.2 Far-Field Beam Shapers: Gratings and Diffusers

Diffusers fall into the second category of beam shaping optics, known as the far-field beam shapers. Far-field diffractive optics shape the beams through the interference of a large number of diffractive orders. These elements impart a defined spatial frequency distribution to the phase of the laser beam, and as the beam propagates, the spatial frequencies in the phase cause the beam to interfere with itself. Typical devices are made up of many very small phase apertures (typical <10 wavelengths in size), so the beam is in the far field almost immediately beyond the optic. This means that the resulting shape of the beam will continue to propagate with the predefined angular divergence as defined by the spatial frequencies in the phase. Figure 8.5 illustrates the extreme differences in the phase of a near-field (single-order) optic and a far-field (multiorder) optic. Figure 8.5a is the phase applied to the input field to generate the shaped beam shown in Figure 8.3, while Figure 8.5b is the phase of a diffuser that projects a square top-hat pattern.

Far-field optics have the advantage of being relatively insensitive to the shape, size, and alignment of the input beam. An input beam that is a TEM_{00} mode will produce a very similar output to an input beam that is a TEM_{01} mode.[4,5] This is due to the multiplicative property of a Fourier transform. The resulting beam of a far-field optic is simply the convolution of the Fourier transform of the input beam and the spatial frequencies of the optic. As we will see later, the energy envelope of the output pattern is dominated by the phase function of the diffuser and not the shape of the input beam.

Since gratings and diffusers are both far-field diffractive optics, they share many characteristics. It is useful to describe a diffuser in terms of a grating due to its familiarity to most readers. In general, a grating is a periodic amplitude and/or phase

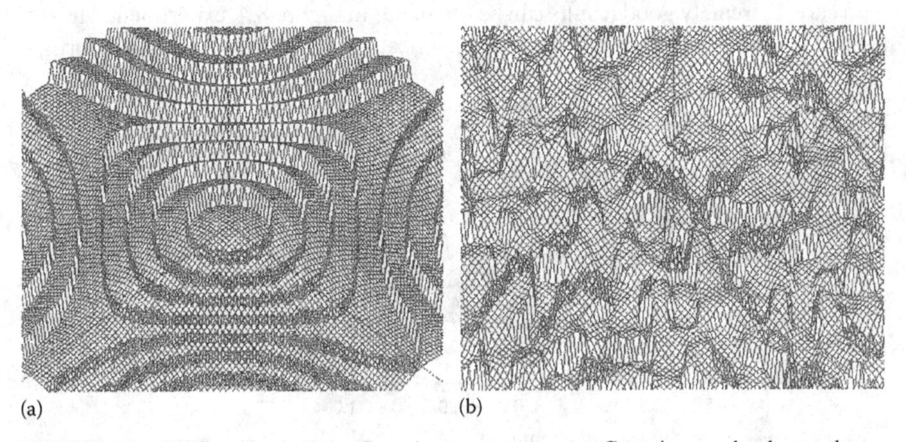

FIGURE 8.5 (a) The phase of the Gaussian to square super-Gaussian top-hat beam shaper shown in Figure 8.3. (b) A portion of the phase of a diffuser that projects a square energy envelope.

structure. For the purposes of this discussion, we will limit a grating to a phase-only structure.

By starting with the differential form of Maxwell's equations, and making simplifying assumptions for a homogeneous medium, one can arrive at the homogeneous wave equation for the electric field[6]:

$$\nabla^2 \mathbf{E} = \mu\varepsilon \frac{\partial^2 \mathbf{E}}{\partial t^2} \tag{8.2}$$

where:

E is the electric field vector

μ and ε are the material property parameters called the permeability and permittivity, respectively

A similar equation exists for the magnetic field. This vector equation can be separated into three scalar equations, one scalar equation for each component of the coordinate system. Using Cartesian coordinates and choosing the scalar equation dependent on the z spatial coordinate, we have

$$\frac{\partial^2 E}{\partial t^2} - \frac{1}{\mu\varepsilon}\frac{\partial^2 E}{\partial z^2} = 0 \tag{8.3}$$

The general form of the solution of Equation 8.3 is[6]

$$E(z,t) = Af_+(\omega_0 t - k_{0z}z) + Bf_-(\omega_0 t - k_{0z}z) \tag{8.4}$$

where:

A and B are constants

ω_0 is the angular frequency with the units of radians/time

k_{0z} is called the propagation constant with the units of radians/length

Equation 8.4 is the solution of the scalar equation (8.3) if

$$\frac{\omega_0}{k_0} \equiv \frac{1}{\sqrt{\mu\varepsilon}} = \upsilon \tag{8.5}$$

where:

υ is the velocity of the light in the medium

The two terms on the right side of Equation 8.4 describe two waves: one traveling in the positive z-direction and the other traveling in the negative z-direction. In general, the wave can travel in any direction. The argument of the first term in Equation 8.4 can be written more generally as $\omega_0 t - \mathbf{k}_0 \cdot \mathbf{r}$ where in Cartesian coordinates

$$\mathbf{k}_0 = k_{0x}\hat{x} + k_{0y}\hat{y} + k_{0z}\hat{z} \tag{8.6}$$

$$\mathbf{r} = x\hat{x} + y\hat{y} + z\hat{z} \tag{8.7}$$

Equation 8.4 can be rewritten as

$$E(x,y,z,t) = Af(\omega_0 t - \mathbf{k}_0 \cdot \mathbf{r}) \tag{8.8}$$

Equation 8.8 is a wave with amplitude A and velocity υ traveling in the $\mathbf{k}_0$ direction. $\mathbf{k}_0$ is called the propagation vector or the wave vector. The direction of the power flow density of a field is equal to the direction of the propagation vector.[4] The magnitude of the propagation vector is given by[4,7]

$$\left|\mathbf{k}_0\right| = k_0 = \frac{\omega_0}{\upsilon} = \frac{2\pi}{\lambda} \tag{8.9}$$

where:
λ is the wavelength of the light in a given material
k_0 is a constant while the light is propagating in the material

A wave described by Equation 8.8 is often referred to as a plane wave. An arbitrary complex electromagnetic field can be analyzed in terms of its Fourier components. The Fourier components of a complex field are simply a series of plane waves traveling in different directions.[5] When analyzing periodic structures such as gratings, it is often advantageous to perform the analysis in the Fourier domain.

Figure 8.6 shows a circle whose radius is k_0. Along the k_x axis is a periodic structure with a grating vector of $\mathbf{K}_g$ whose magnitude is given by

$$k_g = \frac{2\pi}{\Lambda} \tag{8.10}$$

where:
Λ is the period of the grating

$\mathbf{K}_g$ has only an x component and adds to the x component of the propagating wave in discrete multiples. A graphical illustration of this is shown in Figure 8.6. Due to the constraint that the wave vector has a constant magnitude of k_0, we see from the figure that we can graphically determine the direction of the series of plane waves that result from the grating

$$\sin(\theta_t^m) = \frac{(k_{0x} + mk_g)}{k_0} \tag{8.11}$$

which then reduces to the familiar grating equation

$$\sin(\theta_t^m) = \sin(\theta_i) + \frac{m\lambda}{\Lambda} \tag{8.12}$$

where:
θ_i is the incidence angle
θ_t^m is the transmitted angle of a given diffracted order m

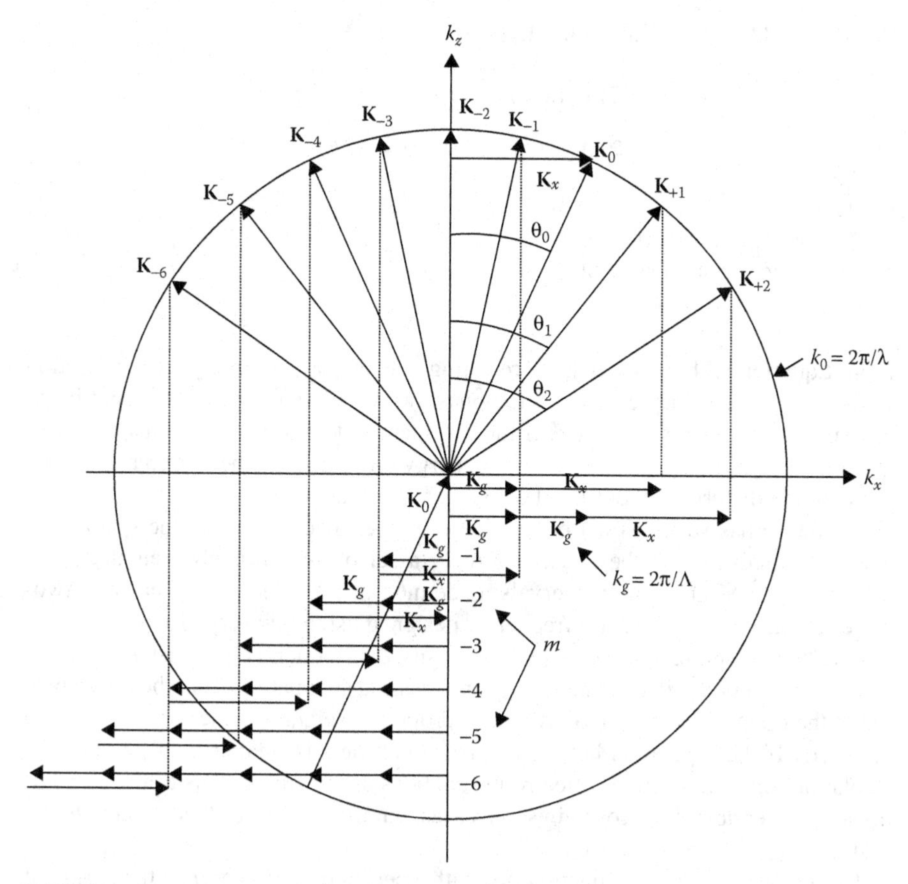

FIGURE 8.6 The wave vector map of light as it transmits through a periodically varying structure such as a grating. The lower portion of the plot shows the x component summations of the undeviated beam ($\mathbf{K}_0$) and the grating vector ($\mathbf{K}_g$). The orders of the grating are the result of an integer number of grating vectors added (or subtracted) to k_0.

Orders (values of m) that require $|\sin(\theta_t^m)| > 1$ are called evanescent orders. The wave vector of an evanescent order has an imaginary z component and thus attenuates exponentially beyond the surface of the grating.[7]

For the "real" orders, the resulting electric field at $z = 0$ is of the form:

$$\mathbf{E}_g(x, z=0) = \mathbf{A}(x,0)e^{j(2\pi/\lambda)(n-1)g(x)} = \mathbf{A}(x,0)P(x) \tag{8.13}$$

where:

n is the index of refraction of the material

$\mathbf{A}(x,0)$ is the amplitude of the input beam

$g(x)$ is a periodic phase function whose height is $\lambda/(n-1)$

In the far field ($z = z'$), Equation 8.13 becomes

$$\begin{aligned}\mathbf{E}_g(k_x,z') &= \Im[\mathbf{E}_g(x,0)] \\ &= \Im[\mathbf{A}(x,0)] * \Im[P(x)] = \mathbf{A}'(k_x,0) * \Im[P(x)]\end{aligned} \tag{8.14}$$

where:

$\Im$ stands for Fourier transform
* is the convolution symbol
$k_x = k_0 x'/z$

From Equation 8.14, we see that the resulting field at $z = z'$ is simply the convolution of the spatial frequency content of the phase with the amplitude of the input beam after propagating a distance of z'. If the divergence of the diffuser (high spatial frequency) is significant, the shape of the energy envelope will be dominated by the phase of the diffuser, rather than the divergence of the input beam.

A grating has very distinct orders due to its periodic structure. The spatial frequency composition of the phase is simply a set of appropriately weighted delta functions spaced at angular intervals as defined in Equation 8.12. One- and two-dimensional phase screens are frequently designed using this very principle and are termed "beam splitters." The far-field intensity distribution is composed of clearly distinct diffraction orders arranged in some desired spot pattern, and the phase profile of the optic is optimized to give the appropriate weighting function to this comb of orders. If the separation between diffraction orders is smaller than the detector resolution, the beam splitter effectively functions as a diffuser. This offers an additional diffuser design approach in some cases, but the discrete orders are not always desirable for certain applications.

If we now add a second function $r(x)$ with a period much larger than the period of $g(x)(\Lambda_r >> \Lambda_g)$ to the phase in Equation 8.13, we see that

$$\mathbf{E}_d(x,z=0) = \mathbf{A}(x,0)e^{j(2\pi/\lambda)(n-1)[g(x)+r(x)]} = \mathbf{A}(x,0)P(x)R(x) \tag{8.15}$$

The resulting field after propagating a distance z' then becomes

$$\mathbf{E}_d(k_x,z') = \Im[\mathbf{E}_d(x,0)] = \mathbf{A}'(k_x,0) * \Im[P(x)] * \Im[R(x)] \tag{8.16}$$

When this second component is added to the phase, the distinct orders become blurred by the spatial frequency components of the function $r(x)$. By choosing Λ_r and Λ_g appropriately, an apparent continuum of spatial frequencies may be obtained, resulting in a "solid" filled region of light in the far field.

Notice that since $\Lambda_r >> \Lambda_g$ the period of the diffuser Λ_d is approximately equal to Λ_r. Thus, if a grating has a large period such that the orders of that grating, as governed by Equation 8.12 (with the period equal to Λ_g), are spaced in such a way that the resulting beams significantly overlap, the angular region will be "solidly" filled with light. Any two coherent beams that overlap will interfere. This interference is the source of the speckle indicative in diffuser patterns. The subject of speckle is covered in detail in Section 8.6.

Also notice the case that as Λ_r becomes so large that the size of the input beam is no longer large enough to sample one full period of the phase, the phase is effectively no longer periodic. At this point, the location of the orders is ill defined, since optically the phase is not periodic. This is a result of the fact that nonperiodic functions have a continuum of frequencies rather than a discrete set of frequencies. However, the phase still has the same spatial frequency spectrum, and thus the envelope of the pattern will remain virtually unchanged. This is of course much like a traditional hologram in that a small piece of a hologram will still produce the same image.

The transition from a periodic to nonperiodic phase function is the essential distinction between a diffuser and a beam splitter. Frequently, it makes sense to design a diffuser pattern somewhat larger than the incident beam and then tile it across the part. This allows the system to have very loose alignment tolerances, and it adds flexibility in wafer-based fabrication of diffusers. With this design approach, the diffuser essentially becomes a type of beam splitter with the only difference being that the diffraction orders are blurred together and are less distinct.

8.2.3 Mathematical Description of a Diffuser

To mathematically describe a diffuser, we first note the shift property of the Fourier transform:

$$A(k_x - k_{x0}) \Leftrightarrow a(x)e^{-jk_{x0}x} = a(x)e^{-j(2\pi/\Lambda)x} \tag{8.17}$$

When designing optics such as diffusers, it is often useful to define things in terms of a discrete Fourier transform. For a calculation grid of dimension D, the smallest frequency increment is $\delta f = 1/D$.[8] Physically, D is the diameter of the input beam or the period of a grating. Thus, any frequency is an integer multiple of δf. For example, define a frequency f_0

$$f_0 = N\delta f = \frac{N}{D} \tag{8.18}$$

or, in terms of the wave number

$$k_{x0} = 2\pi f_0 = 2\pi\frac{N}{D} \tag{8.19}$$

where:

N is an integer

From Equations 8.17 and 8.19, it then follows that

$$\Lambda = \frac{D}{N} \tag{8.20}$$

Substituting Equation 8.20 into Equation 8.12, we find that any discrete spatial frequency can be described as

$$\sin(\theta_t) = \sin(\theta_i) + \frac{N\lambda}{D} \quad (8.21)$$

Recognizing the fact that $D = \delta dM$, where δd is the smallest distance increment and M is the number of data points across the calculation grid, Equation 8.21 becomes

$$\sin(\theta_t) = \sin(\theta_i) + \frac{N\lambda}{\delta dM} \quad (8.22)$$

Finally, solving Equation 8.22 for N, we have

$$N = \left[\sin(\theta_t) - \sin(\theta_i)\right]\frac{\delta dM}{\lambda} \quad (8.23)$$

This equation is useful for computational reasons to calculate a particular grid point number on a discrete Fourier grid to produce a phase function of a given dimension that will bend light of a wavelength λ by an angle θ_t.

As noted in the previous section, when the input beam diameter is smaller than the periodicity of the phase function (D in the above equations), there is a continuum of diffraction orders. However, a discrete Fourier transform operates with discrete physical positions and discrete frequency components, the latter of which are equivalent to the diffraction orders of a beam splitter. It is convenient to refer to the frequency-domain components resulting from the discrete Fourier transform as diffraction orders even if the phase function is not strictly periodic.

8.3 DIFFUSER DESIGN PROCESS

Typically, high-efficiency diffuser elements are phase-only transmissive elements. With antireflective coatings, the transmission through the elements can be assumed to be unity everywhere. The goal of the design process is then to determine the phase function required to generate the desired output intensity distribution. From Equation 8.14, it is clear that there is a Fourier transform relationship between the phase and the output intensity pattern. In fact, there is a long history in the image processing and the optical pattern recognition fields of using only the phase, or even binary versions of the phase, to reconstruct a function. Stark[9] outlines the significance of the phase of the Fourier transform of an image in its reconstruction more extensively than can be discussed in this chapter.

8.3.1 Inverse Fast Fourier Transform Design Approach

It has been demonstrated that the Fourier transform phase can provide a good reconstruction of an image. This is especially true for images with a lot of high-frequency content such as edges. Image reconstruction from the phase of the Fourier transform can be viewed as a high-frequency enhanced filtering process. Flannery and Horner discuss the applications of phase-only and binary phase-only filters to Fourier optical signal processing and pattern recognition.[10]

A naive diffuser design approach would be to take the inverse Fourier transform of a target intensity pattern and only keep the phase term. However, the high-frequency filtering aspect is a key concern. This is illustrated in Figure 8.7. The inverse Fourier transform of a square aperture (which represents a desired square top-hat diffuser pattern) is not well represented only by the phase term; Figure 8.7a). If one throws away the amplitude component and takes the Fourier transform of the phase term only, the resulting pattern is no longer close to the square top-hat pattern. A more complex image, such as the Jenoptik logo in Figure 8.7b, is far better reconstructed

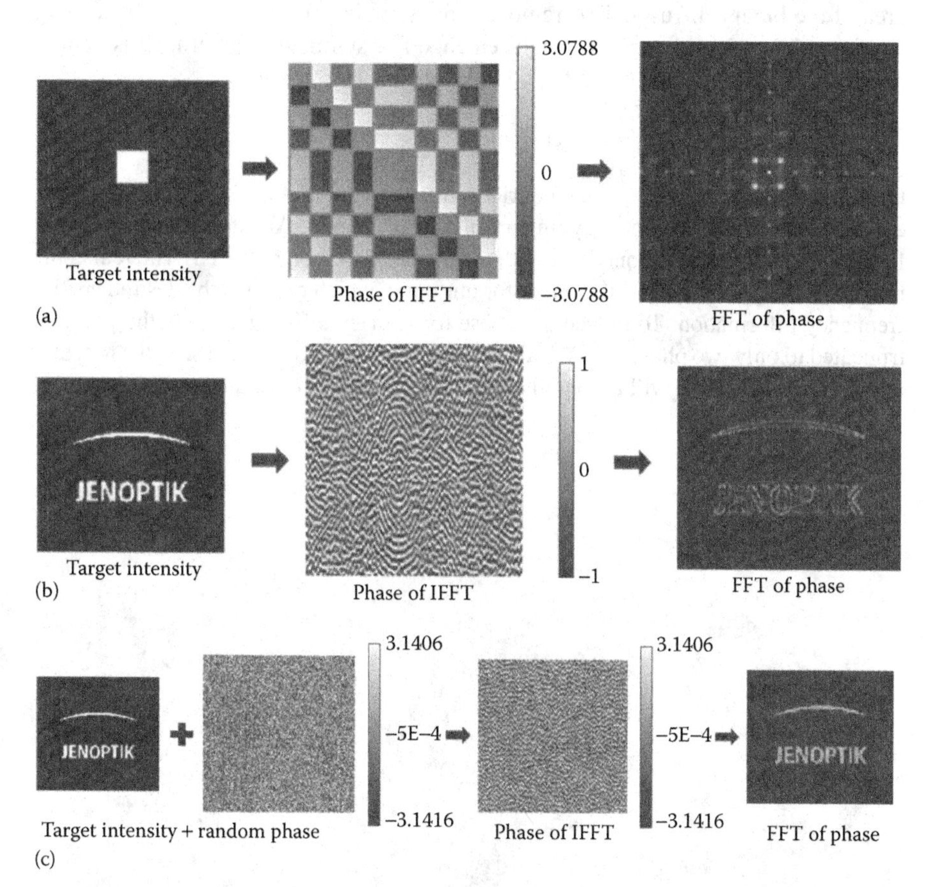

FIGURE 8.7 Reconstruction of images from the phase of the inverse Fourier transform of a target pattern. (a) The phase of the inverse Fourier transforms of a square does not Fourier transform back into a distinct square pattern. The low-frequency content of the target image does not lend itself well to the Fourier reconstruction. (b) A target image with higher frequencies, sharper edges, and more content can be better reconstructed from the phase of its inverse Fourier transform. The low-frequency regions from the larger blocks in the centers of the various shapes are still not well represented in the Fourier transform of the phase. (c) When the target intensity from (b) is given a random phase distribution, its high-frequency nature allows it to be fully reconstructed from the phase of its inverse Fourier transform.

from the Fourier transform phase term alone. But even in this case the low-frequency information is lost. The target pattern can be supplied with high-frequency information by giving it a random phase distribution as in Figure 8.7c. The phase of the inverse Fourier transform of the new pattern allows for a good reconstruction of the original.

The basic design approach is then to randomize the nonzero values of the amplitude between 0 and 1, and randomize the phase between 0 and 2π. The randomization step reduces the output dependence on the input beam. Effectively, the high-frequency random function being multiplied by the desired frequency envelope ensures that the spectral content of the envelope function is distributed over the full area of the binary diffuser. This removes any input beam alignment tolerances and any input beam intensity profile requirements. For symmetric patterns, it is convenient at this point to ensure that the complex grid is Hermitian:

$$f(i,j) = f^*(-i,-j) \tag{8.24}$$

Once this is done, the inverse fast Fourier transform (FFT) of the complex grid is calculated. Due to the conjugate symmetry, the result is real. All of the desired spatial frequency information is contained in the real part of the complex FFT. The real component of the grid is used as the phase for our optic, which contains the desired spatial frequency information. To reduce the phase to a binary diffractive optic, the phase is truncated to only two phase levels: 0 and π. Assigning any positive phase value to π and any negative value to 0 will accomplish this. The process is illustrated in Figure 8.8.

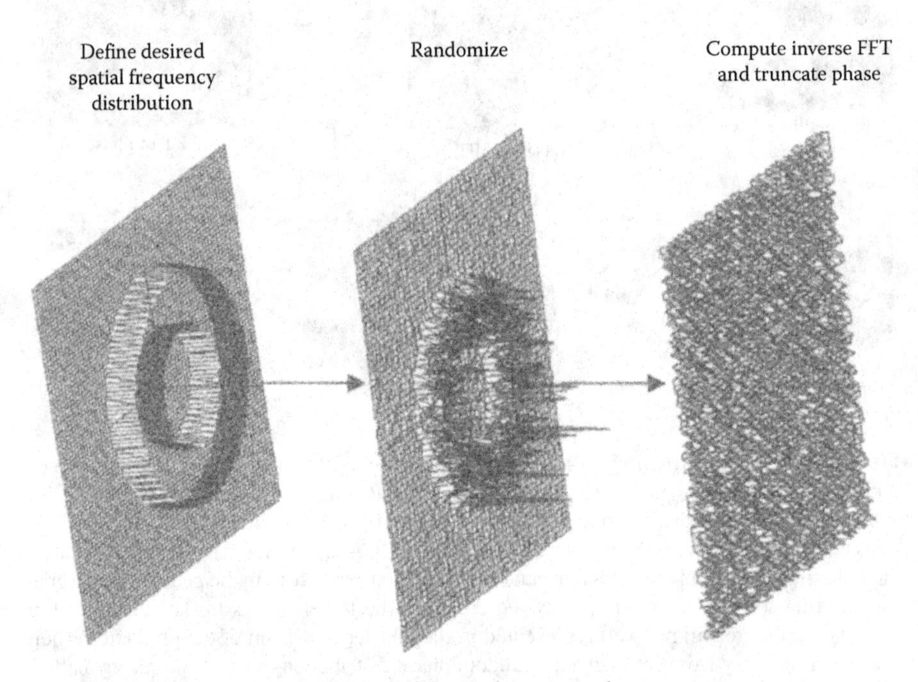

FIGURE 8.8 The steps for a simple diffuser design approach.

For a nonsymmetric pattern, the conjugate-symmetric condition is not possible, which results in a complex field following the inverse Fourier transform. One way to design a more complicated pattern that can still be generated with a binary diffuser is to define the center of the desired pattern far enough off-axis so that it does not overlap its symmetric counterpart. Alternatively, a truly nonsymmetric pattern can be produced by simply using the phase of the resulting inverse Fourier transform as the phase for the diffractive diffuser.

8.3.2 Iterative Fourier Transform Algorithm

A more general numerical design process for diffusers is known as the Gerchberg–Saxton algorithm,[11] or the iterative Fourier transform algorithm (IFTA). A simulation grid is generated where each pixel represents a single diffractive order in the output plane. The grid is initialized with the desired output amplitude and each pixel is given a random phase between 0 and 2π. This provides the high-frequency content necessary to obtain a good reconstruction from the phase term, as noted in Section 8.3.1.

Once this is done, the inverse FFT of the complex grid is calculated. The phase component of the resulting complex field gives us an initial guess for the phase profile of the diffuser required to produce the specified output. At this point, the amplitude should be reset to match the incident beam geometry. If the design is to be completed for plane wave illumination, the amplitude can simply be set to unity.

In addition, the fabrication process must be accounted for at this point in the process. Diffusers are often fabricated through either single or multiple binary etch processes or through a single grayscale etch.[12] The latter case provides a continuous phase profile, so no additional modification to the phase is required. However, if a binary or multilevel phase is intended, the phase at the diffuser plane must be discretized to the appropriate number of phase levels.

Once the phase and amplitude have been appropriately adjusted at the diffuser plane, it is used as the input for a forward FFT to obtain an updated version of the expected output. The amplitude in the output plane is reset to the desired output amplitude, and the inverse FFT of this complex field is taken once again. The entire process iterates a set number of times or until the output intensity converges on the specified pattern. The algorithm is diagrammed in Figure 8.9.

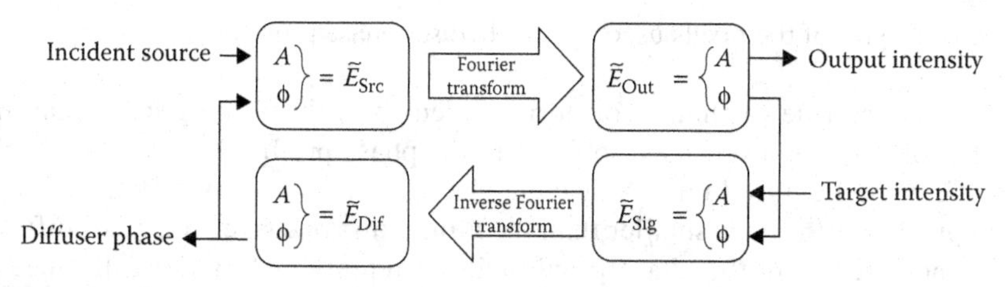

FIGURE 8.9 A flowchart describing the IFTA for diffuser design.

8.3.3 Algorithm Implementation

The diffuser design algorithm is relatively straightforward to program, but there are also commercially available design packages (such as LightTrans VirtualLab) that implement the approach in a very customizable way and can account for other aspects such as fabrication tolerances, minimum feature sizes, and impact on the output pattern caused by the pixel shape.

It is also worth noting that earlier implementations of the algorithm were limited by the available computing power and memory. They required a great deal of care in the choice of the simulation size (a square grid of 2^N pixels to ensure an efficient FFT) and the types of output patterns (a conjugate-symmetric pattern would allow the diffuser phase to be purely real, which could save computation time and memory in an optimized implementation) to obtain a design in a reasonable timeframe. However, modern computers are significantly faster and the algorithm efficiency is now rarely a limitation. Typical diffuser designs can be produced in a matter of minutes or even seconds even when the iterative algorithm is used.

8.3.4 Input Parameters

As with any optical system, a diffuser design starts with a well-defined input and output. The clear aperture, beam profile, and wavelength are the key input parameters, and the required far-field intensity distribution in angle space characterizes the output. One of the design aspects to verify early in the design process is the minimum feature size. This is the smallest spatial frequency expected in the diffuser phase profile. This value is determined by the maximum angular extent of the illuminated output pattern according to

$$d = \frac{\lambda}{\sin(\theta_{\max})} \tag{8.25}$$

From the minimum feature size, one can determine a maximum pixel size. The ratio of minimum feature size to pixel size is a parameter termed the "design freedom." For small divergence angles, the design freedom can also be defined as the ratio of the angular extent of the output design space to the maximum angle of the illuminated output area. For a high-efficiency diffuser pattern, a good minimum value for design freedom is 8:

$$\frac{\lambda}{\delta d \sin(\theta_{\max})} \geq 8 \tag{8.26}$$

where:

δd is the size of the pixels used for the diffuser phase profile

An alternative rule of thumb for design freedom is that it be greater than the number of discrete levels used to fracture the phase profile. For example, if the phase is to be fractured into 16 discrete phase levels, one should target a maximum pixel size 16 times smaller than the minimum zone size so that all diffractive zones present in the phase profile will be represented by the full range of discrete phase levels.

In some cases, the pixel size will be determined by the fabrication process, and the design freedom may be limited by output requirements and fabrication limitations. In such cases, it is still possible to produce a high-quality design, but it is often much less tolerant of variations in the fabrication process.

Once the pixel size is chosen, the simulation grid will also be specified. The number of points is equal to the size of the clear aperture divided by the pixel size:

$$M = \frac{D}{\delta d} \tag{8.27}$$

The output plane must have the same number of pixels as the input plane to allow for the Fourier transform relationship between the two. In the output plane, each order number (ranging from $1 - [M/2]$ to $+[M/2]$) will have a divergence angle as specified in Equation 8.23.

Since the maximum angular extent of the output pattern is known, one can determine the highest order number, N, to be illuminated, which allows one to create the target output amplitude.

The amplitude target, diffuser pattern parameters (pixel size and simulation grid size), design wavelength, and incident beam shape are the required parameters for the diffuser design process. After obtaining the diffuser phase profile, the final step should be to evaluate the optical performance of the structure in relation to fabrication tolerances.

8.4 TOLERANCING AND FABRICATION CONSIDERATIONS

When qualifying a diffuser for a given application, the two primary metrics are diffraction efficiency and uniformity. Diffraction efficiency is defined as the ratio of the energy transmitted into the design region to the total energy incident on the diffuser (note that the absence of antireflective coatings will introduce an immediate efficiency penalty). It can be shown analytically that diffractive lenses can have very high diffraction efficiency (99% for a 16-level kinoform and 95% for an 8-level kinoform), though Swanson showed[13] that this scalar-theory calculation is only true if the diffractive feature sizes are much larger than the wavelength (typically 10–20 times larger). Furthermore, the calculation applies specifically to a single-order diffractive optic. When a large number of diffractive orders is present, maximum efficiency can be significantly lower than expected, particularly when the divergence angle is sufficiently large that the smallest diffractive zones are no longer significantly larger than the wavelength (which begins to occur when the diffuser half-angle divergence exceeds 3°).

The other metric for characterizing diffusers is uniformity. The presence of speckle can make this difficult to fully quantify, and a full analysis requires accurate characterization of the beam profile and coherence. However, one can offer a uniformity value that ignores speckle by treating the diffuser as a beam splitter once again and specifying the uniformity of the diffracted orders.

The uniformity metric essentially describes the degree to which the output relative intensity distribution in the illuminated orders differs from the design. Calculation of uniformity can be a bit complicated for diffuser designs with varying intensity in the illuminated orders. The general form for uniformity error is calculated as follows:

First, determine the orders with the maximum and minimum transmitted intensity as a ratio of the designed intensity for that order:

$$\hat{I}_{max} = \max_{m} \frac{|U_{out}(m)|^2}{|U_{des}(m)|^2} \tag{8.28}$$

$$\hat{I}_{min} = \min_{m} \frac{|U_{out}(m)|^2}{|U_{des}(m)|^2} \tag{8.29}$$

Note that for diffuser patterns designed for uniform transmission in all illuminated orders, $\hat{I}_{max}$ and $\hat{I}_{min}$ and simplify to the maximum and minimum intensity in any illuminated orders. Then, these two orders are used to calculate a peak-to-valley uniformity error, E_{unif}:

$$E_{unif} = \frac{\hat{I}_{max} - \hat{I}_{min}}{\hat{I}_{max} + \hat{I}_{min}} \tag{8.30}$$

When using this formulation for uniformity error, it is common to say that all orders have intensities within $\pm E_{unif}$ percent of the designed relative intensity. For example, if the uniformity for a top-hat diffuser is specified as <5% (ignoring speckle), all orders within the transmitted field will have intensities between 95% and 105% of the median value. There is an alternate formulation that differs from this by a factor of 2 and would define a variation between 95% and 105% as being 10% uniformity error. However, that formulation is better applied to gratings and beam splitters with very few orders.

Uniformity and efficiency can readily be calculated from a Fourier transform simulation of a given diffuser phase profile. However, this only gives a theoretical value for the optical performance of the design. To be useful, the design must be fabricated, and any fabrication process will introduce tolerances that will impact actual performance.

There are many ways to transfer a designed diffractive optic into the actual physical diffractive optic: binary mask(s) photolithography followed by an etch(es),[14] gray-scale mask photolithography followed by an etch, e-beam lithography, direct laser writing, focused ion-beam milling, plastic molding, and embossing to mention a few. The limiting factor in the fabrication of a diffractive optic is the level of accuracy with which the process can reproduce the designed structure. This is generally limited by the resolution of the process which defines a minimum pixel size, δd in Equations 8.23 and 8.26, the smallest feature in the grid.

There are a few basic fabrication tolerances to consider when evaluating a diffuser design:

1. Feature etch depth errors
2. Feature shape errors
3. Misalignment of successive levels in multistep lithography processes
4. A regular grid of pixels introduces its own high-frequency diffraction pattern that will impact the actual output

Each of these issues impacts uniformity and efficiency in a somewhat different way. We consider each of them individually.

8.4.1 Diffuser Feature Depth Errors

The designed diffuser phase profile has phase values ranging from 0 to 2π. Larger phase values correspond to deeper etch depths. The mapping between the phase depth, ϕ, and the etch depth, d, is

$$d = \frac{\lambda}{2\pi(n-1)}\phi \tag{8.31}$$

where:
λ is the design wavelength
n is the refractive index of the material

Note also that if the phase profile is patterned using discrete binary etch steps, producing an m-level stair-step approximation to the continuous phase profile, the maximum phase is $[(m-1)/m]2\pi$.

With some of the more common diffuser fabrication processes, it is easy to miss the target feature depth by as much as a few percent. Simple phase gratings provide an illustrative example to the resulting impact on diffuser performance. In general, deriving analytical expressions for the diffraction efficiency for each order for arbitrary diffuser phase patterns is very much nontrivial. However, the relationship between the phase profile and the relative intensity into the various diffraction orders is a Fourier transform, as discussed in Section 8.2.2, and the Fourier transform can be calculated analytically for a few special cases.

A square-wave phase grating with a phase depth of ϕ has diffraction efficiencies given by the following:

$$\begin{cases} \mathrm{DE}_0 = \cos^2(\phi) \\ \mathrm{DE}_{+1} = \mathrm{DE}_{-1} = \left(\dfrac{2}{\pi}\sin(\phi)\right)^2 \\ \mathrm{DE}_m = 0 & [m \text{ even}] \\ \mathrm{DE}_m = \dfrac{1}{m^2}\mathrm{DE}_{+1} & [m \text{ odd}] \end{cases} \tag{8.32}$$

The summation over all the diffraction efficiencies is 1. Note that all diffraction efficiencies depend on the square of the trigonometric functions, and as the phase depth deviates from $\pi/2$, more of the incident energy is pulled out of the higher orders and shifted to the zero order. This has a direct impact on output uniformity. If the phase depth is approximately 1.0 rad, the energy in the 0, +1, and −1 orders will be the same. However, if the phase depth is reduced by 5% to 0.95 rad, the zero order will contain approximately 10% more energy than either the +1 or −1 order.

Similarly, a sinusoidal phase grating with a phase depth of φ has diffraction efficiencies expressed in terms of Bessel functions:

$$\begin{cases} \mathrm{DE}_0 = J_0^2(\phi) \\ \mathrm{DE}_{+1} = \mathrm{DE}_{-1} = J_1^2(\phi) \end{cases} \tag{8.33}$$

As with the square wave, slight changes in the phase depth can produce significant changes in the output uniformity, and, to a lesser degree, the efficiency.

The diffuser efficiency (the fraction of the incident light that is transmitted into the design orders) is degraded as efficiency increases for higher scattered orders. Additionally, some of the energy loss resulting from the lower energy diffracted into the design orders will instead end up in the zero order, which can result in a spike in the center of the diffuser output pattern.

The full one-wave modulation depth of a diffractive structure is given in terms of the wavelength, λ, and refractive index, n, by

$$d_{2\pi} = \frac{\lambda}{n-1} \tag{8.34}$$

Combining Equations 8.31 and 8.34, the phase depth of a grating can be calculated from its physical depth, d, by the following expression:

$$\phi = \frac{2\pi d}{d_{2\pi}} = \frac{2\pi d(n-1)}{\lambda} = k_0 d(n-1) \tag{8.35}$$

From Equation 8.35, one should additionally note that a change in the incident wavelength affects the output of a given phase profile in a similar way as structure depth errors, and it thus has an equivalent impact to optical performance. In fact, modeling the structure's output over a range of grating depth errors can adequately capture both effects.

In a similar manner, the phase depth is impacted by a variation in the incident angle of the light. The general form of Equation 8.35 for nonnormal illumination is

$$\phi = \frac{2\pi d(n-1)}{\lambda}\cos(\theta) \tag{8.36}$$

This effect, along with variations in refractive index (either due to the use of different materials than was called for by the design or due to induced optical effects in the grating material), can likewise be quantified directly from the feature depth error tolerancing calculations.

8.4.2 Diffuser Feature Shape Errors

Feature shape errors manifest as rounded edges and corners due to lithography resolution limitations and as nonvertical sidewalls resulting from the various fabrication process. These errors are highly dependent on the fabrication methods used, and the actual impact on performance will vary significantly with diffuser feature sizes and incident wavelength. Simulating the impact can be a very intensive process because the shape errors are typically a small fraction of the pixel size. An accurate simulation will require a grid with a resolution at least an order of magnitude smaller than the original design grid.

One approach that can provide a rough estimate for the shape error impact is to assume some convolution kernel that captures the fabrication processes. For example, a Gaussian with a sufficiently small waist radius convolved with the original shape profile will introduce rounding of the corners and will smooth out the sharp vertical transitions. The Fourier transform of the resulting phase profile gives a first-order estimate to the performance impact.

Typically, if the magnitude of the shape errors is much smaller than the minimum zone sizes and the incident wavelength, the performance impact will be small, though it will affect both efficiency and uniformity. Conversely, if the rounding is sufficiently large, it can effectively change the feature depth and also increase the zero order as discussed above and produce a noticeable intensity spike at the center of the pattern.

8.4.3 Alignment Errors in Multistep Fabrication Processes

In multistep lithography processes, such as a binary multimask fabrication sequence, a small pixel size can yield additional errors due to layer-to-layer registration errors. These errors usually manifest themselves as very small peaks and/or valleys in the phase, which can result in a very large number of scattering sites. It is possible to model this scattering, but as with the shape errors, it requires a high-resolution representation of the surface. Since registration tolerances typically produce a random variation within a certain range, it is virtually impossible to predict actual results, especially when there are multiple (more than two) successive fabrication steps. It is possible to use a Monte Carlo approach to model the impact over the alignment tolerance range, but this can be impractical for most diffuser design projects. In reality, as with the shape errors, the actual impact is typically significant primarily in the case of small zone sizes and short wavelengths.

8.4.4 Diffraction from Phase Pixelation

There is one additional fabrication factor missing in Equations 8.13 through 8.16: the pixelated nature of the phase. In many cases, the phase is composed of square pixels on a fixed uniform grid. The very small squares in the phase will manifest themselves in the far field as a large $\mathrm{sinc}^2(x, y)$ function envelope. The desired pattern (e.g., a top-hat function) will be present in the center of the main lobe and repeated at the null points of the $\mathrm{sinc}^2(x, y)$ function as shown in Figure 8.10. The location of the null points are at $\sin(\theta_m) = m\lambda/\delta d$. This problem can become very significant if the desired diffuse pattern contains angles that are large enough to approach these null points. From Equation 8.23, we see that this occurs when

$$N \sim \frac{M}{2} \tag{8.37}$$

Figure 8.11 further illustrates this effect with a practical example of a square top-hat diffuser. In Figure 8.11a, the pixelation is ignored in the output simulation, while the simulated output in Figure 8.11b accounts for pixel size.

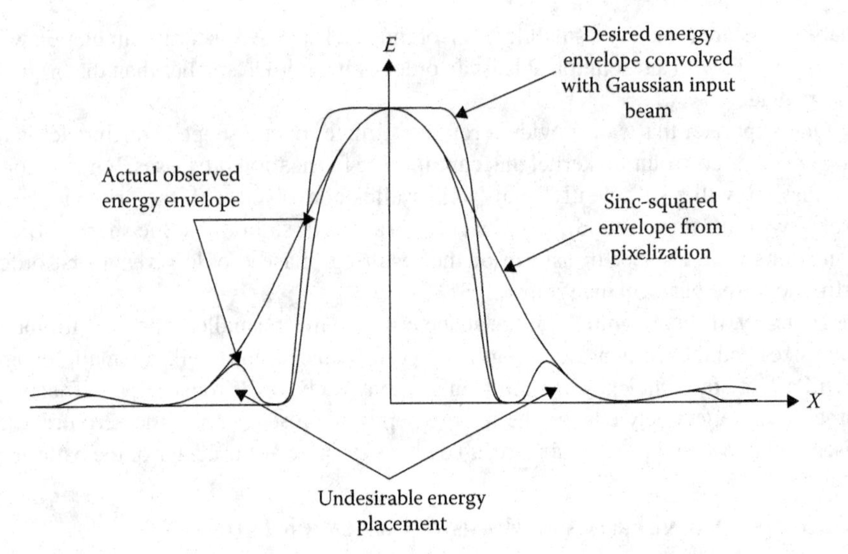

FIGURE 8.10 An illustrative energy envelope plot of the output of a diffuser and the effects of a pixelated phase.

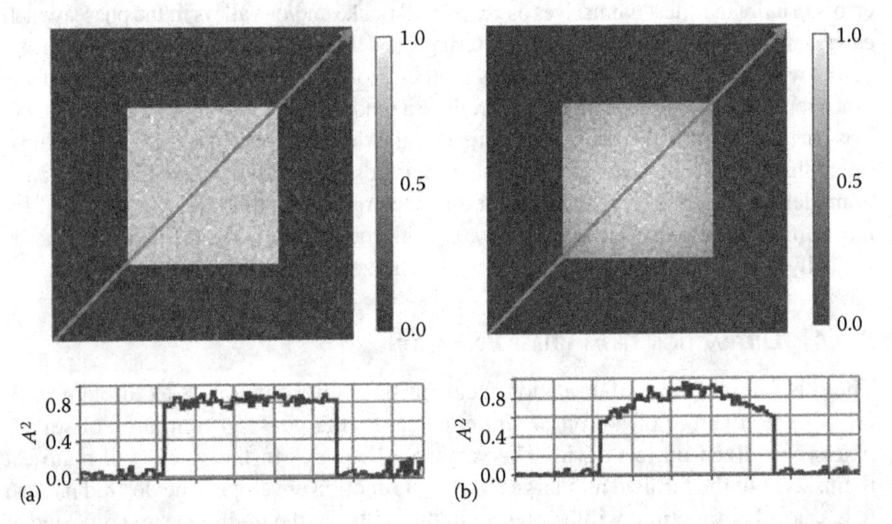

FIGURE 8.11 Empirical results for a simple 8-level top-hat diffuser with a design freedom of 2 illustrating the impact of pixelation. False color 2D image and corner-to-corner cross section are shown for each case. (a) The simulated diffuser output without including the pixel size in the calculation. (b) A simulation of the same diffuser profile with pixelation effects included in the simulation. The sinc^2 intensity envelope is quite evident.

In the example shown in Figure 8.10, N is 50 and M is 64. It is therefore best if $N \ll M/2$, which is one of the reasons for maintaining a large design freedom. For a given divergence angle, decreasing δd is a way to either decrease N or increase M. However, the fabrication method of choice may not be able to support such a reduction in δd. The advantage is that as $\delta d \rightarrow 0$ the sinc$^2(x,y)$ envelope becomes very large and thus less of a factor. In single-step lithography, such as grayscale or direct write methods, the resolution limits actually help minimize the $\text{sinc}^2(x, y)$ envelope effect by poorly representing the pixelated nature of the design grid, producing more of a smooth continuous phase. This is often the phase that a coarse calculation grid is trying to approximate.

An alternative to trying to avoid the pixelation issue is to determine its impact ahead of time and compensate for it in the design process. This is accomplished by dividing the design amplitude pattern by the sinc-squared function resulting from the selected pixel size. The target output will appear to have higher intensity at the outer edges and corners compared to the center, but the resulting intensity pattern will be properly compensated and will provide a better realization of the design goal.

The phase structures of two diffusers are shown in Figure 8.12. In Figure 8.12a $\delta d \sim 2.0$ μm, and in Figure 8.12b $\delta d \sim 0.33$ μm. Both of these diffusers

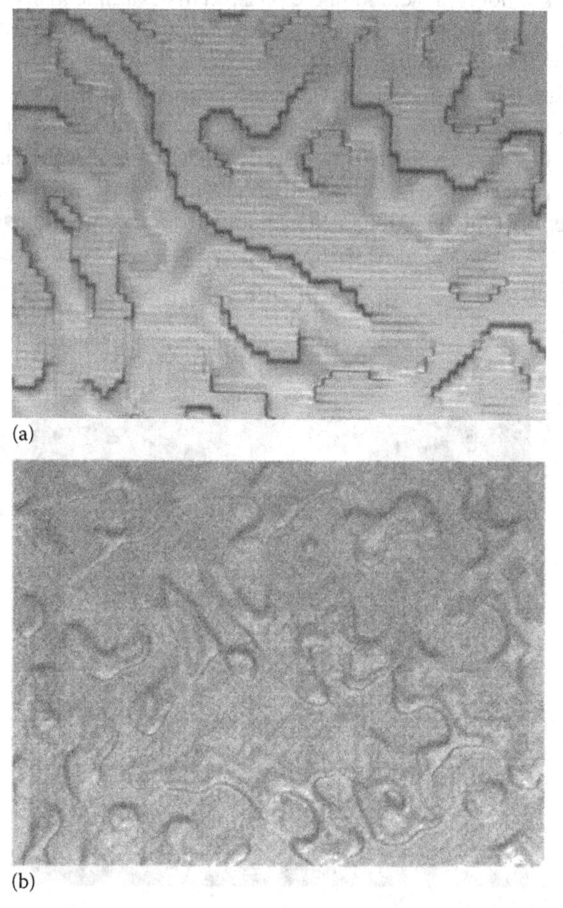

FIGURE 8.12 Pictures of a portion of the phase structures of two diffusers with the same desired output. Both phase functions have 64 phase levels and are in a photoresist layer on a fused silica substrate. (a) A pixelated version with a pixel size of ~2.0 μm. (b) A nonpixelated version. The alternating dark and light contour fringes are the results of thin-film interference from the illuminating source.

were fabricated with grayscale mask photolithography. In this case, the phase structures have 64 phase levels and are only in a photo-resist layer on a fused silica substrate. In the case of Figure 8.12b, one can see that the final result is not pixelated and δd is only a consideration for the design grid. To make them more robust, a reactive ion etch would be used to transfer the surface relief structure into the fused silica.

8.4.5 Example of a Fabricated Diffuser

Figure 8.13 shows the experimental measurement of the output of the diffuser whose phase is shown in Figure 8.12b. The sinc-squared envelope indicative in a pixelated phase is not present in the output from the phase shown in Figure 8.12b since the pixel structures are not represented in the phase. This example also demonstrates the complexity that is available in the desired pattern is almost limitless provided that the fabrication method is appropriate in turning the design into reality.

It is interesting to note that the diffuser shown in Figure 8.12b was also tested in white light and worked fairly well. The face was clear but the letters were blurred. From Equation 8.22, the divergence angle is dependent on the wavelength of light (λ). This causes the pattern to be chromatically blurred. The separation of colors is most evident in the letters which are the farthest points from the optical axis. A blue-green dot at the zero order (optical axis) was present due to the lower diffraction efficiency of the off wavelengths.[14]

FIGURE 8.13 Empirical results of a 632.8 nm laser illuminating the phase of the diffuser pictured in Figure 8.12b. This demonstrates the complexity and clarity that is possible with complex diffusers. Note that some of the fine detail present in the actual output was not faithfully transferred to this figure. For instance, individual hairs on the very top of his head are distinguishable in the live presentation.

8.5 SIMPLE DESIGN EXAMPLE

As a simple illustrative design example, suppose we wish to design a diffuser with a clear aperture of $D = 1.0$ mm that projects a ring of laser light with a wavelength of $\lambda = 0.6328$ μm between 1° and 2°.

Before beginning the design process, we must consider the overall problem. Note that if the incident source is a 1 mm diameter Gaussian, it has a full-angle divergence of 0.046°. This is relevant for estimating speckle, but it also has implications of the output beam. Specifically, if the largest diffraction orders are specified by the design to be at a diffraction angle of 2°, the actual $1/e^2$ full divergence angle of the output will be 4.046°. For this example, we will assume that the roll-off at the edges is acceptable, and this is the case in many applications. However, if the transmitted efficiency within a fixed angular window is a critical design requirement, the input divergence may require significant consideration.

8.5.1 BINARY DIFFUSER DESIGN

The maximum divergence angle is 2°, which means that the minimum feature size from Equation 8.25 is 18 μm. These features are comfortably fabricated with modern diffractive optic fabrication processes. We begin by comparing a simple inverse FFT design approach with the IFTA algorithm and assume a binary diffuser pattern in both cases. Since the clear aperture is specified as 1 mm, Equation 8.23 gives the inner and outer diffraction orders as $N_1 = N(\theta = 1°) = 28$ and $N_2 = N(\theta = 2°) = 55$. Notice that we have to round to the nearest integer. The pixel size for the diffuser phase profile is chosen to be 2 μm, which results in a 500 × 500 point simulation grid. From this information, we now prepare a grid of 500 × 500 points with a width and height of 1.0 mm that is zero everywhere except grid points whose radius falls between 28 and 55.

$$R = \sqrt{i^2 + j^2} \quad 28 \le R \le 55 \tag{8.38}$$

where:

i and j are grid indices that have the range $-250 < i, j < 249$

Figure 8.14 shows the result. This target intensity profile is a sufficient condition with which to start both design algorithms. In many cases, better results are obtained by assuming a uniform plane wave source condition for the optimization process. The resulting phase profiles are shown in Figure 8.15. Figure 8.15a is the binary phase profile resulting from the inverse FFT, while Figure 8.15b is the output phase profile from the IFTA optimization.

To verify our designs, we first assign an arbitrary amplitude function: Gaussian with $\omega_0 = 0.2$ mm. We then simulate an optical Fourier transform assuming a 10 mm focal length Fourier transform lens. The simulated output resulting from the (1) inverse FFT design approach and (2) the IFTA optimization are shown in Figure 8.16. The two phase profiles show markedly similar output, though the IFTA-optimized phase profile offers 58% diffraction efficiency into the ring pattern for this source condition compared to 51% for the phase profile generated by

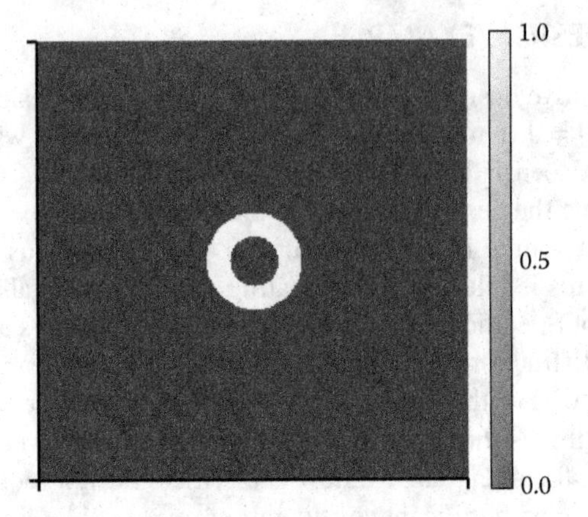

FIGURE 8.14 A plot of the target spatial frequency band intensity profile that is desired in the ring diffuser design example.

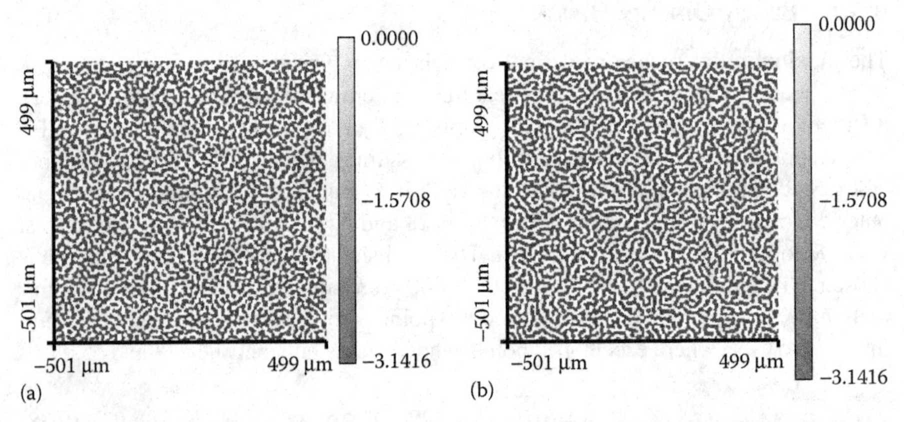

FIGURE 8.15 Designed binary phase profile resulting from (a) inverse FFT method and (b) IFTA optimization.

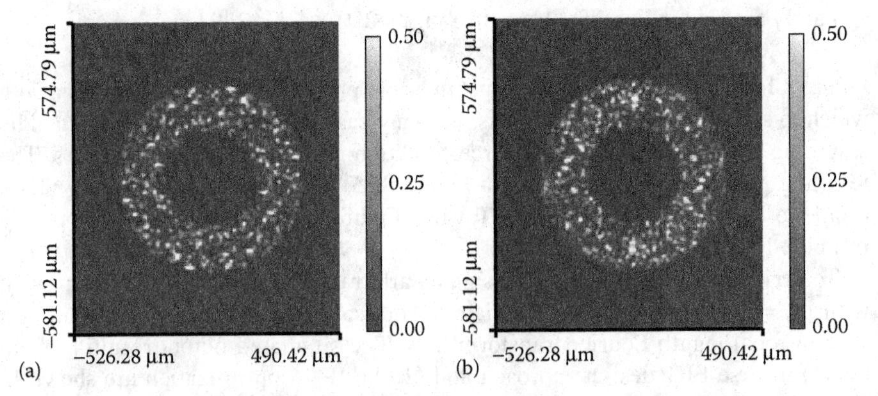

FIGURE 8.16 Simulated output with a 10 mm focal length Fourier transform lens and a 0.2 mm waist radius incident source. Simulation uses the phase function designed by (a) inverse FFT method and (b) IFTA optimization.

the inverse FFT approach. Also, the IFTA-optimized pattern offers a somewhat better signal-to-noise (SNR) ratio and noticeably lower maximum intensity of stray light.

Notice the excellent suppression outside of the ring area in both cases. We know that at the focal plane $r = f\tan(\theta)$, where r is the radius from the optical axis, f is the focal length of the lens, and θ is the angle of the incoming ray. From this, we can verify that the divergence angles of the design are indeed correct. We also note that the output shape is independent of the incident source profile. The uniformity and speckle pattern in the output may be impacted, but the overall shape envelope is determined solely by the phase profile.

8.5.2 Multilevel Diffuser Simulation

For completeness, we also use the IFTA algorithm to generate an 8-level phase profile which, as expected, offers increased efficiency and uniformity. The 8-level phase profile (shown in Figure 8.17a) has a 66% efficiency and similar stray light and SNR levels as the IFTA-optimized binary pattern. The simulated intensity profile (Figure 8.17b) also appears more uniformly filled than the two binary designs (Figure 8.16).

We previously noted the subtleties between beam splitters and diffusers. Figure 8.18 shows the output of the 8-level pattern with different source conditions. In both cases, we still assume the same 10 mm focal length Fourier Transform lens as with the earlier results. In Figure 8.18a, the constant amplitude incident source results in a very uniform output pattern. The diffraction order locations are well defined and have a grid-like shape in the speckle pattern. In the extreme case, with a large incident source covering many periods of the phase profile, each diffraction order becomes a delta function, and the speckle pattern does not appear in the output plane. The output is that of a beam splitter composed of discrete diffraction orders rather than a uniformly illuminated region. On the other hand, as the beam diameter is reduced, as in Figure 8.18b where the diameter is 0.8 mm, the diffractive orders are less well defined and the speckle pattern is much coarser.

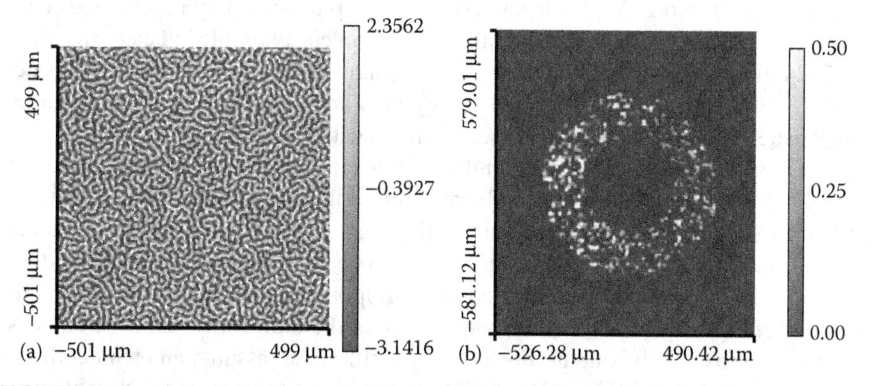

FIGURE 8.17 (a) An 8-level phase profile for the ring diffuser and (b) output intensity distribution with 0.2 mm waist radius source.

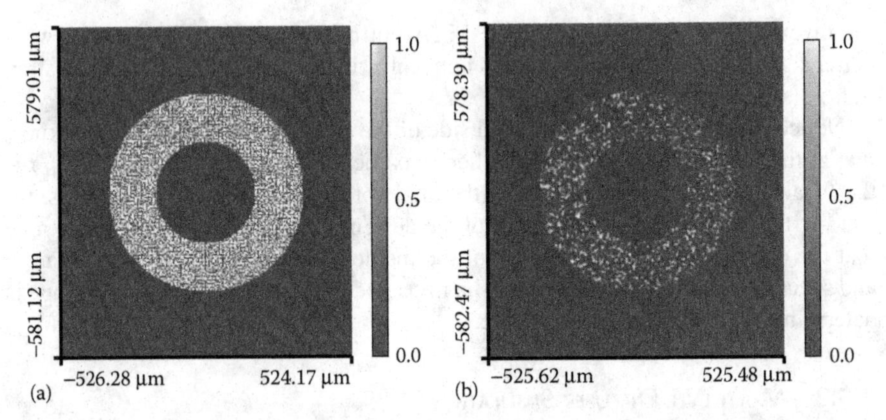

FIGURE 8.18 Simulated intensity plots of the 8-level ring diffuser. Both simulations use a 10 mm focal length Fourier transform lens in the calculations. (a) This simulation assumes an incident uniform source representing an incident beam somewhat larger than the phase function. The phase profile is assumed to be periodic, and the period size determines the diffractive order spacing. (b) This simulation used a Gaussian input beam of diameter 0.8 mm. The phase profile is no longer periodic, and the speckle pattern becomes more evident. Note that the output divergence angles are unaffected by incident beam size.

8.5.3 Diffuser Performance and Tolerancing

The final step in the design process is to tolerance it against anticipated fabrication limitations. As discussed earlier, some of the tolerances required fairly intensive computations to adequately simulate. Others are tied very closely to the exact fabrication method and tools. The one class of tolerances that can be simulated in a fairly straightforward manner is the feature depth or phase depth. Simulations of the diffuser performance over a range of feature depths also provide an immediate approach to quantify the performance variation due to changes in incident wavelength or angle, material variations, and other effects that impact the phase depth of the structure without distorting the individual features relative to each other.

For the ring diffuser example, we are particularly interested in diffraction efficiency, uniformity, and zero-order efficiency. The later will produce a bright spot in the center of the pattern. A quick calculation from the target intensity profile indicates that there are 7024 illuminated diffraction orders in the design. Thus, if the zero-order efficiency exceeds 0.014%, it will contain more energy than any single diffraction order in the target pattern. It is not uncommon to see diffusers with zero-order efficiency as high as 0.5%–1%, so this could become a significant concern in some applications. However, with diffusers, it is typically the combination of all the orders that is more important, and while the zero order may be significantly brighter than any individual diffraction order, its relative brightness is less apparent compared to the integrated output pattern.

Tolerance calculations assume a constant amplitude input field for ease of comparison. This will not fully capture the speckle or the uniformity, but it will at least be representative of the output. For the ring diffuser, we assume an etch depth tolerance of up to ±5% and plot the optical performance in Figure 8.19. Note that the

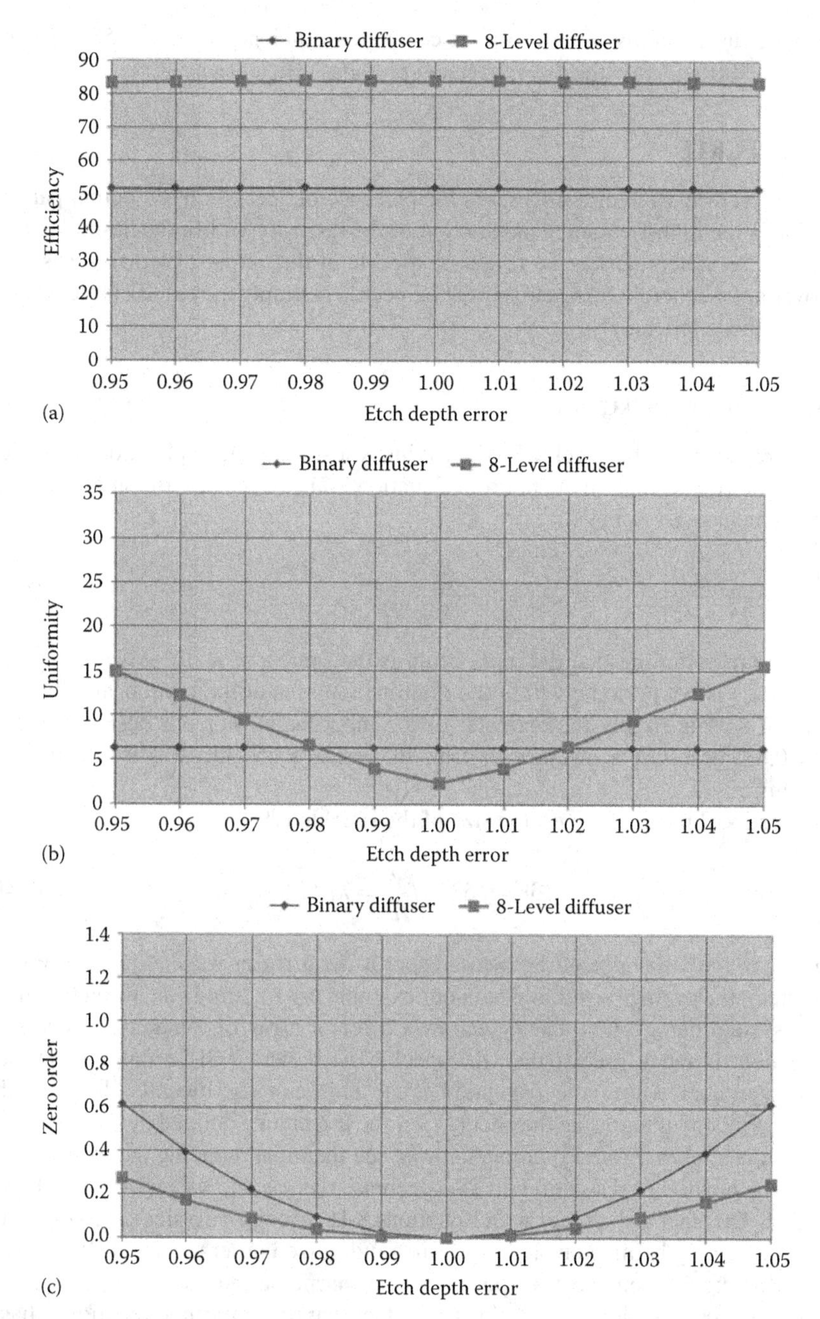

FIGURE 8.19 Fabrication tolerance simulations for the ring diffuser design example. All three simulations are completed over an etch depth range of ±5% from target and assume a constant input field. (a) Simulated diffuser diffraction efficiency for binary and 8-level designs. (b) Simulated uniformity error for binary and 8-level designs. (c) Simulated zero-order efficiency for binary and 8-level designs.

binary diffuser is only able to offer a calculated efficiency of about 50% for this source condition, while the 8-level version provides efficiencies >80%.

8.6 SPECKLE

The biggest drawback of a diffuser is the presence of speckle in the output pattern. Speckle is the high-frequency modulation of the intensity within the desired energy envelope. As noted earlier, the origin of speckle in the output of a diffuser is from the overlap of coherent diffraction orders. Speckle is simply the result of interference between those orders.

8.6.1 Size of Speckle

The average size of the speckle is of great interest since some applications can tolerate speckle if it is small enough. From Equation 8.21, we see that the smallest angular increment is given by[15]

$$\delta\theta \sim \frac{\lambda}{D} \tag{8.39}$$

Here we are assuming that the input beam is the same size as the calculation grid. If the phase is not periodic within the illumination area of the input beam, the D in Equation 8.39 is simply the diameter of the input beam. For our design example, $\delta\theta = 0.63$ mrad. At a distance of 1 m, the average speckle size would then be ~0.63 mm.

At the focal plane of a lens, the size of the speckle is[16]

$$\delta r \sim f\delta\theta \sim f\frac{\lambda}{D} = \lambda f/\# \tag{8.40}$$

which is roughly the size of the focused spot if the diffuser was not present. For an *f*/2 system, the average speckle size in our example is ~1.2 μm. If the resolution limit of the system being illuminated, such as a detector array or a material processing application, is much greater than the speckle size, there will be multiple speckle lobes integrated within the resolution area. This can significantly decrease the speckle effect by integrating the energy over the larger area defined by the resolution limit of the system. From Equation 8.40, we see then that reducing the focused spot size of the nondiffused system will also decrease the speckle size when the diffuser is added. This fact also agrees with Equations 8.11 through 8.16 since the amplitude $A(k_x,z')$ is simply being convolved with the randomized orders of the diffuser. Also notice that the frequency cut-off point is much sharper at the focus of a lens than in a free space propagation. This is due to the fact that the amplitude function A has a smaller diameter at focus. Thus, we see that if the application will allow the diffuse pattern to exist only on a particular plane along the optical axis, the focal plane of a lens generally gives the best results.

The number of speckle lobes within the desired pattern can be seen from Equation 8.23. For a solid pattern, such as a square or a circle, the number of speckle lobes across the diameter or along a side is simply ~$2N$. In our example, there are ~55 − 28 = 27 speckle lobes across the ring's width.

8.6.2 Speckle Reduction

The reduction of the amount of speckle modulation is accomplished by three basic techniques.[17] The first is to illuminate the diffuser with temporally or spatially partially coherent light. Since speckle is caused by coherent interference, then it stands to reason that reducing the coherence of the beam will also reduce the contrast within the speckle pattern. Contrast is defined as[16,17]

$$C = \frac{\sigma_I}{\langle I \rangle} \tag{8.41}$$

where:

σ_I is the variance of the intensity
$\langle I \rangle$ is the mean value of the intensity

For coherent light the contrast is equal to unity. The contrast of M mutually incoherent speckle patterns superimposed with the same wavelength and equal average intensities then becomes[16,17]

$$C = M^{-1/2} \tag{8.42}$$

From the above equation, we see that a laser with many incoherent modes will have lower contrast speckle than a single-mode coherent laser.

The second technique to reduce speckle is to time-average the speckle. This involves physically moving the beam or the diffuser very quickly and integrating the speckle pattern over a short period of time. This can work quite well for applications where the sensor, such as the human eye, a charge-coupled device (CCD) array, or a photographic film, has a finite exposure time. In general, any approach that can be used to break up the spatial coherence of the input can reduce the speckle contrast. One final technique is to spatial filter the speckle by observing the pattern through a finite aperture.

The subject of speckle is usually described in terms of the statistics of the speckle within the pattern. The details of the statistics of the speckle pattern are beyond the scope of this book. Excellent sources for further reading on the subject are found in the works of Hariharan,[16] Dainty,[17] and Goodman.[18]

8.7 APPLICATIONS OF DIFFUSERS

There is a wide variety of applications that demand diffractive diffusers. They are used to homogenize light sources, including broad band light sources in some cases (note that diffractive diffusers function with white light even though they are dispersive). They are used to illuminate specific regions for scanning applications. They are also used in alignment applications where a specific pattern is desired, such as laser targeting systems for firearms, machine tooling and assembly alignment systems, and even for space station to shuttle docking alignment. A grid diffuser pattern can be used to map the topography of a region. Diffusers can accompany illuminators for night vision systems, product marking systems, pen pointers with corporate logos or sports team logos, and laser light shows, and they can increase the viewing angle of a display. Diffusers could even be used in a wireless, free space interoffice communication network to reduce alignment requirements between components.

In general, the applications for diffuser beam shapers as opposed to remapping beam shapers depend on the system limitation and the application requirements. Diffusers should not be used in applications where speckle is not acceptable and cannot be reduced to a tolerable level. They should also not be used when collimation of the shaped pattern is a requirement. However, diffusers offer a better solution for applications where the input beam quality and/or system alignment capabilities are not sufficient for a remapping beam shaper. Factors that affect the input beam quality are the ability to measure the intensity and the phase of the beam, the stability of the beam with time, and the consistency of the beam from laser to laser. Remapping beam shapers can only shape one mode of the laser, and other modes in the input source will produce background noise. Diffusers will shape all modes of the laser, and an increased number of modes will actually lower the contrast of the speckle.

In addition, when one is using remapping beam shapers, any potential instability or variation of the input laser beam must to be evaluated to determine if the impact to the resulting output is acceptable. It is prudent to design the optical system with a spatial filter and methods to adjust the beam to better match the designed input beam. Alignment to the beam shaper is critical, and depending on the specifics of the beam shaper, the errors in alignment often have a multiplicative effect on the errors observed in the output. For example, a 2% translational misalignment in the beam shaper may result in a 10% tilt in the top hat.

Diffusers do not suffer from such requirements and can easily tolerate a range of input source conditions with minimal impact to the optical performance. They are, consequently, quite effective for high-power applications where the laser that is being used has a large number of mutually incoherent modes. Diffusers are relatively simple and customizable structures that offer a wide degree of flexibility for myriad applications.

REFERENCES

1. A.E. Siegman. *Lasers*. Mill Valley, CA: University Science Books, 1986.
2. G. Borek, D.R. Brown, D.M. Brown, and R. Clark. High performance diffractive optics for beam shaping. *SPIE Proceedings*, San Jose, CA, January 1999.
3. D. Brown and A. Kathman. Multi-element diffractive optical designs using evolutionary programming. *SPIE Proceedings*, San Jose, CA, February 1995.
4. H.A. Haus. *Waves and Fields in Optoelectronics*. Englewood Cliffs, NJ: Prentice Hall, 1984.
5. J.W. Goodman. *Introduction to Fourier Optics*. New York: McGraw-Hill, 1968.
6. P.P. Banerjee and T.C. Poon. *Principles of Applied Optics*. Boston, MA: Aksen Associates, 1991.
7. E.G. Loewen and E. Popov. *Diffraction Gratings and Applications*. New York: Marcel Dekker, 1997.
8. A.V. Oppenheim and R.W. Schafer. *Discrete-Time Signal Processing*. Englewood Cliffs, NJ: Prentice Hall, 1989.

9. M.H. Hayes and H. Stark (ed.) *Image Recovery: Theory and Application.* San Diego, CA: Academic Press, 1987.
10. D.L. Flannery and J.L. Horner. Fourier optical signal processors. *Proceedings of the IEEE* 77(10):1511–1527, 1989.
11. R.W. Gerchberg and W.O. Saxton. A practical algorithm for the determination of the phase from image and diffraction plane pictures. *Optik* 35:237–246, 1972.
12. D.M. Brown, D.R. Brown, and J.D. Brown. High performance analog profile diffractive elements. *SPIE Proceedings*, San Jose, CA, January 1999.
13. G.J. Swanson. Binary optics technology: Theoretical limits on the diffraction efficiency of multilevel diffractive optical elements. Massachusetts Institute of Technology, Lincoln Laboratory, Technical Report 914, 1991.
14. G.J. Swanson. Binary optics technology: The theory and design of multi-level diffractive optical elements. Massachusetts Institute of Technology, Lincoln Laboratory, Technical Report 854, 1989.
15. N. George. Speckle. *SPIE Proceedings*, San Diego, CA, July 29–30, 1980.
16. P. Hariharan. *Optical Holography Principles, Techniques, and Applications.* Cambridge, NY: Cambridge University Press, 1996.
17. J.C. Dainty. *Laser Speckle and Related Phenomena.* New York: Springer-Verlag, 1984.
18. J.W. Goodman. *Statistical Optics.* New York: Wiley, 1985.

9 Engineered Microlens Diffusers

Tasso R. M. Sales

CONTENTS

9.1 INTRODUCTION

Optical diffusers are often used to condition raw illumination and make it useful in a variety of applications. Some examples of conditioning include homogenization, controlled spread and distribution, and intensity shaping. Diffusers also help reduce sensitivity to variations in source properties, amplitude or phase, and provide a stable output beam. Common diffusers include ground glass,[1] holographic diffusers,[2] opal glass,[3] and volume diffusers.[4] With the exception of opal glass, a very inefficient Lambertian diffuser, other diffusers are notorious for spreading light with a Gaussian intensity profile in the far field. Therefore, unless one is specifically looking for a Gaussian profile, common diffusers find most applicability in homogenization and spreading, not beam shaping. (Here we will be concerned with surface diffusers so we will not have much to say about volume diffusers, of which opal glass is an example. Narrow-angle volume diffusers also tend to produce Gaussian scatter.) For general beam shaping, until recently diffractive optical elements (DOEs) were the only viable approach,[5] even though these are limited to monochromatic illumination and small divergence angles, unless a strong zero diffraction order can be tolerated.

Microlens arrays (MLAs; Figure 9.1), although not exactly diffusers, are also used in the homogenization of laser beams. Early work[6] suggested periodic MLAs in combination with prismatic elements to generate uniform illumination for holography. Regular MLAs have been combined[7] with a focusing lens and some defocus to obtain a flattened focal profile. A similar approach has been used[8] with an array of

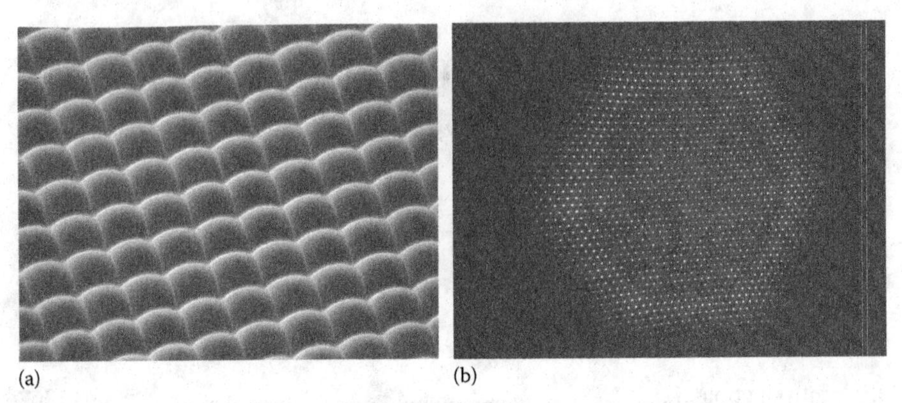

FIGURE 9.1 (a) Square microlens array. (b) Example of a diffraction pattern from a microlens array with hexagonal geometry.

binary diffractive lenses. As a beam shaper, an MLA samples the input beam within the scale of each microlens unit and then overlaps the individual contribution of each element, thus the term "beam integrator."[9] In the far field, because of their periodic nature, MLAs give rise to a characteristic diffraction pattern that results from the regular arrangement of the lenses (see Figure 9.1 for an example). For this reason, in many applications a random diffuser is used in combination with the periodic MLA to wash out the diffraction pattern and minimize color and moiré fringing, even though it may lead to some loss in resolution, particularly for screen applications. MLAs can also only provide limited beam shaping capabilities due to the need for 100% fill factor, if one is to maintain high efficiency and no zero order. Most commercially available MLAs present a spherical profile and are made by thermal reflow[10] of photoresist or direct machining, which further limits their diffusing or beam shaping capabilities. Aspheric profiles can be achieved in some cases by means of alternative processes such as reactive-ion etching.[11] In spite of their limitations, MLAs do provide a working, if not ideal, solution to a problem that common diffusers cannot solve, that is, producing uniform illumination over a defined angular domain.

Diffusers that uniformly distribute illumination over an angular range or at a plane are of great practical interest.[12] Depending on the particularities of the system, different approaches need to be developed to provide uniform illumination without sacrificing efficiency. Lithographic illumination systems often rely on fly's eye MLAs to generate uniform illumination,[13] sometimes in combination with additional diffusers and/or motion to minimize the diffraction artifacts induced by the periodic lens arrays; laser displays make use of optical diffusers[14] for shaping and uniformly light distribution as well as speckle management. Several approaches have been proposed to solve this problem using single-surface diffusers but only limited success can be claimed,[15] due to either the design approach or the lack of an appropriate manufacturing technology.

Illumination confined to a well-defined angular region is termed "band-limited illumination" and a diffuser capable of producing such illumination pattern is known as a band-limited diffuser.[16] The ideal band-limited diffuser[17] spreads all available energy within the desired angular range, the target, only limited by diffraction. Immediately outside the target there is a falloff region where the intensity drops toward zero. Unless intentionally designed in, the amount of energy within the falloff region is mostly dictated by diffraction and is therefore dependent on the structures that define the diffuser. As a result, the ideal band-limited diffuser is one whose falloff region is limited only by diffraction with nearly zero energy outside of it. The total fraction of light scattered within the target region represents the target efficiency. The ideal band-limited diffuser maximizes target efficiency with the only losses coming from Fresnel reflections and diffraction-induced broadening in the falloff region.

Until recently, probably the best example of a band-limited diffuser was provided by the diffractive diffuser.[13] While capable of creating uniform diffuse patterns with general distributions, diffractive elements cannot prevent a certain fraction of the incident illumination from spilling outside of the target. With binary elements, for example, at least 20% of the light is lost to higher diffraction orders outside the target. Continuous-phase diffractives are more efficient but, still, about 5%–10% is lost to higher diffraction orders. (Note that these figures do not include Fresnel losses.) We can therefore say that, while a diffractive diffuser is able to provide uniform illumination, it is not band limited, although the continuous-phase element comes very close. Common diffusers, refractive in nature, are fundamentally free of the intrinsic losses associated with high-order diffraction and zero order and would seem ideal candidates for band-limited behavior. However, the typical examples of ground glass and holographic diffusers generate Gaussian scatter and, therefore, do not have a uniform scatter region or well-defined cutoff for the diffuse light and are thus not band limited.

Over the last decade or so, a novel approach to achieve beam shaping, diffusion, and homogenization based on arrays of randomized microlenses has been developed[18] that combines the homogenization capabilities characteristic of random diffusers, such as ground glass, and the beam shaping capabilities of diffractive elements but without their drawbacks. An engineered diffuser constitutes an ensemble of microstructures, generally microlenses, where each individual element is designed and fabricated to produce a controlled scatter pattern. The ensemble is generated and randomized according to well-defined rules and probability distributions, but the engineered diffuser is best described as a deterministic optical element where the aggregate of microlenses and their individual optical prescriptions are precisely defined in a completely repeatable manner, contrary to a random diffuser which by construction is only repeatable in a statistical sense. The implication is that engineered diffusers enable the control of both energy distribution and intensity profiles. In particular, engineered diffusers have made possible the production of diffusers that come very close to the band-limited ideal. This chapter covers the concept of engineered microlens diffusers describing in some detail design aspects and illustrating performance features. The organization is as follows. Section 9.2 discusses the general concept of engineered diffusers. Section 9.3 covers typical configurations in optical systems. Section 9.4 explores the general design rules and parameters involved in the generation of engineered diffusers. Section 9.5 shows the use of engineered diffusers to efficiently produce controlled intensity profiles. In Section 9.6 we discuss the modeling of engineered diffusers and present a framework for designers to incorporate these elements in commercial ray-tracing software. Section 9.7 summarizes the main results.

9.2 THE CONCEPT OF ENGINEERED DIFFUSERS

An engineered diffuser is a random assembly of microstructures designed to perform some beam shaping function. It differs from other beam shaping approaches in that each microstructure is individually designed and grouped deterministically to create a unique diffuser component. The random nature of the engineered diffuser enables not only the elimination of diffraction artifacts but also the production of very general light distributions with high efficiency, limited only by surface (Fresnel) losses. It also makes the diffuser robust to variability in fabrication parameters and insensitive to input beam conditions, particularly when compared to diffractive elements or MLAs. The precise definition of each microstructure permits the control of intensity profiles, particularly when compared to statistical diffusers, which are generally confined to Gaussian scatter. Some of the features of engineered diffusers are listed below:

- No zero order
  - Contrary to DOEs operating at wavelengths other than the design or MLAs that attempt to produce certain patterns, such as circular diffusion
- Absence of image artifacts
  - Random speckle, as opposed to that seen with MLAs and the inevitable high-order diffraction losses of DOEs
- Absence of color effects and moiré fringing when used as screens
  - See Figure 9.2 for an example.
- Controlled angular spread and light distribution
  - DOEs become more challenging to manufacture as the angular spread becomes wider.
- Insensitivity to input variations
  - DOEs or MLAs with large feature size are more sensitive to variations in amplitude and phase of the input illumination.
- Robustness to fabrication errors
  - Variations on the local properties of the microstructure of random diffusers have a limited effect on performance. Contrarily, lens aberrations or profile variations have a direct impact on the performance of MLAs and DOEs.

The engineered diffuser incorporates the robustness and homogenization capabilities of random diffusers with the beam shaping capabilities of DOEs and MLAs while enabling performance features that cannot be provided by either approach. An engineered diffuser takes into consideration all aspects that define the diffuser in a deterministic fashion by integrating the various available design parameters to achieve a certain beam shaping goal. The process begins with the specification of the desired beam shaping pattern, in terms of both light distribution and intensity profiles in the

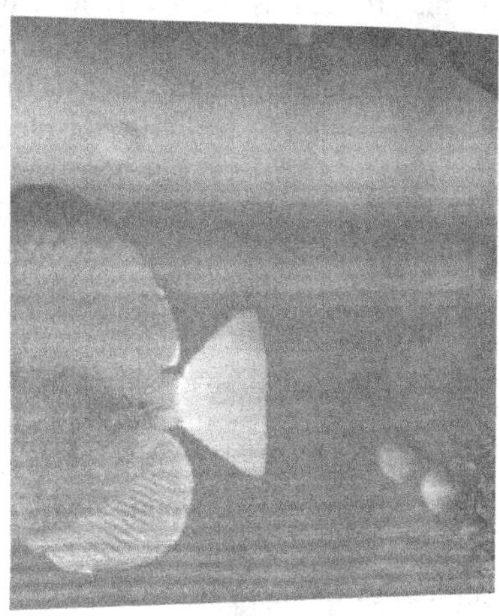

FIGURE 9.2 Image projected through a periodic microlens array (right) and an engineered diffuser (left). Moiré fringing and color artifacts are introduced by the periodic screen, absent with the engineered diffuser.

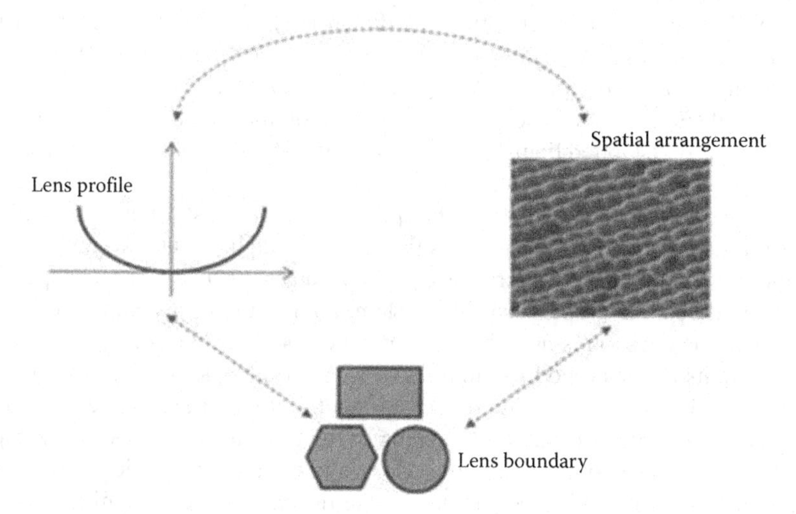

FIGURE 9.3 Elementary components required to create an engineered diffuser: microlens profile, the lens boundary shape, and spatial arrangement of an ensemble of lens elements.

far field. Common distributions include circular, linear, rectangular, and elliptical patterns. Common intensity profiles include flattop and batwing (more intensity toward wider angles compared to the center) for uniformity illuminating flat surfaces. The engineered diffuser is defined by three basic elements (Figure 9.3): the sag profile of the microstructures (typically microlens-based), their boundary shape, and the spatial arrangement of the microstructures. Each element affects diffuser performance and must be taken into account together with the other parameters. Critical to the success of this program is the ability to manufacture microstructures with accurate surface profiles according to a general surface prescription, down to fractions of microns. Some alternative methods include direct machining[19] and lithographic processes.[20]

In Section 9.3, we will consider in more detail some aspects in the design of engineered diffusers, but here we outline how the basic design elements interact and affect performance. As pointed out previously, a periodic distribution of microstructures leads to significant light structure (the diffraction pattern) and color effects (a periodic structure is basically a diffraction grating). Randomization has the advantage of minimizing or eliminating these artifacts but also makes available previously unavailable benefits. First, it introduces a new degree of freedom by allowing microstructures to be combined or placed arbitrarily in controlled proportions. Also, surface slopes can be defined with limited local constraints, which allow general beam shaping to be implemented because the arbitrary slope histograms become possible, at least in principle. The price paid by the additional degrees of freedom is the considerably greater computing power required to design each microstructure and generate the diffuser surfaces. For most applications, however, the current state of the art in computer speed, memory, and storage has proven adequate.

While there is a variety of randomization algorithms, a simple way to gain some understanding into the degrees of freedom as well as the constraints and effects of a random distribution is to start with a one-dimensional periodic MLA and consider ways to randomize it. An example will help to illustrate the concept (Figure 9.4). A one-dimensional periodic MLA is given and the middle element is selected to be modified (Figure 9.4a). The types of modification that can be implemented here include sag profile (dotted line), horizontal shift, vertical shift, and rotations, as indicated by the various arrows. (In two dimensions, one could also consider changes to the microlens boundary.) Of course, the specific type of transformation a particular lens undergoes depends on what one is trying to accomplish, but it seems clear that any modification to a given microlens element changes the histogram of slopes and, therefore, affects the properties of the system, both in the near field and in the far field. Now, the question is how the transformation of a single element affects other elements in its neighborhood. Of all possible transformations, consider the case of a horizontal shift to the right (Figure 9.4b) where the original unmodified microlens is depicted by the dashed line. The shifting of the lens creates on one side a gap in the array and, on the other side, an overlap with a neighboring microlens. A possible approach to deal with the gap is to make the lens larger to eliminate the gap (Figure 9.4c) so that no portion of the input beam is transmitted undeflected. On the other side, where lenses intersect, one might consider several alternative possibilities (Figure 9.4d–f). The profile of the shifted lens might prevail, the profile of the neighbor lens might prevail, or maybe none of these options with some portion of both microlenses prevailing. In either case, the histogram of slopes is modified and, depending on what the beam shaping requirements might be, these transformations need to be compensated or complemented by transformations performed on other lenses in the array. For a small number of lenses, the final goal may be hard to achieve but typically at least hundreds of elements are illuminated and it becomes usually possible to come fairly close to the target requirements.

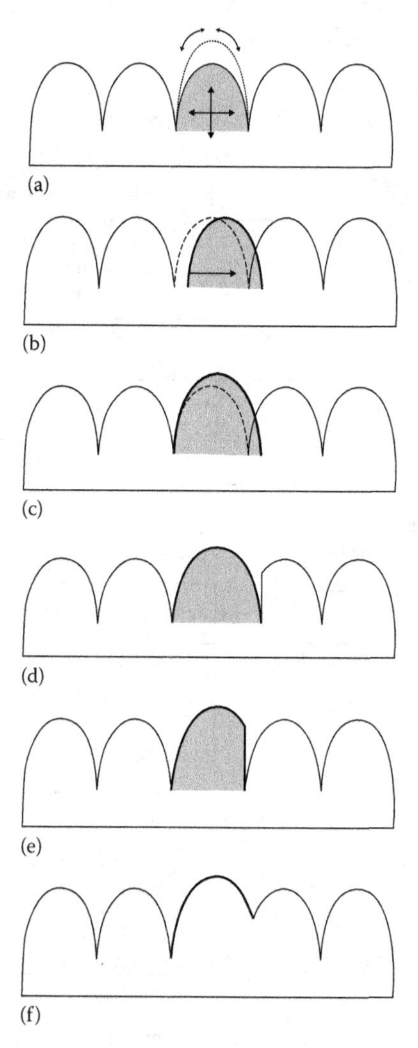

FIGURE 9.4 Example of the randomization of a single lens element (center) in an initially periodic array. (a) Available degrees of freedom include sag profile (dotted line), translations, and rotations. (b) Center element is shifted to the right (dashed line: original microlens). (c) To eliminate gap at the left created by the shift, the microlens diameter, and therefore sag, is increased. (d) Possible approach to deal with the overlap with the lens to the right: the sag of the center lens prevails. (e) Another approach: the sag of the lens to the right prevails. (f) Yet another approach: portions of both lenses are used.

A general algorithm for the design of engineered diffusers is shown in Figure 9.5. The beam shaping requirements determine the appropriate design variables, such as microlens prescription parameters, and rules for intersecting microstructures, as illustrated in Figure 9.4, as the available diffuser space is occupied with scattering elements. The chosen algorithm is thus implemented repeatedly until the diffuser surface is completely defined. An evaluation step then follows with comparison to desired targets, based on ray-tracing or diffraction theory, whichever is appropriate.

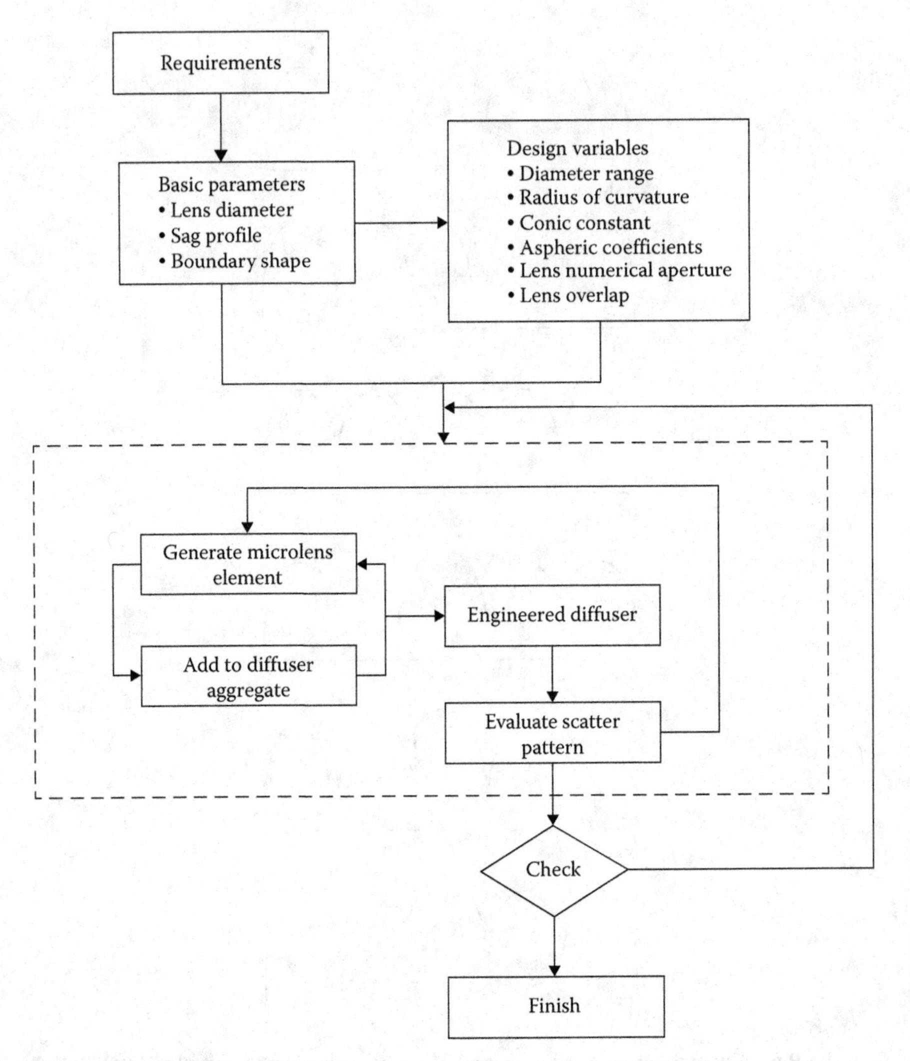

FIGURE 9.5 Basic algorithm for the creation of an engineered diffuser. Once design variables are defined, the main iteration loop consists of generating microlens elements and populating the diffuser aggregate.

This could be the final step in the design process or may require a restart of the design cycle, depending on the results of the evaluation. It is worth noting that in most cases the design parameters and algorithms are not unique, implying a large number of local minima in the space of solutions, but making more difficult the identification of the global minimum solution.

Note that in the approach illustrated above, little attention is paid to the diffuser near-field behavior. The reason is that in most cases, one is interested in energy distribution or intensity profiles in the far field and, therefore, the particular way the microlenses focus the incident beam is inconsequential to the performance of the diffuser, although an analysis of the near field does shed light on the far-field properties of the diffuser, but this is outside the scope of the present discussion. We note, however, that there is one case where the near field is of critical importance, and that is the case of a fly's eye beam shaper[9] where two engineered diffusers are used in tandem with the first array placed near the focus of the second array. In the present discussion, however, we concentrate on single-surface diffusers where near-field behavior is not critical to the design of the engineered diffuser.

In Section 9.3, we briefly review typical operating conditions for engineered diffusers.

9.3 TYPICAL CONFIGURATIONS

The most common mode of operation using engineered diffusers involves a source that provides the input illumination, the engineered diffuser to shape the input beam, and the output beam following propagation through the diffuser, or after reflection, generally in the far field. In the simplest setup, Figure 9.6a, the incident illumination

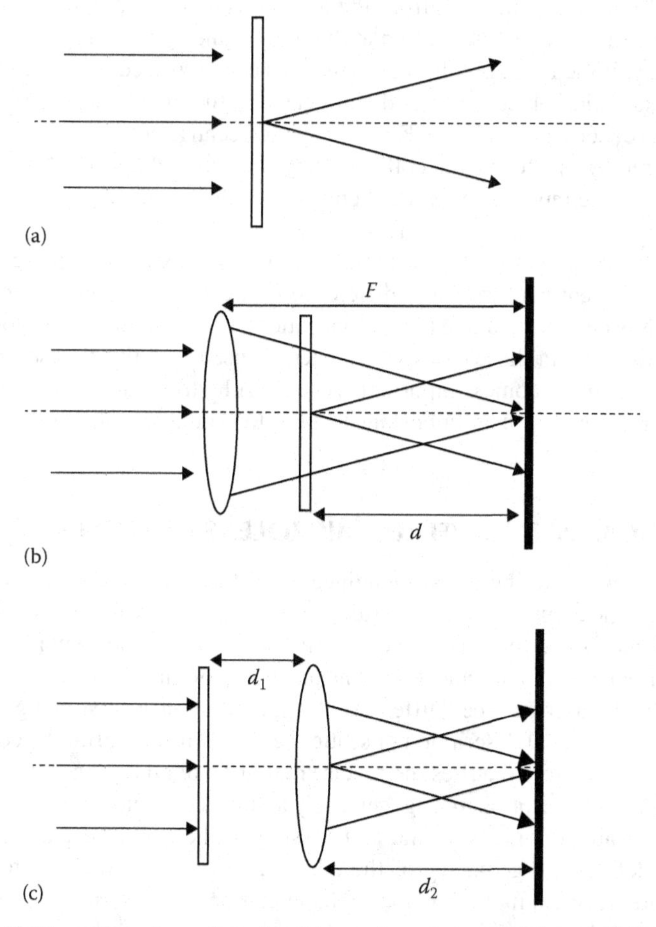

FIGURE 9.6 Typical configurations for the use of engineered diffusers. (a) Free propagation. (b) A lens is used before the diffuser. (c) A lens is used after the diffuser.

hits the substrate from the diffuser side of the substrate, with the scatter pattern being observed at a sufficiently large distance from the diffuser, in the region generally denoted as the far-field or the Fraunhofer region of diffraction.[21] If the diffuser angle is small enough, it is of little consequence if the incident beam hits the diffuser side or the unpatterned side first, but as the angle increases, incidence from the diffuser side is preferable for best efficiency. Also, if the engineered diffuser is used for beam shaping, the input should generally be collimated, particularly if the scatter pattern has sharp features. In some cases, it is possible to take into account the degree of collimation of the source, but generally a collimated input provides the sharpest falloff.

An equivalent, but more compact, configuration is to utilize a focusing lens to bring the far-field pattern to the focal plane of the lens[22] (Figure 9.6b). This geometry allows one to utilize the diffuser over much shorter distances compared to free propagation. For that reason, the diffuser output can also exhibit relatively sharper features. The diffuser is located at distance d from the observation plane, which is at the focal plane, a distance F from the lens. This configuration can also be used to tune the size of the diffuse pattern at the focal plane by varying the separation d. Alternatively, if the input is collimated, the lens can be placed between the diffuser and the observation plane, separated from each by the distances d_1 and d_2, respectively (Figure 9.6c). If $d_1 = d_2 = F$, we have a telecentric configuration[21] where the light scattered by the diffuser is collimated by the lens and delivered parallel to the optical axis, if the input beam is small enough. As the beam size increases, the illumination angles at the base focal plane also increase.

Within the range of validity of the Fraunhofer approximation, these configurations are equivalent with the main difference between them being whether the diffuse pattern is defined in angle (free propagation) or coordinate space (focal plane). Note, however, that in those cases where a lens is used with the diffuser, the quality of the lens design becomes important. A lens with strong aberrations may affect the diffuser output, so it is important to take that into consideration when using this setup.

9.4 DESIGN OF ENGINEERED MICROLENS DIFFUSERS

The elementary scattering units of engineered diffusers are generally microlenses, and to define the entire diffuser structure, we need to specify three essential properties: functional form or prescription of the microlens sag function, geometrical shape of the microlens boundary, and spatial arrangement of an ensemble of microlenses to create the diffuser surface. Differently from random diffusers, like ground glass where only a statistical description of scatter centers is meaningful, the construction of engineered diffusers requires the precise definition of all design parameters.

The surface prescription, or sag, generally assumes the conical form defined by a radius of curvature, conic constant, and possibly aspheric coefficients. The boundary shape defines the geometry of the edges that limit the spatial extent of each scatter center. For example, boundary shape can be circular, square, rectangular, or any general shape or, when considered in an aggregate, combinations of shapes. Finally, the spatial arrangement dictates how scatter centers are placed on the diffuser surface and what happens when lenses overlap, similar to the illustration in Figure 9.4. Each and all of these components that define the engineered diffuser have an effect on its scatter properties and are characterized by probability distribution functions that govern their statistical properties. In what follows, we discuss each component individually and illustrate their significance to the performance of the diffuser.

There are several ways to define the microlens sag function but, for simplicity, we consider the case of a microlens characterized by a radius of curvature R and a conic constant κ, which is given by

$$y = \frac{(r-r_c)^2}{R^2 + \sqrt{R^2 - (\kappa+1)(r-r_c)^2}} + \sum_k a_k (r-r_c)^k \tag{9.1}$$

where:

r designates a coordinate point on a local coordinate system associated with a particular lens element
r_c represents a decenter parameter from the origin
a_k denotes aspheric coefficients

The sag function (Equation 9.1) through its various design parameters is mainly responsible for controlling the far-field intensity profile. For simplicity, we ignore the aspheric coefficients in the present treatment but, depending on the beam shaping requirements, all design parameters may in principle be required to implement some beam shaping functions. To illustrate the basic design rules, we consider the case of the engineered diffuser where each microlens element scatters an input collimated beam with a constant angular spread, thereby fixing the radius of curvature for a certain lens diameter. The remaining design parameters are then diameter, conic constant, and decenter (also microlens boundary in the more general two-dimensional case).

The average microlens diameter defines the diffuser feature size, which in some applications can be a critical design parameter, for example, in illumination systems projected through pixelated devices where the diffuser feature size needs to be several times smaller than the pixel size. An example is the case of projection screens. These are thus the factors to consider in this regard: resolution, total sag, and averaging. Once a certain diameter is selected according to resolution requirements then, to ensure the best uniformity, the largest number of microlenses should be illuminated to provide sufficient averaging, that is, the lens diameter should be small relative to the input beam size. At the same time, given a microlens sag prescription the lens depth decreases as the diameter decreases. Larger microlenses imply deeper sag, whose maximum allowed value is established by the manufacturing process.[20] If the microlens element is so small that the total sag only imparts a phase delay that is on the order of 2π at a certain wavelength of interest, one should expect degradation in performance. In this respect, it is useful to define the phase number

$$M = \frac{y_{\max}}{(\lambda/\Delta n)} \tag{9.2}$$

where:

y_{max} represents the total lens sag in the nomenclature of Equation 9.1
λ is the wavelength under consideration
Δn equals $n(\lambda) - 1$, with n the index of refraction at wavelength λ, for a diffuser in air

The phase number basically expresses the total sag in the language of phase cycles and defines the regime, diffractive or refractive, the microlens operates on: $M = 1$ implies a diffractive element with exactly 2π phase shift. For a microlens to operate in the refractive regime providing achromatic behavior and high target efficiency, the phase number M should be as large as possible. A simple rule of thumb to help decide the minimum feature size or lens diameter to utilize is given by the following relation:

$$D \geq 230M\frac{\lambda}{\theta_0} \tag{9.3}$$

where θ_0 is the half-width diffuser angle in degrees where, to ensure one is safely in the refractive regime, M should be at least 8, preferably more. It should be noted that Equation 9.3 only applies to parabolic profiles and angles no larger than about 20°–30°, strictly speaking. However, it is useful in providing a starting point for more accurate calculations. At this stage, it is instructive to illustrate these concepts with some examples. Consider a single parabolic microlens element, $\kappa = -1$, that spreads an incident collimated beam into a 40° cone, ±20°. The scatter patterns for diameter values 1000, 100, 10, and 1 μm are shown in Figure 9.7. The dotted line represents the diffraction-based calculation, whereas the solid line is the result of ray-tracing or the geometrical optics performance. As expected, under coherent illumination one notes the intensity oscillations, a clear signature for the presence of diffraction effects but without 100% modulation, indicative of single-lens diffraction. The ray-tracing curve is obviously the same independent of the lens size with the geometrical

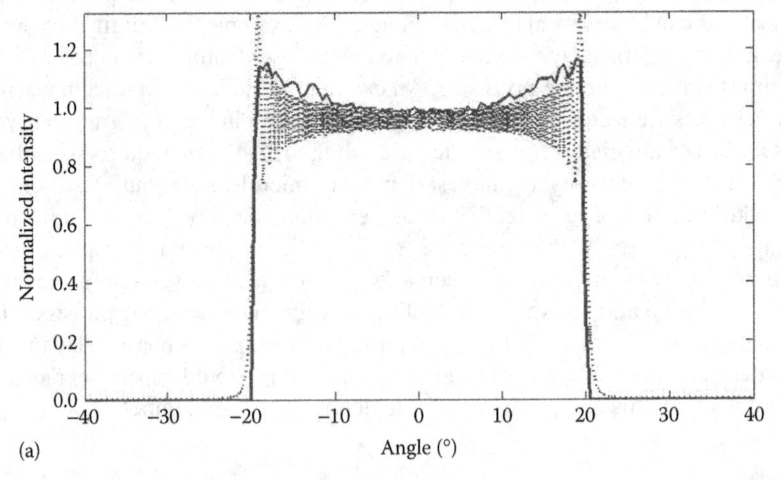

FIGURE 9.7 Scatter pattern due to a single microlens element with diameter (a) 1000 μm,

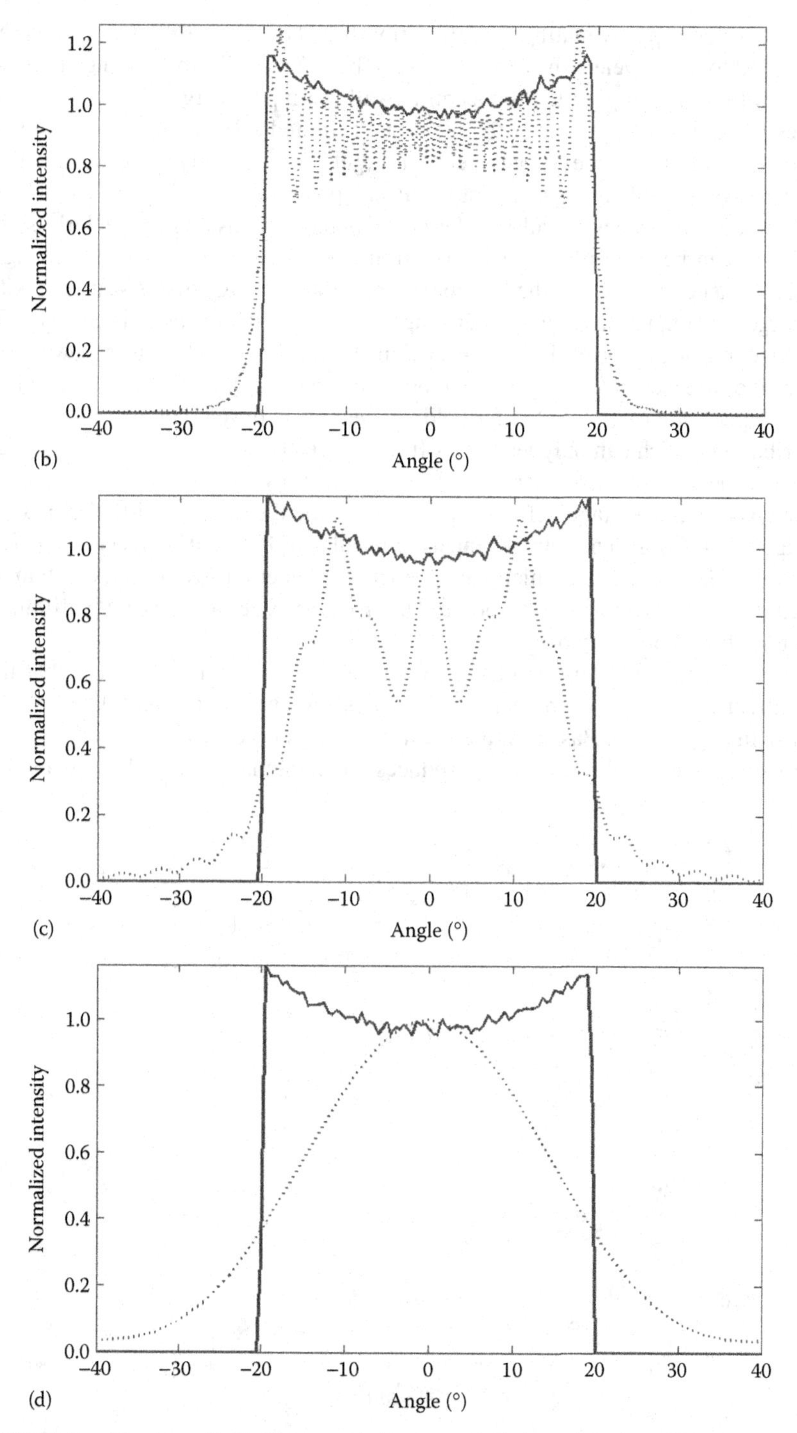

FIGURE 9.7 **(Continued)** (b) 100 μm, (c) 10 μm, and (d) 1 μm based on diffraction (dotted line) and ray tracing (solid line).

optics envelope approximating the diffraction-based profile as the lens size increases compared to the wavelength, assumed here to be 633 nm. The ray-tracing curve does indicate the target region that concentrates all the input energy in the geometrical optics limit. The only losses of the ideal band-limited diffuser are those due to the diffraction falloff outside the geometrical target. Consequently, to maximize target efficiency one should use the largest beam size possible.

The target efficiency calculation based on parabolic lenses is particularly significant as it can be seen as a fundamental limit for microlens-based diffusers. Again, target efficiency is given by the fraction of transmitted energy that is scattered within the target region, usually the angular range over which the intensity is constant. The sag function of the form given by Equation 9.1 can be expanded in a power series where the first element is that of a parabolic lens plus higher order terms. The effect of the higher order terms in the far field is given by a convolution with the parabolic contribution, which can only lead to its further spread. As a result, for an engineered diffuser, the best possible target efficiency is given by the use of parabolic microlenses. The estimated target efficiency for band-limited engineered diffusers is shown in Figure 9.8. Consistent with the intensity profiles of Figure 9.7, higher efficiency is found with larger values of microlens diameter or wider angles, for a fixed diameter. Small-angle diffusers generally require larger feature sizes and large input illumination for efficient beam shaping.

To complete the definition of the elementary microlens unit, we note that there is a direct relation between shape of its boundary and the far-field distribution of energy in angle space. Basically, a circular microlens produces a circular scatter pattern while a rectangular microlens produces a rectangular pattern. This relationship

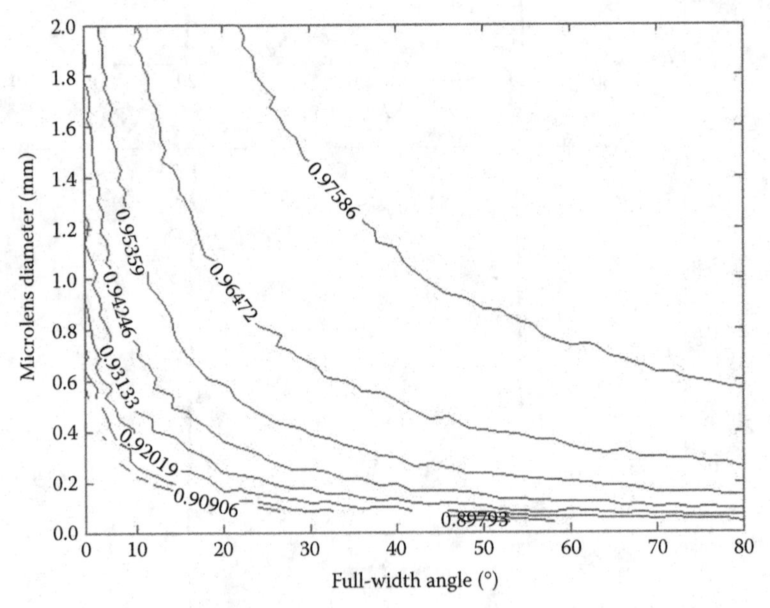

FIGURE 9.8 Target efficiency (fraction of scattered energy within the region of uniform intensity) as a function of full-width scatter angle and microlens diameter.

follows directly from diffraction theory, and while it is possible to violate it and have, for example, a microlens square aperture produce a round scatter pattern, we are currently interested in structures that lead to uniformly distributed scatter patterns and those are typically originated from matching patterns between far-field energy distribution and boundary shape. In other words, microlenses (circular, elliptical, square, rectangular, etc.) are naturally suited to produce scatter patterns (circular, elliptical, square, rectangular, etc.), even though this condition is only sufficient and not necessary.

The conic constant allows one to adjust the intensity profile from flat-top to Gaussian-like profiles (positive values of κ) to batwing profiles (negative values of κ). If a specific dependence of intensity against angle is required that cannot be properly generated by the sag function given by Equation 9.1, one may need to resort to the aspheric coefficients. Examples are shown in Figure 9.9 for a microlens element with $D = 100\ \mu m$, spread angle ±20°, refractive index 1.5, and wavelength 633 nm.

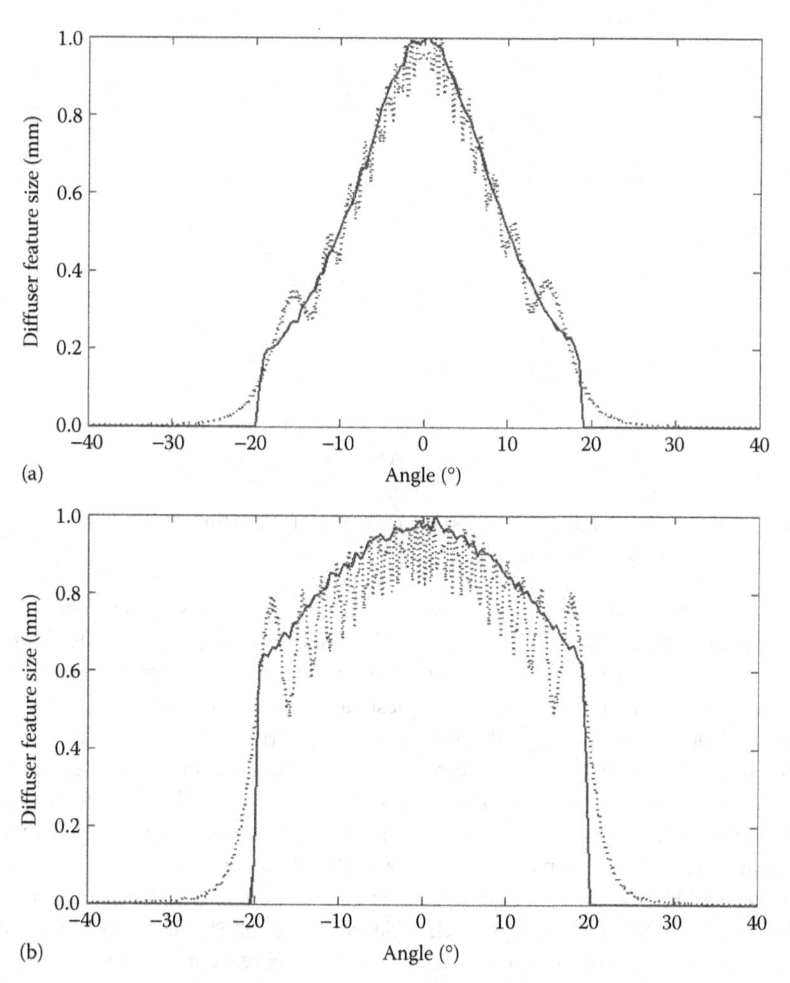

FIGURE 9.9 Effect of conic constant κ on scatter profile for a conic lens: (a) $\kappa = 4$, (b) $\kappa = 0$,

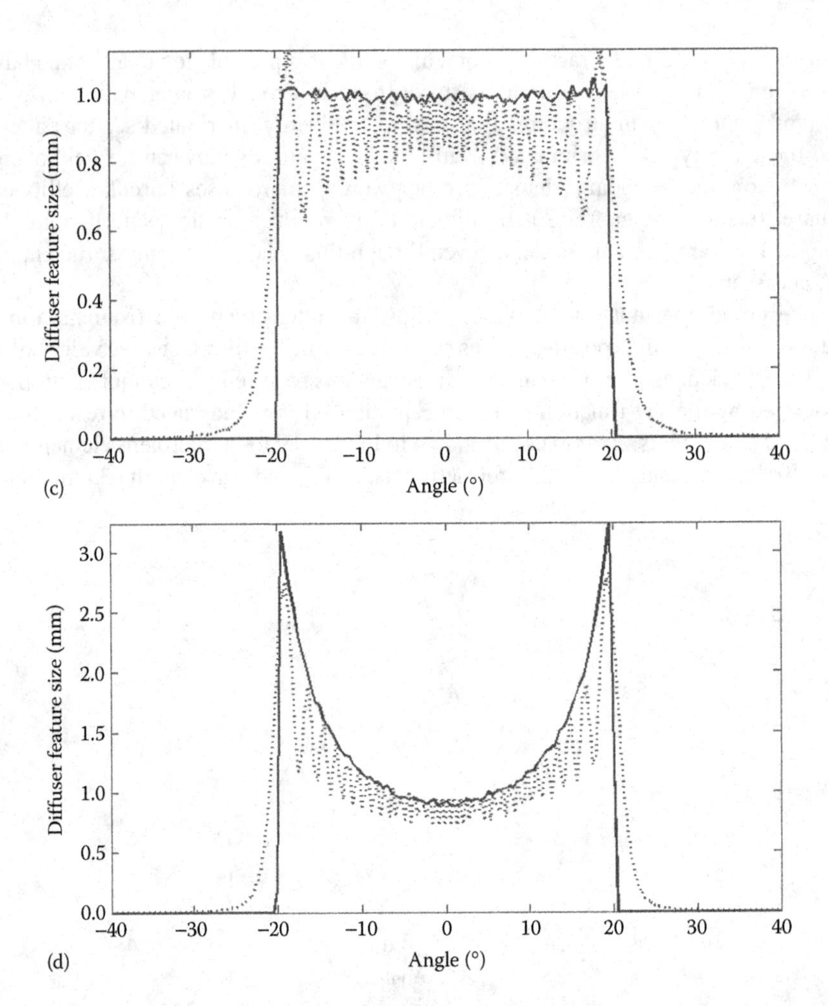

FIGURE 9.9 **(Continued)** (c) $\kappa = -0.8$, and (d) $\kappa = -2$. Dotted line: diffraction-based result. Solid line: ray tracing.

The decenter parameter finds an important role particularly under coherent laser illumination to help minimize intensity artifacts due to diffraction and interference of the various microlenses in the ensemble. The maximum amount of decenter is set as a fraction of the diameter and in the simplest case applied to every microlens element with a uniform distribution, both positive and negative values. Going from a single scatter center to an ensemble of microlenses, one notices the emergence of speckle,[23] that shows up as soon as more than one scatter center are illuminated. Speckle is unavoidable when illuminating any structure whose feature size is smaller than the coherence area of the source. In some applications where detection occurs over an area that includes several modulation cycles, the presence of speckle may not pose problems. For other applications, such as laser projection,[24] speckle is objectionable and measures need to be taken to reduce it to a level where it cannot be perceived by an observer. This is usually accomplished using diffuser motion to average out speckles.

The following example should shed some light on the effect of decenter on uniformity. Consider a single one-dimensional lens element with angular spread ±20°, diameter $D = 100$ μm, conic constant $\kappa = -0.8$, refractive index 1.5, and wavelength 633 nm, similar to previous examples. We now calculate the far-field diffraction pattern due to a combination of two lenses only, one with decenter $r_c = 0$ and another with $r_c = 0.05D$. The results are shown in Figure 9.10 for incoherent and coherent addition of the diffraction patterns. In either case, one notices an effective reduction in the total range of intensity fluctuations within some range of the scatter pattern. In the case of coherence addition of complex amplitudes, speckle is averaged over 1° intervals for this particular example. As more lenses are added and the decenter parameter is randomly applied to the various lenses in the array, an overall improvement in uniformity is observed over the entire angular region of interest. The effect is illustrated with

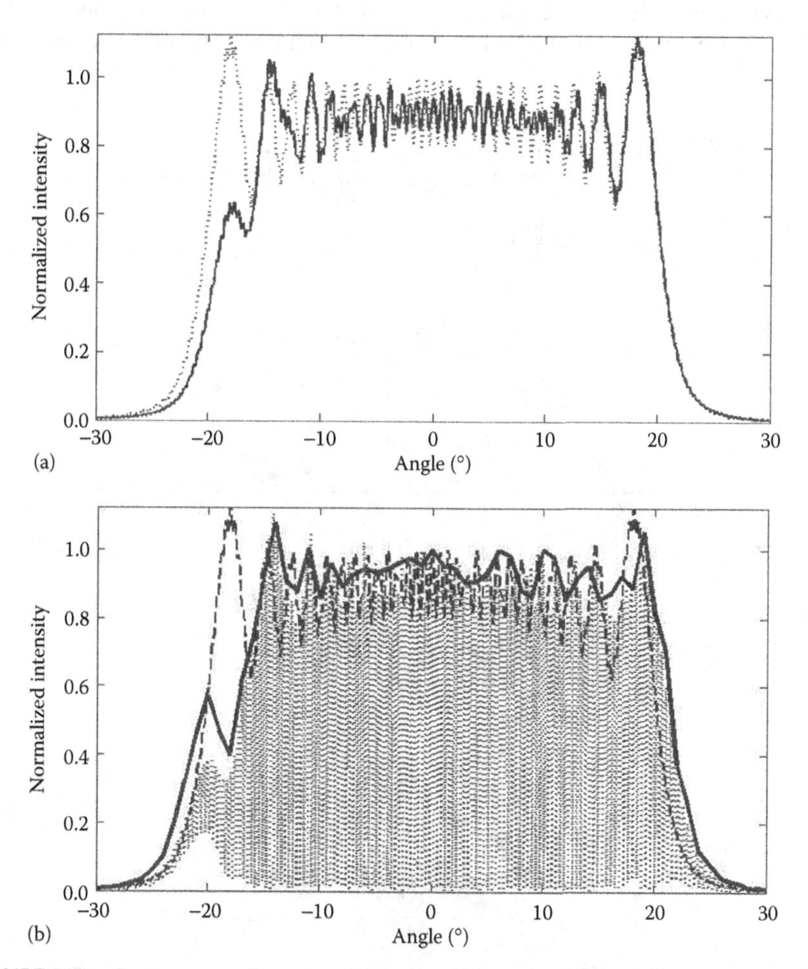

FIGURE 9.10 Scatter pattern from the combination of two microlenses of diameter 100 μm and decenter values 0 μm and 5 μm decenter resulting from (a) incoherent sum of intensities (dotted line: 0 μm, solid line: 5 μm) and (b) coherent sum of complex amplitudes (dashed line: 0 μm; solid line: 5 μm, total intensity with speckle; dashed line: 5 μm, speckle averaged over 0.5°).

an engineered diffuser with microlens diameter in the range of 90–110 μm and same parameters as above. The detector angular range is assumed to be 0.5°, over which speckle is averaged. The results are shown in Figure 9.11, where the dotted line indicates the diffraction-based speckled diffuse calculation, whereas the thick black curve is the result of ray tracing. The thin black curve represents the speckle pattern averaged in steps of 0.5°. The decenter helps minimize the strong fluctuations due to diffraction and brings uniformity closer to the geometrical optics performance.

Once the design parameters are established, the engineered diffuser is constructed by randomly populating a grid with the microlens elements. In its simplest form, the diameter is taken from a range of values, typically within 10%–20% from the average, with a uniformly distributed probability for the spatial distribution of microlenses. Overlap between microlenses as the ensemble is created can be dealt with in a variety of ways similar to the scheme illustrated in Figure 9.4. The process continues until the whole grid is populated with microstructures. A scanning electron microscope (SEM)

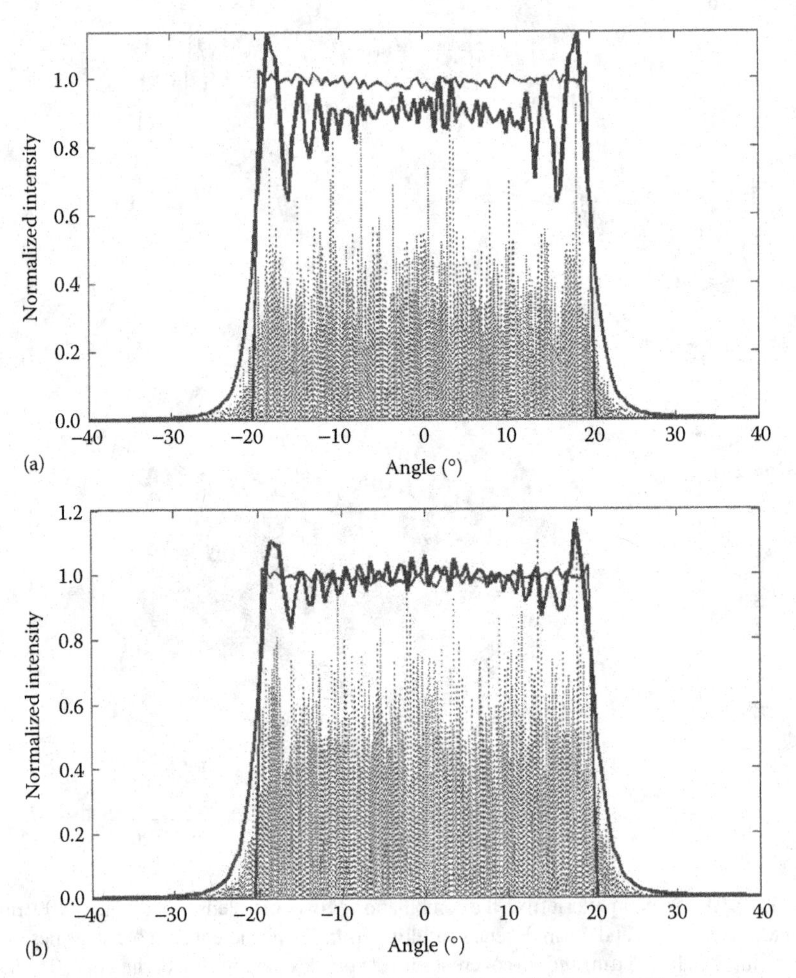

FIGURE 9.11 Effect of decenter values (a) 0%, (b) 4%, (c) 5%, and (d) 6% on scatter profile.

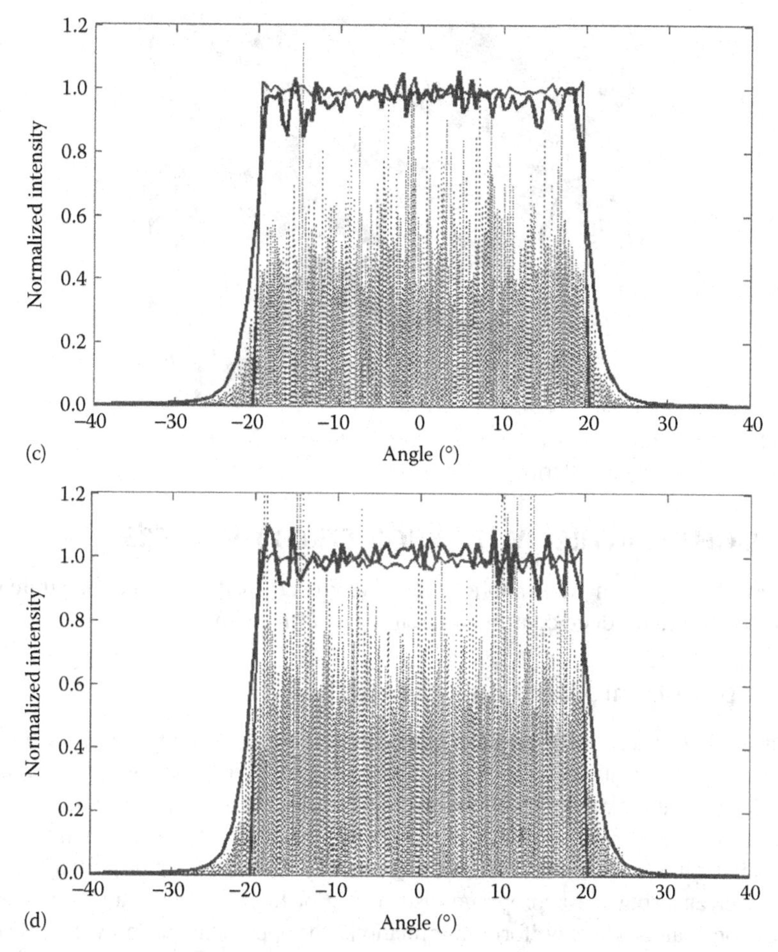

FIGURE 9.11 **(Continued)** Dotted line: diffraction-based intensity profile; thick black curve: diffraction-based profile with speckle averaged over 0.5°; thin black curve: geometrical-optics intensity profile.

picture of an engineered diffuser surface structure that produces a circular pattern is shown in Figure 9.12. For best uniformity, a large number of microlenses should be illuminated to provide sufficient averaging, otherwise one might see strong intensity fluctuations. As the size of the input beam increases and more microstructures are illuminated, uniformity eventually plateaus at an asymptotic value that depends on the diffuser design and systems parameters such as source properties and size of speckle averaging detector. If the detection area is on the order of the speckle size,[25] given by λ/B radians, with λ being the wavelength and the B the input beam size, uniformity will be limited by the coherence area of the source.[26] If the detection area is sufficiently large to allow averaging of a large enough number of speckles, uniformity will be limited by the diffuser design. As is not uncommon, there is always a compromise that needs to be considered between maximum efficiency and uniformity, depending on the beam shaping requirements and operating conditions.

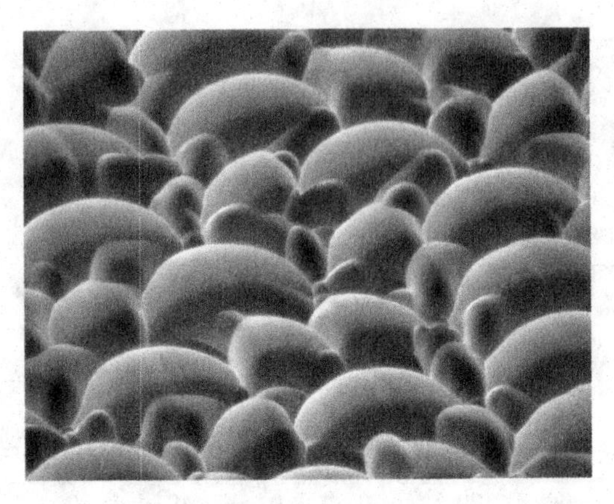

FIGURE 9.12 Scanning electron microscope picture of an engineered microlens diffuser that produces a circular pattern.

9.5 LIGHT CONTROL WITH ENGINEERED DIFFUSERS

We now explore two notable applications of engineered diffusers that illustrate their unique ability to produce controlled intensity patterns with high efficiency.

9.5.1 BAND-LIMITED DIFFUSERS

Engineered microlens diffusers have proven to be naturally suited to generate uniform illumination. Alternative concepts have also been proposed[27] such as a distribution of linear facets that are randomly combined to spread light over a specific angular range. Each facet scatters into a specific direction and by imposing an upper limit to the slope angles in the ensemble, one can theoretically guarantee band-limited behavior. By further ensuring the appropriate distribution of facets over the angular range of interest, one can produce uniform illumination. The approach does provide, at least in principle, band-limited behavior and some experimental demonstration can be found in the literature.[15] However, because each linear facet corresponds to a specific angular direction, this type of diffuser requires a large input beam size to sample enough facets so that uniform illumination could result over some specified angular range. Making the linear facets sufficiently small minimizes the requirements for a large beam but runs into manufacturing and diffraction issues. Engineered diffusers sidestep the issue completely and offer a natural way to generate uniform illumination. Examples are shown in Figure 9.13 for an 80° and a 30° circular diffuser. In each case, the measured intensity is best fit by a super-Lorentzian function of the form

$$I(\theta) = \frac{1}{1 + \left|\frac{2\theta}{w}\right|^p} \tag{9.4}$$

where:

w is the full width at half maximum

p is the power of the super-Lorentzian curve

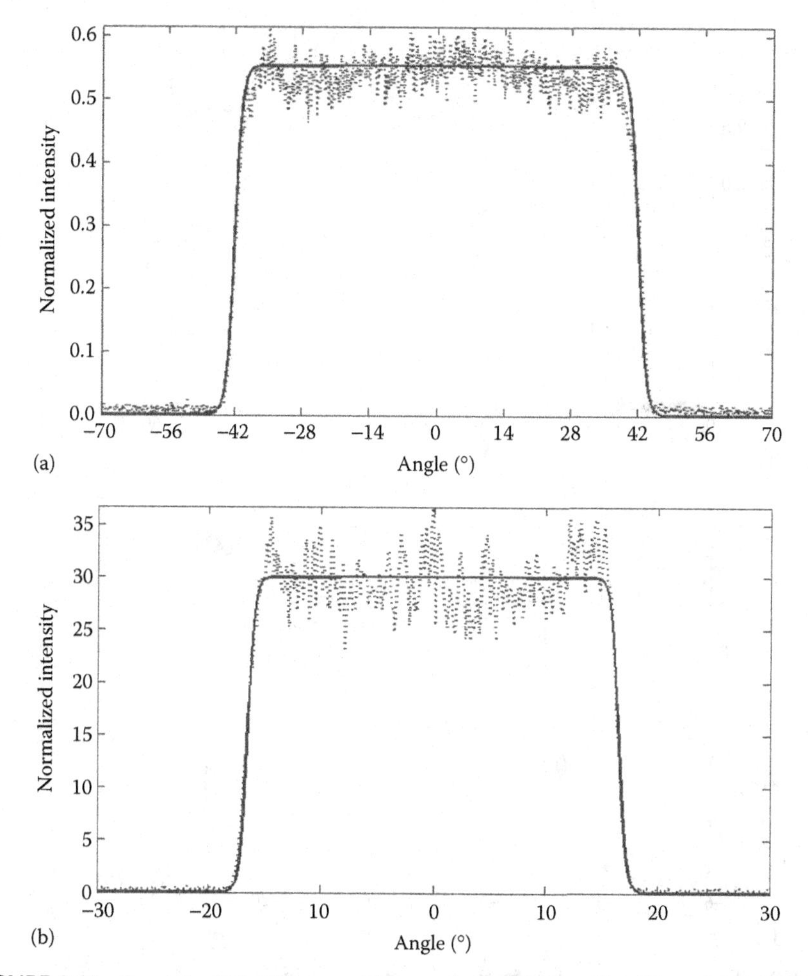

FIGURE 9.13 Band-limited diffusers with super-Lorentzian profiles (Equation 9.1) (a) 80° full width, $p = 60$, and (b) 30° full width, $p = 50$. Dotted line: measured data, solid line: fit.

The larger the power p, the sharper the intensity falloff at the edges of the scatter and the higher the target efficiency.

Other profiles such as those commonly referred to as "batwing" where the intensity increases away from the origin are also possible. This type of profile is needed, for example, if one wants to uniformly illuminate a flat surface. Two examples are shown in Figure 9.14 where the best fit takes the form of an inverse cosine power function versus angle, which is given by the relation:

$$I(\theta) = \begin{cases} \cos^{-p}\theta, & |\theta| \le \theta_0 \\ 0, & |\theta| > \theta_0 \end{cases} \tag{9.5}$$

with p being a real number indicating how fast the intensity increases at the edge compared to the center of the pattern, over the angular range defined by θ_0. The case

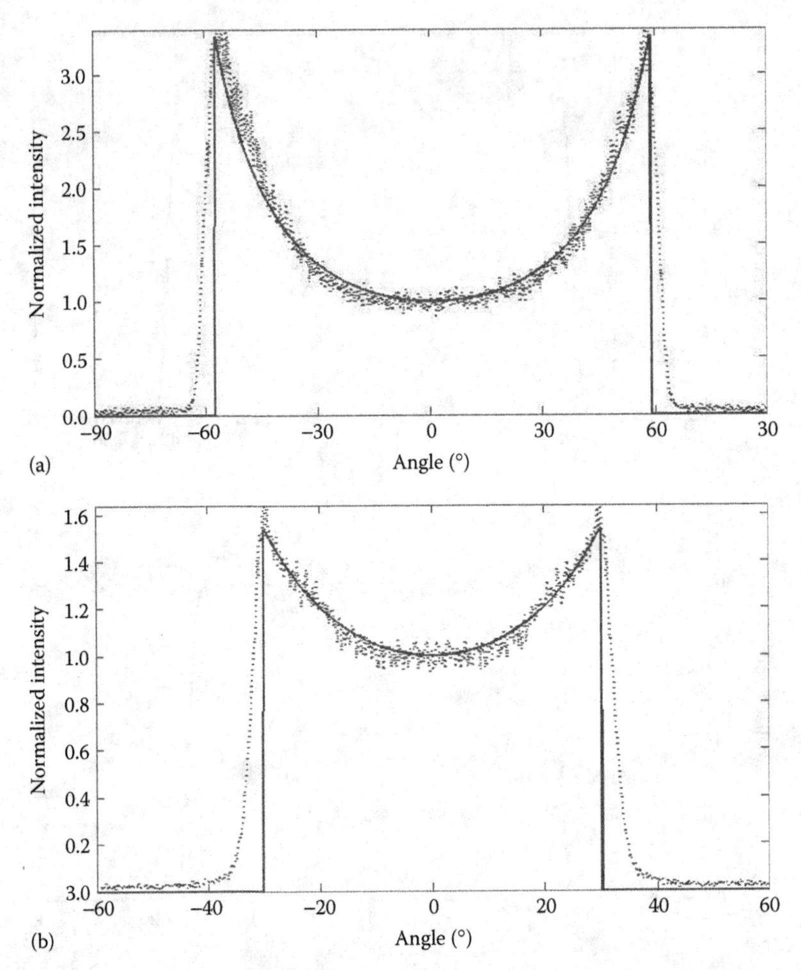

FIGURE 9.14 Band-limited diffusers with inverse cosine profiles (Equation 9.2) (a) 106° full width, $p = 1.9$, and (b) 60° full width, $p = 3$. Dotted line: measured data, solid line: fit.

where $p < 0$ is also possible as well as alternative functional forms, not limited to cosine or Lorentzian forms. In any case, the engineered diffuser can be designed to match general intensity profiles, either in angle space or at a flat target.

9.5.2 High-Efficiency Lambertian Illumination

Lambertian diffusers are often used for radiometric calibration purposes and in backlight display systems. The basic feature of a Lambertian scatterer is that the radiance is a constant,[28] independent of the observation angle with an intensity that depends on scatter angle as a cosine function over the whole half-hemisphere.

One can find Lambertian diffusers that work in reflection or transmission. Reflective diffusers[29] are generally made of materials such as Teflon, barium sulfate, or magnesium oxide and operate typically in the UV, visible, and near-infrared regions of the spectrum. The prototypical transmissive Lambertian diffuser is opal

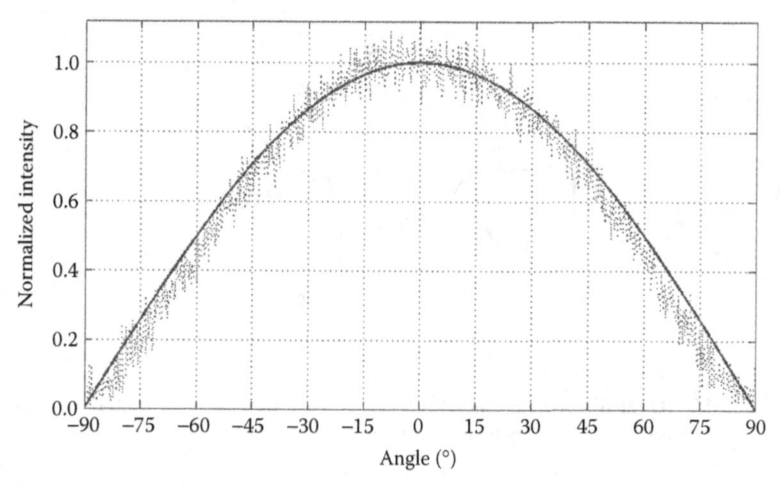

FIGURE 9.15 Scatter from opal glass (dashed line) and cosine fit (solid line). Transmission efficiency measured at 20%.

glass,[3] characterized by a distribution of small particulates embedded within a solid matrix. The scatter from opal glass is very nearly Lambertian (Figure 9.15), but because of the strong scatter angles and the volumetric nature of the diffuser transmission efficiency is fairly low (around 20%). Not much work has been done to improve the efficiency of transmission Lambertian diffusers. A concept has been theoretically proposed[30] based on a rotationally symmetric diffuser to achieve Lambertian scatter, which in practice would have the disadvantage of requiring alignment with the input beam. We show how the use of engineered microlens diffusers enables the design and manufacturing of transmission Lambertian scatter with high efficiency and applicability over a broad spectral region.

The main difficulty in implementing a single-surface Lambertian diffuser in transmission is the controlled fabrication of the steep slopes it requires to spread rays over wide angles within the full 180° hemisphere. For a material with index 1.5, for instance, continuously varying slope angles up to about 83° are necessary—not easy to fabricate with current lithographic methods. In a high-index material, such as silicon, with a refractive index of about 3.4 the maximum slope angle is about 24°, posing no difficulty for lithographic fabrication. However, high-index materials are generally restricted to infrared applications. A possible approach would be to consider a double-sided solution, where a diffuser surface is applied to both sides of a substrate. The idea behind the double-sided diffuser is to reduce the slope requirements on any given surface by having two diffuser surfaces working in conjunction to produce the Lambertian intensity profile. Although at first this might seem a valid approach, it turns out that a Lambertian diffuser cannot be implemented using the double-sided component. In the first diffuser, light propagates from a low incidence medium (typically air) into the material and all rays make it into the substrate, with a small percentage lost to Fresnel reflections. In the second diffuser, however, the incident medium has higher index compared to the output medium and that severely limits the cone of rays that are transmitted to the output medium. An application of

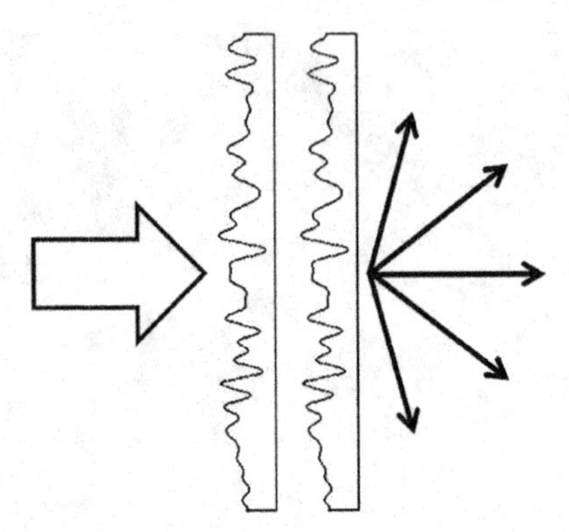

FIGURE 9.16 Dual-diffuser geometry for generating Lambertian scatter: two identical engineered diffusers separated by a small air gap.

Snell's law to the double-sided diffuser shows that this configuration cannot generate rays over the full hemisphere necessary for Lambertian scatter. Consequently, the double-sided diffuser cannot be used to create a Lambertian diffuser.

The concept we have implemented consists of a dual-diffuser geometry where the idea is to use the compounding effect of having diffusers in series to attain the wide angles required to produce Lambertian scatter with individual diffusers that can be manufactured with current lithographic capabilities. In this scheme, two identical diffusers are used, as illustrated in Figure 9.16, with a small air gap between them. The source faces the patterned side of the first diffuser while the patterned side of the second diffuser faces the flat side of the first. In this manner, the only sources of loss are those due to Fresnel reflections, thus minimizing internal losses. Finally, the two diffusers are space-invariant making alignment unnecessary and significantly simplifying assembly.

The problem now becomes that of determining the diffuser structure that, when combined in the fashion shown in Figure 9.16, produces the desired cosine intensity distribution. The elementary treatment of the diffraction problem involves expressing the far-field complex amplitude as the Fourier transform of the transmission functions of each diffuser, or a convolution operation.[31] However, because of the wide angles involved a more precise formulation is required, notwithstanding the fact that the inverse problem is probably much too difficult to solve, either analytically or numerically. We have opted for an alternative approach based on a ray-tracing algorithm that dodges the wide-angle limitation of Fourier optics to optimize the single-diffuser scatter. The result, interestingly, is that the single-diffuser solution can be expressed by

$$I(\theta) = \begin{cases} \cos\left(\dfrac{90}{\theta_0}\theta\right), & |\theta| \leq \theta_0 \\ 0, & |\theta| > \theta_0 \end{cases} \tag{9.6}$$

where:

θ_0 is a constant dependent on the diffuser material, 64° in the present case of index 1.5

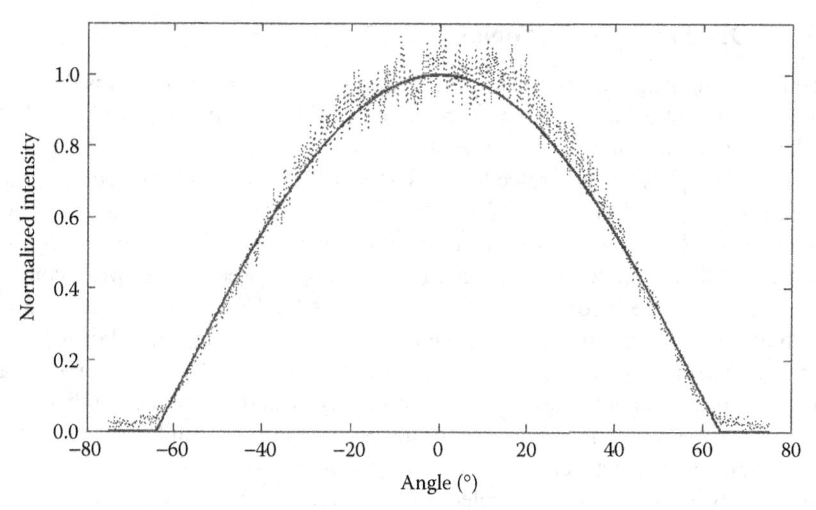

FIGURE 9.17 Single-diffuser profile required for the dual-diffuser solution to producing Lambertian scatter.

A plot of the single-diffuser measured intensity profile is shown in Figure 9.17.

Two diffusers were assembled as illustrated in Figure 9.16 and measured with a helium–neon laser, 633 nm. The measured scatter pattern is shown in Figure 9.18 showing close match to the target cosine profile. The measured transmission of the two-diffuser assembly was just over 70%, a factor of about 3.5 times higher than opal glass.

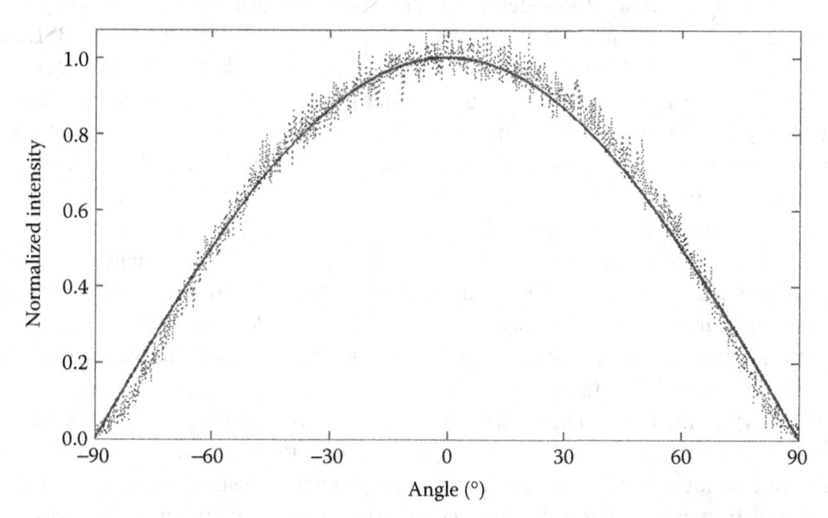

FIGURE 9.18 Lambertian cosine intensity profile produced with engineered microlens diffusers.

9.6 MODELING OF ENGINEERED DIFFUSERS

A common challenge facing the implementation of diffusers in general is the ability to reliably model these components in commercial ray-tracing software. While the fundamental nature of the diffuser scattering properties can only be captured by an accurate description of its surface features, most often one has limited access, if any, to that information. Common diffusers, such as opal glass, ground glass, and holographic, are only known in a statistical sense and therefore direct surface modeling is not at all used in practice, only ray-based or metrology-based models. Most commercial design software incorporates standard models for Gaussian diffusers and also the ability to use measured intensity data to model scatter properties. For deterministic beam shaping elements, such as engineered diffusers, diffractive elements, and MLAs, there is complete knowledge of the surface structure but even in these cases it is generally not practical, possible, or necessary really, to implement the intricacies of their relief structure. A clear exception is that of MLAs that, because of their periodic nature, are routinely modeled in commercial programs. The same, however, cannot be said of engineered diffusers and diffractive elements, mainly because of the large amount of data and complexity required to represent the surface.

There are two situations where modeling of the actual surface would be desirable: near-field behavior and coherent illumination. The diffuser models encountered in most commercial software are generally limited to far-field behavior and, since most are incoherent and ray based, an accurate model of the surface is not needed. Consequently, one must resort to the construction of models that allow one to reliably implement features of interest while sacrificing other features less relevant to the problem at hand. As long as the model provides a sufficiently accurate description of the diffuser behavior in the domain of interest, one can rely on the model for performance analysis and predictive purposes. Here we are concerned with implementing a modeling technique that allows some analysis of an engineered diffuser in commercial design software.

A common approach to modeling general diffusers, available in most programs, is based on the concept of bidirectional scattering distribution function (BSDF),[32] which describes the diffuser response as a function of incident angle. To determine the BSDF, a diffuser sample is measured under plane wave illumination for intensity versus angle at various angles of incidence. The resulting data are then compiled and imported into the optical design software, where some sort of interpolation is used for a complete picture of the model. While the BSDF approach is useful, it does require a significant amount of metrology, which can be time consuming and expensive.

We describe a simpler approach that requires a single measurement of intensity versus angle at normal incidence. The measured data is used to calculate a surface profile that can be used in the ray tracing model. The calculated surface sag is such that the intensity profile resulting from the ray tracing matches the measured data. The reconstruction of the surface profile from the scatter data is carried out using methods well known to the beam shaping literature.[9] Having the surface profile that generates the desired light distribution allows implementation in optical design software. We will not consider such implementation into any specific design software but present a general framework that a designer could adapt to the unique environment of the program she or he uses.

In the geometrical optics limit, the basic element responsible for changing the direction of propagation of a ray is the surface slope between two media of different indices of refraction. As long as one is not looking close enough to the diffuser surface, a convenient model is to represent the diffuser by the slope distribution or histogram that generates its scatter properties. The usefulness of this approach stands on the fact that the slope distribution for a given functional form of intensity versus angle is unique, even though its representation through a sag profile is not. But again, that should not matter as long as the observation point is far enough from the diffuser surface.

As an example, consider a flat-top diffuser that spreads light within a $\pm 10°$ range. An obvious way to represent this distribution is by means of a sequence of linear slopes covering the desired angular range, sort of like a Fresnel lens with each linear facet angle adjusted so that the complete set of facet slopes produces the uniform intensity over the $\pm 10°$ angular range. Another approach, maybe not as obvious but is just as effective, is to use a parabolic lens.[33] Other more elaborate representations, but of dubious usefulness, are also possible. The point to be made, though, is that each representation produces the desired intensity profile equally well even though their near-field properties are completely distinct.

The nonuniqueness of the representation problem is the reason why a single measurement of intensity versus angle cannot, strictly speaking, provide a complete model to be used in a general circumstance, at least not without certain assumptions. In the above case, for example, while both the Fresnel lens and the parabolic lens provide similar performance at normal incidence, they do not at nonzero angles of incidence. For the cases we are interested in, however, we will assume that the diffuser structure is composed of an aggregate of microlenses randomly arranged with a minimum of sag discontinuities, like those in a Fresnel or diffractive lens, in which case a representation valid at a certain incidence angle would also be valid at other angles of incidence, at least as long as angles of incidence are not so wide that a significant fraction of rays undergo more than a single refraction with each surface.

Under these assumptions, then, the approach to the problem becomes that of calculating the aspheric microlens element that reproduces the intensity distribution, the equivalent microlens profile (EMP), measured at normal incidence in the majority of cases. Fortunately, there are several known formalisms that provide solution to this otherwise difficult problem.[9,34] Figure 9.19 shows a diagrammatic illustration of the method under consideration. A certain diffuser (Figure 9.19a) characterized by a feature size D_1 produces an intensity profile of known functional form, either analytical, numerical, or measured, in the far field under normal incidence. Associated with the intensity profile and feature size D_2, there is associated an aspheric lens element that under normal incidence produces the same intensity profile in the far field. In other words, the single element surface profile is described by the same histogram of slopes as the whole diffuser. The EMP lens can be then replicated in the optical design program to create an MLA, thus defining a surface with the same histogram as the original diffuser. Under the assumption of incoherent ray tracing, the equivalent array model reproduces the desired diffuser intensity profile as long as one is not too close to the diffuser surface. Feature size effects are accommodated by setting the size of the microlens elements to correspond to the typical feature size of the diffuser, $D_2 = D_1$. This can be important if the input beam presents nonuniformity significant

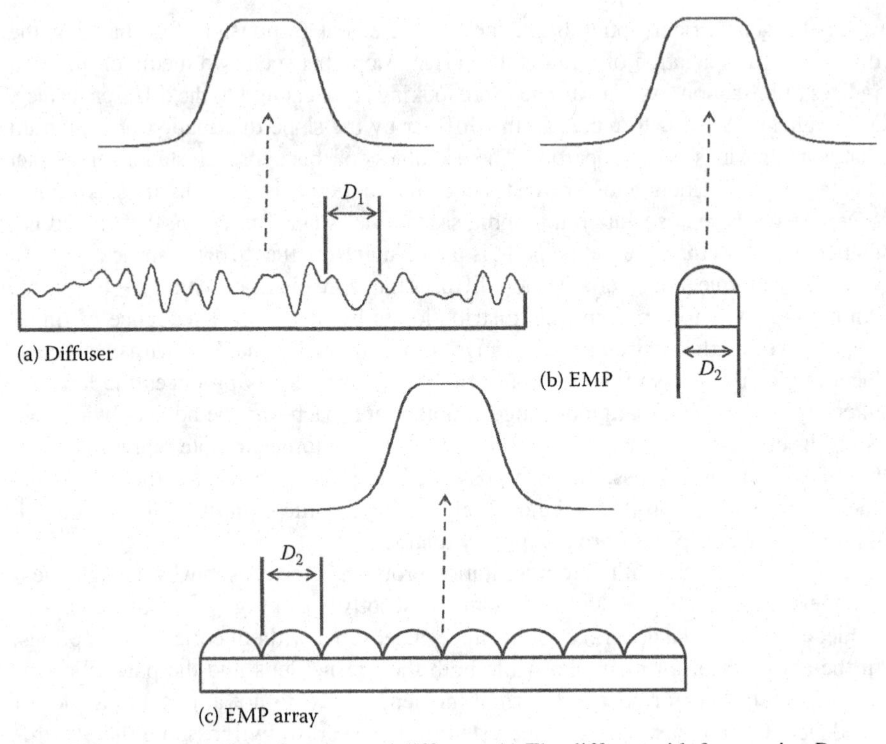

FIGURE 9.19 Modeling of engineered diffusers. (a) The diffuser with feature size D_1 generates a certain scatter pattern. (b) The EMP with feature size D_2 has the same histogram of slopes as the diffuser. (c) Array of EMP lenses that produce the same scatter pattern as the diffuser in the far field.

enough to warrant the higher degree of detail, otherwise the exact feature size used in the model is not particularly critical. The calculated EMP is typically expressed as a numerical table of coordinate versus sag value or expressed as a set of aspheric coefficients, whichever case is most convenient for the program being used. Most design programs have the means to define optical surfaces using either method.

Engineered diffusers that produce rectangular, square, or linear patterns can generally be modeled directly with rectangular or linear arrays with 100% fill factor. Modeling of engineered diffusers that produce a circular scatter pattern is of particular importance due to their ubiquity in many optical applications. However, since the diffuser surface cannot be completely covered by circular apertures, one needs to consider other methods. One possible approach is to utilize a hexagonal MLA to approximate the circular scatter pattern. Another approach is to consider an array of circular lens elements either in a square or hexagonal array with the space between lens elements being defined by a material that completely absorbs incident rays, as illustrated in Figure 9.20. The hexagonal arrangement provides a better sampling of the input beam and the suitable approach to implement will depend on details of the application and the capability of the design software being used to implement the desired geometry. General geometries would need to rely on a unit cell comprised of a combination of more complex microlens boundary shapes and regions of absorbing material.

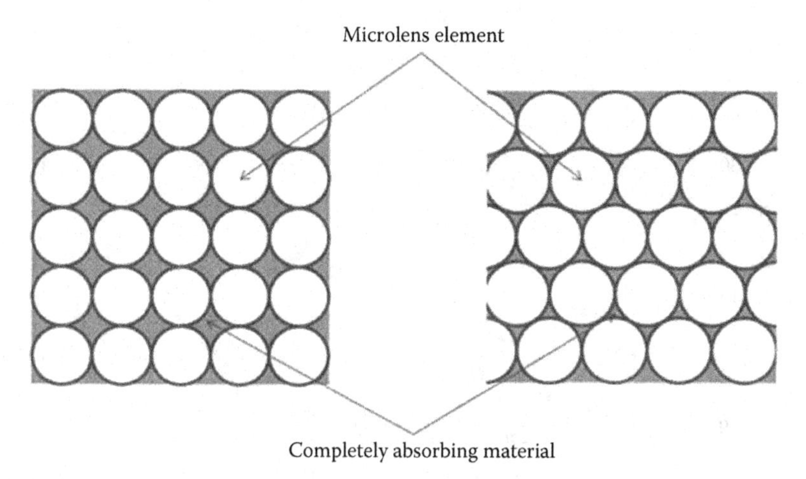

FIGURE 9.20 Alternative periodic geometries to model engineered diffusers that generate circular scatter patterns.

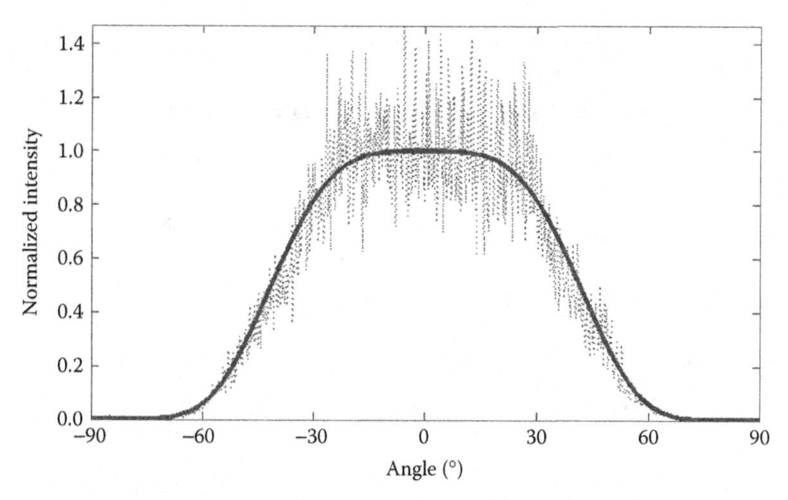

FIGURE 9.21 Measured intensity from a linear engineered diffuser (dotted line) and super-Gaussian fit (solid line).

As an example, of the above method, consider a one-dimensional engineered diffuser that produces the intensity profile illustrated in Figure 9.21. The measured data (dotted line) were taken under laser illumination at 633 nm with the diffuser index of refraction 1.56 at this wavelength. Because the source is coherent, intensity fluctuations due to speckle are observed. The incoherent intensity profile (solid line) is well described by a super-Gaussian function of the form:

$$I(\theta) = \exp\left[-\log(2)\left|\frac{2x}{w}\right|^{p}\right] \tag{9.7}$$

where the full width spread at half maximum $w = 83°$ and $p = 3.8$.

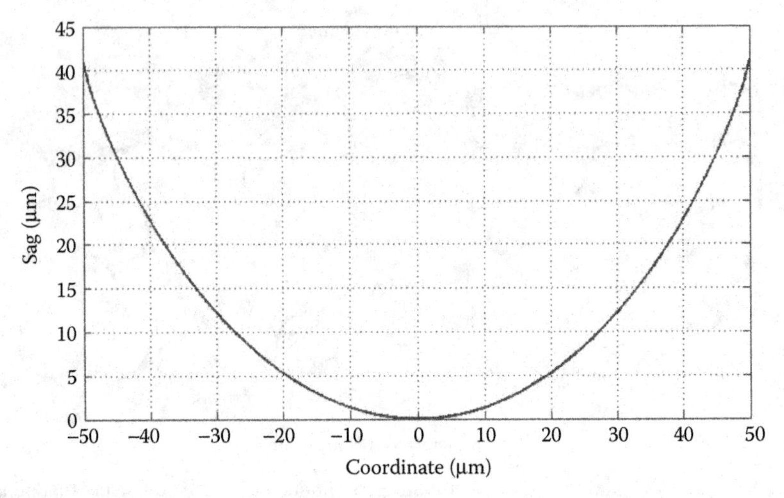

FIGURE 9.22 EMP that produces the diffuse pattern of Figure 9.21.

TABLE 9.1
Sag vs. Coordinate for Linear Diffuser EMP

Coordinate	Sag
0	0
5	0.3286
10	1.3059
15	2.9455
20	5.2727
25	8.3296
30	12.1840
35	16.9472
40	22.8179
45	30.2223
50	41.2284

The aspheric lens profile that reproduces the incoherent envelope, EMP, can be calculated[9,31] as shown in Figure 9.22 assuming a diameter of 100 μm. Total depth for the EMP element is just over 41 μm. The data, symmetric around the origin, are also presented in the sag table shown in Table 9.1.

Alternatively, the sag can be expressed in terms of an aspheric expansion, usable in most programs, in the following form:

$$s(r) = \sum_k a_k r^k \tag{9.8}$$

where the sag function s is written as a polynomial expansion with a finite number of terms. The coefficients associated with the sag profile assuming 20 terms in

TABLE 9.2
Aspheric Coefficient Representation of the Linear Diffuser EMP

k	a_k
0	0
2	1.28E–02
4	3.27E–06
6	−1.78E–08
8	5.52E–11
10	−9.48E–14
12	9.76E–17
14	−6.15E–20
16	2.33E–23
18	−4.85E–27
20	4.28E–31

the expansion are shown in Table 9.2 (odd terms are zero since the sag profile is an even function).

The EMP shown in Figure 9.22 and expressed through the aspheric coefficients of Table 9.2 can be used to trace rays and calculate the expected behavior under different angles of incidence and compare the results to actual measurements. Figure 9.23 shows the results of ray tracing through the EMP lens at normal incidence. The next plots show results of the ray tracing under different values of angles of incidence, showing good agreement between the ray-tracing results and the measured data, except of course for the fluctuations due to speckle.

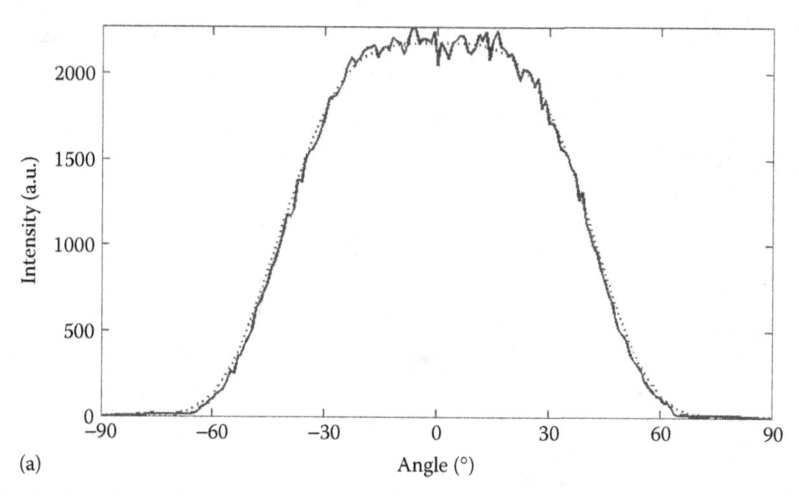

FIGURE 9.23 Comparison of intensity profiles from ray tracing (solid curve) and measured intensity (dotted curve) at various angles of incidence through an engineered diffuser: (a) normal incidence (for this plot only the dotted line is the target super-Gaussian), (b) 5°, (c) 10°, and (d) 15°.

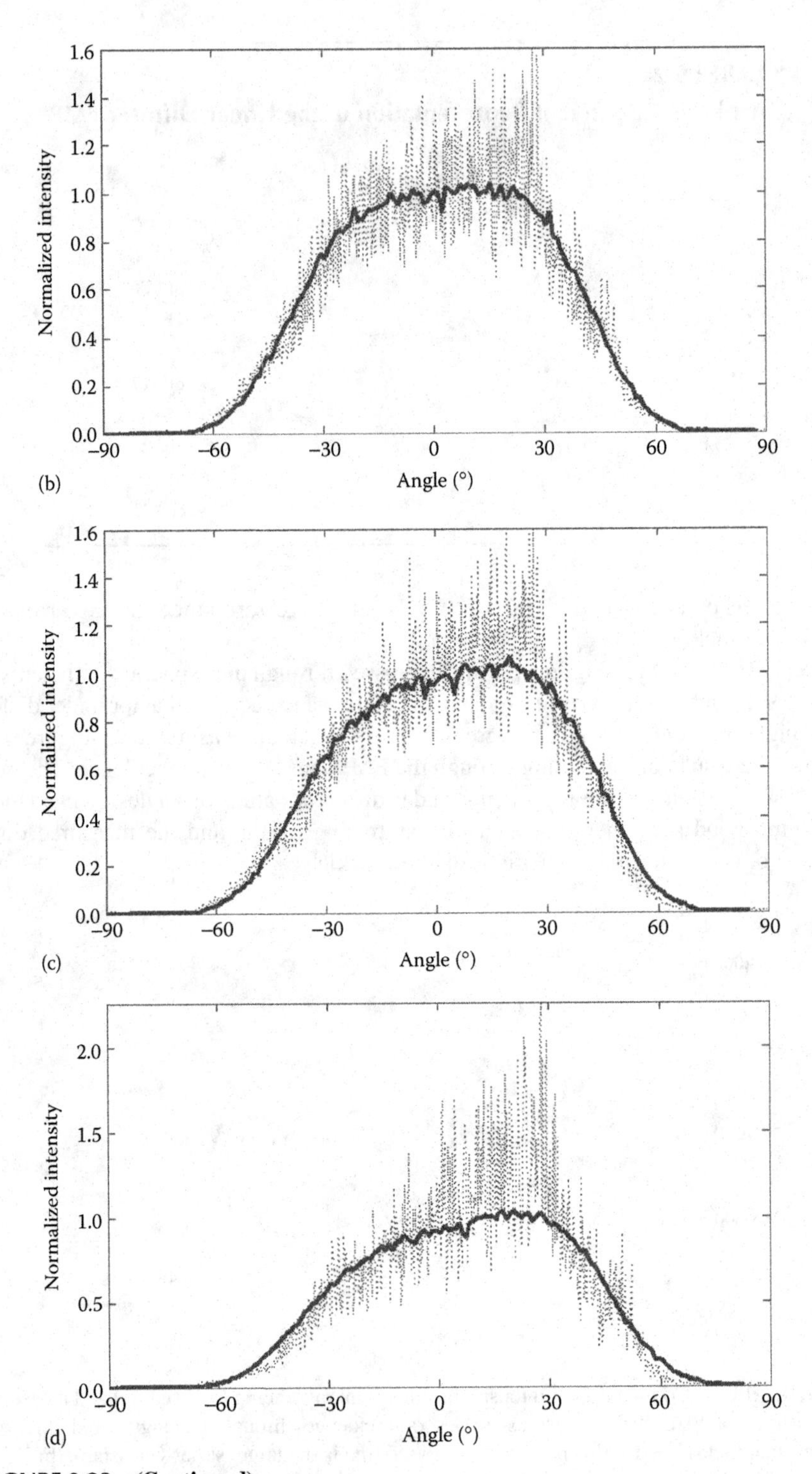

FIGURE 9.23 (Continued)

While additional work is required to more thoroughly understand the range of applicability of the present method, it can certainly provide a starting point in the modeling of engineered diffusers. In fact, it can be used with any single-surface diffuser, deterministic or random, by calculating the EMP sag function and implementing it as an MLA in ray-tracing programs. It is interesting to note that the approach also allows one to incorporate in a ray-tracing setting other effects that cannot be captured in a purely geometrical-optics model such as the broadening of the scatter due to diffraction (or whatever artifacts the particular diffuser introduces) that leads to a finite falloff region in its scatter. Consider, for example, the case of a parabolic lens, conic constant $\kappa = -1$, with diameter 50 μm and radius of curvature 150 μm. The far-field intensity profile for this lens element is shown in Figure 9.24 (solid line). Ray tracing through this same element provides a completely different result (dotted line), particularly missing the energy falloff outside the center region of uniform intensity. Modeling of the parabolic element or array in a ray-tracing program yields the result given by the dotted line, significantly misrepresenting the fraction of energy within the center region.

The incoherent envelop of the intensity profile can be fit with super-Lorentzian curves with full width at half maximum w in the following form:

$$I_{\text{fit}}(\theta) = \begin{cases} \dfrac{1}{1+\left|\dfrac{2\theta}{w}\right|^{p_1}}, & |\theta| \le w \\[2ex] \dfrac{1}{1+\left|\dfrac{2\theta}{w}\right|^{p_2}}, & |\theta| > w \end{cases} \tag{9.9}$$

where:

p_1 and p_2 represent the power of the Lorentzian curves

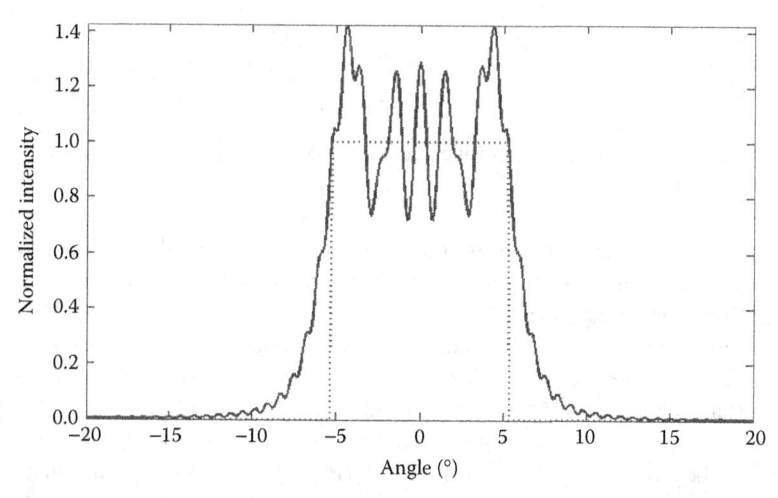

FIGURE 9.24 Intensity profile due to a parabolic lens with diameter $D = 50$ μm and radius of curvature $R = 150$ μm from diffraction theory (solid line) and ray tracing (dotted line).

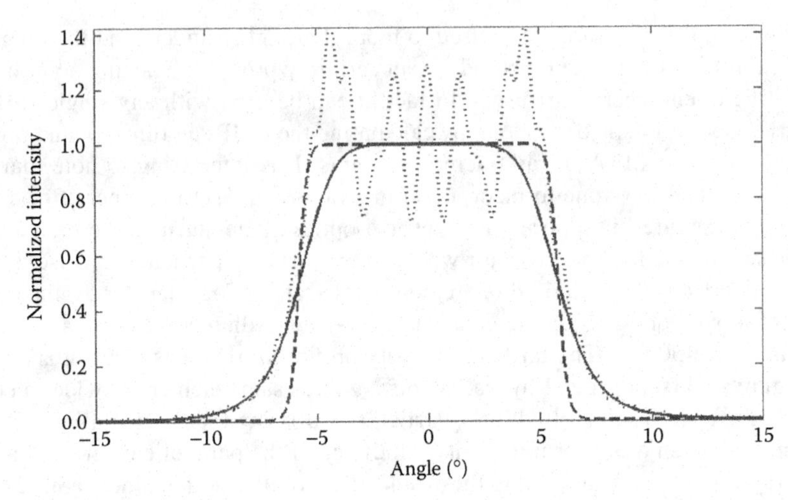

FIGURE 9.25 Fit of intensity profile (dotted line) with super-Lorentzian profiles: outer portion power 7 (solid line) and center portion power 30 (dashed line).

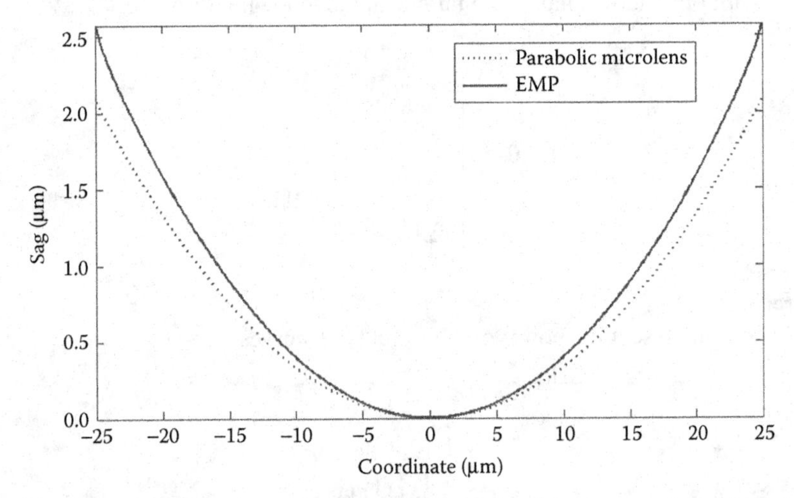

FIGURE 9.26 Calculated EMP of super-Lorentzian example and comparison with original parabolic profile.

For the curve in question the fit is shown in Figure 9.25. The dashed line provides the fit over the center portion with $p_1 = 30$ and the solid line represents the fit over the falloff region where $p_2 = 7$.

The calculated EMP associated with the fit function given by Equation 9.6 and the above parameters is shown in Figure 9.26. The EMP is naturally a little deeper than the original parabolic element as it spreads light over wider angles. Aspheric coefficients are also given for this particular profile in Table 9.3. The output of the ray tracing through the EMP (Figure 9.27) provides a much better model of the diffraction-induced broadening even in the context of a ray-tracing model.

TABLE 9.3
Aspheric Coefficients of Super-Lorentzian EMP Example

k	a_k
2	4.02E–03
4	2.07E–07
6	–2.44E–08
8	3.83E–10
10	–2.90E–12
12	1.25E–14
14	–3.24E–17
16	4.96E–20
18	–4.15E–23
20	1.46E–26
22	4.02E–03

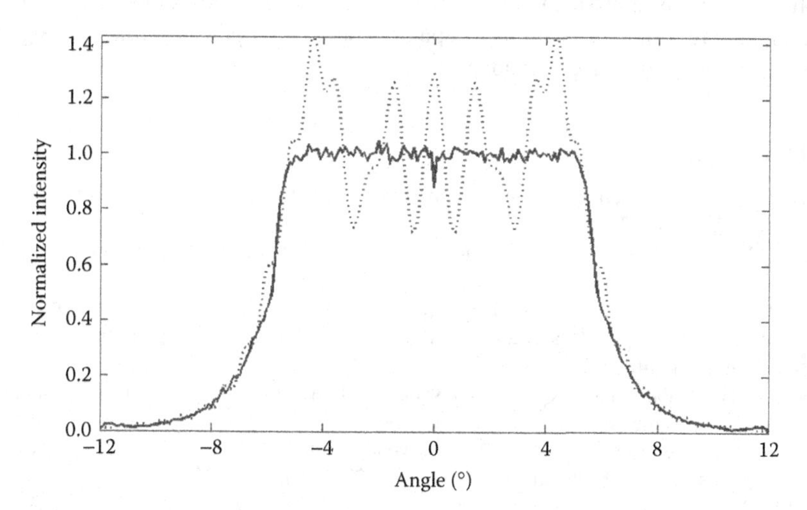

FIGURE 9.27 Calculated intensity through EMP (solid line) for super-Lorentzian example.

9.7 SUMMARY

Engineered microlens diffusers provide a degree of beam shaping capability that is beyond what can be achieved by common statistical diffusers and without the limitations of diffractive elements or periodic MLAs. At the same time, the engineered diffusers share the robustness of random diffusers and the deterministic nature of diffractives. We described some essential concepts that define engineered diffusers in terms of their basic microlens elements and their assembly to create a fully randomized diffuser surface. The typical design unit is a conic microlens characterized by a number of design parameters that control intensity profiles, energy distribution, uniformity, and efficiency. We have shown that engineered diffusers are well suited to produce band-limited illumination where essentially all of the transmitted energy is confined within a certain angular range. Common random diffusers spread light with a Gaussian intensity profile which, by definition, is not band limited. Engineered diffusers, however, are able to produce band-limited illumination with controlled intensity profile to fit different requirements. Interestingly, though, nearly ideal band-limited performance is only part of the general capabilities of these diffusers and more general distributions are feasible, from the usual Gaussian through uniform and Lambertian scatter. The refractive nature of engineered microlens diffusers means achromatic behavior with either laser or light-emitting diode (LED) sources.

An approach to the modeling of engineered diffusers has been described that allows the implementation of general intensity profiles into ray-tracing software by using the capability of most programs to trace rays through MLAs. The idea is based on the calculation of an aspheric lens with a histogram of slopes that matches that of the engineered diffuser, as reflected on a model or measured intensity function versus angle. The aspheric element is thus called the EMP, which generates the same intensity distribution as the diffuser. Generally, the EMP sag function is given in numerical form or by means of a set of aspheric coefficients that fit the numerical profile. Using this approach we can also, to some degree, incorporate into the model effects that are usually outside the scope of geometrical optics ray-tracing programs such as diffraction-induced broadening.

REFERENCES

1. P. Kuttner, "Modulation transfer functions of ground-glass screens," *Appl. Opt.* 11, 2024–2027, 1972.
2. S. L. Yeh, "A study of light scattered by surface-relief holographic diffusers," *Opt. Commun.* 264, 1–8, 2006.
3. B. A. Brice and M. Halwer, "Determination of the diffuse transmittance of opal glass and the use of opal glass as a standard diffuser in light-scattering photometers," *J. Opt. Soc. Am.* 44, 340–340, 1954.
4. J. F. Goldenberg and T. Stewart McKechnie, "Diffraction analysis of bulk diffusers for projection-screen applications," *J. Opt. Soc. Am. A* 2, 2337–2347, 1985.
5. J. S. Liu and M. R. Taghizadeh, "Iterative algorithm for the design of diffractive phase elements for laser beam shaping," *Opt. Lett.* 27, 1463–1465, 2002.
6. W. J. Dallas, "Deterministic diffusers for holography," *Appl. Opt.* 12, 1179–1187, 1973.
7. X. Deng, X. Liang, Z. Chen, W. Yu, and R. Ma, "Uniform illumination of large targets using a lens array," *Appl. Opt.* 25, 377–381, 1986.
8. R. M. Stevenson, M. J. Norman, T. H. Bett, D. A. Pepler, C. N. Danson, and I. N. Ross, "Binary-phase zone plates for the generations of uniform focal profiles," *Opt. Lett.* 19, 363–365, 1994.
9. F. Dickey and S. Holswade, *Laser Beam Shaping: Theory and Techniques*, Marcel Dekker, New York, 2000.
10. M. C. Hutley, "Refractive lenslet arrays," in *Micro-Optics: Elements, Systems and Applications*, H. Peter Herzig (ed.), Taylor & Francis, London, 1997.
11. J. A. Mucha, D. W. Hess, and E. S. Aydill, "Plasma etching," in *Introduction to Microlithography*, L. F. Thompson, C. G. Willson, and M. J. Bowden (eds.), The American Chemical Society, Washington, DC, 1994.
12. V. Kettunen and H. P. Herzig, "Applications of diffractive optics and micro-optics in lithography," in *Encyclopedia of Modern Optics*, R. D. Guenther, D. G. Steel, and L. Bayvel (eds.), Elsevier, Oxford, 2004, pp. 281–290.

13. F. M. Dickey, S. C. Holswade, and D. L. Shealy, *Laser Beam Shaping Applications*, Taylor & Francis, Boca Raton, FL, 2006.
14. D. J. Schertler and N. George, "Uniform scattering patterns from grating—Diffuser cascades for display applications," *Appl. Opt.* 38, 291–303, 1999.
15. E. R. Méndez, E. Efrén García-Guerrero, H. M. Escamilla, A. A. Maradudin, T. A. Leskova, and A. V. Shchegrov, "Photofabrication of random achromatic optical diffusers for uniform illumination," *Appl. Opt.* 40, 1098–1108, 2001.
16. E. R. Méndez, E. E. García-Guerrero, T. A. Leskova, A. A. Maradudin, J. Muñoz-Lopez, and I. Simonsen, "The design of one-dimensional random surfaces with specified scattering properties," *Appl. Phys. Lett.* 81, 798–800, 2002.
17. T. R. M. Sales, "True bandlimited diffusers," in *Latin America Optics and Photonics Conference*, OSA Technical Digest (CD), Optical Society of America, 2010 paper PDPTuK4, Recife, Brazil, September 27–30, 2010.
18. T. R. M. Sales, "Structured microlens arrays for beam shaping," *Opt. Eng.* 42, 3084, 2003.
19. A. Y. Yi and L. Li, "Design and fabrication of a microlens array by use of a slow tool servo," *Opt. Lett.* 30, 1707–1709, 2005.
20. T. R. M. Sales and G. Michael Morris, "Laser writing advances micro-optics fabrication," *Laser Focus World*, January 2008.
21. M. Born and E. Wolf, *Principles of Optics*, 7th edition, Cambridge University Press, Cambridge, 1999.
22. J. D. Gaskill, *Linear Systems, Fourier Transforms, and Optics*, John Wiley & Sons, New York, 1978.
23. J. W. Goodman, *Speckle Phenomena in Optics*, Roberts & Company, Englewood, CO, 2007.
24. Y. Kuratomi, K. Sekiya, H. Satoh, T. Tomiyama, T. Kawakami, B. Katagiri, Y. Suzuki, and T. Uchida, "Speckle reduction mechanism in laser rear projection displays using a small moving diffuser," *J. Opt. Soc. Am. A* 27, 1812–1817, 2010.
25. N. George, "Speckle," *Opt. News* 2(1), 14–20, 1976.
26. J. W. Goodman, *Statistical Optics*, Wiley, New York, 1985.
27. E. R. Méndez, T. A. Leskova, A. A. Maradudin, and J. Muñoz-Lopez, "Design of two-dimensional random surfaces with specified scattering properties," *Opt. Lett.* 29, 2917, 2004.
28. R. W. Boyd, *Radiometry and the Detection of Optical Radiation*, Wiley, New York, 1983.
29. F. Grum and G. W. Luckey, "Optical sphere paint and a working standard of reflectance," *Appl. Opt.* 7, 2289–2294, 1968.
30. A. A. Maradudin, T. A. Leskova, and E. R. Méndez, "Two-dimensional random surfaces that act as circular diffusers," *Opt. Lett.* 28, 72–74, 2003.
31. J. W. Goodman, *Introduction to Fourier Optics*, McGraw-Hill, New York, 1996.
32. F. B. Leloup, S. Forment, P. Dutré, M. R. Pointer, and P. Hanselaer, "Design of an instrument for measuring the spectral bidirectional scatter distribution function," *Appl. Opt.* 47, 5454–5467, 2008.
33. C. N. Kurtz, H. O. Hoadley, and J. J. DePalma, "Design and synthesis of random phase diffusers," *J. Opt. Soc. Am. A* 63, 1080, 1973.
34. J. Rubinstein and G. Wolansky, "Intensity control with a free-form lens," *J. Opt. Soc. Am. A* 24, 463–469, 2007.

10 Multi-Aperture Beam Integration Systems

Daniel M. Brown, Fred M. Dickey, and Louis S. Weichman

CONTENTS

10.1 INTRODUCTION

Various high-power laser applications, such as laser heat processing, cutting, marking, photolithography, and fiber injection, require a laser irradiance that is substantially uniform on a target over a specified area at a fixed longitudinal distance from the source. The irradiance pattern may be circular, hexagonal, rectangular, ring shaped, or practically any other shape that can be defined by the boundary of an aperture. If the laser beam mode is well defined and constant in time, then the beam shaping (field mapping) methods discussed in Chapter 2, Chapters 5 through 7, and the near-field beam shapers discussed briefly in Chapter 8 can be used. The field mapping approach discussed in these chapters is also applicable if the output is required to be collimated. However, in cases where the laser modes are unknown or change with time, and collimation is not required, a multifaceted or multi-aperture beam integrator may be more desirable. This approach to beam shaping is especially suitable to excimer lasers[1,2] and other multimode laser beams, laser diode arrays,[3,4] or other light sources with highly irregular irradiance distributions.[5]

A multi-aperture integrator system basically consists of two components: (1) a subaperture array component consisting of one or more lenslet arrays which segments the entrance pupil or cross section of the beam into an array of beamlets and applies a phase aberration to each beamlet and (2) a beam integrator or focusing component that overlaps the beamlets from each subaperture at the target plane. These elements can be refractive, reflective, or diffractive. Generally, the subaperture elements all have the same shape and phase function to simplify fabrication, but varying their phase and shape within the array can provide greater irradiance uniformity in the target plane as shown by Pepler et al.[6] and others.[7] The target is located at the focal point of the primary focusing element where the chief rays of each subaperture intersect. Thus, the amplitude of the irradiance distribution on the target is a Fourier transform of the incoming wavefront aberrated by the lenslet array.[8] Although this chapter primarily addresses the multi-aperture beam integration problem from the standpoint of refractive optics, the concepts and analysis are directly applicable to reflective optics. Dickey and O'Neil[5] treat multifaceted reflective systems in considerable detail.

All beam integrators can be loosely divided into two categories: diffracting and imaging. A simple diffracting beam integrator (also called a nonimaging integrator[9]) is illustrated in Figure 10.1, consisting of a single lenslet array and a positive primary lens. The target irradiance is the sum of defocused diffraction spots [point spread functions (PSFs)] of an on-axis object point at infinity (assuming a collimated input wavefront). If the source is spatially coherent over the lenslet aperture, or can be defined by a single field point from a ray optics point of view, the diffraction spot will closely replicate the shape of the subaperture with diffraction rings (determined

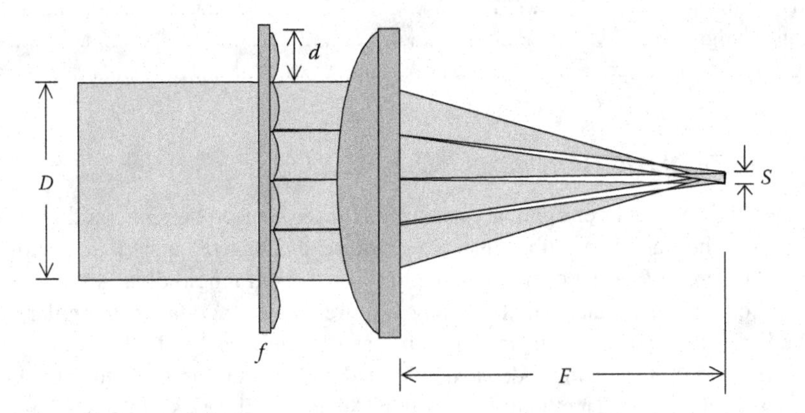

FIGURE 10.1 Diffracting multi-aperture beam integrator concept.

by the degree of defocus and other aberrations) superimposed. The defocus is caused by the additional optical power of the subaperture.

The diffracting beam integrator is based on the assumption that the output is the superposition of the diffraction fields of the beamlet apertures. The diffraction field is obtained using the Fresnel integral. If the beam is not spatially coherent over each beamlet aperture, a more complicated integral is required and, generally, one would not be able to obtain a reasonable replica of the lenslet aperture. For example, a spatially incoherent field is approximated by a Lambertian source that radiates over a large angle and would not produce a localized irradiance distribution at the output plane.

Figure 10.2 illustrates an imaging multi-aperture beam integrator. This type of integrator is especially appropriate for spatially incoherent sources. From a ray optics perspective, these sources produce a wavefront incident over a range of field angles on the lenslet apertures. The first lenslet array segments the beam as before and focuses the beamlets onto a second lenslet array. That is, each lenslet in the first array is designed to confine the incident optical radiation within the corresponding aperture in the second array. A second lenslet array, separated from the first by a distance equal to the focal length of the secondary lenslets, together with the primary focusing lens forms a real image of the subapertures of the first lenslet array on the target plane. The primary lens overlaps these subaperture images at the target to form one integrated image of the subapertures of the first array element. Reimaging the lenslet apertures mitigates the diffraction effects of the integrator in Figure 10.1.

In this chapter, we discuss the theory and design of diffracting and imaging multi-aperture beam integration systems. We show how the subaperture shape and phase function determines the irradiance pattern in the image plane. We discuss the diffraction effects and the interference between the subapertures.

In Section 10.2, we outline the basic theory and design considerations for multifaceted beam integrators. A number of different optical configurations exist for multifaceted beam integrators. Minimizing the interference effects produced by the finite number of subapertures and making the irradiance uniform in the target pattern is the design goal of these systems. We discuss these problems in this section. Multi-aperture beam shaping design methods are discussed in Section 10.3.

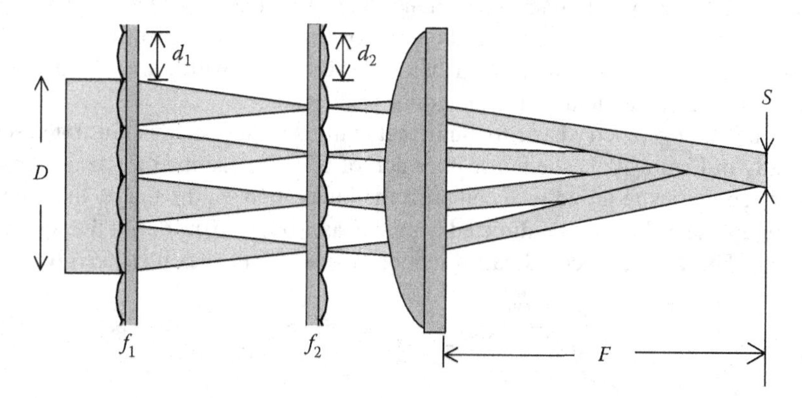

FIGURE 10.2 Imaging multi-aperture beam integrator concept.

In this section, we show how to use geometric ray-tracing codes, such as ZEMAX,* to design multifaceted beam integrators. The effects of geometric aberrations are discussed. Fabrication considerations are discussed in Section 10.4, and applications and experimental data are presented in Section 10.5.

Throughout this chapter, we use the following fundamental design parameters to describe multi-aperture beam integration systems:

D—diameter of input beam at multi-aperture integrator
d—diameter of subaperture or lenslet
F—focal length of primary lens
f—focal length of array lenslet
S—diameter of target spot
λ—wavelength
k—$2\pi/\lambda$
R_w—radius of curvature of wavefront
R_0—radius of curvature of reference sphere centered on target

All other parameters will be defined as they are introduced.

10.2 THEORY

A major assumption in the multifaceted approach to beam integration is that the laser beam does not have a time-varying divergence that is significant over the distance required to accomplish the integration, that is, only the irradiance fluctuates with time. This divergence requirement corresponds to a slowly varying, nearly uniform phase across each subaperture. This condition is required to guarantee a good overlap of the beamlets in the output plane. Further, it is required that the input beam has a high degree of spatial coherence over each facet. If this is not the case, the diffraction pattern of the beamlets will be dominated by the coherence function, not the aperture function defining the beamlets.

The analysis and design of laser beam integrators should include the effects of averaging, diffraction, interference, and imaging. With multifaceted integrators, it is primarily the averaging process that is used to produce the desired irradiance distribution. Diffraction and interference tend to produce undesired fluctuations in the irradiance distribution. Imaging can be used to control diffraction effects as well as the size (scale) of the irradiance distribution. Aberrations, which are inherent in the imaging process, tend to degrade integrator performance.

The beamlet geometry basic to multifaceted mirror integrators is illustrated schematically in Figure 10.3. The figure does not, of course, describe the various lens or mirror geometries required to accomplish the integration. In the figure, the array at the left represents beamlets redirected from the laser beam to overlap at the square on the right. The x_0- and y_0-coordinate direction cosines for the beamlets are given by

$$\alpha_m = \frac{x_{1m}}{\sqrt{x_{1m}^2 + y_{1n}^2 + z^2}}, \qquad \beta_n = \frac{y_{1n}}{\sqrt{x_{1m}^2 + y_{1n}^2 + z^2}} \tag{10.1}$$

* ZEMAX is a commercial lens design software sold by Focus Software, Inc., Tucson, AZ.

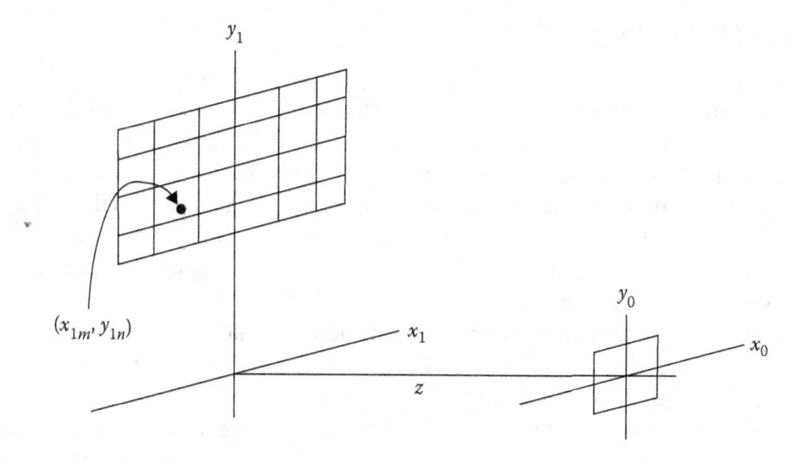

FIGURE 10.3 Beam integrator geometry.

The integrated optical field is the sum of the diffracted fields associated with the individual beamlets. The diffraction integral should be developed with respect to the limiting apertures that produce the beamlets. Also, imaging can be included as generalized diffraction. An assumption appropriate to multifaceted integrators is that the optical field amplitude (or equivalently, irradiance) is approximately uniform over each limiting aperture forming the beamlets. Also, the angle between beamlets should be kept small to reduce the size of the following optics and associated aberrations.

There are four major assumptions in the development of diffracting beam integrators. We will list them here and discuss their impact:

1. The input beam amplitude (or equivalently irradiance) is approximately uniform over each subaperture. This allows for the output to be the superposition of the diffraction patterns of the beamlet defining apertures. It is expected that small deviations will average out in the output plane. That is, the errors associated with a particular aperture will not dominate.
2. The phase across each subaperture is uniform. The discussion in assumption 1 applies in this case also. In addition, a linear phase across a subaperture results in a redirection of the beamlets, causing a misalignment in the output.
3. The input beam divergence does not vary significantly with time. Generally, an input beam divergence will result in a nonoverlapping of the beams in the target plane. This can be corrected in many cases with correction optics in the input beam. However, a time-varying divergence would negate the possibility of correction.
4. The input beam field should be spatially coherent over each subaperture. This is inherent in assumption 1 since the diffraction patterns are assumed to be described by a Fresnel integral.

The imaging integrator does not require assumption 4 since it does not necessarily require that the output pattern be described by a diffraction integral.

10.2.1 Diffraction Considerations

The basic problem of multi-aperture beam integration is to map the input fields in the input apertures in Figure 10.3 (x_1–y_1 plane) into the desired irradiance in the output plane (x_0–y_0 plane). It is assumed that the irradiance in the input plane is relatively uniform (constant) over each aperture. This assumption leads to the requirement of small apertures; however, there is a limit to how small one would make the apertures. This is the averaging problem discussed in Section 10.2.3. In addition, since the beamlets are superimposed in the output plane, small deviations from uniformity of the beam irradiance over the apertures should average out.

The theory of mapping a uniform input irradiance into a uniform output irradiance is developed in Appendix A of Chapter 5. The basic optical layout of a system for accomplishing this mapping, for each subaperture, using a diffracting multi-aperture beam integrator is illustrated in Figure 10.1. The system consists of a lenslet array and a primary lens. The target plane is located at the focal point of the primary lens. A collimated beam of diameter D is incident on an array of lenslets, each of diameter d and focal length f, which segments the beam into multiple beamlets. The primary lens of focal length F overlaps the beamlets, bringing the chief rays of each beamlet to a common focus at the back focal point of the primary objective where the integrated irradiance pattern is formed. The primary lens produces a field distribution at the focal plane that is proportional to the Fourier transform of the product of the functions representing the input beam and the lenslet array.[8]

It is shown in Appendix A of Chapter 5 using the method of stationary phase that the optical element that effects the mapping is a quadratic phase element, that is, a simple lens. The analysis is done in one dimension, but it can be extended to two dimensions. The phase of the optical element is given by

$$\Phi\left(\frac{x}{d}\right) = \beta\left(\frac{x}{d}\right)^2 \tag{10.2}$$

The stationary phase solution includes a parameter β that is a measure of the quality of the solution. This parameter, given by

$$\beta = \frac{\pi d S}{\lambda F} \tag{10.3}$$

has the same form as the Fresnel number, differing only by a constant factor. Note that β is a dimensionless constant. The significance of β is discussed in considerable detail in Chapters 2 and 5. In Chapters 2 and 5, β is shown to be related to the mathematical uncertainty principle. Although the numerical coefficient in front of the factor, $dS/\lambda F$, may vary with the problem, the main result is that control of the shape of the beam cannot be maintained if β is too small. Further, it can be seen from the form of Equation 10.2 that if β is fixed, the solution for the phase of the lenslet is fixed.

It can be shown, using diffraction theory (see Section 5.2.5), that if the phase function representing each lenslet is an even function, the negative of the phase also gives the same output irradiance. If the lenslets in the array are positive, as shown in

Figure 10.1, each beamlet will have a focus ahead of the focal point of the primary lens. If they are negative, the individual beamlets will have either a real focus after the focal point of the primary lens or a virtual focus ahead of the lenslets, depending on the relative optical powers of the lenslets and the primary lens. The numerical aperture (NA) of the beamlets and the distance between beamlet foci and target plane determine the spot size, S. It can be shown from paraxial geometrical optics that the spot size S on the target is equal to the focal length F of the primary lens divided by the f-number of the subaperture lenslets:

$$S = \frac{F}{f/d} \tag{10.4}$$

This result is also obtained using diffraction theory and Fourier optics. It is apparent from Equation 10.4 that the lenslet diameter can be varied over the array as long as its focal length is also varied proportionately such that the f-number (f/d) remains constant. Also, one can change the focal length F of the primary lens to scale the diameter S of the target spot and not change β.

The diffraction pattern of a single subaperture determines the shape of the spot on the target. The Fresnel number is useful for estimating this irradiance distribution on the target. The Fresnel number equals the number of half-waves of optical path difference (OPD) and is approximately given by

$$\text{Fresnel} = 2 \times \text{OPD} = \frac{r^2}{\lambda}\left(\frac{1}{R_w} - \frac{1}{R_0}\right) \tag{10.5}$$

where:

- R_w is the radius of curvature of the wavefront
- R_0 is the radius of curvature of a reference sphere centered on the observation point
- r is the radial coordinate in the subaperture

For a uniformly illuminated aperture, the Fresnel number also equals the number of peaks in the aberrated PSF cross section. Even integer Fresnel numbers have an on-axis minimum in the diffraction PSF. Odd integer Fresnel numbers have an on-axis peak intensity in the PSF. This is shown in Figure 10.4a and b. Generally, the Fresnel number is a measure of the complexity of the diffraction pattern; the number of maxima increases with increasing Fresnel number while the depth of the modulation decreases with increasing Fresnel number.

Assuming a collimated beam incident onto the lenslet array, the optical powers of the lenslets and the primary lens combine to produce a spherical wavefront converging with a radius of curvature of R_w. Substituting $1/R_w = 1/f + 1/F$, $1/R_0 = 1/F$, and $d = 2r$ into Equation 10.5 gives the Fresnel number in terms of lenslet parameters:

$$\text{Fresnel} = \frac{d}{4\lambda(f/d)} \tag{10.6}$$

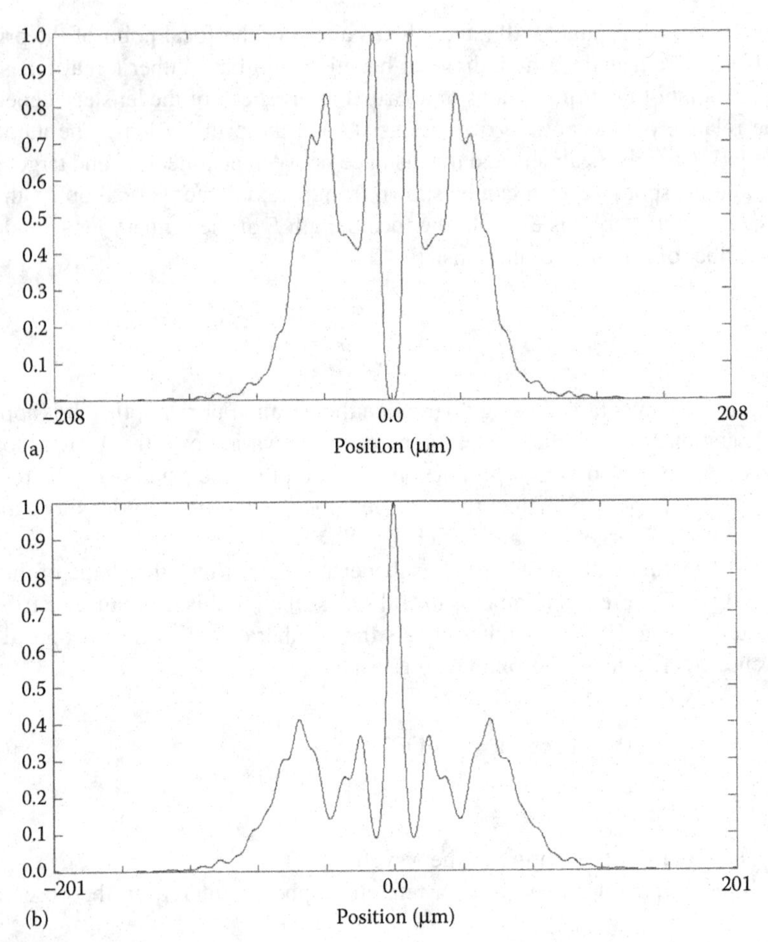

FIGURE 10.4 (a) Diffraction patterns of on-axis subaperture pupil functions as a function of Fresnel number 2 waves defocus (Fresnel no. = 4); (b) 2.5 waves defocus (Fresnel no. = 5).

Alternatively, substituting Equation 10.4 into Equation 10.6, the Fresnel number can be written in terms of the target spot size and primary lens focal length:

$$\text{Fresnel} = \frac{dS}{4\lambda F} \tag{10.7}$$

Since $\lambda F/d$ is proportional to the width of the PSF of a single subaperture, Equation 10.7 shows that the Fresnel number is proportional to the number of PSFs across the target pattern. The lower the Fresnel number, the more rounded the target pattern becomes. Note that Equation 10.7 is directly proportional to Equation 10.3.

10.2.2 Interference Effects

The output irradiance in Figure 10.3 is the superposition of the diffraction patterns of the input aperture fields. Depending on the degree of coherence of the source, the output irradiance will contain a component of an interference or speckle pattern. For these conditions, the integrated irradiance of the coherent component is adequately described by

$$I(x,y) = \left| \sum_{0,0}^{M,N} A_{mn} \exp\{i[k(\alpha_m x + \beta_n y) + \theta_{mn}]\} \right|^2 |F(x,y)|^2 \tag{10.8}$$

where:

α_m and β_n are the direction cosines defined in Equation 10.1
θ_{mn} is the phase of the beamlet
A_{mn} is the amplitude of the beamlet field
the function $F(x,y)$ is the diffraction integral of the beamlet-limiting aperture and the Fourier transform of the aperture function for the optical configuration in Figure 10.1

The first factor in Equation 10.8 describes the averaging and interference effects of the integrator. The interference effect is a result of the sum of linear (in x and y) phase terms, which can be viewed as a Fourier series. The spatial period of the resulting interference pattern is given by

$$P = \frac{\lambda}{\alpha} \tag{10.9}$$

where:

α is the angle between adjacent beamlets

For a spatially coherent source, the interference pattern will generally result in large fluctuations in the integrated irradiance. The only practical way to negate the effects of interference is to choose a sufficiently large value for α so that the interference pattern is too fine to be resolved in the application. If this is done, the effective integrated irradiance will be the local average of the irradiance in Equation 10.8. It is easy to show that the averaged irradiance is

$$I(x,y) = \sum_{0,0}^{M,N} |A_{mn}|^2 |F(x,y)|^2 \tag{10.10}$$

This result represents the ideal performance of a multifaceted beam integrator. Note that this result does not depend on θ_{mn}, the relative phase of the beamlets. The effects of the diffraction term are discussed with respect to specific configurations in Section 10.5.

The above results can be obtained for the system in Figure 10.1 using Fourier transform theory. Since the target is located at the focus of the primary lens, the irradiance pattern on the target is simply the magnitude squared of the Fourier transform of the pupil function modified by the lenslet array (assuming a spatially coherent source). If all the lenslets are identical, the aberrated pupil function is approximately the convolution of a two-dimensional (2D) delta function array (array of lenslet centers) and the lenslet pupil function, all multiplied by the laser beam amplitude profile. The Fourier transform is then the product of the Fourier transform of the delta function array (properly scaled) and the aberrated PSF of the subaperture, convolved with the PSF of the laser beam focused by the primary lens. Mathematically, the field just past the lenslet array is given by

$$E = E_b(x,y)[\Psi(x,y) * A_{\delta\delta}(x,y)] \tag{10.11}$$

where:

$E_b(x,y)$ is the field of the laser beam
$\Psi(x,y)$ is the lenslet pupil function
$A_{\delta\delta}(x,y)$ is the delta function array
$*$ denotes the convolution

The irradiance at the target plane is proportional to the magnitude squared of the Fourier transform of E:

$$I \propto |\Im[E]|^2 = \left|\tilde{E}_b * [\tilde{\Psi}(x,y)\tilde{A}_{\delta\delta}(x,y)]\right|^2 \tag{10.12}$$

where:

the small tilde (˜) denotes the Fourier transform operation

Since the Fourier transform of a periodic 2D delta function array is a 2D delta function array, the function, $\tilde{A}_{\delta\delta}(x,y)$, is the source of the interference effects.[10] The subaperture PSF, $\tilde{\Psi}(x,y)$, represents the diffraction effects. These diffraction and interference effects produce undesirable fluctuations in the irradiance distribution.

For a rectangular aperture array, the spatial period of the interference pattern is given by the ratio of the wavelength to the sine of the angle between the adjacent beamlets:

$$\text{Period} = \frac{\lambda}{\sin\theta} \tag{10.13}$$

Since sin θ approximately equals the lenslet spacing divided by the focal length of the primary lens, the interference periodicity is also given by

$$\text{Period} = \frac{\lambda F}{d} \tag{10.14}$$

Equations 10.13 and 10.14 are obtainable from $\tilde{A}_{\delta\delta}(x,y)$ scaled to the focal plane of the primary lens.

10.2.3 Averaging

The averaging aspect of multi-aperture beam integrating systems consists of making trade-offs between lenslet aperture size and β or, equivalently, the Fresnel number. Increasing the lenslet aperture size, d, increases β which reduces diffraction effects.

However, averaging is reduced. What is desired is to have d as small as possible while maintaining adequate β.

Assuming the intensity distribution within each subaperture is relatively uniform; the superposition of all the subapertures on the target plane will give a relatively uniform irradiance distribution (except for diffraction and interference effects). The assumption of uniform intensity within a subaperture is of course more valid the smaller the diameter of the subaperture. However, holding the f/d constant in order to maintain a spot size in accordance with Equation 10.4, Equation 10.6 shows the Fresnel number decreases linearly with subaperture diameter, which results in fewer peaks in the diffraction pattern with a correspondingly greater depth of modulation and a more gradual rolloff on the edges of the pattern at the target. This may require expanding the input beam in order to make the subaperture diameters larger or reducing the focal length of the primary lens. Varying the lenslet diameter according to Equation 10.6 across the lenslet array can also be used to superimpose different diffraction patterns such that the peaks of one pattern fall into the valleys of another. Varying the subaperture diameters by integer multiples allows 100% fill factor to be maintained. This is the approach taken by Pepler et al.[6]

10.2.4 Coherence Effects

As discussed in Section 10.2.2, multi-aperture beam integration systems will generally exhibit a degree of interference or speckle. The amount of speckle is determined by the spatial coherence of the source. A spatially incoherent source will not produce an interference/speckle pattern, and a spatially coherent source will produce the maximum interference/speckle. The results for intermediate cases will depend on the spatial coherence function representing the source. A general formulation of the problem in terms of coherence theory is quite involved and beyond the scope of this chapter. However, we will give a simplified formulation that will illustrate the major aspects of the problem.

The multi-aperture beam integrator can be viewed as a multibeam interferometer since all of the beamlets are superimposed at the target plane. The superposition of any two of the beamlets mimics Young's experiment. The visibility of fringes in Young's experiment is the definition of spatial coherence. In the following text, we will give only the rudiments of coherence theory needed to develop the simplified model of beam integration systems. For the basics of coherence theory, the authors recommend the book *Statistical Optics* by J. W. Goodman.[11]

Assuming a quasimonochromatic source, a generally good assumption for lasers, the spatial coherence of a laser is adequately described by the mutual intensity

$$J_{12} = \langle u(P_1,t)u^*(P_2,t)\rangle \tag{10.15}$$

where:

$u(P_1,t)$ is the analytic signal describing the optical field

P_1 and P_2 are points in the plane in which the coherence of the beam is being represented

$\langle\cdot\cdot\rangle$ denotes a time average

The mutual intensity is a correlation function when the functional dependence is an explicit function of the difference in coordinates, that is, $J_{12}(P_1,P_2) = J_{12}(P_2 - P_1)$. Note that by definition the irradiance (frequently called intensity) of the optical field is obtained for $P_1 = P_2$ as

$$I(P_1) = J_{12}(P_1 - P_1) \tag{10.16}$$

The complex coherence factor is defined as the normalized mutual intensity

$$\mu_{12}(P_1, P_2) = \frac{J_{12}(P_1, P_2)}{[I(P_1)I(P_2)]^{1/2}} \tag{10.17}$$

When one produces an interference pattern by combining radiation from points P_1 and P_2 in a Young's interferometer configuration, J_{12} may be regarded as the phasor amplitude of the spatial sinusoidal fringe pattern (on axis), whereas μ_{12} is the normalized fringe pattern. The complex coherence factor has the property:

$$0 \le |\mu_{12}| \le 1 \tag{10.18}$$

When $\mu_{12} = 0$, there are no interference fringes, and the two optical fields are mutually incoherent. When $\mu_{12} = 1$, the two optical fields are perfectly correlated and are mutually coherent. For intermediate values of μ_{12}, the fields are partially coherent.

The mutual intensity J_{12} can be computed at an output plane given J_{12} in the input plane using a generalized Van Cittert–Zernike theorem[12] and the relation

$$J_t(P_1, P_2) = t(P_1)t^*(P_2)J_i(P_1, P_2) \tag{10.19}$$

Equation 10.19 relates the mutual intensity transmitted by the object with transmission t to incident mutual intensity. In Equation 10.19, the numerical subscripts have been dropped and replaced by i and t, which represent the incident and transmitted mutual intensity, respectively. The generalized Van Cittert–Zernike theorem is a fourth-order integral over four variables. To develop such an integral for the system shown in Figure 10.1 would be very difficult and may not be very enlightening. For this reason, we will make a simplifying assumption that will illustrate the basic concepts. The basic assumption that we will use is that the field is mutually coherent over each subaperture in Figure 10.1, and will generally be partially coherent over greater distances. The assumption that the field is mutually coherent over each subaperture is fundamental to the performance of diffracting multi-aperture beam integration systems (Section 10.2). This assumption can be reduced for the case of imaging integrators. The following analysis will apply specifically to diffracting beam integrators.

With this assumption, we can approximate the field at the output plane as the sum of the coherent diffraction from each aperture:

$$E(x,y) = \sum_{0,0}^{M,N} (A_{mn} \exp\{i[k(\alpha_m x + \beta_n y) + \theta_{mn}]\})F(x,y) \tag{10.20}$$

The functions and variables in Equation 10.20 are defined following Equation 10.8. Assuming a degree of partial coherence between the beamlets, the coherence aspects of the problem are contained in the correlation between the amplitudes, A_{mn}. Given this, we can write the intensity at the target plane as

$$I(x,y) = \left\langle \sum_{0,0}^{M,N} (A_{mn} \exp\{i[k(\alpha_m x + \beta_n y) + \theta_{mn}]\}) F(x,y) \times \sum_{0,0}^{M,N} (A_{pq}^* \exp\{-i[k(\alpha_p x + \beta_q y) + \theta_{pq}]\}) F^*(x,y) \right\rangle \tag{10.21}$$

Noting that the time average only involves the amplitudes, the time average can be written as

$$I(x,y) = \sum_{0,0}^{M,N} |A_{mn}|^2 |F(x,y)|^2 + \sum_{m,n\neq}^{M,N} \sum_{p,q}^{M,N} \Big[\langle A_{mn} A_{pq}^* \rangle \exp(i\{k[(\alpha_m - \alpha_p)x + (\beta_n - \beta_q)y] + \theta_{mn} - \theta_{pq}\}) \Big] |F(x,y)^2| \tag{10.22}$$

Noting the time average bracket can be interpreted as a mutual intensity, we can write the last equation as

$$I(x,y) = \sum_{0,0}^{M,N} |A_{mn}|^2 |F(x,y)|^2 + \sum_{m,n\neq}^{M,N} \sum_{p,q}^{M,N} \Big[J_{mn,pq} \exp(i\{k[(\alpha_m - \alpha_p)x + (\beta_n - \beta_q)y] + \theta_{mn} - \theta_{pq}\}) \Big] |F(x,y)|^2 \tag{10.23}$$

Using Equation 10.17, the last equation can be written in terms of the complex coherence factor as

$$I(x,y) = \sum_{0,0}^{M,N} |A_{mn}|^2 |F(x,y)|^2 + \sum_{m,n\neq}^{M,N} \sum_{p,q}^{M,N} \sqrt{I_{mn} I_{pq}} \Big[\mu_{mn,pq} \exp(i\{k[(\alpha_m - \alpha_p)x + (\beta_n - \beta_q)y] + \theta_{mn} - \theta_{pq}\}) \Big] |F(x,y)|^2 \tag{10.24}$$

The reader should note that the subscripts refer to aperture elements in the lenslet array and the intensities, I_{mn}, are assumed constant over each aperture. With this in mind, the reader can see that the second term in Equation 10.24 is responsible for the interference (speckle) effects, and the first term is the sum of the irradiances at the output associated with each aperture. Clearly, if the fields in each aperture are mutually incoherent, $\mu_{mn,pq} = 0$, we have

$$I(x,y) = \sum_{0,0}^{M,N} |A_{mn}|^2 |F(x,y)|^2 \tag{10.25}$$

which is just Equation 10.10. When all of the aperture fields are mutually coherent, $\mu_{mn,pq} = 1$, we have maximum interference, and the exact form is dominated by the I_{mn} and θ_{mn}. In all other cases, the interference pattern will be influenced by the correlation between the various aperture fields, $\mu_{mn,pq}$, as well as I_{mn} and θ_{mn}. In all cases, the envelope of the irradiance pattern is given by the function, $|F(x,y)|^2$.

To estimate the effect of the coherence of the input laser beam on the interference pattern in the output using Equation 10.24, one needs to estimate the complex coherence factor. The complex coherence factor can be measured by repeating Young's experiment for pinhole pairs with different spacings between the holes. This is generally a tedious process. It is not uncommon, in practice, to design a prototype beam integration system for a given laser and experimentally measure its performance as part of the engineering design process. Experimental evaluation of the effects of spatial coherence is presented in Section 10.5. This section presents experimental data for spatially coherent (between subapertures) beams, partially coherent beams, and spatially incoherent beams.

10.2.5 Imaging Integrators

Diffracting integrators (Figure 10.1) are restricted to sources with a relatively high degree of spatial coherence over each subaperture. In addition, a change in the angle of incidence of the incoming light on the array causes a lateral shift in the irradiance spot on the target. As spatial coherence decreases or the angular spread or field angle of the incident light increases, the irradiance blurs to a spot larger than that given by Equation 10.4. This problem is eliminated with an imaging integrator, illustrated in Figure 10.2. The angular spread or field angle of the incident light can be quite large and still maintain overlap of the beamlets on the target plane. The imaging integrator can also offer improved integrator performance when the beam is collimated with a high degree of spatial coherence by reducing the diffraction effects. This effect is discussed in detail for reflective systems in Section 10.5.

An imaging integrator requires two lenslet arrays. The first lenslet array segments the input beam into multiple beamlets and directs these onto the second lenslet array. The first lenslet array serves to produce an intermediate image plane on or near the second lenslet array so that the second lenslet array can reimage the subapertures of the first array onto the target plane. The magnification of the subaperture images is given by the ratio of the focal lengths of the primary integrator lens and the second lenslet array. Thus, the spot size is again given by Equation 10.4 with slight modification that the second array lenslet focal length and first array lenslet diameter are used:

$$S = \frac{F}{f_2/d_1} \tag{10.26}$$

The second lenslet array serves as a field lens to redirect the off-axis chief rays back toward the optical axis. If the distance between the two lenslet arrays equals the focal length of the first array, the integrator can receive incident light over a maximum field angle equal to the second array subaperture semidiameter divided by the focal length of the first array:

$$\theta = \frac{d_2}{2f_1} \tag{10.27}$$

The simplest form of the second lenslet array element incorporates identical lenslets with mutually parallel optical axes. This is not a necessary requirement, for example, if a continuously variable tilt or decenter is applied to each of the lenslets as a function of position in the array, the system can be made into a zoom integrator with a continuously adjustable spot size. Zoom integrators are discussed further in Section 10.3.2 and several configurations are reported in the work of Dickey and O'Neil.[5]

10.2.6 Channel Integrators

Another approach to beam integration is the channel integrator. This section is taken verbatim from *Laser Beam Shaping Applications*.[12] As first proposed, a channel integrator is a reflective cylinder with a rectangular cross section.[13] Other cross sections are possible. The channel integrator concept can be easily explained in one dimension. These ideas can then be extended to two dimensions.

A one-dimensional (1D) schematic of the channel integrator is shown in Figure 10.5. In the figure, the two solid lines labeled 1 and 2 represent the channel integrator. The input beam is focused, with focal length *F*, on the center of the front face of the integrator. The integrator aperture size is *S*. To understand the integrator, consider the input rays centered on the optical axis that are bounded by an aperture of width *d*. This bundle of rays just fills the output aperture of the integrator. The ray bundle just above this one with the same aperture width *d* will also fill the output aperture after a reflection. This can be readily visualized by repeating virtual integrator walls, shown by the dashed lines. These virtual walls define the reflections and the input ray bundles that fill the output aperture. It can be seen that successive apertures with size *d* in the input are geometrically projected with inversion on the output aperture. Thus, this configuration is equivalent to the multi-aperture arrays discussed above. Although they are not addressing "laser" beam shaping, Chen et al.[14] provide interesting analysis that is applicable to channel integrators.

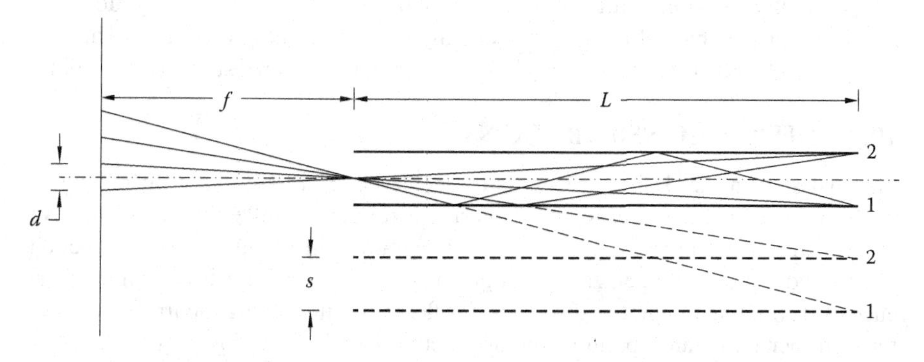

FIGURE 10.5 Schematic of the channel integrator.

The above argument can readily be extended to channel integrators with square or rectangular cross sections. It appears that channel integrators can be made with cross sections that can both tile a plane and meet the condition that each edge is a line of reflection. This clearly eliminates circular (or elliptical) cross sections. Using a different argument, channel integrators with circular cross sections can also be eliminated by considering the diagram that results from rotating Figure 10.5 about the optical axis. In this case, one can show that there are concentric rings that are mapped onto the ring defining the outer aperture of the integrator. One can also show that there are interspersed rings that map to a point at the center of the output aperture. The result is that the input irradiance mapping is not uniform over the output aperture. A complete analysis of the channel integrator with respect to possible aperture shapes would involve tiling and group theory and is beyond the scope of this chapter.

For the channel integrator, Equation 10.4 relating the integrator (input) aperture size to the output spot size is given by

$$d = \frac{FS}{L} \tag{10.28}$$

Equation 10.14 for the interference pattern period is also valid in this case since it depends on the angle between adjacent beamlets. The parameter β (derived in Appendix A of Chapter 5) is not directly applicable for the channel integrator. It is suggested that the related Fresnel number[15] be used for β in this case. The Fresnel number for the channel integrator is

$$N_F = \frac{d^2 L}{\lambda F(L+F)} \tag{10.29}$$

This equation for the Fresnel number is obtained by including the phase function representing the lens in the Fresnel integral defining the propagation of the beam. Requiring a large Fresnel number implies that L should be large or F should be small.

As discussed above, the channel integrator is equivalent to the multi-aperture beam integrator that comprises refractive lenslet arrays or multi-aperture reflective systems. There are a couple of disadvantages associated with channel integrators. One major disadvantage is that they tend to be lossy due to the multiple reflections involved. Another disadvantage is the complexity of needing, in most cases, to add a second lens to relay (image) the output pattern onto the working surface. There is the possibility of eliminating the first lens by tapering the channel integrator. One advantage of the channel integrator is the high-power handling capability. This is a result of the fact that the channel can be made of metal structures suitable for cooling.

10.3 DESIGN CONSIDERATIONS

The first step in the design process is to decide between building a diffracting or imaging beam integrator, and whether to use refractive, reflective, or diffractive components. Generally, an imaging beam integrator will produce lower diffraction effects and better homogenization, particularly for sources with low spatial coherence. A diffracting integrator allows greater flexibility in shaping the irradiance spot through aberrations and aperture flipping (Section 10.3.3). Imaging integrators generally introduce more complexity since there are more optical elements and associated alignment issues. Equation 10.4 is used to determine the first-order parameters of the lens elements of a diffracting integrator. The first-order parameters for imaging integrators are obtained using Equations 10.26 and 10.27.

Equation 10.3 or 10.7 is used to ensure that the parameters in a first-order design result in β or Fresnel number values required for a good beam shaping result. Depending on requirements, the Fresnel number should be at least 3.0 and the β parameter should be at least 40. The period of the interference pattern is obtained using Equation 10.14. Generally, it is desirable for the period of the interference to be small to reduce the interference effects. Source spatial incoherence also reduces the contrast of the interference pattern. Various software programs, either geometrical ray tracing or wave propagation, can be used to evaluate and optimize the lens aberrations to achieve the desired integration and smoothing.

10.3.1 Diffracting Integrator Layout

The optical layout of a basic multi-aperture beam integrator is shown in Figure 10.6. For clarity, only the rays for two outer lenslets are drawn. A collimated beam from a laser source is incident from the left. A lenslet array breaks up the incoming beam into an array of beamlets which are then overlapped at the target by a primary integrator lens. Either positive or negative lenslets (or a combination of the two) may be used in the array. Positive lenslets will produce a real beamlet focus ahead of the target plane (as shown in the figure). Negative lenslets will have a virtual beamlet focus either ahead of the lenslet array or behind the target plane, depending on the relative optical powers of the lenslets and primary lens. In the simplest configuration, all the lenslets are identical. Note that the spacing between the lenslet array and the primary integrator lens is not critical to first order.

The target plane is located at the focal point of the primary lens. The positive primary lens focuses the chief rays of each beamlet to a common point on the target, thus overlapping the defocused beamlets at the target. If the lenses are sufficiently free of aberrations, the spot formed at the focal point of the primary will replicate the lenslet aperture. Square lenslet apertures will form a square irradiance pattern, and circular lenslet apertures will form a circular irradiance pattern. If the lenslet has positive optical power, the spot will be an inverted replica of the aperture. If it has negative optical power, the spot will be an upright replica of the aperture. This is the basis for the aperture flipping technique discussed in Section 10.3.3. As shown in Section 10.3.4, aberrations can be added to the lenslets to significantly distort the irradiance pattern into almost any arbitrary shape.

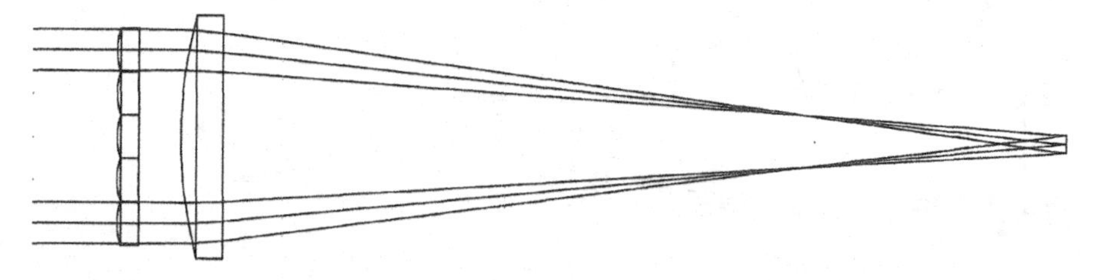

FIGURE 10.6 Optical layout of a refractive diffracting beam integrator.

10.3.2 Imaging Integrator Layout

Imaging integrators are well suited to sources with a low degree of spatial coherence.[16] Their drawback is the loss of flexibility in irradiance pattern shaping through lenslet aberrations. Figure 10.7 shows an optical layout for a simple imaging integrator with an extended source. For clarity, only the rays for two outer lenslets are drawn. The first lenslet array segments the input into multiple beamlets and focuses the beamlets onto the corresponding lenslets of the second array. To minimize stray light outside the target area, the beamlet diameters must not be greater than the lenslet clear apertures at the second array. Minimum spot size on the second array elements occurs when the array separation distance equals the first array lenslet focal length, in which case Equation 10.27 applies. Each element of the second lenslet array combined with the primary lens forms a relay lens that produces a real image of the pupils (of the first array lenslets) at the target plane. The geometry of each relay lens is such that the pupil images are superimposed at the output plane.

To make a zoom imaging integrator that allows continuous adjustment of the target spot diameter, one simply needs to replace the single primary integrator lens with a multielement variable focal length camera lens.

Alternatively, one can add to each lenslet in the second array a tilt that is proportional to the lenslet's radial distance from the system optical axis. This is equivalent to superimposing the primary integrator lens onto the lenslet array and thus eliminating one element. Figure 10.8 shows such a zoom imaging integrator.

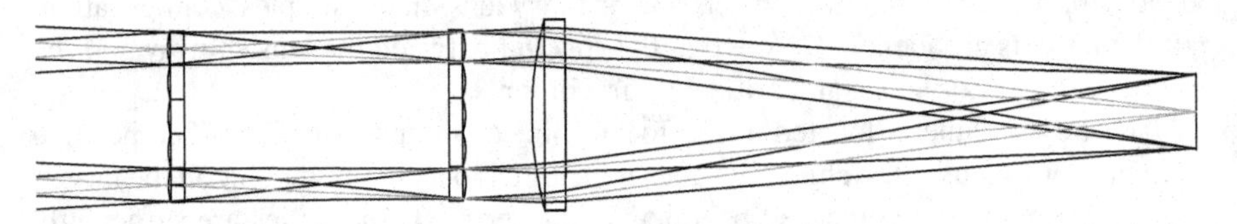

FIGURE 10.7 Optical layout of a refractive imaging integrator.

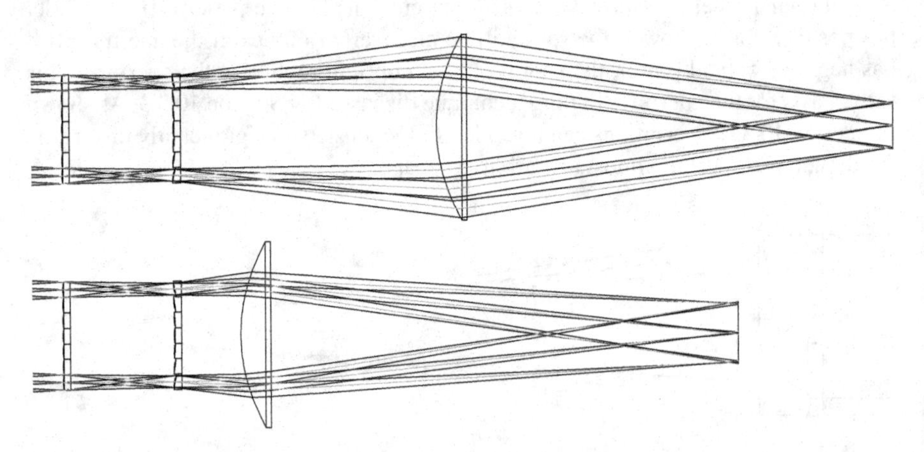

FIGURE 10.8 Zoom imaging integrator for variable spot size.

10.3.3 Subaperture Shape

The fact that the spot has the same shape as the subaperture allows almost any spot shape to be produced. Usually, the designer wishes to stack the lenslets with 100% fill factor in order to maximize the energy on the image plane. Lenslet aperture shapes that are easily stacked with 100% fill factor include square, rectangular, and hexagonal. If a diffracting integrator is used, one can form a triangular irradiance pattern with 100% fill factor at the lenslet array. If a triangular subdivision of a square, rectangle, or hexagon results in only two different triangles that are inverted images of each other, then the sign of the phase function can be flipped for inverted apertures resulting in a single integrated image of one of the triangles.

An optical layout of an aperture-flipped diffracting integrator is illustrated in Figure 10.9. As shown in Figure 10.10, the lenslet apertures are equilateral triangles

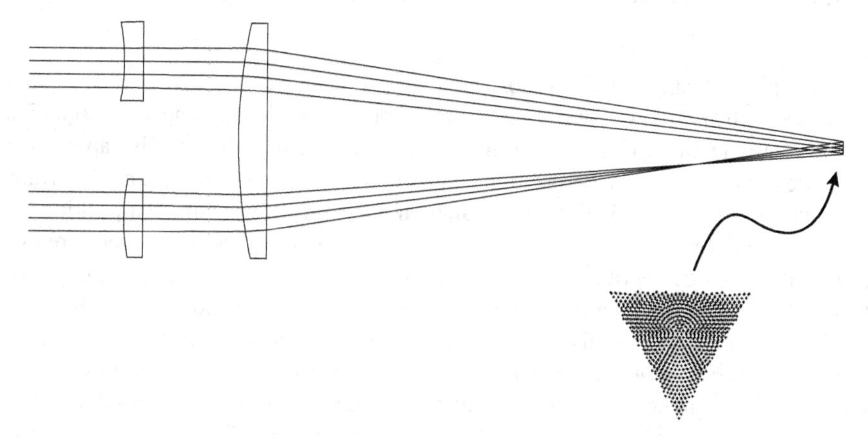

FIGURE 10.9 Diffracting integrator incorporating aperture flipping.

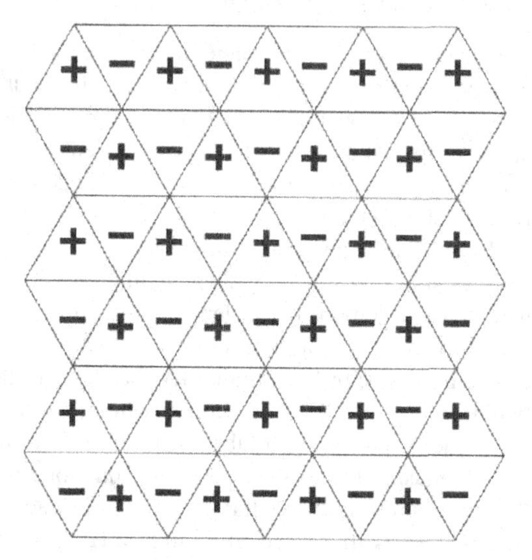

FIGURE 10.10 Triangular apertures in a hexagonal packing with sign-flipped optical powers.

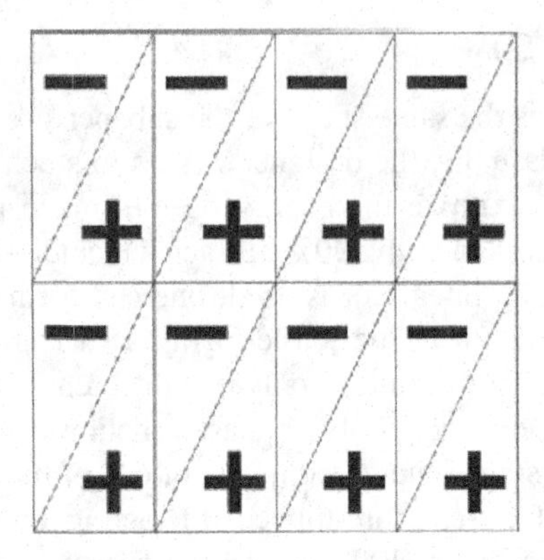

FIGURE 10.11 Triangular apertures in a rectangular packing with sign-flipped optical powers.

in a hexagonal packing. Triangles oriented with apex up have positive optical powers. Triangles with apex down have negative optical powers but of the same magnitude. The positive apex-up apertures become inverted at the target but the negative apex-down apertures are not. Their superposition results in a single apex-down triangular irradiance pattern. Figure 10.11 illustrates triangular apertures in a rectangular packing.

Aperture flipping is practical only with diffracting integrators. To use aperture flipping with imaging integrators would require at least a third lenslet array, adding greatly to system complexity. In order to incorporate aperture flipping and maintain 100% fill factor at the lenslet array, the subdivision of the basic square, rectangle, or hexagon shape must result in two subapertures which are inverted images of each other. The phase functions in these two apertures have the same magnitudes but opposite signs.

10.3.4 Lens Phase Functions and Aberrations

Once the first-order layout and aperture shape are determined, aberrations in the lens elements can be adjusted to fine-tune the irradiance distribution on the target. Aberrations in both the lenslet array and the primary integrator lens affect the irradiance uniformity on the target. Aberrations in the primary lens (e.g., spherical aberration) will result in a lateral displacement between the overlapped beamlet diffraction patterns at the target which will tend to compensate for the diffraction-induced irradiance modulation. Equivalently, a slight defocus of the target from the ideal focal point produces a similar effect, as shown by Deng et al.[17]

The phase function of the lenslet array elements affects the diffraction pattern or image of the lenslet apertures. For diffracting integrators, nonquadratic lenslet phase functions can be used to significantly modify the shape of the irradiance pattern. This is particularly useful in laser machining where one desires to shape the laser irradiance to the clear aperture of a fabrication mask in order to increase the laser power through the mask.[18] For example, a phase function that is a linear function of radius, $\phi = \alpha r$, where r is the radial coordinate in each subaperture, will produce a ring pattern. This is illustrated in Figure 10.12. Horizontal and vertical rectangular subapertures with decentered lens functions can be used to form hollow

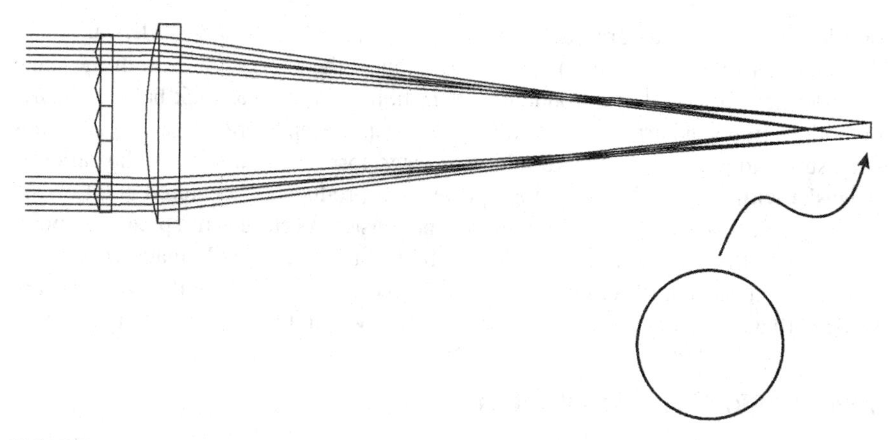

FIGURE 10.12 Diffracting integrator forming a ring pattern on target.

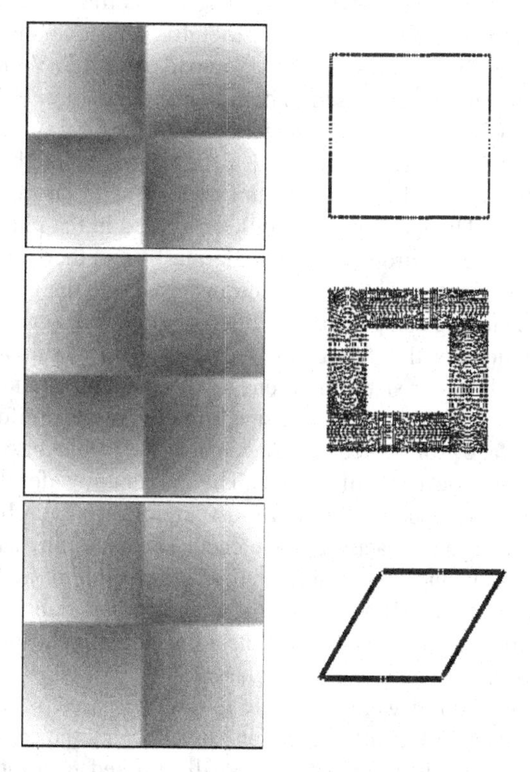

FIGURE 10.13 Lenslet surface profiles (represented as grayscale levels) and resulting irradiance patterns: hollow square, thick hollow square, and hollow trapezoid represent irradiance patterns.

square patterns. Rooftop prisms can be superimposed onto the lenslets, giving a linear phase tilt in one axis, to produce two parallel bars in the irradiance pattern.[19] Alternatively, lenslets can be decentered to produce multiple spots on the target.[20]

Figure 10.13 illustrates three different lenslet surface profiles and their corresponding irradiance patterns on the target (determined by ray tracing). The surface profiles, represented as grayscale patterns, are similar to the first order but differ in the higher order (aberration) terms. Variation of the aberrations within the same basic lenslet profile allows producing a thin hollow square, a thick hollow square, or hollow trapezoid irradiance patterns. The square aperture of each lenslet has been subdivided into four square subapertures to form the four sides of the patterns. A lenslet with only tilt but no optical power will produce a decentered point on the target. Optical power in one direction in the lenslet aperture will spread the point in the same direction on the target. The hollow patterns are easily made with cylindrical lenslets with a tilt or decenter term superimposed. Thick hollow patterns are made with a small anamorphic term or optical power in the orthogonal direction.

10.4 FABRICATION CONSIDERATIONS

The key element in multi-aperture beam integrators and most difficult element to fabricate is the lenslet array. This element can be refractive, reflective, diffractive–transmissive, or diffractive–reflective. Metallic reflective anamorphic or non-rotationally symmetric elements can be diamond turned.[21] Various methods of fabricating microlenses are discussed in detail in the literature;[22,23] so we will only briefly outline two of the more common technologies for fabricating these elements. Photolithographic technology can be used to fabricate any of the above types of lenslet arrays. Photoresist of appropriate thickness is spun on the glass wafer and hardened by baking. The lens surface profile is formed in the photoresist by exposure to UV light through chrome or grayscale masks and then development of the photoresist. The lens surface profile is then transferred into the glass by reactive ion milling or plasma etching using primarily fluorine or chlorine gas.

Inherent limitations of this fabrication technology vary from vendor to vendor, but generally include wafer size limitations, wafer material limitations, maximum etch depth limitations, minimum feature size limitations, etch uniformity across the wafer, and lens surface profile accuracy. Acceptable wafer thickness typically ranges from about 300 μm to several millimeters. The maximum wafer diameter that can be easily processed is about 150 mm. Almost any material can be ion milled but only a few materials can be reactively ion etched. The common material choices for reactive ion etching include fused silica, silicon, germanium, and ZnSe. Jenoptik Optical Systems (formerly MEMS Optical) in Huntsville, Alabama, has developed etch chemistries for many other materials, including gallium phosphide, zinc sulfide, chalcogenide glasses, and other common glasses.

Etch selectivity (ratio of wafer etch rate to photoresist etch rate) can be varied over a limited range with reactive ion etching by changing the etchant gas mixture. Selectivity for the ion milling process is usually limited to a ratio of about one. Limits on selectivity and maximum workable photoresist thickness place a limit on the maximum etch depth or lens surface sag. The increase in surface roughness with increasing etch depth also limits maximum etch depth. The maximum etch depth for fused silica is about 20 μm. Silicon, due to its higher etch rate, can be etched much deeper to 60 μm or more. The above numbers are loose approximations and vary from vendor to vendor, but they provide the reader a general idea of the types of fabrication limitations involved.

Photoresist heating and reflow is a common inexpensive method of fabricating microlenses. A single binary chrome mask is used to produce pillars of photoresist that are subsequently reflowed into a lenslet surface by heating. The profile is then etched into the glass wafer by reactive ion etching or ion milling. A few drawbacks of this method are noted here. First, it is difficult to get 100% fill factor (ratio of lens surface area to wafer surface area) with this method. A sufficiently large gap must exist between adjacent lenses to prevent the merging of photoresist from separate lenses. Second, the reflow method cannot produce negative lens elements. Third, the lens elements tend to have focal lengths that are too short for beam integrators.

The alternative grayscale mask fabrication method solves the above problems as can be seen in Figure 10.14. The drawback of grayscale mask technology is greater process development and cost. The inherent nonlinearities of this process often require iterative corrections to the grayscale mask to accurately produce the desired lenslet surface profile.

Reliable surface profilometry equipment, whether contact or optical, is essential for fabricating microlenses. Contact surface profilometers that do not raster scan the lenslet profile, but rather make only a single scan (cross section) of the lens, can be difficult to use as the stylus must scan through the vertex of the lens in order to obtain an accurate measurement of the surface profile. Optical profilometers, such as interferometric microscopes, can capture the entire three-dimensional (3D) surface profile of the lens. However, optical profilometers designed to measure flat surfaces can introduce erroneous spherical aberration into the measurements when measuring the steep surface curvatures of microlenses. A Fizeau interferometer can give

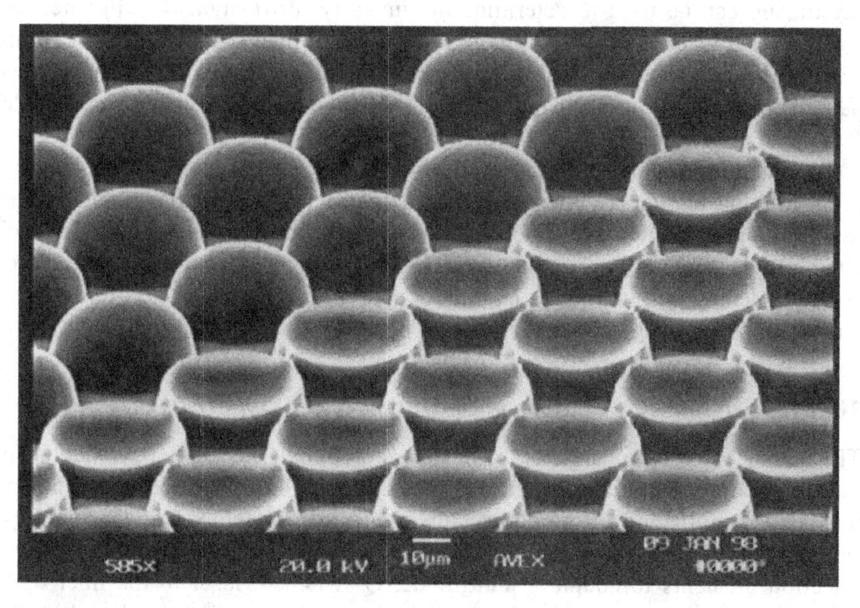

FIGURE 10.14 Array of positive and negative elements fabricated by grayscale mask technology.

better results for surface profiles that are close to spherical. A ball lens or metal microsphere can be used to check a particular instrument to determine if it introduces erroneous aberrations into the measurements.

10.5 APPLICATIONS AND EXPERIMENTAL EVALUATION

Numerous applications exist for multi-aperture beam integrators. A few of these applications include laser heat processing (including medical and dental applications), laser machining, product marking, laser diode array integration for laser pumping and fiber injection, photolithographic mask aligners and steppers, and fiber injection systems. We show experimental results of a fiber injection application below.

The light sources of photolithographic steppers and contact mask aligners are typically high-pressure mercury or xenon arc lamps that supply the required high-intensity UV irradiance. The energy produced by these highly compact and relatively noisy arc sources must be uniformly distributed over the area of the photolithographic mask. A highly uniform irradiance at the mask plane is particularly critical for grayscale mask processes. Multi-aperture beam integrators have been successfully used on photolithographic equipment to homogenize arc sources for many years. Examples of such beam integrators are described in patents by Mori and Komatsuda[24] and Komatsuda et al.[25]

Arc sources can be modeled as a series of small concentric ellipsoids, located near the cathode tip, whose radiance decreases with increasing size of the ellipsoids. Most of the radiant power originates from an ellipsoidal region less than 1 mm in diameter near the cathode tip. Although these sources are often referred to as "point sources," their finite size still results in a finite angular distribution of intensity in the collimated beam. Thus, arc sources are only partially spatially coherent. Nonsequential ray tracing can be used to determine the intensity distribution of collimated arc sources for a given collimator system. Due to the limited spatial coherence of arc sources, imaging beam integrators are ideally suited for forming uniform irradiance patterns with these sources. The Van Cittert–Zernike theorem can be used to model the spatial coherence of collimated arc sources.

A related but slightly different application of beam integrators is the combining of the multiple emitters of laser diode arrays to form a single irradiance pattern. For example, a laser printer application might require the magnified line images of each emitter to be superimposed on the target. Instead of segmenting a collimated source input, the lenslet array reimages multiple sources at finite conjugates. An integrator lens overlaps the images at the target.

10.5.1 Experimental Evaluation of Diffracting Beam Integrator

The theory and effectiveness of the design approach discussed in this chapter for the diffracting or nonimaging beam integrator can be illustrated using experimental data collected on a compact fiber injection system.[26] For this application, a single lenslet array and a plano–convex lens were employed as the fiber injection elements to couple a multimode, Q-switched, laser to the fiber optic transmission system. Characterization of the intensity profiles produced by the diffracting beam integrator using various laser sources and primary lens focal

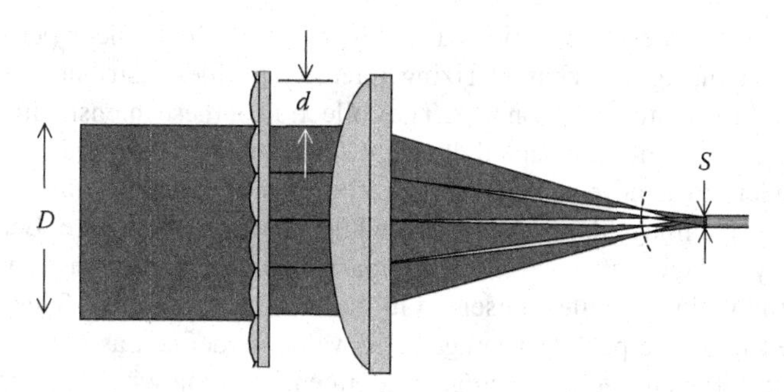

FIGURE 10.15 Optical architecture and design specifications for the diffracting beam integrator and fiber injection system.

lengths illustrate the diffraction, interference, and averaging effects discussed in Section 10.2.

The optical components and their specifications that form the basis of the fiber injection system are shown in Figure 10.15. This system comprises a multi-aperture, refractive, lenslet element fabricated by Jenoptik Optical Systems and a primary injection lens that overlaps or integrates the beamlets from each subaperture at the lens focal plane. Given the set of specifications shown in Figure 10.15 and the design equations presented earlier in this chapter, the functional injection parameters are calculated and summarized for the reader in Table 10.1.

Input Source	**Primary Injection Lens**
Wavelength: 1061 nm	Type: Plano/convex
Diameter (D): 5.0 mm	Material: Fused silica
Divergence: 2 mR (full angle)	Focal length (F): 17.1 mm
Lenslet Array	**Integrator Output**
Fill factor: >98% (hexagonal)	Spot size (S): 0.31 mm
Subaperture (d): 1.25 mm	Fiber size (core): 0.365 mm
Focal length (f): 68.95 mm	Fiber NA: 0.22

TABLE 10.1
Calculated Injection Parameters for the Diffracting Beam Integrator

Equation	Calculated Value	Measured Value
$\beta = \dfrac{\pi d S}{\lambda F}$ (10.3)	67	N/A
$S = \dfrac{F}{f/d}$ (10.4)	310 μm	353 μm
$\text{Fresnel} = \dfrac{dS}{4\lambda F}$ (10.7)	5.34	Modeled
$\text{Period} = \dfrac{\lambda F}{d}$ (10.14)	14.3	16.3 μm

Verification of the design equations and illustration of the lenslet's performance are based on capturing and characterizing intensity profiles distributed along the optical axis of the primary injection lens. The collection of these intensity images was accomplished using a charge-coupled device (CCD) camera* configured with a suitable objective lens to achieve an approximate 10× magnification. Postprocessing and manipulation of the image data was performed using analysis software, Beamcode.† With this analysis package, a qualitative comparison of the peak intensity value can be made for the various profiles presented in the sections that follow. This figure of merit, referred to as the peak-to-average (*P*/*A*) value, is defined as the ratio of the peak pixel intensity count to the average pixel intensity count within a user-defined diameter. Subsequently, qualitative comparisons of intensity profiles are made by maintaining a consistent user-defined diameter for all cases in which the peak and average pixel counts are calculated. (All *P*/*A* values presented herein are based on a 365 μm diameter positioned at the centroid of the imaged intensity profile.) With this definition, a perfect flat-top intensity profile extending over the entire 365 μm diameter would have a *P*/*A* value of unity—the theoretical ideal for most fiber injection applications.

10.5.1.1 Diffraction and Interference Effects

The theory of the effects of laser coherence on the performance of the diffracting beam integrators discussed in Section 10.2 can be illustrated by observing the intensity distribution located at the focal plane of the primary injection lens when such a system is illuminated by a variety of laser sources—representing varying degrees of spatial coherence. As discussed in Section 10.2.1, the diffraction pattern of a single aperture of the lenslet array determines the shape of the spot on the target, and the Fresnel number (or β) yields a general measure of the diffraction structure or deviation from the geometric ideal (i.e., flat-top profile). In addition, it was shown that both diffraction and interference effects play an integral role in determining the ultimate intensity distribution. For example, a source that is characterized as "highly spatially coherent" produces an intensity pattern dominated by interference effects characterized by large intensity fluctuations (or spikes) with a well-defined periodicity at the lens focal plane—while sources having a "lesser degree" of spatial coherence exhibit a mixture of both interference and diffraction effects.

Three different laser sources were used to "separate" and better illustrate the effects of diffraction and interference on the ultimate performance of the diffracting beam integrator described in Figure 10.15 and Table 10.1. Although each laser source that we evaluated presents a different spatial intensity pattern at the input to the lenslet element, it is the spatial coherence of the source that appears to dominate the resulting profiles at the injection lens focal plane.[26]

Using a multimode, Q-switched, Nd:YAG laser (Laser Photonics Model YQL-102), diffraction effects are clearly distinguishable in the intensity profiles shown in Figure 10.16a. Visible in the intensity profiles are the shape of the lenslet elements (hexagonal) and the expected diffraction pattern. With this laser source,

* COHU 4800 Camera active picture pixel size—horizontal: 23.0 μm; vertical: 27.0 μm.

† It is an optical beam diagnostic system providing real-time analysis of captured intensity profiles.

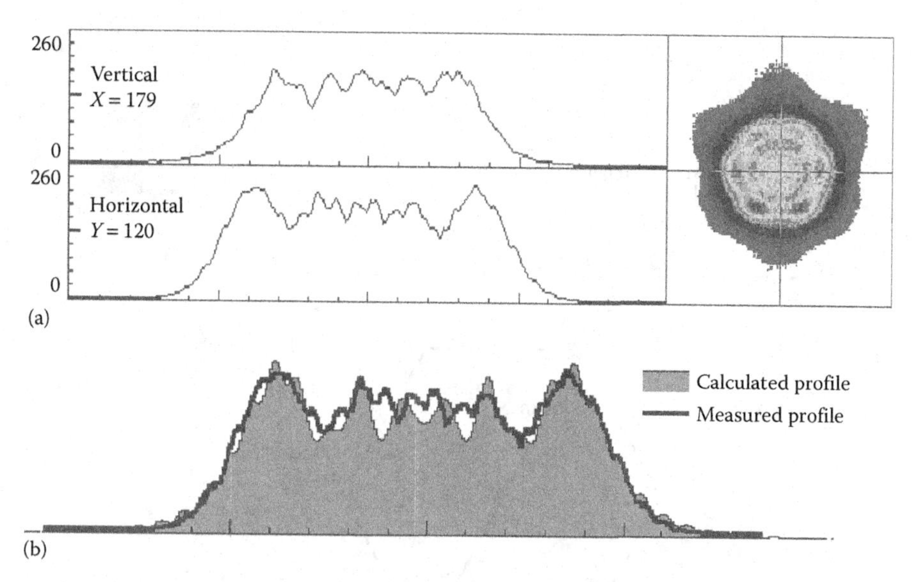

FIGURE 10.16 (a) Measured intensity profile at the focal plane of the diffracting beam integrator with a low spatially coherent source ($P/A = 2.50$, source laser: Laser Photonics YQL-102). (b) Comparison of measured and calculated intensity profile at the focal plane of the diffracting integrator.

lenslet diffraction dominates the intensity structure and the effects of the Fresnel number can be experimentally observed and analytically verified. Overlaying the calculated 1D diffraction profile with the experimental results displayed in Figure 10.16b reveals excellent agreement with the measured data. This suggests that the structure observed is indeed dominated by lenslet diffraction effects.

At the other performance extreme, a CW Cr:Nd:GSGG, TEM_{00} laser (AMOCO Model 1061-40P) was used to illustrate the interference effects resulting from a source that can be characterized as highly spatially coherent. As expected, the intensity profile shown in Figure 10.17a displays both diffraction and interference effects. The hexagonal shape of the lenslet elements and a slowly varying intensity modulation earlier attributed to diffraction effects (Figure 10.16b) are apparent; however, the intensity profile is clearly dominated by narrow spikes—indicative of interference effects. Confirmation that these features are a consequence of interference effects can be made by comparing the calculated and measured periodicity of this structure (Table 10.1). It should be noted that Equations 10.9, 10.13, and 10.14 derived for the periodicity of the interference pattern are based on paraxial approximations and assume a 1D lenslet pattern. The calculated intensity profile and its periodicity can be further studied to include the effects of both the hexagonal lenslet array structure and system aberrations using more advanced optical modeling software. Analytical results, obtained using an optical modeling package, Advanced System Analysis Program (ASAP),* shown in Figure 10.17b, reveal the effects of the lenslet geometry

* ASAP is an optical software program for geometrical and physical modeling. It is a trademark of Breault Research Corporation.

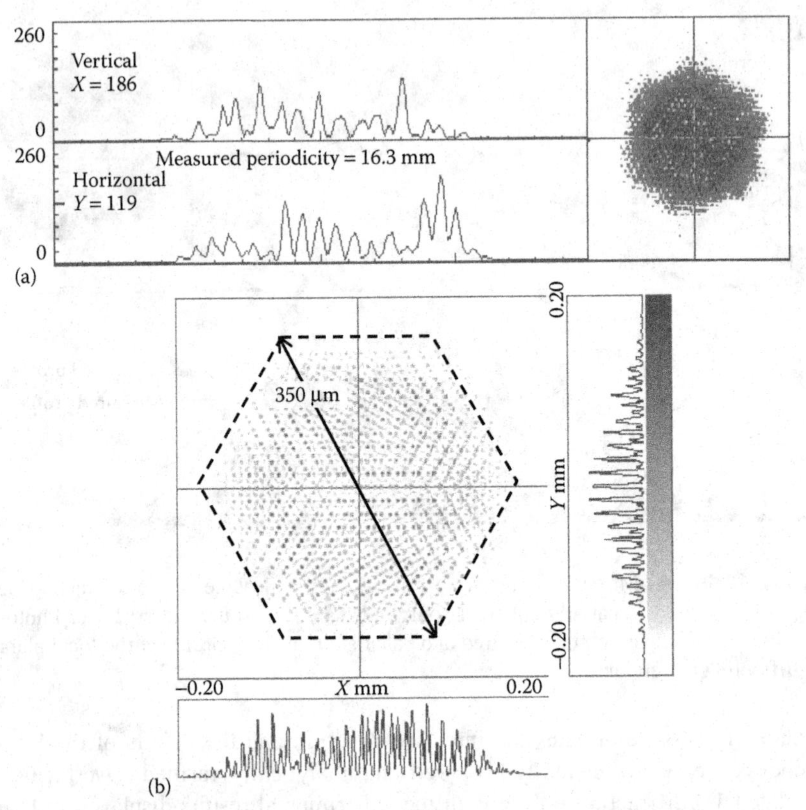

FIGURE 10.17 (a) Measured intensity profile at the focal plane of the diffracting integrator with a high spatially coherent source (*P*/*A* = 8.08, source laser: AMOCO 1061-40P). (b) Calculated intensity profile at the focal plane of the diffracting beam integrator using the ASAP modeling software (period = 14.2 μm).

and system aberrations on the intensity profile recorded at the primary lens focal plane. Assuming a perfectly coherent source, the model displays the expected 2D dependence for both the interference periodicity (14.2 μm) and intensity.

For the condition of partial spatial coherence, a third laser source was evaluated against the diffracting beam integrator described in Figure 10.15. In this case, the source was a "home-built," multimode, Q-switched, Cr:Nd:GSGG laser. (This same laser was used extensively to characterize other performance parameters of the beam integrator further described in the works of Weichman et al.[26]) Illumination of the diffracting beam integrator with this source produces the intensity profile shown in Figure 10.18 at the primary lens focal plane. Close observation of this profile illustrates effects of both diffraction and interference. The spot geometry and Fresnel structure displayed earlier in Figure 10.16 are visible; however, superimposed on the diffraction pattern are narrow, high-intensity "spikes." Comparing the periodicity of the high-frequency structure observed in Figure 10.18 with that shown in Figure 10.17 implies that these features are indeed generated by interference effects. Moreover, comparing the *P*/*A* pixel response from Figures 10.16

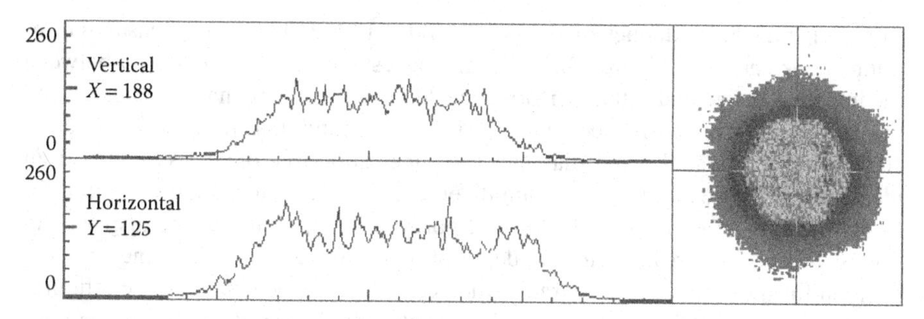

FIGURE 10.18 Measured intensity profile at the focal plane of the diffracting beam integrator with a partial spatially coherent source (P/A = 3.19, source laser: Q-switched, Cr:Nd:GSGG).

through 10.18 (2.50, 8.08, and 3.19, respectively) further suggests that for this beam shaping configuration, interference effects will be a major contributor in determining the localized peak intensity at the intended target.[26]

10.5.1.2 Spot Diameter and Averaging

Characterization of the intensity profiles collected along the optical axis of the diffracting beam integrator described in Figure 10.15 yields a location and estimate for the minimum beam diameter and further describes how the profile evolves past the lens focal plane. The spot size as a function of distance along the optical axis is plotted in Figure 10.19. In this case, a functional description of the spot size is defined as the diameter in which 98% of the energy is contained. As expected, the minimum spot size is achieved at the focal plane of the injection lens and that the growth in the beam diameter behind this plane is described by the paraxial approximation given by the input beam diameter and the primary lens focal length (i.e., *f*-number).

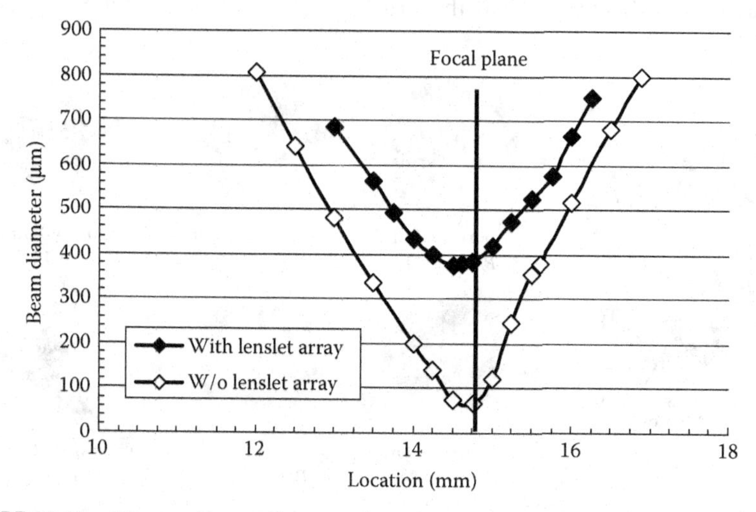

FIGURE 10.19 Measured beam diameter along the optical axis with and without the lenslet array (injection NA = 0.17).

Comparing the beam diameter calculated in Table 10.1 (310 m) to the measured minimum value shown in Figure 10.19 (353 m) suggests reasonable agreement between the intended design and actual performance. Differences between the calculated and measured results are easily accounted for in those assumptions inherent in Equation 10.4 for the calculated beam diameter (i.e., aberration-free optical system and an ideal source with essential zero beam divergence) and the functional definition for beam diameter presented earlier. Another potentially important feature displayed in Figure 10.19 is the greater effective depth of focus around the focal plane or target location for the diffracting integrator when compared to the simple lens configuration. The importance of this characteristic could become significant when the integrator is incorporated into an assembly where alignment or position insensitivity is considered desirable (i.e., fiber injection, photolithography, laser drilling, etc.). This characteristic will be discussed in more detail in the context of the compact fiber injection system presented in Section 10.5.2.

The averaging aspect of the diffracting beam integrators and the subsequent insensitivity of the intensity distribution at the target plane to spatial perturbations of the input source is a highly desirable characteristic of many beam shaping systems. Once again, intensity profiles collected at the focal plane using various laser sources and different input beam diameters have been used to illustrate this feature. Applying various apertures to the input source and hence effectively exposing different near-field features to the lenslet array can yield some insight into the sensitivity of the diffracting beam integrator to these changes. The results of such an experimental characterization using the Q-switched, Cr:Nd:GSGG laser described earlier are shown in Figure 10.20. The contour sequence displayed in Figure 10.20 reveals little change in the output intensity profile (or *P*/*A* value) recorded at the lens focal plane until the input beam diameter is comparable in size to the diameter of a single lenslet element on the array. When this occurs, the primary injection lens no longer "overlaps" or integrates inputs from multiple lenslet elements. In the limit of illuminating a single lenslet, this effectively yields the spatial profile that would result using just the

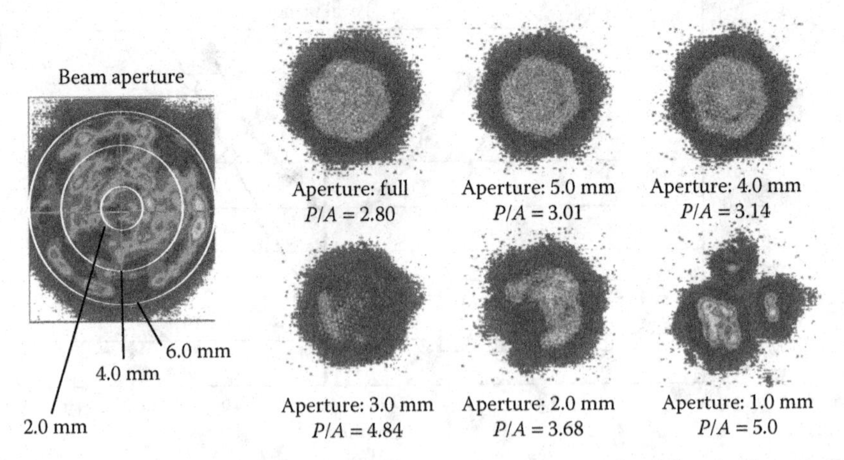

FIGURE 10.20 Measured intensity profiles at the focal plane of the diffracting beam splitter with various apertures at the source laser.

primary injection lens slightly defocused from its focal plane. (This slight "defocus" is a result of the nonzero optical power provided by the lenslet element.)

10.5.2 Compact Fiber Injection System

Various system applications exist in which the advantages offered by the multi-aperture integration systems discussed in the chapter can readily be applied. Specifically, a number of medical and industrial laser applications exist in which the limiting factor for performance or additional capability resides in the transmission of the optical power (or energy) from the source to the target. Typical transmission systems for these applications include the use of fiber optics or "open air" designs requiring numerous optical elements to direct and "reshape" the laser radiation along the intended path to the target. Depending on the specific application and configuration (i.e., target accessibility, laser wavelength, peak power, etc.), the use of fiber optics is often the preferred transmission system. Although the use of optical fibers provides the system designer and the end user effective control over a number of application parameters (i.e., beam diameter, spatial profile, target accessibility, etc.), it is often limited by the maximum power level that can be reliably injected into and ultimately transmitted to the target.

Maximizing power throughput and minimizing fiber damage require a thorough understanding of fiber damage mechanisms and the control of a number of fiber injection criteria.[27,28] In contrast to the "simple" injection lens, effective control of a number of these parameters is provided by the diffracting beam integrator discussed in Section 10.5.1 and shown in Figure 10.15.

Performance characterization of a compact fiber injection system featuring the diffracting beam integration approach, conducted at Sandia National Laboratories,* Albuquerque, New Mexico, showed that geometrical and mechanical constraints were significant factors in determining the optical architecture employed to effectively integrate and optically couple a miniaturized, Q-switched laser source with the desired fiber optic transmission system. Mechanically constrained in the overall length of the fiber injection and alignment system to less than 25 mm, more complex beam shaping techniques were quickly abandoned and the more traditional simple injection lens approach was initially evaluated. However, as Figure 10.21 summarizes, the simple lens approach for high-power applications imposes a number of functional limitations on the power or energy that can be reliably injected and transmitted by the optical fiber. With a single injection lens element, the ultimate limitations and performance variability are in practice a combination of the low air breakdown threshold and the strong interdependence of the laser source characteristics and injection alignment to the peak optical fluence incident on the fiber face. It is worth noting that these issues are further exasperated by the use of short focal length lenses and high-peak-power, multimode lasers—both conditions inherent in this application.

---

* Sandia is a multiprogram laboratory operated by Lockheed Martin for the US Department of Energy under Contract DE-AC04-94AL85000.

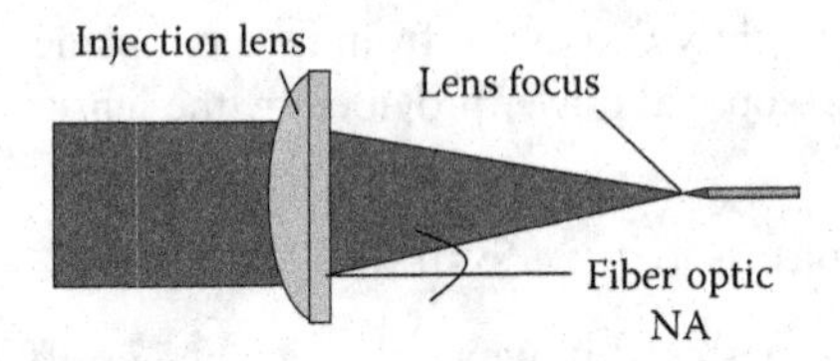

FIGURE 10.21 Fiber injection issues using the simple plano–convex injection lens.

• Low air breakdown thresholds—*all* energy is focused at *single* foci. • Spatial profile at the fiber is strongly *dependent* on input laser and alignment. • Beam diameter at the fiber face is *strong* function of lens displacement.	• NA of injection system is *limited* by the need to avoid air breakdown. • Exit profile and potential for damage internal to the fiber *dependent* on the initial mode power distribution (MPD) incident on the entrance face.

Effective control of a number of those fiber injection issues shown in Figure 10.21 is provided by the diffracting beam integrator evaluated and discussed in Section 10.2. Beyond those results presented earlier illustrating the effects of diffraction, interference, and spatial averaging on the intensity profile at the target (or fiber) plane, a brief performance comparison of the simple injection lens and the diffracting beam integrator is discussed in the following material.

With the simple injection lens approach, the conditions imposed to avoid air breakdown and provide "adequate" filling of the fiber NA are diametrically opposed. Avoiding air breakdown with the simple lens implies controlling the minimum spot size and hence the lens focal length. Unfortunately, as the lens focal length is increased to accommodate a higher threshold breakdown, the injected entrance angle or NA to the fiber is correspondingly decreased. In contrast, the diffracting integrator distributes the input laser energy over a larger cross-sectional area and into multiple foci representing the number of active lenslet elements. Consequently, eliminating air breakdown allows the designer the freedom to select the primary lens focal length to better match the acceptance angle of the optical fiber. However, as discussed in Section 10.2.1, as the focal length of the primary lens is varied, the *f*-number of the lenslets must be correspondingly adjusted to maintain the desired spot size per Equation 10.4. Maintaining the spot size is achieved at the expense of changing β or the Fresnel number defined in Equations 10.3 and 10.7, respectively. It is the interdependence of the lens focal length and β that must be optimized given the specifics of the desired injection geometry.

Another important characteristic provided by the diffracting beam integrator in the evaluated fiber injection system is the large depth of field discussed in Section 10.5 (Figure 10.19) and the alignment insensitivity to the input source that is a result of the lenslet's spatial averaging behavior (Figure 10.20). The large depth of field enables the designer to use a very simple mechanical mount that is required to provide only gross adjustment capabilities along the lens optical axis.

A more troublesome design issue with fiber injection systems and a parameter substantially relaxed by the use of the diffracting beam integrator is the alignment

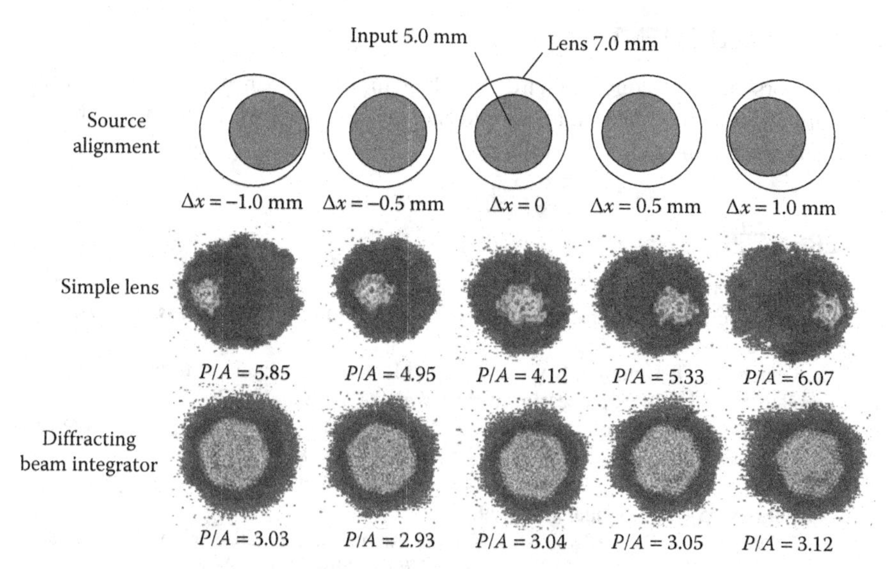

FIGURE 10.22 Measured intensity profiles at the fiber face with lateral misalignment of the source.

or centralization of the input beam on the primary injection lens. As shown in Figure 10.22, when the input beam is displaced across the injection lens without the lenslet array, the spatial intensity profile incident at the fiber face is strongly affected by system aberrations—resulting in higher peak fluences physically decentered from the fiber core. In contrast, when the lenslet array is added, the spatial profile (or *P*/*A* value) remains unchanged over a wide range of lateral misalignment. Once again insensitivity to alignment significantly relaxes the mechanical requirements for orienting the laser source to the fiber injection system.

10.6 SUMMARY

In this chapter, we have presented the theory and design of multi-aperture beam integration systems. These systems are especially applicable to the shaping of multimode laser beams that are characterized by an irregular irradiance pattern that frequently varies with time. The major assumptions applicable to the design and analysis of a multi-aperture beam integration system are stated explicitly.

As discussed in Chapters 2 and 5, β, or equivalently the Fresnel number, is again an important parameter in determining the performance of the beam shaping system. The basic concepts and equations needed for system design are developed. Diffraction and interference effects associated with multi-aperture beam integration systems are treated in detail. The impact of the input beam spatial coherence on beam integrator performance is analyzed.

Experimental data illustrating the effects of diffraction, interference, and spatial coherence on beam integrator performance are presented. Finally, we discuss the design and present data for a diffractive multi-aperture beam integration system for optical fiber injection.

ACKNOWLEDGMENT

The authors acknowledge Richard N. Shagam for Advanced System Analysis Program (ASAP) modeling analysis and Stephen E. Yao and Jeremy D. Brown for their help with manuscript preparation.

REFERENCES

1. H.J. Kahlert, U. Sarbach, B. Burghardt, and B. Klimt. Excimer laser illumination and imaging optics for controlled microstructure generation. *Proceedings of SPIE* 1835: 110–112, 1992.
2. L. Unnebrink, T. Henning, E.W. Kreutz, and R. Poprawe. Excimer laser beam homogenization for materials processing. *Proceedings of SPIE* 3573: 126–129, 1998.
3. J.R. Leger and W.C. Goltsos. Method and apparatus for efficient concentration of light from laser diode arrays. US Patent No. 5787107, July 28, 1998.
4. J. Endriz. Diode laser source with concurrently driven light emitting segments. US Patent No. 5594752, January 14, 1997.
5. F.M. Dickey and B.D. O'Neil. Multifaceted laser beam integrators: General formulation and design concepts. *Optical Engineering* 27(11): 999–1007, 1988.
6. D.A. Pepler C.N. Danson, I.N. Ross, S. Rivers, and S. Edwards. Binary-phase Fresnel zone plate arrays for high-power laser beam smoothing. *Proceedings of SPIE* 2404: 258–265, 1995.
7. K. Kamon. Fly-eye lens device and lighting system including same. US Patent No. 5251067, October 5, 1993.
8. J.W. Goodman. *Introduction to Fourier Optics*. New York: McGraw-Hill, 1968.
9. L. Unnebrink, T. Henning, E.W. Kreutz, and R. Poprawe. Optical system design for excimer laser materials processing. *Proceedings of SPIE* 3779: 413–422, 1999.
10. K.F. Cheung and R.J. Marks II. Imaging sampling below the Nyquist density without aliasing. *Journal of the Optical Society of America A* 7(1): 92–105, 1990.
11. J.W. Goodman. *Statistical Optics*. New York: Wiley, 1985.
12. F.M. Dickey, S.C. Holswade, and D.L. Shealy (eds.) *Laser Beam Shaping Applications*. Boca Raton, FL: CRC Press, 2005.
13. J.M. Geary. Channel integrator for laser beam uniformity on target. *Optical Engineering* 27: 972–977, 1988.
14. M.M. Chen, J.B. Berkowitz-Mattuck, and P.E. Glaser. The use of a kaleidoscope to obtain uniform flux over a large area in a solar or arc imaging furnace. *Applied Optics* 2: 265–271, 1963.
15. B.E.A. Saleh and M.C. Teich. *Fundamentals of Photonics*. New York: Wiley, 1991.
16. T. Henning, L. Unnebrink, and M. Scholl. UV laser beam shaping by multifaceted beam integrators: Fundamental principles and advanced design concepts. *Proceedings of SPIE* 2703: 62–73, 1996.
17. X. Deng, X. Liang, Z. Chen, W. Yu, and R. Ma. Uniform illumination of large targets using a lens array. *Applied Optics* 25(3): 377–381, 1986.
18. J. Bernges, L. Unnebrink, T. Henning, E.W. Kreutz, and R. Poprawe. Mask adapted beam shaping for material processing with eximer laser radiation. *Proceedings of SPIE* 3573: 108–111, 1998.
19. B. Burghardt, H.J. Kahlert, and U. Sarbach. Device for homogenizing a light beam. US Patent No. 5414559, May 9, 1995.
20. H.J. Kahlert and B. Burghardt. Optical apparatus for the homogenization of laser radiation and the generation of several lighting fields. US Patent No. 5796521, August 18, 1998.

21. T. Henning, M. Scholl, L. Unnebrink, U. Habich, R. Lebert, and G. Herziger. Beam shaping for laser materials processing with non-rotationally symmetric optical elements. *Proceedings of SPIE* 3209: 126–129, 1996.
22. S. Sinzinger and J. Jahns. *Microoptics*. Weinheim, Germany: Wiley-VCH, 1999.
23. M. Kufner and S. Kufner. *Micro-Optics and Lithography*. Brussels, Belgium: VUB Press, 1997.
24. T. Mori and H. Komatsuda. Optical integrator and projection exposure apparatus using the same. US Patent No. 5594526, January 14, 1997.
25. H. Komatsuda, H. Hirose, and T. Kikuchi. Illumination device with allowable error amount of telecentricity on the surface of the object to be illuminated and exposure apparatus using the same. US Patent No. 5594587, January 14, 1997.
26. L. Weichman, F. Dickey, and R. Shagam. Beam shaping element for compact fiber injection systems. *Proceedings of SPIE* 3929: 176, 2000.
27. R. Setchell. An optimized fiber delivery system for Q-switched, Nd:YAG lasers. *Proceedings of SPIE* 2966: 608, 1997.
28. R. Setchell. End-face preparation methods for high intensity fiber applications. *Proceedings of SPIE* 3578: 743, 1998.

11 Axicon Ring Generation Systems

Fred M. Dickey, Carlos López-Mariscal, and Daniel M. Brown

CONTENTS

11.1 INTRODUCTION

The production of a ring of light has applications in laser material processing and machining, particle manipulation (optical tweezers), and corneal surgery. It is well known that axicons (conical lenses) can produce light ring patterns. An axicon combined with a focusing lens produces a ring pattern at the focal plane of the lens. A combination of a positive (convex) and negative (concave) axicon pair can be used to achieve a variable (zoom) ring diameter. Negative axicons are expensive and more difficult to obtain. Fortunately, a zoom system can be achieved by using two positive axicons. Axicons were introduced by McLeod[1] as early as 1954 as an optical element. The ring forming properties of axicons are discussed in detail in the excellent papers by Belanger and Rioux[2] and Rioux et al.[3] They present a zoom system using two axicons, one positive and one negative. The axicon depth of focus is treated by Lit and Tremblay.[4] Shi et al.[5] use an axicon structured lens to obtain a large depth of focus. This is related to the subject of nondiffracting beams. Axicons are used to generate Bessel-like beams.[6] Zeng et al.[7–9] treat the application of axicons to optical trepanning in considerable detail. Goncharov et al.[10] give a design for an axicon system that produces a line focus.

This chapter discusses the use of axicons to generate laser light ring patterns. The next section reviews what might be considered the fundamental axicon systems for ring pattern generation. Section 11.3 gives the theory, based on geometrical optics, of a zoom system using only positive axicons. The diffractive theoretical model behind this design is outlined in Section 11.4, while a numerical simulation of the model is presented in Section 11.5. A method for producing a rectangular line pattern is discussed in Section 11.6. Although the method in Section 11.6 does not use axicons, the irradiance patterns and applications are similar to those of axicon systems, so it is included in this chapter. Both variable ring (axicon) and rectangular line image forming systems have the potential to reduce the amount of material vaporization in certain types of laser machining applications as compared to circular and square top-hat systems. A summary of the chapter is provided in Section 11.7.

11.2 THE FUNDAMENTAL AXICON SYSTEMS

A lens axicon combination forms an annular ring focus. Due to the rotational symmetry, it is easy to develop designs using geometrical optics in one dimension. This section and Section 11.3 are taken in part from the paper by Dickey and Conner.[11] The basic design equations given by Rioux et al.[3] are used in this section.

11.2.1 Radial Focusing

A configuration for forming a radial focus, annular ring using a positive axicon is shown in Figure 11.1, where it is assumed that the axicon is placed next to the lens. In two dimensions, the combination of a lens and axicon forms two converging beams that cross after the axicon. An annular ring focus is produced because of the rotational symmetry of the system. Similar results are obtained for a negative (concave) axicon, except that the two converging beams do not cross. For both positive and negative axicon systems, the radius of the ring, R, is given by

$$R \cong (n-1)\alpha F \tag{11.1}$$

where:

α is the base angle of the axicon
n is the index of refraction
F is the focal length of the lens

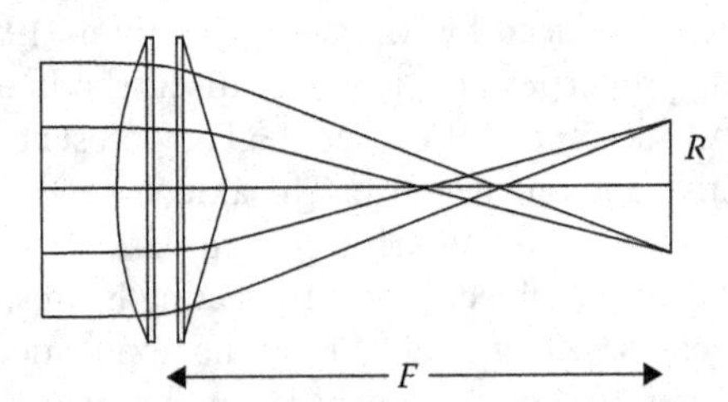

FIGURE 11.1 Annular ring formation with a focusing lens and a positive axicon after the lens.

Equation 11.1 is good if the base angle of the axicon is less than 10°.

The distribution of light at the focal plane of a lens is the Fourier transform of the input light pattern, except for a phase factor that depends on the position of the object. The phase factor is a constant if the object (axicon in our case) is placed at a focal length before the lens. Thus, the ring pattern produced by the axicon is the Fourier transform of the phase function of the axicon, which can be written as

$$p = e^{i\gamma\sqrt{x^2+y^2}} = e^{-i\gamma r} \tag{11.2}$$

where:

γ is a scale factor

Goodman[13] has shown that the Fourier transform relation exists if the object is placed after the lens, but the size of the Fourier transform scales linearly with the distance d from the focal plane. In this case, Equation 11.1 becomes

$$R \cong (n-1)\alpha d \tag{11.3}$$

Thus, the system in Figure 11.1 offers some zoom capability. However, the Fourier transform relation breaks down as the axicon approaches the focal plane (d is small). Also, one could not go to a zero radius ring without having the axicon at the focal plane, which is not practical for most applications.

The axicon can also be placed before the lens as shown in Figure 11.2. This configuration produces a ring pattern with the properties described above, except that the ring pattern does not change with the separation distance between the lens and the axicon. Based on Fourier optics,[9] the axicon can be placed in an arbitrary distance before the lens.

Belanger and Rioux[2] have developed an analytical expression for the field produced by an axicon at the focal plane of a lens for an input Gaussian beam as the complex sum of hypergeometric functions. However, the result is very close to a Gaussian function of the radial distance, with the width of the ring given by

$$2\Delta r = 3.3\left(\frac{\lambda F}{\pi W}\right) \tag{11.4}$$

where:

W is the root-mean-square input beam radius

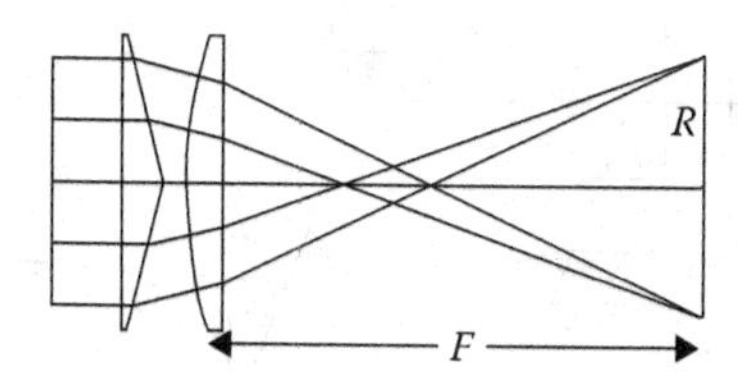

FIGURE 11.2 Annular ring formation with a focusing lens and a positive axicon before the lens.

It is interesting to note that this result is close to the focal spot size for a focused beam of half the input size. Belanger and Rioux[12] also give a solution for a uniform input beam. The solution in the case of a uniform input beam is also a complex sum of hypergeometric functions. In this case, the width of the ring is given by

$$2\Delta r = 7.2\left(\frac{\lambda F}{\pi W}\right) \tag{11.5}$$

The radial profile of the ring is close to an Airy pattern.

11.2.2 Positive and Negative Axicon Zoom Systems

Rioux et al.[3] present a configuration using a positive and a negative axicon as shown in Figure 11.3. They used two matching axicons, one positive and one negative, with equal base angles. This is an ideal annular ring zoom system, except for the problem of producing negative axicons. Negative axicons are very expensive and difficult to obtain. The zoom range goes from a radius of zero to a maximum radius limited by the diameter of the second axicon. The ring radius goes to zero as the separation between the two axicons approaches zero. The axicons do not have to have equal base angles. In this case, the radius of the ring would be a double-valued function of the separation distance between the axicons if the base angle of the second axicon is greater than that of the first axicon. If the base angle for the first is greater than that of the second, the ring radius goes from some minimum to its maximum value. This introduces a complexity that is not needed for most applications. They give the following formula for the radius as a function of the separation of the axicons:

$$R = \frac{d(n-1)}{1-\alpha} \tag{11.6}$$

For a collimated input beam, if the axicon pair is placed before the lens, a ring pattern is produced, but the radius of the ring does not change regardless of the separation of the axicons, or their distance from the focusing lens. The zoom capability is lost in this case. It might be noted that in the case that both axicons have the same base angle and are brought into contact, the two axicons become a window and no

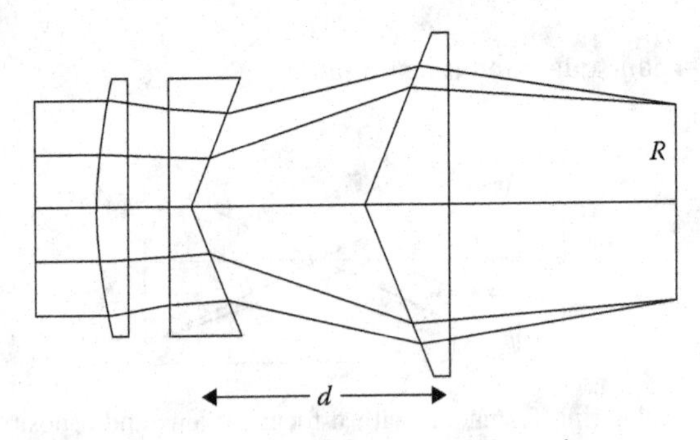

FIGURE 11.3 A zoom system with a positive and negative axicon.

ring is produced. A ring is produced, but the ring radius depends only on the base angles of the axicons. If the negative axicon is placed before and the positive axicon is placed after the lens, a zoom capability is maintained, and it can be achieved by moving either axicon. However, in this case, there is a limit on how small of a ring radius can be obtained.

11.3 POSITIVE AXICON ZOOM SYSTEM

A zoom system having the capability of the system shown in Figure 11.3 can be designed using two positive axicons, one before and one after the lens, as shown in Figure 11.4. In this configuration, the first axicon appears to form two intersecting beamlets, while the second axicon redirects the beamlets to produce the desired annular pattern. The distance, *d*, between the first axicon and the lens must be large enough that the beamlets completely cross without overlap before the second axicon, otherwise the distance is part of the design. When this condition is violated, a pattern of two concentric rings is obtained; the irradiance of each one of which varies with the distance *d*.

The design for this configuration is more complicated than that for the system shown in Figure 11.3. The cone base angles are chosen to give the desired range for the ring radius. There is a degree of flexibility that allows one to generally develop a design from commercially available axicons. If the axicons have equal base angles, the system performs similar to the configuration in Figure 11.3, except that the ring radius zooms from some minimum to a maximum as the second axicon is moved to the right. Although it is not necessary, it is probably best to design the system so that the minimum ring radius is obtained with the second axicon next to the lens. This would allow for the position of the second axicon not having to be too close to the output plane.

The design of a system based on Figure 11.4 has a large degree of flexibility. One can juggle the axicon base angles, the focal length of the lens, and the distances between the elements to come up with a workable system. Generally, a large *F*-number system is desirable, as long as the desired width of the ring is obtained. In the following text, we give the equations to be solved to accomplish a first-order design. The equations trace the central ray of the beamlets through the system, beginning with the beamlet exiting the lower half of the first prism. Before giving

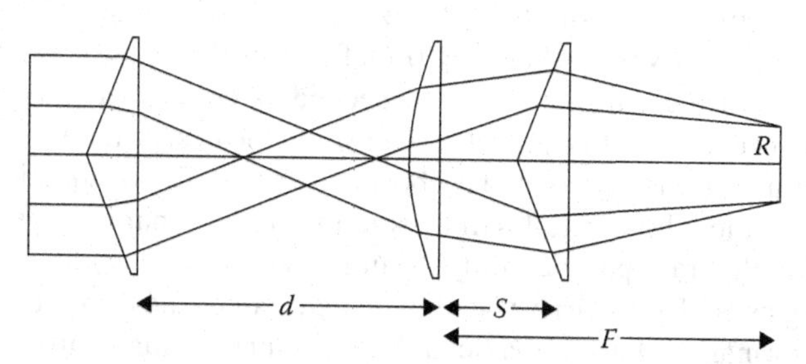

FIGURE 11.4 A zoom system with two positive axicons.

the equations, we need to define distances and angles not shown in the figure. The following angles are measured counterclockwise from the horizontal:

θ_1—the angle the beam deflected by the first axicon
θ_2—the angle of the beamlet exiting the lens
θ_3—the angle the beam deflected by the second axicon
z_1—the distance from the lens that the center ray of each beamlet crosses the optical axis
z_2—the distance from the lens that the beamlets separate after crossing each other
D—the effective diameter of the input beam

The radius of the output ring is given by the following equations:

$$R = F\tan\theta_1 + (F - S)\tan\theta_3 \tag{11.7}$$

$$\theta_1 = -\alpha_1 + \sin^{-1}(n\sin\alpha_1) \tag{11.8}$$

$$\theta_2 = \tan^{1}\left[\frac{(F - z_1)\tan\theta_1}{F}\right] \tag{11.9}$$

$$\theta_3 = -\theta_2 + \alpha_2 - \sin^{-1}[n^2 - \sin^2\theta_2]^{1/2}\sin\alpha_2 - \cos\alpha_2\sin\theta_2 \tag{11.10}$$

$$z_1 = d - \frac{D}{4\tan\theta_1} \tag{11.11}$$

$$z_2 = d - \frac{D}{2\tan\theta_1} \tag{11.12}$$

Equation 11.7 has a simple interpretation. The first term is equivalent to Equation 11.1; it gives the size of the ring produced by the first axicon and the lens combination. The second term describes the redirection of the rays to size the output ring. The two axicons are coupled in the design due to the fact that θ_3 is dependent on θ_1.

One approach to a first-order design is to use Equation 11.1 to select a base angle for the first axicon for a ring radius greater than the maximum desired radius. Then, iterations of the above equations are used to determine the second axicon base angle that gives the desired range of zoom of the ring radius. The ring radius obtains its minimum value when $s = 0$ (defined in the figure). We suggest that this position be the design position of the second axicon for the maximum radius. As this is the usual case, it is desirable to develop the design for a large *F*-number system to minimize aberrations. The beam diameter, *D*, in Equation 11.11 is a source of aberration; however, it will be small for large *F*-number systems. This is due to the fact that rays parallel to the beamlet central ray exit the lens at slightly different angles and thus arrive at the second axicon at different angles. The detailed performance of a zoom axicon design can be evaluated using the diffraction theory in the following section.

11.4 POSITIVE AXICON SYSTEM DIFFRACTION THEORY

In this section, an analysis of the two positive axicons system using Fourier optics is presented. The layout of the system is shown in Figure 11.5 with the planes of every optical component labeled as they will be referred to in what follows. The separation between the planes of the first axicon P_1 and the lens P_L is given by $a = s$, and the separation between the second axicon and the focal plane of the lens is given by $b = s - f$. The purpose of the following analysis is to find an expression for the transverse field distribution at P_L as a function of the system's optical parameters.

Consider first an optical subsystem that consists of only the thin lens L and the axicon at plane P_2 as shown in Figure 11.6. Let the transmission function of the axicon be g_2 and its base angle α_2. The field produced at P_f under uniform illumination of the lens is given by the Fraunhofer integral[13]:

$$U_f = \frac{1}{i\lambda b}\exp\left[i\frac{k}{2b}\left(x_f^2 + y_f^2\right)\right]\left(\frac{f}{b}\right)$$

$$\int\int_{-\infty}^{\infty} g_2(x_2,y_2)\mathrm{circ}\left[\sqrt{\left(\frac{x_2 f}{bR}\right)^2 + \left(\frac{y_2 f}{bR}\right)^2}\right]\exp\left[-i\frac{2}{\lambda b}(x_2x_f + y_2y_f)\right]\mathrm{d}x_2\mathrm{d}y_2 \tag{11.13}$$

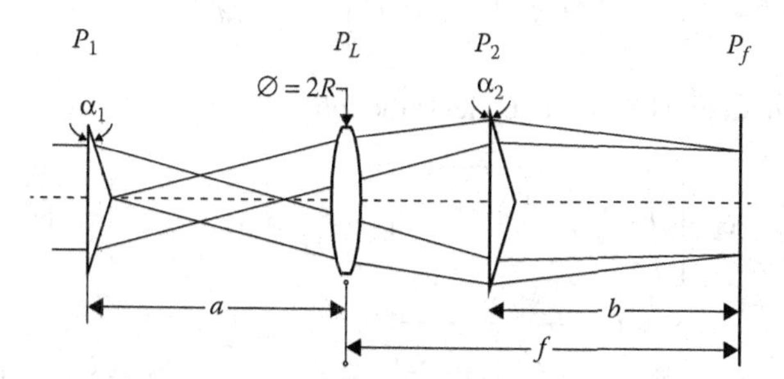

FIGURE 11.5 Two positive axicons system layout and parameters.

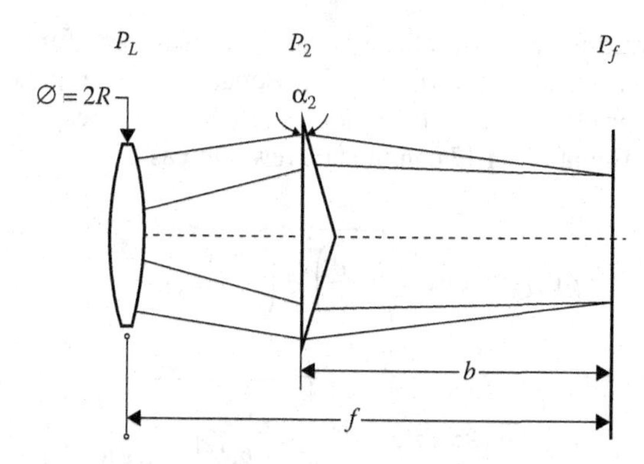

FIGURE 11.6 Axicon–lens subsystem from the system in Figure 11.5.

where:

λ is the wavelength
R is the radius of the lens
(x_f, y_f) and (x_2, y_2) are the transverse coordinates at planes P_f and P_L, respectively

Let also

$$\psi_z = \psi(x,y,z) = \frac{1}{i\lambda z}\exp\left[i\frac{k}{2z}(x^2+y^2)\right]\left(\frac{f}{z}\right) \tag{11.14}$$

so that Equation 11.13 can now be written as

$$U_f = \psi_b \iint_{-\infty}^{\infty} g_2(x_2, y_2)\mathrm{circ}\left[\sqrt{\left(\frac{x_2 f}{bR}\right)^2 + \left(\frac{y_2 f}{bR}\right)^2}\right]\exp\left[-i\frac{2\pi}{\lambda b}(x_2 x_f + y_2 y_f)\right]\mathrm{d}x_2\mathrm{d}y_2 \tag{11.15}$$

Consider now a hypothetical thin optical element with transmission function $g_2'(x_L, y_L)$ located at the principal plane of lens P_L that produces the same field as described by Equation 11.15 under uniform illumination so that

$$U_f = \psi_f \iint_{-\infty}^{\infty} g_2'(x_L, y_L)\mathrm{circ}\left[\sqrt{\left(\frac{x_L}{R}\right)^2 + \left(\frac{y_L}{R}\right)^2}\right]\exp\left[-i\frac{2\pi}{\lambda f}(x_L x_f + y_L y_f)\right]\mathrm{d}x_L\mathrm{d}y_L \tag{11.16}$$

Equating Equations 11.5 and 11.6 yields the following equation:

$$\begin{aligned}&\psi_f \iint_{-\infty}^{\infty} g_2'(x_L, y_L)\mathrm{circ}\left[\sqrt{\left(\frac{x_L}{R}\right)^2 + \left(\frac{y_L}{R}\right)^2}\right]\exp\left[-i\frac{2\pi}{\lambda f}(x_L x_f + y_L y_f)\right]\mathrm{d}x_L\mathrm{d}y_L = \\ &\psi_b \iint_{-\infty}^{\infty} g_2(x_2, y_2)\mathrm{circ}\left[\sqrt{\left(\frac{x_2 f}{bR}\right)^2 + \left(\frac{y_2 f}{bR}\right)^2}\right]\exp\left[-i\frac{2\pi}{\lambda b}(x_2 x_f + y_2 y_f)\right]\mathrm{d}x_2\mathrm{d}y_2\end{aligned} \tag{11.17}$$

which can be solved for g_2' to find the equivalent transmission function of the hypothetical object. Each one of the integrals in Equation 11.17 represents the Fourier transform of the product of a transmission function and its respective pupil. Using Fourier notation, Equation 11.17 can thus be rewritten as

$$\begin{aligned}&\Im\left\{g_2'(x_L, y_L)\mathrm{circ}\left[\sqrt{\left(\frac{x_L}{R}\right)^2 + \left(\frac{y_L}{R}\right)^2}\right]\right\} = \\ &\frac{\psi_b}{\psi_f}\Im\left\{g_2(x_2, y_2)\mathrm{circ}\left[\sqrt{\left(\frac{x_2 f}{bR}\right)^2 + \left(\frac{y_2 f}{bR}\right)^2}\right]\right\}\end{aligned} \tag{11.18}$$

Applying the inverse Fourier transform on both sides yields

$$g_2'(x_L,y_L)\mathrm{circ}\left[\sqrt{\left(\frac{x_L}{R}\right)^2+\left(\frac{y_L}{R}\right)^2}\right]=$$
$$\mathfrak{I}^{-1}\left\{\frac{\psi_b}{\psi_f}\right\}*\left\{g_2(x_2,y_2)\mathrm{circ}\left[\sqrt{\left(\frac{x_2 f}{bR}\right)^2+\left(\frac{y_2 f}{bR}\right)^2}\right]\right\} \tag{11.19}$$

The right-hand side of Equation 11.18 represents the Fresnel diffraction integral of the input field as seen at the plane of the axicon. The phase of the factor ψ_b/ψ_f is a transverse spherical phase modulation proportional to the separation $f-b$ between the lens and the axicon:

$$\frac{\psi_b}{\psi_f}=\left(\frac{f}{b}\right)^2\exp\left[\mathrm{i}\frac{k}{2bf}(x^2+y^2)(f-b)\right] \tag{11.20}$$

When the axicon is placed very close to the lens, $\psi_b/\psi_f\approx 1$ and this phase modulation vanishes.

Consider now the subsystem shown in Figure 11.7. The field produced at P_L by the axicon under uniform illumination is given by the Fresnel integral

$$U_L=\frac{\exp(ika)}{i\lambda a}\int\int_{-\infty}^{\infty}g_1(x_1,y_1)\exp\left\{-\mathrm{i}\frac{k}{2a}\left[(x_L-x_1)^2+(y_L-y_1)^2\right]\right\}\mathrm{d}x_1\mathrm{d}y_1 \tag{11.21}$$

which can be expressed as the convolution of $g_1(x_1,y_1)$ and the Fresnel kernel

$$U_L=g_1(x,y)*\frac{\exp(ika)}{i\lambda a}\exp\left[-\mathrm{i}\frac{k}{2a}(x^2+y^2)\right]$$
$$U_L=g_1(x,y)*h(x,y) \tag{11.22}$$

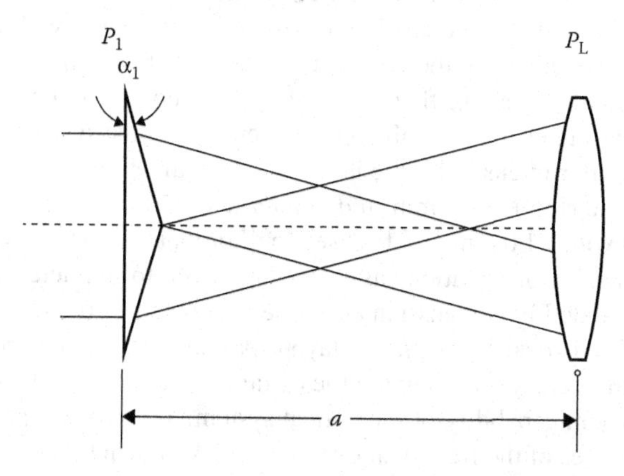

FIGURE 11.7 Lens–axicon subsystem from the system in Figure 11.5.

The field produced by the two axicon and lens system at P_f is proportional to the Fourier transform of the field at P_L, which is in turn given by the product of U_L and the left-hand side of Equation 11.19. Explicitly,

$$U(x_f, y_f) = \Im\left\{U_L(x_L, y_L) g_2'(x_L, y_L) \mathrm{circ}\left[\sqrt{\left(\frac{x_L}{R}\right)^2 + \left(\frac{y_L}{R}\right)^2}\right]\right\} \quad (11.23)$$

or using the convolution theorem

$$U(x_f, y_f) = \Im[U_L(x_L, y_L)] * \Im\left\{g_2'(x_L, y_L) \mathrm{circ}\left[\sqrt{\left(\frac{x_L}{R}\right)^2 + \left(\frac{y_L}{R}\right)^2}\right]\right\} \quad (11.24)$$

The output field is thus given by the convolution of two Fourier transforms. Equation 11.24 can now be readily evaluated by substituting the corresponding transmission function for each axicon:

$$g_{1,2}(x, y) = \exp\left[-i\frac{2\pi}{\lambda}(n_{1,2} - 1)\tan\alpha_{1,2}\sqrt{x^2 + y^2}\right] \quad (11.25)$$

where:

$n_{1,2}$ are the refractive indices of the axicons

It must be noted that the approach used in this analysis is applicable to other systems composed of a thin lens and any two optical elements of arbitrary transmission functions g_1 and g_2, not necessarily axicons, placed in front and behind the lens, respectively.

11.5 NUMERICAL EVALUATION OF THE DIFFRACTION EQUATION

Using Equation 11.24, the spatial properties of the intensity distribution produced by the optical system can now be numerically calculated. An advantageous feature of Equation 11.24 is that the convolution can be implemented with high efficiency as a point-wise product using the convolution theorem. Evaluation of this product is less computationally intensive than calculating the output using successive Fresnel integrals considering each one of the optical elements in the system, for example.

A typical annular intensity profile is shown in Figure 11.8 along with a plot of its cross section. The ring is 1.83 mm in diameter and 25 μm in width. In this case, a Gaussian beam with a 1.0 mm waist is used to illuminate the first axicon. Notice that the resulting intensity distribution has a very low width-to-diameter ratio. The annular profile is bounded by a Gaussian envelope as a result of the illumination beam. For a lens of fixed focal length *f*, the layout parameters that determine the mean width and the diameter of the ring are the distances *a* and *b*. The plot in Figure 11.9 illustrates the zoom capability of the optical system. The annular profiles produced for a range of values of the fractional distance *b*/*f* are shown to decrease in diameter with an increase in *b*, therefore providing a mechanism to vary the diameter of the

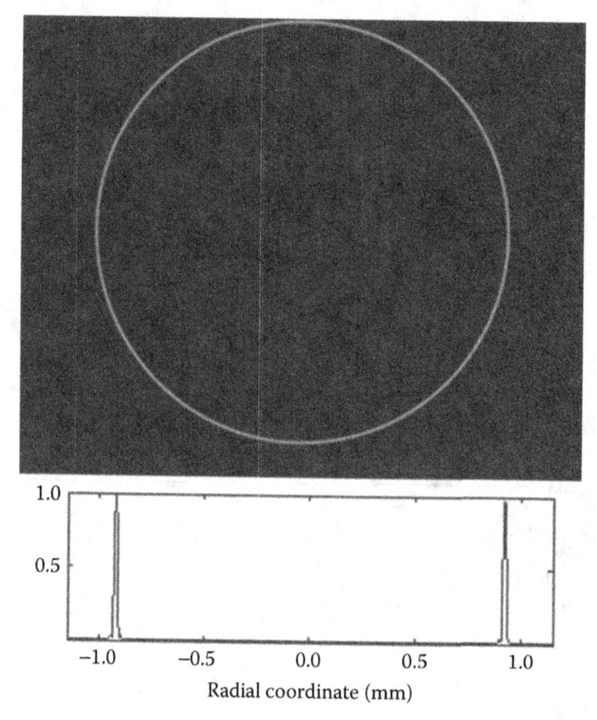

FIGURE 11.8 Typical annular intensity profile as evaluated by Equation 11.24.

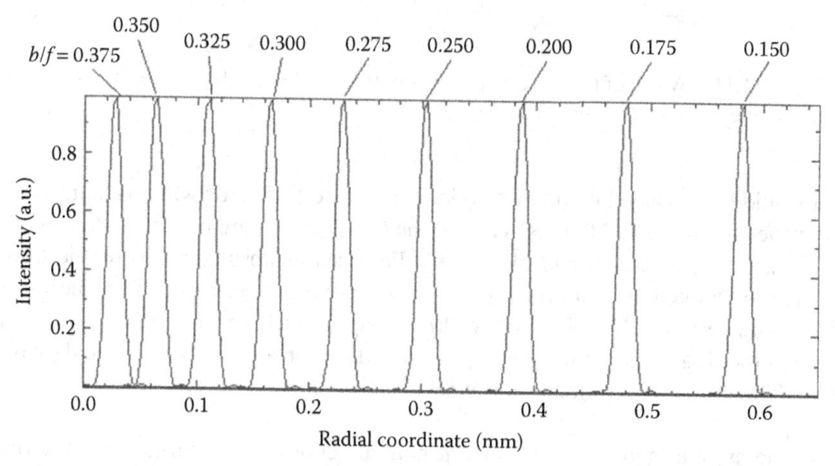

FIGURE 11.9 Annular intensity profiles at P_f for varying distance b as a fraction of f.

annulus continuously as intended by design. Figure 11.10 shows the ring diameters for the range b/f = [0.15–0.40]. In the particular calculations shown in Figures 11.9 and 11.10, the wavelength is 775 nm, f = 15 mm, $\alpha_1 = 2°$, and $\alpha_2 = 5°$.

The widths of the annular patterns in Figure 11.9 are shown in Figure 11.11 as a function of ring diameter. Equation 11.4 predicts a ring width of 24.4 μm for the ring

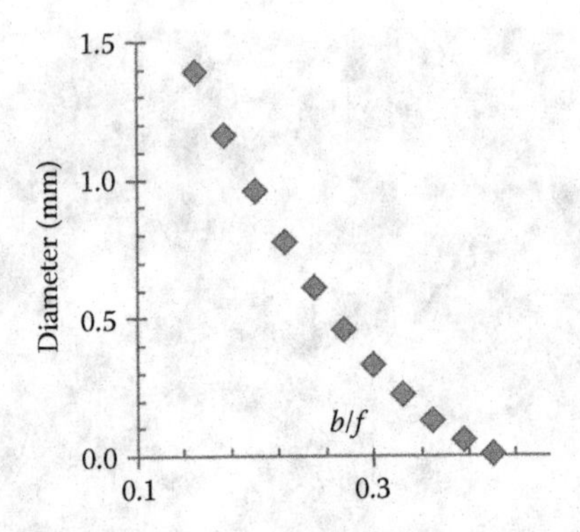

FIGURE 11.10 Ring diameters as a function of *b*/*f*.

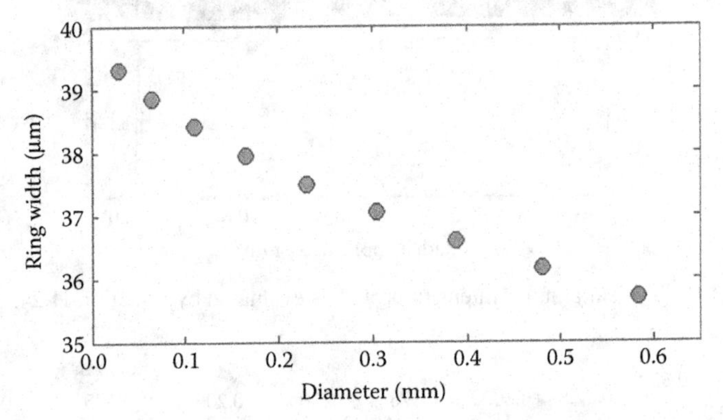

FIGURE 11.11 Width of the annular patterns in Figure 11.9 as a function of their diameter. Note that the vertical scale shows a range of only 5 µm.

that would be produced by the first axicon in Figure 11.5. This width is smaller than the values in Figure 11.11. This is due to the fact that the ring pattern is described as a correlation function instead of a simple Fourier transform. The correlation function can be looked at as an interaction between the two axicons, or as aberrations discussed in Section 11.3. The ring widths in Figure 11.11 are approaching a smaller value as the ring diameter increases [the second axicon approaches the focal (output) plan]. The effect (aberrations) of the second axicon decreases as it approaches the focal plane.

In the system layout, the distance a must be chosen to be larger than the overlapping distance of the rays originating on the top and bottom halves of the first axicon, as mentioned in Section 11.3. Notice that the minimum allowed value of a is dependent on the value of ω, and the spatial extent of the field at P_L can exceed the physical aperture of the lens for large values of a. For this reason, it is best to adjust the zoom factor by continuously varying b and leaving a fixed for a fixed value of the illuminating aperture. Violation of this condition results in the double ring pattern shown in Figure 11.12.

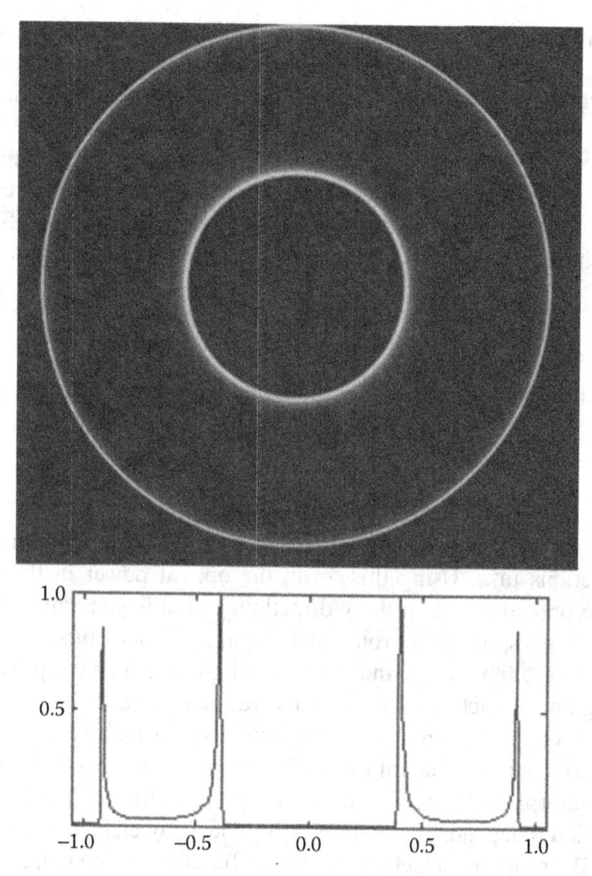

FIGURE 11.12 Double ring intensity pattern for an insufficient first axicon to lens separation.

Additional design considerations of the two positive axicons system thus include calculating the transverse size of the intensity pattern at P_L prior to selecting a lens diameter. Likewise, the extent of the input at the second axicon must be considered in order to choose an adequate axicon diameter.

11.6 CONTINUOUSLY VARIABLE RECTANGULAR LINE GENERATOR

Multi-aperture beam integration systems were discussed in Chapter 10. In this section, we show how to combine a multi-aperture beam integrator with highly anamorphic lenslet arrays to generate a continuously variable rectangular line pattern. The aspect ratio and size of this rectangular line pattern are varied by small lateral displacements (in x- and y-directions) of the lenslet arrays. Such patterns are useful for very flexible laser machining and marking tools. This method was first disclosed in the US Patent 6909553,[14] and subsequently in a paper by D. M. Brown.[15] The method is a generalization of the Alvarez lens, shown in the US Patents 3305294[16] and 3507565,[17] where Alvarez showed that a pair of antisymmetric, highly anamorphic, refractive lens elements can be combined to produce a spherical lens with variable optical power. The variable power is obtained by translating the elements laterally in opposite directions. The problem lies in the fact that these highly anamorphic lenses are difficult to fabricate. However, Wallace[18] points out that using diffractive optical elements solves the fabrication problem and opens up new opportunities for these types of lenses. We expand upon these ideas to apply diffractive Alvarez-like lenslets to the problem of laser beam shaping and show how to design the diffractive phase function to achieve the desired continuously variable complex pattern on the target.

11.6.1 Theory

The functional form Alvarez gives for the refractive element thickness is

$$t = A\left(xy^2 + \frac{1}{3}x^3\right) + Dx + E \tag{11.26}$$

which allows the optical power to be varied by translating matching elements in opposite directions in x. Using this form, the optical power in the x-direction is coupled to the optical power in the y-direction, which is generally desirable if one wishes to produce a spherical or rotationally symmetric lens function. If an array of these elements were fabricated, and the source light was made to pass only through the proper regions of each lenslet pair, the arrays could be used in a multi-aperture beam integrator configuration to form a continuously variable spot size on the target.

However, using a more general form of the element thickness (or phase function), a variable anamorphic lens can be produced in which the x- and y-direction powers can be varied independently by translating the array elements in both the x- and y-directions. By properly selecting the phase function, orthogonal optical powers and tilts can be continuously varied and combined in a beam integrator system to produce complex polygon line patterns. Three or more arrays can also be used to provide even greater degrees of freedom in forming complex image patterns, but we confine our discussion here to using two movable arrays.

One generally desires to use all of the available laser light (or other collimated light source) in the generated pattern. However, only a small fraction of the clear aperture or overlap region of each lenslet pair can be used to modify the wavefront and form the image pattern in order to allow space for translating. This requires that the input laser beam be segmented into an array of separated beamlets so that each beamlet passes only through the overlap regions and not the invalid regions between neighboring lenslet pairs. This can be accomplished with a stationary array of afocal telescopes, or beamlet reducers, placed ahead of the movable arrays, which segments the entrance beam with 100% fill factor. Laterally, translating the lenslet arrays about these beamlets and then combining the beamlets with an integrator lens produces the variable irradiance pattern with nearly all the beam energy. The optical designer should consider the edge blurring effects resulting from the increase in angular divergence exiting the beam reducer as a result of the optical invariant. Alternatively, a diffractive fan-out grating could be used to replicate the laser beam into a 2D array in angle space.

To produce the four line segments of the rectangular pattern, four unique lenslet phase functions are required in the array. A lenslet in one array is paired with another lenslet in the second array element to produce the phase function. The paired lenslets have equal and opposite phase distortions such that when the lateral translation of the arrays is zero in x and y, the combination produces zero phase distortion. Translation of the arrays in opposite directions in x and y varies the combined phase distortion resulting in varying the length and position of the line segments of the rectangular image. This group of four lenslet pairs is replicated over the entire area of the arrays in order to achieve beam homogenization.

The following derivation of the phase aberration function assumes that the two movable arrays are placed in sufficient near proximity to each other so that wavefront propagation between the two arrays can be ignored. Diffractive optical elements make this much easier to achieve. When the lenslet arrays are placed in a diffracting beam integrator configuration (Section 10.3.1), a beamlet from a lenslet with no optical power (or defocus) will form a point focus on the image plane. If the lenslet pair has optical power in the x-direction, forming a cylindrical lens, a horizontal line image is formed. A vertical line image is formed if the lenslet pair has only power in the y-direction.

The phase or surface profile of each lenslet in the movable arrays is described by the general polynomial of the form:

$$\Phi_1 = \sum_{i=0}^{m} \sum_{j=0}^{n} a_{i,j} x^i y^j \tag{11.27}$$

$$\Phi_2 = \sum_{i=0}^{m} \sum_{j=0}^{n} b_{i,j} x^i y^i \tag{11.28}$$

where:

Φ_1 is the phase function (or surface profile) of the lenslet in the first movable array
Φ_2 is the phase function (or surface profile) of the paired lenslet in the second movable array

We assume an insignificant propagation distance between a pair of lenslets so that $\Phi = \Phi_1 + \Phi_2$.

In general, when $a_{i,j} \neq b_{i,j}$, very complex shapes are formed, which we briefly discuss here by showing how each term in the polynomial distorts the phase. But our primary concern in this section is with simple rectangular line images where $a_{i,j} = -b_{i,j}$ and only two of the polynomial terms are used. The piston phase term, $a_{0,0}$, is ignored in this discussion as it only contributes interference effects with adjacent lenslets that can be considered as a separate issue from the image shape generated by the phase function. The first term in the polynomial is linear in x:

$$\Phi_{1x} = a_{1,0} x \tag{11.29}$$

$$\Phi_{2x} = b_{1,0} x \tag{11.30}$$

Assuming opposite translations of equal magnitude for the two arrays, we make the substitutions:

$$x \to x+\xi$$
$$y \to y+\eta \tag{11.31}$$

for the first array, and

$$x \to x-\xi$$
$$y \to y-\eta \tag{11.32}$$

for the second array, where ξ and η are the lateral translations of the arrays in the x- and y-directions, respectively. The wavefront aberration contribution for the lenslet pair due to the first polynomial term is

$$\Phi_x = a_{1,0}(x+\xi)+b_{1,0}(x-\xi) \tag{11.33}$$

which is a wavefront tilt in the x-direction. The displacement on the image plane is proportional to the gradient of the phase function. Taking the gradient, $\vec{\nabla}\Phi_x = (a_{1,0}+b_{1,0})i$, shows a constant displacement in the x-direction at the image plane, independent of lenslet array translation (ξ). The magnitude of the displacement depends on the relative magnitudes of the coefficients and the focal length of the integrator lens. Similarly, the linear term in y results in a constant displacement in the y-direction.

The wavefront aberration contribution due to the first quadratic term, x^2, can be written as

$$\Phi_{xx} = a_{2,0}(x+\xi)^2+b_{2,0}(x-\xi)^2 \tag{11.34}$$

with its gradient given by

$$\vec{\nabla}\Phi_{xx} = [(2a_{2,0}+2b_{2,0})x+(2a_{2,0}-2b_{2,0})\xi]i \tag{11.35}$$

With the condition $a_{2,0} = b_{2,0}$, the second term above becomes zero and the first term gives defocus or a constant optical power in the x-direction. This would form a horizontal line image at the focal point of the integrator lens. With the condition $a_{2,0} = -b_{2,0}$, the first term becomes zero and the second term shows a linearly varying displacement in the x-direction on the image plane, proportional to the array lateral translation distance, ξ. With $|a_{2,0}| \neq |b_{2,0}|$, the lenslet pair provides both constant optical power in the x-direction and variable lateral translation in the x-direction. This would translate a horizontal line of constant length in the horizontal direction with translation of the arrays in the horizontal direction. Similar results apply to the y^2 term in the vertical direction.

The wavefront aberration contribution due to the cross term, xy, is given by

$$\Phi_{xy} = a_{1,1}(x+\xi)(y+\eta)+b_{1,1}(x-\xi)(y-\eta) \tag{11.36}$$

with its gradient given by

$$\vec{\nabla}\Phi_{xy} = \left[(a_{1,1} + b_{1,1})y + (a_{1,1} - b_{1,1})\eta\right]\vec{i} + \left[(a_{1,1} + b_{1,1})x + (a_{1,1} - b_{1,1})\xi\right]\vec{j} \quad (11.37)$$

With the condition $a_{1,1} = b_{1,1}$, the lenslet pair gives defocus or constant optical power in a direction 45° to the x- and y-axes and an optical power of the opposite sign in the orthogonal direction, producing astigmatism. However, since a negative and a positive lens of the same optical power produce the same spot size at the focus of the integrator lens, this case is indistinguishable from a rotationally symmetric lens with the same optical power in both x- and y-directions. With $a_{1,1} = -b_{1,1}$, a vertical translation of the arrays results in a horizontal (x-direction) displacement at the image plane, and a horizontal translation of the arrays results in a vertical displacement at the image plane. With $|a_{1,1}| \neq |b_{1,1}|$, the constant optical power (defocus) and variable displacement result.

The wavefront aberration contribution due to the first cubic phase term, x^3, is given by

$$\Phi_{xxx} = a_{3,0}(x + \xi)^3 + a_{3,0}(x - \xi)^3 \quad (11.38)$$

with its gradient given by

$$\vec{\nabla}\Phi_{xxx} = \left[(3a_{3,0} + 3b_{3,0})x^2 + (6a_{3,0} - 6b_{3,0})\xi x + (3a_{3,0} + 3b_{3,0})\xi^2\right]i \quad (11.39)$$

The first term in the gradient is a constant coma-like aberration in the x-direction. The second term is a linearly varying defocus in the x-direction. The third term is a one-sided displacement varying quadratically in ξ. With $a_{3,0} = b_{3,0}$, this phase term gives a constant coma-like aberration that can be displaced in the x-direction with translation of the arrays. With $a_{3,0} = -b_{3,0}$, this phase term gives a linearly varying defocus in the x-direction proportional to array translation distance. With $|a_{3,0}| \neq |b_{3,0}|$, a combination of coma, defocus, and displacement in the x-direction result. The fourth cubic phase term, y^3, produces similar results in the y-direction.

The aberration contribution for third cubic term, xy^2, along with its gradient is given by

$$\Phi_{xyy} = a_{1,2}(x + \xi)(y + \eta)^2 + b_{1,2}(x - \xi)(y - \eta)^2 \quad (11.40)$$

and

$$\begin{aligned} \vec{\nabla}\Phi_{xyy} = \vec{i}&\left[(a_{1,2} + b_{1,2})y^2 + (2a_{1,2} - 2b_{1,2})\eta y + (a_{1,2} + b_{1,2})\eta^2\right] \\ &+ \vec{j}\left[(2a_{1,2} + 2b_{1,2})xy + (2a_{1,2} - 2b_{1,2})\eta x\right. \\ &\left. + (2a_{1,2} - 2b_{1,2})\xi y + (2a_{1,2} + 2b_{1,2})\xi\eta\right] \end{aligned} \quad (11.41)$$

This term produces a constant coma-like aberration for $a_{1,2} = b_{1,2}$, which can be displaced horizontally and vertically with array translation. When this cubic term is combined with the first cubic term, x^3, and $a_{3,0} = b_{3,0} = a_{1,2} = b_{1,2}$, the result is conventional coma with the axis of symmetry in the x-direction. The coma patch remains constant, but is laterally displaced on the image plane with the square of the array translation distances. The displacement is one sided in x and two sided in y.

With $a_{1,2} = -b_{1,2}$, a translation of the arrays in the x-direction produces a linearly varying defocus in the y-direction, resulting in a vertical line at the image plane. Translation of the arrays in the y-direction produces defocus in both the x- and y-directions. Combined translations of the arrays in both x and y result in an astigmatic patch with its axes rotated with respect to the x–y axes. With $|a_{1,2}| \neq |b_{1,2}|$, a combination of displacement, astigmatic-like aberrations, and coma-like aberrations are produced. Similar results are obtained in the orthogonal directions for the second cubic term, x^2y, which can also be combined with the y^3 terms to produce conventional coma with a vertical axis of symmetry.

The wavefront aberration contribution and its gradient for the fourth-order term, x^4, are given by

$$\Phi_{xxxx} = (a_{4,0} + b_{4,0})(x^4 + 6\xi^2x^2 + \xi^4) + (a_{4,0} - b_{4,0})(4\xi x^3 + 4\xi x^3) \tag{11.42}$$

and

$$\vec{\nabla}\Phi_{xxx} = \vec{i}\left[(a_{4,0} + b_{4,0})(4x^3 + 12\xi^2 x) + (a_{4,0} - b_{4,0})(12\xi x^2 + 4\xi^3)\right] \tag{11.43}$$

With $a_{4,0} = b_{4,0}$, this term produces a constant spherical aberration in the x-direction and an x-direction defocus that varies quadratically with x-direction array translation. With $a_{4,0} = -b_{4,0}$, this term results in a coma-like aberration and displacement in the x-direction. This process can be carried to higher order terms, but the above is sufficient to provide a great deal of flexibility in designing continuously variable complex image-forming beam integrators. We use only two of the above terms, the quadratic term of Equation 11.34 and the cubic term of Equation 11.38, to form the continuously variable rectangular line image. An example of this is shown in the following section.

11.6.2 Example

Figure 11.13 shows an example of beam integrator consisting of a stationary afocal beam reducer array with 2 mm² lenslets and two movable arrays that translate in both the x- and y-directions. The stationary array forms an array of Galilean telescopes that collects the incoming light with 100% fill factor and segments the laser beam into multiple spatially separated beamlets. The beam diameters are sufficiently reduced in order to prevent vignetting at the following movable arrays. In this example, the stationary afocal array reduces 2 to 1.2 mm² beamlets, allowing a maximum translation distance of 0.4 mm for the movable arrays. The two movable arrays have diffractive anamorphic lenses patterned on the surfaces facing each other. The two movable arrays have a group of four unique lenslets replicated over the arrays. Each lenslet in the group produces one of the four line segments of the rectangular line image.

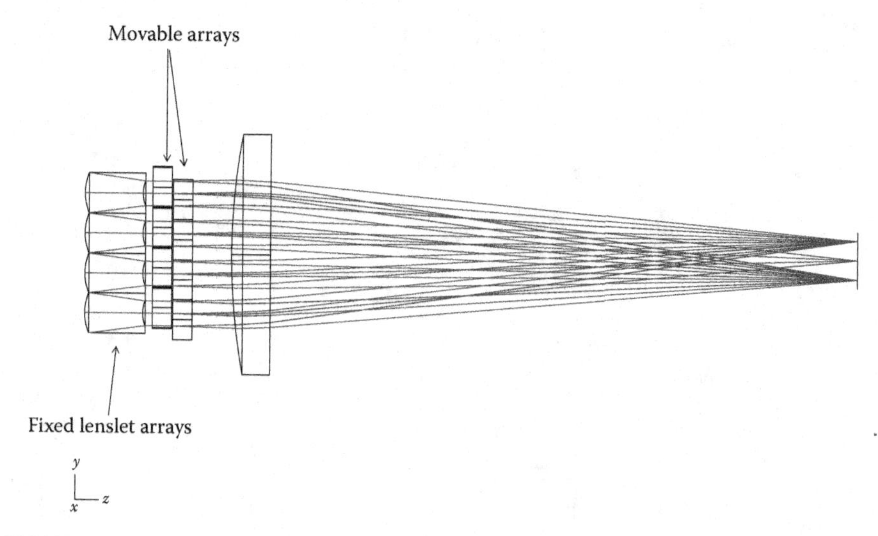

FIGURE 11.13 Layout showing stationary beam reducer and two movable arrays.

TABLE 11.1
Coefficients for Example Variable Rectangular Line Image

	Lens 1	Lens 2	Lens 3	Lens 4
A	0	0	256	−256
B	140	140	0	0
C	256	−256	0	0
D	0	0	140	140

The general phase functions for each pair of lenslets, required to produce the rectangular line image, are given by

$$\Phi_1 = Ax^2 + Bx^3 + Cy^2 + Dy^3 \quad (11.44)$$

$$\Phi_2 = -Ax^2 - Bx^3 - Cy^2 - Dy^3 \quad (11.45)$$

The four coefficient values (A, B, C, and D) are easily optimized in a lens design program for the specific optical layout. The coefficient values are adjusted so that the line segments intersect at the corners and the desired sensitivity with lateral displacement of the arrays is achieved. The coefficient values given in Table 11.1 are optimized for the layout of Figure 11.13 and will produce a continuously variable rectangular line image. The cubic terms produce a linearly varying defocus or line image of varying length. The quadratic terms produce a linearly varying displacement from the center of the image, moving the segments closer or further from the center as the rectangle shrinks or grows. The image changes with lateral displacements of the two movable arrays, which are moved in opposite directions by equal amounts. Figure 11.14 shows

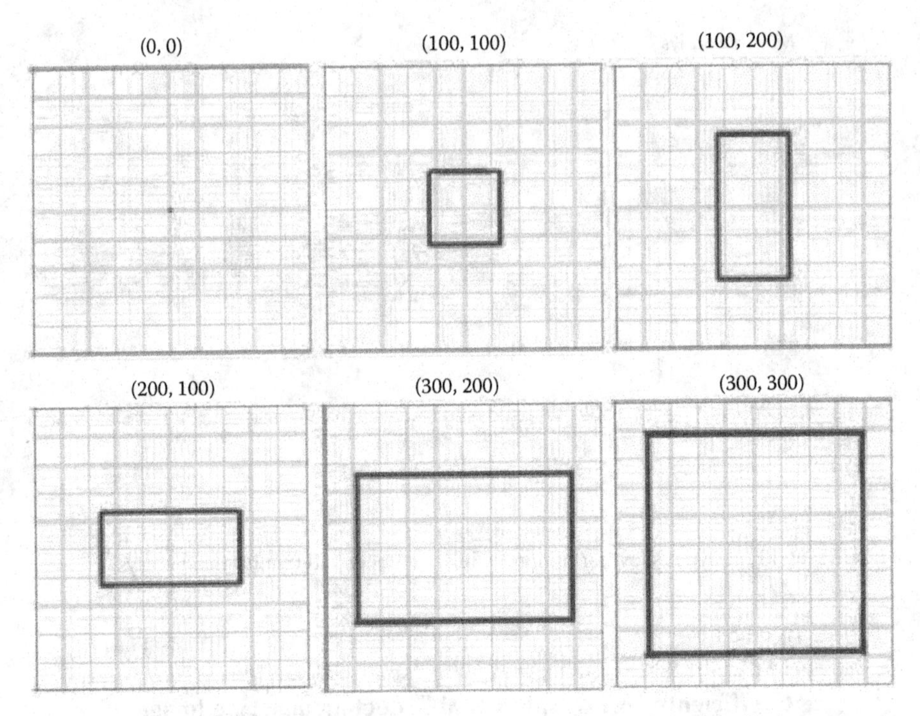

FIGURE 11.14 Spot diagrams of rectangular image for various lateral displacements in x and y.

a sequence of spot diagrams of the rectangular image for various lateral translations of the movable arrays. (Diffraction effects are not included in spot diagrams.) The x- and y-displacements (in microns) for these six spot diagrams are (0, 0), (100, 100), (100, 200), (200, 100), (300, 200), and (300, 300). Note that zero displacement in both x and y produces a single focused point at the image center.

In this example, the lenslet arrays are square packed on a 2 mm pitch. The beamlets are about 1.2 mm^2 after the reducer. This allows a maximum translation distance of 0.4 mm in either the x- or y-direction. The greater the beam reduction, the more the arrays are allowed to translate. However, the size of the image is also proportional to the diameter of the beams after reduction. Also, the beam reduction factor affects the blurring of the line image. The sensitivity of the system, or the rate of image increase with translation distance, is affected by the amount of beam reduction as well as the magnitudes of the phase coefficients.

This example shows that multi-aperture beam integrators need not be confined to only producing stationary simple spots. It also demonstrates how to design a multi-aperture beam integrator system that allows continuous variation of the shape, magnification, or aspect ratio of a target image by means of small lateral translations of two anamorphic lenslet arrays in a beam integrator configuration. Since the lateral translations of the arrays are small (on the order of tens or hundreds of microns), the arrays can be moved with motor-driven micrometers providing real-time variation of the spot shape and great flexibility to a laser machining or marking tool.

Diffractive optical elements allow paired lenslets to be placed in close proximity to each other, simplifying the design and fabrication process.

In principle, refractive nonsymmetric anamorphic lenslets can be fabricated using grayscale photolithographic processes. However, previous efforts to do so have proven very difficult because the grayscale process is highly nonlinear. Many more fabrication iterations are required for highly anamorphic surface profiles than for simple rotationally symmetric profiles. This fabrication problem is solved using diffractive lenslets. MEMS Optical, Huntsville, Alabama (now Jenoptic Optical Systems), successfully manufactured diffractive anamorphic lenslet arrays based on the example shown here. But attempts to fabricate them as refractive anamorphic lenslets were not as successful. Most laser machining applications can tolerate the higher order scattering from the diffractives. But if this is a problem, an off-axis system can be designed by adding a constant tilt to one of the diffractive arrays, separating the various orders at the final image plane so that the extraneous orders can be adequately blocked. An example of an off-axis beam integrator is shown in the work of Brown.[19]

11.7 SUMMARY

This chapter has presented a review of the classical ring forming properties of axicons, including a zoom system using a positive and negative axicon combination. The design of a zoom system using two positive axicons is developed. As part of this development, a unique diffraction theory is given for the case of a system with two elements, one before the Fourier transform lens and one after the lens. This diffraction theory is then used as a basis for calculation of the output patterns produced by the positive axicon zoom system. A design is developed for producing rectangular line patterns. This design uses a multi-aperture beam integrator system with highly anamorphic lenslet arrays.

REFERENCES

1. JH McLeod, "Axicon: A new type of optical element," *Journal of the Optical Society of America*, 44(8), 592–597, 1954.
2. PA Belanger and M Rioux, "Ring pattern of a lens-axicon doublet illuminated by a Gaussian beam," *Applied Optics*, 17(7), 1080–1088, 1978.
3. N Rioux, R Tremblay, and PA Belanger, "Linear, annular, and radial focusing with axicons and applications to laser machining," *Applied Optics*, 17(10), 1532–1536, 1978.
4. JWY Lit and R Tremblay, "Focal depth of a transmitting axicon," *Journal of the Optical Society of America*, 63(4), 445–449, 1973.
5. L Shi, X Dong, Q Deng, Y Lu, Y Ye, and C Du, "Design and characterization of an axicon structured lens," *Optical Engineering*, 50(6), 063001, 2011.
6. V Belyi, A Forbes, N Kazak, N Khilo, and P Ropot, "Bessel-like beams with z-dependent cone angles," *Optics Express*, 18(3), 1966–1973, 2010.
7. D Zeng, WP Latham, PF Jacobs, and A Kar, "Annular beam analysis of an optical trepanning system," *Proceedings of SPIE*, 5876OL, 184–193, 2005.
8. D Zeng, WP Latham, and A Kar, "Design and characteristic analysis of shaping optics for optical trepanning system," *Proceedings of SPIE*, 5876OM, 194–202, 2005.

9. D Zeng, WP Latham, and A Kar, "Optical trepanning with a refractive axicon lens system," *Proceedings of SPIE*, 6290, 6290OJ, 2006.
10. AV Goncharov, A Burvall, and C Dainty, "Systematic design of an anastigmatic lens axicon," *Applied Optics*, 46(24), 6076–6080, 2007.
11. FM Dickey and JD Conner, "Annular ring zoom system using two positive axicons," *Proceedings of SPIE*, 8130, 8130OB, 2011.
12. PA Belanger and M Rioux, "Diffraction ring pattern at the focal plane of a spherical lens-axicon doublet," *Canadian Journal of Physics*, 54, 1774–1780, 1976.
13. JW Goodman, *Introduction to Fourier Optics*, 3rd edn. New York: McGraw-Hill, 1996.
14. DM Brown, "Multi-aperture beam integrator/method producing a continuously variable complex image," US Patent No. 6909553, June 2005.
15. DM Brown, "Multi-aperture beam integrator producing continuously variable complex images," *Proceedings of SPIE*, 5175, 130–138, 2003.
16. LW Alvarez, "Two-element variable-power spherical lens," US Patent No. 3305294, February 1967.
17. LW Alvarez and WE Humphrey, "Variable-power lens and system," US Patent No. 3507565, April 1970.
18. J Wallace, "Alvarez lens enters the real world," *Laser Focus World*, pp. 15–16, March 2000.
19. DM Brown, "Variable ring beam integrators for product marking and machining," *Proceedings of SPIE*, 4443, 159–165, 2001.

12 Current Technology of Beam Profile Measurement

Kevin D. Kirkham and Carlos B. Roundy

CONTENTS

12.1 INTRODUCTION

As explained in Chapters 2 through 10, laser beam shaping is a process whereby the irradiance of the laser beam is changed along its cross section. In some cases, the laser beam is shaped so that it is uniform or top hat. In other cases, it is given a different shape such as a Gaussian or super-Gaussian. For laser beam shaping to be effective, it is necessary to be able to measure the degree to which the irradiance pattern or beam profile has been modified by the shaping medium. In some cases, the beam shaping requires a specific input beam. For example, in many cases, the input beam must be Gaussian for the shaping to create an undistorted top-hat beam. In this case, the profile of the input beam must be measured to assure that it is close enough to the desired Gaussian distribution. If the input beam does not have the proper profile, measurements will inform the user that adjustments to the source must be made before attempting to perform laser beam shaping. Therefore, laser beam profile analysis is an essential part of effective laser beam shaping.

Section 12.2 discusses laser beam properties and the general need for beam profile analysis [1–10]. Section 12.3 reviews the art of laser beam profile analysis and historic methods. Section 12.4 introduces the use of electronic cameras for measuring the laser beam profile and includes descriptions of other instrumentation useful in beam profile measurement. Sections 12.5 and 12.6 discuss accommodations necessary for the use of cameras [11,12] and Section 12.7 considers the information that can be obtained by simply viewing the beam profile. In Section 12.8, quantitative measurements and their significance are explained. Sections 12.9 through 12.11 discuss other electronic measurement techniques that are used to provide a more complete understanding of a laser's performance including propagation characteristics and wavefront analysis [13,14]. Chapter 10 also goes into the need for signal processing to enhance the accuracy of electronic laser beam measurements. The chapter closes with a summary in Section 12.12.

12.2 LASER BEAM PROPERTIES

12.2.1 Unique Laser Beam Characteristics

Laser beams produce light with many characteristics that are unique to this type of illumination. Some of the characteristics that make laser beams unique are listed in Table 12.1. For example, the monochromatic nature of a laser beam means that it is typically a single wavelength with very little light at wavelengths other than the central peak. The temporal characteristic of a laser beam covers a wide range from a continuous wave (CW) to an extremely short pulse, providing very high-peak-power densities. The coherence of a laser enables it to travel in a narrow beam with a small angle of divergence or spread. This allows a user to precisely define the area illuminated by the laser beam at any distance from the source. Because of this coherence, a laser beam can also be focused to a very small and intense spot of light in a highly concentrated area. This ability to be concentrated makes the laser beam useful for many applications in physics, chemistry, medicine, and industry. Finally, a laser beam has a unique irradiance profile that offers significant opportunities for its application. The beam profile is the cross-sectional pattern of irradiance at a typically orthogonal plane along the propagation axis of the beam.

12.2.2 Significance of the Beam Profile

Since the beam profile is a description of the two-dimensional (2D) energy density distribution at a plane perpendicular to the laser's propagation, information regarding the concentration and the collimation of the laser is evident in the beam profile. Propagation characteristics of the beam through space can be easily determined by observing a series of beam profile measurements along its path. Figure 12.1 shows a number of typical laser beam profiles illustrating the variety that can exist. Since such a variety may exist, it is essential to measure the profile in any application where energy distribution affects the performance of the laser or its intended purpose.

TABLE 12.1
Unique Characteristics of a Laser Beam

Monochromatic (single wavelength)
Temporal (continuous wave to femtosecond pulses)
Coherence (consistent phase between all light elements)
Highly concentrated (focusable to extremely small spots)
Beam irradiance profile (unique spatial power or energy distribution)

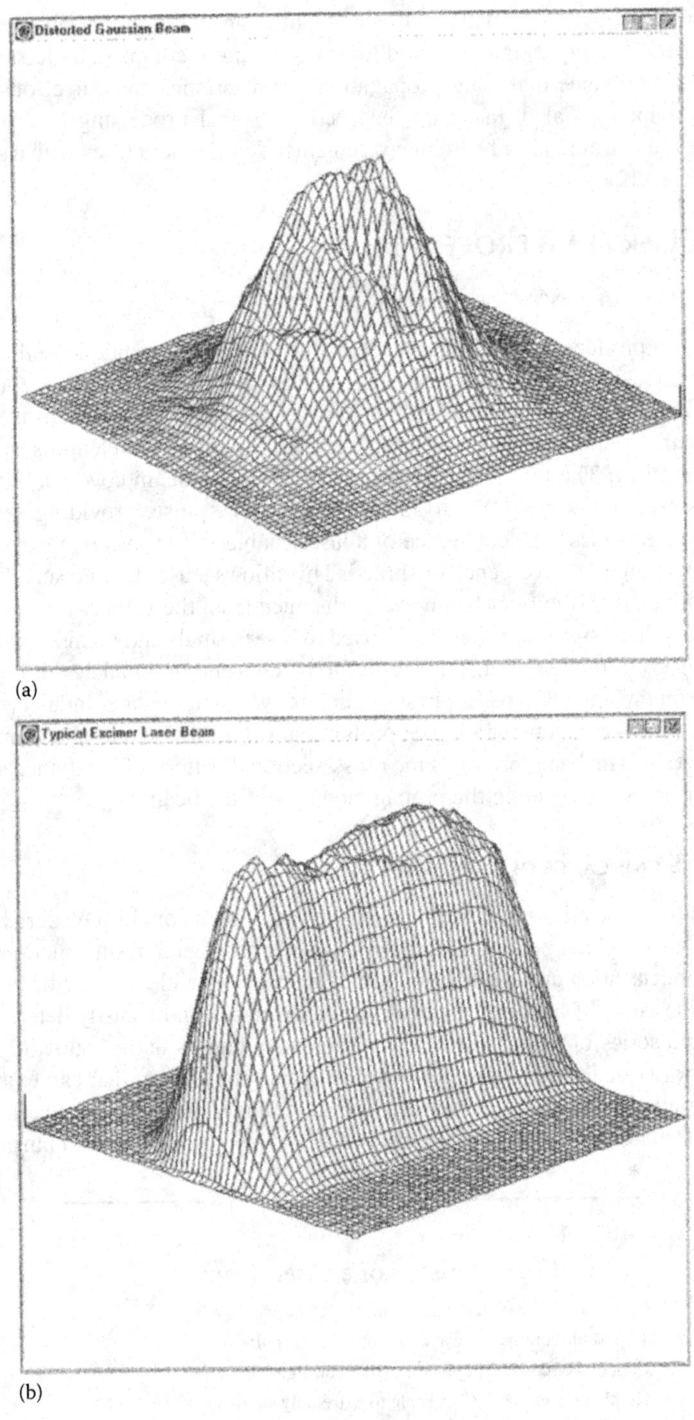

FIGURE 12.1 Various laser beam profiles: (a) HeNe; (b) excimer; (c) nitrogen ring laser.

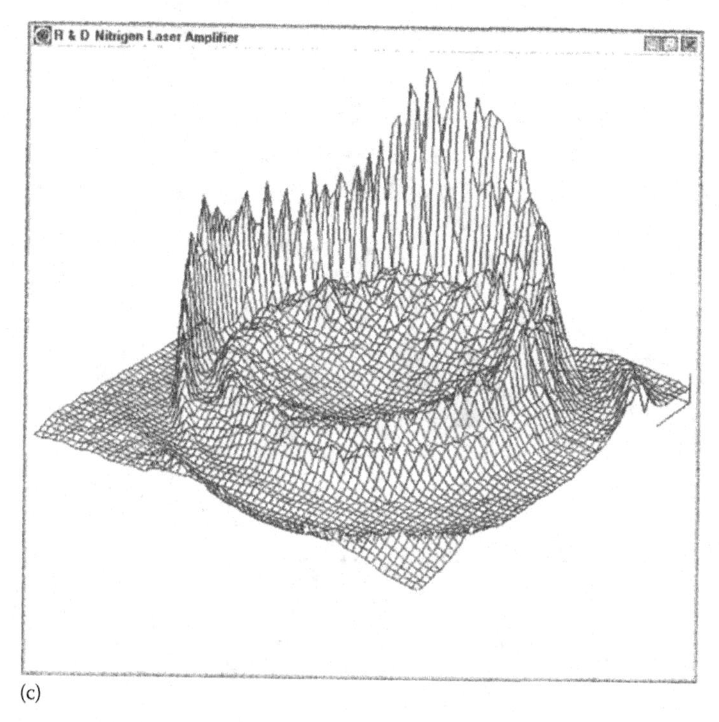

(c)

FIGURE 12.1 **(Continued)**

Examples of two different types of ideal laser beams for different purposes are a Gaussian and a top-hat beam. A true Gaussian or single-mode beam allows the highest concentration of focused light, whereas a top-hat beam allows for very uniform distribution of the energy across a given area. These two idealized beams are shown in Figure 12.2.

12.2.3 Effects of Distorted Beam Profiles

Lasers rarely exhibit the most uniform irradiance profile. Sometimes Gaussian beams are highly structured, and often-intended top-hat beams are nonuniform across the top, or may be tilted in energy density from one side to the other. Figure 12.3 illustrates some real-world examples of distorted beam profiles. For example, in Figure 12.3a, the highly structured beam would not focus nearly as well as the ideal Gaussian beam. The tilted top-hat beam in Figure 12.3b would not give uniform illumination as intended, and could cause poor performance in the process for which it is being applied.

12.2.3.1 Scientific Applications

The significance of distorted beam profiles varies with the application. In scientific applications, nonlinear processes are typically proportional to the irradiance squared

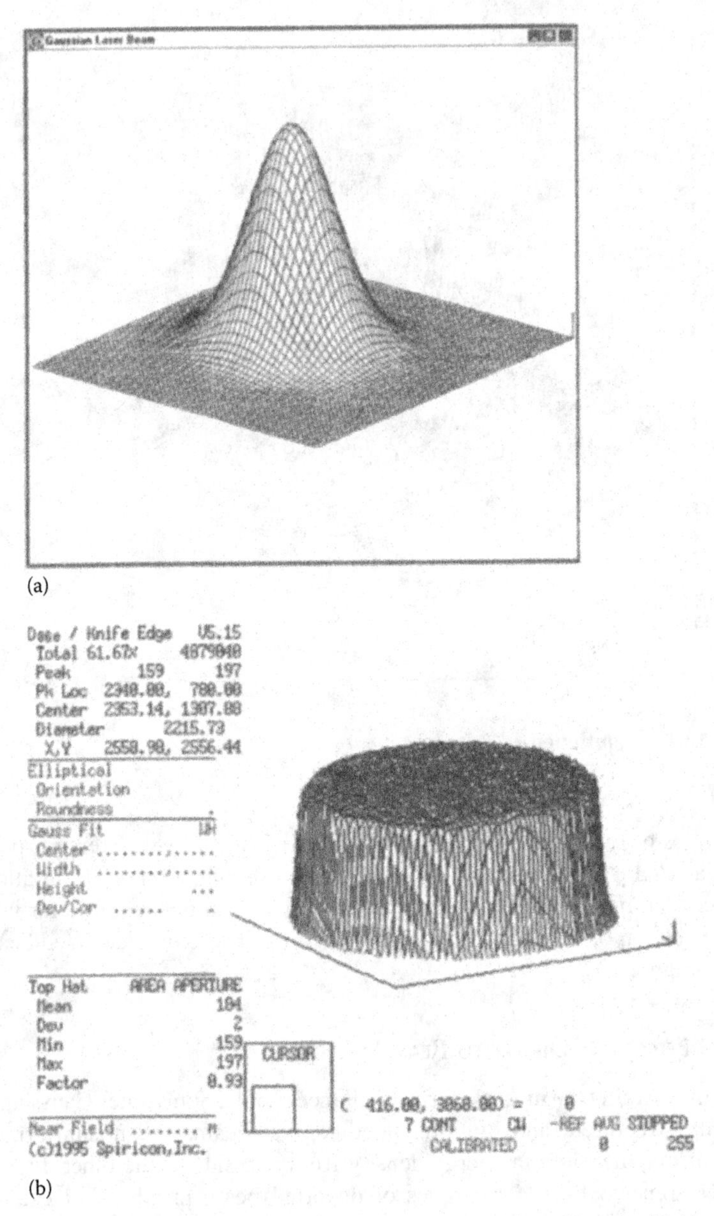

FIGURE 12.2 (a) Ideal Gaussian beam for highest concentration of energy; (b) ideal top-hat beam for uniform laser illumination.

or cubed. Thus, a non-Gaussian profile may have peak energy as low as 50% of what a Gaussian beam would have under the same conditions of total power or energy. Therefore, the nonlinear process may deteriorate to 25% or 12% of what is expected. This is a 300%–700% error on an experiment that should be accurate to within ±5%. Figure 12.4 shows beam profiles of a Cr:LiSAF oscillator with subsequent amplifier outputs when the amplifier is properly aligned and when it is not.

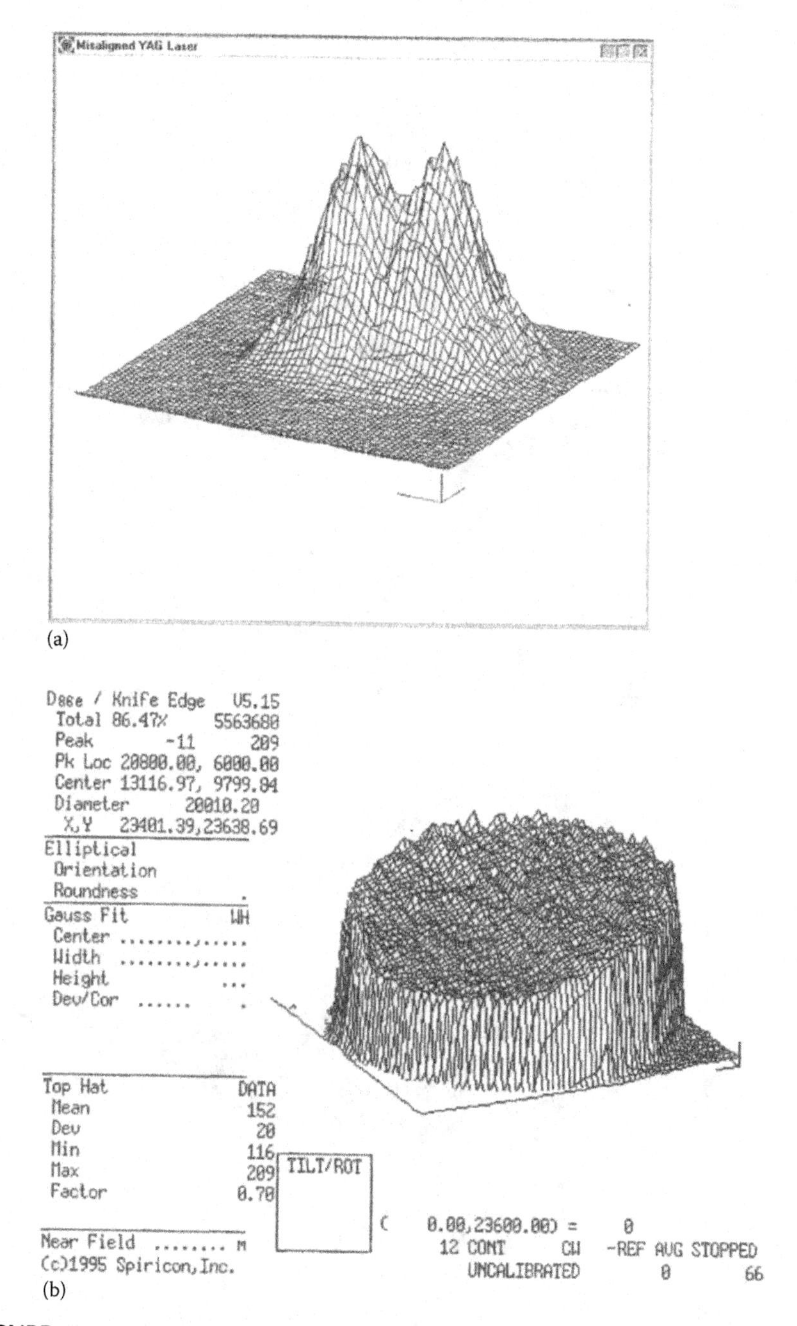

FIGURE 12.3 (a) Highly structured would-be Gaussian beam; (b) tilted or nonuniform top-hat beam.

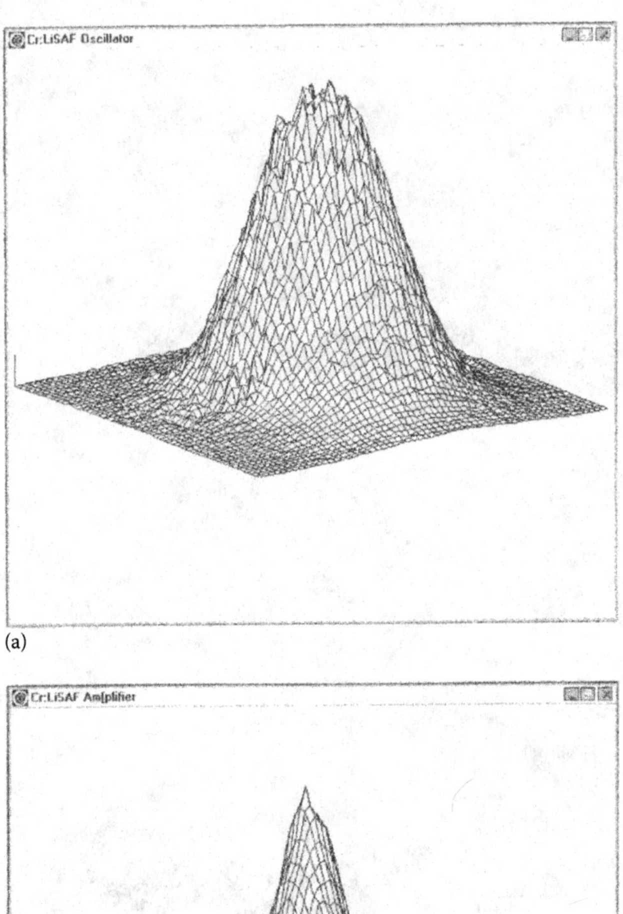

(a)

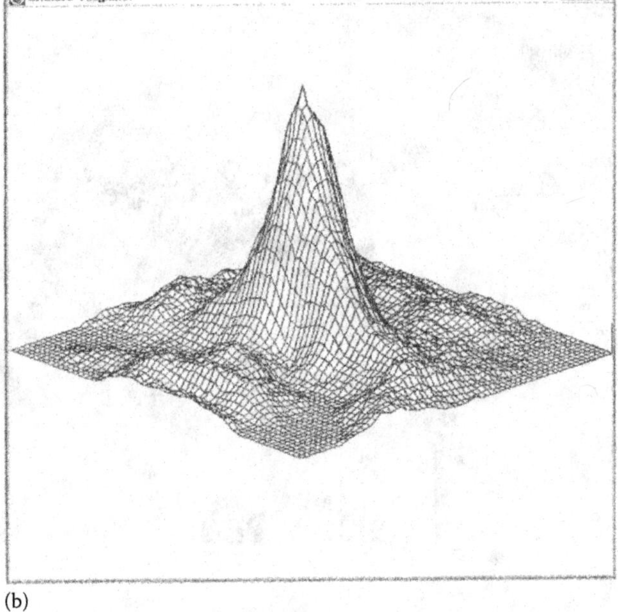

(b)

FIGURE 12.4 (a) Cr:LiSAF laser oscillator; (b) Cr:LiSAF laser with well-aligned amplifier; (c) Cr:LiSAF laser with misaligned amplifier.

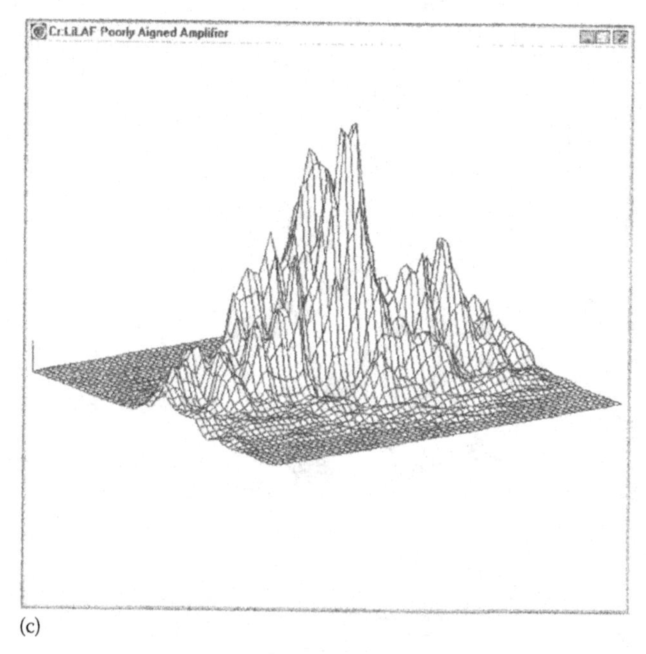

(c)

FIGURE 12.4 **(Continued)**

12.2.3.2 Instruments Using Lasers

Instruments using lasers, such as printers, fiber optic communications, and cell cytometry, require a high degree of control of the laser light to accomplish the intended task. The uniformity, pointing direction, pointing stability, and mode pattern of a typical laser diode used in instruments can be dramatically deteriorated by misalignment of the collimating optics or mounting, causing the instrument not to perform as expected. For example, Figure 12.5 illustrates a collimated laser diode beam being focused into a single-mode fiber optic. In Figure 12.5b, the z-axis of the focused laser diode is poorly aligned to the fiber, and much of the energy is fed into the cladding rather than to the inner fiber. Thus, the efficiency of the fiber as an energy conduit is significantly degraded. Energy in the central lobe is greatly reduced due to misalignment. In Figure 12.5c, the z-axis is adjusted slightly and the major portion of the laser beam is coupled into the core of the fiber optic; thus, the fiber more efficiently transports optical power.

12.2.3.3 Medical Applications

There are many medical applications of lasers [15]. One of these is laser-assisted *in situ* Keratomileusis (LASIK) [16], in which one technique, a top-hat beam is used to make corrections to the optical properties of the cornea. If the homogenizer producing the top hat is out of alignment and there is a 50% tilt in the top-hat intensity pattern, the correction to the eye may be four diopters on one side of an iris, with only two diopters on the opposite side. The top-hat beam in Figure 12.2b would give expected results, whereas the tilted or nonuniform beam in Figure 12.3b would cause

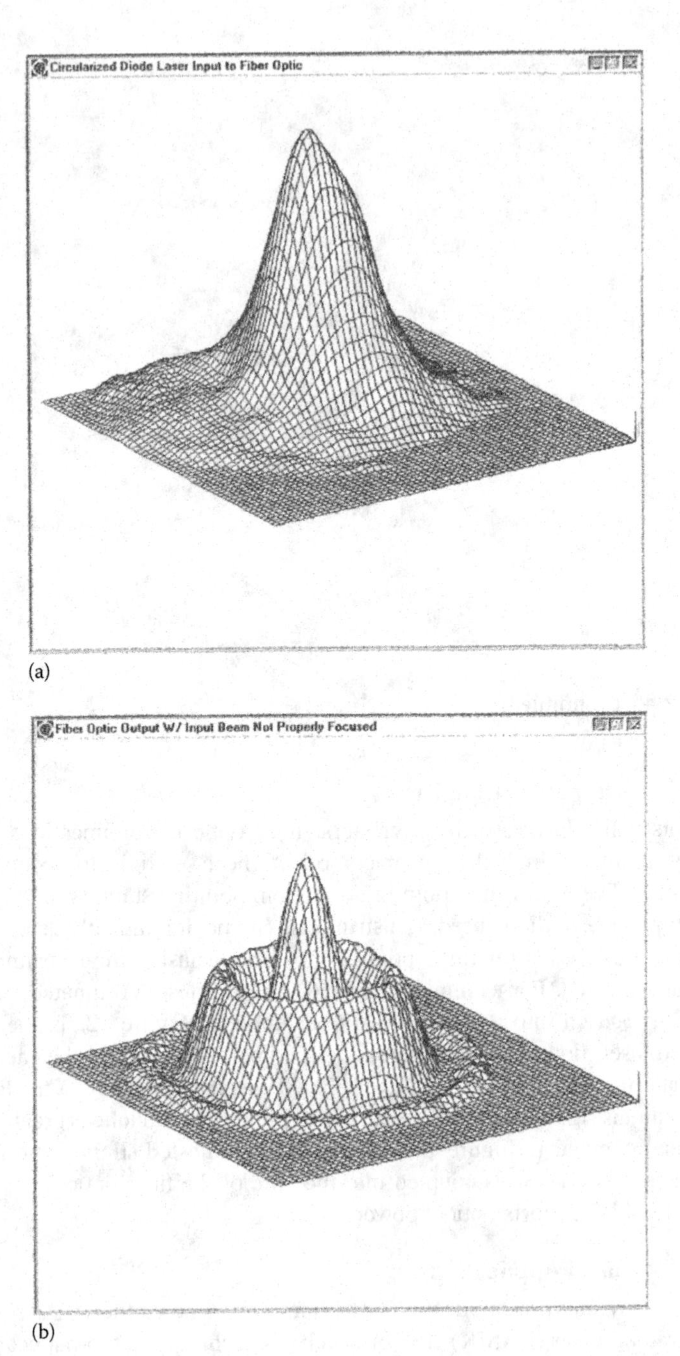

FIGURE 12.5 (a) Collimated laser diode beam; (b) fiber output with diode poorly coupled into fiber optic; (c) fiber output with diode well coupled into fiber optic.

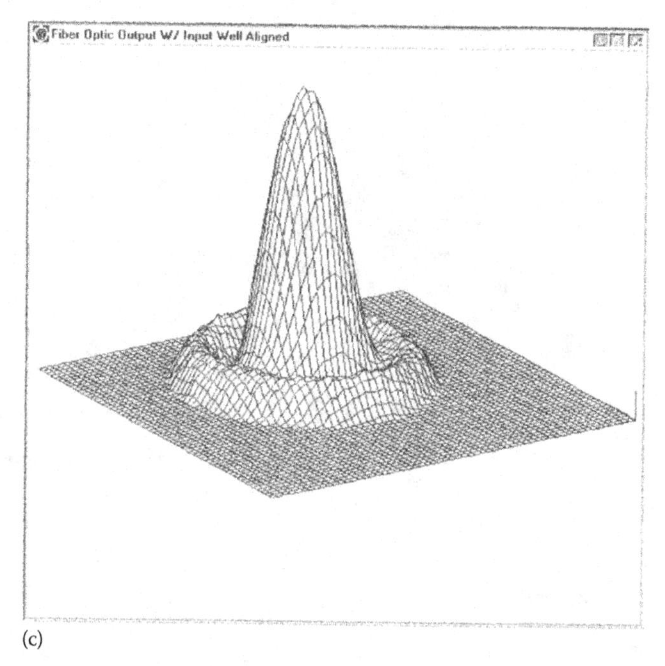

(c)

FIGURE 12.5 **(Continued)**

severe difficulty in the surgical outcome. The flatness or uniformity of the laser beam top hat is also critical in the removal of port wine stains, wrinkle remediation, and in other cosmetic surgery.

Tissue cutting and welding requires an extremely well-controlled irradiance density to accomplish the task properly. Finally, many medical applications, such as photodynamic therapy, use fiber optic delivery systems, and the efficiency of these systems is strongly affected by the initial alignment of the laser beam into the fiber, as shown in Figure 12.5.

12.2.3.4 Industrial Applications

In industrial laser applications [17–19], most high-power Nd:YAG lasers and some CO_2 lasers produce multimode beams. The cutting, welding, and drilling efficiencies of these lasers are directly related to the beam profile. For example, an Nd:YAG laser with a double peak can cause one cut width in the x-direction and a different cut width in the y-direction. A beam with a poor profile can result in percussive laser hole drilling of a different size than expected, and a weld that is not as strong or as precise as necessary, for instance, to create a hermetic seal.

Figure 12.6 shows the beam profile of a poorly aligned and a well-aligned CO_2 laser cavity. An industrial laser shop was using CO_2 lasers for scribing ceramic wafers before dicing them into individual pieces. Most of the lasers in the machine shop gave good results. However, one laser gave inconsistent results, which caused very erratic breaking of the ceramic which produced excessive scrap. The laser had been measured by nonelectronic mode burns in wooden tongue depressors, which

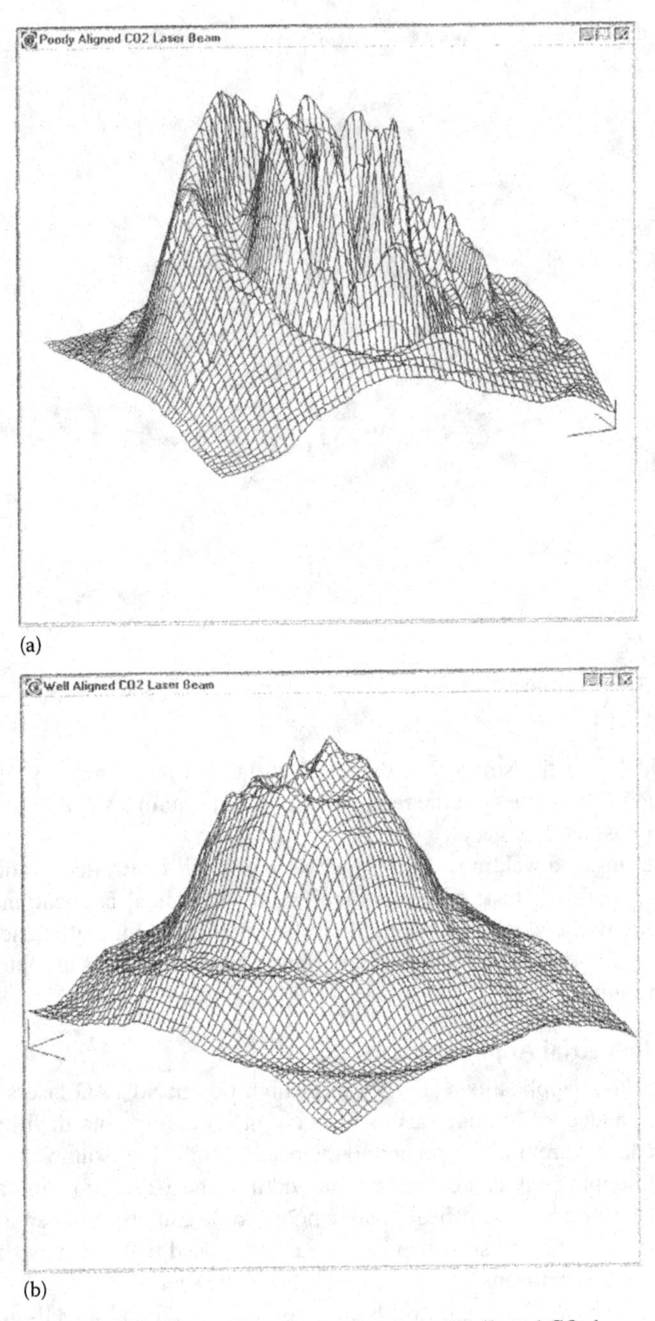

FIGURE 12.6 (a) Poorly aligned CO_2 laser cavity; (b) well-aligned CO_2 laser cavity.

gave the appearance that the laser had an appropriate output. However, when the laser was measured with an electronic pyroelectric camera, the structure in the laser beam of Figure 12.6a became immediately evident. When the technician saw this structure, he adjusted the laser cavity mirrors, and within a short time, using the visual feedback of the electronic beam profiling system, the beam was improved to that shown in Figure 12.6b. It was observed that the beam profile of Figure 12.6b was similar to that of the other lasers that were operating efficiently in this shop. Once the problematic laser was tuned to the beam profile of Figure 12.6b, it gave consistent results.

12.3 LASER BEAM PROFILE MEASUREMENT METHODS

12.3.1 Nonelectronic Methods

There are a number of nonelectronic methods of laser beam profile measurement that have been used since lasers were invented. The first of these is observance of a visible laser beam reflected from a wall or other object. This is by far the simplest, least expensive, and most widely used method of observing a laser beam profile. One problem with this method is that the human eye is logarithmic, and can see many orders of magnitude difference in light irradiance. Even though it is logarithmic, the eye can only distinguish 8–12 shades of gray. Thus, it is nearly impossible for a visual inspection of a laser beam to provide anything more than the most basic qualitative measurement of the beam size and shape. A beam width measurement by eye may have as much as 100% error. Figure 12.7a is a photograph of a HeNe laser beam being reflected off the wall.

Photographic film or burn paper has even less dynamic range than the human eye. Figure 12.7b shows a very intense beam at the center, but a very large amount of structure far out from the center. This structure, which one might mistake as part of the laser beam, is less than 1% of the total energy. The eye and the film are clearly able to discern this low energy. However, the eye may not distinguish structure in a laser beam with less than a 20% magnitude variation.

Burn paper and exposed photographic film are often used for making beam profile observations. Figure 12.7b illustrates thermal paper having been illuminated by a laser beam. The burn paper may have a dynamic range of only 3, unburned paper, brown, and blackened paper. Skilled, experienced operators can make more critical burn paper patterns, and obtain a dynamic range of 5 or more. The main objection to this manual method is that the spot size is highly subjective to the integration time on the paper and the experience of the operator. With longer exposures the center may not change, but the width of the darkened area could change 50% or more.

Wooden tongue depressors and burn spots on metal plates are used as methods similar to burn paper. Sometimes the depth of the burn gives additional insight into the laser irradiance pattern. Experienced operators learn to read these burn spots to determine which beam tuning gives a specific result in a specific application. They might be tuning to one burn spot type for cutting, and a different one for drilling holes. However, this measurement system is archaic, crude, nonquantitative, and subject to the capability and experience of the operator. Therefore, burn patterns are dependent on the experience and interpretation of the operator and thus unreliable.

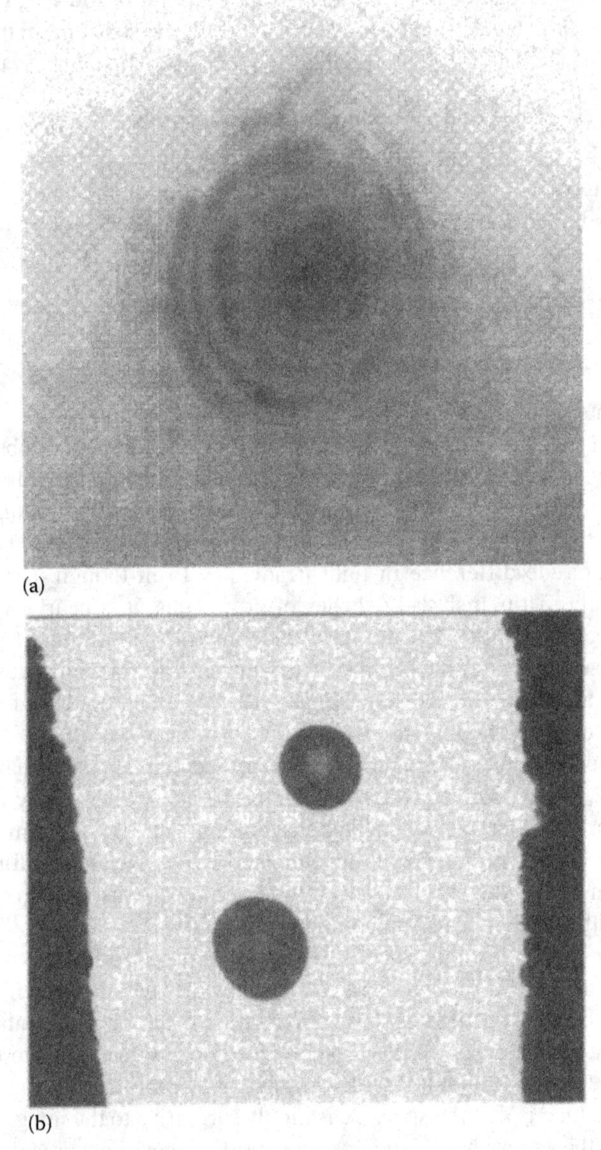

(a)

(b)

FIGURE 12.7 (a) Reflected laser beam; (b) laser beam burn spots.

Fluorescing plates or cards permit the laser operator to see laser beams by converting UV or IR radiation into visible light where they can be seen by the human eye. Fluorescing plates have limited dynamic range. This permits the determination of the relative spot size but not the distribution of intensities of light within the spot. Similar dynamic range problems are encountered when viewing reflected beams.

Acrylic mode burns such as those shown in Figure 12.8 provide quite representative beam profiles of CO_2 lasers. The depth and detailed pattern of the acrylic burns

FIGURE 12.8 Laser beam acrylic mode burn.

clearly shows the irradiance profile of the beam, and it is often possible to see mode structure. This gives an excellent visual interpretation of the beam profile. However, the acrylic mode burns are not presented in real time, which makes it very cumbersome to tune the laser. They also do not enable one to see short-term fluctuations in the laser beam that can be problematic in some CO_2 lasers. It is possible that fumes from the burning acrylic may form plasma or a smoke plume at the center of the hole. This material can block the incoming CO_2 beam, thus distorting the pattern of the melt. Unless care is taken to have a fan to extract the fumes, the acrylic mode burn will contain structure that is nonexistent in the beam. Fumes from burning acrylic are toxic to humans, and care must be taken to exhaust these fumes outside of the work area.

12.3.2 Electronic Measurement Methods

For electronic laser beam profile analysis, it is typically necessary to attenuate the laser beam to match the range of the sensor. The degree of attenuation required depends on two factors. The first is the irradiance of the laser beam being measured. The second is the sensitivity or dynamic range of the beam profile sensor. Figure 12.9 shows a typical setup for the case where a significant amount of attenuation is required for the sensor to operate in the range of its linear response.

Typically, when measuring the beam profile of a high-power laser, that is, in excess of 5–50 W, the beam has enough energy to damage most sensors that might be placed in the beam path. Therefore, the first element of Figure 12.9, the beam sampling assembly, is typically used. It should be noted that there are some beam profiling sensors that can be placed directly into the path of a high-power laser to 10 kW.

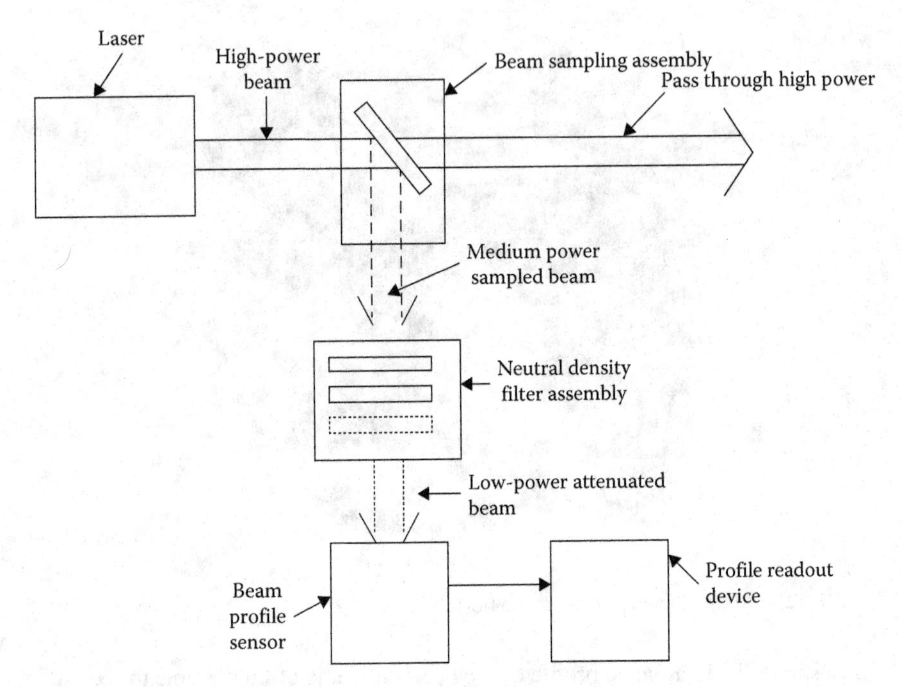

FIGURE 12.9 Optical setup for electronic laser beam analysis.

For mechanical scanning instruments, the beam splitter assembly is usually sufficient to reduce the signal from high-power lasers to the level that is acceptable. If the original laser beam is in the range of 50 W/cm² or less, mechanical scanning instruments can measure the beam directly without using beam splitting or sampling optics. Mechanical scanning instruments are able to be placed directly in the path of medium power beams because the scanning device also acts to reflect a large percentage of the incident laser light. In the case of a scanning slit-based system, the drum blocks light from the sensor during most of the duty cycle of the sensor. The rotating drum either absorbs or reflects the incident laser except when the microscopic slit, which is patterned on to the drum, rotates through the beam. In this case, only a very small fraction of the laser power is placed on the sensing element. For this type of beam profile sensor when excess laser power is placed on the sensor, the typical point of failure is the mechanical slit and not the detector element.

For camera-based beam analyzers, beam splitters alone may not provide sufficient attenuation to reduce the beam power to less than the saturation level of the camera sensor. In this case, additional attenuation filters are placed directly in front of the camera sensor to reduce the power density to a level acceptable by the camera. In some cases, the beam power, even after reflection from one sampling surface, is too high and would burn neutral density (ND) filters or cause the filters to distort the beam profile. In this case, a second reflecting surface is used to further reduce the incident power before impinging on the ND filter set. This is described in more detail in Section 12.4.

ND filter systems can be adjusted over a very wide dynamic range, as much as from ND0 to ND10 (or transmission of 1–10^{-10}). These ND filter assemblies are discussed in more detail in Section 12.5.2.

The fourth item in Figure 12.9, the beam profile sensor, can be a mechanical scanning device, described in the next section, or a charge-coupled device (CCD) or other type of 2D sensor array-based camera, described in the following section. The beam profile display device consists of a dedicated monitor or a PC style computer and monitor for providing an electronic image of the beam profile.

12.3.2.1 Mechanical Scanning Instruments

One of the first methods of measuring laser beams electronically employed a mechanical scanning device. This usually consisted of a rotating drum containing a knife-edge, slit, or pinhole that moves in front of a single-element detector. This method provided excellent resolution, typically to less than 1 μm. The limit of resolution of this type of device is determined by the size of the slit or pinhole and the sampling frequency of the electronics that define the location of the mechanical scanning device. These devices can be used directly in the beam of low- to medium-power lasers with little or no attenuation because only a small part of the beam is impinging on the detector element at any one time. The rotating drum reflects the beam away from the detector the majority of the time.

Mechanical scanning methods work with CW lasers and high-frequency pulsed lasers. They do not work with pulsed lasers with pulse repetition frequencies of less than ~20 kHz. They have a limited number of axes for measurement, usually two, and integrate the beam along those axes. Thus, they only give detailed information about the structure of the beam in the direction of travel of the edge, slit, or pinhole. These beam profile instruments are appropriate for work in the visible, UV, and IR regions by using different types of single element detectors to sense the intensity of the laser radiation. Software has been developed which provides useful beam profile displays, as well as detailed quantitative measurements describing the spatial characteristics of the laser that is being measured.

Figure 12.10a illustrates a commercial version of the knife-edge scanning slit beam profiling instrument. Figure 12.10b shows a typical Windows computer display. Figure 12.10c illustrates a typical mechanical diagram of a scanning slit beam profiler. Figure 12.11a is a photograph of a seven-axis scanning blade system. Figure 12.11b illustrates a typical Windows display from this seven-axis system. The mechanical layout of the multiaxis profiler is similar to Figure 12.10c. The angles of the knife-edges are varied so that the beam profile data come from multiple axes.

One variation of the rotating drum is the scanning slit or pinhole system, which is used for measuring the propagation characteristics of the beam under test. This system includes a lens mounted in front of the drum [20]. The lens is mounted on a moving stage under software control. By moving the lens in the beam, a series of measurements can be made that enables calculation of M^2. (A more detailed discussion of M^2 will be provided in Section 12.9.) A photograph of this M^2 measuring instrument and readout is given in Figure 12.12a. A mechanical layout of the instrument is shown in Figure 12.12b.

Another mechanical scanning system consists of a rotating needle that is placed directly in the beam. This needle or waveguide has a very small opening allowing

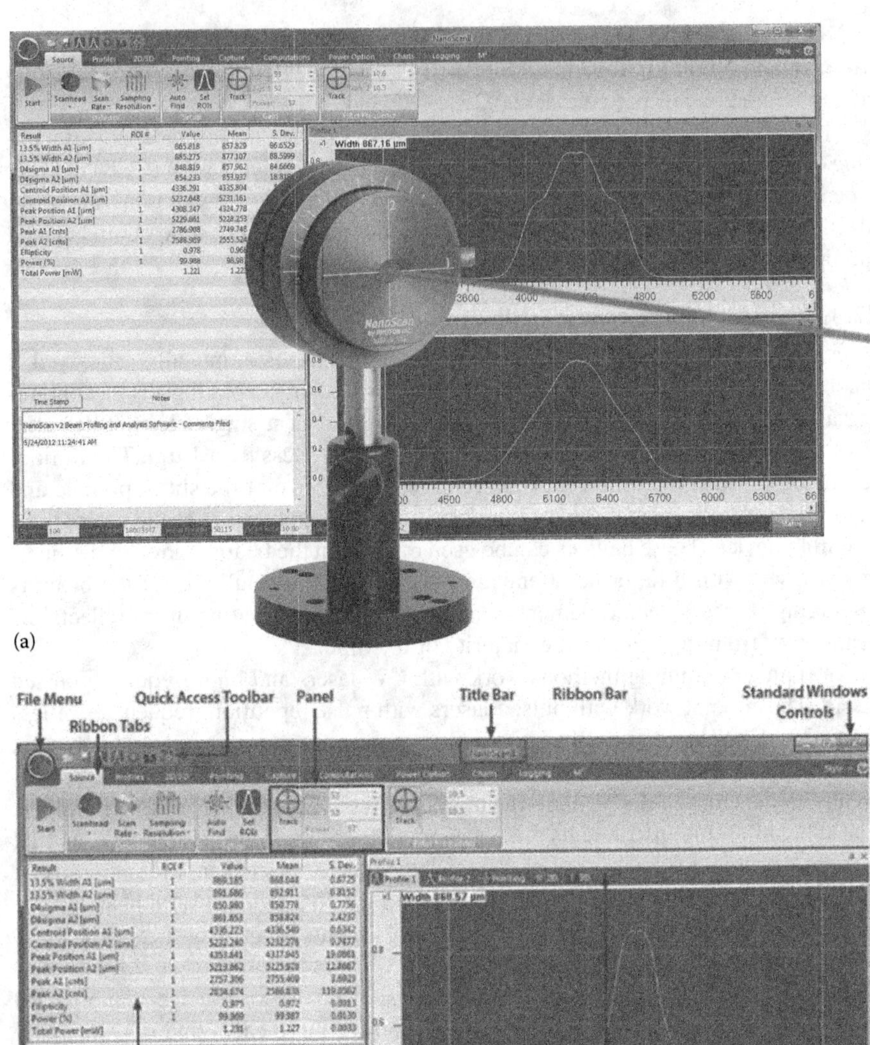

FIGURE 12.10 **(See color insert.)** (a) Commercial knife scanner; (b) Windows PC display; (c) mechanical diagram of scanning slit or knife-edge beam profiler.

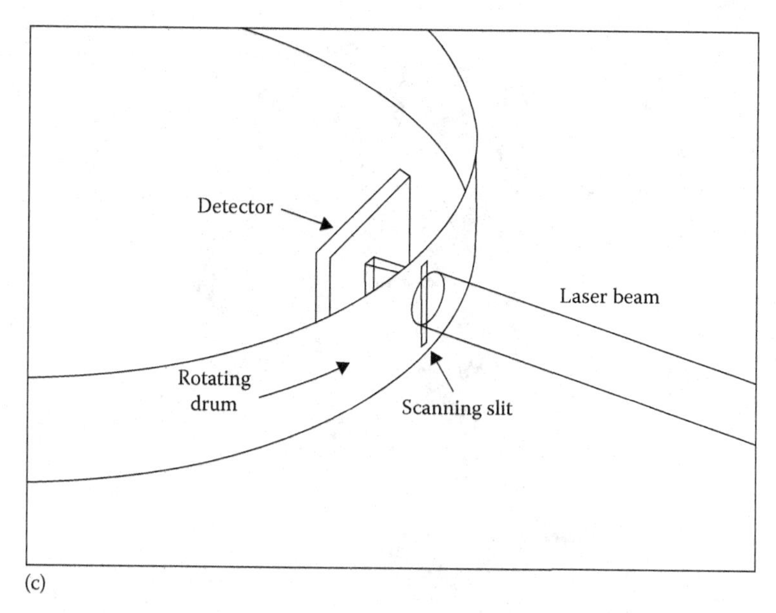

FIGURE 12.10 **(Continued)**

only a portion of the beam to enter the needle. A 45° mirror at the bottom of the needle reflects the sampled part of the beam to a single-element detector. The needle is both rotated in the beam and axially moved in and out of the beam to sample it in a complete, 2D manner. The advantage of a rotating needle system is that it can be placed directly in the beam of high-power industrial lasers, including Nd:YAG, fiber, and CO_2. In addition, the step size of the mechanical translation of the needle can be made extremely small for measuring focused laser beams, or large for measuring unfocused beams. It has, however, some characteristics of the rotating drum mentioned earlier. Specifically, it is not very useful for pulsed lasers because time is required to scan the beam. More scanning time is required than the duration of the laser pulse. Rotating needle systems provide extensive computer processing of the signal to produce displays of the measured beam profile and quantitative calculations describing the laser spot size and location.

12.3.2.2 Camera-Based Systems

Cameras are used to provide simultaneous, 2D laser beam profile measurements. They work with both pulsed and CW lasers. Silicon-based cameras operate in the UV region to the near IR to 1.1 μm, which is useful with a great majority of lasers. In addition, there are other types of cameras that operate in the x-ray and UV regions, as well as cameras that cover the IR wavelengths from 1.0 μm to >400 μm. One limitation to the use of cameras as a sensor for measuring laser beam profiles is that the spatial resolution is limited by the pixel size of the cameras. In CCD cameras, this is roughly 5 μm, and for most IR cameras, the size is somewhat larger. However, a focused spot can be reimaged with lenses to provide a larger spot for viewing on the camera, which can provide resolution down to approximately 1 μm

(a)

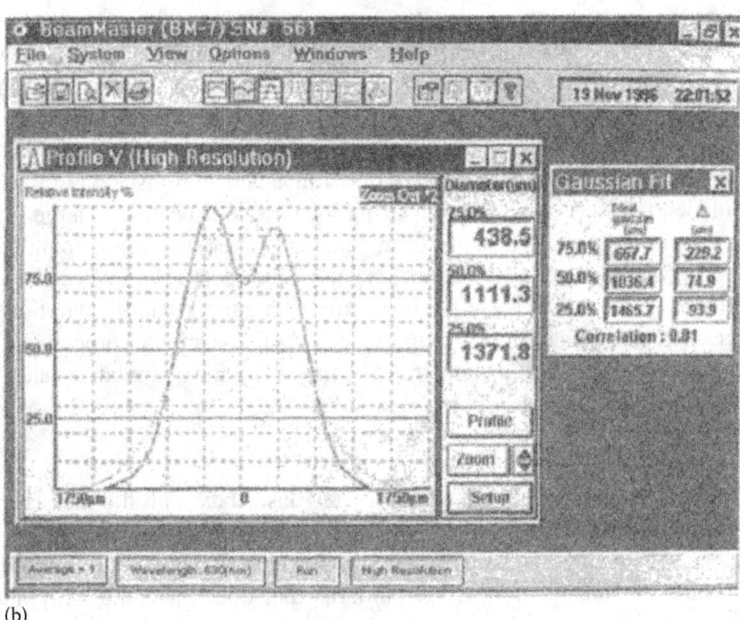

(b)

FIGURE 12.11 (a) Seven-axis knife-edge instrument; (b) typical readout from seven-axis system.

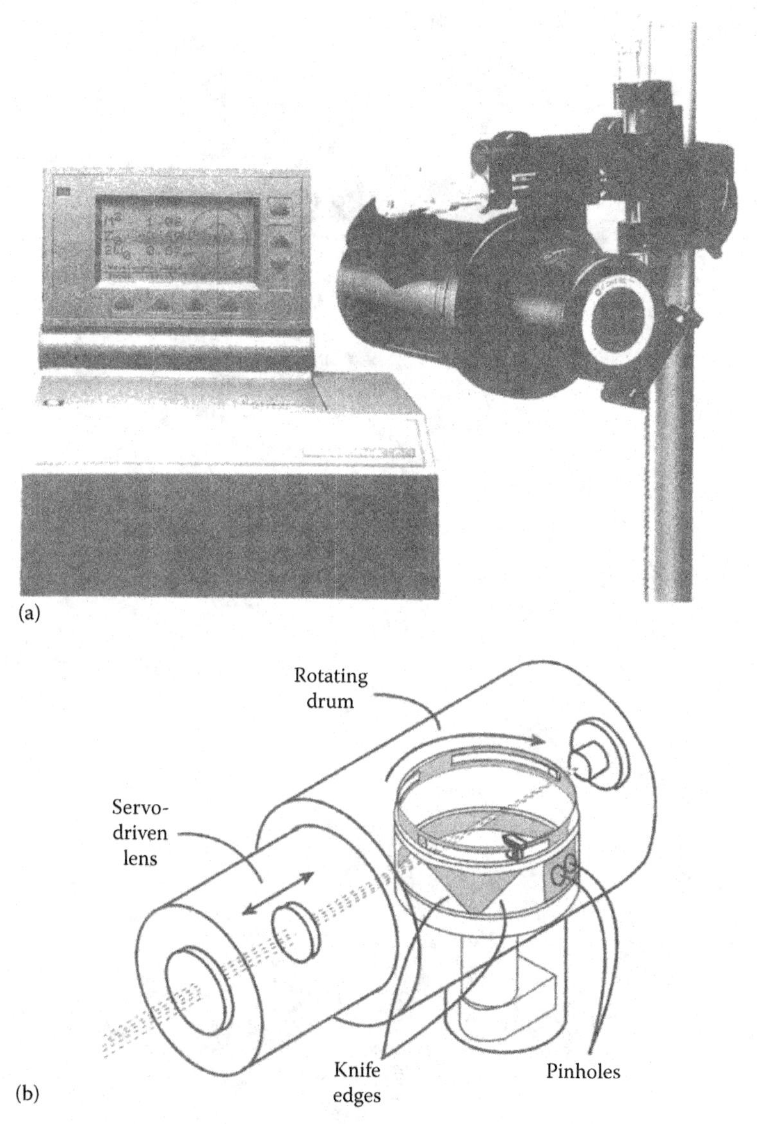

FIGURE 12.12 (a) M^2 measuring instrument and readout; (b) mechanical diagram of M^2 measuring instrument.

or less depending on the laser wavelength and the optics that are employed. The resolution of camera systems using beam reimaging or magnification optics is limited by diffraction of the optics. Modern cameras provide easy computer connection through universal serial bus (USB), Firewire (IEEE1394), or gigabit Ethernet (GigE) interfaces that are already available on most PCs. Current commercial beam profiling software provides illuminating two- and three-dimensional (3D) beam displays as shown in Figure 12.13. Modern beam profiling systems also provide sophisticated

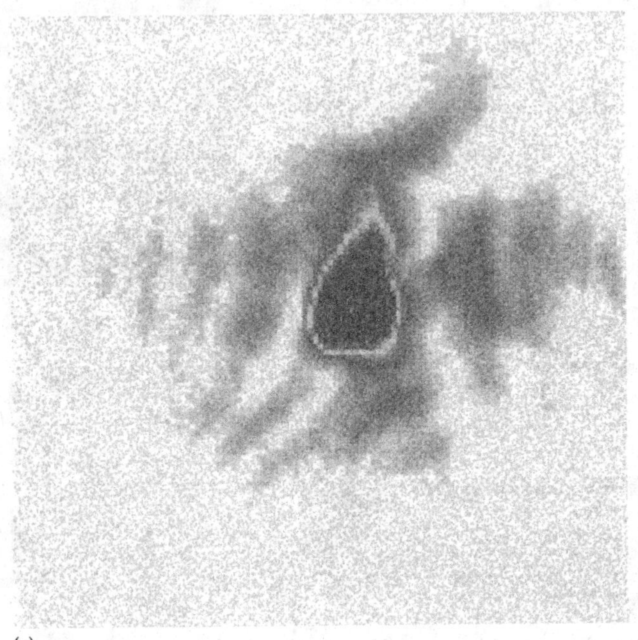

(a)

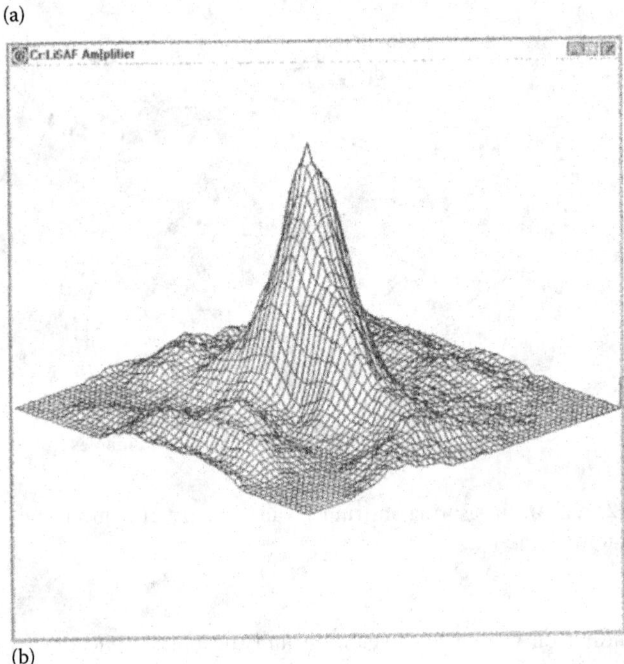

(b)

FIGURE 12.13 Highly structured laser beam measured with a CCD camera and shown in both 2D (a) and 3D (b) views.

spatial analysis of the beam profile. A drawback of camera-based systems is that the cameras are extremely sensitive relative to the energy in most laser beams. Thus, nearly all lasers must be significantly attenuated before the cameras can be used for beam profile analysis.

12.4 CAMERA-BASED INSTRUMENT DESCRIPTIONS

Complete instrumentation for a camera-based beam profiling system includes the camera sensor, PC, and software for making quantitative calculations and displaying the graphical representations of the beam profile as well as the computational results. Optics is almost always needed to split off the part of the beam, and attenuate the beam before going into the camera. Often the beam is either too large or too small, and optics or other techniques must be used to size the beam appropriately.

12.4.1 COMPUTERS

Modern computers used for laser beam profiling permit the imaging and analysis of the beam at frame rates as fast as 30 Hz with camera matrices as large as 1200 × 1600 pixels. The performance of these PC-based systems is enhanced by the ease of use of USB 2.0, Firewire, and GigE interfaces.

Measuring parameters of a laser beam presents many unique challenges that do not exist when measuring other items that are commonly measured with machine vision systems. As renowned laser scientist Tom Johnston once said, "Measuring the width of a laser beam is like trying to measure the size of a cotton ball with a caliper." The difficulty in measuring laser beam widths comes from the fact that laser beams never cut off succinctly, but almost always have energy that extends out into the wings where very low light levels can mimic camera noise. Thus, processing of very low-level optical signals becomes critical in the measurement of laser beams. Therefore, sensors with high dynamic range are necessary to accurately perform the task.

12.4.2 BEAM ANALYSIS SOFTWARE

Laser beam analysis systems run on the current operating system, 64-bit Windows 7 and 8, and 32-bit Windows XP as well as legacy operating systems and non-Windows environments. Sophisticated signal processing of large data matrices provides very detailed views of the laser as well as accurate quantitative beam profile measurements. Software must also control the sensor and thus be able to manipulate the camera baseline, exposure, and gain settings. In addition to providing beam displays and calculations, beam analysis software may also interface to other sensor systems such as laser power and energy measurement devices, wavelength, and pulse width measurement devices to name a few. The capabilities of the software to provide intuitive beam profile displays are described in more detail in Section 12.7, and the quantitative calculations made by these products are provided in Section 12.8.

12.4.3 Cameras Used in Beam Profile Measurement

Various types of cameras are used in laser beam profile analysis. Each of these has advantages and disadvantages for various applications. The most common type of camera used for laser beam diagnostics incorporates a silicon-based sensor. These cameras consist of at least two types: CCD and complementary metal–oxide–semiconductor (CMOS). Silicon-based cameras cover the wavelength range from 190 nm to 1.1 μm when the typical glass window and IR cutoff filter are removed. Such optics would otherwise attenuate or eliminate the UV and IR portions of the usable spectrum. CMOS and CCD cameras are fairly inexpensive because they cover the UV, visible, and near-IR regions these cameras are used for the measurement of many lasers. Also in use are cameras with InGaAs matrix sensors. InGaAs cameras cover the range from visible to 1.7 μm and typically have pixels two to five times larger than CCD or CMOS cameras. Pyroelectric solid-state cameras are also used for IR lasers with wavelengths longer than 1 μm. These cameras can have matrix sizes of 124 × 124 and 320 × 320 pixels with 100–80 μm elements. Pyroelectric cameras also operate in the 13–350 nm soft x-ray to UV spectral range. Solid-state cameras based on 2D pyroelectric arrays have been provided for many years. Recently, higher resolution models have become available.

12.4.3.1 Characteristics of Cameras

One of the initial camera technologies used for beam profile analysis was a charge injection device (CID). CID cameras are very versatile in that they have *X*/*Y* readout rather than a sequential readout, and can thus be programmed to read out only a part of the camera matrix. This enables them to operate at high frame rates. In addition, CID cameras are being coated with phosphors that make them especially sensitive for UV and x-ray. CID cameras are currently produced by Thermo Fisher Scientific, Liverpool, NY.

CCD and CMOS cameras are the most common types of cameras used in beam profile analysis. In addition to standard CCD and CMOS cameras, there are CCD and CMOS cameras with coolers attached to the sensor. Active cooling reduces the noise of the camera, and allows greater signal-to-noise ratio for measuring laser beams. A typical uncooled CCD/CMOS has a 10–16-bit dynamic range, whereas cooled cameras can be obtained with 16–24-bit signal-to-noise ratio and higher. Cooled cameras can be as much as a factor of 5–20 times more expensive than uncooled cameras.

Another type of solid-state camera useful for profiling lasers in the far IR is a camera based on a micro-bolometer array sensor. These IR bolometers are designed for thermal imaging, but nevertheless could be useful for long-wavelength laser beam analysis. In addition to IR bolometers, there are ferroelectric cameras. This camera is also designed for thermal imaging in the 8–12 μm region, but is potentially useful for laser beam analysis.

Finally, there are cooled IR cameras that could be used for laser beam analysis. This includes cameras with sensors made from materials such as indium antimonide (InSb) and mercury cadmium telluride (HgCdTe). These cooled IR cameras require significantly more cooling than cooled CCDs, and typically use liquid nitrogen as the cooling mechanism. In addition, they are relatively expensive, costing between 20 and 50 times as much as CCD cameras, and 2–4 times as much as solid-state pyroelectric cameras. A drawback of these cooled IR cameras is that they are extremely sensitive. This means they require additional attenuation over and above what would be required for uncooled, solid-state pyroelectric cameras.

12.4.3.2 Characteristics of Cameras Relevant to Beam Profile Analysis

There are a number of characteristics to evaluate in choosing one camera over another, or in specifying a given type of camera for use in laser beam profile analysis. The most significant characteristic is the wavelength response of the camera. This was alluded in Section 12.4.3.1. For example, CCD and CMOS cameras are the most useful cameras for the UV, visible, and near-IR spectra. A second essential factor is that the sensor on the camera be windowless to eliminate possible fringes caused by reflections of the two window surfaces interfering with each other. This is the same effect as that of an etalon. Alternatively, if a window is required, then the window should be configured to minimize these interference effects. This can be done by anti-reflection (AR) coating the window for the wavelength of use, wedging the surfaces of the window and angling the window relative to the sensor array, or making the window of bulk absorbing ND filter, which attenuates the reflection from the second surface going back and interfering with the incoming irradiation of the first surface. The dynamic range of the camera is another factor for serious consideration.

Another useful feature to consider in choosing a camera is the availability of an electronic shutter or exposure control. This allows the cameras to integrate light only during a short time, for example, 1/1000 s. This feature can enable the camera to select a single laser pulse out of a kilohertz pulse train. In some cases, exposure control can be used to make the camera less sensitive and in effect attenuate the laser beam. Multiple frame integration or exposure control also enables very low-level light signals to be accumulated on the camera, thus obtaining an image with higher signal-to-noise ratio.

Fill factor should be considered in the choice of a camera. Normally, CCDs and most other cameras have a relatively high fill factor, and thus do not lose signal in between the active parts of the pixel. Cameras with low fill factor may not accurately measure all of the spatial characteristics of a laser.

Linearity of the camera output is another very important factor to be considered. Most solid-state cameras have nonlinearity of less than 1% over the specified dynamic range of the camera, which enables accurate beam profile measurements.

A useful feature of some cameras is that they can be triggered externally. This enables a trigger pulse from the laser to synchronize the camera to the laser. Another feature in beam profiling systems provides an electronic signal such that the laser can be synchronized to the frame rate of the camera. When these synchronization methods are impossible, the laser and camera may run asynchronously, the user takes a slight chance that the camera will be in a reset mode when the laser pulse arrives. However, this typically occurs less than 3% of the time with most CCD-type cameras.

Camera sensitivity is another consideration. Almost all of the silicon-based cameras, such as CCD and CMOS, have similar sensitivities. The solid-state uncooled pyroelectric cameras are about 6 orders of magnitude less sensitive than CCDs, and thus require less attenuation than these camera types.

12.4.3.3 CCD-Type Cameras

CCD cameras are the most common type of cameras used in beam profile analysis. There are very inexpensive CCDs typically used in webcams and consumer applications such as cell phones. These CCDs typically have limited signal-to-noise ratios and thus are not very suitable for laser beam profile analysis. Industrial grade CCD cameras have electronics designed to mask or correct bad or inoperative pixels that may exist. Bad pixel correction provides high fidelity images thus making it easier for beam analysis software to process the signal.

There are basically two types of CCD camera technology currently in use: frame transfer and interline transfer. In frame transfer cameras, there is only one sensor site for both fields of the signal frame. Thus on a pulsed laser, since there is only one cell, this cell is read out during the first field, and no signal remains for the second field. Thus, frame transfer cameras have only one-half the spatial resolution for pulsed lasers that they do for CW lasers. Some frame transfer CCD cameras have been shown to have signal response beyond the normal 1.1 μm cutoff of silicon sensors, out to 1.3 μm, even though the sensitivity is typically 1000 times less than it is at 0.9 μm. Thus, the dynamic range of the camera is reduced when used in the wavelength of range >1250 nm.

Interline transfer sensors have individual pixels for each field of the camera frame. They maintain twice the resolution of frame transfer cameras with pulsed lasers. An interline transfer camera can pick out a single pulse from pulse rates up to 10 kHz with a 1/10,000 s shutter speed. Interline transfer cameras typically have higher speed shutters than with frame transfer cameras. However, a problem exists with interline transfer cameras in that the readout electronics are typically on the rear of the silicon wafer behind the sensor cells. For IR lasers with wavelengths approaching 1.06 μm, the absorption of all the radiation does not occur in the sensor cells on the front, and some of the radiation is absorbed in the transfer electronics on the rear of the cell. This absorption of radiation creates a ghost image in the beam, which distorts the view of the beam profile. Even more significantly, it greatly distorts measurements on the beam, since this ghost image appears as low-level energy stripe that transits the center or most intense portion of the beam. Thus, interline transfer cameras are recommended for pulsed lasers in which the wavelength is less than 1 μm. Frame transfer cameras are recommended for YAG lasers at 1.06 μm, even though on pulsed lasers they have only half the resolution as interline transfer cameras.

12.4.3.4 Pyroelectric Solid-State Cameras

Pyroelectric solid-state cameras [21] have been developed that cover the wavelength range from 1.1 μm to >400 μm. These cameras have a very reliable and linear output. Pyroelectric cameras interface to ubiquitous computer ports such as USB, Firewire, and GigE much as do CCD cameras and provide the same viewing and numerical capability. However, pyroelectric cameras have a somewhat lower spatial resolution of 80 μm per pixel and 320 × 320 matrices.

Figure 12.14 shows a pyroelectric camera with the output of a CO_2 laser displayed on a computer monitor. The false color image is provided by laser beam analysis software.

FIGURE 12.14 **(See color insert.)** Pyroelectric camera video graphics array (VGA) output of CO_2 laser.

Pyroelectric solid-state cameras work well with pulsed and CW laser radiation. However, it is necessary that the camera be triggered from the pulsed laser to synchronize the scanning. The reason for this is that the pyroelectric sensor is a thermal sensor, and after the signal is read out from the heating radiation pulse, the heated area of the sensor cools and generates a signal of the opposite polarity. It is necessary to read out this negative polarity signal and reset the sensor before the next laser pulse. This is done by having the camera synchronize to an electronic pulse that is coincident with the optical laser pulse. Camera electronics calculates the interval between pulses, and then adjusts the resetting scan to occur just before the next laser pulse. Depending on the pulse rate of the laser—single shot, very low frequency below 5 Hz, intermediate frequency between 5 and 50 Hz, or high frequency between 50 Hz and 1 kHz—the electronic resetting operation is performed differently in the pyroelectric camera.

For CW operation, the sensor must be mechanically chopped to provide an alternating heating and cooling cycle. This typically is done with a 50% duty cycle between heating and cooling, and is normally performed by a rotating chopper blade. The chopper blade is usually incorporated into the camera. The camera readout is then triggered to read out each row from the camera just as the blade crosses that row of pixels. In this manner, every row of pixels in the pyroelectric sensor has the same integration time, is read out immediately after being covered or uncovered, and thus gives optimum uniformity of the signal.

12.5 LASER BEAM ATTENUATION

Laser beam profile measurements are made on lasers that vary from <1 mW to >10 kW average powers. This typically corresponds to a power density of $<10^{-1}$ W/cm^2 to $>10^5$ W/cm^2. A CCD camera typically saturates at a power density in the range of 10^{-7} W/cm^2. Solid-state pyroelectric cameras typically saturate at ~1 W/cm^2. Thus, the necessary attenuation arranged for CCD cameras varies from 10^5 to 10^{12}. For pyroelectric cameras, the attenuation range is a little more modest from about 10^4 to as much as 10^{11}. Initial attenuation of high average power laser beams is usually performed by one of the two methods. The first is using a beam splitter to pick off a small percentage of a beam, allowing the main part to pass through the beam splitter. The second method is in-line attenuation in which the beam is reduced in power by the absorption of ND filters.

12.5.1 Beam Pickoff

The first step in attenuating a high-power laser beam is to pick off or sample a small percentage of the beam from the main beam. This must be done without affecting the beam profile of the sampled beam. There are basically three ways to perform this pickoff. The most common is to have a beam splitter that is mostly transmitting and partially reflecting. The beam splitter is typically put in the beam at 45°, so that a small percentage of the beam is reflected at 90° to the incident beam. However, this beam sampling surface can be placed at any angle, and there is an advantage to placing the pickoff surface nearly perpendicular to the beam so that the reflection becomes less polarization sensitive.

Another type of pickoff is to use a mostly reflecting and partially transmitting surface. In this case, the surface is placed in the beam at an angle to reflect the majority of the beam, and then transmit a small part through the surface to be measured by the beam analyzer.

A third method of beam pickoff is to use a diffraction grating. This can be either a reflecting or a transmitting type. In the transmitting type of diffraction grating, the beam is typically incident upon the grating perpendicular to the surface, and most of the beam passes directly through the diffraction grating. However, a small percentage of the beam is transmitted at an angle offset from the output angle of the main pass through beam. The portion diffracted typically has multiple modes, whereas, for example, 1% of the beam may transmit at, for example, 15° from the emitting main beam. Second-order diffraction may be 0.01% at 30°, and even a third-order beam may be 10^{-6} of the input beam at 45°. The angle and the diffraction percent depend on the manufacturing characteristics of the diffraction grating, as well as the wavelength of the beam incident upon the grating.

A reflection type of diffraction grating works in a similar manner, except that the beam incident on the diffraction grating is at an angle other than perpendicular. For example, instead of at 90° to the plane of the grating, it may be 30° from normal incidence. The main reflected beam is then reflected at 30° from normal incidence in the opposite direction. Now the first-, second-, and third-order beams are reflected at angles other than the angle of the main reflection. As in the case of the transmitted reflection beam, these refracted and attenuated beams may be 5°, 10°, or 15° away from the main reflected beam, and are typically 1%, 0.01%, or 10^{-6} of the main beam, and so on. The diffracted low-intensity beam maintains all of the beam profile characteristics of the main beam. This small intensity beam can then be used for beam analysis.

The most commonly used beam pickoff surface is made of fused silica and is usually used at a 45° angle to the incoming beam. If the fused silica is not AR coated, it reflects an average of 4% of the beam per surface. However, at 45° the material becomes polarization sensitive, and one polarization is reflected at about 2%, and the other can be as high as 8%–10%. Thus the reflected sample beam does not truly represent the incoming laser beam. This problem can be solved by placing a second surface in the path of the initially sampled beam, but angled in a perpendicular plane that reflects the two polarizations in the opposite way to the first surface. That is, the first surface may reflect the beam 90° in the horizontal, and the second surface reflects 90° in the vertical. After two such reflections, the sampled beam once again has the same characteristics as the initial beam.

The quartz sampling plates have two configurations. One is a wedge so that the back surface of the quartz reflects at a different angle than the front surface. This keeps the reflections from the two surfaces separate, thus keeping them from interfering with each other. The other configuration uses a very thick, flat quartz plate placed at 45° such that the reflection from the back surface is displaced sufficiently far from the front surface reflection so that it does not overlap and interfere. Flat pickoffs have the advantage that the throughput beam, while being slightly displaced in position from the input beam, exits the flat at the same angle as the entrance and is not distorted. Figure 12.15a shows a commercially available quartz reflecting device using a thick flat as the reflector. The mechanical layout is shown in Figure 12.15b. With the wedge, the exit beam is displaced in position and angle, as well as being slightly elongated. Thus if a beam pickoff were to be used in process and left permanently in place, the flat would have superior characteristics to a wedge.

Figure 12.16a shows a commercially available attenuation device using a wedge as the reflecting mechanism. Figure 12.16b shows the mechanical layout of the device,

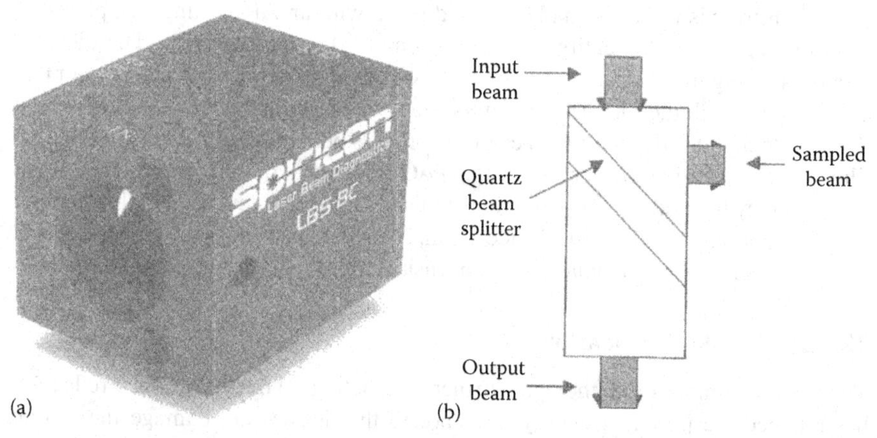

FIGURE 12.15 (a) Beam sampling device; (b) beam sampling mechanical layout.

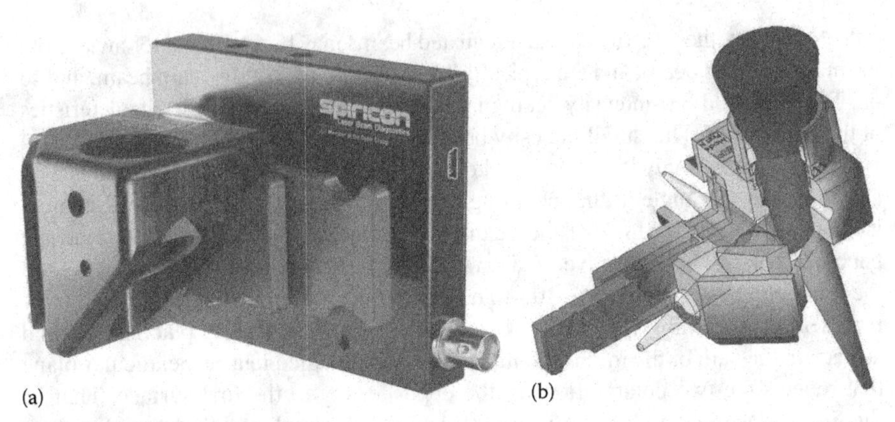

(a) (b)

FIGURE 12.16 **(See color insert.)** (a) Combined beam splitter and ND filter holder; (b) mechanical diagram of combined beam splitter and ND filter holder.

with indents for mounting the wedge on the right side and slots for ND filters on the left side. For IR lasers, ZnSe commonly replaces the quartz as the bulk reflecting material. ZnSe can be AR coated for the specific wavelength of interest, and achieve reflection lower than 1% per surface. Such coating may also be designed so that when placed at 45°, the reflection is polarization insensitive.

An advantage of reflecting gratings is that they can be made from metal, and then the rear cooled with water to enable them to withstand very high powers. Some gratings are of a transmitting type, either from quartz for visible radiation or from ZnSe for IR radiation, and have the advantage that the main beam continues along the same path as its entrance.

Sometimes a thin pellicle of 10–50 μm thick is used for a beam sampler. This is so that the rear reflection is so close to the front side that the interference effects can sometimes be negligible. However, a pellicle as thin as 10 μm may still create interference fringes that could be resolved with a 4 μm pixel camera.

Finally, for industrial Nd:YAG lasers a good pickoff scheme is the use of the dichroic mirror that is normally employed as a turning mirror for the laser. This dichroic mirror is typically made of fused silica with an AR coating and placed at 45° to reflect nearly the entire 1.06 μm beam at 90° from the input. The dichroic mirror is configured so that visible light passes through the filter, so an operator can either see through the filter to the work surface or a camera can be mounted behind the filter to monitor the work surface of the process being performed. These dichroic filters transmit a small percentage of the YAG laser beam directly through the filter, so that it may be used as a sampling mechanism. Dichroic filters used in this manner are polarization sensitive, and so once again, two filters must be placed at 90° to each other in order to obtain a polarization insensitive representation of the input beam.

12.5.2 In-Line Attenuation

There are a number of methods of further attenuating a laser beam once reflection has reduced the power or energy low enough that it does not damage the in-line attenuators. One type of in-line attenuators consists of fused silica plate with a metalized reflecting surface coating. These ND filters made of fused silica are particularly useful for UV radiation. However, when the multiple surface reflecting ND filters are used in conjunction with each other, there is the danger of causing reflections between the multiple surfaces. These reflections can cause interference fringes, which can significantly distort the transmitted beam profile. This problem can be somewhat alleviated by tilting the filters so that the reflected beams are not similar and thus do not interfere with each other.

Another type of surface-reflecting ND filter consists of circular variable filters, in which the attenuation varies around the surface of a circular disk. This type of filter is very useful for single-element detectors, but is not very useful for beam profilers, in that the attenuation is continuously varying, and therefore will attenuate one portion of the beam more than the other.

The more common in-line attenuation filters for beam profile analysis consist of bulk absorbing ND filters. Bulk absorbing filters are usually made of BK7 glass impregnated with an absorbing material. The range of attenuation achievable with these filters varies from an optical density (OD) of 0.1 to >4. The OD number is defined by

$$\text{OD} = \log\left(\frac{1}{T}\right) \tag{12.1}$$

where:

T is the transmission ratio of the output divided by the input

The term ND is somewhat of a misnomer as the OD is strongly dependent on wavelength.

Since the absorption is within the material, there is very little danger of reflection from one surface reflecting back and interfering with the reflection from the other surface. Nevertheless, when two filters are stacked together, the back surface of one and the front surface of the next need to be slightly angled so that interference of the surface reflections do not distort the attenuated beam. Bulk absorbing filters are useful for the entire visible and near-IR spectrum. However, they begin to cut off at about 380 nm; thus, they are not useful for UV lasers. They also tend to change their attenuation characteristics at 900 nm in the IR, and then cut off completely between 2 and 2.5 μm.

Bulk absorbing ND filters are commercially available in a number of forms. One common form is simply a flat plate, up to 2 inch (50 mm) square, which can be placed one after another. Figure 12.16 shows 2 inch square ND filter flat plates used to attenuate a beam. This same instrument could also accommodate surface reflecting filters for the UV. A second commercially available type is to have individual round filters mounted on a wheel so that the wheel can be turned, enabling the user to change attenuation simply by rotating the wheel. Often these wheels can be stacked one behind another, so that multiple ND filters can be selected. Typical transmissive filters have an OD range between 0 and 10^{-4}. Figure 12.17 shows a commercially available rotating wheel ND filter set.

A third type of bulk absorbing ND filter consists of two filters made in the form of wedges. An individual wedge would be like a circular variable filter and attenuate

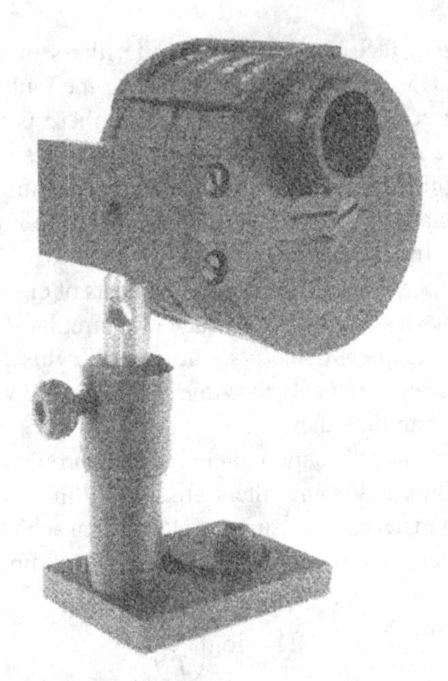

FIGURE 12.17 Rotating wheel ND filter set.

more on one side than another. However, an opposing wedge is placed behind the first wedge, and the entire beam passes through the same amount of attenuating material. These wedges enable a user to make continuous changes in attenuation in small increments, which can be very convenient. However, in some instances beam distortion has been observed from these filters. The amount of distortion is somewhat based on the beam diameter relative to the size of the bulk absorbing wedges.

None of the in-line filters discussed thus far is useful for IR lasers beyond 2 μm. It turns out that for CO_2 lasers, CaF_2 flats are a very useful attenuating filter. A 1 mm thick CaF_2 plate absorbs roughly 50% of 10.6 μm radiation impinging upon it. Thus, by stacking CaF_2 flats, CO_2 lasers can be attenuated so that the signal is reduced in fine increments to within the range of the IR camera.

Cross polarizers are popular for in-line attenuation of laser beams. However, it would be difficult to assure that the cross polarizers are attenuating each polarization of the beam identically. Therefore, they are not commonly used in beam profile analysis, even though they work very well to attenuate beams for single-element power measurement.

Finally, the last method of attenuating a laser beam is to allow the beam to impinge upon a scattering surface. The beam must first be attenuated by beam sampling, so it does not burn or damage the scattering surface. Once the beam impinges upon the scattering surface, the camera can use a lens to image the reflection of the scattering surface. The image reflection is typically very representative of the beam profile. A problem that can exist is that speckle always occurs from scattering surfaces. Speckle is a situation in which the roughness of the scattering causes interference to create both bright and dark spots in the image reflection. Having the scattering surface move at a rate faster than the camera integration, frame rate can solve this problem. There exists a commercial product called a "speckle eater," which is simply a scattering surface mounted to a small vibrating motor. An advantage of imaging scattered beam reflection is that absorption-type filters behind the camera lens can be used for attenuation to achieve a fine degree of beam irradiance reduction.

12.6 BEAM SIZE

Laser beams typically vary from a few microns in beam width to over 50 mm, and in some applications, much larger. Focused laser beam spots can be as small as 1 μm in width. Since camera pixels, at the smallest, are approximately 4 μm, cameras are not very useful for measuring the smallest focused spots. In addition, typical commercial grade cameras have an overall sensitive area of roughly 6–36 mm, with 36 mm being the size of the largest area CCD cameras. Thus, there are many cases when the beam is too small or too large to be measured directly with a camera.

Very small beams can be measured by mechanical scanning devices or optically enlarged for measurement with a 2D array such as a CCD or CMOS camera. Both the rotating drum and the rotating needle systems can measure small beams. However, these systems have a problem in that they do not work with low repetition rate pulsed lasers, and do not give instantaneous whole beam analysis.

Many times when looking at a semiconductor laser such as a laser diode, a microscope objective is used to focus the emitter surface on to the camera. This technique provides a measurement of the spatial output of the emitter surface.

An indirect method of measuring a small focused spot is to allow the beam to go through focus and use another lens to collimate the beam. A third lens with a long focal length then refocuses the beam to a spot size that can be resolved by the camera pixels. If the beam is not a tightly focused spot, but rather a long waist, a beam expander can perform the same function to increase the size of the beam. Finally, a small focused spot can be scattered from either a reflecting or transmitting surface, and imaged with a camera lens. Difficulty with this technique is that if the spot is very small, it is difficult to obtain scattering surfaces with structure small enough to accurately scatter the beam, rather than simply reflect off one of the facets of the scattering medium. Magnification of the beam can also be achieved in the computer display, however, at the loss of resolution. An example of the magnification capability in the computer software is shown in Figure 12.18.

When the beam is too large for the camera, the first solution is to use a beam expander in reverse. A beam expander can typically give beam reductions in the order of 10 to 1. Thus, a 5 cm beam could be reduced to 5 mm, which would fit nicely on a CCD camera. A second method with large beams is to use large area sensors. This is limited to approximately 35 × 25 mm for large area or multiple sensor cameras.

Finally, the most common method of viewing very large beams is to reflect the beam from a scattering surface and image the beam with a lens. A scale can be used to calibrate the pixel pitch of the lens/camera system. This is the same technique as is used for beam attenuation, but now the primary purpose is to be able to image a large area beam, rather than attenuate a large amount of energy. Techniques described above can be used to minimize speckle when it reduces the fidelity of the beam profile measurement.

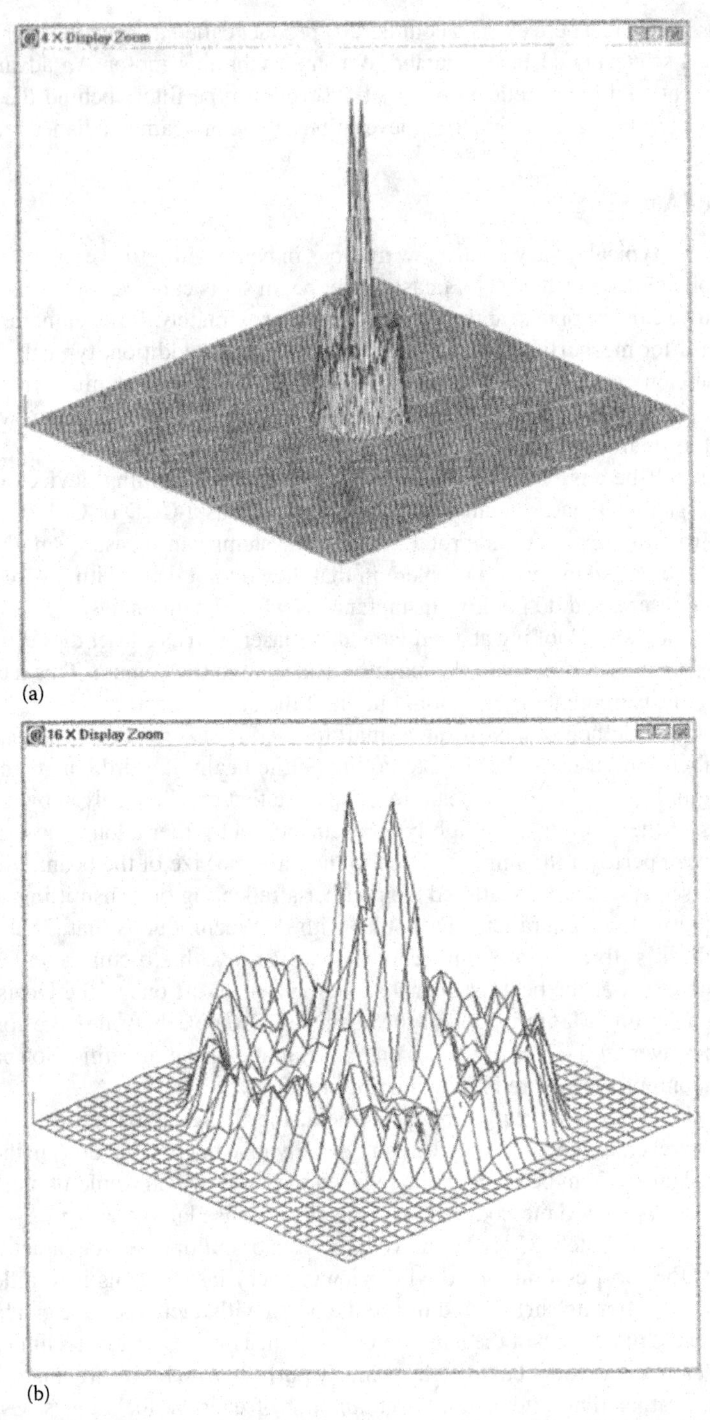

FIGURE 12.18 Focused laser diode beam shown at (a) 4× computer magnification and (b) 16× magnification.

12.7 VIEWING BEAM PROFILES

A tremendous amount of information can be gained about the beam profile simply by being able to clearly see it on a computer screen. Mode structure and distortion of the beam are immediately recognized. Examples are a Gaussian beam distorted into an elliptical shape, or the introduction of spurious multimode beams into the main beam. The beam splitting up into multiple spots or clipping of the beam on an edge of the transport system is immediately seen. In top-hat beams, an electronic display can show hot and cold spots in the top hat, as well as distortion in the vertical sides or walls of the beam.

12.7.1 2D Beam Profile Displays

A 2D view of the beam enables the user to see the entire beam simultaneously. A false color or a grayscale plot is given which enables the user to tell intuitively where the hot and cold spots are within the beam. Cross sections through the beam, located either manually or automatically at some part of the beam, introduce displays of beam irradiance in the vertical and/or horizontal axis, which help interpret the 2D display. Figure 12.19 shows the 2D display of a beam profile with the cross-sectional vertical displays drawn through the peak of the beam. The cross-sectional

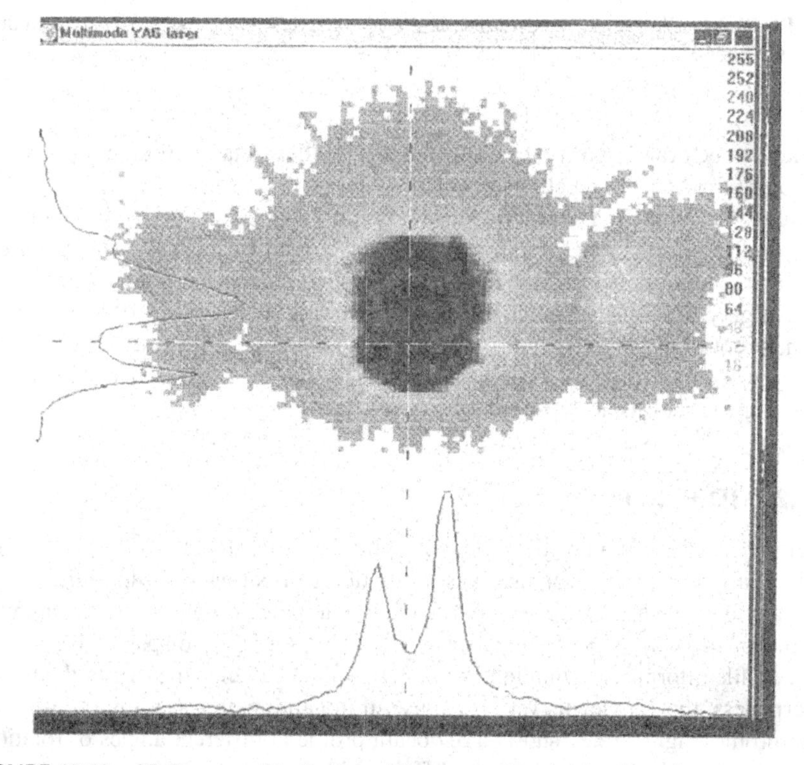

FIGURE 12.19 2D Beam profile display with cross section on the *x*/*y*–axis.

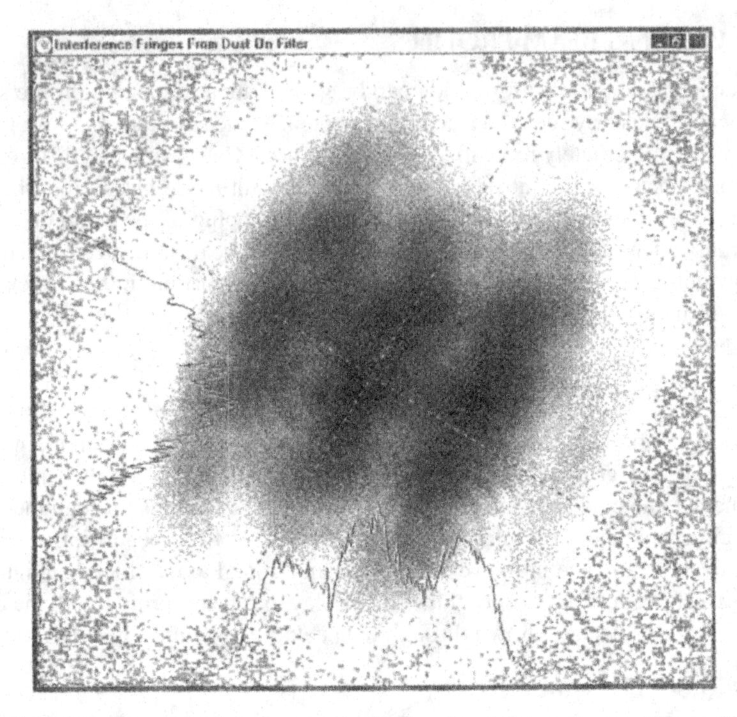

FIGURE 12.20 2D grayscale beam profile display showing interference fringes from dust on an ND filter used to attenuate the beam.

profiles can be drawn at any part of the beam as well as rotated from the *x*/*y*-axis to the major/minor axis of an elliptical beam.

Sometimes false color schemes can have a significant effect in providing intuitive beam profile information. Oftentimes a grayscale image can show patterns in the beam that color does not impart. Figure 12.20 shows a 2D beam profile display created as a grayscale image. Notice the interference rings that show up dramatically in the shades of gray. If this were in color, one might not notice the interference rings. These interference rings are reflections from small specks of dust on one of the transmissive optics such as an ND filter.

12.7.2 3D Beam Profile Displays

A 3D view of the beam profile renders a higher level of intuition of what the beam profile really looks like. The user has the option of rotating and tilting the beam, changing the resolution and color, and so on, to maximize his ability to obtain intuitive information from the beam display. However, while 2D displays give all the beam profile information simultaneously, 3D displays can hide the rear of the beam. Nevertheless, the 3D view is very often useful in gaining greater intuition from the beam profile. Figure 12.21 shows a 3D beam profile at different angles of rotation, illustrating how the beam looks from different sides.

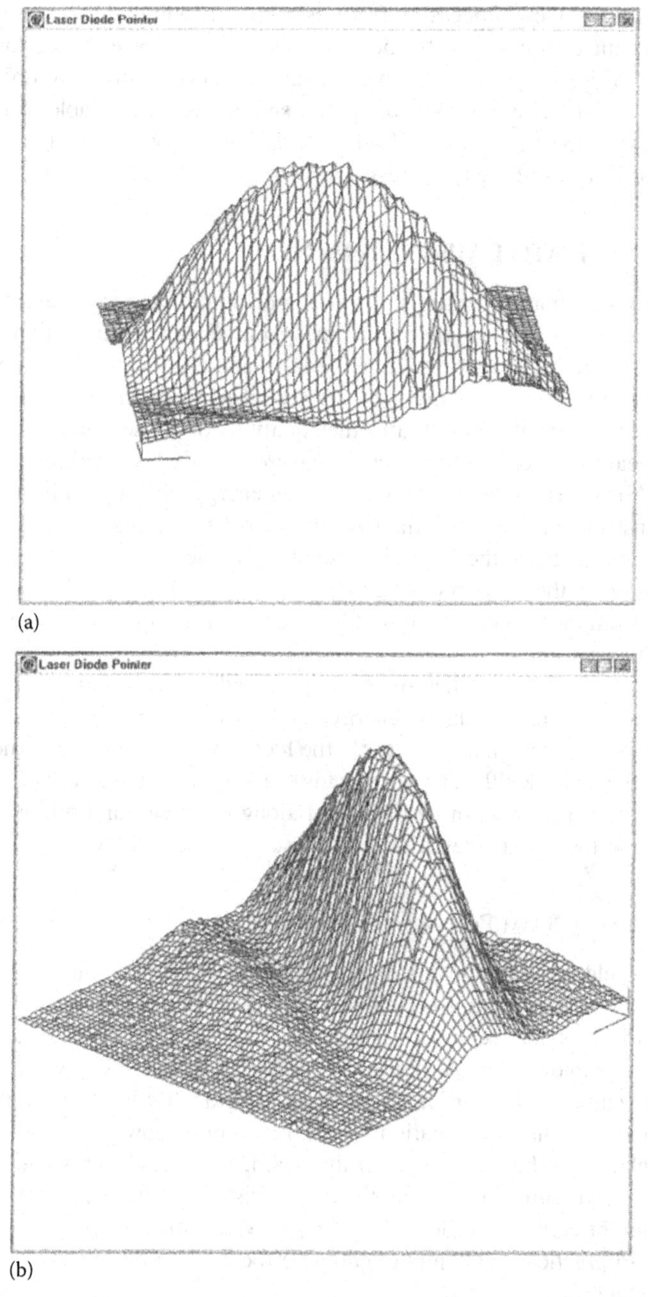

FIGURE 12.21 3D Beam profile shown at two rotation angles: (a) 135°; (b) 225°.

There are many view processing features now available that assist in enhancing the intuition gained from seeing the beam. These include choices of resolution in the 3D display, that is, the number of lines displayed. Adjacent pixel summing, called convolution kernel, Gaussian blur, or spatial smoothing, is available to reduce the signal-to-noise ratio of a beam profile. This technique enables a user to more clearly see the major features of the beam, especially in the presence of noise.

12.8 QUANTITATIVE MEASUREMENTS

One of the most important features of modern beam profilers is the ability to make very accurate measurements of the beam spatial characteristics. Two important laser characteristics, the wavelength of the laser and the temporal pulse width, are unrelated to beam profile measurement and are measured by instruments other than beam profile instruments. Nearly all other qualities of a laser beam are related to the spatial beam profile. One important measurement that is not directly measured by beam profile instruments is the total power or energy, which must be measured by a separate instrument. However, the total power or energy can be measured with a power or energy meter at the same time that the sampled beam profile is measured. When this is done, the beam profiling system can be calibrated to the total power or energy, and from then on the beam profiler is able to track the total power or energy of the laser under test.

Characteristics of a laser that are directly related to beam profile measurements include the pulse-to-pulse relative energy, as discussed in Section 12.8.1, the peak power or energy, the location of the peak, the location of the centroid of the power or energy, and the beam width. The beam width can be measured either on an *x*/*y*-axis or, for an elliptical beam, can be measured along the major and minor axis of the ellipse. Each of these characteristics is discussed in Section 12.8.

12.8.1 Relative Beam Power or Energy

Cameras are seldom able to give a direct measurement of the total energy or power in a laser beam. The camera is typically placed at the end of a long chain of attenuation so that it does not see the total beam directly. Since this optical sampling and attenuation is employed for the purpose of reducing the energy down to the usable range of the camera, and can be as much as a factor of 10^{11}, it is not practical to calibrate each element of attenuation. Thus, the absolute power fed to the camera is unknown relative to the total power of the beam. Second, cameras and attenuation optics do not have uniform wavelength characteristics. Therefore, there would have to be a different calibration factor for every wavelength of laser that is measured. It would be impractical to attempt to calibrate the beam profiling system as a function of wavelength.

Nevertheless, as described earlier, a power meter can be used in the direct beam path to measure the total power while a fraction of the beam's total power is split off and sent to the camera. After correcting for the power lost or ratio of the beam splitter, the total energy or power measured by the meter can then be entered into the beam analysis instruments software. From then on the measured beam profile fluence map represents the total power or energy contained within the beam. This is especially useful with camera as they "see" the entire 2D beam distribution, and thus can provide a measurement of the total beam power or energy as accurately as a power or energy meter within the dynamic range of the sensor.

12.8.2 Peak Power or Energy

Peak power or energy is a relatively easy measurement that is derived from the total power. Since the total power on a camera is a summation of the irradiance on each pixel, it becomes relatively easy to determine what part of this total power is contained within each pixel, and thus the energy on the pixel with the highest power is derived in software. This is a useful measurement in that it tells whether there are hot spots in the beam and the magnitude of these hot spots. This can be particularly useful when the laser power or energy is approaching the damage threshold of optics through which the beam must pass. A hot spot in the beam could cause damage even when power averaged over the area of the beam is well below the damage threshold.

12.8.3 Peak Pixel Location

When the software in the beam analyzer finds the magnitude of the pixel with the highest irradiance, it can also provide the location of this pixel. This may be useful to track the stability of the hot spot or peak irradiance, and determine whether or not this highest irradiance is stable or is moving back and forth across the beam. The actual peak irradiance location is seldom useful in telling where the majority of the energy of the beam is located, however.

12.8.4 Beam Centroid Location

Quite often, more significant than the peak pixel location is the location of the centroid of the beam. The centroid is defined as the center of mass or first moment of the laser beam, and is described in the following equation:

$$X = \frac{\iint xE(x,y,z)\,dx\,dy}{\iint E(x,y,z)\,dx\,dy} \tag{12.2a}$$

$$Y = \frac{\iint yE(x,y,z)\,dx\,dy}{\iint E(x,y,z)\,dx\,dy} \tag{12.2b}$$

The centroid of the beam can be more significant than the peak pixel because it is independent of hot spots in the beam and is not as strongly affected by noise in the measurement system. This is where the energy center is located. Pointing stability of a beam is measured by doing statistical analysis on the centroid location rather than the peak pixel. Pointing stability provides significance in showing the spatial stability of the laser beam.

The significance of the beam centroid can be very important in alignment of laser beams. This is true in optical trains, on research tables, and in industrial laser applications where it is important to know that the beam is positioned correctly in the optics. It is also significant in aligning lenses to laser diodes to collimate the beam. The beam centroid must also be accurately known when aligning beams into fiber optics. Many beam shaping systems require alignment of the beam, usually the centroid, to the shaping optics.

12.8.5 Beam Width

One of the most fundamental laser profile measurements is the beam width. It is a measurement of primary significance because it affects many other beam parameters. For example, the beam width gives the size of the beam at the point where measured. This can be significant in terms of the size of the elements that are in the optical train. Measurement of beam width is the critical part of measuring divergence of laser beams, which is significant in predicting what size the beam will be at some other point in the optical train. Beam divergence also predicts the fluence of a beam at any point along its propagation. The beam width is critical for the performance of most nonintegrating beam shaping systems. Statistical measurement of the width of the beam is also a significant factor in determining the stability of the laser output. Finally, measurement of the beam width is essential in calculating the M^2 of the laser. This is an important characteristic of laser beams that will be discussed later in this section. Even though fundamental and important, the beam width is sometimes a very difficult measurement to perform accurately.

12.8.5.1 Considerations in Accurate Beam Width Measurement

A number of characteristics of a sensor used for beam profile measurement must be carefully considered and accounted for to accurately measure laser beam width. Among these considerations is the signal-to-noise ratio, that is, the magnitude of the beam relative to the background noise in the sensor. The amount of attenuation used or the sensitivity of the sensor is usually adjusted to enable the peak to be as near to saturation as possible without overdriving or saturating the sensor. If the beam is small, a significant number of pixels or data points must be utilized or the measurement accuracy will suffer. To assure accurate beam width measurements, it is important that a minimum of 10 pixels is covered by the laser beam. If a scanning slit sensor is used, a beam that is four times the slit width must be considered as the minimum resolvable spot size.

The camera baseline offset is another factor that must be accurately controlled. Because the energy of a laser does not abruptly go to zero, but trails off to a width roughly four times the standard deviation, or twice the $1/e^2$ width, there is a lot of low-power energy that must be accounted for in accurately measuring the width of the beam [22–24]. The proportion of energy in a Gaussian beam is 68% in $\pm 1\sigma$, 95% in $\pm 2\sigma$, and 99.7% in $\pm 3\sigma$. Nevertheless, experiments performed by the author have shown that as an aperture cuts off the beam at less than $\pm 4\sigma$ the measured beam width begins to decrease. Correct and incorrect baseline controls are illustrated in Figure 12.22a–c. In Figure 12.22a, the baseline is set too low, and the digitizer cuts

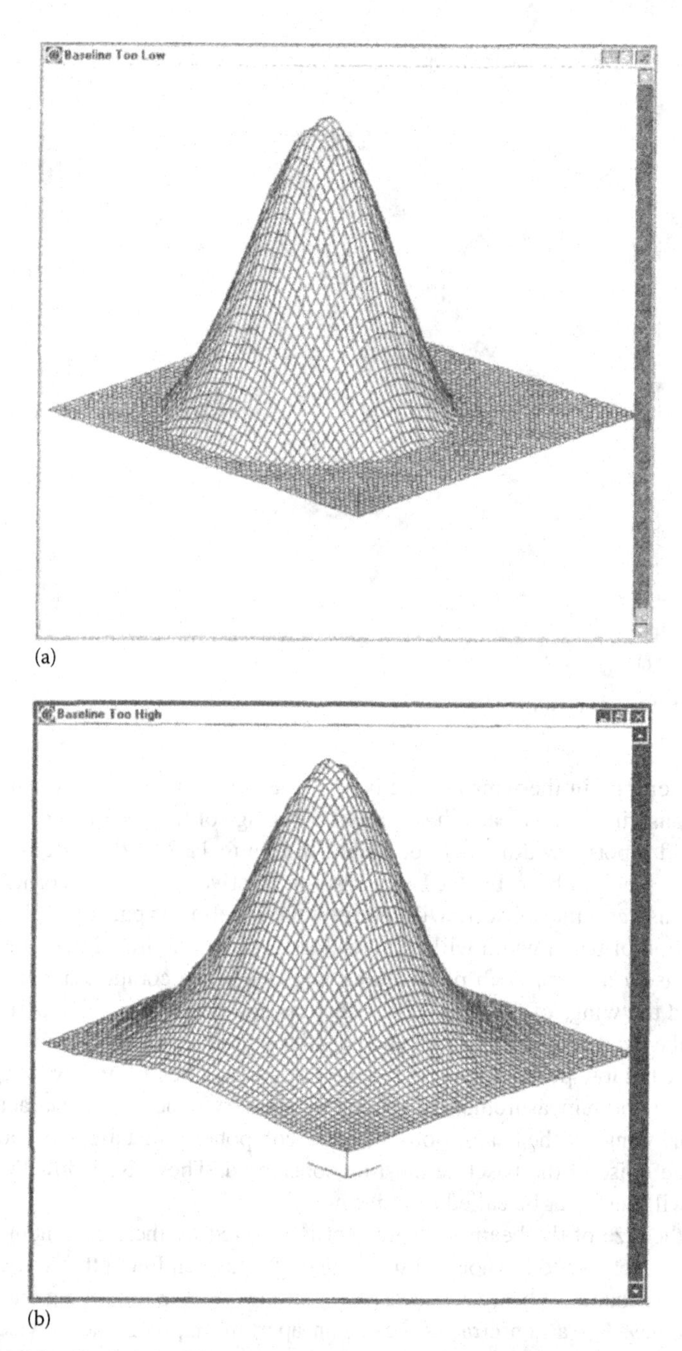

FIGURE 12.22 Camera baseline set: (a) too low; (b) too high; (c) precisely at zero. (Low baseline shows beam rising out of a flat background, which would cause a beam width calculation too small. High baseline would cause a beam width measurement too large.)

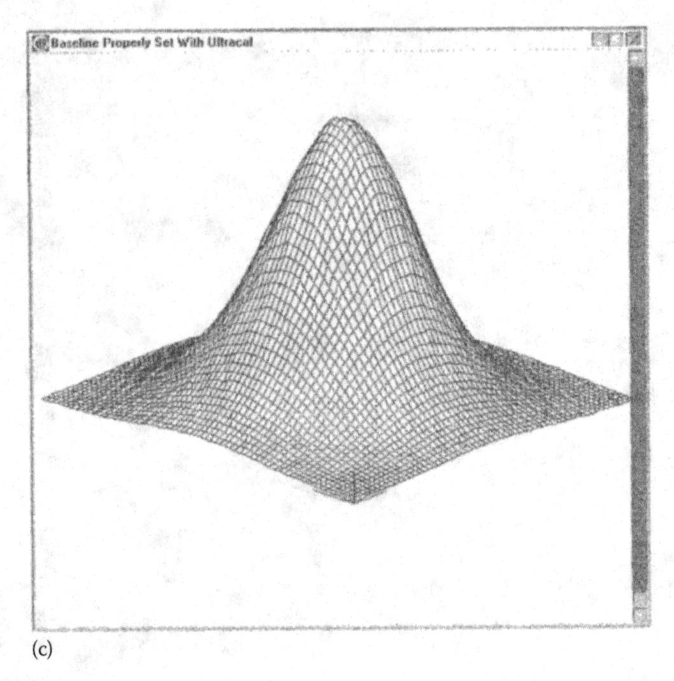

(c)

FIGURE 12.22 **(Continued)**

off all the energy in the wings of the beam. The beam is seen to rise out of a flat, noiseless baseline. This means that without the wings of the laser beam, a measurement would report a width much too small. In Figure 12.22b, the baseline offset is too high, as seen by observing the beam baseline relative to the small corner defining mark. In this case, the software will interpret the baseline as part of the laser beam. A calculation of beam width will be much too large. In Figure 12.22c, the baseline is set precisely at zero. Both positive and negative noise components are retained out beyond the wings of the beam where there is no beam energy. The software will interpret the average of the positive and negative signals as nearly zero.

Because the low-power energy in the wings of a laser beam can have a significant effect on the width measurement, it becomes necessary to be able to characterize the noise in the wings of the beam. Both the noise components that are above and below the average noise in the baseline must be considered. The noise below the average baseline will hereafter be called negative noise.

Since the size of the beam measurement is affected by the total amount of laser beam energy relative to the noise of the camera, it has been found that software apertures placed around the beam can have a very strong effect in improving the measurement accuracy. For a nonrefracted beam, an aperture approximately twice the $1/e^2$ width of the beam can be placed around the beam and all noise outside the aperture can be excluded from the calculation. This greatly improves the relative signal-to-noise ratio when small beams are being measured in a large camera field. Finally, the measurement algorithm that is used to measure the beam width can have a notable effect on the accuracy and significance of the measurement.

12.8.5.2 Beam Width Definitions

There are various traditional definitions of beam width, which may or may not contribute to knowing what the beam will do when focused or propagated into space. Some of these include measurements of the width at a percentage of the peak; full width/half maximum, which would be 50% of peak, or a percentage of energy; the $1/e^2$ width encircles 13.5% of the beam's energy. Software equivalent knife-edge measurements are also used as a means of determining the beam width. Finally, a more important definition of beam width is called the second moment [25].

The software equivalent knife-edge measurement and the second moment measurement are becoming the most widely accepted means of measuring the size of laser beams. Both measurements are independent of nulls or structure within the beam. When the knife-edge measurement is performed, it can do an excellent job of approximating a second moment measurement [26] but is dependent on the mode structure of the beam. The knife-edge measurement with a camera is simply a software algorithm simulating the motion of an actual moving knife-edge. One advantage of cameras over actual mechanical scanning knife-edges is that the software can quickly find the major and minor axes of an elliptical beam, and perform the knife-edge measurement along these axes without having to actually reposition the mechanical knife edge.

12.8.5.3 Second Moment Beam Width Measurements

Recent International Organization for Standardization (ISO) standards [26–29] have defined a second moment beam width, abbreviated $D4\sigma$, which, for many cases, gives the most realistic measure of the actual beam width. The equation for the second moment beam width is given in Equation 12.3. Equation 12.3 is an integral of the irradiance of the beam multiplied by the square of the distance from the centroid of the beam, and then divided by the integrated irradiance of the beam. This equation is called the second moment due to the analogy to the second moment of mechanics, and is abbreviated $D4\sigma$ because it is the diameter at $\pm 2\sigma$ which is $\pm 1/e^2$ for Gaussian beams. This second moment definition of a beam width enables a user to accurately predict what will happen to the beam as it propagates, what is its real divergence, and the size of the spot when the beam is focused.

$$D4\sigma_x = 4\left(\frac{\iint (x-X)^2 E(x,y,z)\ dx\ dy}{\iint E(x,y,z)\ dx\ dy}\right)^{1/2} \tag{12.3a}$$

$$D4\sigma_y = 4\left(\frac{\iint (y-Y)^2 E(x,y,z)\ dx\ dy}{\iint E(x,y,z)\ dx\ dy}\right)^{1/2} \tag{12.3b}$$

where:

$x - X$ and $y - Y$ are the distances to the centroid coordinates X and Y, respectively

Sometimes there are conditions of laser beams in which the second moment measurement is not an appropriate measurement to make. This is particularly true when there are optical elements in the beam smaller than twice the $1/e^2$ widths that cause diffraction of part of the energy in the beam. This diffraction will put energy further out into the wings of the beam, which when measured by the second moment method, will cause a measurement of the beam width much larger than is significant for the central portion of the beam. In Equation 12.3, the $(x - X)^2$ term overemphasizes small signals far from the centroid. This requires judgment on the part of users as to whether or not measurement of this diffracted energy is significant for their application. If the diffracted energy, which typically diverges more rapidly than the central lobe, is not significant, it is possible to place a physical or software aperture around the main lobe of the beam and make second moment measurements only within this aperture, and disregard the energy in the wings. However, if the application is dependent on the total amount of energy, and it is important to know that part of this energy is diffracted, one would want to place this aperture such that it includes all the beam energy in making the calculation.

Second moment beam width measurements are somewhat difficult to make with CCD cameras because camera noise out in the wings of the beam is multiplied by $(x - X)^2$ producing a large error component. Also any offset or shading of the camera in the wings of the beam causes very large errors because these small energy numbers are multiplied by $(x - X)^2$. For example, Figure 12.23a and b illustrates the difficulty of making second moment measurements. These figures are from theoretical calculations based on creating a perfect Gaussian beam, adding random noise to the mathematically derived beam, and then using beam width measurement algorithms to calculate the beam size. In Figure 12.23a, it is seen that a knife-edge measurement can measure a beam of 64 pixels in a 512 field with only 3% error. However, using second moment measurement and random camera noise, the beam width error rises to over 60%. For this reason, a few years ago, theoreticians believed that it was not possible to make an accurate second moment beam width measurement with a commercial grade CCD camera. However, as shown in Figure 12.23b, using a knife-edge can initially calculate a relatively accurate beam width. Then by placing a 2× software aperture around the beam, the second moment measurement can make very accurate beam width calculations down to a beam containing as few as 13 pixels.

In the following comparisons (Figures 12.24 through 12.26), measurements were made to determine the effect on beam width measurement accuracy of various parameters. Since there is currently no "traceable standard beam width," the beam was first measured under the most ideal conditions. This includes a large beam of high intensity and using 2× apertures and negative noise components. Then as measurement conditions are changed, the "error" is calculated as the percentage change in measured beam width from the measurement made under the ideal conditions. All measurements in Figures 12.24 through 12.26 were made on the same beam and in the same time frame.

Figure 12.24a and b illustrates the measured experimental accuracy of making second moment beam width measurements with and without a 2× software aperture. Figure 12.24a illustrates the accuracy versus the irradiance of the peak pixel on the camera. Notice that with the 2× software aperture around the beam, the irradiance can be reduced to as low as 16 counts out of 256, or roughly 5% of saturation and

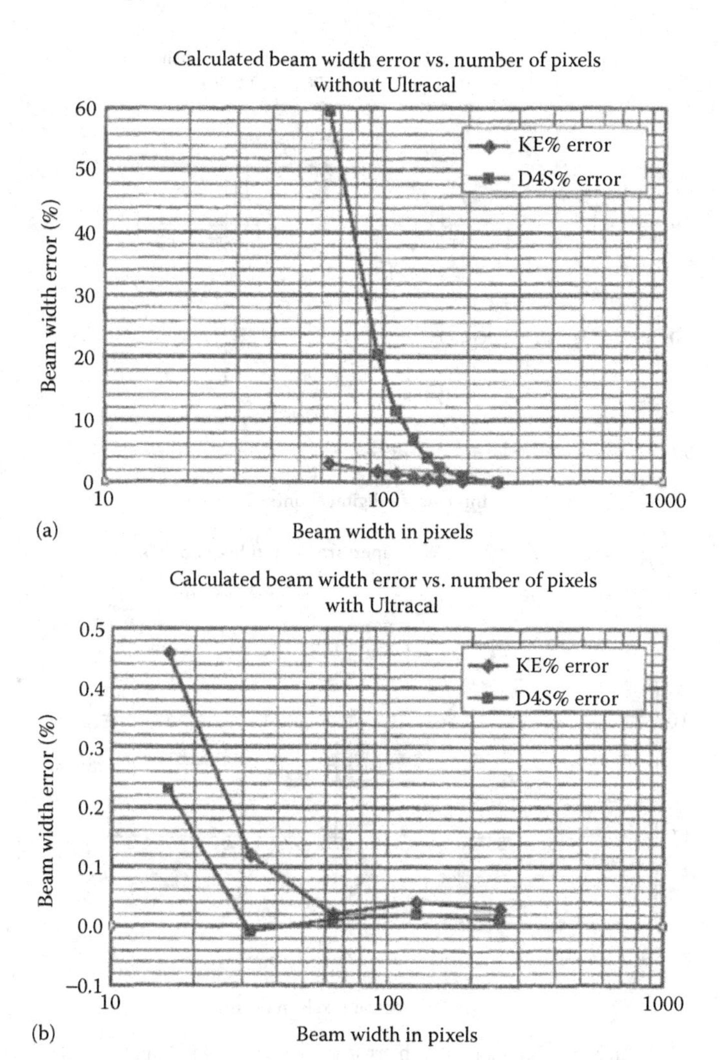

FIGURE 12.23 (a) Simulated beam width error vs. number of pixels without a 2× aperture; (b) simulated beam width error vs. number of pixels with a 2× aperture.

the beam width measurement error is still only about 3%. Without an aperture, the beam width measurement error is in the 3%–5% range, regardless of the irradiance of the beam. In Figure 12.24b, it is shown that the number of pixels in the beam can be reduced to about 3 × 3 pixels before the beam width error measurement rises to 3%. Without an aperture in the beam, the beam width error is always in the 3%–5% range, and at 3 pixels the error rises to over 60%.

Other conditions that are necessary to accurately measure the second moment beam width include accurate baseline control. This is done by having the software perform a multiple frame average of each individual pixel in the camera while the camera is not illuminated. This baseline is then subtracted from the signal when the laser is being measured. This baseline subtraction eliminates not only the total offset of the baseline

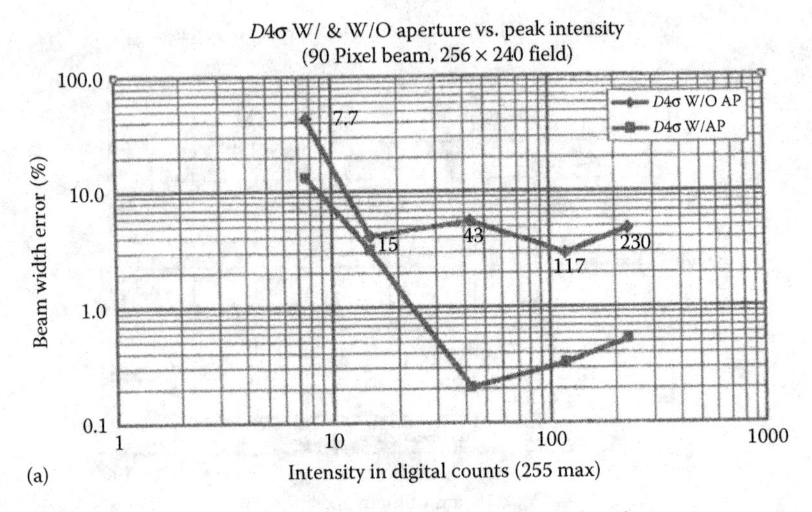

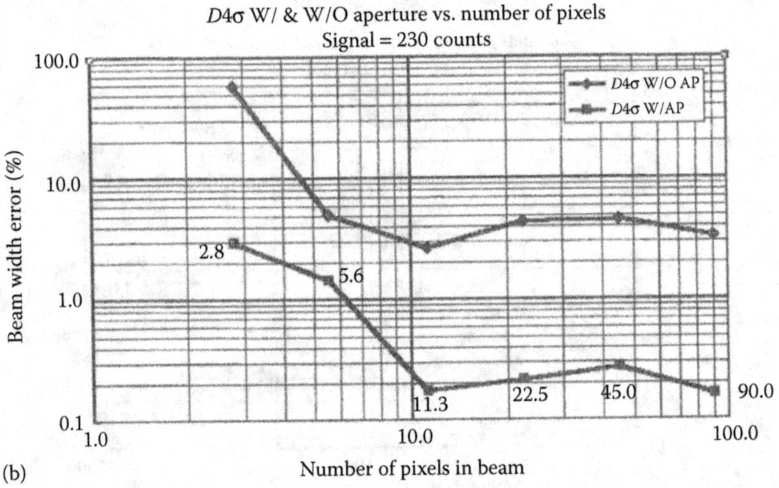

FIGURE 12.24 (a) Measured beam width error vs. irradiance with and without a 2× aperture; (b) measured beam width error vs. number of pixels with and without a 2× aperture.

but also any shading in the camera. Shading is defined as the offset in the baseline not being uniform across the camera, but varying from one side to the other.

In addition to accurate baseline control and 2× software apertures mentioned in the previous paragraph, it is also very important to maintain the negative numbers derived from background subtraction as described previously. Figure 12.25a and b illustrates measurements made on an actual laser beam to determine the relative accuracy of making beam width measurements using both second moment and knife edge under varying conditions. Figure 12.25a and b illustrates the ability of the second moment algorithm to accurately measure beam width with and without negative numbers in the baseline. Notice that in Figure 12.25a where the beam is reduced in irradiance, the beam can be as low as 15 counts or 5% of saturation, with only 3% error. However, without negative numbers in the baseline, at 15 counts the beam

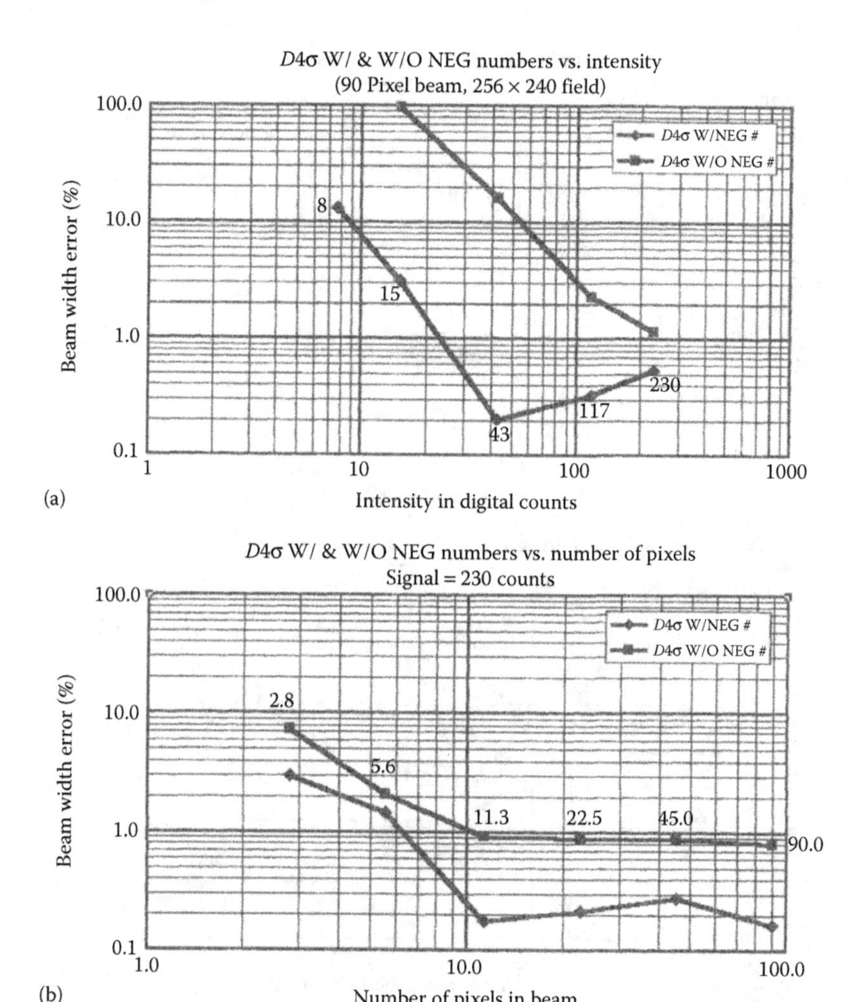

FIGURE 12.25 (a) Measured beam width error vs. irradiance with and without negative baseline numbers; (b) measured beam width error vs. number of pixels with and without negative baseline numbers.

width error is 100%. Figure 12.25b illustrates the ability to accurately measure beam width as a function of the number of pixels in the beam with and without negative numbers in the baseline. Having the negative numbers improves the accuracy by about a factor of 5 for larger beams. At 3 × 3 pixels, the accuracy is 3% with negative numbers, and just over 7% without negative numbers.

Figure 12.26a and b illustrates the measurement accuracy of the second moment beam width method compared to the accuracy of the knife-edge algorithm. In Figure 12.26a, the measurements are compared with the irradiance of the beam. In Figure 12.26b, they are compared with the number of pixels in the beam. In both cases, an aperture and negative numbers are used. Note that both second moment and knife edge have approximately the same measurement accuracy in these conditions.

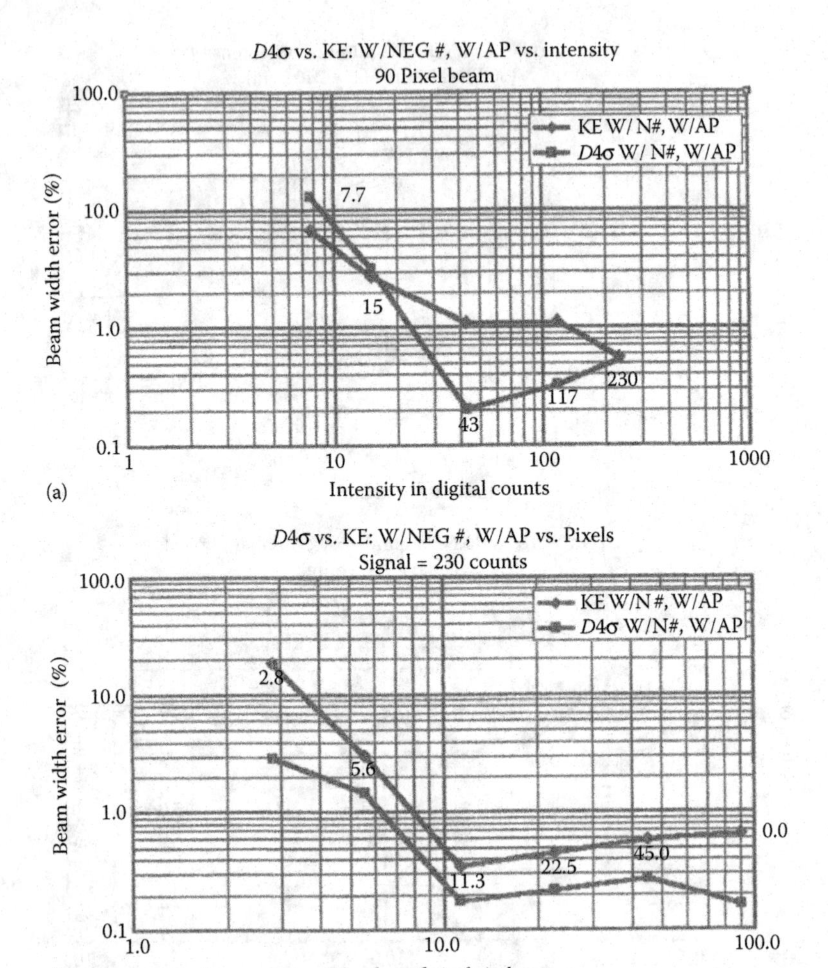

FIGURE 12.26 (a) $D4\sigma$ vs. knife-edge (KE) beam width error vs. irradiance (with aperture and negative numbers); (b) $D4\sigma$ vs. KE beam width error vs. number of pixels (with aperture and negative numbers).

12.8.6 Other Important Beam Profile Measurements

12.8.6.1 Beam Ellipticity

With camera-based beam profiling systems, it is relatively simple for the software to measure the ellipticity of laser beams. The software typically finds the major axis of a beam, and then sets the minor axis perpendicular to the major axis. Once the major axis is found, the angle that the major axis deviates, typically from the x-axis, is given, and the ratio of the major to minor axis widths is calculated. This is an extremely useful measurement in laser beam alignment. It is particularly useful in aligning lenses to laser diodes, which are highly elliptical. Typically a special lens is used with laser diodes to circularize the beam. The alignment of this lens to the diode is extremely critical. With mechanical scanning systems, it is very cumbersome to find the major and minor axes. Whereas with a camera-based system the entire beam profile is obtained in every frame of the camera, so the ellipticity can be found instantaneously. This makes it extremely rapid to do beam and component alignment in real time.

Another important reason for knowing the ellipticity of the laser beam is in industrial applications. Typically, if the beam becomes elliptical, a laser used for cutting irregular shapes will have a different cut width in one axis than in the other. By measuring the ellipticity and correcting it when it goes beyond the acceptable limits, industrial users can eliminate creating scrap materials.

12.8.6.2 Gaussian Fit

In many cases, the desired beam irradiance profile is a Gaussian beam with its irradiance at any point in the x–y plane corresponding to Equation 12.4. There are a number of ways to perform a fit of the real beam to the Gaussian equation. One of these is to minimize the deviation, which is defined in Equation 12.5. This fit can be either along an x/y-axis, a major/minor axis, or performed over the entire laser beam. Being performed over the entire beam is useful in that it means that any energy off-axis contributes to determining how well the beam fits a perfect Gaussian. In addition to these equations, the actual data of beam profile irradiance can be exported to a spreadsheet and users can perform the calculations according to their own method.

The Gaussian equation is as follows:

$$J = J_0 e^{-2[(x-\bar{x}/w_x)^2+(y-\bar{y}/w_y)^2]} + A \tag{12.4}$$

where:

J is the amplitude at the point (x, y)
J_0 is the amplitude at the Gaussian center
x is the x location of the pixel
$\bar{x}$ is the x location of the Gaussian center
w_x is the horizontal radius at $1/e^2$ of energy
y is the y location of the pixel
$\bar{y}$ is the y location of the Gaussian center
w_y is the vertical radius at $1/e^2$ of energy
A is the offset

Minimization of the deviation can be performed by varying the parameters of Equation 12.4 using the spreadsheet solve feature. A is an offset term that is set to zero, that is, disregarded in beam analyzers, because as stated earlier, the background is carefully set to zero. The definition of the deviation is

$$\sigma = \sqrt{\frac{\sum (Z-s)^2}{n-2}} \tag{12.5}$$

where:

σ is the standard deviation
Z is the pixel irradiance
s is the Gaussian surface irradiance
n is the number of pixels

Gaussian fit as a measure of the quality of a laser beam is becoming less important. It has been shown that a multimode beam with the right combination of modes can look Gaussian (Equation 12.26), and can very closely fit to a Gaussian curve. Nevertheless, the beam has many modes and is far from true TEM_{00} mode. A multimode will not follow the propagation laws of a perfect Gaussian beam, and a user can be misled by the Gaussian fit. Instead, the parameter M^2 has become more popular as representing the reality of how close the beam is to a true TEM_{00} Gaussian. The parameter M^2 will be discussed in more detail in Section 12.9.

12.8.6.3 Top-Hat Measurement

Many real beams are intended to be top hat. Some of the earlier chapters on beam shaping discuss how to obtain uniform top hats from Gaussian and other input beams. A top-hat beam is useful in many applications where the irradiance should be uniform over a given cross section. Applications include medical processes such as wine spot removal and photorefractive keratotomy in which a uniform portion of the cornea of the eye is removed. Industrial applications in which a top hat is useful include cleaning of surfaces and marking.

Camera-based systems enable easy and accurate measurements of top-hat beams. The software is programmed to calculate and display the average irradiance or the mean across the top hat, the standard deviation of the variations from the mean, and the standard deviation divided by the mean, which gives a percentage of the flatness or top-hat uniformity. Also the minimum and maximum can be provided, which give additional information about the relative flatness of the beam. The top-hat factor [30] is a way to give a quantitative and intuitive measure of how flat a top-hat beam is. (The equations are given in Reference [30].) A typically square beam would have a top-hat factor of 1. A Gaussian beam has a top-hat factor of 0.5. Therefore, most beams will fall somewhere between 0.5 and 1. In addition to measuring the flatness of the top hat, the software can also calculate the top-hat area and the size or width of the top-hat beam. Figure 12.27a shows a typical top-hat beam, and Figure 12.27b shows the typical calculations.

12.8.6.4 Divergence Measurement

Divergence is an important characteristic of laser beams. It gives the angle at which the beam is diverging from a perfectly collimated parallel beam. It is important because the lower the divergence, the longer the beam will remain at a given diameter. Typically, when low divergence is necessary, a beam is often expanded to a large width, and then the divergence of this large width beam is smaller. Nevertheless, beam divergence by itself does not provide the true characteristics of a beam, since as just mentioned; simply expanding the beam to a larger waist can change it. This will be explored in more detail in Section 12.9.

12.8.6.5 Statistical Measurement

Statistics on all measurements can provide information on long-term stability of the laser beam. A typical example of statistical measurement is shown in Figure 12.28. This figure illustrates a number of basic measurements possible from the software, along with the statistics provided by sampling 20 calculations to determine

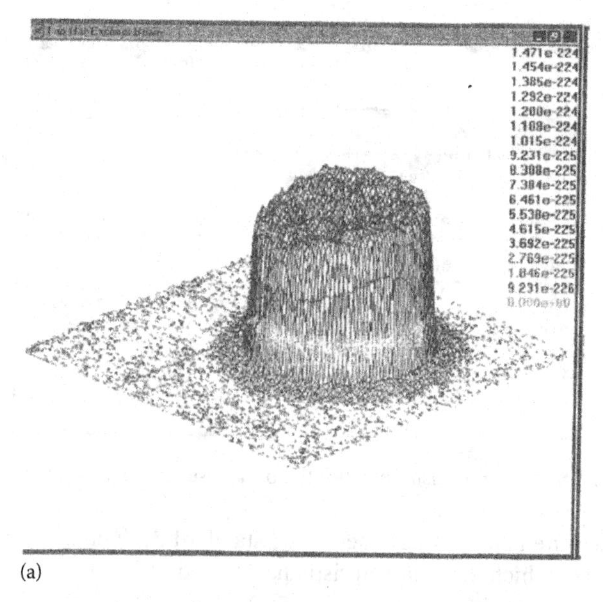

(a)

Frame 9

	Current	Units
—Quantitative—90/10 Knife Edge		
Total	2,.02,6e-,225	mw
% Above Clip	84.16	%
Peak	8.481e-225	mw/cm²
Min	-5.769e-227	mw/cm²
Peak Loc X	8.300e+03	um
Peak Loc Y	6.500e+03	um
Centroid X	7.035e+03	um
Centroid Y	6.765e+03	um
Width X	6.371e+03	um
Width Y	6.328e+03	um
Diameter	6.349e+03	um
—Top Hat—Data		
Mean	7.761e-225	mw/cm²
StdDev	3.903e-226	mw/cm²
SD/Mean	5.03	%
Min	6.808e-225	mw/cm²
Max	8.481e-225	mw/cm²
Factor	0.823	
Effective Area	2.197e+07	um²
Effective Diam	5.289e+03	um
—Divergence—Focal Length		
Divergence X	6.371e+00	mrad
Divergence Y	6.328e+00	mrad

(b)

FIGURE 12.27 (a) Typical top-hat beam; (b) top-hat calculations.

the beam stability. Statistics can be performed in a large variety of ways. For example, software can be arranged so that only one measurement is made out of every few hundred frames, then statistics are calculated on thousands of such frames. This enables one to track the stability of a laser with respect to time, temperature fluctuations, or other characteristics of interest. Statistics typically provide the mean

Frame 20

	Current	Mean	Deviation	Minimum	Maximum	Units
—Statistics						
Samples	20	20	20	20	20	
—Quantitative—90/10 Knife Edge						
Total	5.002e+05	5.100e+05	5.094e+05	5.088e+04	2.006e+06	mw
% Above Clip	86.76	87.19	.94	86.50	90.68	%
Peak	1.490e+02	1.458e+02	6.172e+01	4.600e+01	2.450e+02	mw/cm²
Min	-1.200e+01	-8.050e+00	5.463e+00	-1.500e+01	0.000e+00	mw/cm²
Peak Loc X	5.300e+03	6.680e+03	1.263e+03	4.000e+03	8.400e+03	um
Peak Loc Y	8.100e+03	7.530e+03	1.584e+03	5.200e+03	1.060e+04	um
Centroid X	6.584e+03	6.734e+03	9.008e+02	5.484e+03	8.534e+03	um
Centroid Y	5.980e+03	6.456e+03	7.703e+02	5.478e+03	8.287e+03	um
Width X	1.005e+04	8.464e+03	2.209e+03	5.186e+03	1.403e+04	um
Width Y	1.183e+04	1.059e+04	1.615e+03	7.608e+03	1.330e+04	um
Diameter	1.094e+04	9.527e+03	1.670e+03	6.890e+03	1.367e+04	um

FIGURE 12.28 Statistical measurement of the basic laser beam parameters.

or average measurement of a parameter, the standard deviation, and the minimum and maximum to which that characteristic has drifted.

12.8.6.6 Pass/Fail Measurements

Figure 12.29 shows a typical dialog for a pass/fail measurement. Essentially, all of the quantitative measurements being made on the laser beam can have pass/fail limits set in one of these dialog boxes. Thus, for example, if the centroid location is critical in a manufacturing or other environment, a maximum radius from a given position can be set. The software can then be programmed to provide an alarm if the parameter of interest drifts outside the limits. This feature can be used in many environments, including industrial, instrument design, laser stability and design, and others.

12.9 M^2 MEASUREMENTS

M^2, or the factor $k = 1/M^2$, has become increasingly important in recent years in describing the focusability of a laser beam [31–43]. In many applications, especially those in which a single mode or TEM_{00} beam is the desired profile, M^2 is the most important characteristic describing the focusability of the beam. Figure 12.30 illustrates the essential features of the concept of M^2 as defined by Equation 12.6a and b. As shown in Figure 12.30, if a given input beam of width D_{in} is focused by a lens, the focused spot size and divergence can be readily predicted. If the input beam is a pure TEM_{00}, the spot size equals a minimum defined by Equation 12.6a and d_{00} in Figure 12.30. However, if the input beam D_{in} is composed of modes other than pure TEM_{00}, the beam will focus to a larger spot size, namely, M^2 times larger than the minimum, as shown mathematically in Equation 12.6b, and d_0 in Figure 12.30. In addition to defining the minimum spot size, M^2 also predicts the divergence of the beam after the focused spot. Specifically, the real beam will diverge M^2 times faster than an equivalent TEM_{00} beam of the same width. Figure 12.30 illustrates what happens to a beam after going through a focusing lens. The same principles apply if no lens is involved. The beam will diverge more rapidly by a factor of M^2 than if it were

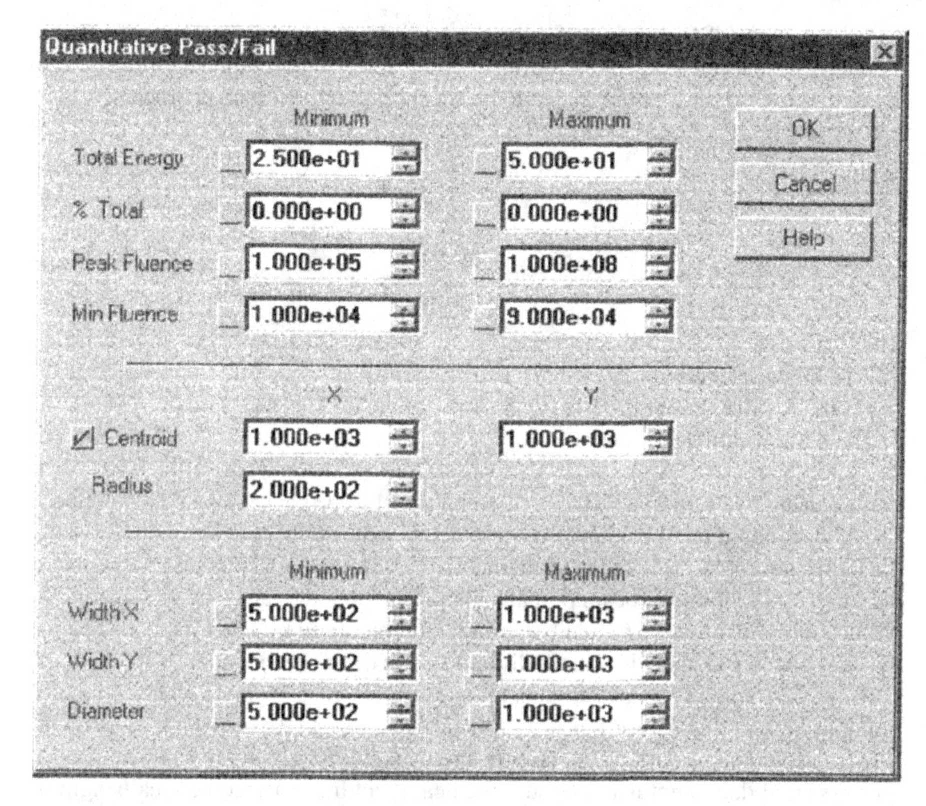

FIGURE 12.29 Pass/fail dialog box.

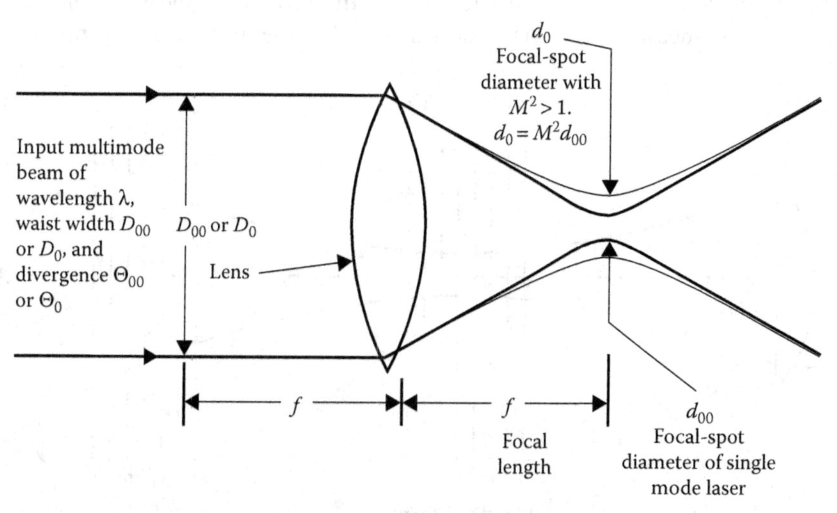

FIGURE 12.30 Curve showing M^2 (characteristics and equations relating M^2 to the beam-focused spot size).

true TEM_{00}. The ISO definition for the focusability of a laser beam uses M^2 as the fundamental parameter. This is useful in applications that require the beam to be focused to the smallest possible spot to be successful in the intended process.

$$d_{00} = \frac{4\lambda f}{\pi D_{in}} \tag{12.6a}$$

$$d_0 = \frac{M^4 4\lambda f}{\pi D_{in}} \tag{12.6b}$$

where:

λ is the wavelength
f is the focal length of the lens
D_{in} is the width of the input beam

In measuring and depicting M^2, it is essential that the correct beam width be defined. The ISO standard and beam propagation theory indicate that the second moment is the most relevant beam width measurement in defining M^2. Only the second moment measurement follows the beam propagation laws so that the future beam size will be predicted by Equation 12.6a and b. Beam width measured by other methods may or may not give the expected width in different parts of the beam path.

M^2 is as simple as the measurement of a beam profile at a single plane. Typically multiple measurements are made as shown in Figure 12.31 in which an artificial waist is generated by passing the laser beam through a lens with known focal length. One essential data point is to measure the beam width exactly at the focal length of the lens. This gives one way to measure the divergence of the beam. Other measurements are made near the focal length of a lens to find the width of the beam and the position at the smallest point. In addition, measurements are made beyond the Rayleigh range of the beam waist to confirm the divergence measurement. With these multiple measurements, one can then calculate the divergence and minimum

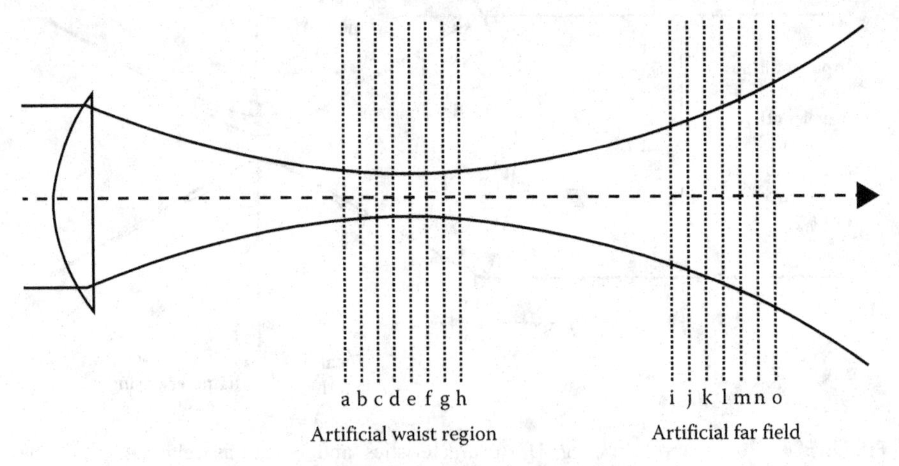

FIGURE 12.31 Multiple measurements made to measure M^2.

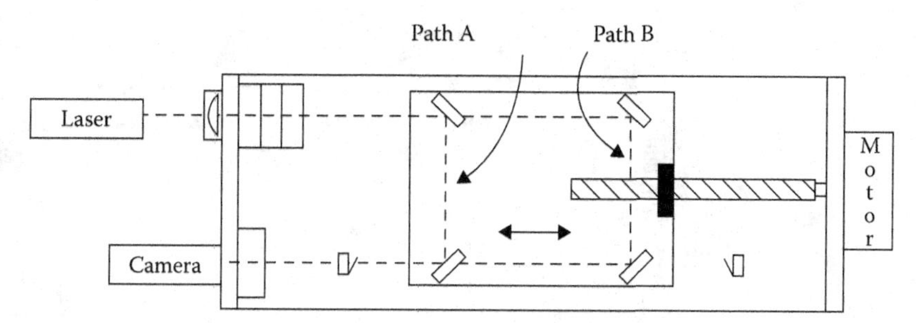

FIGURE 12.32 Instrument with fixed position lens for measuring M^2.

spot size, and then going backward through Equation 12.6b, one can find the M^2 of the input beam.

The measurement shown in Figure 12.31 can be made in a number of ways. In one commercial instrument, shown in Figure 12.12, a detector is placed behind a rotating drum with knife edges, and then the lens is moved in the beam to effectively enable the measurement of the multiple spots without having to move the detector. This instrument works extremely well as long as the motion of the lens is in a relatively collimated part of the laser beam.

The ISO method [28] for measuring M^2 requires the lens to be in a fixed position. Beam width measurements are made at multiple detector positions along the beam's propagation as shown in Figure 12.31. This can be done by placing a lens on a rail and then moving the camera along the rail through the waist and into the far-field region. There are commercial instruments that perform this measurement automatically without having to manually position the camera along the rail. One of these is shown in Figure 12.32, in which the lens and the camera are fixed, but folding mirrors are mounted on a translation table, and moved back and forth to provide the changing path length of the beam.

A typical readout of an M^2 measurement is shown in Figure 12.33. In this case, a collimated laser diode was measured, which gave a much greater divergence in the x-axis than in the y-axis. The steep V curve displayed is the x-axis of the beam coming to a focus following the lens. The more gradual curve is the focus of the less divergent y-axis. Notice that while for most of the range the x-axis has a wider beam width, at focus the x-axis focuses smaller than the y-axis. In addition, the x-axis M^2 was 1.46, whereas the y-axis M^2 was only 1.10. The M^2 reported in the numbers section is calculated from the measurements of the beam width at the focal length, the minimum width, and the divergence in the far field according to the equations in the ISO standard.

One of the difficulties of accurately measuring M^2 is that precise beam width measurements are required. This is one of the reasons that so much effort has been made to define the second moment beam width, and create algorithms to accurately make this measurement. Another difficulty in measuring this beam width is that the irradiance at the beam focus is much greater than it is far from the Rayleigh length. This necessitates that the measurement instrument operates over a wide signal dynamic range. ND filters may be used to increase the dynamic range of the beam measurement system. An alternative exists with cameras or detectors that have extremely wide

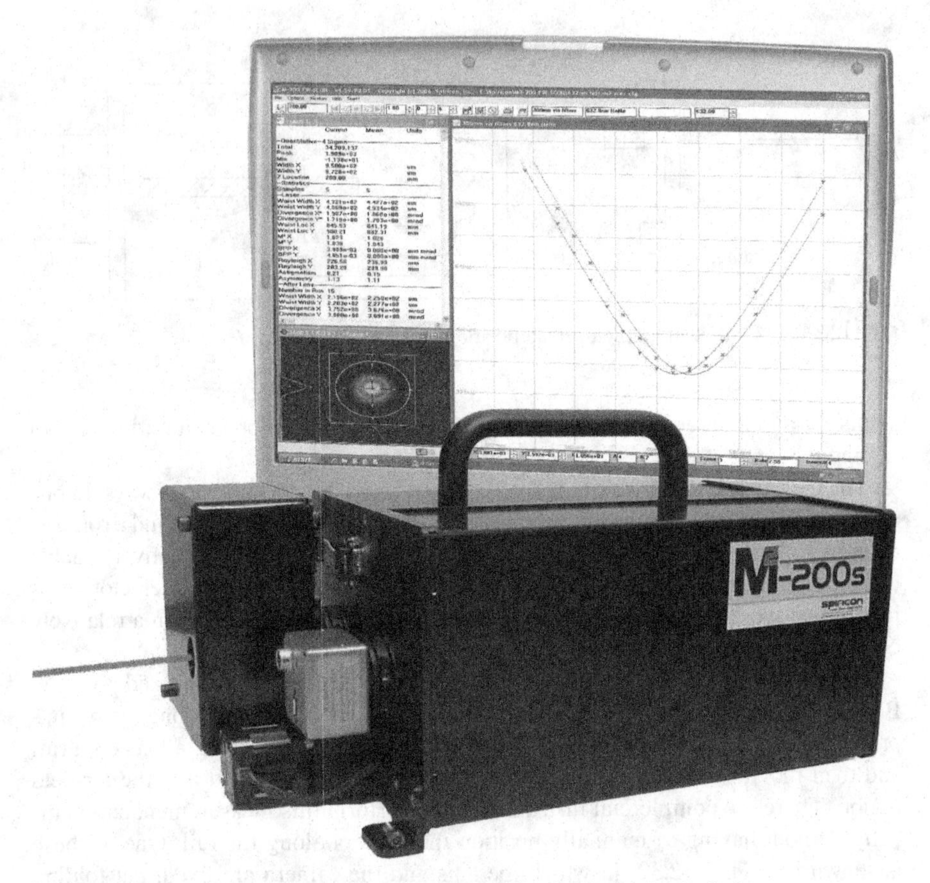

FIGURE 12.33 **(See color insert.)** M^2 measurement display and calculation.

dynamic range so that sufficient signal-to-noise ratio is obtained when the irradiance is low yet does not saturate the detector near the focused waist where the irradiance is much higher.

There are some cases when M^2 is not a significant measure of the quality of a laser beam. For example, top-hat beams for surface processing typically have a very large M^2, and M^2 is not at all relevant to the quality of the beam. Nevertheless, for many applications in nonlinear optics, industrial laser processing, and many others, the smallest possible beam with the M^2 closest to 1 is the ideal. Some top-hat beam shapers are designed for an input Gaussian beam and then the M^2 of the input beam should be very close to 1, and the beam widths should closely match the design width. This was discussed in more detail in Chapter 3.

12.10 SIGNAL PROCESSING

Careful analysis of the camera baseline pixel values, including proper treatment of both positive and negative going noise, enables significant improvement in beam profile measurement accuracy. Figure 12.34 shows a HeNe laser beam at near

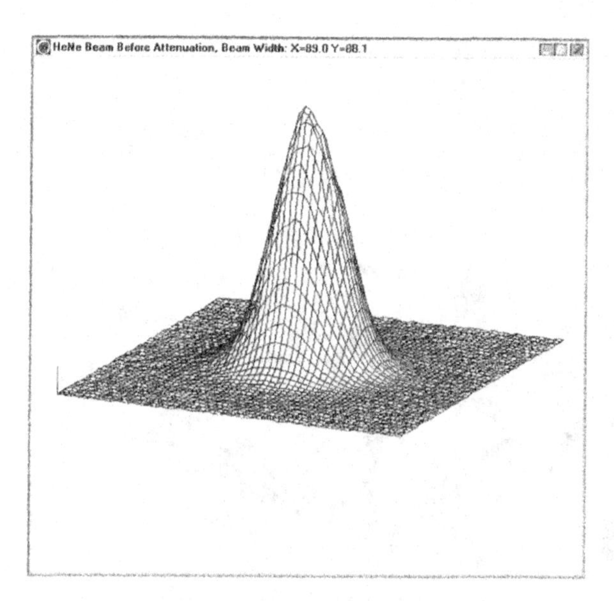

FIGURE 12.34 HeNe laser beam used in signal processing experiment.

saturation of a CCD camera. This beam was then blocked, and signal summing of 256 frames was performed to determine the noise distribution under summing conditions. This noise is shown in three dimensions in Figure 12.35a. The darker components of noise at the bottom of the distribution are the negative-going components. With accurate baseline control and treatment of negative noise components, Figure 12.35b shows that the distribution of the noise is roughly Gaussian, and is centered at zero. This is what would be hoped for from summing random noise in any measurement system.

The laser beam in Figure 12.34 was then measured after passing through an ND2 filter, which attenuated it by a factor of roughly 100. At this point, the laser beam was completely buried in the random noise for each single frame. Hundred frames of signal were summed, and the signal rose out of the noise as shown in Figure 12.36a. In this case, the signal sums as the number of frames, whereas the noise sums roughly as the square root of the number of frames, thus the signal-to-noise ratio is improved by approximately the square root of the number of frames summed. Note that this is possible only when negative noise components are used in determining the mean baseline value of each pixel. Otherwise, if negative components are clipped at zero, the noise will sum with a positive DC offset. Figure 12.36b shows the beam profile of Figure 12.36a when adjacent pixels in a 4×4 matrix are summed together. Notice significant noise cancellation producing a much clearer view of the beam profile. Figure 12.36c shows a similar way of providing a clearer beam profile picture by using convolution to average out the noise in the background. In all three cases of Figure 12.36, the beam width measurement, from the measurement of the beam in Figure 12.34, was in error by only about 5%–7%. This is quite impressive for a beam that started out buried in noise.

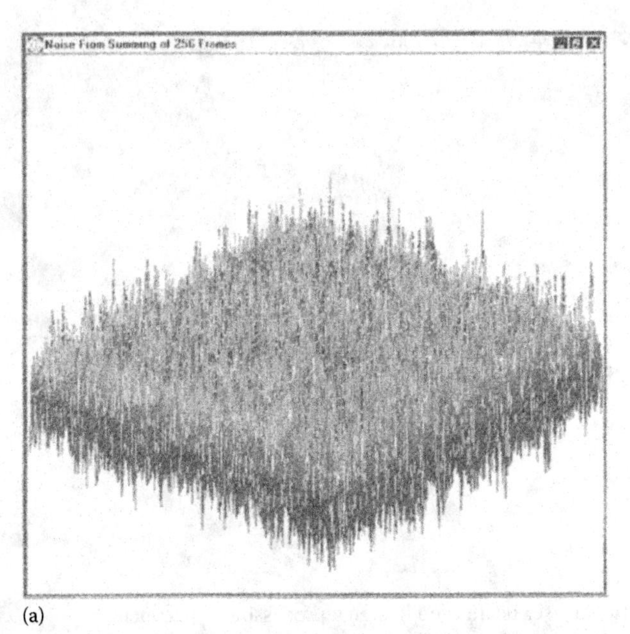

(a)

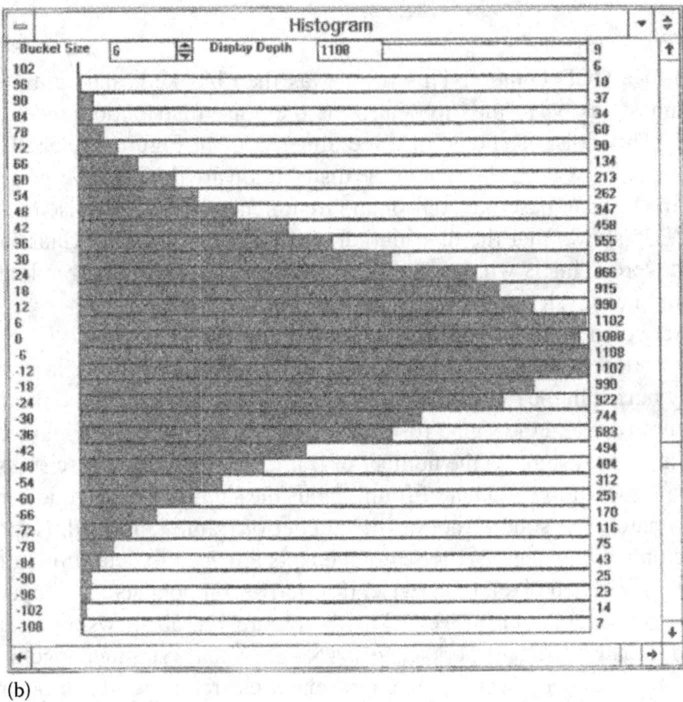

(b)

FIGURE 12.35 (a) CCD camera noise after a sum of 256 frames; (b) distribution of noise shown in (a).

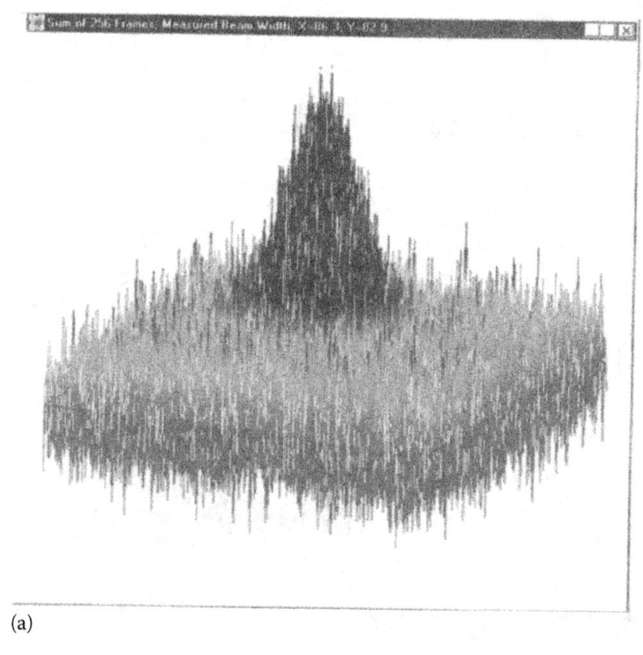

(a)

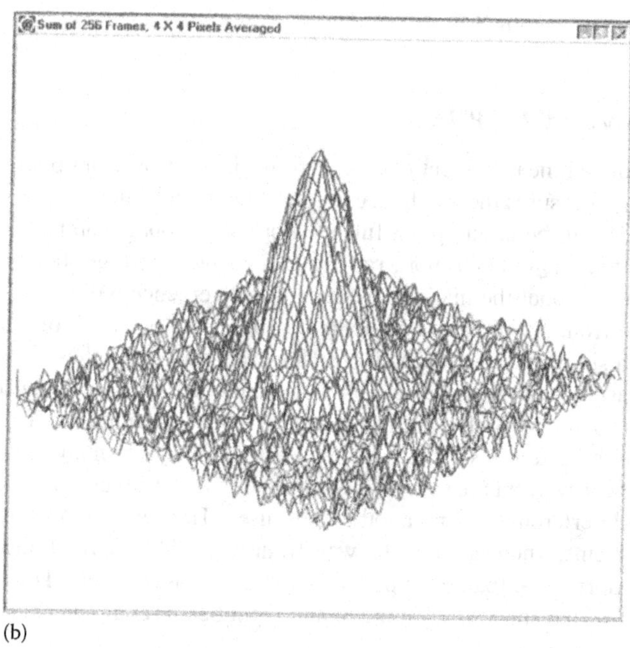

(b)

FIGURE 12.36 (a) Beam of Figure 12.34 after attenuation of about 100 and summed for 256 frames; (b) beam of (a) with summing of pixels in a 4 × 4 matrix; (c) beam of (a) with convolution over a 7 × 7 array.

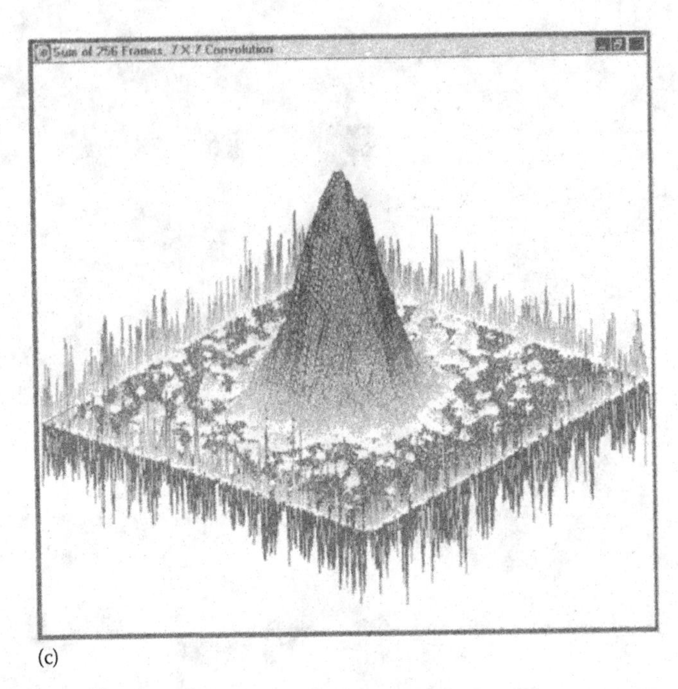

(c)

FIGURE 12.36 (Continued)

12.11 WAVEFRONT PHASE

A more advanced measurement of a laser beam is the wavefront phase. The beam profile simply measures the irradiance at a selected plane, but does not predict what the irradiance will be at any point further along the propagation path. A measurement of M^2 tells how much more rapidly a beam will diverge, but does not give any information about the manner in which this divergence will occur. A measurement of wavefront phase gives additional details of the beam distortion than those reported as a simple number such as M^2. However, wavefront phase is a more complicated measurement to make, as well as to make use of. It is likely that as users become more sophisticated, wavefront phase will become an increasingly important measurement related to laser beam characteristics. There are numerous methods of measuring the wavefront phase for which commercial instruments are available. One is to use an interferometer, and another is to use a Hartman array. For some beam shaping problems, knowledge of the wavefront phase is important and can impart additional information than a simple beam profile measurement. (This is discussed in Chapter 3 with respect to collimation and input beam requirements.)

12.12 SUMMARY

Beam shaping generally requires beam profile measurement. This is required on the input beam to make sure that it has the proper characteristics. It is also required on the output beam to make sure the beam shaping mechanism is operating properly.

Mechanical scanning instruments can provide single axis profiles, which are sufficient in many cases.

Electronic measurements of laser beams using CCD and other solid-state cameras' sensors yield very detailed information on both the input beam and the output beam. Using such beam profilers, scientists and users in beam shaping and many other fields of lasers are able to greatly enhance the success of their endeavors. Giving an accurate view of the beam profile and making precise measurements of beam parameters, such as beam width and other characteristics, provide the ability to properly condition the input beam and measure the shaped output beam.

REFERENCES

1. Darchuk J. Beam profilers beat laser-tuning process. *Laser Focus World* 205–212, May 1991.
2. Forrest G. Measure for measure (letters). *Laser Focus World* 55, September 1994.
3. Langhans L. Measure for measure (letters). *Laser Focus World* 55, September 1994.
4. Roundy CB. A beam profiler that stands alone. *Lasers & Optronics* 81, June 1990.
5. Roundy CB. The importance of beam profile. *Physics World* 65–66, July 1990.
6. Roundy CB. Instrumentation for laser beam profile measurement Industrial. *Laser Review* 5–9, March 1994.
7. Roundy CB. So, who needs beam diagnostics? *Lasers & Optronics* 19–22, April 1994.
8. Roundy CB. Seeing is believing with visual laser-beam diagnostics. *Laser Focus World* 117–119, July 1994.
9. Roundy CB. Measure for measure (letters). *Laser Focus World* 55, September 1994.
10. Roundy CB. Practical applications of laser beam profiling. *Lasers & Optronics* 21, April 1994.
11. Roundy CB. Electronic beam diagnostics evaluate laser performance. *Laser Focus World* 119–125, May 1996.
12. Roundy CB. PC-based laser analyzers: New uses require improved devices. *Photonics Spectra* 97–98, January 1997.
13. Sasnett MW. Propagation of multimode laser beams: The M^2 factor. In *The Physics and Technology of Laser Resonators*, DR Hall and PE Jackson, eds., New York: Adam Hilger, 1989, pp. 132–142.
14. Siegman AF. *Lasers*. Mill Valley, CA: University Science Books, 1986.
15. Carts YV. Excimer-laser work spurs UV beam-profiler development. *Laser Focus World* 21:24–30, August 1989.
16. Roundy CB. Laser-assisted radialeratotomy. *Photonics Spectra* 122, October 1994.
17. Roundy CB. Applying beam profiling to industrial lasers. *Lasers & Optronics*, supplement to *Metalworking Digest* 5, August 1996.
18. Sasnett MW. Beam geometry data helps maintain and improve laser processes (Part 1). *Industrial Laser Review* 9–13, August 1993.
19. Sasnett MW. Beam geometry data helps maintain and improve laser processes (Part 2). *Industrial Laser Review* 15–16, May 1994.
20. Sasnett MW. Characterization of laser beam propagation. *Coherent Mode-Master Technical Notes*, 1990.
21. Roundy CB. Pyroelectric arrays make beam imaging easy. *Lasers and Applications* 55–60, January 1982.
22. Roundy CB. Digital imaging produces fast and accurate beam diagnostics. *Laser Focus World* 117, October 1993.

23. Roundy CB, Slobodzian GE, Jensen K, and Ririe D. Digital signal processing of CCD camera signals for laser beam diagnostics applications. *Electro Optics* 11, November 1993.
24. Roundy CB. 12-Bit accuracy with an 8-bit digitizer. *NASA Tech Briefs* 55H, December 1996.
25. Jones RD and Scott TR. Error propagation in laser beam spatial parameters. *Optical and Quantum Electronics* 26, 25–34, 1994.
26. Siegman AE, Sasnett MW, and Johnston, TF, Jr. Choice of clip level for beam width measurements using knife-edge techniques. *IEEE Journal of Quantum Electronics* 27(4):1098–1104, 1991.
27. Fleischer J. Laser beam width, divergence, and propagation factor: Status and experience with the draft standard. *Proceedings of SPIE* 1414:2–11, 1991.
28. International Organization for Standardization. Test methods for laser beam parameters: Beam widths, divergence angle and beam propagation factor. Document ISO/11146, 1993.
29. Sasnett MW, Johnston TF, Jr., Siegman AE, Fleischer J, Wright D, Austin L, and Whitehouse D. Toward an ISO beam geometry standard. *Laser Focus World* 53, September 1994.
30. Klauminzer G and Abele C. Excimer lasers need specifications for beam uniformity. *Laser Focus World* 153–158, May 1991.
31. Belanger PA. Beam propagation and the ABCD ray matrices. *Optics Letters* 16:196–198, 1991.
32. Borghi R and Santarsiero M. Modal decomposition of partially coherent flat-topped beams produced by multimode lasers. *Optics Letters* 23:313–315, 1998.
33. Chapple PB. Beam waist and M^2 measurement using a finite slit. *Optical Engineering* 33:2461–2466, 1994.
34. Herman RM and Wiggins TA. Rayleigh range and the M^2 factor for Bessel–Gauss beams. *Applied Optics* 37:3398–3400, 1998.
35. Johnston TF, Jr. M-squared concept characterizes beam quality. *Laser Focus World* 173–183, 1990.
36. Johnston TF, Jr. Beam propagation (M^2) measurement made as easy as it gets: The four-cuts method. *Applied Optics* 37:4840–4850, 1998.
37. Lawrence GN. Proposed international standard for laser-beam quality falls short. *Laser Focus World* 109–114, 1994.
38. Sasnett MW and Johnston TF, Jr. Beam characterization and measurement of propagation attributes. *Proceedings of SPIE* 1414:21–32, 1991.
39. Siegman AE. New developments in laser resonators. *Proceedings of SPIE* 122:2–14, 1990.
40. Siegman AE. New development in laser resonators. *Presented at Conference on Laser Resonators SPIE/OE LASE*, Los Angeles, CA, January 1990.
41. Siegman AE. Conference on lasers and electro-optics. Conference on Lasers and Electro-Optics, and International Quantum Electronics, Anaheim, CA, May 1990.
42. Siegman AE. Output beam propagation and beam quality from a multimode stable-cavity laser. *IEEE Journal of Quantum Electricity* 29:1212–1217, 1993.
43. Woodward W. A new standard for beam quality analysis. *Photonics Spectra* 24:139–142, 1990.

13 Classical (Nonlaser) Methods

David L. Shealy

CONTENTS

13.1 INTRODUCTION

In this chapter, the design and analysis of nonlaser optical systems used for beam shaping are discussed. Geometrical optics is used to evaluate the irradiance throughout the optical system. A method based on differential equations is presented for evaluating the contour of an optical surface (mirror or lens) that will transform a given input beam profile into a specified output beam profile or irradiance distribution over a detector surface. Nonlaser beam shaping differs from laser beam shaping in two major ways. First, it deals with more general sources, such as Lambertian sources, line sources, or light-emitting diodes (LEDs). Second, it is only concerned with providing a certain irradiance distribution at a particular surface. It is not concerned with propagating a beam beyond that surface or form of the irradiance distribution at intermediate points.

Early thoughts of beam shaping in nonlaser systems can be traced to before the days of Archimedes and his burning glass [1,2], where optics was reported to concentrate on—to increase the power density of—solar radiation. The literature is rich with reports of various optical systems used as solar collectors [3–9]. Welford and Winston [10] have presented a good accounting of nonimaging (nonfocusing) optics used as solar collectors, including an ideal light collector [11,12], which concentrates a beam by the maximum amount allowed by phase space considerations. Burkhard and Shealy [13] have used a differential equation method to design a reflecting surface, which distributes the irradiance over a receiver surface in a prescribed manner. McDermit [14] and Horton and McDermit [15] presented a generalized technique for designing a rotationally symmetric reflective solar collector, which can heat the collector surface in a prescribed manner. Beam shaping has also been used in optoelectronics to achieve a maximum power transfer between a micro-optics light source and an optical fiber [16,17], in radiative heat transfer [18–20], in illumination applications [21–24], and for reflector synthesis [25,26].

In Section 13.2, ray tracing and the flux flow equation are discussed within the context of design and analysis of nonlaser beam profile shaping. For analysis, the flux flow equation can be used to compute the irradiance over any surface within an optical system. For design, the flux flow equation has been inverted to give a differential equation for the shape of one optical surface of the system, when the input and output beam profiles are known. Section 13.3 discusses the application of this design method for a point and Lambertian source of radiation.

13.2 THEORY OF NONLASER BEAM PROFILE SHAPING

For nonlaser systems, a typical beam shaping system design goal is to illuminate a detector or substrate surface with a specified irradiance distribution. For laser-based systems, Section 6.2.1 discusses that shaping an irradiance profile can be achieved by using the energy balance condition to determine the geometrical shape of one optical surface [lens, mirror, or gradient index (GRIN) optics] of the system. In addition, laser beam shaping applications often seek to retain the output irradiance profile as the beam propagates (a collimated beam). This can be achieved by requiring that the system has a constant optical path length between the input and output surfaces, as discussed in Section 6.2.2. This propagation constraint is not needed in nonlaser applications. The following theory, then, extends the laser analysis of Chapter 6 to the more general sources found in nonlaser systems.

Ray tracing [27] is widely used to simulate the performance of both imaging and nonimaging optical systems. By assigning to each incoming ray equal energy density, and by counting the number of rays crossing a unit of area within the optical system, the irradiance can be computed throughout the optical system. Kock [28] reports a method to simplify photo-radiometric calculations of optical systems by using a reference sphere and ray tracing. The flux flow equation [29,30] offers an alternate approach for evaluating the irradiance within an optical system. The flux flow equation along with the ray trace equations is used to monitor the change in size of an element of area of a bundle of rays [31] as the wavefront propagates through the optical system. The flux flow equation depends on the beam parameters and the shape/orientation of the optical surfaces and allows the irradiance to be computed along a ray path as it propagates through an optical system. The flux flow equation can also be considered as a differential equation of the optical surface contour, which can be solved if the input and output beam profiles are known. In the next section, conservation of energy within a bundle of rays is used to derive two alternate expressions of the flux flow equation. Then, in Section 13.2.2, the flux flow equation is used to formulate a method based on differential equations for design of nonlaser beam shaping systems.

13.2.1 Irradiance (Illuminance) Analysis with the Flux Flow Equation

In this section, a formula is derived for the flux density (irradiance or illuminance, which is the energy per unit area per unit time) of a ray passing through an optical system. This formula, which has been labeled the flux flow equation in the literature, can be expressed as the ratio of the products of the principal radii of curvature* of the wavefront as it approaches and leaves an optical surface. The principal radii of curvature and torsion of the incident and reflected wavefront are related to similar quantities of the reflecting surface using a generalization of the Coddington equations [32]—a procedure also known as generalized ray tracing [33].

Assume that the flux density incident upon an optical interface surface s along the direction of a particular ray is denoted by $\sigma(\mathbf{r})$. If i is the angle of incidence of a ray upon the element of surface area $\mathrm{d}a$ on s, the total flux incident upon $\mathrm{d}a$ is given by

$$\mathrm{d}F = \sigma(\mathbf{r}) \cos i \,\mathrm{d}a \tag{13.1}$$

The radiation is reflected (or refracted) to the element of area $\mathrm{d}A$ on surface S as shown in Figure 13.1. Then, the flux density over the surface S is given by

$$\frac{\mathrm{d}F}{\mathrm{d}S} \equiv E_{\mathrm{d}s-\mathrm{d}S} = \sigma(\mathbf{r})\tau(\mathbf{r})\cos i(\mathbf{r})\left(\frac{\mathrm{d}a}{\mathrm{d}A}\right) \tag{13.2}$$

where:

$\tau(\mathbf{r})$ is the reflection or transmission coefficient at the point $\mathbf{r}$ on surface s

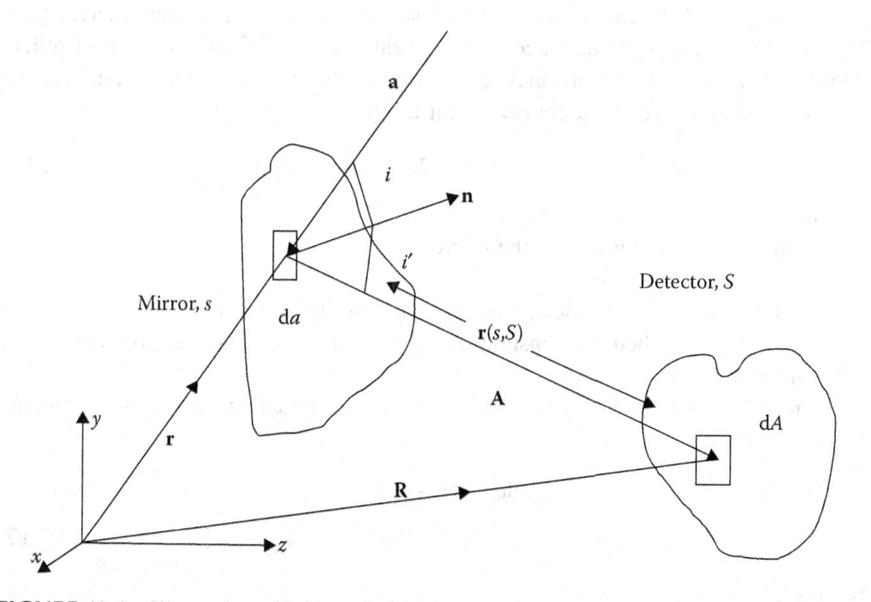

FIGURE 13.1 Illustration of beam reflected from surface s to surface S.

* See Appendix A for a discussion of the principal radii of curvature of a surface and other basic concepts from differential geometry and the theory of surfaces.

If the radiation incident upon surface S is reflected or refracted to another surface, the process of evaluating the flux density along a ray path within an optical system can be generalized to an arbitrary number of surfaces [29]. The problem now is to evaluate the ratio of elements of surface areas ($\mathrm{d}a/\mathrm{d}A$) along a ray path.

The ray trace equations (Equation 6.43) can be regarded as a coordinate transformation between the elements of surface area $\mathrm{d}a$ and $\mathrm{d}A$. Then, the Jacobian, J, of this transformation enables the connection $\mathrm{d}A = J\,\mathrm{d}a$. This approach has been used by Shealy et al. [19] and Shealy and Burkhard [30]. A simpler approach follows by recognizing that the flux is carried by the wavefront [32]. From conservation of energy along a bundle of rays (Equation 6.35), an element of area of the wavefront must be related to an element of area on the optical surfaces s and S. Equation 13.2 may then be replaced by either a quadratic equation in the distance $\mathbf{r}(s,S)$ between $\mathrm{d}a$ and $\mathrm{d}A$ along the ray path or an equation involving the wavefront elements of area before and after reflection (or refraction), which in turn may be replaced by an expression involving wavefront curvatures.

The position vector of $\mathrm{d}A$ along the ray after it leaves $\mathrm{d}a$ is given by

$$\mathbf{R} = \mathbf{r}(u,v) + \mathbf{r}(s,S)\mathbf{A}(u,v) \tag{13.3}$$

where:

$$\mathbf{r}(u,v) = \hat{\mathbf{i}}\,x(u,v) + \hat{\mathbf{j}}\,y(u,v) + \hat{\mathbf{k}}\,z(u,v) \tag{13.4}$$

is the equation of the mirror surface s expressed in terms of the parameters (u,v). The quantity $r(s,S)$ is the distance along the ray measured from the point of reflection $\mathrm{d}a$ on surface s to $\mathrm{d}A$ on surface S. $\mathbf{A}(u,v)$ is the unit vector along the reflected ray and is related to the incident ray vector $\mathbf{a}$ by

$$\mathbf{A} = \mathbf{a} - 2\mathbf{n}(\mathbf{a}\cdot\mathbf{n}) \tag{13.5}$$

where:
$\mathbf{n}$ is the unit normal vector to the mirror

Equation 13.3 can also be viewed as an equation of the reflected wavefront in the vicinity of the ray, when the constant phase condition as measured from the source is considered.

To map a bundle of rays continuously across an optical surface s, the following relations hold:

$$\mathrm{d}a \cos i = \mathrm{d}W(s) \tag{13.6}$$

$$\mathrm{d}a \cos i' = \mathrm{d}W'(s) \tag{13.7}$$

where:
i' is the angle of reflection (or refraction)
$\mathrm{d}W(s)$ is an element of area on the wavefront incident upon $\mathrm{d}a$
$\mathrm{d}W'(s)$ is an element of area on the wavefront reflected from $\mathrm{d}a$

At surface S, similar relations hold:

$$dA \cos I = dW(S) \quad (13.8)$$

$$dA \cos I' = dW' \quad (13.9)$$

Now, derivations for two alternate expressions of the flux flow equation will be presented, which involve evaluating the ratio (da/dA)—the first case—in terms of the partial derivatives of the direction cosines of the reflected ray vector **A** and the equation of the mirror surface s. In the second case, the flux flow equation is expressed in terms of the principal curvatures of the incident and reflected wavefront at S and s, respectively.

13.2.1.1 Flux Flow Equation: First Case

The first expression for the flux flow equation is obtained by evaluating the ratio (da/dA) from the equation of the surface and the ray trace equation relating da to dA. An element area on the surface s is equal to the magnitude of the cross product of the independent surface–tangent vectors:

$$da = |\mathbf{r}_u \times \mathbf{r}_v|\, du dv = \sqrt{g}\, du dv \quad (13.10)$$

where:

$$g = g_{uu} g_{vv} - g_{uv}^2 \quad (13.11)$$

$$g_{uu} = \mathbf{r}_u \cdot \mathbf{r}_u, \quad gv_{vv} = \mathbf{r}_v \cdot \mathbf{r}_v, \quad g_{uv} = \mathbf{r}_u \cdot \mathbf{r}_v \quad (13.12)$$

$$\mathbf{r}_u \equiv \left[\frac{\partial \mathbf{r}(u,v)}{\partial u}\right] \quad \text{and} \quad \mathbf{r}_v \equiv \left[\frac{\partial \mathbf{r}(u,v)}{\partial v}\right] \quad (13.13)$$

See Appendix A for derivation of Equation 13.10. From Equation 13.8, the element of area dA can be expressed in terms of an element of area $dW(S)$ on the wavefront incident upon S as follows:

$$dA = \frac{dW(S)}{\cos I} \quad (13.14)$$

Evaluating $dW(S)$ in terms of the coordinates (u,v) in a similar manner as in Equation 13.10 leads to

$$dW(S) = \mathbf{A} \cdot \left[\frac{\partial \mathbf{R}(u,v)}{\partial u}\right] \times \left[\frac{\partial \mathbf{R}(u,v)}{\partial v}\right] du dv \quad (13.15)$$

where the magnitude of an element of area on the wavefront is obtained by projecting the vector cross product along the direction of the ray vector **A**, which is also normal to the wavefront. Using Equation 13.3 to simplify Equation 13.15 leads to the following:

$$\begin{aligned} dW(S) = \{&\mathbf{A} \cdot [\mathbf{r}_u \times \mathbf{r}_v] + r(s,S)\mathbf{A} \cdot [\mathbf{r}_u \times \mathbf{A}_v + \mathbf{A}_u \times \mathbf{r}_v] \\ &+ r^2(s,S)\mathbf{A} \cdot [\mathbf{A}_u \times \mathbf{A}_v]\} du dv \end{aligned} \quad (13.16)$$

where:

the subscripts (u,v) of vector $\mathbf{r}$ or $\mathbf{A}$ represent the partial derivatives with respect to u or v, as defined in Equation 13.13

Substituting Equation 13.16 into Equation 13.14 and using this result with Equation 13.10 leads to the following expression for the $\mathrm{d}a/\mathrm{d}A$ ratio:

$$\frac{\mathrm{d}a}{\mathrm{d}A} = \frac{\cos I}{\{\mathbf{A}\cdot[\mathbf{r}_u \times \mathbf{r}_v] + r(s,S)\mathbf{A}\cdot[\mathbf{r}_u \times \mathbf{A}_v + \mathbf{A}_u \times \mathbf{r}_v] + r^2(s,S)\mathbf{A}\cdot[\mathbf{A}_u \times \mathbf{A}_v]\}/\sqrt{g}} \tag{13.17}$$

Define the denominator of Equation 13.17 to be $L(s)$:

$$L(s) \equiv L_0(1) + r(s,S)L_1(1) + r^2(s,S)L_2(1) \tag{13.18}$$

where:

$$L_0(1) = \frac{\mathbf{A}\cdot[\mathbf{r}_u \times \mathbf{r}_v]}{\sqrt{g}} = \cos i' \tag{13.19}$$

$$L_1(1) = \frac{\mathbf{A}\cdot[\mathbf{r}_u \times \mathbf{A}_v + \mathbf{A}_u \times \mathbf{r}_v]}{\sqrt{g}} \tag{13.20}$$

$$L_2(1) = \frac{\mathbf{A}\cdot[\mathbf{A}_u \times \mathbf{A}_v]}{\sqrt{g}} \tag{13.21}$$

Substituting Equation 13.17 into Equation 13.2 gives the following expression for the irradiance over surface S:

$$E_{\mathrm{d}s-\mathrm{d}S} = \frac{\sigma(\mathbf{r})\tau(\mathbf{r})\cos i \cos I}{L(s)} \tag{13.22}$$

with an immediate generalization to n surfaces [29]. Equation 13.22 has been called the flux flow equation. So far, only the conservation of energy within a bundle of rays has been used to derive the flux flow equation. The law of reflection is introduced at each optical surface when evaluating L. Detailed discussions of using the flux flow equation to compute irradiance distributions for reflective and refractive optical systems have been reported in Refs. [34,35]. For example, Figure 13.2 shows the contours of equal irradiance over a plane for light reflected from a paraboloid; Figure 13.3 shows the contours of equal irradiance over a plane for light reflected from a cone; and Figure 13.4 shows the contours of equal irradiance over a plane for light refracted by a convex-plano lens.

Explicit expressions for the terms L_0, L_1, and L_2 in Equation 13.22 have been reported in the literature for reflection or refraction within multi-interface optical systems [29]. In case of reflection of collimated light from a mirror, these coefficients are given by

$$L_0(1) = \cos i' \tag{13.23}$$

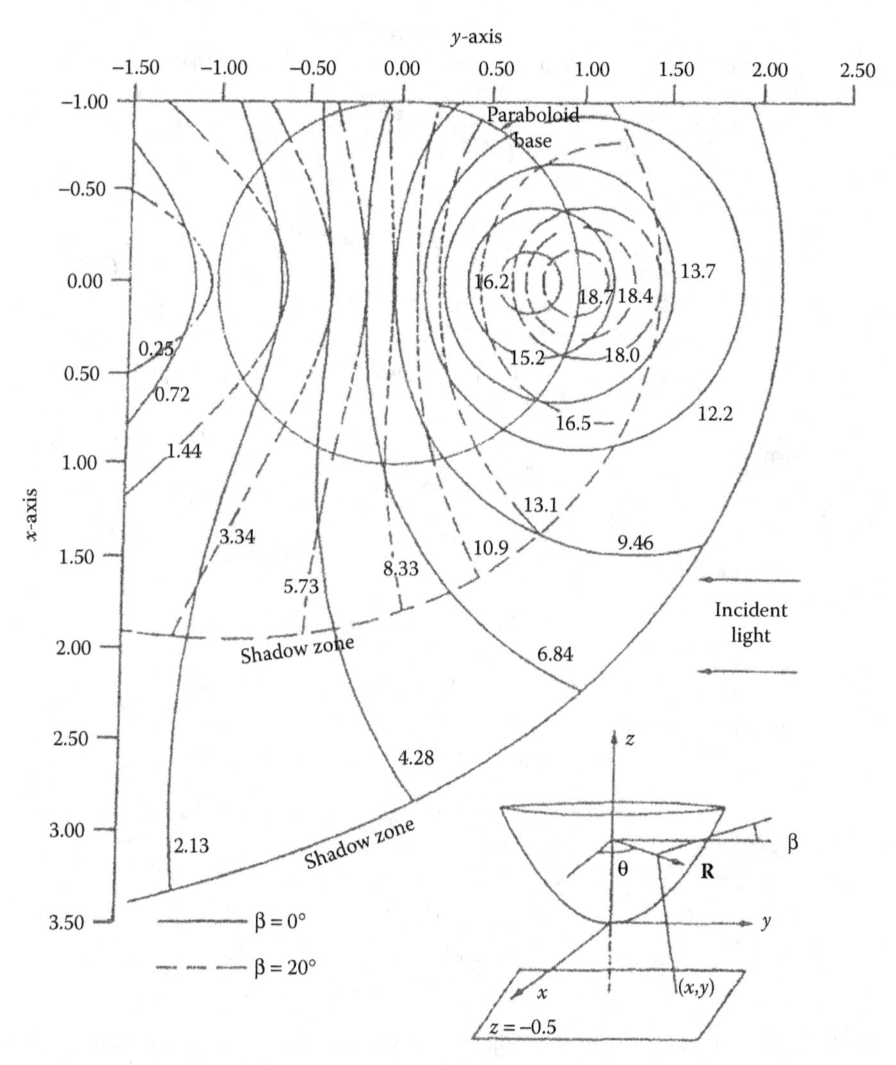

FIGURE 13.2 Contours of equal irradiance on a plane for light reflected from a paraboloid. Flux density values associated with each contour represent the percentage of light incident upon a specific region of the receiver plane. The reflection coefficient of the mirror was assumed to be 1. (From Shealy, D.L. and Burkhard, D.G., *International Journal of Heat and Mass Transfer*, 16, 281–291, 1973. With permission.)

$$L_1(1) = 4H\cos^2 i + 2K_n \sin^2 i \tag{13.24}$$

$$L_2(1) = 4K\cos^2 i' \tag{13.25}$$

where:

H, K, and K_n are the mean, Gaussian, and normal curvatures, respectively, of the mirror

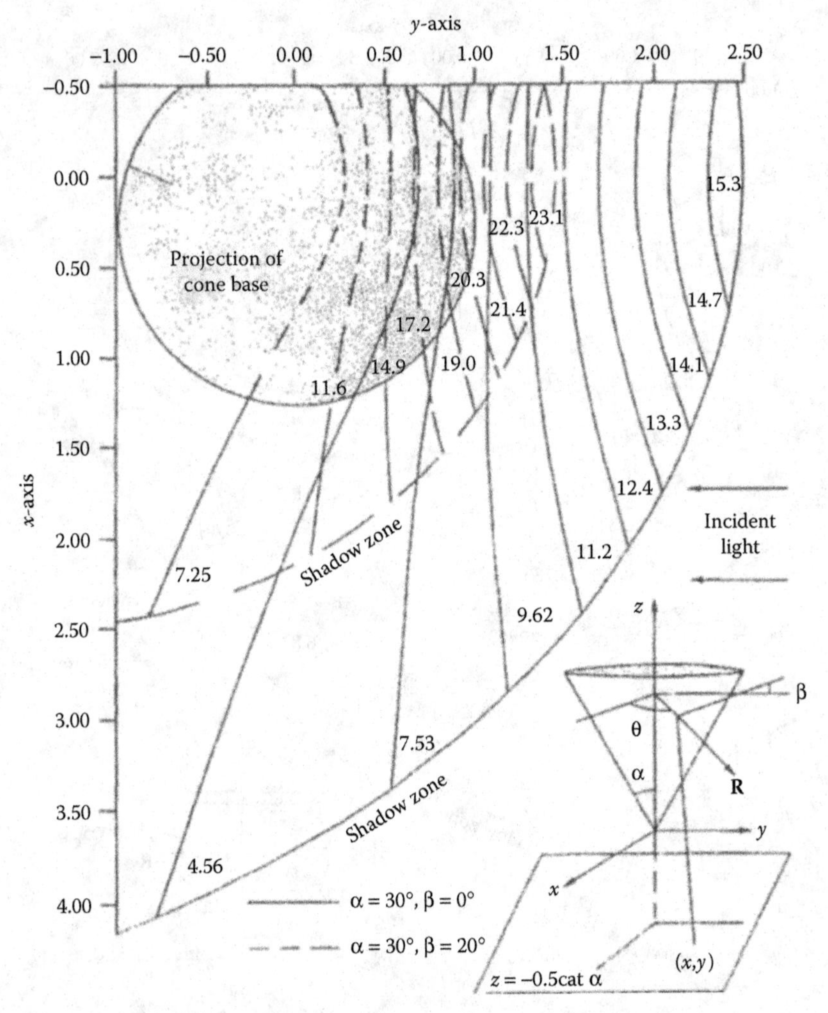

FIGURE 13.3 Contours of equal irradiance on a plane for light reflected from a cone. Flux density values associated with each contour represent the percentage of light incident upon a specific region of the receiver plane. The reflection coefficient of the mirror was assumed to be 1. (From Shealy, D.L. and Burkhard, D.G., *International Journal of Heat and Mass Transfer*, 16, 281–291, 1973. With permission.)

It can be expressed in terms of the equation of the mirror surface and the direction of the incident light using the following expressions:

$$K = \frac{b_{uu}b_{vv} - b_{uv}^2}{g} \tag{13.26}$$

$$H = \frac{g_{uu}b_{vv} - 2g_{uv}b_{uv} + g_{vv}b_{uu}}{2g} \tag{13.27}$$

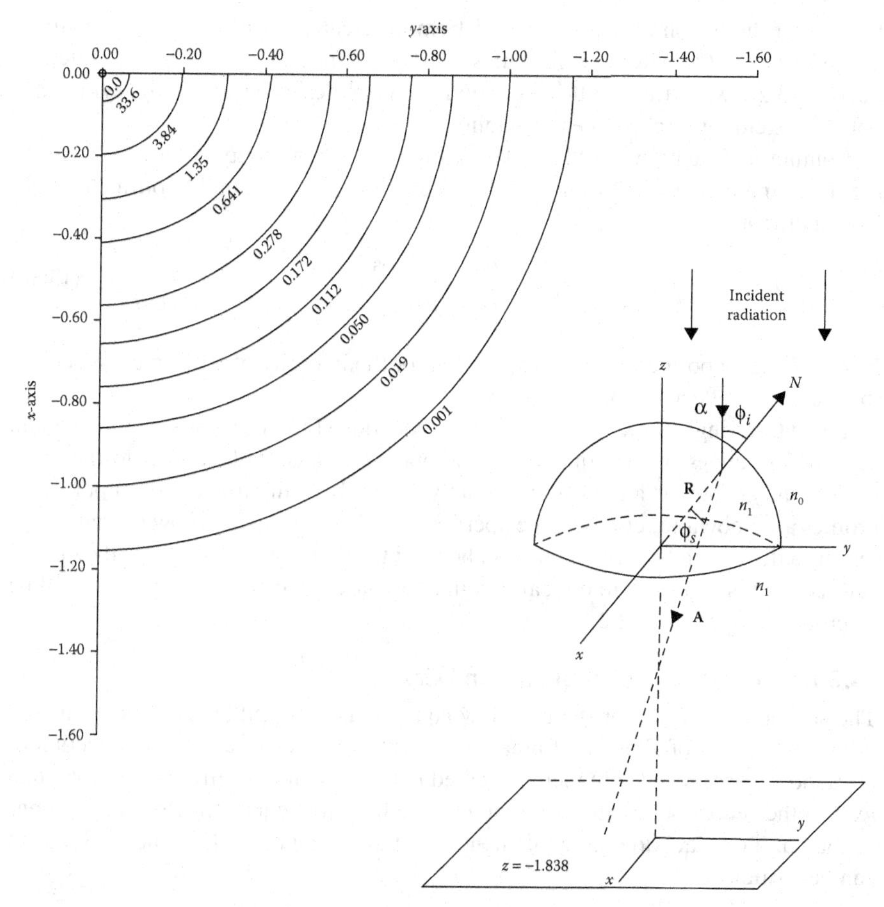

FIGURE 13.4 Contours of equal irradiance on a plane for light reflected from a plano-spherical lens of index of refraction 1.544. Flux density values associated with each contour represent the percentage of light incident upon a specific region of the receiver plane. The transmission coefficient of the lens was assumed to be 1. (From Burkhard, D.G. and Shealy, D.L., *Journal of the Optical Society of America*, 63, 299–304, 1973. With permission.)

$$K_n \sin^2 i = \sum_{i,j=u}^{v} a^i a^j b_{ij} \tag{13.28}$$

$$b_{uv} = g^{-1/2} \left| \left(\frac{\partial \mathbf{r}}{\partial u} \right) \times \left(\frac{\partial \mathbf{r}}{\partial v} \right) \cdot \left(\frac{\partial^2 \mathbf{r}}{\partial u \partial v} \right) \right| \tag{13.29}$$

$$a^i = g^{i1} \mathbf{a} \cdot \left(\frac{\partial \mathbf{r}}{\partial u} \right) + g^{i2} \mathbf{a} \cdot \left(\frac{\partial \mathbf{r}}{\partial v} \right) \tag{13.30}$$

$$g^{uu} = \frac{g_{vv}}{g}; \quad g^{vv} = \frac{g_{uu}}{g}; \quad g^{uv} = -\frac{g_{uv}}{g} \tag{13.31}$$

For a brief discussion of some concepts from differential geometry, see Appendix A or Reference [36]. Alternate formulas for these curvatures will be presented in Section 13.2.2 when the flux flow equation is used to design a rotationally symmetric optical system for nonlaser beam shaping.

Combining Equations 13.18 and 13.23 through 13.25 with Equation 13.22 leads to the following expression for the flux flow equation for reflection of collimated light from a mirror:

$$E_{ds-dS} = \frac{\sigma(\mathbf{r})\rho(\mathbf{r})\cos i \cos I}{\left|\cos i' + 2r(s,S)[2H\cos^2 i + K_n \sin^2 i] + 4r^2(s,S)K\cos i'\right|} \tag{13.32}$$

Fock [37] has reported a similar expression for the intensity of a beam cross section that has been reflected from a surface.

It is interesting to note that the flux flow equation (13.22) depends on the first and second derivatives of the equation of the optical surface with respect to the coordinates (u,v). Thus, it follows conceptually that if both the input and output beam profiles are known functions of the aperture coordinates, the flux flow equation represents a differential equation that can be used to determine the shape of the optical surface s. This approach to optical design of nonlaser beam shaping systems will be discussed in Section 13.2.2.

13.2.1.2 Flux Flow Equation: Second Case

The second expression for the flux flow equation is obtained by expressing the da/dA ratio in terms of element of area on the wavefront before and after reflection or refraction. These results are then expressed in terms of the wavefront curvatures that lead to the second alternate expression of the flux flow equation. Both expressions of the flux flow equation are equivalent [32]. Using Equations 13.7 and 13.8, da/dA can be written as

$$\frac{da}{dA} = \frac{\cos I}{\cos i'}\frac{dW'(s)}{dW(S)} \tag{13.33}$$

Then, from Equation 13.2, the flux density over surface S can be written as

$$E_{ds-dS} = \sigma(\mathbf{r})\tau(\mathbf{r})\frac{\cos i \cos I}{\cos i'}\frac{dW'(s)}{dW(S)} \tag{13.34}$$

We will show (justify) that an element of area on the wavefront can be expressed in terms of the principal radii of curvature of the surface. The principal radii of curvature of an element of area on a surface are the maximum and minimum curvatures of the surface at that point.* They are found by taking a plane through the surface normal at a point and rotating it around the normal. The intersection of the plane and the surface forms curves.

* See Appendix A for a discussion of some concepts from differential geometry which may be helpful to the reader in better understanding the physical meaning of the different curvatures of a surface discussed in this chapter.

Born and Wolf [31] have shown that there are two focal (imaging) points for each point on a wavefront. The caustic surface has also been defined [32] to be the loci of the focal (imaging) points of an optical system. The irradiance, as computed from the flux flow equation (e.g., see Equation 13.22), is infinite on the caustic surface, which occurs in geometrical optics when $dA = 0$. Points on the caustic surface for rays reflected from a mirror surface s are computed by setting $L(s)$ in Equation 13.22 equal to zero and solving the resulting quadratic equation for $r(s,S)$:

$$L(s) = L_0(1) + r(s,S)L_1(1) + r^2(s,S)L_2(1) = 0 \tag{13.35}$$

The two roots of Equation 13.35 are labeled $r_1(s)$ and $r_2(s)$ and represent the distance from the point of reflection to one of the focal (caustic) points on the ray from da. Stavroudis and Fronczek [38] have shown that the caustic points of a wavefront are the principal radii of curvature of the wavefront. Therefore, $r_1(s)$ and $r_2(s)$ are the principal radii of curvature of the wavefront as it leaves da on surface s. Solving the quadratic equation (13.35) for the distance from ds to the caustic surface gives

$$r_1(s) = \frac{-L_1(1) + \sqrt{L_1^2(1) - 4L_0(1)L_2(1)}}{2L_2(1)} \tag{13.36}$$

$$r_2(s) = \frac{-L_1(1) - \sqrt{L_1^2(1) - 4L_0(1)L_2(1)}}{2L_2(1)} \tag{13.37}$$

where:

the + and – of the radical in Equations 13.36 and 13.37, respectively, were arbitrarily assigned to $r_1(r_2)$

Recognizing that $r(s,S)$ is the distance along ray path from da to dA, it has been shown in Refs. [23,32] that the principal radii of curvature of the wavefront as it reaches dA are given by

$$\begin{aligned} r_1(S) &= r_1(s) - r(s,S) \\ r_2(S) &= r_2(s) - r(s,S) \end{aligned} \tag{13.38}$$

where the optics sign convention [39] for the radii of curvature* has been used in Equation 13.38. Further, it has been shown in Refs. [23,32] that

$$\frac{dW'(s)}{dW(S)} = \frac{r_1(s)r_2(s)}{r_1(S)r_2(S)} \tag{13.39}$$

* The radius of curvature of a surface is positive if the center of curvature of the surface is located to the right of the vertex of the surface with respect to the optical axis, when the light is traveling from the left to the right. Otherwise, the radius of curvature of a surface is negative.

which permits Equation 13.34 to be written in terms of wavefront principal radii of curvature:

$$E_{ds\text{-}dS} = \sigma(\mathbf{r})\tau(\mathbf{r})\frac{\cos i \cos I}{\cos i'}\frac{r_1(s)r_2(s)}{r_1(S)r_2(S)} \tag{13.40}$$

The generalized ray trace equations (13.32 and 13.33) are used to compute the principal radii of curvature of a reflected or refracted wavefront in terms of the curvatures and torsion of the incident wavefront and optical interface. References [23,32] provide additional details for using the flux flow equation to evaluate irradiance distributions over surfaces of an optical system.

13.2.2 Optical Design of Nonlaser Illumination Systems

As noted in Section 13.2.1, when the input beam profile and irradiance over a receiver surface are given, the flux flow equation can be viewed as a second-order differential equation that can be solved for the contour of one optical surface. This approach to optical design will be discussed in Section 13.2.2.1. Equivalently, the energy balance condition and ray trace equation can be used to obtain a first-order differential equation for the contour of one surface in the system, as discussed in Section 13.2.2.2.

13.2.2.1 Using the Flux Flow Equation

As noted in Section 13.2.1, the flux flow equation (13.22 or 13.40) depends on the first and second derivatives of the surface equation $\mathbf{r}(u,v)$ of the mirror as well as on the direction of incident radiation and $\sigma(\mathbf{r})$. Therefore, the flux flow equation may be considered as a differential equation for the shape of the mirror. If both the input and output beam profiles are given along with the geometrical surface parameters (boundary conditions), the resulting differential equation can be solved for the shape of the mirror.

To illustrate this approach for design of a nonlaser beam profile shaping system, consider a collimated beam with irradiance profile $\sigma(\mathbf{r})$ incident upon a rotationally symmetric mirror shown in Figures 13.5 and 13.6.

The Fresnel reflection losses are not considered in this design approach, that is, put $\tau(\mathbf{r}) = 1$ in the flux flow equation. Assume that the equation of the mirror surface can be written as

$$\mathbf{r}(r,\phi) = \hat{\mathbf{i}}r\cos\phi + \hat{\mathbf{j}}r\sin\phi + \hat{\mathbf{k}}z(r) \tag{13.41}$$

where:

- (r,ϕ) are the polar coordinates in the x–y plane
- $(\hat{\mathbf{i}},\hat{\mathbf{j}},\hat{\mathbf{k}})$ are the Cartesian unit vectors
- the term $z(r)$ is an unknown function to be determined by the solution of the flux flow differential equation to be written out below

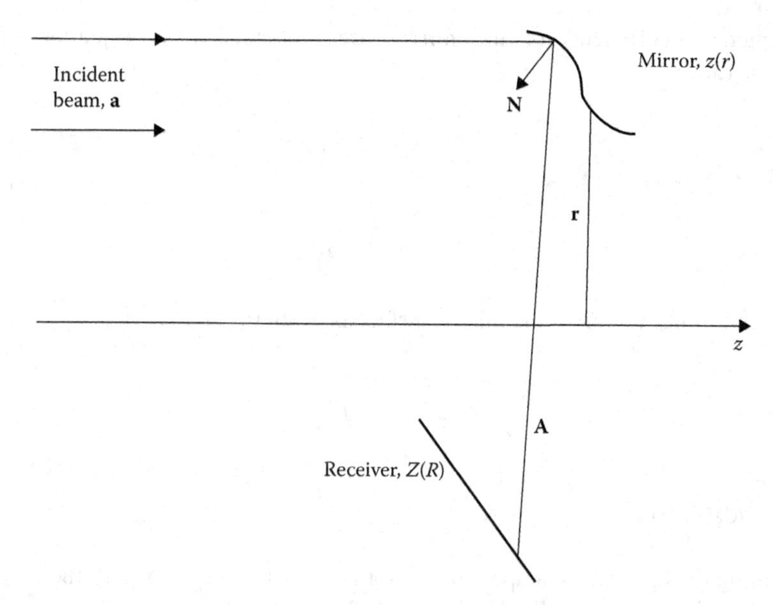

FIGURE 13.5 Collimated beam incident upon mirror and reflected to detector.

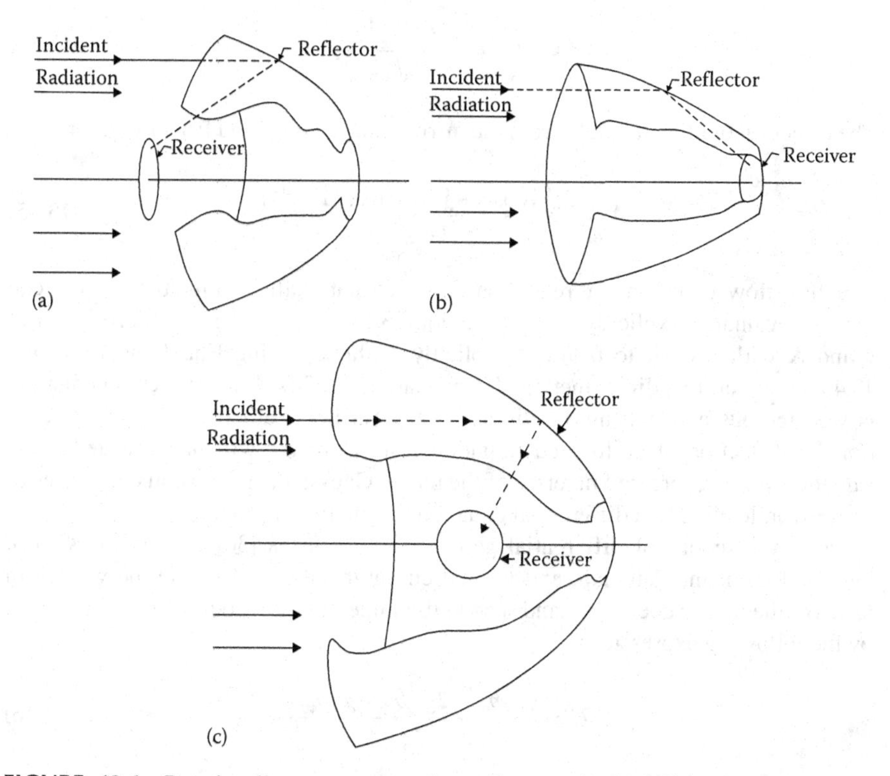

FIGURE 13.6 Rotationally symmetric solar collectors: (a) back-lighted configuration; (b) direct-lighted configuration; (c) spherical detector. (From Burkhard, D.G. and Shealy, D.L., *Solar Energy*, 17, 221–227, 1975. With permission.)

The metric coefficients of the mirror are computed from Equations 13.10 through 13.13:

$$\begin{aligned} g_{rr} &= 1+z'^2 \\ g_{\phi\phi} &= r^2 \\ g_{r\phi} &= 0 \\ g &= r^2(1+z'^2) \end{aligned} \tag{13.42}$$

The unit normal vector on the mirror surface is given by

$$\mathbf{n} = \frac{\mathbf{r}_r \times \mathbf{r}_\phi}{\sqrt{g}} = \frac{-\hat{\mathbf{i}}\, z'\cos\phi - \hat{\mathbf{j}}\, z'\sin\phi + \hat{\mathbf{k}}}{\sqrt{1+z'^2}} \tag{13.43}$$

where:
  $z' = (dz(r)/dr)$

Assuming the light incident upon the mirror is along the z-axis ($\mathbf{a} = \hat{\mathbf{k}}$), the cosine of the angle of incidence on the mirror is given by

$$\cos i = \mathbf{a}\cdot\mathbf{n} = \frac{1}{\sqrt{1+z'^2}} \tag{13.44}$$

The direction of reflected light from the mirror can be computed from Equation 13.5:

$$\mathbf{A} = \frac{2z'(\hat{\mathbf{i}}\cos\phi + \hat{\mathbf{j}}\sin\phi) - \hat{\mathbf{k}}(1-z'^2)}{1+z'^2} \tag{13.45}$$

The flux flow equation for reflection from a rotationally symmetric mirror can now be evaluated explicitly using Equation 13.17 where the partial derivatives of **r** and **A** with respect to (r,ϕ) are explicitly evaluated using Equations 13.41 and 13.45. However, this direct method for evaluating the flux flow differential equation is very tedious and will not be discussed any further. Rather, the flux flow equation for reflection of collimated radiation from a mirror will be evaluated using Equation 13.32, expressed in terms of the mean, Gaussian, and normal curvatures of the mirror, leading to a differential equation for the mirror surface.

From the theory of differential geometry of surfaces [36] and previous work [30,32], the mean, Gaussian, and normal curvature of a surface can be written in terms of the metric coefficient and second fundamental form coefficients of a surface by the following expressions:

$$H = \frac{(g_{rr}b_{\phi\phi} - 2g_{r\phi}b_{r\phi} + g_{rr}b_{\phi\phi})}{2g} \tag{13.46}$$

$$K = \frac{b_{rr}b_{\phi\phi} - (b_{r\phi})^2}{g} \tag{13.47}$$

$$K_n \sin^2 i = \sum_{i,j=r}^{\varphi} a^i a^j b_{ij} \tag{13.48}$$

where:

$$b_{ij} = \frac{1}{\sqrt{g}}\left\{\mathbf{r}_i \times \mathbf{r}_j \cdot \left[\frac{\partial^2 \mathbf{r}(r,\phi)}{\partial i \partial j}\right]\right\} \quad \text{for } i,j = r \text{ or } \phi \tag{13.49}$$

$$a^i = g^{ir}\mathbf{a}\cdot\mathbf{r}_r + g^{i\phi}\mathbf{a}\cdot\mathbf{r}_\phi \tag{13.50}$$

$$g^{rr} = \frac{g_{\phi\phi}}{g}, \quad g^{\phi\phi} = \frac{g_{rr}}{g}, \quad g^{r\phi} = \frac{-g_{r\phi}}{g} \tag{13.51}$$

or explicitly for the application shown in Figures 13.5 and 13.6 using surface equations (13.41 and 13.42) for the metric coefficients

$$\begin{aligned} b_{rr} &= \frac{(\partial^2 z''/\partial r^2)}{\sqrt{1+z'^2}}, \quad b_{\phi\phi} = \frac{rz'}{\sqrt{1+z'^2}}, \quad b_{r\phi} = 0, \\ g^{rr} &= \frac{1}{(1+z'^2)}, \quad g^{\phi\phi} = \frac{1}{r^2}, \quad g^{r\phi} = 0 \end{aligned} \tag{13.52}$$

$$a^r = \frac{z'}{1+z'^2}, \quad a^\phi = 0 \tag{13.53}$$

where:
$z'' \equiv (\partial^2 z/\partial r^2)$

The mean, Gaussian, and normal curvatures of the surface are given by the following expressions:

$$H = \frac{1}{2}\left[\frac{z' + rz''}{r(1+z'^2)^{3/2}}\right] \tag{13.54}$$

$$K = \frac{z'z''}{r(1+z'^2)^2} \tag{13.55}$$

$$K_n \sin^2 i = \frac{z'^2 z''}{(1+z'^2)^{5/2}} \tag{13.56}$$

Then, the flux flow equation for collimated light reflected from a rotationally symmetric mirror can be explicitly written from Equation 13.32 in the following form:

$$E_{ds-dS} = \frac{\sigma(\mathbf{r})\cos i \cos I}{\cos i\left\{1 + 2r(s,S)\left[\frac{(z'/r) + z''}{(1+z'^2)^2}\right] + 4r^2(s,S)\left[\frac{(z'/r) + z''}{(1+z'^2)^2}\right]\right\}} \tag{13.57}$$

or, after factoring the denominator,

$$E_{ds\text{-}dS} = \frac{\sigma(\mathbf{r})\cos I}{\left\{1 + \left[\frac{2r(s,S)z'}{r(1+z'^2)}\right]\right\}\left\{1 + \left[\frac{2r(s,S)z''}{(1+z'^2)}\right]\right\}} \tag{13.58}$$

Using the following relationships

$$r(s,S) = \frac{R-r}{A_r} = R - r\frac{1+z'^2}{2z'} \tag{13.59}$$

$$\cos I = \mathbf{A}\cdot\mathbf{N} = \frac{\left[\left(\frac{2z'}{1+z'^2}\right)\left(\frac{dZ}{dR}\right) + \left(\frac{1-z'^2}{1+z'^2}\right)\right]}{\sqrt{1+\left(\frac{dZ}{dR}\right)^2}} \tag{13.60}$$

where:

N is the unit normal vector to the receiver surface S

$Z(R)$ specifies the shape of the receiver surface S

Then, the flux flow equation (13.58) can be written as a second-order differential equation for the mirror surface:

$$\frac{z''}{z'} = \frac{1}{R-r}\left\{\left(\frac{\sigma(r)}{E_{ds\text{-}dS}}\right)\left(\frac{r}{R}\right)\frac{\left[(2z'/1+z'^2)(dZ/dR) + (1-z'^2/1-z'^2)\right]}{\sqrt{1+(dZ/dR)^2}} - 1\right\} \tag{13.61}$$

The above equation is equivalent to Equation 6.98 and the results of McDermit [14] and McDermit and Horton [18] (Equation 3.14 of Reference [14] and Equation 13 of Reference [18]). When appropriate boundary conditions are given, Equation 13.61 can be solved for the shape of the mirror surface that will illuminate the receiver surface S with a prescribed irradiance for a given source profile.

13.2.2.2 Using the Conservation of Energy Condition

Instead of using the second-order differential equation (13.61) for evaluating the contour of the beam shaping optics, the energy balance equation (6.35) can be integrated and combined with the ray trace equations to obtain a first-order differential equation for the reflecting surface. This approach has been used by Schruben [21] to design a mirror that illuminates its aperture with a specified distribution. This approach is also equivalent to using the flux flow equation to obtain a second-order differential equation (13.61) of the mirror surface.

Consider the rotationally symmetric geometry shown in Figures 13.5 and 13.6. The radiation is incident upon reflector surface s with equation $z = z(r)$. The equation of the receiver surface S is $Z = Z(R)$. The flux density $\sigma(r)$ is incident upon a circular ring about the z-axis of area $2\pi r\, dr$ and is reflected to a circular ring on the receiver surface S of area dA. The equation of the receiver surface S can be written in terms of the polar and radial coordinates $[\Phi, R]$:

$$\mathbf{R}(\Phi,R) = \hat{\mathbf{i}} + R\cos\Phi + \hat{\mathbf{j}}R\sin\Phi + \hat{\mathbf{k}}Z(R) \tag{13.62}$$

Then, applying Equations 13.10 and 13.42 to the receiver surface S, a rotationally symmetric element of area on the receiver surface can be written as

$$\mathrm{d}A = \int_0^{2\pi} G^{1/2}\mathrm{d}\Phi\,\mathrm{d}R = 2\pi R\sqrt{1+\left(\frac{\mathrm{d}Z}{\mathrm{d}R}\right)^2}\,\mathrm{d}R \tag{13.63}$$

where:

$$\begin{aligned} G &= G_{RR}G_{\Phi\Phi} - G_{R\Phi}^2 = R^2\left[1+\left(\frac{\partial Z(R)}{\partial R}\right)^2\right] \\ G_{RR} &= 1 + \left[\frac{\partial Z(R)}{\partial R}\right]^2 \\ G_{\Phi\Phi} &= R^2; \quad G_{R\Phi} = 0 \end{aligned} \tag{13.64}$$

The element of area dA in Equation 13.63 represents a circular ring on the receiver surface as illustrated in Figure 13.7.

When the receiver surface is a disk, as shown in Figure 13.6a and b, Z = constant and d$A = 2\pi R\,\mathrm{d}R$. For a spherical receiver, as shown in Figure 13.6c,

$$\mathrm{d}A = 2\pi\left[1+\left(\frac{R^2C^2}{1-C^2R^2}\right)\right]^{1/2}\mathrm{d}R = \frac{2\pi R\mathrm{d}R}{\sqrt{1-C^2R^2}} \tag{13.65}$$

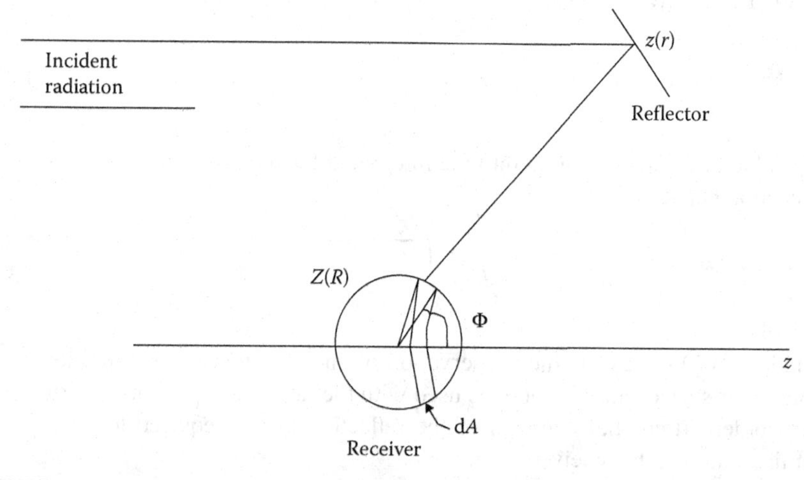

FIGURE 13.7 Rotationally symmetric beam shaping mirror and spherical detector.

when the equation of the spherical receiver surface is written in a form similar to that given in Equation 6.46:

$$Z(R) = \frac{CR^2}{1+\sqrt{1-C^2R^2}}$$
$$R^2 = X^2 + Y^2 \tag{13.66}$$
$$C \equiv \frac{1}{\text{Radius of spherical surface}}$$

The energy balance equation (6.10) is

$$ER\sqrt{1+\left(\frac{dZ}{dR}\right)^2}\,dR = \sigma r dr \tag{13.67}$$

where:

E gives the irradiance on the receiver surface

When the receiver surface equation is specified, both sides of Equation 13.67 can be integrated. E may be an arbitrary function of position on S, but it must have an adjustable parameter so that conservation of energy is satisfied between the input beam and the receiver surface. For a flat receiver, $Z = Z_0$, conservation of energy between the input beam and the receiver surface is given by

$$\int_{R_0}^{R} E\, RdR = \int_{r_0}^{r} \sigma r dr \tag{13.68}$$

For uniform irradiance over a flat receiver with $R_0 = 0$, r_0, r_f, $R_f \neq 0$, integrating Equation 13.68 gives

$$E_0 = \sigma\frac{(r_f^2 - r_0^2)}{R_f^2} \tag{13.69}$$

This value of E_0 is substituted into the integrated form of Equation 13.68 to yield a connection between r and R:

$$R = \sqrt{\frac{\sigma(r^2 - r_0^2)}{E_0}} \tag{13.70}$$

Equation 13.70 represents the conservation of energy between the input beam and the receiver surface and can now be used with the ray trace equations to write down a first-order differential equation of the reflector surface required to provide this illumination over the receiver.

A unit vector along the reflected light for this geometry is given by Equation 13.45. Then the ray trace equation between the reflector and the receiver in the r–z plane is given by

$$\frac{R-r}{Z_0-z}=\frac{2z'}{z'^2-1} \tag{13.71}$$

Equation 13.71 can be solved as a quadratic equation for z' to give

$$\begin{aligned} z' &= \left(\frac{z-Z_0}{r-R}\right) \pm \sqrt{1+\left(\frac{z-Z_0}{r-R}\right)^2} \\ &= \left[\frac{z-Z_0}{r-\sqrt{\sigma(r^2-r_0^2)/E_0}}\right] \pm \sqrt{1+\left[\frac{z-Z_0}{r-\sqrt{\sigma(r^2-r_0^2)/E_0}}\right]^2} \end{aligned} \tag{13.72}$$

where:

the $\pm$ sign is used when the concave side of mirror is oriented toward the positive or negative z-direction

R as a function of r is given by Equation 13.70

Equation 13.72 can be solved numerically to determine the shape of the reflector. The initial conditions are $z = z_0$ when $r = r_0$. Reference [13] contains several solar collectors designed by the solution of Equation 13.72 for different initial conditions.

In this section, two methods of designing nonlaser beam profile shaping optical systems have been discussed. These methods are equivalent and are based on the application of conservation of energy between the input beam and the receiver surface, and on the ray trace equations between the reflector and the receiver. These methods for designing nonlaser beam shaping systems are generally applicable to all forms of incident radiation. In Section 13.3, the optical design method using the conservation of energy condition will be extended to include a point and line source of radiation for heating and illumination applications.

13.3 APPLICATION TO POINT AND LAMBERTIAN SOURCE

In Section 13.2, the principle of conservation of energy and the ray trace equations were used to obtain a first- and a second-order differential equation used in optical design of beam shaping systems to illuminate a receiver with a prescribed irradiance when the input beam was collimated. However, for many nonlaser beam shaping applications associated with heating or illumination, it is important to consider the finite size and location of the source of radiation. Schruben [21] reported a differential equation-based design of an illumination system for a reflector and small

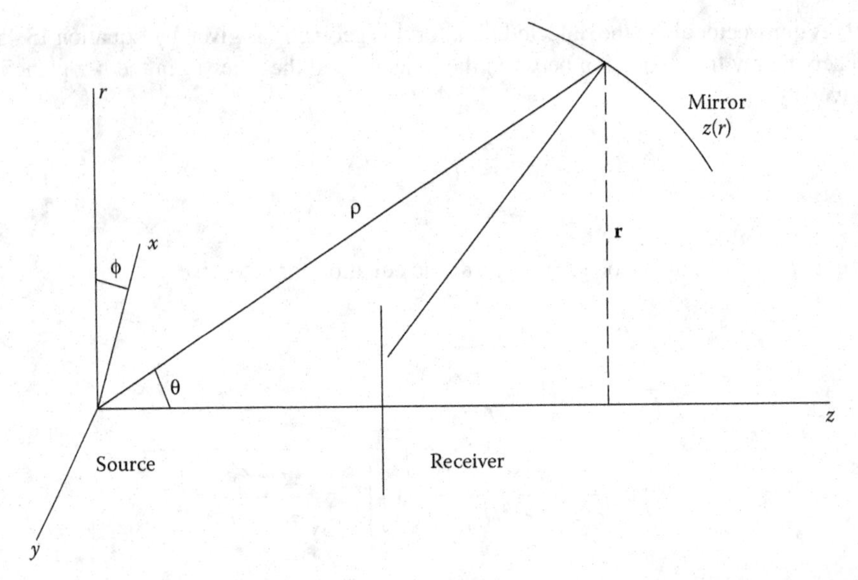

FIGURE 13.8 Geometrical configuration of point source, mirror, and receiver surface.

Lambertian source. Burkhard and Shealy [22] reported an optical design method for shaping a mirror to transform a point or line source of light into a prescribed irradiance over a receiver surface. In this section, the first-order differential equation* discussed in Section 13.2.2 will be revised to use with point or small line (Lambertian) source of light.

Consider the geometrical configuration of point source, mirror, and receiver surface shown in Figure 13.8, where the mirror and receiver surfaces have rotational symmetry around the z-axis.

The current analysis will take into account the fact that the incident radiation is not collimated, or its wavefront is not planar. This means that the direction of the incident light from the source will vary over the surface of the optics, which differs from the beam shaping applications addressed in Section 13.2.2.2 and shown in Figures 13.5 and 13.6 where the direction of incident light over optics was constant. A unit vector along an incident ray upon the mirror is given by

$$\mathbf{a} = \hat{\mathbf{r}} \sin\theta + \hat{\mathbf{k}} \cos\theta \tag{13.73}$$

where:

$\hat{\mathbf{r}}(= \hat{\mathbf{i}} \cos\phi + \hat{\mathbf{j}} \sin\phi)$ is the radial polar unit vector in the x–y plane

It will be helpful to understand clearly that all the variables used in the present analysis (φ,θ,ρ) are the conventional spherical coordinates, where the z-axis is also the optical (symmetry) axis. Since the beam shaping optics has rotational symmetry about the optical (z-) axis, it is convenient to use the r–z plane polar coordinates

* For a more complete discussion of differential geometry, the reader is referred to one or more of the comprehensive books in the literature on this topic, such as Reference [36].

(ρ,θ) to solve for the shape of the mirror surface. The slope of the mirror in the r–z plane is given by

$$z' \equiv \frac{dz}{dr} = \frac{(\rho' \cos\theta - \rho' \sin\theta)}{(\rho' \sin\theta + \rho' \cos\theta)} \tag{13.74}$$

where:

$\rho' \equiv d\rho/d\theta$
$z = \rho \cos\theta$
$r = \rho \sin\theta$

These variables are shown in Figure 13.8. From Equation 13.43, a unit normal vector of the mirror can be written in terms of the coordinates (ρ,θ) as

$$\mathbf{n} = \frac{-z'\hat{\mathbf{r}} + \hat{\mathbf{k}}}{\sqrt{1+z'^2}} = \frac{-\hat{\mathbf{r}}(\rho' \cos\theta - \rho \sin\theta) + \hat{\mathbf{k}}(\rho' \sin\theta + \rho \cos\theta)}{\sqrt{\rho'^2 + \rho^2}} \tag{13.75}$$

The direction of the reflected ray **A** can be evaluated from Equation 13.5 to give

$$\mathbf{A} = \frac{\hat{\mathbf{r}}(\rho'^2 \sin\theta + 2\rho\rho' \cos\theta - \rho^2 \sin\theta)}{\rho'^2 + \rho^2} + \frac{\hat{\mathbf{k}}(\rho'^2 \cos\theta + 2\rho\rho' \sin\theta - \rho^2 \cos\theta)}{\rho'^2 + \rho^2} \tag{13.76}$$

Then, the ray trace equation in the r–z plane, connecting the point (r,z) on the reflector to the point (R,Z) on the receiver, is given by

$$\frac{R - r}{Z(R) - z(r)} = \frac{A_r}{A_z} = \frac{\rho'^2 \sin\theta + 2\rho\rho' \cos\theta - \rho^2 \sin\theta}{\rho'^2 \cos\theta + 2\rho\rho' \sin\theta - \rho^2 \cos\theta} \tag{13.77}$$

For a planar receiver surface, $Z = Z_0$ (a constant), Equation 13.77 can be written as a first-order differential equation for the reflector surface:

$$\rho' = \frac{\left[\rho(R\sin\theta + Z_0 \cos\theta - \rho) \pm \rho\sqrt{(-R\sin\theta - Z_0\cos\theta + \rho)^2 + (R\cos\theta - Z_0 \sin\theta)^2}\right]}{\rho\cos\theta \; - \; Z_0 \sin\theta} \tag{13.78}$$

where:

the $\pm$ sign in Equation 13.78 is chosen to ensure that ρ' is positive or negative as required by the geometrical configuration of source, reflector, and receiver shown in Figure 13.6

In subsequent calculations in this chapter, the positive root of Equation 13.78 will be used.

In order to solve Equation 13.78, the energy balance equation must be used to obtain an expression for $R(\rho,\theta)$. For a flat receiver, the conservation of energy condition (6.35) becomes

$$\int_0^R 2\pi E(R) R \, dR = \int_{\theta_0}^{\theta} I(\theta) 2\pi r^2 \sin\theta \, d\theta \tag{13.79}$$

where:

$I(\theta)$ is the intensity of the source

the reflectance of mirror has been assumed to be equal to unity

If the source is a small Lambertian source [39] along the z-axis, then $I(r,\theta) = I_0 \sin\theta/r^2$, where I_0 is a constant. If the optics is such that the direct illumination from source to receiver can be ignored, Equation 13.79 can be integrated to obtain the constant E_0, which ensures conservation of energy for this system. For uniform irradiance over the receiver, $E = E_0$, integrating Equation 13.79 over the full beam [$\theta \in (\theta_0,\theta_m)$ and $R \in (0,R_m)$] leads to the following result:

$$E_0 = \left(\frac{I_0}{R_m^2}\right)\left[\theta_m - \theta_0 + \left(\frac{\sin 2\theta_0 - \sin 2\theta_m}{2}\right)\right] \tag{13.80}$$

where:

R_m and θ_m are the maximum values of R and θ, respectively

Similarly, for an isotropic point source, uniform irradiance of the receiver leads to the following expression for E:

$$E_0 = 2I_0 \frac{\cos\theta_0 - \cos\theta_m}{R_m^2} \tag{13.81}$$

The constants I_0, θ_0, and θ_m can be chosen to give the desired value of E_0.

For backlighting configuration (Figure 13.6a), integrating Equation 13.79 leads to an expression for $R(\theta)$ that can be used to integrate Equation 13.78 to determine the shape of mirror required for a specific beam shaping application. For a Lambertian source,

$$R = \sqrt{\frac{I_0}{2E_0}\left[2(\theta - \theta_0) - (\sin 2\theta - \sin 2\theta_0)\right]} \tag{13.82}$$

and for an isotropic point source,

$$R = \sqrt{\frac{2I_0}{E_0}(\cos\theta_0 - \cos\theta)} \tag{13.83}$$

Equation 13.78 can now be integrated to obtain the shape of the reflector that will transform point source light into uniform irradiance on the back of a detector. Figure 13.9a is a scaled drawing of a point source, mirror, and back-lighted detector of an example solution of this differential equation for the optical design of nonlaser beam shaping systems. Similar calculations could also be done for a Lambertian source.

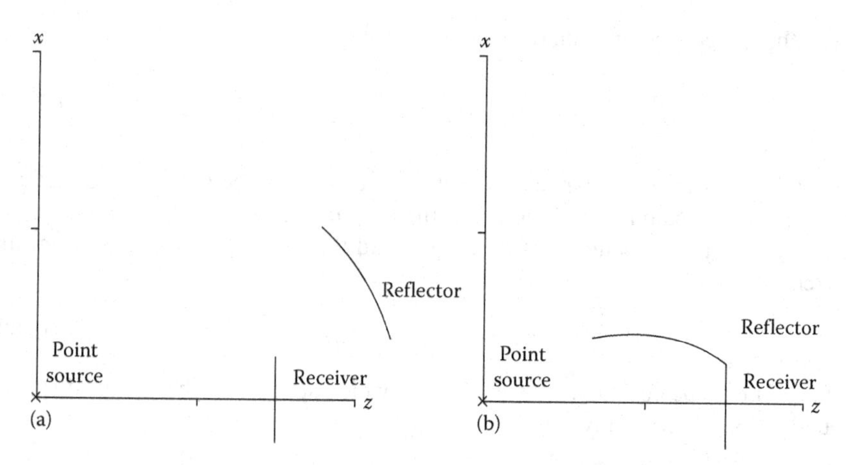

FIGURE 13.9 Uniform illumination of a flat disk by a rotationally symmetric reflector for point source light incident upon (a) back-lighted and (b) direct-lighted configurations. (From Burkhard, D.G. and Shealy, D.L., *Solar Energy*, 17, 221–227, 1975. With permission.)

When the optical configuration (Figure 13.6b) allows direct illumination of the receiver from the source as well as from the reflector, the analysis leading to Equations 13.80 through 13.83 needs to take into account the direct illumination. The irradiance, E_{direct}, directly incident upon receiver from source, as shown in Figure 13.6b, is given by

$$E_{\text{direct}} = \sigma \cos \chi \tag{13.84}$$

where:
- σ is the flux density from source evaluated at the receiver
- χ is the angle between the ray and the normal to the receiver

For a Lambertian source [39],

$$\sigma = \frac{I_0 \sin \chi}{r_0^2} \tag{13.85}$$

where:
- r_0 is the distance from the source to a point on the receiver

From the geometry of a direct illumination system, the following relations hold:

$$\begin{aligned} r_0 &= \sqrt{R^2 + Z_0^2} \\ \sin \chi &= \frac{R}{r_0} \\ \cos \chi &= \frac{Z_0}{r_0} \end{aligned} \tag{13.86}$$

Then, the direct irradiance on receiver is given by

$$E_{\text{direct}} = \frac{RZ_0 I_0}{(R^2 + Z_0^2)^2} \tag{13.87}$$

To obtain a uniform irradiance (E_0) over the receiver surface when both direct and reflected light are considered, note that the total irradiance $E_{\text{total}}(R)$ at the receiver is the sum of the reflected light $E_{\text{reflected}}(R)$ and the direct light $E_{\text{direct}}(R)$ from the source:

$$E_{\text{total}}(R) = E_{\text{direct}}(R) + E_{\text{reflected}}(R) \tag{13.88}$$

If $E_{\text{total}}(R)$ is to be constant, then $E_{\text{reflected}}(R)$, which appears on the left-hand side of Equation 13.79, will be given by

$$E_{\text{reflected}} = E_{\text{total}} - E_{\text{direct}} \tag{13.89}$$

and the constant E_{total} needs to satisfy the following integral equation:

$$\int_0^{R_m} 2\pi \left[E_{\text{total}} - \frac{RZ_0 I_0}{(R^2 + Z_0^2)^2} \right] R\mathrm{d}R = \int_{\theta_0}^{\theta_m} I_0 2\pi \sin^2\theta \, \mathrm{d}\theta \tag{13.90}$$

After carrying out integrals and solving for E_{total} in Equation 13.90, one has

$$E_{\text{total}} = \frac{I_0}{R_m^2} \left[(\theta_m - \theta_0) - \frac{1}{2}(\sin 2\theta_m - \sin 2\theta_0) + \tan^{-1}\left(\frac{R_m}{Z_0} \right) - \frac{(R_m/Z_0)}{1 + (R_m/Z_0)^2} \right] \tag{13.91}$$

The above equation expresses the total energy leaving the sources that is intercepted by the mirror and the receiver. The value of E_{total} from Equation 13.91 is substituted into Equation 13.90 with upper limits of R and θ to solve for $R(\rho,\theta)$ and subsequently solve the differential equation (13.78) numerically for $\rho(\theta)$, which is the shape of a mirror surface [13,21]. Figure 13.9b is a scaled drawing of point source and reflector that will uniformly illuminate receiver plane when taking both direct and reflected light into account. This optical design method has been extended in Reference [22] to include multiple point sources and continuous line sources along the symmetry axis when solving for the shape of a mirror that will uniformly illuminate a detector.

13.4 CONCLUSION

The design and analysis of nonlaser beam shaping systems has been discussed in this chapter. The ray trace equations and the principle of conservation of energy within a bundle of rays have been used to derive several alternate forms of the flux flow equation, Equations 13.22 and 13.40. Equation 13.22 is useful when computing the irradiance distribution for a collimated incident beam. Equation 13.40 is useful when computing the irradiance distribution for cases when the incident beam wavefront is not planar and, subsequently, when the incident beam wavefront curvatures vary over the beam shaping optics, which happens for point or extended sources near the optics.

The flux flow equation can be used to monitor the irradiance along a ray path as it propagates through the optical system. When the input and output beam profiles are known, the shape of a single surface is determined by a differential energy balance equation. Specific examples of using this optical design method are presented for collimated, point, and Lambertian sources of radiation.

REFERENCES

1. AC Claus. On Archimedes' burning glass. *Applied Optics* 12(10): A14, 1973.
2. ON Stavroudis. Comments On: On Archimedes' burning glass. *Applied Optics* 12(10): A16, 1973.
3. MH Cobble. Theoretical concentration for solar furnaces. *Solar Energy* 5(2): 61, 1961.
4. O Kamada. Theoretical concentration and attainable temperature in solar furnaces. *Solar Energy* 9(1): 39–17, 1965.
5. K-E Hassan and MF El-Refaie. Theoretical performance of cylindrical parabolic solar concentrators. *Solar Energy* 15: 219–244, 1973.
6. RA Zakhidov and AA Vainer. Distribution of radiation produced by a paraboloidal concentrator. *Geliotekhnika* (Applied Solar Energy) 10(3): 34–40, 1974.
7. WS Duff, GF Lameiro, and GOG Löf. Parametric performance and cost models for solar concentrators. *Solar Energy* 17: 47–58, 1975.
8. GL Strobel and DG Burkhard. Irradiance for skew rays incident upon trough-like solar collector of arbitrary shape. *Solar Energy* 20: 25–27, 1978.
9. RF Jones Jr. Collection properties of generalized light concentrators. *Journal of the Optical Society of America* 67(11): 1594–1598, 1977.
10. WT Welford and R Winston. *High Collection Nonimaging Optics*. New York: Academic Press, 1989.
11. R Winston. Principles of solar concentrators of a novel design. *Solar Energy* 16: 89–95, 1974.
12. R Winston and H Hinterberger. Principles of cylindrical concentrators for solar energy. *Solar Energy* 17: 255–258, 1975.
13. DG Burkhard and DL Shealy. Design of reflectors which will distribute sunlight in a specified manner. *Solar Energy* 17(1): 221–227, 1975.
14. JH McDermit. Curved reflective surfaces for obtaining prescribed irradiation distributions. PhD dissertation, University of Mississippi, Oxford, MS, 1972.
15. TE Horton and JH McDermit. Optical design of solar concentrators. *Journal of Energy* 4(1): 4–9, 1980.
16. DL Shealy and HM Berg. Simulation of optical coupling from surface emitting LEDs. *Applied Optics* 22(11): 1722–1730, 1983.
17. DR Gabardi and DL Shealy. Coupling of domed light-emitting diodes with a multimode step-index optical fiber. *Applied Optics* 25(19): 3435–3442, 1986.
18. JH McDermit and TE Horton. Reflective optics for obtaining prescribed irradiative distributions from collimated sources. *Applied Optics* 13(6): 1444–1450, 1974.
19. DG Burkhard, DL Shealy, and RU Sexl. Specular reflection of heat radiation from an arbitrary reflector surface to an arbitrary receiver surface. *International Journal of Heat and Mass Transfer* 16: 271–280, 1973.
20. DG Burkhard and DL Shealy. View function in generalized curvilinear coordinates for specular reflection of radiation from a curved surface. *International Journal of Heat and Mass Transfer* 16: 1492–1496, 1973.
21. JS Schruben. Analysis of rotationally symmetric reflectors for illuminating systems. *Journal of the Optical Society of America* 64(1): 55–58, 1974.
22. DG Burkhard and DL Shealy. Specular aspheric surface to obtain a specified irradiance from discrete or continuous line source radiation: Design. *Applied Optics* 14(6): 1279–1284, 1975.

23. DG Burkhard and DL Shealy. A different approach to lighting and imaging: Formulas for flux density, exact lens and mirror equations, and caustic surfaces in terms of the differential geometry of surfaces. *Proceedings of SPIE* 692: 248–272, 1986.
24. JM Gordon and A Rabl. Reflectors for uniform far-field irradiance: Fundamental limits and example of an axisymmetric solution. *Applied Optics* 37(1): 44–47, 1998.
25. BS Westcott and AP Norris. Reflector synthesis for generalized far-fields. *Journal of Physics A: Mathematical and General* 8(4): 521–532, 1975.
26. F Brickell and BS Westcott. Reflector design for two-variable beam shaping in the hyperbolic case. *Journal of Physics A: Mathematical and General* 9(1): 113–128, 1976.
27. P Mouroulis and J Macdonald. *Geometrical Optics and Optical Design*. New York: Oxford University Press, 1997.
28. DG Kock. Simplified irradiance/illuminance calculations in optical systems. *International Symposium on Optical Design*, Berlin, Germany, September 14, 1992.
29. DL Shealy and DG Burkhard. Analytical illuminance calculation in a multi-interface optical system. *Optica Acta* 22(6): 485–501, 1975.
30. DL Shealy and DG Burkhard. Flux density for ray propagation in discrete index media expressed in terms of the intrinsic geometry of the deflecting surface. *Optica Acta* 20(4): 287–301, 1973.
31. M Born and E Wolf. *Principles of Optics*. 5th edn. New York: Pergamon Press, 1975.
32. DG Burkhard and DL Shealy. Simplified formula for the illuminance in an optical system. *Applied Optics* 20(5): 897–909, 1981.
33. ON Stavroudis. *The Optics of Rays, Wavefronts, and Caustics*. New York: Academic Press, 1972.
34. DL Shealy and DG Burkhard. Heat flux contours on a plane for parallel radiation specularly reflected from a cone, a hemisphere and a paraboloid. *International Journal of Heat and Mass Transfer* 16: 281–291, 1973.
35. DG Burkhard and DL Shealy. Flux density for ray propagation in geometrical optics. *Journal of the Optical Society of America* 63(3): 299–304, 1973.
36. E Kreyszig. *Introduction to Differential Geometry*. Toronto, ON: University of Toronto Press, 1968.
37. VA Fock. *Electromagnetic Diffraction and Propagation Problems*. New York: Pergamon Press, 1965.
38. ON Stavroudis and RC Fronczek. Caustic surfaces and the structure of the geometrical image. *Journal of the Optical Society of America* 66: 795–800, 1976.
39. MV Klein. *Optics*. New York: Wiley, 1970.

APPENDIX A

SUMMARY OF SOME CONCEPTS AND RESULTS FROM DIFFERENTIAL GEOMETRY*

In general, the equation of a surface is a constraint equation between three coordinates such that when two coordinates are given the third coordinate will indeed

* Reference [19] contains a formula for the flux flow equation applied to reflection (or refraction) of point source light to illuminate a receiver surface. This formula for the flux flow equation could be used to derive a second-order differential equation for design of a mirror for beam shaping of point source light. However, it is more straightforward to use the two first-order differential equations resulting from application of conservation of energy and the ray trace equations than the second-order differential equation obtained from the flux flow equation as part of the optical design of a mirror used with point or extended source of light.

lie on the surface. The equation of the surface can be represented by the following symbolic equation:

$$w = w(u,v) \tag{A.1}$$

where:

$[u,v,w]$ are curvilinear coordinates that themselves are defined in terms of the Cartesian coordinates $[x,y,z]$:

$$\begin{aligned} x &= x(u,v,w) \\ y &= y(u,v,w) \\ z &= z(u,v,w) \end{aligned} \tag{A.2}$$

Combining the equation of the surface (A.l) and the relationship between the Cartesian coordinates $[x,y,z]$ of a point and the curvilinear coordinates $[u,v,w]$ of the same point (Equation A.2), one obtains a parametric representation of the surface:

$$\begin{aligned} x &= x(u,v) \\ y &= y(u,v) \\ z &= z(u,v) \end{aligned} \tag{A.3}$$

where:

$[u,v]$ are the curvilinear coordinates of the surface that can also be considered as surface parameters

It will be convenient to write Equation A.3 as the vector equation:

$$\mathbf{r} = \mathbf{r}(u,v) \tag{A.4}$$

where:

the Cartesian components of the vector $\mathbf{r}(\equiv \hat{\mathbf{i}}x + \hat{\mathbf{j}}y + \hat{\mathbf{k}}z)$ are given by Equation A.3

The two vectors $\mathbf{r}_u \equiv \partial\mathbf{r}/\partial v$ and $\mathbf{r}_v \equiv \partial\mathbf{r}/\partial v$, which are tangent to the u-parameter curve and the v-parameter curve on the surface, are two linearly independent vectors in the tangent plane of the surface at a specified point on the surface, and thus, any vector in the tangent plane can be written as a linear combination of the vectors $[\mathbf{r}_u,\mathbf{r}_v]$. The unit normal vector to the surface can then be written as the vector cross product of $[\mathbf{r}_u,\mathbf{r}_v]$:

$$\mathbf{n} = \frac{\mathbf{r}_u \times \mathbf{r}_v}{|\mathbf{r}_u \times \mathbf{r}_v|} \tag{A.5}$$

The measurement of lengths on the surface terms of the coefficients is conveniently expressed in

$$g_{uu} \equiv \mathbf{r}_u \cdot \mathbf{r}_u = \left(\frac{\partial x}{\partial u}\right)^2 + \left(\frac{\partial y}{\partial u}\right)^2 + \left(\frac{\partial z}{\partial u}\right)^2$$

$$g_{vv} \equiv \mathbf{r}_v \cdot \mathbf{r}_v = \left(\frac{\partial x}{\partial v}\right)^2 + \left(\frac{\partial y}{\partial v}\right)^2 + \left(\frac{\partial z}{\partial v}\right)^2 \tag{A.6}$$

$$g_{uv} \equiv \mathbf{r}_u \cdot \mathbf{r}_v = \left(\frac{\partial x}{\partial u}\right)\left(\frac{\partial x}{\partial v}\right) + \left(\frac{\partial y}{\partial u}\right)\left(\frac{\partial y}{\partial v}\right) + \left(\frac{\partial z}{\partial u}\right)\left(\frac{\partial z}{\partial v}\right)$$

$$g_{vu} \equiv \mathbf{r}_v \cdot \mathbf{r}_u = g_{uv}$$

or by

$$g_{jk} = \mathbf{r}_j \cdot \mathbf{r}_k \tag{A.7}$$

where it is understood in Equation A.7 that (j,k) may each take on the value of u or v.

The coefficients g_{uv} transform like a second rank symmetric tensor and are called the metric coefficients of the surface or the coefficients of the first fundamental form of the surface. The first fundamental form of the surface is a quadratic expression for the differential arc length of a curve on the surface and is given by

$$ds^2 \equiv \mathbf{dr} \cdot \mathbf{dr} = \left(\frac{\partial \mathbf{r}}{\partial u} du + \frac{\partial \mathbf{r}}{\partial v} dv\right) \cdot \left(\frac{\partial \mathbf{r}}{\partial u} du + \frac{\partial \mathbf{r}}{\partial v} dv\right)$$

$$= g_{uu}(du)^2 + 2g_{uv}dudv + g_{vv}(dv)^2 \tag{A.8}$$

$$= \sum_{j,k=u}^{v} g_{jk} d(j)\, d(k) = g_{jk} d(j) d(k)$$

where the simplified (last) expression uses the summation convention, which means that if a given letter appears twice on the same side of an equation, then the summation of that letter must be carried out. For surfaces, the summation will be over the two curvilinear coordinates u and v.

The determinant of the metric coefficients is given by

$$g \equiv \begin{vmatrix} g_{uu} & g_{uv} \\ g_{uv} & g_{vv} \end{vmatrix} = g_{uu}g_{vv} - (g_{uv})^2 \tag{A.9}$$

By direct computation, one sees that the determinant g is identically equal to the magnitude square of the normal vector $\mathbf{r}_u \times \mathbf{r}_v$:

$$g = \left|\mathbf{r}_u \times \mathbf{r}_v\right|^2 \tag{A.10}$$

In terms of g, one can write the Cartesian components of $\mathbf{n}$:

$$n_x = g^{-1/2}\left(\frac{\partial y}{\partial u}\frac{\partial z}{\partial v} - \frac{\partial z}{\partial v}\frac{\partial z}{\partial u}\right)$$

$$n_y = g^{-1/2}\left(\frac{\partial z}{\partial u}\frac{\partial x}{\partial v} - \frac{\partial z}{\partial v}\frac{\partial u}{\partial u}\right) \quad \text{(A.11)}$$

$$n_z = g^{-1/2}\left(\frac{\partial x}{\partial u}\frac{\partial y}{\partial v} - \frac{\partial x}{\partial v}\frac{\partial y}{\partial u}\right)$$

An expression for the element of area, da, on the surface in terms of the differentials du and dv is found by separately varying the position vector $\mathbf{r}(u,v)$ of a point on the surface by an amount du and dv, and then taking the cross product of the resulting differential vectors:

$$da = \left|\frac{d\mathbf{r}}{du}du \times \frac{d\mathbf{r}}{dv}dv\right| = \left|\mathbf{r}_u \times \mathbf{r}_v\right| dudv = g^{1/2}dudv \quad \text{(A.12)}$$

We have seen that a knowledge of the metric coefficients g_{jk} is sufficient for calculating lengths and areas on a surface; however, they do not uniquely determine a surface. In order to fully characterize a surface in terms of the radii of curvature, for example, it is necessary to introduce another quadratic form in the coordinate differentials du and dv, which is usually referred to as the second fundamental form of a surface and is given by

$$b_{uu}(du)^2 + 2b_{uv}dudv + b_{vv}(dv)^2 \quad \text{(A.13a)}$$

or simply by

$$b_{jk}d(j)d(k) \quad \text{(A.13b)}$$

where the summation convention is implied. The second fundamental form coefficients, b_{jk}, will transform like a second rank symmetric tensor and are given by

$$b_{uu} = \mathbf{r}_{uu}\cdot\mathbf{n} = n_x\frac{\partial^2 x}{\partial u^2} + n_y\frac{\partial^2 y}{\partial u^2}n_z\frac{\partial^2 z}{\partial u^2}$$

$$b_{uv} = \mathbf{r}_{uv}\cdot\mathbf{n} = n_x\frac{\partial^2 x}{\partial u\partial v} + n_y\frac{\partial^2 y}{\partial u\partial v}n_z\frac{\partial^2 z}{\partial u\partial v} \quad \text{(A.14)}$$

$$b_{vu} = \mathbf{r}_{vu}\cdot\mathbf{n} = b_{uv}$$

$$b_{vv} = \mathbf{r}_{vv}\cdot\mathbf{n} = n_x\frac{\partial^2 x}{\partial v^2} + n_y\frac{\partial^2 y}{\partial v^2} + n_z\frac{\partial^2 z}{\partial v^2}$$

or when $\mathbf{r}_{jk} \equiv \partial^2 r/\partial(j)\,\partial(k)$

$$b_{jk} = \mathbf{r}_{jk}\cdot\mathbf{n};\ (j,k = u,v) \quad \text{(A.15)}$$

The determinant of the coefficients b_{jk} is given by

$$b = b_{uu}b_{vv} - (b_{uv})^2 \tag{A.16}$$

Since the unit normal $\mathbf{n}$ to the surface is given by

$$\mathbf{n} = \frac{\mathbf{r}_u \times \mathbf{r}_v}{\sqrt{g}} \tag{A.17}$$

the second fundamental form coefficients, b_{jk}, can be written explicitly as

$$\begin{aligned} b_{uu} &= \frac{\mathbf{r}_u \times \mathbf{r}_v \cdot \mathbf{r}_{uu}}{\sqrt{g}} \\ b_{uv} &= \frac{\mathbf{r}_u \times \mathbf{r}_v \cdot \mathbf{r}_{uv}}{\sqrt{g}} \\ b_{vv} &= \frac{\mathbf{r}_u \times \mathbf{r}_v \cdot \mathbf{r}_{vv}}{\sqrt{g}} \end{aligned} \tag{A.18}$$

or by

$$b_{jk} = \frac{\mathbf{r}_u \times \mathbf{r}_v \cdot \mathbf{r}_{jk}}{\sqrt{g}} \tag{A.19}$$

Since $\mathbf{r}_j$ is a vector in the tangent plane

$$\mathbf{r}_j \cdot \mathbf{n} = 0 \tag{A.20}$$

And therefore, one has after taking the partial derivative of Equation A.20 with respect to the variable k (= u or v)

$$\mathbf{r}_{jk} \cdot \mathbf{n} + \mathbf{r}_j \cdot \mathbf{n}_k = 0 \tag{A.21}$$

Hence, the second fundamental form coefficients b_{jk} given by Equation A.15 can also be written as

$$b_{jk} = -\mathbf{r}_j \cdot \mathbf{n}_k \tag{A.22}$$

One can now write the second fundamental form as

$$b_{jk}\, \mathrm{d}(j)\mathrm{d}(k) = -\mathrm{d}\mathbf{r} \cdot \mathrm{d}\mathbf{n} \tag{A.23}$$

One can associate the second fundamental form with the distance from the tangent plane of a point on the surface to an adjacent point on the surface. This can be verified if one draws a tangent plane at a point P on the surface whose position vector is

$\mathbf{r}(u,v)$. Then, the distance δ to an adjacent point P' whose position vector is $r(u + \Delta u, \Delta v)$ will be given by

$$\delta = \Delta r \cdot \mathbf{n} \tag{A.24}$$

where:

$\mathbf{n}$ is the unit normal to the surface at the point P and $\Delta \mathbf{r}$ is given by

$$\Delta \mathbf{r} = r(u+\Delta u, v+\Delta v) - r(u,v)$$

or, expanding in a Taylor series,

$$\Delta \mathbf{r} = (\mathbf{r}_u \Delta u + \mathbf{r}_v \Delta v) + \tfrac{1}{2}\left[\mathbf{r}_{uu}(\Delta_u)^2 + 2\mathbf{r}_{uv}\Delta u \Delta v + \mathbf{r}_{vv}(\Delta v)^2 + \cdots\right] \tag{A.25}$$

Hence, one has to second order in displacement

$$\begin{aligned} \delta &= \Delta \mathbf{r} \cdot \mathbf{n} \\ &= \tfrac{1}{2}\left[\mathbf{r}_{uu} \cdot \mathbf{n}(\Delta u)^2 + 2\mathbf{r}_{uv} \cdot \mathbf{n}\Delta u \Delta v + \mathbf{r}_{vv} \cdot \mathbf{n}(\Delta v)^2\right] \end{aligned}$$

or in the limit of small displacement

$$\begin{aligned} \delta &= \tfrac{1}{2}\left[b_{uu}(\mathrm{d}u)^2 + 2b_{uv}\mathrm{d}u\mathrm{d}v + b_{vv}(\mathrm{d}v)^2\right] \\ &= \tfrac{1}{2} b_{jk}\mathrm{d}(j)\mathrm{d}(k), \quad (j,k \text{ summed over } u,v) \end{aligned} \tag{A.26}$$

It is interesting to note that the coefficients b_{jk} are strictly functionally dependent on the properties of the surface at the point P, whereas the location of the adjacent point P' is uniquely characterized by the displacement du, dv. Thus, the interpretation of the second fundamental form as being proportional to the distance between P' and the tangent plane to the surface at P seems to be reasonable.

We shall now be interested in deriving expressions for suitable measures of the curvature of a surface. The concept of curvature of a surface is given meaning in terms of the curvature of an arbitrary curve C on the surface S, which is represented by

$$u = u(s), \quad v = v(s) \tag{A.27}$$

where:

s is the arc length of C

For such a curve C on S, one could not expect the unit principal normal $\mathbf{p}$ of the curve C, which is given by

$$\mathbf{p} = \frac{1}{K}\frac{\mathrm{d}^2\mathbf{r}\left[u(s),v(s)\right]}{\mathrm{d}s^2} \tag{A.28}$$

where:

K is the curvature of C to lie along the unit normal to the surface

On the contrary, **p** and **n** will make a nonzero angle η that will be a function of both the curve *C* and the surface *S*. The cosine of η is given by

$$\cos\eta = p\cdot n \tag{A.29}$$

The differentiation of **r**[*u*(*s*), *v*(*s*)] with respect to *s* appearing in Equation A.28 is given by

$$\begin{aligned}
\frac{\mathrm{d}\mathbf{r}}{\mathrm{d}s} &= \frac{\partial\mathbf{r}}{\partial u}\frac{\mathrm{d}u}{\mathrm{d}s}+\frac{\partial\mathbf{r}}{\partial v}\frac{\mathrm{d}v}{\mathrm{d}s}=\mathbf{r}_j\frac{\mathrm{d}(j)}{\mathrm{d}s}\\
\frac{\mathrm{d}^2\mathbf{r}}{\mathrm{d}s^2} &= \frac{\mathrm{d}}{\mathrm{d}s}\left(\frac{\partial\mathbf{r}}{\partial u}\right)\frac{\mathrm{d}u}{\mathrm{d}s}+\frac{\partial\mathbf{r}}{\partial u}\frac{\mathrm{d}^2u}{\mathrm{d}s^2}+\frac{\mathrm{d}}{\mathrm{d}s}\left(\frac{\partial\mathbf{r}}{\partial v}\right)\frac{\mathrm{d}v}{\mathrm{d}s}+\frac{\partial\mathbf{r}}{\partial u}\frac{\mathrm{d}^2v}{\mathrm{d}s^2}\\
&= \left(\frac{\partial^2\mathbf{r}}{\partial u^2}\frac{\mathrm{d}u}{\mathrm{d}s}+\frac{\partial^2\mathbf{r}}{\partial u\partial v}\frac{\mathrm{d}v}{\mathrm{d}s}\right)\frac{\mathrm{d}u}{\mathrm{d}s}+\frac{\partial\mathbf{r}}{\partial u}\frac{\mathrm{d}^2u}{\mathrm{d}s^2}\\
&= \left(\frac{\partial^2\mathbf{r}}{\partial u\partial v}\frac{\mathrm{d}u}{\mathrm{d}s}+\frac{\partial^2\mathbf{r}}{\partial v^2}\frac{\mathrm{d}v}{\mathrm{d}s}\right)\frac{\mathrm{d}v}{\mathrm{d}s}+\frac{\partial\mathbf{r}}{\partial v}\frac{\mathrm{d}^2v}{\mathrm{d}s^2}\\
&= \mathbf{r}_{jk}\frac{\mathrm{d}(j)}{\mathrm{d}s}\frac{\mathrm{d}(k)}{\mathrm{d}s}+\mathbf{r}_j\frac{\mathrm{d}^2(j)}{\mathrm{d}s^2}
\end{aligned} \tag{A.30}$$

Combining Equations A.28 through A.30 and making use of Equation A.30, one finds

$$K\cos\eta = \mathbf{r}_{jk}\cdot\mathbf{n}\frac{\mathrm{d}(j)}{\mathrm{d}s}\frac{\mathrm{d}(k)}{\mathrm{d}s} \tag{A.31}$$

However, we have already defined $\mathbf{r}_{jk}\cdot\mathbf{n}$ as being the coefficients b_{jk} of the second fundamental form, and from Equation A.8 we have identified $\mathrm{d}s^2$ with the first fundamental form $g_{lm}\mathrm{d}(l)\mathrm{d}(m)$. Therefore, Equation A.31 becomes

$$K\cos\eta = \frac{b_{jk}\mathrm{d}(j)\mathrm{d}(k)}{g_{lm}\mathrm{d}(l)\mathrm{d}(m)} \tag{A.32}$$

It is interesting to note that the right-hand side of Equation A.32 is only a function of the point (*u*,*v*) on the surface and the direction (d*u*/d*v*) of the curve passing through that point. Thus, at a given point *P* on the surface *S*, if we fix the tangent to the curve, then the right-hand side of Equation A.32 is a constant, which shall be denoted by K_n:

$$K_n = \frac{b_{jk}\mathrm{d}(j)\mathrm{d}(k)}{g_{lm}\mathrm{d}(l)\mathrm{d}(m)} \tag{A.33}$$

K_n is called the normal curvature of the surface *S* at the point *P*. From Equations A.32 and A.33,

$$K\cos\eta = K_n \tag{A.34}$$

If $\eta = 0$, $K = K_n$, and if $\eta = \pi$, $K = -K_n$. Hence $|K_n|$ is the curvature of the intersection of the surface S and the plane that passes through both the tangent to the curve on the surface and the normal to the surface at P. Such a curve will be called a normal section of S. One may introduce the idea of the radii of curvature of a normal section by putting $R = 1/K_n$ in Equation A.33.

We are now interested in obtaining an expression for suitable measures of the curvature of a surface in terms of the coefficients g_{jk} and b_{jk}. We shall see that these measures can be expressed in terms of the two primary curvatures of a surface. Writing out Equation A.33 explicitly, one has

$$K_n = \frac{b_{uu}(du)^2 + 2b_{uv}dudv + b_{vv}(dv)^2}{g_{uu}(du)^2 + 2g_{uv}dudv + g_{vv}(dv)^2} \tag{A.35a}$$

or, in terms of the direction $q \equiv du/dv$,

$$K_n = \frac{b_{uu}q^2 + 2b_{uv}q + b_{vv}}{g_{uu}q^2 + 2g_{uv}q + g} \tag{A.35b}$$

which can be written as

$$(b_{uu} - K_n g_{uu})q^2 + 2(b_{uv} - K_n g_{uv})q + (b_{vv} - K_n g_{vv}) = 0 \tag{A.36}$$

The curvature K_n in Equation A.36 is a function of q. Therefore, if one differentiates Equation A.36 with respect to q and makes use of the condition for an extremum of K_n, namely, $dK_n/dq = 0$, one obtains

$$(b_{uu} - K_n g_{uu})q + (b_{uv} - K_n g_{uv}) = 0 \tag{A.37}$$

In order to solve for the explicit values of the extrema of the normal curvature, one must eliminate q between Equations A.36 and A.37 and solve for K_n. One obtains

$$(b_{vv} - K_n g_{vv})(b_{uu} - K_n g_{uu}) - (b_{uv} - K_n g_{uv})^2 = 0$$

or

$$K_n^2 + \frac{1}{g}(g_{uu}b_{vv} - 2g_{uv}b_{uv} + g_{vv}b_{uu})K_n + \frac{b}{g} = 0 \tag{A.38}$$

The principal curvatures K_1 and K_2, which are the extrema of K_n, will be the solutions of Equation A.38. It then follows that

$$(K_n - K_1)(K_n - K_2) = K_n^2 - (K_1 + K_2)K_n + K_1K_2 = 0 \tag{A.39}$$

where the coefficient of K_n in Equation A.38 is put equal to $(K_1 + K_2)$. It is conventional to denote the product of K_1K_2 by K, and by comparing Equations A.38 and A.39, one can write

$$K \equiv K_1K_2 = \frac{b}{g} \tag{A.40}$$

K is called the Gaussian curvature. In spite of the fact that the Gaussian curvature given by Equation A.40 appears to depend on both the first and second fundamental forms, it can be shown that K depends only on the first fundamental form coefficients and their first and second derivatives.

The arithmetic mean $(K_1 + K_2)/2$ of the principal curvatures is called the mean curvature of the surface and is denoted by H. From Equations A.38 and A.39, one sees

$$H = \tfrac{1}{2}(K_1 + K_2) = \frac{1}{2g}(g_{uu}b_{vv} - 2g_{uv}b_{uv} + g_{vv}b_{uu}) \tag{A.41}$$

The Gaussian curvature and the mean curvature are useful expressions of the curvature of a surface in terms of the coefficients g_{jk} and b_{lm}. As seen from Equations A.40 and A.41, a knowledge of H and K determines the principal curvatures K_1 and K_2 of a surface, which themselves are extrema of the normal curvature K_n. Furthermore, by the Euler theorem of differential geometry, one can express the curvature of a normal section in an arbitrary direction in terms of the two principal curvatures K_1 and K_2 and the angle between the direction of the curve and the direction of the curve with the curvature K_1.

This brief discussion of the theory of surfaces is intended only to highlight some of the key ideas of differential geometry used in Chapter 13 as well as to present an accessible reference to derivations of some of the results that we have used in this chapter. For a more complete discussion of differential geometry with application to the theory of surfaces, see Reference [36], for example.

Index

Note: Locators "*f*" and "*t*" denote figures and tables in the text

C

D

E

F

O

P

Q

R

W

X

Z